The following sections feature extra practice problems posted on the Internet. They can be downloaded to your disk or hard drive from the *Intermediate Algebra: Concepts and Applications*, Fifth Edition, home page, which is located at **http://hepg.awl.com/be/inter_5**. For your convenience, these exercises are also printed in the *Instructor's Resource Guide*. Please contact your instructor for assistance.

INTERMEDIATE
Algebra
CONCEPTS AND APPLICATIONS

Marvin L. Bittinger

Indiana University–Purdue University at Indianapolis

David J. Ellenbogen

Community College of Vermont

FIFTH EDITION

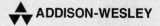
ADDISON-WESLEY

An imprint of Addison Wesley Longman, Inc.

Reading, Massachusetts • Menlo Park, California • New York • Harlow, England
Don Mills, Ontario • Sydney • Mexico City • Madrid • Amsterdam

Publisher	Jason A. Jordan
Associate Editor	Christine Poolos
Production Supervisors	Kathleen Manley and Ron Hampton
Text Design	Geri Davis/The Davis Group, Inc.
Editorial and Production Services	Martha Morong/Quadrata, Inc.
Art Editor	Janet Theurer
Marketing Managers	Ben Rivera and Liz O'Neil
Illustrators	Scientific Illustrators and Jim Bryant
Compositor	The Beacon Group
Cover Designer	Barbara Atkinson
Cover Photograph	Eric Neurath/Stock, Boston/PNI
Manufacturing Supervisor	Ralph Mattivello

PHOTO CREDITS

1, Comstock **40,** Bob Daemmrich/Stock Boston **65,** David Muench/Tony Stone Images **86,** Alexander Tsiaras/Stock Boston **115,** Art Wolfe/Tony Stone Images **131,** Donald Williams Wood Floors **197,** David A. Meardon/Silver Images **205,** Corbis/Bettmann **236,** Comstock **247,** Bert Blokjuis/Tony Stone Images **251,** Simon Wilkinson/The Image Bank **311,** Gale Zucker/Stock Boston **375,** Steve Dunwell/The Image Bank **433,** Randall Hyman/Stock Boston **456,** Reuters/Ira Strickstein/Archive Photos **477,** National Park Service **513,** Ken Cole/Animals Animals **558,** Comstock **561,** AP/Wide World Photos **567,** From *Classic Baseball Cards*, by Bert Randolph Sugar, copyright © 1977 by Dover Publishing, Inc. **568,** AP/Wide World Photos **577,** Felix St. Clair Renard/The Image Bank **617,** Bonnie Kamin/PhotoEdit

Library of Congress Cataloging-in-Publication Data

Bittinger, Marvin L.

Intermediate algebra: concepts and applications/Marvin L. Bittinger, David Ellenbogen,—5th ed.

p. cm.

Includes index.

ISBN 0-201-84750-7.—ISBN 0-201-41732-4 (teacher's ed.)

1. Algebra. I. Ellenbogen, David. II. Title.

QA154.2.B54 1997

512.9—dc20 96-36430

CIP

Reprinted with corrections, April 1999

4 5 6 7 8 9 10—RNV—0099

To Monroe and Zachary

Contents

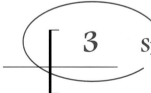

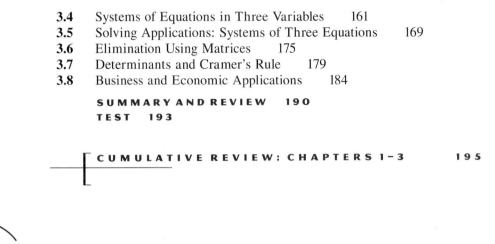

4   Inequalities and Problem Solving   197

5   Polynomials and Polynomial Functions   247

6 Rational Expressions, Equations, and Functions   311

7 Exponents and Radicals   375

8   Quadratic Functions and Equations   433

9   Exponential and Logarithmic Functions   513

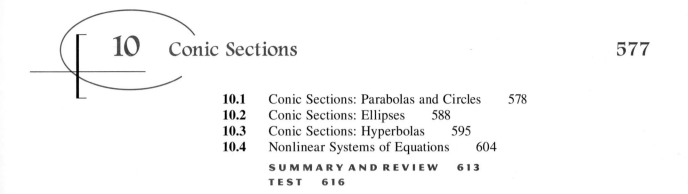

Preface

*A*ppropriate for a one-term course in intermediate algebra, this text is intended for those students who have completed a first course in algebra. It is the second of three texts in an algebra series that also includes *Elementary Algebra*: *Concepts and Applications*, Fifth Edition, by Bittinger/Ellenbogen and *Elementary and Intermediate Algebra*: *Concepts and Applications*, *A Combined Approach*, Second Edition, by Bittinger/Ellenbogen/Johnson. *Intermediate Algebra*: *Concepts and Applications*, Fifth Edition, is a significant revision of the Fourth Edition with respect to design, contents, pedagogy, and an expanded supplements package. This series is designed to prepare students for any mathematics course at the college algebra level.

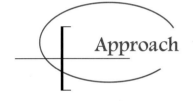

Approach

Our approach is designed to help today's students both learn and retain mathematical concepts. The goal of this revision was to address the major challenges for teachers of developmental mathematics courses that we have seen emerging during the 1990s. The first challenge is to prepare students of developmental mathematics for the transition from "skills-oriented" elementary and intermediate algebra courses to more "concept-oriented" college-level mathematics courses. The second is to teach these same students critical thinking skills: to reason mathematically, to communicate mathematically, and to solve mathematical problems. The third challenge is to reduce the amount of content overlap between elementary algebra and intermediate algebra texts.

Following are some aspects of the approach that we have used in this revision to help meet the challenges we all face teaching developmental mathematics.

Problem Solving

One distinguishing feature of our approach is our treatment of and emphasis on problem solving. We use problem solving and applications to motivate the material wherever possible, and we include real-life applications and problem-solving techniques throughout the text. Problem solving not only encourages students to think about how mathematics can be used, it helps to prepare them for more advanced material in later courses.

- In Chapter 1, we introduce the five-step process for solving problems: (1) Familiarize, (2) Translate, (3) Carry out, (4) Check, and (5) State the answer. These steps are introduced in Chapter 1 and used consistently throughout the text whenever we encounter a problem-solving situation. Repeated use of this problem-solving strategy gives students a sense that they have a starting point for any type of problem they encounter, and frees them to focus on the mathematics necessary to successfully translate the problem situation. In this edition (see pages 29–31 and 153), estimation has been used more consistently to help with the Familiarize and Check steps.

Functions and Graphing

To retain skills and to apply them at a more conceptual level in later courses, students must have an intuitive understanding of the material. A visual interpretation of mathematical concepts can provide this type of understanding to those students with a visual, rather than symbolic, orientation.

- We introduce functions and graphing in Chapter 2, substantially earlier than in many intermediate algebra texts. Functions and graphs then appear throughout the text to enrich the coverage of other topics. For instance, examples of polynomial and rational functions are introduced along with polynomial and rational expressions and equations in Chapters 5 and 6. (See pages 250 and 341.) Increased familiarity and practice with functions and graphing techniques make students more comfortable with these important topics when they move on to later courses.
- We have more than doubled the number of Technology Connections in the Fifth Edition. This optional feature allows students to use a graphing calculator or computer to help visualize concepts, and provides additional opportunities for students to see the usefulness of functions and graphing. (See pages 71, 89, and 137.)

Applications

Interesting applications of mathematics help motivate both students and instructors. Solving applied problems gives students the opportunity to see their conceptual understanding put to use in a real way. In the Fifth Edition of *Intermediate Algebra: Concepts and Applications*, not only have we increased the number of applications by more than 60 percent, but we have also included a wide variety of real-data applications throughout the text. Over 80 percent of our applications are new. Art has also been integrated into the applications and exercises to aid the student in visualizing the mathematics. (See pages 34, 97, and 347.)

Content

Many intermediate algebra texts contain a substantial review of elementary algebra topics. This can lull students into complacency and prevent instructors from covering the intermediate algebra topics required in later courses.

- By introducing graphing and functions in Chapter 2, we present students with "intermediate algebra" topics almost immediately. These topics reappear throughout the text to give students familiarity and practice with concepts that will be critical in later courses.
- Systems of equations are introduced in Chapter 3 to provide students with a valuable problem-solving tool. Students can then translate problem situations into systems of equations throughout the remainder of the text. This approach provides a useful alternative to always translating problems into equations in which only one variable is used.

Pedagogy

Skill Maintenance Exercises.   Retention of skills is critical to the future success of our students. To this end, nearly every exercise set includes carefully chosen exercises that review skills and concepts from preceding chapters of the text, often in preparation for the next section. In this edition, we have increased the number of skill maintenance exercises by about 50%.

Cumulative Review.   After every three chapters, and at the end of the text, we have also included a Cumulative Review, which reviews skills and concepts from all preceding chapters of the text. (See pages 195, 373, 574, and 653.)

Synthesis Exercises.   Each exercise set ends with a set of synthesis exercises. These problems can offer opportunities for students to synthesize skills and concepts from earlier sections with the present material, or they can provide students with deeper insights into the current topic. Synthesis exercises are generally more challenging than those in the main body of the exercise set. We have increased the number of synthesis exercises by 10% for the Fifth Edition. (See pages 10, 148, 168, and 226.)

Writing Exercises.   Nearly every set of synthesis exercises begins with four writing exercises, which nearly doubles the number found in the Fourth Edition. All writing exercises are marked with a maze icon (◆). These exercises are usually not as difficult as other synthesis exercises, but require written answers that aid in student comprehension, critical thinking, and conceptualization. Because some instructors may collect answers to writing exercises, and because more than one answer may be correct, answers to writing exercises are not listed at the back of the text. (See pages 86, 108, and 234.)

What's New in the Fifth Edition?

We have rewritten many key topics in response to user and reviewer feedback and have made significant improvements in design, art, and pedagogy. Detailed information about the content changes is available in the form of a

Conversion Guide. Please ask your local Addison Wesley Longman sales consultant for more information. Following is a list of the major changes in this revision.

New Design

- The new design is more open and readable. We have increased the type size in many features to emphasize key topics. Pedagogical use of color has been increased to make it easier to see how examples unfold.
- The entire art program is new for this edition. We have ensured the accuracy of the graphical art through the use of computer-generated graphs. Color in the graphical art is used pedagogically and precisely to help the student visualize the mathematics. (See pages 103 and 455.)

Technology Connections

- The optional Technology Connections, which appear throughout the text (see pages 71, 89, and 137), integrate technology, increase the understanding of concepts through visualization, encourage exploration, and motivate discovery learning. Due to user demand, we have more than doubled the number of Technology Connections in the Fifth Edition. Optional Technology Connection exercises appear in many exercise sets and are marked with a grapher icon (![grapher icon]). (See pages 74, 100, and 337.)

Collaborative Corners

- In today's professional world, teamwork is essential. We have included optional Collaborative Corner features throughout the text to allow students to work in groups to solve problems. There is an average of three Collaborative Corner activities per chapter, each one appearing after the appropriate section's exercise set. (See pages 36, 149, and 485.) Additional Collaborative Corner activities and suggestions for directing collaborative learning appear in the *Printed Test Bank/Instructor's Resource Guide*.

World Wide Web Integration

- The World Wide Web is a powerful resource that reaches more and more people every day. In an effort to get students more involved in using this resource, we have enhanced every chapter opener to include a World Wide Web address **http://hepg.awl.com/be/inter_5**. Students can go to this page on the World Wide Web to further explore the subject matter of the chapter opening application. Selected exercise sets also have additional practice-problem worksheets that can be downloaded over the Internet. These exercise sets are listed opposite the inside front cover.

Content Changes

A variety of content changes have been made throughout the text. Some of the more significant changes are listed below.

- In Chapter 1, Section 1.1 now includes order of operations to facilitate work in Section 1.2 and beyond. Sections 1.5 and 1.6 have been reorganized for better flow.

- In Chapter 2, Section 2.2 now includes more discussion of visualizing domain and range on a graph. Section 2.3 now includes reading solutions from graphs, and the definition of slope now precedes slope–intercept equations and is covered in more detail. Both of these changes encourage students to use their visualization skills and to read and interpret graphs.
- In Chapter 4, Section 4.2 now includes more material on domains of functions in order to better prepare students for Chapters 6 and 7. Section 4.3 now includes more graphing, for increased visualization. This change will also aid students in relating inequalities to equations.
- In Chapter 5, Section 5.8, we make greater use of function notation and provide more work with domain. This will better prepare students for Chapters 6 and 7 and for college algebra.
- Chapter 6 has many changes. Sections 6.1 and 6.2 have been reorganized to clarify simplifying rational expressions and more work with domain has been added. The reorganization of this traditional problem area for students should improve comprehension. Sections 6.4 and 6.6 make greater use of function notation and graphs. This is a more visual approach to the material and better prepares students for college algebra.
- Chapter 7 has been extensively revised. Section 7.1 now has more function notation and more work with domains of radical functions. Sections 7.3–7.6 include additional discussion of functions and represent a streamlining of Sections 7.3–7.7 of the Fourth Edition.
- In Chapter 8, Section 8.1, we have included more work with quadratic functions and more exercises with imaginary solutions to better prepare students for college algebra. Section 8.4 now includes more on writing equations from solutions to build mathematical understanding. Section 8.10 has been streamlined for increased clarity.
- In Chapter 10, Section 10.2 now includes ellipses not centered on the origin and Section 10.3 is dedicated to hyperbolas. These changes give the student a more complete picture and will better prepare them for college algebra.

Supplements for the Instructor

Instructor's Edition

The *Instructor's Edition* is a specially bound version of the student text with worked-out solutions to the even-numbered exercises and answers to the odd-numbered exercises at the back of the text.

Instructor's Solutions Manual

The *Instructor's Solutions Manual* contains worked-out solutions to all exercises in the exercise sets.

Printed Test Bank/Instructor's Resource Guide
by Donna DeSpain

This supplement contains the following:

- Extra practice problems with more sections covered than in the Fourth Edition
- Black-line masters of grids and number lines for transparency masters or test preparation
- A videotape index and section cross references to the tutorial software packages available with this text
- Additional collaborative learning activities and suggestions
- A syllabus conversion guide from the Fourth Edition to the Fifth Edition

The test bank portion contains the following:

- Six alternative free-response test forms for each chapter
- Two multiple-choice versions of each chapter test
- Eight final examinations: three with questions organized by chapter, three with questions organized by type, and two with multiple-choice questions

All test forms have been completely rewritten.

TestGen—EQ

TestGen—EQ is a computerized test generator that allows instructors to select test questions manually or randomly from selected topics. The test questions are algorithm-driven so that regenerated number values maintain problem types and provide a large number of test items in both multiple-choice and open-ended formats for one or more test forms. Test items can be viewed on screen, and the built-in question editor lets instructors modify existing questions or add new questions that include pictures, graphs, math symbols, and variable text and numbers. Instructors can also customize both the look and content of test banks and tests. Test questions are easily transferred from the test bank to a test and can be sorted, searched, and displayed in various ways. Available in both Windows and Macintosh versions, TestGen—EQ is free to qualifying adopters.

QuizMaster—EQ

QuizMaster—EQ enables instructors to create and save tests and quizzes using TestGen—EQ so students can take them on a computer network. Instructors can set preferences for how and when tests are administered. QuizMaster—EQ automatically grades the exams and allows the instructor to view or print a variety of reports for individual students, classes, or courses. This software is available for both Windows and Macintosh and is fully networkable. QuizMaster—EQ is free to qualifying adopters.

InterAct Math Plus

This software, available for both Windows and Macintosh, combines course management and on-line testing with the features of InterAct Math Tutorial Software (see *InterAct Math Tutorial* under Supplements for the Student) to create an invaluable teaching resource. Contact your local Addison Wesley Longman sales consultant for a demonstration.

Supplements for the Student

Student's Solutions Manual by Judith A. Penna

This manual contains completely worked-out solutions with step-by-step annotations for all the odd-numbered exercises in the exercise sets. Solution processes match those illustrated in the text. The manual also contains answers for all the even-numbered exercises in the text.

Videotapes

Developed especially for the Bittinger/Ellenbogen texts, these videotapes feature an engaging team of lecturers presenting material from every section of the text in an interactive format. The lecturers' presentations support an approach that emphasizes visualization and problem solving. The videotapes are free to qualifying adopters.

Video Manual by Janina Udrys

Designed to be used with the Bittinger/Ellenbogen videotapes in distance-learning situations, this manual includes many examples and art pieces from the video series. The manual also includes additional problems specifically developed for distant learners.

InterAct Math Tutorial Software

InterAct Math Tutorial Software includes exercises that are linked one-to-one with the odd-numbered exercises in the textbook. Every exercise is accompanied by an example and an interactive guided solution designed to involve students in the solution process and to help them identify precisely where they are having trouble. In addition, the software recognizes common student errors and provides students with appropriate customized feedback. Available for both Windows or Macintosh and fully networkable, Interact Math Tutorial Software is free to qualifying adopters.

World Wide Web Supplement http://hepg.awl.com/be/inter_5

This on-line supplement contains links to web sites related to the chapter openers in the text. It also contains extra practice-problem worksheets for selected text sections. Students can download these worksheets and use them for practice, as needed. Text sections that correlate with on-line worksheets are listed opposite the inside front cover.

Acknowledgments

No book can be produced without a team of professionals who take pride in their work and are willing to put in long hours. Barbara Johnson, in particular, deserves special thanks for her work as development editor. Barbara's tireless attention to detail and her many fine suggestions have contributed immeasurably to the quality of this text. Laurie A. Hurley and Irene Doo also deserve deeply felt thank you's for their careful accuracy checks and well-thought-out suggestions. Judy Penna's work in preparing the *Student's Solution Manual*, the *Instructor's Solution Manual*, and the indexes, amounts to an inspection of the text that goes beyond the call of duty, and for which we are extremely grateful. We are also indebted to Kim McDowell, Patty Schwarzkopf, and Doris Lewis for their careful checking of answers and to Chris Burditt for his many fine ideas that appear in our Collaborative Corners.

Jason Jordan, the sponsoring editor, displayed both wisdom and restraint in providing direction when it was needed and allowing us to work independently as much as possible. Martha Morong, of Quadrata, Inc., provided editorial and production services that set the standard for the industry. Janet Theurer, of Theurer Briggs Design, performed outstanding work as art editor. George and Brian Morris of Scientific Illustrators generated the graphs, charts, and many of the illustrations. Their work is always precise and attractive. The many hand-drawn illustrations appear thanks to Jim Bryant, an artist with true mathematical sensibilities. Finally, a special thank you to Christine Poolos for coordinating reviews, tracking down information, and managing so many of the day-to-day details—always with a pleasant demeanor.

In addition, we thank the students at Community College of Vermont and the following professors for their thoughtful reviews and insightful comments.

Prerevision Diary Reviewers (Fourth Edition)

Omar DeWitt, *University of New Mexico*
Robert Grant, *Mesa Community College*
Gary Long, *Mt. San Antonio College*
Linda Reist, *Macomb County Community College*

Telephone Survey Participants

Paul Blankenship, *Lexington Community College*
Dawn Burke, *University of New Hampshire*
Chuy Carreon, *Mesa Community College*

Antonio Ciro, *University of Delaware*
David Demic, *Tyler Junior College*
Barbara Duncan, *Hillsborough Community College*
Carol Lawrence, *North Carolina State University*
Patty Schwarzkopf, *University of Delaware*
Rick Serpone, *Mesa Community College*
Mahendra Singhal, *University of Wisconsin—Parkside*
Lois Speg, *Delaware Technical Community College*
Carl Thomas, *Napa Valley College*
Gerry Tolimante, *Mesa Community College*
William Waters, *North Carolina State University*
Larry Watson, *North Carolina State University*

Manuscript Reviewers

Diane Blansett, *Delta State University*
Connie Buller, *Metropolitan Community College*
Joseph Conrad, *Solano Community College*
Sally Copeland, *Johnson County Community College*
Ben Cornelius, *Oregon Institute of Technology*
Linda Crabtree, *Metropolitan Community College*
Sharon Hamsa, *Longview Community College*
Lonnie Hass, *North Dakota State University*
Robert Hickey, *Kishwaukee College*
Louise Hoover, *Clark College*
Judy Kasabian, *El Camino College*
Marie Long, *Pennsylvania State University—DuBois*
Jann McInness, *Florida Community College at Jacksonville*
Jane Murphy, *Middlesex Community College*
Paul O'Heron, *Broome Community College*
Mark Omodt, *Anoka Ramsey Community College*
Mark Parker, *Shoreline Community College*
Marjorie Stinespring, *Chicago State University*
June Strohm, *Pennsylvania State University—DuBois*
Sharon Testone, *Onondaga Community College*
Kathy Tomsich, *Anoka Ramsey Community College*
Marilyn Treder, *Rochester Community College*
Bettie Truitt, *Black Hawk Community College*
Terry Wyberg, *University of Minnesota*

Finally, a special thank you to all those who so generously agreed to discuss their professional uses for mathematics in our chapter openers. These dedicated people, none of whom we knew prior to writing this text, all share a desire to make math more meaningful to students. We cannot imagine a finer set of role models.

M.L.B.
D.J.E.

Feature
Walkthrough

The following six pages show you how to use *Intermediate Algebra* to maximize understanding while making studying easier.

- **Chapter Opener:** Each chapter opens with a list of the sections covered and a real-life application that helps show how mathematics is needed when solving real problems. Real data is often used in these applications, as well as in many other exercises and examples. For those students with access to the internet, the authors have equipped every chapter opener with a World Wide Web address, **http://hepg.awl.com/be/inter_5**. At this site, students can further explore the subject matter of the chapter opening application.

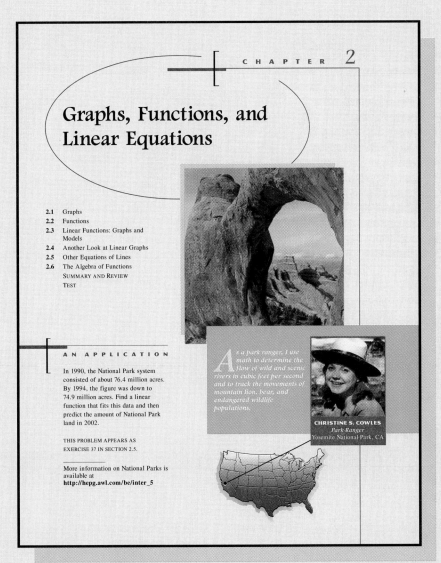

CHAPTER 2

Graphs, Functions, and Linear Equations

2.1 Graphs
2.2 Functions
2.3 Linear Functions: Graphs and Models
2.4 Another Look at Linear Graphs
2.5 Other Equations of Lines
2.6 The Algebra of Functions
SUMMARY AND REVIEW
TEST

AN APPLICATION

In 1990, the National Park system consisted of about 76.4 million acres. By 1994, the figure was down to 74.9 million acres. Find a linear function that fits this data and then predict the amount of National Park land in 2002.

THIS PROBLEM APPEARS AS
EXERCISE 37 IN SECTION 2.5.

More information on National Parks is available at
http://hepg.awl.com/be/inter_5

As a park ranger, I use math to determine the flow of wild and scenic rivers in cubic feet per second and to track the movements of mountain lion, bear, and endangered wildlife populations.

CHRISTINE S. COWLES
Park Ranger
Yosemite National Park, CA

• One distinguishing feature of the authors' approach is their **Five-Step Process for Solving Problems:** 1) Familiarize, 2) Translate, 3) Carry Out, 4) Check, and 5) State the Answer. These steps are used throughout the text whenever a problem-solving situation arises. Use of this problem-solving process gives students a sense that they have a starting point for any problem they encounter, and helps them focus on the mathematics necessary to translate the problem situation.

Problem Solving

At this point, our study of algebra is just beginning. Thus we have few algebraic tools with which to work problems. As the number of tools in our algebraic "toolbox" increases, so will the difficulty of the problems being solved. For now our problems may seem simple; however, to gain practice with the problem-solving process, you should try to use all five steps. Later some steps may be shortened or combined.

EXAMPLE 3

Purchasing.   Elka pays $1187.20 for a computer. If the price paid includes a 6% sales tax, what is the price of the computer itself?

SOLUTION

1. Familiarize.   Familiarize yourself with the problem. Note that tax is calculated from, and then added to, the computer's price. We let

C = the computer's price.

Let's guess that the computer's price is $1000. To check the guess, we calculate the amount of tax, $(0.06)(\$1000) = \$60$, and add it to $1000:

$$(0.06)(\$1000) + \$1000 = \$60 + \$1000$$
$$= \$1060. \qquad \text{\small $1060 \neq \$1187.20$}$$

Our guess was wrong, but it was useful. The manner in which we checked the guess will guide us in the next step.

2. Translate.   Translate the problem to mathematical language. Our guess leads us to the following translation:

6% of the computer's price	plus	the computer's price	is	the price with sales tax.
↓	↓	↓	↓	↓
$(0.06)C$	$+$	C		$\$1187.20$

3. Carry out.   Carry out some mathematical manipulation:

$$0.06C + 1C = 1187.20$$
$$1.06C = 1187.20 \qquad \text{\small Combining like terms}$$
$$\frac{1}{1.06} \cdot 1.06C = \frac{1}{1.06} \cdot 1187.20 \qquad \text{\small Using the multiplication principle}$$
$$C = 1120.$$

4. Check.   Check the answer in the original problem. To do this, note that the tax on a computer costing $1120 would be $(0.06)(\$1120) = \$67.20$. When this is added to $1120, we have

$$\$67.20 + \$1120, \text{ or } \$1187.20.$$

We see that $1120 checks in the original problem.

5. State.   State the answer clearly. The computer itself costs $1120.

- **Applications:** In this edition the authors have increased the number of applications and have also included real data applications that come from a wide variety of disciplines and fields, such as natural and social sciences, business and economics, and health care. Over 80% of the applications are new. Art has also been integrated to aid the student in visualizing the many applications that appear in the examples and exercises.

earth's distance from the sun, and L is some fixed distance, is used in calculating when lunar eclipses occur. Solve for D.

SOLUTION We first clear fractions by multiplying by the LCD, which is $D - d$:

$$(D - d)L = (D - d)\frac{dR}{D - d}$$

$$(D - d)L = dR.$$

We do *not* multiply the factors on the left since we wish to get D all alone. Instead we multiply both sides by $1/L$ and then add d:

$$D - d = \frac{dR}{L} \qquad \text{Multiplying on both sides by } \frac{1}{L}$$

$$D = \frac{dR}{L} + d. \qquad \text{Adding } d \text{ on both sides}$$

We now have D all alone on one side of the equation. Since D does not appear on the other side, we have solved the formula for D.

EXAMPLE 3 *Acoustics (the Doppler Effect).* The formula

$$f = \frac{sg}{s + v}$$

is used to determine the frequency f of a sound that is moving at velocity v toward a listener who hears the sound as frequency g. Here s is the speed of sound in a particular medium. Solve for s.

SOLUTION We first clear fractions by multiplying by the LCD, $s + v$:

$$f \cdot (s + v) = \frac{sg}{s + v}(s + v)$$

$$fs + fv = sg. \qquad \text{Here, because } s \text{ does appear on both sides,}$$
$$\text{we do distribute on the left side.}$$

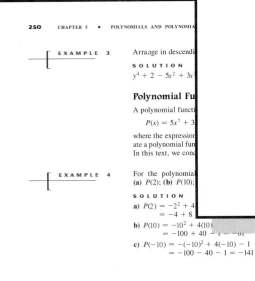

EXAMPLE 3 Arrange in descendi

SOLUTION
$y^4 + 2 - 5x^2 + 3x$

Polynomial Fu

A polynomial functi

$$P(x) = 5x^7 + 3$$

where the expression ate a polynomial fun In this text, we cond

EXAMPLE 4 For the polynomial
(a) $P(2)$; (b) $P(10)$;

SOLUTION
a) $P(2) = -2^2 + 4$
$\quad = -4 + 8$

b) $P(10) = -10^2 + 4(10)$
$\quad = -100 + 40 - 1 = -61$

c) $P(-10) = -(-10)^2 + 4(-10) - 1$
$\quad = -100 - 40 - 1 = -141$

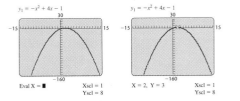

TECHNOLOGY CONNECTION 5.1A   A visual way to evaluate a polynomial function is to enter and graph the function as y_1. If a CALC menu exists, selecting VALUE allows us to enter the value of x in which we are interested. The corresponding y-value and point on the graph then appear.

1. Use this approach to check Examples 4(b) and 4(c).
2. What advantage(s) does this approach offer over use of the TRACE feature?

$y_1 = -x^2 + 4x - 1$
30
$-15 \quad 15$
-160
Eval X = ■   Xscl = 1   Yscl = 8

$y_1 = -x^2 + 4x - 1$
30
$-15 \quad 15$
-160
X = 2, Y = 3   Xscl = 1   Yscl = 8

- **Technology Connection Features:** Due to user demand the authors have more than doubled the number of Technology Connection features in this edition. Optional Technology Connection exercises also appear in many exercise sets and are marked with a grapher icon (⌁).

- **New Design:** The new design is more open and readable. With an entirely new art program for this edition, generous amounts of color-coded technical and situational art appear throughout the text to enhance student understanding of an example or exercise and to aid in the visualization of mathematical concepts.

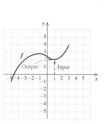

b) The domain of the function is the set of all x-values that are in the graph. These extend from -5 to 3 and can be viewed as the curve's shadow, or *projection*, on the x-axis. Thus the domain is $\{x \mid -5 \leq x \leq 3\}$.

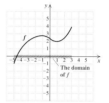

c) To determine what member of the domain is paired with 1, we locate 1 on the vertical axis. (See the graph on the left below.) From there we look left and right to the graph of f to find any points for which 1 is the second co-ordinate. One such point exists, $(-4, 1)$. We observe that -4 is the only element of the domain paired with 1.

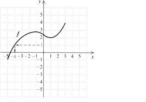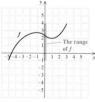

d) The range of the function is the set of all y-values that are in the graph. (See the graph on the right above.) These extend from -1 to 4 and can be viewed as the curve's projection on the y-axis. Thus the range is $\{y \mid -1 \leq y \leq 4\}$.

---

87. Assume that the graphs of $y_1 = -\frac{1}{2}x + 5$, $y_2 = x - 1$, and $y_3 = 2x - 3$ are as shown below. Solve each inequality, referring only to the figure.

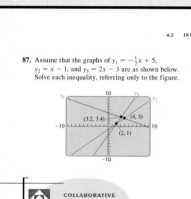

COLLABORATIVE
C·O·R·N·E·R

Focus: Inequalities and problem solving

Time: 10–15 minutes

Group size: 2

In the United States, the amount of solid waste being recycled is slowly catching up to the amount being generated. In 1991, each person generated, on average, 4.3 lb of solid waste every day, of which 0.8 lb was recycled. In 1994, each person generated, on average, 4.4 lb of solid waste, of which 1.0 lb was recycled. (*Source:* Characterization of Municipal Solid Waste in the United States: 1995 Update, Executive Summary, United States Environmental Protection Agency, March 1996.)

ACTIVITY

Assume that the amount of solid waste being generated and the amount being recycled are both increasing linearly. One group member should find a linear function w for which $w(t)$ represents the number of pounds of waste generated per person per day t years after 1991. The other group member should find a linear function r for which $r(t)$ represents the number of pounds recycled per person per day t years after 1991. Finally, working together, the group should determine those years for which the amount recycled will meet or exceed the amount generated.

4.2 Intersections, Unions, and Compound Inequalities

Intersections of Sets and Conjunctions of Sentences • Unions of Sets and Disjunctions of Sentences • More on Domains of Functions

We now consider **compound inequalities**—that is, sentences formed by two or more inequalities, joined by the word *and* or the word *or*.

- **Collaborative Corner Features:** The authors now include optional Collaborative Corner features throughout the text that allow students to work in groups to solve problems or to perform specially designed activities. There is an average of three Collaborative Corners per chapter, each one appearing after the appropriate exercise set.

- **Skill Maintenance and Synthesis Exercises:** To help students retain material, nearly every exercise set includes Skill Maintenance Exercises that review material from earlier sections of the book. The number of Skill Maintenance Exercises in this edition has been increased by about 50%. Synthesis Exercises offer students opportunities to combine skills and concepts from earlier sections with newer material, or to gain deeper insights into the current topic. The number of synthesis exercises has been increased by 10%.

---

48. $\dfrac{x^2 - 6x + 9}{12 - 4x} \cdot \dfrac{x^6 - 9x^4}{x^3 - 3x^2}$

49. $\dfrac{x^2 - 2x - 35}{2x^3 - 3x^2} \cdot \dfrac{4x^3 - 9x}{7x - 49}$

50. $\dfrac{y^2 - 10y + 9}{y^2 - 1} \cdot \dfrac{y + 4}{y^2 - 5y - 36}$

51. $\dfrac{c^3 + 8}{c^5 - 4c^3} \cdot \dfrac{c^6 - 4c^5 + 4c^4}{c^2 - 2c + 4}$

52. $\dfrac{x^3 - 27}{x^4 - 9x^2} \cdot \dfrac{x^5 - 6x^4 + 9x^3}{x^2 + 3x + 9}$

53. $\dfrac{a^3 - b^3}{3a^2 + 9ab + 6b^2} \cdot \dfrac{a^2 + 2ab + b^2}{a^2 - b^2}$

54. $\dfrac{x^3 + y^3}{x^2 + 2xy - 3y^2} \cdot \dfrac{x^2 - y^2}{3x^2 + 6xy + 3y^2}$

55. $\dfrac{4x^2 - 9y^2}{8x^3 - 27y^3} \cdot \dfrac{4x^2 + 6xy + 9y^2}{4x^2 + 12xy + 9y^2}$

56. $\dfrac{3x^2 - 3y^2}{27x^3 - 8y^3} \cdot \dfrac{6x^2 + 5xy - 6y^2}{6x^2 + 12xy + 6y^2}$

Divide and simplify.

57. $\dfrac{9x^5}{8y^2} \div \dfrac{3x}{16y^9}$

58. $\dfrac{16a^7}{3b^5} \div \dfrac{8a^3}{6b}$

59. $\dfrac{6x + 12}{x^8} \div \dfrac{x + 2}{x^3}$

60. $\dfrac{3y + 15}{y^7} \div \dfrac{y + 5}{y^2}$

61. $\dfrac{x^2 - 4}{x^3} \div \dfrac{x^5 - 2x^4}{x + 4}$

62. $\dfrac{y^2 - 9}{y^2} \div \dfrac{y^5 + 3y^4}{y + 2}$

63. $\dfrac{25x^2 - 4}{x^2 - 9} \div \dfrac{2 - 5x}{x + 3}$

64. $\dfrac{4a^2 - 1}{a^2} \div \dfrac{2a - 1}{a^2}$

74. $\dfrac{x^3 + 8y^3}{2x^2 + 5xy + 2y^2} \div \dfrac{x^3 - 2x^2y + 4xy^2}{8x^2 - 2y^2}$

SKILL MAINTENANCE

75. Solve by substitution:
$$3x + y = 13,$$
$$x = y + 1.$$

76. Evaluate: $\begin{vmatrix} 3 & -2 \\ 4 & 7 \end{vmatrix}$.

77. Solve: $\frac{2}{3}(3x - 4) = 8$.

78. A concert committee needs to take in $4000 from ticket sales in order to break even. If a total of 400 tickets is to be sold at full price and 200 tickets sold at half price, how should the tickets be priced?

SYNTHESIS

79. ◆ Is it possible to understand how to simplify rational expressions without first understanding how to multiply rational expressions? Why or why not?

80. ◆ Nancy *incorrectly* simplifies $\dfrac{x + 2}{x}$ as
$$\dfrac{x + 2}{x} = \dfrac{x + 2}{x} = 1 + 2 = 3.$$
She insists this is correct because it checked when x was replaced by 1. Explain her misconception.

81. ◆ Tony *incorrectly* argues that since
$$\dfrac{a^2 - 4}{a - 2} = \dfrac{a^2}{a} + \dfrac{-4}{-2} = a + 2,$$
it follows that
$$\dfrac{x^2 + 9}{x + 1} = \dfrac{x^2}{x} + \dfrac{9}{1} = x + 9.$$
Explain his misconception.

82. Let
$$g(x) = \dfrac{2x + 3}{4x - 1}.$$
Determine each of the following.
a) $g(x + h)$
b) $g(2x - 2) \cdot g(x)$
c) $g\left(\frac{1}{2}x + 1\right) \cdot g(x)$

83. Graph the function given by
$$f(x) = \dfrac{x^2 - 9}{x - 3}.$$
(*Hint*: Determine the domain of f and simplify.)

---

43. $y + 3 = 5x,$
$3x - y = -2$

44. $y + 8 = -6x,$
$-2x + y = 5$

45. $f(x) = 3x + 9,$
$2y = 6x - 2$

46. $f(x) = -7x - 9,$
$-3y = 21x + 7$

Write an equation of the line containing the specified point and parallel to the indicated line.

47. $(3, 7),\ x + 2y = 6$

48. $(0, 3),\ 3x - y = 7$

49. $(2, -1),\ 5x - 7y = 8$

50. $(-4, -5),\ 2x + y = -3$

51. $(-6, 2),\ 3x - 9y = 2$

52. $(-7, 0),\ 5x + 2y = 6$

53. $(-3, -2),\ 3x + 2y = -7$

54. $(-4, 3),\ 6x - 5y = 4$

Without graphing, tell whether the graphs of each pair of equations are perpendicular.

55. $f(x) = 4x - 5,$
$4y = 8 - x$

56. $2x - 5y = -3,$
$2x + 5y = 4$

57. $x + 2y = 5,$
$2x + 4y = 8$

58. $y = -x + 7,$
$f(x) = x + 3$

Write an equation of the line containing the specified point and perpendicular to the indicated line.

59. $(2, 5),\ 2x + y = -3$

60. $(4, 0),\ x - 3y = 0$

61. $(3, -2),\ 3x + 4y = 5$

62. $(-3, -5),\ 5x - 2y = 4$

63. $(0, 9),\ 2x + 5y = 7$

64. $(-3, -4),\ -3x + 6y = 2$

65. $(-4, -7),\ 3x - 5y = 6$

66. $(-4, 5),\ 7x - 2y = 1$

SKILL MAINTENANCE

67. The price of a radio, including 5% sales tax, is $36.75. Find the price of the radio before the tax was added.

68. A basketball team increases its score by 7 points in each of three consecutive games. If the team scored a total of 228 points in all three games, what was its score in the first game?

69. 15% of what number is 12.4?

70. Subtract: $-\frac{7}{4} - \left(-\frac{2}{3}\right)$.

SYNTHESIS

71. ◆ The total number of reported cases of AIDS in the United States grew from 372 in 1981 to 100,000 in 1989 and 200,000 in 1992. Can a linear function be used to predict the number of cases in 2004? Why or why not?

72. ◆ The information in Exercises 34 and 36 indicates that when suppliers charge $8 per pound for coffee, the supply will not meet the demand. How could suppliers determine a price for which their supply will exactly meet the demand?

73. ◆ Explain why the negative slope in Exercise 31 is "good," whereas the negative slope in Exercise 37 is "bad."

74. ◆ A firm offers its entering employees a starting salary with a guaranteed 7% increase each year. Can a linear function be used to express the yearly salary as a function of the number of years an employee has worked? Why or why not?

For Exercises 75–78, assume that a linear equation models each situation.

75. *Depreciation of a computer.* Gina's computer cost $2500 new. The computer's value had dropped to $2150 after 5 mos. What will the computer be worth after 8 mos?

76. *Cellular phone charges.* Rick's cellular phone cost him $70 to purchase and activate. After 4 mos, his total cost for the phone is $190. Predict Rick's total cost for the phone after 9 mos.

77. *Operating expenses.* The total cost for operating Ming's Wings was $7500 after 4 mos and $9250 after 7 mos. Predict the total cost after 10 mos.

78. *Depreciation of a printer.* After 6 mos of use, the value of Pearl's printer had dropped to $900. After 8 mos, the value had gone down to $750. How much did the printer cost when new?

79. *Temperature conversion.* Water freezes at 32° Fahrenheit and at 0° Celsius. Water boils at 212°F and at 100°C. What Celsius temperature corresponds to a room temperature of 70°F?

80. For a linear function f, $f(-1) = 3$ and $f(2) = 4$.
a) Find an equation for f.
b) Find $f(3)$.
c) Find a such that $f(a) = 100$.

- **Writing Exercises:** These exercises are generally less challenging than other synthesis exercises, but require written answers that aid in student comprehension, critical thinking, and communication skills. All writing exercises are marked with a maze icon (◆). In response to user feedback, the number of writing exercises has been doubled.

• Each chapter ends with a **Summary** and **Review** of key terms, important properties and formulas, review exercises that include Skill Maintenance, Synthesis and Writing Exercises, and a chapter test.

SUMMARY AND REVIEW: CHAPTER 4   **243**

SUMMARY AND REVIEW
4

KEY TERMS

Inequality, p. 198
Solution, p. 198
Solution set, p. 198
Set-builder notation, p. 199
Interval notation, p. 199
Open interval, p. 199
Closed interval, p. 199

Half-open interval, p. 199
Compound inequality, p. 209
Intersection, p. 210
Conjunction, p. 210
Union, p. 212
Disjunction, p. 213
Absolute value, p. 219

Linear inequality, p. 227
Related equation, p. 227
Objective function, p. 236
Constraint, p. 236
Linear programming, p. 237
Feasible region, p. 237

IMPORTANT PROPERTIES AND FORMULAS

The Addition Principle for Inequalities

For any real numbers a, b, and c:

$a < b$ is equivalent to $a + c < b + c$;
$a > b$ is equivalent to $a + c > b + c$.

Similar statements hold for $\leq$ and $\geq$.

Set union:

$$A \cup B = \{x \mid x \text{ is in } A \text{ or in } B, \text{ or both}\}$$

For any real numbers a and b with $a < b$, $a < x$ and $x < b$ can be abbreviated $a < x < b$. Intersection corresponds to "and"; union corresponds to "or."

$|x| = x$ if $x \geq 0$; $|x| = -x$ if $x < 0$.

The Absolute Value Principles for Equations and Inequalities

For any positive number p and any algebraic expression X:

a) The solutions of $|X| = p$ are those numbers that satisfy $X = -p$ or $X = p$.
b) The solutions of $|X| < p$ are those numbers that satisfy $-p < X < p$.
c) The solutions of $|X| > p$ are those numbers that satisfy $X < -p$ or $p < X$.

If $|X| = 0$, then $X = 0$. If p is negative, then $|X| = p$ has no solution.

CUMULATIVE REVIEW: CHAPTERS 1–6   **373**

CUMULATIVE REVIEW
1-6

1. Evaluate
$$\frac{2x - y^2}{x + y}$$
for $x = 3$ and $y = -4$.

2. Convert to scientific notation: 5,760,000,000.

3. Determine the slope and the y-intercept for the line given by $7x - 4y = 12$.

4. Find an equation for the line that passes through the points $(-1, 7)$ and $(2, -3)$.

5. Solve the system
$$5x - 2y = -23,$$
$$3x + 4y = 7.$$

6. Solve the system
$$-3x + 4y + z = -5,$$
$$x - 3y - z = 6,$$
$$2x + 3y + 5z = -8.$$

7. Luigi's Pizzeria donated 45 pizzas for a charity event. Small pizzas sold for $7.00 each and large pizzas for $10.00 each. The total amount of funds raised from the sale of the pizzas was $402. How many of each size pizza were donated?

8. The sum of three numbers is 20. The first number is 3 less than twice the third number. The second number minus the third number is -7. What are the numbers?

9. If
$$f(x) = \frac{x - 2}{x - 5},$$
find **(a)** $f(3)$ and **(b)** the domain of f.

Solve.

10. $8x = 1 + 16x^2$

11. $625 = 49y^2$

12. $20 > 2 - 6x$

13. $\frac{1}{3}x - \frac{1}{5} \geq \frac{1}{5}x - \frac{1}{3}$

14. $-8 < x + 2 < 15$

15. $3x - 2 < -6 \text{ or } x + 3 > 9$

16. $|x| > 6.4$

17. $|4x - 1| \leq 14$

18. $\frac{2}{n} - \frac{7}{n} = 3$

19. $\frac{6}{x - 5} = \frac{2}{2x}$

20. $\frac{3x}{x - 2} - \frac{6}{x + 2} = \frac{24}{x^2 - 4}$

21. $\frac{3x^2}{x + 2} + \frac{5x - 22}{x - 2} = \frac{-48}{x^2 - 4}$

22. Let $f(x) = |3x - 5|$. Find all values of x for which $f(x) = 2$.

23. Write the domain of f using interval notation if $f(x) = \sqrt{x - 7}$.

24. Solve $5m - 3n = 4m + 12$ for n.

25. Solve $P = \frac{3a}{a + b}$ for a.

Graph on a plane.

26. $4x \geq 5y + 20$ **27.** $y < -2$

Perform the indicated operations and simplify.

28. $(2x^2 - 3x + 1) + (6x - 3x^3 + 7x^2 - 4)$

29. $(5x^3y^2)(-3xy^2)$

30. $(3a + b - 2c) - (-4b + 3c - 2a)$

31. $(5x^2 - 2x + 1)(3x^2 + x - 2)$

32. $(2x^2 - y)^2$

33. $(2x^2 - y)(2x^2 + y)$

34. $(-5m^3n^2 - 3mn^3) + (-4m^3n^2 + 4m^3n^2) - (2mn^3 - 3m^2n^2)$

35. $\frac{y^2 - 36}{2y + 8} \cdot \frac{y + 4}{y + 6}$

36. $\frac{x^4 - 1}{x^2 - x - 2} \div \frac{x^2 + 1}{x - 2}$

37. $\frac{5ab}{a^2 - b^2} + \frac{a + b}{a - b}$

38. $\frac{2}{m + 1} + \frac{3}{m - 5} - \frac{m^2 - 1}{m^2 - 4m - 5}$

39. $y - \frac{2}{3y}$

40. Simplify: $\dfrac{\frac{1}{x} - \frac{1}{y}}{x + y}$.

41. Divide: $(9x^3 + 5x^2 + 2) \div (x + 2)$.

• After every three chapters, and at the end of the text, the authors include a **Cumulative Review**, which reviews skills and concepts from all preceding chapters of the text.

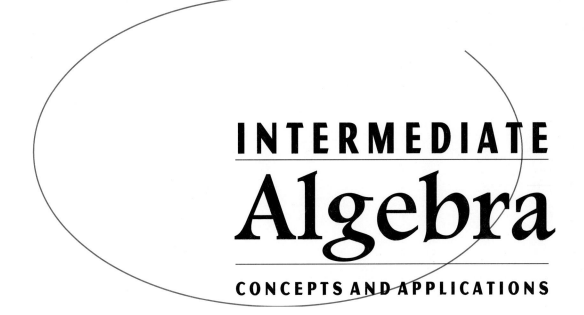

INTERMEDIATE
Algebra
CONCEPTS AND APPLICATIONS

Algebra and Problem Solving

AN APPLICATION

A jeweler suspects that a "gold" medal is not solid gold. The density of gold is 19.3 grams per cubic centimeter (g/cm^3) and the medal is 0.5 cm thick with a radius of 3 cm. If the medal is really gold, how much should it weigh?

THIS PROBLEM APPEARS AS EXAMPLE 5 IN SECTION 1.5.

More information on gold is available at **http://hepg.awl.com/be/inter_5**

All that glitters is not gold. Some things that appear to be gold are just base metal. A knowledge of basic mathematics will help a person be a more educated consumer of precious metals.

MARK M. COHEN
Gold Refiner
Philadelphia, PA

The principal theme of this text is problem solving in algebra. An overall strategy for solving problems is presented in Section 1.4. Additional and increasing emphasis on problem solving appears throughout the book.

This chapter begins with a short review of algebraic symbolism and properties of numbers. As you will see, the manipulations of algebra, such as simplifying expressions and solving equations, are based on the properties of numbers.

1.1   Some Basics of Algebra

Algebraic Expressions and Their Use • Translating to Algebraic Expressions • Evaluating Algebraic Expressions • Sets of Numbers

This section introduces some important concepts of algebra. We will examine different types of numbers and certain expressions that arise in problem solving.

Algebraic Expressions and Their Use

We are all familiar with expressions like

$$73 + 21, \qquad 18 \times 34, \qquad 9 - 5, \quad \text{and} \quad \frac{21}{34}.$$

In algebra, we use these as well as expressions like

$$x + 21, \qquad l \cdot w, \qquad 9 - s, \quad \text{and} \quad \frac{d}{t}.$$

When a letter is used to stand for various numbers, it is called a **variable**. If a letter represents one particular number, it is called a **constant**. Let d = the number of hours in a day. Then d is a constant. If t = the number of hours that a jet has been flying, then t is a variable since t changes as the flight progresses.

An **algebraic expression** consists of variables, numbers, and operation signs. All of the expressions above are examples of algebraic expressions. When an equals sign is placed between two expressions, an **equation** is formed.

Algebraic expressions and equations arise frequently in problem-solving situations. Suppose, for example, that we want to determine how much the price paid by recyclers for a ton of waste paper increased from 1994 to 1995.*

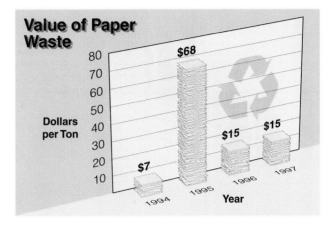

By using x to represent the increase in price, we can form an equation:

Price per ton of waste paper in 1994	plus	increase in price per ton	is	price per ton of waste paper in 1995
↓	↓	↓	↓	↓
7	+	x	=	68.

To find a **solution**, we can subtract 7 on both sides of the equation:

$$x = 68 - 7$$
$$x = 61.$$

We see that the price increased $61 per ton from 1994 to 1995.

Translating to Algebraic Expressions

To translate problems to equations, we need to know that certain words correspond to certain symbols:

KEY WORDS

Addition	Subtraction	Multiplication	Division
add	subtract	multiply	divide
sum	difference	product	divided by
plus	minus	times	quotient
increased by	decreased by	twice	ratio
more than	less than	of	per

*Source: Chittenden Solid Waste District, Williston VT, 1996 telephone interview.

Phrase	Algebraic Expression
Five *more than* some number	$n + 5$
Half *of* a number	$\dfrac{1}{2}t$ or $\dfrac{t}{2}$
Five *more than* three *times* some number	$3p + 5$
The *difference* of two numbers	$x - y$
Six *less than* the *product* of two numbers	$rs - 6$
Seventy-six percent *of* some number	$0.76z$ or $\dfrac{76}{100}z$

Note that expressions like rs represent products and can also be written as $r \cdot s$, $r \times s$, or $(r)(s)$. The multipliers r and s are also called **factors**.

EXAMPLE 1   Translate to an algebraic expression:

Five less than forty-three percent of the quotient of two numbers.

SOLUTION   We let r and s represent the two numbers.

$$(0.43) \cdot \frac{r}{s} - 5$$

Five less than forty-three percent of the quotient of two numbers

Evaluating Algebraic Expressions

When we replace a variable by a number, we say that we are **substituting** for the variable. This process is called **evaluating the expression.**

EXAMPLE 2   Evaluate the expression $3xy + z$ for $x = 2$, $y = 5$, and $z = 7$.

SOLUTION   We substitute and carry out the multiplication and addition:

$$\begin{aligned} 3xy + z &= 3 \cdot 2 \cdot 5 + 7 \\ &= 30 + 7 \\ &= 37. \end{aligned}$$

Geometric formulas are often evaluated. In the next example, we use the formula for the area A of a triangle with a base of length b and a height of length h:

$$A = \tfrac{1}{2} \cdot b \cdot h.$$

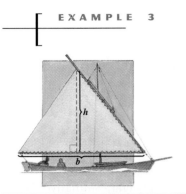

EXAMPLE 3   The base of a triangular sail is 8 m and the height is 6.4 m. Find the area of the sail.

SOLUTION   We substitute 8 for b and 6.4 for h and multiply:

$$\frac{1}{2} \cdot b \cdot h = \frac{1}{2} \cdot 8 \cdot 6.4$$
$$= 25.6 \text{ square meters (sq m)}.$$

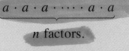

Before evaluating other algebraic expressions, we need to develop *exponential notation*. Many different kinds of numbers can be used as *exponents*. Here we establish the meaning of a^n when n is a counting number, 1, 2, 3, ...

EXPONENTIAL NOTATION

The expression a^n, in which n is a counting number, means

$$\underbrace{a \cdot a \cdot a \cdot \cdots \cdot a \cdot a}_{n \text{ factors.}}$$

In a^n, a is called the *base* and n is the *exponent,* or *power.*
The symbol a^1 means a.

The expression a^n is read "a raised to the nth power" or simply "a to the nth." We read s^2 as "s-squared" and x^3 as "x-cubed." This terminology comes from the fact that the area of a square of side s is $s \cdot s = s^2$ and the volume of a cube of side x is $x \cdot x \cdot x = x^3$.

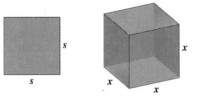

Exponential notation tells us that 5^2 means $5 \cdot 5$, or 25, but what does $1 + 2 \cdot 5^2$ mean? If we add 1 and 2 and multiply by 25, we get 75. If we multiply 2 times 25 and add 1, we get 51. A third possibility is to square $2 \cdot 5$ to get 100 and then add 1. The following convention indicates that only the second of these approaches is correct.

RULES FOR ORDER OF OPERATIONS

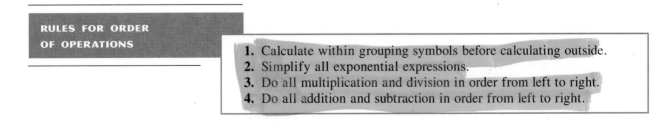

1. Calculate within grouping symbols before calculating outside.
2. Simplify all exponential expressions.
3. Do all multiplication and division in order from left to right.
4. Do all addition and subtraction in order from left to right.

E X A M P L E   4

Evaluate $5 + 2(a - 1)^2$ for $a = 4$.

S O L U T I O N

$$
\begin{aligned}
5 + 2(a - 1)^2 &= 5 + 2(4 - 1)^2 && \text{Substituting} \\
&= 5 + 2(3)^2 && \text{Working within parentheses first} \\
&= 5 + 2(9) && \text{Simplifying } 3^2 \\
&= 5 + 18 && \text{Multiplying} \\
&= 23 && \text{Adding}
\end{aligned}
$$

Step (3) in the rules for order of operations tells us to divide before we multiply when division appears first, reading left to right. Similarly, if subtraction appears before addition, we subtract before we add.

E X A M P L E   5

Evaluate $9 - x^3 + 6 \div 2y^2$ for $x = 2$ and $y = 5$.

S O L U T I O N

$$
\begin{aligned}
9 - x^3 + 6 \div 2y^2 &= 9 - 2^3 + 6 \div 2(5)^2 && \text{Substituting} \\
&= 9 - 8 + 6 \div 2 \cdot 25 && \text{Simplifying } 2^3 \text{ and } 5^2 \\
&= 9 - 8 + 3 \cdot 25 && \text{Dividing} \\
&= 9 - 8 + 75 && \text{Multiplying} \\
&= 1 + 75 && \text{Subtracting} \\
&= 76 && \text{Adding}
\end{aligned}
$$

Sets of Numbers

When evaluating algebraic expressions, and in problem solving in general, we often must examine the *type* of numbers used. For example, if a formula is used to determine an optimal class size, any fractional results must be rounded off, since it is impossible to have a fractional part of a student. Three frequently used sets of numbers are listed below.

NATURAL NUMBERS, WHOLE NUMBERS, AND INTEGERS

Natural Numbers (or Counting Numbers)
Those numbers used for counting: $\{1, 2, 3, \ldots\}$

Whole Numbers
The set of natural numbers with 0 included: $\{0, 1, 2, 3, \ldots\}$

Integers
The set of all whole numbers and their opposites:

$$\{\ldots, -4, -3, -2, -1, 0, 1, 2, 3, 4, \ldots\}$$

The dots mean that the pattern continues without end.

The integers correspond to the points on a number line as follows:

$$\begin{array}{c} \longleftarrow \bullet\;\bullet\;\bullet\;\bullet\;\bullet\;\bullet\;\bullet\;\bullet\;\bullet\;\bullet\;\bullet\;\bullet\;\bullet\;\bullet\;\bullet \longrightarrow \\ {\small -7\;\;-6\;\;-5\;\;-4\;\;-3\;\;-2\;\;-1\;\;\;0\;\;\;1\;\;\;2\;\;\;3\;\;\;4\;\;\;5\;\;\;6\;\;\;7} \end{array}$$

To fill in the rest of the points on our number line, we must describe two more sets of numbers. To do so, we must first discuss set notation.

The set containing the numbers -2, 1, and 3 can be written $\{-2, 1, 3\}$. This method of writing a set is known as **roster notation.** Roster notation was used for the sets listed above. A second type of set notation, **set-builder notation,** specifies conditions under which a number is in the set. The following example of set-builder notation is read as shown:

$$\{x \mid x \text{ is an odd number less than 5}\}$$

"The set of all x"   such that   "x is an odd number less than 5"

EXAMPLE 6   Using both roster notation and set-builder notation, name the set consisting of the first four even natural numbers.

SOLUTION

Using roster notation: $\{2, 4, 6, 8\}$

Using set-builder notation: $\{n \mid n \text{ is an even number between 1 and 9}\}$

The symbol $\in$ is used to indicate that an element belongs to a set. Thus if $A = \{2, 4, 6, 8\}$, we can write $4 \in A$ to indicate that 4 *is an element of A*. We can also write $5 \notin A$ to indicate that 5 *is not an element of A*.

EXAMPLE 7   Classify the statement $8 \in \{x \mid x \text{ is an integer}\}$ as true or false.

SOLUTION   Since 8 *is* an integer, the statement is true. In other words, since 8 is an integer, it belongs to the set of all integers.

With set-builder notation, we can describe the set of all *rational numbers*.

RATIONAL NUMBERS

Numbers that can be expressed as an integer divided by a nonzero integer are called *rational numbers*:

$$\left\{ \frac{p}{q} \mid p \text{ is an integer, } q \text{ is an integer, and } q \neq 0 \right\}.$$

Rational numbers can be written using fractional or decimal notation.

Fractional notation uses symbolism like the following:

$$\frac{5}{8}, \quad \frac{12}{-7}, \quad \frac{-17}{15}, \quad -\frac{9}{7}, \quad \frac{39}{1}, \quad \frac{0}{6}.$$

In *decimal notation,* rational numbers either *terminate* or *repeat.*

EXAMPLE 8   When written in decimal form, does each of the following numbers terminate or repeat? **(a)** $\frac{5}{8}$; **(b)** $\frac{6}{11}$.

SOLUTION

a) Since $\frac{5}{8}$ means $5 \div 8$, we perform long division to find that $\frac{5}{8} = 0.625$, a decimal that ends, or terminates. Thus $\frac{5}{8}$ can be written as a terminating decimal.

b) Using long division, we find that $6 \div 11 = 0.5454\ldots$, so we can write $\frac{6}{11}$ as a repeating decimal. Repeating decimal notation can be abbreviated by writing a bar over the repeating part—in this case, $0.\overline{54}$.

Many numbers, like π, $\sqrt{2}$, and $-\sqrt{15}$, can be only approximated by rational numbers. For example, $\sqrt{2}$ is the number for which $\sqrt{2} \cdot \sqrt{2} = 2$. A calculator's representation of $\sqrt{2}$ as 1.414213562 is an approximation since $(1.414213562)^2$ is not exactly 2.

To see that $\sqrt{2}$ is a "real" point on the number line, recall that when a right triangle has two legs of length 1 unit, the remaining side has length $\sqrt{2}$ units. Thus we can "measure" $\sqrt{2}$ and locate it precisely on a number line.

Numbers like π, $\sqrt{2}$, and $-\sqrt{15}$ are said to be **irrational**. Decimal notation for irrational numbers neither terminates nor repeats.

The set of all rational numbers, combined with the set of all irrational numbers, gives us the set of all **real numbers.**

REAL NUMBERS

Numbers that are either rational or irrational are called *real numbers*:

$$\{x \mid x \text{ is rational or } x \text{ is irrational}\}.$$

Every point on the number line represents some real number and every real number is represented by a point on the number line.

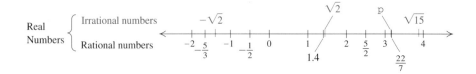

The following figure shows the relationships among various kinds of numbers.

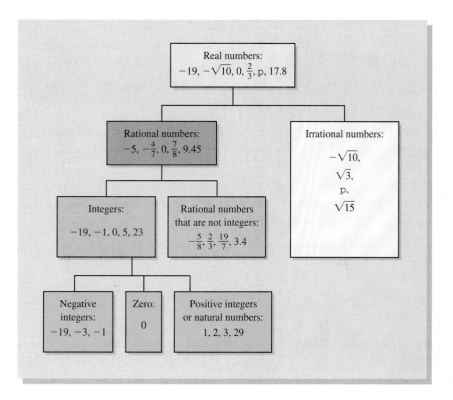

When all members of one set are found in a second set, the first set is a **subset** of the second set. Thus if $A = \{2, 4, 6\}$ and $B = \{1, 2, 4, 5, 6\}$, we write $A \subseteq B$ to indicate that *A is a subset of B*. Similarly, if $\mathbb{N}$ represents the set of all natural numbers and $\mathbb{Z}$ the set of all integers, we can write $\mathbb{N} \subseteq \mathbb{Z}$. Additional statements can be made using other sets in the diagram above.

EXERCISE SET
1.1

Use mathematical symbols to translate each phrase.

1. Seven more than some number

2. Two less than some number

3. Twelve times a number

4. Twice a number

5. Sixty-five percent of some number

6. Thirty-nine percent of some number

7. Nine less than twice a number

8. Four more than half of a number

9. Eight more than ten percent of some number

10. Five less than six percent of some number

11. One less than the difference of two numbers

12. Two more than the product of two numbers

13. Ninety miles per every four gallons of gas

14. One hundred words per every sixty seconds

Evaluate each expression using the values provided.

15. $4x - y$, for $x = 3$ and $y = 2$

16. $3a + b$, for $a = 5$ and $b = 4$

17. $2c \div 3b$, for $b = 4$ and $c = 6$

18. $3z \div 2y$, for $y = 1$ and $z = 6$

19. $25 - r^2 + s$, for $r = 3$ and $s = 7$

20. $n^3 - 2 + p$, for $n = 2$ and $p = 5$

21. $3n^2p + 2p^4$, for $n = 5$ and $p = 3$

22. $2a^3b - 2b^2$, for $a = 3$ and $b = 7$

23. $5x \div (2 + x - y)$, for $x = 6$ and $y = 2$

24. $3(m + 2n) \div m$, for $m = 7$ and $n = 0$

25. $29 - (a - b)^2$, for $a = 7$ and $b = 2$

26. $15 + (2x - y)^2$, for $x = 4$ and $y = 1$

27. $m + n(5 + n^2)$, for $m = 15$ and $n = 3$

28. $a^2 - 3(a - b)$, for $a = 7$ and $b = 4$

Find the area of a triangular window with the given base and height.

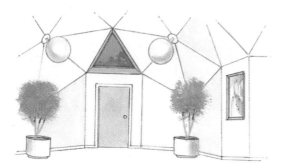

29. Base = 5 ft, height = 7 ft

30. Base = 2.9 m, height = 2.1 m

31. Base = 4 m, height = 3.2 m

32. Base = 4.6 ft, height = 4 ft

Use roster notation to name each set.

33. The set of all vowels in the alphabet

34. The set of all days of the week

35. The set of all odd natural numbers

36. The set of all even natural numbers

37. The set of all natural numbers that are multiples of 7

38. The set of all natural numbers that are multiples of 10

Use set-builder notation to name each set.

39. The set of all odd numbers between 10 and 30

40. The set of all multiples of 4 between 22 and 45

41. $\{0, 1, 2, 3, 4\}$

42. $\{-3, -2, -1, 0, 1, 2\}$

43. The set of all multiples of 5 between 7 and 79

44. The set of all even numbers between 9 and 99

Classify each statement as true or false. The following sets are used:

$\mathbb{N}$ = the set of natural numbers;

$\mathbb{W}$ = the set of whole numbers;

$\mathbb{Z}$ = the set of integers;

$\mathbb{Q}$ = the set of rational numbers;

$\mathbb{H}$ = the set of irrational numbers;

$\mathbb{R}$ = the set of real numbers.

45. $7.3 \in \mathbb{N}$ **46.** $8 \in \mathbb{N}$

47. $\mathbb{N} \subseteq \mathbb{W}$ **48.** $\mathbb{W} \subseteq \mathbb{Z}$

49. $\sqrt{8} \in \mathbb{Q}$ **50.** $4.1 \in \mathbb{H}$

51. $\mathbb{H} \subseteq \mathbb{R}$ **52.** $\sqrt{10} \in \mathbb{R}$

53. $4.3 \notin \mathbb{Z}$ **54.** $\mathbb{Z} \nsubseteq \mathbb{N}$

55. $\mathbb{Q} \subseteq \mathbb{R}$ **56.** $\mathbb{Q} \subseteq \mathbb{Z}$

SYNTHESIS

To the student and the instructor: *The synthesis exercises in each section are designed to challenge students to extend the concepts or skills studied in that section. Some synthesis exercises will require the assimilation of skills and concepts from several sections.*

Writing exercises, denoted by ◈ *, should be answered using one or more complete English sentences. In nearly every section, several writing exercises appear as the first synthesis exercises. These exercises are not as challenging as those exercises appearing later in the exercise set and can be assigned to students who might not otherwise attempt synthesis exercises. Because many writing exercises have a variety of correct answers, solutions are not listed in the answer section.*

57. ◈ On a quiz, Francesca answers $6 \in \mathbb{Z}$ while Jacob writes $\{6\} \in \mathbb{Z}$. Jacob's answer does not receive full credit while Francesca's does. Why?

58. ◈ Explain the difference between rational numbers and irrational numbers.

59. ◈ What advantage does set-builder notation have over roster notation?

60. ◈ Werner insists that $15 - 4 + 1 \div 2 \cdot 3$ is 2. What error is he making?

Translate to an algebraic expression.

61. The product of the sum of two numbers and their difference

62. Three times the sum of two numbers

63. Half of the difference of two numbers

64. The quotient of the difference of two numbers and their sum

Use roster notation to write each set.

65. The set of all whole numbers that are not natural numbers

66. The set of all integers that are not whole numbers

67. $\{x \mid x = 5n, n \text{ is a natural number}\}$

68. $\{x \mid x = 3n, n \text{ is a natural number}\}$

69. $\{x \mid x = 2n, n \text{ is an integer}\}$

70. $\{x \mid x = 2n + 1, n \text{ is a whole number}\}$

71. Draw a right triangle that could be used to measure $\sqrt{13}$ units.

1.2 Operations and Properties of Real Numbers

Absolute Value • Inequalities • Addition, Subtraction, and Opposites • Multiplication, Division, and Reciprocals • The Commutative, Associative, and Distributive Laws

In this section, we review how real numbers are added, subtracted, multiplied, and divided. We also study important rules for manipulating algebraic expressions.

The following discussions of absolute value and inequalities will help us when we begin to add real numbers.

Absolute Value

It is convenient to have a notation that represents a number's distance from zero on the number line.

ABSOLUTE VALUE

We write $|a|$, read "the absolute value of a," to represent the number of units that a is from zero.

EXAMPLE 1   Find the absolute value: **(a)** $|-3|$; **(b)** $|2.5|$; **(c)** $|0|$.

SOLUTION

a) $|-3| = 3$ -3 is 3 units from 0.

b) $|2.5| = 2.5$ 2.5 is 2.5 units from 0.

c) $|0| = 0$ 0 is 0 units from itself.

Note that whereas the absolute value of a nonnegative number is the number itself, the absolute value of a negative number is its opposite.

Inequalities

We need a notation to indicate how any two real numbers compare with each other. For any two numbers on the number line, the one to the left is said to be less than, or smaller than, the one to the right. The symbol $<$ means "is less than" and the symbol $>$ means "is greater than." The symbol $\leq$ means "is less than or equal to" and the symbol $\geq$ means "is greater than or equal to." These symbols are used to form **inequalities**.

In the figure below, note that although $|-6| > |-1|$, we have $-6 < -1$ since -6 is to the left of -1.

EXAMPLE   2

Write out the meaning of each inequality and determine whether it is a true statement: **(a)** $-7 < -2$; **(b)** $4 > -1$; **(c)** $-3 \geq -2$; **(d)** $5 \leq 6$; **(e)** $6 \leq 6$.

SOLUTION

Inequality	Meaning
a) $-7 < -2$	"-7 is less than -2," a true statement since -7 is to the left of -2.
b) $4 > -1$	"4 is greater than -1," a true statement.
c) $-3 \geq -2$	"-3 is greater than or equal to -2," a false statement since -3 is to the left of -2.
d) $5 \leq 6$	"5 is less than or equal to 6." Because $5 < 6$ is true, $5 \leq 6$ is true.
e) $6 \leq 6$	"6 is less than or equal to 6." Because $6 = 6$ is true, $6 \leq 6$ is true.

Addition, Subtraction, and Opposites

We are now ready to review the addition of real numbers.

ADDITION OF TWO REAL NUMBERS

1. *Positive numbers*: Add the numbers. The result is positive.
2. *Negative numbers*: Add absolute values. Make the answer negative.
3. *A negative and a positive number*: If the numbers have the same absolute value, the answer is 0. Otherwise, subtract the smaller absolute value from the larger one:
   a) If the positive number is further from 0, make the answer positive.
   b) If the negative number is further from 0, make the answer negative.
4. *One number is zero*: The sum is the other number.

EXAMPLE 3

Add: **(a)** $-9 + (-5)$; **(b)** $-3.2 + 9.7$; **(c)** $-\frac{3}{4} + \frac{1}{3}$.

SOLUTION

a) $-9 + (-5)$ — We add the absolute values, getting 14. The answer is *negative*, -14.

b) $-3.2 + 9.7$ — The absolute values are 3.2 and 9.7. Subtract 3.2 from 9.7 to get 6.5. The positive number is further from 0, so the answer is *positive*, 6.5.

c) $-\frac{3}{4} + \frac{1}{3} = -\frac{9}{12} + \frac{4}{12}$ — The absolute values are $\frac{9}{12}$ and $\frac{4}{12}$. Subtract to get $\frac{5}{12}$. The negative number is further from 0, so the answer is *negative*, $-\frac{5}{12}$.

When numbers like 7 and -7 are added, the result is 0. Such numbers are called **opposites**, or **additive inverses,** of one another.

THE LAW OF OPPOSITES

For any two numbers a and $-a$,

$$a + (-a) = 0.$$

(When opposites are added, their sum is 0.)

EXAMPLE 4

Find the opposite: **(a)** -17.5; **(b)** $\frac{4}{5}$; **(c)** 0.

SOLUTION

a) The opposite of -17.5 is 17.5 because $-17.5 + 17.5 = 0$.

b) The opposite of $\frac{4}{5}$ is $-\frac{4}{5}$ because $\frac{4}{5} + \left(-\frac{4}{5}\right) = 0$.

c) The opposite of 0 is 0 because $0 + 0 = 0$.

To name the opposite, we use the symbol "$-$" and read the symbolism $-a$ as "the opposite of a."

Note that $-a$ does not necessarily denote a negative number. In fact, when a is *negative*, $-a$ is *positive*.

EXAMPLE 5

Find $-x$ for the following: **(a)** $x = -2$; **(b)** $x = \frac{3}{4}$.

SOLUTION

a) If $x = -2$, then $-x = -(-2) = 2$.   The opposite of -2 is 2.

b) If $x = \frac{3}{4}$, then $-x = -\frac{3}{4}$.   The opposite of $\frac{3}{4}$ is $-\frac{3}{4}$.

Using the notation of opposites, we can formally define absolute value.

ABSOLUTE VALUE

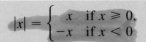

$$|x| = \begin{cases} x & \text{if } x \geq 0, \\ -x & \text{if } x < 0 \end{cases}$$

(When x is nonnegative, the absolute value of x is x. When x is negative, the absolute value of x is the opposite of x. Thus, $|x|$ is never negative.)

A negative number is said to have a negative "sign" and a positive number a positive "sign." To subtract, we can add an opposite. Thus we sometimes say that we "change the sign of the number being subtracted and then add."

EXAMPLE 6   Subtract: **(a)** $5 - 9$; **(b)** $-1.2 - (-3.7)$; **(c)** $-\frac{4}{5} - \frac{2}{3}$.

SOLUTION

a) $5 - 9 = 5 + (-9)$ Change the sign and add.
$\qquad\quad = -4$

b) $-1.2 - (-3.7) = -1.2 + 3.7$ Instead of *subtracting* -3.7, we *add* 3.7.
$\qquad\qquad\qquad\quad = 2.5$

c) $-\frac{4}{5} - \frac{2}{3} = -\frac{4}{5} + \left(-\frac{2}{3}\right)$
$\qquad\qquad = -\frac{12}{15} + \left(-\frac{10}{15}\right)$ Finding a common denominator
$\qquad\qquad = -\frac{22}{15}$

Multiplication, Division, and Reciprocals

Multiplication of real numbers can be regarded as repeated addition or as repeated subtraction that begins at 0. For example,

$$3 \cdot (-2) = 0 + (-2) + (-2) + (-2) = -6$$

and

$$(-2)(-5) = 0 - (-5) - (-5) = 0 + 5 + 5 = 10.$$

MULTIPLICATION OF TWO REAL NUMBERS

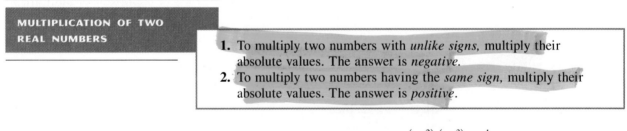

1. To multiply two numbers with *unlike signs*, multiply their absolute values. The answer is *negative*.
2. To multiply two numbers having the *same sign*, multiply their absolute values. The answer is *positive*.

Thus we have $(-4)9 = -36$ and $\left(-\frac{2}{3}\right)\left(-\frac{3}{8}\right) = \frac{1}{4}$.

To divide, recall that the quotient $a \div b$ (also denoted a/b) is that number c for which $c \cdot b = a$. For example, $10 \div (-2) = -5$ since $(-5)(-2) = 10$;

$(-12) \div 3 = -4$ since $(-4)3 = -12$; and $-18 \div (-6) = 3$ since $3(-6) = -18$. Thus the rules for division are just like those for multiplication.

DIVISION OF TWO REAL NUMBERS

1. To divide two numbers with *unlike signs,* divide their absolute values. The answer is *negative.*
2. To divide two numbers having the *same sign,* divide their absolute values. The answer is *positive.*

Thus we have

$$\frac{-45}{-15} = 3 \quad \text{and} \quad \frac{20}{-4} = -5.$$

Note that since

$$\frac{-8}{2} = \frac{8}{-2} = -\frac{8}{2} = -4,$$

we have the following generalization.

THE SIGN OF A FRACTION

For any number a and any nonzero number b,

$$\frac{-a}{b} = \frac{a}{-b} = -\frac{a}{b}.$$

Recall that

$$\frac{a}{b} = \frac{a}{1} \cdot \frac{1}{b} = a \cdot \frac{1}{b}.$$

That is, if we prefer, we can multiply by $1/b$ rather than divide by b. Provided that b is not 0, the numbers b and $1/b$ are called **reciprocals,** or **multiplicative inverses,** of each other.

THE LAW OF RECIPROCALS

For any two numbers a and $1/a$ ($a \neq 0$),

$$a \cdot \frac{1}{a} = 1.$$

(When reciprocals are multiplied, their product is 1.)

EXAMPLE 7   Find the reciprocal: **(a)** $\frac{7}{8}$; **(b)** $-\frac{3}{4}$; **(c)** -8.

SOLUTION

a) The reciprocal of $\frac{7}{8}$ is $\frac{8}{7}$ because $\frac{7}{8} \cdot \frac{8}{7} = 1$.

b) The reciprocal of $-\frac{3}{4}$ is $-\frac{4}{3}$.

c) The reciprocal of -8 is $\frac{1}{-8}$, or $-\frac{1}{8}$.

To divide, we can multiply by a reciprocal. We sometimes say that we "invert and multiply."

EXAMPLE 8   Divide: **(a)** $-\frac{1}{4} \div \frac{3}{5}$; **(b)** $-\frac{6}{7} \div (-10)$.

SOLUTION

a) $-\frac{1}{4} \div \frac{3}{5} = -\frac{1}{4} \cdot \frac{5}{3}$ "Inverting" $\frac{3}{5}$ and changing division to multiplication

$= -\frac{5}{12}$

b) $-\frac{6}{7} \div (-10) = -\frac{6}{7} \cdot \left(-\frac{1}{10}\right) = \frac{6}{70}$, or $\frac{3}{35}$

Thus far, we have never divided by 0 or, equivalently, had a denominator of 0. There is a reason for this. Suppose 5 were divided by 0. The answer would have to be a number that, when multiplied by 0, gave 5. But any number times 0 is 0. Thus we cannot divide 5 or any other nonzero number by 0.

What if we divide 0 by 0? In this case, our solution would need to be some number that, when multiplied by 0, gave 0. But then *any* number would work as a solution to $0 \div 0$. This could lead to contradictions so we agree to exclude division of 0 by 0 also.

DIVISION BY ZERO

We never divide by 0. If asked to divide a nonzero number by 0, we say that the answer is *undefined*. If asked to divide 0 by 0, we say that the answer is *indeterminate*.

The rules for order of operations discussed in Section 1.1 apply to *all* real numbers, regardless of their signs.

EXAMPLE 9   Simplify: $7 - 5^2 + 6 \div 2(-5)^2$.

SOLUTION

$$7 - 5^2 + 6 \div 2(-5)^2 = 7 - 25 + 6 \div 2 \cdot 25 \qquad \text{Simplifying } 5^2 \text{ and } (-5)^2$$
$$= 7 - 25 + 3 \cdot 25 \qquad \text{Dividing}$$
$$= 7 - 25 + 75 \qquad \text{Multiplying}$$
$$= -18 + 75 \qquad \text{Subtracting}$$
$$= 57 \qquad \text{Adding}$$

Besides the usual grouping symbols—parentheses, brackets, and braces—a fraction bar, absolute value symbol, or radical sign ($\sqrt{}$) may indicate groupings.

EXAMPLE 10

Calculate: $\dfrac{12|7 - 9| + 4 \cdot 5}{(-3)^4 + 2^3}$.

SOLUTION   We simplify the numerator and the denominator and divide the results:

$$\frac{12|7 - 9| + 4 \cdot 5}{(-3)^4 + 2^3} = \frac{12|-2| + 4 \cdot 5}{81 + 8}$$

$$= \frac{12(2) + 20}{89}$$

$$= \frac{44}{89}. \qquad \text{Multiplying and adding}$$

The Commutative, Associative, and Distributive Laws

As was the case with the numbers used in arithmetic, when a pair of real numbers are added or multiplied, the order in which the numbers are written does not affect the result.

THE COMMUTATIVE LAWS

For any real numbers a and b,

$a + b = b + a$; $a \cdot b = b \cdot a$.
(for Addition)   (for Multiplication)

The commutative laws provide one way of writing *equivalent expressions*.

EQUIVALENT EXPRESSIONS

Two expressions that have the same value for all possible replacements are called *equivalent expressions*.

EXAMPLE 11

Use a commutative law to write an expression equivalent to $7x + 9$.

SOLUTION   Using the commutative law of addition, we have

$7x + 9 = 9 + 7x$.

We can also use the commutative law of multiplication to write

$7x + 9 = x7 + 9$.

The expressions $7x + 9$, $9 + 7x$, and $x7 + 9$ are all equivalent. They name the same number for any replacement of x.

The *associative laws* also enable us to form equivalent expressions.

THE ASSOCIATIVE LAWS

For any real numbers a, b, and c,

$$a + (b + c) = (a + b) + c; \qquad a \cdot (b \cdot c) = (a \cdot b) \cdot c.$$
(for Addition) $\qquad\qquad\qquad$ (for Multiplication)

EXAMPLE 12   Write an expression equivalent to $(3x + 7y) + 9z$, using the associative law of addition.

SOLUTION   We have

$$(3x + 7y) + 9z = 3x + (7y + 9z).$$

The expressions $(3x + 7y) + 9z$ and $3x + (7y + 9z)$ are equivalent. They name the same number for any replacements of x, y, and z.

EXAMPLE 13   Use the commutative and associative laws to write an expression equivalent to

$$\frac{5}{x} \cdot (yz).$$

SOLUTION   Answers may vary. We use the associative law first, then the commutative law.

$$\frac{5}{x} \cdot (yz) = \left(\frac{5}{x} \cdot y \right) \cdot z \qquad \text{Using the associative law of multiplication}$$

$$= \left(y \cdot \frac{5}{x} \right) \cdot z \qquad \text{Using the commutative law in the parentheses}$$

The *distributive law* that follows provides another way of forming equivalent expressions. In essence, the distributive law allows us to rewrite the *product* of a and $b + c$ as the *sum* of ab and ac.

THE DISTRIBUTIVE LAW

For any numbers a, b, and c,

$$a(b + c) = ab + ac.$$

EXAMPLE 14   Obtain an expression equivalent to $5x(y + 4)$ by multiplying.

SOLUTION   We use the distributive law to get

$$5x(y + 4) = 5xy + 5x \cdot 4 \qquad \text{Using the distributive law}$$

$$= 5xy + 5 \cdot 4 \cdot x \qquad \text{Using the commutative law of multiplication}$$

$$= 5xy + 20x. \qquad \text{Simplifying}$$

The expressions $5x(y + 4)$ and $5xy + 20x$ are equivalent. They name the same number for any replacements of x and y.   ⬤

When we do the opposite of what we did in Example 14, we say that we are **factoring** an expression. This allows us to rewrite a sum as a product.

EXAMPLE 15   Obtain an expression equivalent to $3x - 6$ by factoring.

SOLUTION   We use the distributive law to get

$$3x - 6 = 3(x - 2).$$ ⬤

In Example 15, since the product of 3 and $x - 2$ is $3x - 6$, we say that 3 and $x - 2$ are **factors** of $3x - 6$. Thus "factor" can act as a noun or as a verb.

EXERCISE SET 1.2

Find each absolute value.

1. $|-8|$
2. $|-7|$
3. $|9|$
4. $|12|$
5. $|-6.2|$
6. $|-7.9|$
7. $|0|$
8. $\left|3\frac{3}{4}\right|$
9. $\left|1\frac{7}{8}\right|$
10. $|0.91|$
11. $|-4.21|$
12. $|-5.309|$

Write the meaning of each inequality, and determine whether it is a true statement.

13. $-8 \leqslant -2$
14. $-1 \leqslant -5$
15. $-7 > 1$
16. $7 \geqslant -2$
17. $3 \geqslant -5$
18. $9 \leqslant 9$
19. $-9 < -4$
20. $7 \geqslant -8$
21. $-4 \geqslant -4$
22. $2 < 2$
23. $-5 < -5$
24. $-2 > -12$

Add.

25. $5 + 12$
26. $9 + 7$
27. $-4 + (-7)$
28. $-8 + (-3)$
29. $-5.9 + 2.7$
30. $-1.9 + 7.3$
31. $\frac{2}{7} + \left(-\frac{3}{5}\right)$
32. $\frac{3}{8} + \left(-\frac{2}{5}\right)$
33. $-4.9 + (-3.6)$
34. $-2.1 + (-7.5)$
35. $-\frac{1}{9} + \frac{2}{3}$
36. $-\frac{1}{2} + \frac{4}{5}$

37. $0 + (-4.5)$
38. $-3.19 + 0$
39. $-7.24 + 7.24$
40. $-9.46 + 9.46$
41. $15.9 + (-22.3)$
42. $21.7 + (-28.3)$

Find the opposite, or additive inverse.

43. 7.29
44. 5.43
45. $-4\frac{1}{3}$
46. $2\frac{3}{5}$
47. 0
48. $-2\frac{3}{4}$

Find $-x$ for each of the following.

49. $x = 7$
50. $x = 3$
51. $x = -2.7$
52. $x = -1.9$
53. $x = 1.79$
54. $x = 3.14$
55. $x = 0$
56. $x = -1$

Subtract.

57. $9 - 7$
58. $8 - 3$
59. $4 - 9$
60. $3 - 10$
61. $-6 - (-10)$
62. $-3 - (-9)$
63. $-4 - 13$
64. $-7 - 8$
65. $2.7 - 5.8$
66. $3.7 - 4.2$
67. $-\frac{3}{5} - \frac{1}{2}$
68. $-\frac{2}{3} - \frac{1}{5}$
69. $-3.9 - (-6.8)$
70. $-5.4 - (-4.3)$
71. $0 - (-7.9)$
72. $0 - 5.3$

Multiply.

73. $(-4)7$

74. $(-5)9$

75. $(-3)(-8)$

76. $(-7)(-8)$

77. $(4.2)(-5)$

78. $(3.5)(-8)$

79. $\frac{3}{7}(-1)$

80. $-1 \cdot \frac{2}{5}$

81. $(-17.45) \cdot 0$

82. 15.2×0

83. $(-3.2) \times (-1.7)$

84. $(1.9) \cdot (4.3)$

Divide.

85. $\frac{-10}{-2}$

86. $\frac{-15}{-3}$

87. $\frac{-100}{20}$

88. $\frac{-50}{5}$

89. $\frac{73}{-1}$

90. $\frac{-62}{1}$

91. $\frac{0}{-7}$

92. $\frac{0}{-11}$

Find the reciprocal, or multiplicative inverse.

93. 5

94. 3

95. -9

96. -7

97. $\frac{2}{3}$

98. $\frac{4}{7}$

99. $-\frac{3}{11}$

100. $-\frac{7}{3}$

Divide.

101. $\frac{2}{3} \div \frac{4}{5}$

102. $\frac{2}{7} \div \frac{6}{5}$

103. $-\frac{3}{5} \div \frac{1}{2}$

104. $\left(-\frac{4}{7}\right) \div \frac{1}{3}$

105. $\left(-\frac{2}{9}\right) \div (-8)$

106. $\left(-\frac{2}{11}\right) \div (-6)$

107. $\frac{12}{7} \div (-1)$

108. $\left(-\frac{2}{7}\right) \div (-1)$

Calculate using the rules for order of operations.

109. $12 - (9 - 3 \cdot 2^3)$

110. $19 - (4 + 2 \cdot 3^2)$

111. $\dfrac{5 \cdot 2 - 4^2}{27 - 2^4}$

112. $\dfrac{7 \cdot 3 - 5^2}{9 + 4 \cdot 2}$

113. $\dfrac{3^4 - (5 - 3)^4}{1 - 2^3}$

114. $\dfrac{4^3 - (7 - 4)^2}{3^2 - 7}$

115. $5^3 - [2(4^2 - 3^2 - 6)]^3$

116. $7^2 - [3(5^2 - 4^2 - 7)]^2$

117. $|2^2 - 7|^3 + 1$

118. $|-2 - 3| \cdot 4^2 - 1$

119. $30 - (-5)^2 + 15 \div (-3) \cdot 2$

120. $55 - (-9 + 2)^2 + 18 \div 6 \cdot (-2)$

121. $12 - \sqrt{7 - 3} + 4 \div 3 \cdot 2^3$

122. $15 - 1 + \sqrt{5 + 2^2} \div 6 + (-1)^3$

Write an equivalent expression using a commutative law. Answers may vary.

123. $3x + 8y$

124. $ab + 9$

125. $(7x)y$

126. $-9(ab)$

Write an equivalent expression using an associative law.

127. $(3x)y$

128. $-7(ab)$

129. $x + (2y + 5)$

130. $(3y + 4) + 10$

Write an equivalent expression using the distributive law.

131. $3(a + 1)$

132. $8(x + 1)$

133. $4(x - y)$

134. $9(a - b)$

135. $-5(2a + 3b)$

136. $-2(3c + 5d)$

137. $2a(b - c + d)$

138. $5x(y - z + w)$

Find an equivalent expression by factoring.

139. $5x + 5y$

140. $7a + 7b$

141. $3p - 9$

142. $12x - 3$

143. $7x - 21y$

144. $6y - 9$

145. $2x - 2y + 2z$

146. $3x + 3y - 3z$

SKILL MAINTENANCE

To the student and the instructor: *Exercises included for Skill Maintenance review skills previously studied in the text. You can expect such exercises in every exercise set. Often these serve as preparation for an upcoming section.*

147. Translate to an algebraic expression:

Five less than seventy percent of a number.

148. Translate to an algebraic expression:

Two more than half a number.

SYNTHESIS

149. ◈ Describe in your own words a method for determining the sign of the sum of a positive number and a negative number.

150. ◈ What is the difference between the associative law of multiplication and the distributive law?

151. ◈ Explain in your own words why 7/0 is undefined.

152. ◈ Write a sentence in which the word "factor" appears once as a verb and once as a noun.

Insert one pair of parentheses to convert each false statement into a true statement.

153. $3 - 8^2 + 9 = 34$

154. $2 \cdot 7 + 3^2 \cdot 5 = 104$

155. $5 \cdot 2^3 \div 3 - 4^4 = 40$

156. $2 - 7 \cdot 2^2 + 9 = -11$

157. Find the greatest value of a for which $|a| \geqslant 6.2$ and $a < 0$.

158. Use the commutative, associative, and distributive laws to show that $5(a + bc)$ is equivalent to $c(b5) + a5$. Use only one law in each step of your work.

1.3   Solving Equations

Equivalent Equations   •   The Addition and Multiplication Principles   •
Combining Like Terms   •   Types of Equations

Solving equations is essential for problem solving in algebra. In this section, we review and practice solving simple equations.

Equivalent Equations

In Section 1.1, we saw that the solution of $7 + x = 68$ is 61. That is, when x is replaced by 61, the equation $7 + x = 68$ is a true statement. Although this solution may seem obvious, it is important to know how to find such a solution using the principles of algebra. These principles can produce *equivalent equations* from which solutions can be easily found.

EQUIVALENT EQUATIONS

Two equations are said to be *equivalent* if they have the same solution(s).

EXAMPLE 1   Determine whether $2x = 6$ and $10x = 30$ are equivalent equations.

SOLUTION   The equation $2x = 6$ is true only when x is 3. Similarly, $10x = 30$ is true only when x is 3. Since both equations have the same solution, they are equivalent.

EXAMPLE 2   Determine whether $x + 4 = 9$ and $x = 5$ are equivalent equations.

SOLUTION   Each equation has only one solution, the number 5. Thus the equations are equivalent.

EXAMPLE 3   Determine whether $3x = 4x$ and $3/x = 4/x$ are equivalent equations.

SOLUTION   When x is replaced by 0, neither $3/x$ nor $4/x$ is defined, so 0 is *not* a solution of $3/x = 4/x$. Since 0 *is* a solution of $3x = 4x$, the equations are not equivalent.

The Addition and Multiplication Principles

Suppose that a and b represent the same number and that some number c is added to a. If c is also added to b, we will get two equal sums, since a and b

are the same number. The same is true if we multiply both a and b by c. In this manner, we can produce equivalent equations.

THE ADDITION AND MULTIPLICATION PRINCIPLES FOR EQUATIONS

For any real numbers a, b, and c:
a) $a = b$ is equivalent to $a + c = b + c$;
b) $a = b$ is equivalent to $a \cdot c = b \cdot c$, provided $c \neq 0$.

EXAMPLE 4

Solve: $y - 4.7 = 13.9$.

SOLUTION

$$y - 4.7 = 13.9$$
$$y - 4.7 + 4.7 = 13.9 + 4.7 \qquad \text{Using the addition principle; adding 4.7}$$
$$y + 0 = 13.9 + 4.7 \qquad \text{The law of opposites}$$
$$y = 18.6$$

Check:

$$\frac{y - 4.7 = 13.9}{18.6 - 4.7 \ ? \ 13.9} \qquad \text{Substituting 18.6 for } y$$
$$13.9 \ | \ 13.9 \ \text{TRUE}$$

The solution is 18.6.

In Example 4, why did we add 4.7 on both sides? Because we wanted the variable y alone on one side of the equation. Adding 4.7 gave us $y + 0$, or just y, on the left side. This led to the equivalent equation, $y = 18.6$, from which the solution, 18.6, is immediately apparent.

EXAMPLE 5

Solve: $\frac{2}{5}x = \frac{9}{10}$.

SOLUTION

$$\frac{2}{5}x = \frac{9}{10}$$
$$\frac{5}{2} \cdot \frac{2}{5}x = \frac{5}{2} \cdot \frac{9}{10} \qquad \text{Using the multiplication principle, we multiply by } \frac{5}{2}, \text{ the reciprocal of } \frac{2}{5}.$$
$$1x = \frac{45}{20} \qquad \text{The law of reciprocals}$$
$$x = \frac{9}{4} \qquad \text{Simplifying}$$

The check is left to the student. The solution is $\frac{9}{4}$.

In Example 5, why did we multiply by $\frac{5}{2}$? Because we wanted x alone on one side of the equation. When we multiplied by $\frac{5}{2}$, we got $1x$, or just x, on the left side. This led to the equivalent equation $x = \frac{9}{4}$, from which the solution, $\frac{9}{4}$, was obvious.

There is no need for a subtraction or division principle since subtraction can be regarded as adding opposites and division can be regarded as multiplying by reciprocals.

Combining Like Terms

In an expression like $8a^5 + 17 + 4/b + (-6a^3b)$, the parts that are separated by addition signs are called *terms*. A **term** is a number, a variable, a product of numbers and/or variables, or a quotient of numbers and/or variables. Thus, $8a^5$, 17, $4/b$, and $-6a^3b$ are terms in $8a^5 + 17 + 4/b + (-6a^3b)$. When terms have variable factors that are exactly the same, we refer to those terms as **like**, or **similar**, **terms**. Thus, $3x^2y$ and $-7x^2y$ are similar terms, but $3x^2y$ and $4xy^2$ are not. We can often simplify expressions by **combining**, or **collecting, like terms.**

E X A M P L E   6   Combine like terms: $3a + 5a^2 + 4a + a^2$.

S O L U T I O N

$$3a + 5a^2 + 4a + a^2 = 3a + 4a + 5a^2 + a^2 \qquad \text{Using the commutative law}$$
$$= (3 + 4)a + (5 + 1)a^2 \qquad \text{Using the distributive law.}$$
$$\text{Note that } a^2 = 1a^2.$$
$$= 7a + 6a^2$$

Sometimes we must use the distributive law to remove grouping symbols before combining like terms.

E X A M P L E   7   Simplify: $3x + 2[4 + 5(x + 2y)]$.

S O L U T I O N

$$3x + 2[4 + 5(x + 2y)] = 3x + 2[4 + 5x + 10y] \qquad \text{Using the distributive law}$$
$$= 3x + 8 + 10x + 20y \qquad \text{Using the distributive law}$$
$$= 13x + 8 + 20y \qquad \text{Combining like terms}$$

The product of a number and -1 is its opposite, or additive inverse. For example,

$$-1 \cdot 8 = -8 \qquad \text{(the opposite of 8).}$$

Thus we have $-8 = -1 \cdot 8$, and in general, $-x = -1 \cdot x$. We can use this fact along with the distributive law when parentheses are preceded by a negative sign or subtraction.

E X A M P L E   8   Simplify $-(a - b)$, using multiplication by -1.

S O L U T I O N   We have

$$-(a - b) = -1 \cdot (a - b) \qquad \text{Replacing } - \text{ with multiplication by } -1$$
$$= -1 \cdot a - (-1) \cdot b \qquad \text{Using the distributive law}$$
$$= -a - (-b) \qquad \text{Replacing } -1 \cdot a \text{ by } -a \text{ and } (-1) \cdot b \text{ by } -b$$
$$= -a + b, \text{ or } b - a. \qquad \text{Try to go directly to this step.}$$

The expressions $-(a - b)$ and $b - a$ are equivalent. They name the same number for all replacements of a and b.

Example 8 illustrates a useful shortcut worth remembering:

The opposite of $a - b$ is $-a + b$, or $b - a$.

EXAMPLE 9

Simplify: $9x - 5y - (5x + y - 7)$.

SOLUTION

$$9x - 5y - (5x + y - 7) = 9x - 5y - 5x - y + 7 \qquad \text{Using the distributive law}$$
$$= 4x - 6y + 7 \qquad \text{Combining like terms}$$

In Examples 1–5, we studied equivalent *equations* whereas Examples 6–9 dealt with forming equivalent *expressions*. These ideas are merged when we form an equivalent equation by replacing part of an equation with an equivalent expression.

EXAMPLE 10

Solve: $5x - 2(x - 5) = 7x - 2$.

SOLUTION

$$5x - 2(x - 5) = 7x - 2$$
$$5x - 2x + 10 = 7x - 2 \qquad \text{Using the distributive law}$$
$$3x + 10 = 7x - 2 \qquad \text{Combining like terms}$$
$$3x + 10 - 3x = 7x - 2 - 3x \qquad \text{Using the addition principle; adding } -3x \text{ or subtracting } 3x \text{ on both sides}$$
$$10 = 4x - 2 \qquad \text{Combining like terms}$$
$$10 + 2 = 4x - 2 + 2 \qquad \text{Using the addition principle}$$
$$12 = 4x \qquad \text{Simplifying}$$
$$\tfrac{1}{4} \cdot 12 = \tfrac{1}{4} \cdot 4x \qquad \text{Using the multiplication principle; multiplying by } \tfrac{1}{4} \text{ or dividing by 4 on both sides}$$
$$3 = x \qquad \text{The law of reciprocals; simplifying}$$

Check:

$$\begin{array}{c|c}
\multicolumn{2}{c}{5x - 2(x - 5) = 7x - 2} \\
\hline
5 \cdot 3 - 2(3 - 5) \ ? \ 7 \cdot 3 - 2 & \\
15 - 2(-2) & 21 - 2 \\
15 + 4 & 19 \\
19 & 19 \qquad \text{TRUE}
\end{array}$$

The solution is 3.

Types of Equations

In Examples 4, 5, and 10, we solved *linear equations*. A **linear equation** in one variable—say, x—is an equation equivalent to one of the form $ax = b$ with a and b constants and $a \neq 0$.

Don't rush to solve linear equations in your head. Work neatly, keeping in mind that the number of steps in a solution is less important than producing a simpler, yet equivalent, equation in each step.

Any equation falls into one of three categories. An **identity** is an equation that is true for all replacements that can be used on both sides of the equation (for example, $x + 3 = 2 + x + 1$ or $3a - 5 = 3a - 5$). A **contradiction** is an equation, like $n + 5 = n + 7$, that is *never* true. A **conditional equation,** like $2x + 5 = 17$, is sometimes true and sometimes false, depending on what the replacement of x is. Most of the equations examined in this text are conditional.

EXAMPLE   11   Solve each of the following equations and classify the equation as an identity, a contradiction, or a conditional equation.

a) $2x + 7 = 7(x + 1) - 5x$
b) $3x - 5 = 3(x - 2) + 4$
c) $3 - 8x = 5 - 7x$

SOLUTION

a) $2x + 7 = 7(x + 1) - 5x$

$\qquad 2x + 7 = 7x + 7 - 5x \qquad$ Using the distributive law

$\qquad 2x + 7 = 2x + 7 \qquad$ Combining like terms

The equation $2x + 7 = 2x + 7$ is true regardless of what x is replaced by, so all real numbers are solutions. Note that $2x + 7 = 2x + 7$ is equivalent to $2x = 2x$, $7 = 7$, or $0 = 0$. All real numbers are solutions and the equation is an identity.

b)
$$3x - 5 = 3(x - 2) + 4$$
$$3x - 5 = 3x - 6 + 4 \qquad \text{Using the distributive law}$$
$$3x - 5 = 3x - 2$$
$$-3x + 3x - 5 = -3x + 3x - 2 \qquad \text{Using the addition principle}$$
$$-5 = -2$$

Since our original equation is equivalent to $-5 = -2$, which is false for any choice of x, there is no solution to this problem. There is no choice of x that will solve the original equation. The equation is a contradiction.

c)
$$3 - 8x = 5 - 7x$$
$$3 - 8x + 7x = 5 - 7x + 7x \qquad \text{Using the addition principle}$$
$$3 - x = 5 \qquad \text{Simplifying}$$
$$-3 + 3 - x = -3 + 5 \qquad \text{Using the addition principle}$$
$$-x = 2 \qquad \text{Simplifying}$$
$$x = \frac{2}{-1}, \text{ or } -2 \qquad \text{Dividing by } -1 \text{ or multiplying by } \frac{1}{-1} \text{ on both sides}$$

There is one solution, -2. For other choices of x, the equation is false. This equation is conditional since it can be true or false, depending on the replacement for x.

We will sometimes refer to the set of solutions, or **solution set,** of a particular equation. Thus the solution set for Example 11(c) is $\{-2\}$. The solution set for Example 11(a) is simply $\mathbb{R}$, the set of all real numbers, and the solution set for Example 11(b) is the **empty set,** denoted $\varnothing$ or $\{\ \}$. As its name suggests, the empty set is the set containing no elements.

EXERCISE SET

1.3

Determine whether the two equations in each pair are equivalent.

1. $3x = 15$ and $2x = 10$

2. $5x = 20$ and $15x = 60$

3. $x + 5 = 11$ and $3x = 18$

4. $x - 3 = 7$ and $3x = 24$

5. $13 - x = 4$ and $2x = 20$

6. $3x - 4 = 8$ and $3x = 12$

7. $5x = 2x$ and $\dfrac{4}{x} = 3$

8. $6 = 2x$ and $5 = \dfrac{2}{3 - x}$

Solve. Be sure to check.

9. $x - 5.2 = 9.4$ **10.** $y + 4.3 = 11.2$

11. $9y = 72$ **12.** $7x = 63$

13. $4x - 12 = 60$ **14.** $4x - 6 = 70$

15. $5y + 3 = 28$ **16.** $7t + 11 = 74$

17. $2y - 11 = 37$ **18.** $3x - 13 = 29$

Simplify by combining like terms. Use the distributive law as needed.

19. $4a + 5a$ **20.** $9x + 3x$

21. $7rt - 9rt$ **22.** $3ab + 7ab$

23. $8x^2 + x^2$ **24.** $7a^2 + a^2$

25. $12a - a$ **26.** $15x - x$

27. $t - 9t$ **28.** $x - 6x$

29. $5x - 3x + 8x$ **30.** $3x - 11x + 2x$

31. $5x - 2x^2 + 3x$ **32.** $9a - 5a^2 + 4a$

33. $3a + 5a^2 - a + 4a^2$

34. $9x + 2x^3 + 5x - 6x^2$

35. $4x - 7 + 18x + 25$

36. $13p + 5 - 4p + 7$

37. $-7t^2 + 3t + 5t^3 - t^3 + 2t^2 - t$

38. $-9n + 8n^2 + n^3 - 2n^2 - 3n + 4n^3$

39. $a - (2a + 5)$ **40.** $x - (5x + 9)$

41. $4m - (3m - 1)$ **42.** $5a - (4a - 3)$

43. $3d - 7 - (5 - 2d)$ **44.** $8x - 9 - (7 - 5x)$

45. $-2(x + 3) - 5(x - 4)$

46. $-9(y + 7) - 6(y - 3)$

47. $5x - 7(2x - 3)$

48. $8y - 4(5y - 6)$

49. $9a - [7 - 5(7a - 3)]$

50. $12b - [9 - 7(5b - 6)]$

51. $5\{-2a + 3[4 - 2(3a + 5)]\}$

52. $7\{-7x + 8[5 - 3(4x + 6)]\}$

53. $2y + \{7[3(2y - 5) - (8y + 7)] + 9\}$

54. $7b - \{6[4(3b - 7) - (9b + 10)] + 11\}$

Solve. Be sure to check.

55. $5x + 2x = 56$ **56.** $3x + 7x = 120$

57. $9y - 7y = 42$ **58.** $8t - 3t = 65$

59. $-6y - 10y = -32$ **60.** $-9y - 5y = 28$

61. $2(x + 6) = 8x$ **62.** $3(y + 5) = 8y$

63. $80 = 10(3t + 2)$ **64.** $27 = 9(5y - 2)$

65. $180(n - 2) = 900$ **66.** $210(x - 3) = 840$

67. $5y - (2y - 10) = 25$

68. $8x - (3x - 5) = 40$

69. $7y - 1 = 23 - 5y$

70. $15x + 20 = 8x - 22$

71. $\frac{1}{5} + \frac{3}{10}x = \frac{4}{5}$

72. $-\frac{5}{2}x + \frac{1}{2} = -18$

73. $0.9y - 0.7 = 4.2$

74. $0.8t - 0.3t = 6.5$

75. $5r - 2 + 3r = 2r + 6 - 4r$

76. $5m - 17 - 2m = 6m - 1 - m$

77. $\frac{1}{8}(16y + 8) - 17 = -\frac{1}{4}(8y - 16)$

78. $\frac{1}{6}(12t + 48) - 20 = -\frac{1}{8}(24t - 144)$

79. $5 + 2(x - 3) = 2[5 - 4(x + 2)]$

80. $3[2 - 4(x - 1)] = 3 - 4(x + 2)$

Find each solution set. Then classify each equation as a conditional equation, an identity, or a contradiction.

81. $4x - 2x - 2 = 2x$

82. $2x + 4 + x = 4 + 3x$

83. $2 + 9x = 3(3x + 1) - 1$

84. $4 + 7x = 7(x + 1)$

85. $-8x + 5 = 5 - 10x$

86. $-8x + 5 = 5 - 8x$

87. $2\{9 - 3[-2x - 4]\} = 12x + 42$

88. $3\{7 - 2[7x - 4]\} = -40x + 45$

SKILL MAINTENANCE

89. Write the set consisting of the positive integers less than 10, using both roster notation and set-builder notation.

90. Write the set consisting of the negative integers greater than -9, using both roster notation and set-builder notation.

SYNTHESIS

91. ◈ Explain the difference between equivalent expressions and equivalent equations.

92. ◈ Explain how the distributive and commutative laws can be used to rewrite $3x + 6y + 4x + 2y$ as $7x + 8y$.

93. ◈ Explain how an identity can be easily altered so that it becomes a contradiction.

94. ◈ As the first step in solving
$$2x + 5 = -3,$$
a student multiplies by $\frac{1}{2}$ on both sides. Is this incorrect? Why or why not?

Solve and check. The symbol ▦ indicates an exercise designed to be solved with a calculator.

95. ▦ $4.23x - 17.898 = -1.65x - 42.454$

96. ▦ $-0.00458y + 1.7787 = 13.002y - 1.005$

97. $x - \{3x - [2x - (5x - (7x - 1))]\} = x + 7$

98. $3x - \{5x - [7x - (4x - (3x + 1))]\} = 3x + 5$

99. $17 - 3\{5 + 2[x - 2]\} + 4\{x - 3(x + 7)\}$
$\qquad\qquad = 9\{x + 3[2 + 3(4 - x)]\}$

100. $23 - 2\{4 + 3[x - 1]\} + 5\{x - 2(x + 3)\}$
$\qquad\qquad = 7\{x - 2[5 - (2x + 3)]\}$

1.4 Introduction to Problem Solving

The Five-Step Strategy • Problem Solving

We now begin to study and practice the "art" of problem solving. Although we are interested mainly in using algebra to solve problems, much of what we say here applies to solving all kinds of problems.

What do we mean by a *problem*? Perhaps you've already used algebra to solve some "real-world" problems. What procedure did you use? Was there anything in your approach that could be used to solve problems of a more general nature? These are some questions that we will answer in this section.

In this text, we do not restrict the use of the word "problem" to computational situations involving arithmetic or algebra, such as $589 + 437 = a$ or $3x + 5x = 9$. We mean instead some question to which we wish to find an answer. Perhaps this can best be illustrated with some sample problems:

1. Can I afford to buy a new car?
2. If I exercise twice a week and eat 3000 calories a day, will I lose weight?
3. Do I have enough time to take 4 courses while working 20 hours a week?
4. My fishing boat travels 12 km/h in still water. How long will it take me to cruise 25 km upstream if the river's current is 3 km/h?

Although these problems are all different, there are some similarities. While there are no rules for problem solving, a general *strategy* can be used.

The Five-Step Strategy

Since you have already studied some algebra, you have had some experience with problem solving. The following steps constitute a strategy that you may already have used and a good strategy for problem solving in general.

FIVE STEPS FOR PROBLEM SOLVING WITH ALGEBRA

1. *Familiarize* yourself with the problem situation.
2. *Translate* to mathematical language.
3. *Carry out* some mathematical manipulation.
4. *Check* your possible answer in the original problem.
5. *State* the answer clearly.

Of the five steps, probably the most important is the first: becoming familiar with the problem situation. Here are some hints for familiarization.

THE FIRST STEP IN PROBLEM SOLVING WITH ALGEBRA

Familiarize yourself with a problem situation.

1. If a problem is given in words, read it carefully.
2. Reread the problem, perhaps aloud. Verbalize the problem to yourself.
3. List the information given and restate the question being asked. Select a variable or variables to represent any unknown(s) and clearly state what each variable represents. Be descriptive! For example, let t = time, in seconds; p = Paul's weight, in kilograms; and so on.
4. Find further information. Look up formulas or definitions with which you are not familiar. (Geometric formulas appear at the very end of this text.) Consult an expert in the field or a reference librarian.
5. Create a table using both variables and known information. Look for possible patterns.
6. Make and label a drawing.
7. Estimate or guess an answer and check to see if it's correct.

EXAMPLE 1

How might you familiarize yourself with the situation of Problem 1: "Can I afford to buy a new car?"

SOLUTION   Clearly more information is needed to solve this problem. You might:

a) Estimate the cost of various cars in which you are interested.
b) Examine what your savings are and how your income is budgeted.
c) Find out what payment plans are available.

When enough information is known, it might be wise to make a chart or table to help you reach an answer.

EXAMPLE 2

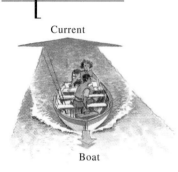

Current

Boat

How might you familiarize yourself with the situation of Problem 4: "How long will it take the boat to cruise 25 km upstream?"

SOLUTION   First read the question *very* carefully. This may even involve speaking aloud. You may need to reread the problem several times to fully understand what information is given and what information is required. A sketch is often helpful.

In this case, a table can be constructed to list all relevant information.

Distance to be Traveled	25 km
Speed of Boat in Still Water	12 km/h
Speed of Current	3 km/h
Speed of Boat Upstream	?
Time Required	?

As a next step in the familiarization process, we should determine, possibly with the aid of outside references, what relationships exist among the various quantities in the problem. With some effort it can be learned that the current's speed should be subtracted from the boat's speed in still water to determine the speed of the boat going upstream. We might consult a physics book to note that

Distance = Speed × Time.

We rewrite part of the table, letting $t =$ the time, in hours, required for the boat to cruise 25 km upstream.

Distance to be Traveled	25 km
Speed of Boat Upstream	$12 - 3 = 9$ km/h
Time Required	t

At this point we might try a guess. Suppose the boat traveled upstream for 2 hr. The boat would have then traveled

$$9 \quad \times \quad 2 \quad = \quad 18 \text{ km.}$$
$$\uparrow \qquad \uparrow \qquad\quad \uparrow$$
$$\text{Speed} \times \text{Time} = \text{Distance}$$

Since $18 \neq 25$, our guess is wrong. Still, examining how we checked our guess gives added insight into how the problem might translate to an equation. We might note that a better guess, when multiplied by 9, would yield a number closer to 25.

The second step in problem solving is to translate the situation to mathematical language. In algebra, this often means forming an equation.

THE SECOND STEP IN PROBLEM SOLVING WITH ALGEBRA

Translate the situation of the problem to mathematical language. In some cases, translation can be done by writing an algebraic expression, but most problems in this text are best solved by translating to an equation.

In the third step of our process, we work with the results of the first two steps. Often this will require us to use the algebra that we have studied.

THE THIRD STEP IN PROBLEM SOLVING WITH ALGEBRA

Carry out some mathematical manipulation. If you have translated to an equation, this means to solve the equation.

To properly complete the problem-solving process, we should always **check** our solution and then **state** the solution in a clear and precise manner. Normally our check consists of returning to the original problem and determining whether all its conditions have been satisfied. If our answer checks, we write a complete English sentence to state what the solution is. Our five steps are listed again below. Try to apply them regularly in your work in mathematics.

FIVE STEPS FOR PROBLEM SOLVING WITH ALGEBRA

1. *Familiarize* yourself with the problem situation.
2. *Translate* to mathematical language.
3. *Carry out* some mathematical manipulation.
4. *Check* your possible answer in the original problem.
5. *State* the answer clearly.

Problem Solving

At this point, our study of algebra is just beginning. Thus we have few algebraic tools with which to work problems. As the number of tools in our algebraic "toolbox" increases, so will the difficulty of the problems being solved. For now our problems may seem simple; however, to gain practice with the problem-solving process, you should try to use all five steps. Later some steps may be shortened or combined.

EXAMPLE 3

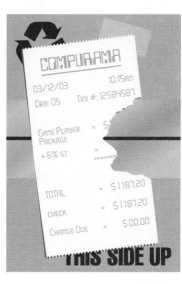

Purchasing. Elka pays $1187.20 for a computer. If the price paid includes a 6% sales tax, what is the price of the computer itself?

SOLUTION

1. **Familiarize.** Familiarize yourself with the problem. Note that tax is calculated from, and then added to, the computer's price. We let

   C = the computer's price.

   Let's guess that the computer's price is $1000. To check the guess, we calculate the amount of tax, $(0.06)(\$1000) = \$60$, and add it to $1000:

   $$(0.06)(\$1000) + \$1000 = \$60 + \$1000$$
   $$= \$1060. \qquad \$1060 \neq \$1187.20$$

   Our guess was wrong, but it was useful. The manner in which we checked the guess will guide us in the next step.

2. **Translate.** Translate the problem to mathematical language. Our guess leads us to the following translation:

6% of the computer's price	plus	the computer's price	is	the price with sales tax.
$\downarrow$	$\downarrow$	$\downarrow$	$\downarrow$	$\downarrow$
$(0.06)C$	$+$	C	$=$	$\$1187.20$

3. **Carry out.** Carry out some mathematical manipulation:

   $$0.06C + 1C = 1187.20$$
   $$1.06C = 1187.20 \qquad \text{Combining like terms}$$
   $$\frac{1}{1.06} \cdot 1.06C = \frac{1}{1.06} \cdot 1187.20 \qquad \text{Using the multiplication principle}$$
   $$C = 1120.$$

4. **Check.** Check the answer in the original problem. To do this, note that the tax on a computer costing $1120 would be $(0.06)(\$1120) = \$67.20$. When this is added to $1120, we have

   $67.20 + 1120, or $1187.20.

   We see that $1120 checks in the original problem.

5. **State.** State the answer clearly. The computer itself costs $1120.

EXAMPLE 4

Framing.   A piece of wood trim 100 in. long is to be cut into two pieces, and those pieces are each to be cut into four pieces to form a square frame. The length of a side of one square is to be $1\frac{1}{2}$ times the length of a side of the other. How should the wood be cut?

SOLUTION

1. **Familiarize.**   Note that the *perimeter* of (distance around) each square is four times the length of a side. Furthermore, if s is used to represent the length of a side of the smaller square, then $\left(1\frac{1}{2}\right)s$ will represent the length of a side of the larger square. Finally, note that the two perimeters must add up to 100 in.

   Perimeter of a square $= 4 \cdot$ length of a side

2. **Translate.**   Rewording the problem can help us translate:

   Rewording: The perimeter of one square   plus   the perimeter of the other   is   100 in.

   Translating: $4s$ $+$ $4\left(1\frac{1}{2}s\right)$ $=$ 100

3. **Carry out.**   We solve the equation:

   $4s + 4\left(1\frac{1}{2}s\right) = 100$

   $4s + 6s = 100$ Simplifying

   $10s = 100$ Combining like terms

   $s = \dfrac{1}{10} \cdot 100$ Multiplying by $\frac{1}{10}$ on both sides

   $s = 10.$ Simplifying

4. **Check.**   If 10 is the length of the smaller side, then $\left(1\frac{1}{2}\right)(10) = 15$ is the length of the larger side. The two perimeters would then be

   $4 \cdot 10 = 40$ and $4 \cdot 15 = 60.$

   Since $40 + 60 = 100$, our answer checks.

5. **State.**   The wood should be cut into two pieces, one 40 in. long and the other 60 in. long. Each piece should then be quartered to form the frames.

> 100 in.

s

$1\frac{1}{2}s$

EXAMPLE 5

Three numbers are such that the second is 6 less than 3 times the first and the third is 2 more than $\frac{2}{3}$ the first. The sum of the three numbers is 150. Find the largest of the three numbers.

SOLUTION   We proceed according to the five-step process.

1. **Familiarize.**   Three numbers are involved, and we want to find the largest. We list the information in a table, letting x represent the first number.

First Number	x
Second Number	6 less than 3 times the first
Third Number	2 more than $\frac{2}{3}$ the first

First + Second + Third = 150

Try to check a guess at this point. We will proceed to the next step.

2. **Translate.**   Because we wish to write an equation in just one variable, we need to express the second and third numbers using x ("in terms of x"). To do so, we expand the table:

First Number	x	x
Second Number	6 less than 3 times the first	$3x - 6$
Third Number	2 more than $\frac{2}{3}$ the first	$\frac{2}{3}x + 2$

We know that the sum is 150. Substituting, we obtain an equation:

$$\text{First} + \text{second} + \text{third} = 150.$$
$$x + (3x - 6) + \left(\tfrac{2}{3}x + 2\right) = 150$$

3. **Carry out.**   We solve the equation:

$$x + 3x - 6 + \tfrac{2}{3}x + 2 = 150 \qquad \text{Leaving off unnecessary parentheses}$$

$$\left(4 + \tfrac{2}{3}\right)x - 4 = 150 \qquad \text{Combining like terms}$$

$$\tfrac{14}{3}x - 4 = 150$$

$$\tfrac{14}{3}x = 154 \qquad \text{Adding 4 on both sides}$$

$$x = \tfrac{3}{14} \cdot 154 \qquad \text{Multiplying on both sides by } \tfrac{3}{14}$$

$$x = 33. \qquad \text{Remember, } x \text{ represents the first number.}$$

Going back to the table, we can find the other two numbers:

Second: $3x - 6 = 3 \cdot 33 - 6 = 93$;
Third: $\frac{2}{3}x + 2 = \frac{2}{3} \cdot 33 + 2 = 24$.

4. **Check.**   We return to the original problem. There are three numbers: 33, 93, and 24. Is the second number 6 less than 3 times the first?

$$3 \times 33 - 6 = 99 - 6 = 93$$

The answer is *yes*.

   Is the third number 2 more than $\frac{2}{3}$ the first?

$$\tfrac{2}{3} \times 33 + 2 = 22 + 2 = 24$$

The answer is *yes*.

Is the sum of the three numbers 150?

$$33 + 93 + 24 = 150$$

The answer is *yes*. The numbers do check.

5. State. The problem asks us to find the largest number, so the answer is: "The largest of the three numbers is 93." ——●

> **CAUTION!** In Example 5, although the equation $x = 33$ enabled us to find the largest number, 93, the number 33 was *not* the solution to the problem. By carefully labeling our variable in the first step of problem solving, we may avoid the temptation of thinking that our variable always represents the solution of the problem.

EXERCISE SET

1.4

For each problem, familiarize yourself with the situation. Then translate to mathematical language. You need not actually solve the problem; just carry out the first two steps of the five-step strategy. You will be asked to complete some of the solutions as Exercises 31–38.

1. The sum of two numbers is 65. One of the numbers is 7 more than the other. What are the numbers?

2. The sum of two numbers is 83. One of the numbers is 11 more than the other. What are the numbers?

3. *Moving sidewalks.* The moving sidewalk in O'Hare Airport is 300 ft long and moves at a rate of 5 ft/sec. If Alida walks at a rate of 4 ft/sec, how long will it take her to walk the length of the moving sidewalk?

4 ft/sec

5 ft/sec

4. *Swimming.* Fran swims at a rate of 5 km/h in still water. The Lazy River flows at a rate of 2.3 km/h. How long will it take Fran to swim 1.8 km upstream?

5. *Flight into a headwind.* An airplane traveling 390 km/h in still air encounters a 65-km/h headwind. How long will it take the plane to travel 725 km into the wind?

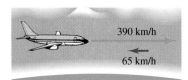

390 km/h

65 km/h

6. *Boating.* A paddleboat moves at a rate of 14 km/h in still water. How long will it take the boat to travel 56 km downstream if the river's current moves at a rate of 7 km/h?

7. *Angles in a triangle.* The degree measures of the angles in a triangle are three consecutive integers. Find the measures of the angles.

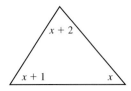

$x + 2$

$x + 1$ x

8. *Pricing.* The Sound Connection prices its blank audiotapes by raising the wholesale price 50% and adding 25 cents. What must a tape's wholesale price be if the tape is to sell for $1.99?

9. *Cruising altitude.*   A commercial jet has been instructed to climb from its present altitude of 8000 ft to a cruising altitude of 29,000 ft. If the plane ascends at a rate of 3500 ft/min, how long will it take to reach the cruising altitude?

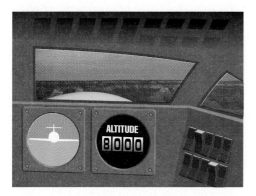

10. A piece of wire 10 m long is to be cut into two pieces, one of them $\frac{2}{3}$ as long as the other. How should the wire be cut?

11. *Angles in a triangle.*   One angle of a triangle is 3 times as great as a second angle. The third angle measures 12° less than twice the second angle. Find the measures of the angles.

12. *Angles in a triangle.*   One angle of a triangle is 4 times as great as a second angle. The third angle measures 5° more than twice the second angle. Find the measures of the angles.

13. Find three consecutive odd integers such that the sum of the first, 2 times the second, and 3 times the third is 70.

14. Find two consecutive even integers such that 2 times the first plus 3 times the second is 76.

15. A piece of wire 100 cm long is to be cut into two pieces, each to be bent to make a square. The length of a side of one square is to be twice the length of a side of the other. How should the wire be cut?

16. A piece of wire 100 cm long is to be cut into two pieces, and those pieces are each to be bent to make a square. The area of one square is to be 144 cm² greater than that of the other. How should the wire be cut? (Remember: Do not solve.)

17. Three numbers are such that the second is 6 less than 3 times the first, and the third is 2 more than $\frac{2}{3}$ of the second. The sum of the three numbers is 172. Find the largest number.

18. *Pricing.*   Whitney's Appliances is having a sale on 13 TV sets. They are displayed in order of increasing price from left to right. The price of each set differs by $20 from either set next to it. For the price of the set at the extreme right, a customer can buy both the second and seventh sets. What is the price of the least expensive set?

19. *Test scores.*   Deirdre's scores on five tests are 93, 89, 72, 80, and 96. What must the score be on her next test so that the average will be 88?

20. *Population growth.*   The population of Newcastle grew 12% for each of three consecutive years. At the end of that time, the population was 50,577. Find Newcastle's population at the start of the three-year period.

Solve each problem. Use all five problem-solving steps.

21. The product of two numbers is 125. If one number is 50, find the other number.

22. The number 38.2 is less than some number by 12.1. What is the number?

23. The number 173.5 is greater than a certain number by 16.8. What is the number?

24. The number 128 is 0.4 of what number?

25. The number 456 is $\frac{1}{3}$ of what number?

26. The quotient of two numbers is 12.3. If the divisor is 4, find the other number.

27. One number exceeds another by 12. The sum of the numbers is 114. What is the larger number?

28. One number is less than another by 65. The sum of the numbers is 92. What is the smaller number?

29. A rectangle's length is twice its width and its perimeter is 21 m. Find the dimensions of the rectangle.

30. A rectangle's width is one-third its length and its perimeter is 32 m. Find the dimensions of the rectangle.

31. Solve the problem of Exercise 4.

32. Solve the problem of Exercise 3.

33. Solve the problem of Exercise 14.

34. Solve the problem of Exercise 13.

35. Solve the problem of Exercise 10.

36. Solve the problem of Exercise 9.

37. Solve the problem of Exercise 8.

38. Solve the problem of Exercise 15.

SKILL MAINTENANCE

Solve.

39. $3[2x - (5 + 4x)] = 5 - 7x$

40. $5[3x - (2 + 4x)] = 9 - 4x$

41. $4 + 2(a - 7) = 2(a - 5)$

42. $5 + 3(a - 2) = 3(4 + a)$

SYNTHESIS

43. ◆ Write a problem for a classmate to solve. Devise the problem so that the solution is "The material should be cut into two pieces, one 30 cm long and the other 45 cm long."

44. ◆ Write a problem for a classmate to solve. Devise the problem so that the solution is "The first angle is 40°, the second angle is 50°, and the third angle is 90°."

45. ◆ How can a guess or estimate help prepare you for the *Translate* step when solving problems?

46. ◆ Why is it important to check the solution from step 3 (*Carry out*) in the original wording of the problem being solved?

47. *Test scores.* Tico's scores on four tests are 83, 91, 78, and 81. How many points above the average must Tico score on the next test in order to raise his average 2 points?

48. *Geometry.* The height and sides of a triangle are four consecutive integers. The height is the first integer, and the base is the fourth integer. The perimeter of the triangle is 42 in. Find the area of the triangle.

49. ▦ *Home prices.* Panduski's real estate prices increased 6% from 1994 to 1995 and 2% from 1995 to 1996. From 1996 to 1997, prices dropped 1%. If a house sold for $117,743 in 1997, what was its worth in 1994? (Round to the nearest dollar.)

50. *Adjusted wages.* Blanche's salary is reduced $n\%$ during a period of financial difficulty. By what number should her salary be multiplied in order to bring it back to where it was before the reduction?

COLLABORATIVE
C◆O◆R◆N◆E◆R

Focus: Problem solving

Time: 15 minutes

Group size: 5

Suppose that two of the five members in each group are celebrating birthdays and the entire group goes out to lunch. Suppose further that each member whose birthday it is gets treated to his or her lunch by the other *four* members. Finally, suppose that all meals cost the same amount and that the total bill is $40.00.*

ACTIVITY

1. Determine, as a group, how much each group member should pay for the lunch described above. Then explain how this determination was made.

2. Compare the results and methods used for part (1) with those of the other groups in the class.

*This activity was inspired by "The Birthday-Lunch Problem," *Mathematics Teaching in the Middle School,* vol. 2, no. 1, September–October 1996, pp. 40–42.

1.5   Formulas, Models, and Geometry

Solving Formulas   •   Mathematical Models

A **formula** is an equation that uses letters to represent a relationship between two or more quantities. For example, in Section 1.4, we made use of the formula $P = 4s$, where P represents the perimeter of a square and s the length of a side. Other formulas that you may recall from geometry are $A = \pi r^2$ (for the area A of a circle of radius r), $C = \pi d$ (for the circumference C of a circle of diameter d), and $A = b \cdot h$ (for the area A of a parallelogram of height h and base length b).* A more complete list of geometric formulas appears at the very end of this text.

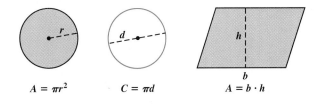

$$A = \pi r^2 \qquad\qquad C = \pi d \qquad\qquad A = b \cdot h$$

Solving Formulas

Suppose we remember the area and the width of a rectangular room and want to find the length. To do so, we could "solve" the formula $A = l \cdot w$ (Area = Length · Width) for l, using the same principles that we use for solving equations.

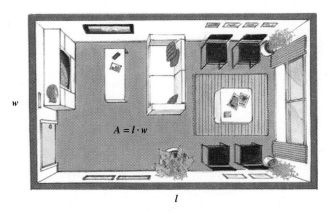

$$A = l \cdot w$$

---

*The Greek letter π, read "pi," is *approximately* 3.14159265358979323846264. Often 3.14 or 22/7 is used to approximate π when a calculator with a π key is unavailable.

EXAMPLE 1

Area of a Rectangle.   Solve the formula $A = l \cdot w$ for l.

SOLUTION

$$A = l \cdot w \qquad \text{We want this letter alone.}$$

$$A \cdot \frac{1}{w} = l \cdot w \cdot \frac{1}{w} \qquad \text{Multiplying by } \frac{1}{w} \text{ on both sides}$$

$$\frac{A}{w} = l \qquad \text{Simplifying}$$

Thus to find the length of a rectangular room, we can divide the area of the room by its width. Were we to do this calculation for a variety of rectangular rooms, the formula $l = A/w$ would be more useful than repeatedly substituting into $A = l \cdot w$.

EXAMPLE 2

Simple Interest.   The formula $I = Prt$ is used to determine the simple interest, I, earned when P dollars is invested for t years at an interest rate r. Solve this formula for t.

SOLUTION

$$I = Prt \qquad \text{We want this letter alone.}$$

$$\frac{1}{Pr} \cdot I = \frac{1}{Pr} \cdot Prt \qquad \text{Multiplying by } \frac{1}{Pr} \text{ on both sides}$$

$$\frac{I}{Pr} = t \qquad \text{Simplifying}$$

EXAMPLE 3

Area of a Trapezoid.   A trapezoid is a geometric shape with four sides, exactly two of which, the bases, are parallel to each other. The formula for calculating the area A of a trapezoid with bases b_1 and b_2 (read "b sub one" and "b sub two") and height h is given by

$$A = \frac{h}{2}(b_1 + b_2),$$

where the *subscripts* 1 and 2 distinguish one base from the other. Solve for b_1.

SOLUTION

$$A = \frac{h}{2}(b_1 + b_2)$$

$$\frac{2}{h} \cdot A = \frac{2}{h} \cdot \frac{h}{2}(b_1 + b_2) \qquad \text{Multiplying by } \frac{2}{h} \text{ on both sides}$$

$$\frac{2A}{h} = b_1 + b_2 \qquad \text{Simplifying. The right side is "cleared" of fractions.}$$

$$\frac{2A}{h} - b_2 = b_1 \qquad \text{Adding } -b_2 \text{ on both sides}$$

The similarities between solving formulas and solving equations can be seen below. In (a), we solve as we did before; in (b), we do not carry out all calculations; and in (c), we cannot carry out all calculations because the numbers are unknown.

a)
$$9 = \frac{3}{2}(x + 5)$$
$$\frac{2}{3} \cdot 9 = \frac{2}{3} \cdot \frac{3}{2}(x + 5)$$
$$6 = x + 5$$
$$1 = x$$

b)
$$9 = \frac{3}{2}(x + 5)$$
$$\frac{2}{3} \cdot 9 = \frac{2}{3} \cdot \frac{3}{2}(x + 5)$$
$$\frac{2 \cdot 9}{3} = x + 5$$
$$\frac{2 \cdot 9}{3} - 5 = x$$

c)
$$A = \frac{h}{2}(b_1 + b_2)$$
$$\frac{2}{h} \cdot A = \frac{2}{h} \cdot \frac{h}{2}(b_1 + b_2)$$
$$\frac{2A}{h} = b_1 + b_2$$
$$\frac{2A}{h} - b_2 = b_1$$

EXAMPLE 4

Simple Interest. The formula $A = P + Prt$ gives the amount A that a principal of P dollars will be worth in t years when invested at simple interest rate r. Solve the formula for P.

SOLUTION

$$A = P + Prt \qquad \text{We want this letter alone.}$$
$$A = P(1 + rt) \qquad \text{Factoring (using the distributive law)}$$
$$A \cdot \frac{1}{1 + rt} = P(1 + rt) \cdot \frac{1}{1 + rt} \qquad \text{Multiplying by } \frac{1}{1 + rt} \text{ on both sides}$$
$$\frac{A}{1 + rt} = P \qquad \text{Simplifying}$$

This last equation can be used to determine how much should be invested at interest rate r in order to have A dollars t years later.

Note in Example 4 that the factoring enabled us to write P once rather than twice. This is comparable to combining like terms when solving an equation like $16 = x + 7x$.

You may find the following summary useful.

TO SOLVE A FORMULA FOR A GIVEN LETTER:

1. Multiply on both sides to clear fractions or decimals, if that is needed.
2. Combine like terms on each side where convenient.
3. Get all terms with the letter being solved for on one side of the equation and all other terms on the other side, using the addition principle.
4. Combine like terms again, if necessary. This may require factoring.
5. Solve for the letter in question, using the multiplication principle.

Mathematical Models

The above formulas from geometry and economics are examples of *mathematical models*. A **mathematical model** can be a formula, or set of formulas, developed to represent a real-world situation. In problem solving, a mathematical model is formed in the *Translate* step.

EXAMPLE 5

Density. A jeweler suspects that a "gold" medal is not solid gold. The density of gold is 19.3 grams per cubic centimeter (g/cm³) and the medal is 0.5 cm thick with a radius of 3 cm. If the medal is gold, how much should it weigh?

SOLUTION

1. **Familiarize.** In a science book, we can learn that density depends on mass and volume and that, in this setting, mass means weight. A formula for the volume of a right circular cylinder appears at the very end of this text.

2. **Translate.** We need to use two formulas:

$$D = \frac{m}{V} \quad \text{and} \quad V = \pi r^2 h,$$

where D is the density, m is the mass, V is the volume, r is the radius, and h is the height of a right circular cylinder. Since we need a model relating mass to the measurements of the medal, we solve for m and then substitute for V:

$$D = \frac{m}{V}$$

$$V \cdot D = V \cdot \frac{m}{V} \qquad \text{Multiplying by } V$$

$$V \cdot D = m \qquad \text{Simplifying}$$

$$\pi r^2 h \cdot D = m. \qquad \text{Substituting}$$

3. **Carry out.** The model $m = \pi r^2 hD$ can be used to find the mass of any right circular cylinder for which the dimensions and the density are known:

$$m = \pi r^2 hD$$
$$= \pi(3)^2(0.5)(19.3) \qquad \text{Substituting}$$
$$\approx 272.847322. \qquad \text{Using a calculator with a } \pi \text{ key}$$

4. **Check.** To check, we could repeat the calculations. We might also check the model by using it to predict the mass of an object of known weight, like a coin.

5. **State.** The medal, if it is really gold, should weigh about 273 grams.

Solve.

1. $d = rt$, for t (a distance formula)

2. $d = rt$, for r

3. $F = ma$, for a (a physics formula)

4. $A = lw$, for w (an area formula)

5. $W = EI$, for E (an electricity formula)

6. $W = EI$, for I

7. $V = lwh$, for h (a volume formula)

8. $I = Prt$, for r (a formula for interest)

9. $L = \dfrac{k}{d^2}$, for k (a formula for intensity)

10. $L = \dfrac{k}{d^2}$, for d^2

(*Hint:* Multiply by d^2 to "clear" the fraction.)

11. $G = w + 150n$, for n
(a formula for the gross weight of a bus)

12. $P = b + 0.5t$, for t (a formula for parking prices)

13. $2w + 2h + l = p$, for l
(a formula used when shipping boxes)

14. $g = \dfrac{km_1m_2}{d^2}$, for d^2 (Newton's law of gravitation)

15. $Ax + By = C$, for y (a formula for graphing lines)

16. $P = 2l + 2w$, for l (a perimeter formula)

17. $C = \frac{5}{9}(F - 32)$, for F (a temperature formula)

18. $T = \frac{3}{10}(I - 12{,}000)$, for I (a tax formula)

19. $V = \frac{4}{3}\pi r^3$, for r^3 (a volume formula)

20. $V = \frac{4}{3}\pi r^3$, for π

21. $A = \dfrac{h}{2}(b_1 + b_2)$, for b_2 (an area formula)

22. $A = \dfrac{h}{2}(b_1 + b_2)$, for h (an area formula)

23. $F = \dfrac{mv^2}{r}$, for m (a physics formula)

24. $F = \dfrac{mv^2}{r}$, for v^2

25. $A = \dfrac{q_1 + q_2 + q_3}{n}$, for n (a formula for averaging)

26. $r = \dfrac{s + t}{d}$, for d

27. $v = \dfrac{d_2 - d_1}{t}$, for t (a physics formula)

28. $v = \dfrac{s_2 - s_1}{m}$, for m

29. $v = \dfrac{d_2 - d_1}{t}$, for d_1

30. $v = \dfrac{s_2 - s_1}{m}$, for s_1

31. $r = m + mnp$, for m

32. $p = x - xyz$, for x

33. $y = ab - ac^2$, for a

34. $d = mn - mp^3$, for m

35. *Banking.*   Yvonne plans to buy a one-year certificate of deposit (CD) that earns 7% simple interest. If she needs the CD to earn \$110, how much should Yvonne invest?

36. *Investing.*   Janos has \$1300 to invest for 6 months. If he needs the money to earn \$39 in that time, at what rate of simple interest must Janos invest?

37. *Geometry.*   The area of a parallelogram is 78 cm^2. The base of the figure is 13 cm. What is the height?

38. *Geometry.*   The area of a parallelogram is 72 cm^2. The height of the figure is 6 cm. How long is the base?

39. ▦ *Weight of salt.*   The density of salt is 2.16 g/cm^3 (grams per cubic centimeter). An empty cardboard salt canister weighs 28 g, is 13.6 cm tall, and has a 4-cm radius. How much will a filled canister weigh? Use the model developed in Example 5.

40. ▦ *Weight of aluminum.*   The density of aluminum is 2.7 g/cm^3. If the medal in Example 5 were made of aluminum instead of gold, how much less would it have weighed?

Projected birth weight.   *Ultrasonic images of 29-week-old fetuses can be used to predict weight. One model, de-*

veloped by Thurnau,* is $P = 9.337da - 299$; *a second model, developed by Weiner,[†] is $P = 94.593c + 34.227a - 2134.616$. For both formulas, P represents the estimated fetal weight in grams, d the diameter of the fetal head in centimeters, c the circumference of the fetal head in centimeters, and a the circumference of the fetal abdomen in centimeters.*

41. 🖩 Use Thurnau's model to estimate the diameter of a fetus' head at 29 weeks when the estimated weight is 1614 g and the circumference of the fetal abdomen is 24.1 cm.

42. 🖩 Use Weiner's model to estimate the circumference of a fetus' head at 29 weeks when the estimated weight is 1277 g and the circumference of the fetal abdomen is 23.4 cm.

43. *Gardening.* A garden is being constructed in the shape of a trapezoid. The dimensions are as shown in the figure. The unknown dimension is to be such that the area of the garden is 90 ft². Find that unknown dimension.

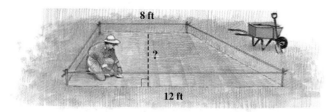

44. *Fencing.* A rectangular garden is being constructed, and 76 ft of fencing is available. The width of the garden is to be 13 ft. What should the length be, in order to use just 76 ft of fence?

45. *Investing.* Bok Lum Chan is going to invest $1600 at simple interest at 9%. How long will it take for the investment to be worth $2608?

46. Rik is going to invest $950 at simple interest at 7%. How long will it take for his investment to be worth $1349?

Waiting time. *In an effort to minimize waiting time for patients at a doctor's office without increasing a physician's idle time, Michael Goiten of Massachusetts General Hospital has developed a model. Goiten suggests that the interval time I, in minutes, between scheduled* appointments be related to the total number of minutes T that a physican spends with patients in a day and the number of scheduled appointments N according to the formula $I = 1.08(T/N)$.*

47. A doctor insists on an interval time of 20 min and must be able to schedule 25 appointments a day. According to Goiten's model, how many hours a day should the doctor be prepared to spend with patients?

48. A doctor determines that she has a total of 8 hr a day to see patients. If she insists on an interval time of 15 min, according to Goiten's model, how many appointments should she make in one day?

SKILL MAINTENANCE

49. What percent of 5800 is 4176?

50. Simplify: $-5a + 9b - (3a - 4b)$.

51. Subtract: $-72.5 - (-14.06)$.

52. Simplify: $(-5)^3$.

SYNTHESIS

53. ◈ Predictions made using the models of Exercises 41 and 42 are often off by as much as 10%. Does that mean the models should be discarded? Why or why not?

54. ◈ Give an example of a mathematical model you have seen and explain what it was used for.

55. ◈ Is every rectangle a trapezoid? Explain why or why not.

56. ◈ Which would you expect to have the greater density, and why: cork or steel?

57. 🖩 The density of a penny is 8.93 g/cm³. The mass of a roll of pennies is 177.6 g. If the diameter of a penny is 1.85 cm, how tall is a roll of pennies?

58. 🖩 The density of copper is 8.93 g/cm³. How long must a copper wire be if it is 1 cm thick and has a mass of 4280 g?

Solve.

59. $s = v_i t + \frac{1}{2}at^2$, for a

60. $A = 4lw + w^2$, for l

61. $\dfrac{P_1 V_1}{T_1} = \dfrac{P_2 V_2}{T_2}$, for T_2

62. $\dfrac{P_1 V_1}{T_1} = \dfrac{P_2 V_2}{T_2}$, for T_1

63. $\dfrac{b}{a - b} = c$, for b

64. $m = \dfrac{(d/e)}{(e/f)}$, for d

*Thurnau, G. R., R. K. Tamura, R. E. Sabbagha, et al. *Am. J. Obstet Gynecol* 1983; **145**:557.

[†]Weiner, C. P., R. E. Sabbagha, N. Vaisrub, et al. *Obstet Gynecol* 1985; **65**:812.

1.6 Properties of Exponents

The Product and Quotient Rules   •   The Zero Exponent   •
Negative Integers as Exponents   •   Raising Powers to Powers   •
Raising a Product or a Quotient to a Power

In Section 1.1, we discussed how whole-number exponents are used. We now develop rules for manipulating exponents and define what zero and negative integers will mean as exponents.

The Product and Quotient Rules

Observe that the expression $x^3 \cdot x^4$ can be rewritten as follows:

$$x^3 \cdot x^4 = \underbrace{x \cdot x \cdot x}_{3 \text{ factors}} \cdot \underbrace{x \cdot x \cdot x \cdot x}_{4 \text{ factors}}$$

$$= \underbrace{x \cdot x \cdot x \cdot x \cdot x \cdot x \cdot x}_{7 \text{ factors}}$$

$$= x^7.$$

This result is generalized in the *product rule*.

**MULTIPLYING WITH
LIKE BASES:
THE PRODUCT RULE**

For any number a and any positive integers m and n,

$$a^m \cdot a^n = a^{m+n}.$$

(When multiplying with exponential notation, if the bases are the same, keep the base and add the exponents.)

EXAMPLE 1   Multiply and simplify: **(a)** $m^5 \cdot m^7$; **(b)** $(5a^2b^3)(3a^4b^5)$.

SOLUTION

a) $m^5 \cdot m^7 = m^{5+7} = m^{12}$

b) $(5a^2b^3)(3a^4b^5) = 5 \cdot 3 \cdot a^2 \cdot a^4 \cdot b^3 \cdot b^5$ Using the associative and commutative laws

$$= 15a^{2+4}b^{3+5}$$ Multiplying; using the product rule

$$= 15a^6b^8$$

We next simplify a quotient:

$$\frac{x^8}{x^3} = \frac{x \cdot x \cdot x \cdot x \cdot x \cdot x \cdot x \cdot x}{x \cdot x \cdot x} \qquad \text{Using the definition of exponential notation}$$

$$= \frac{x \cdot x \cdot x}{x \cdot x \cdot x} \cdot x \cdot x \cdot x \cdot x \cdot x$$

$$= x \cdot x \cdot x \cdot x \cdot x \qquad \text{Removing a factor equal to 1: } \frac{x \cdot x \cdot x}{x \cdot x \cdot x} = 1$$

$$= x^5.$$

The generalization of this result is the *quotient rule*.

DIVIDING WITH LIKE BASES: THE QUOTIENT RULE

For any nonzero number a and any positive integers m and n, $m > n$,

$$\frac{a^m}{a^n} = a^{m-n}.$$

(When dividing with exponential notation, if the bases are the same, keep the base and subtract the exponent of the denominator from the exponent of the numerator.)

EXAMPLE 2

Divide and simplify: **(a)** $\dfrac{r^9}{r^3}$; **(b)** $\dfrac{10x^{11}y^5}{2x^4y^3}$.

SOLUTION

a) $\dfrac{r^9}{r^3} = r^{9-3} = r^6 \qquad$ Using the quotient rule

b) $\dfrac{10x^{11}y^5}{2x^4y^3} = 5 \cdot x^{11-4} \cdot y^{5-3} \qquad$ Dividing; using the quotient rule

$$= 5x^7y^2$$

The Zero Exponent

Suppose now that the bases in the numerator and the denominator are both raised to the same power:

$$\frac{x^5}{x^5} = 1 \quad \text{or} \quad \frac{8^3}{8^3} = 1.$$

These results follow from the fact that any (nonzero) expression, divided by itself, is equal to 1. On the other hand, were we to subtract exponents, we would obtain

$$\frac{x^5}{x^5} = x^{5-5} = x^0 \quad \text{or} \quad \frac{8^3}{8^3} = 8^{3-3} = 8^0.$$

Thus, in order to continue subtracting exponents when dividing like bases raised to the same power, we must have $x^0 = 1$ and $8^0 = 1$. This leads to the following definition.

THE ZERO EXPONENT

For any real number a, $a \neq 0$,

$$a^0 = 1.$$

(Any nonzero number raised to the zero power is 1.)

E X A M P L E   3

Evaluate each of the following for $x = 2.9$: **(a)** x^0; **(b)** $-x^0$; **(c)** $(-x)^0$.

S O L U T I O N

a) $x^0 = 2.9^0 = 1$ Using the definition of 0 as an exponent
b) $-x^0 = -2.9^0 = -1$ The exponent is used before the negative sign.
c) $(-x)^0 = (-2.9)^0 = 1$ The negative sign is used before the exponent.

Parts (b) and (c) of Example 3 illustrate an important result:

Since $-a^n$ means $-1 \cdot a^n$,   in general, $-a^n \neq (-a)^n$.

Negative Integers as Exponents

Later in this text we will explain what numbers like $\frac{2}{9}$ or $\sqrt{2}$ mean as exponents. Until then, integer exponents will suffice.

To develop a definition for negative integer exponents, we simplify $5^3/5^7$ two ways. First we proceed as in arithmetic:

$$\frac{5^3}{5^7} = \frac{5 \cdot 5 \cdot 5}{5 \cdot 5 \cdot 5 \cdot 5 \cdot 5 \cdot 5 \cdot 5} = \frac{5 \cdot 5 \cdot 5 \cdot 1}{5 \cdot 5 \cdot 5 \cdot 5 \cdot 5 \cdot 5 \cdot 5}$$

$$= \frac{5 \cdot 5 \cdot 5}{5 \cdot 5 \cdot 5} \cdot \frac{1}{5 \cdot 5 \cdot 5 \cdot 5}$$

$$= \frac{1}{5^4}.$$

Were we to apply the quotient rule, we would have

$$\frac{5^3}{5^7} = 5^{3-7}$$

$$= 5^{-4}.$$

These two expressions for $5^3/5^7$ suggest that

$$5^{-4} = \frac{1}{5^4}.$$

This leads to the definition of negative exponents.

NEGATIVE EXPONENTS

For any real number a that is nonzero and any integer n,

$$a^{-n} = \frac{1}{a^n}.$$

(The numbers a^{-n} and a^n are reciprocals of each other.)

The definitions above preserve the following pattern:

$4^3 = 4 \cdot 4 \cdot 4,$

$4^2 = 4 \cdot 4,$ Dividing by 4 on both sides

$4^1 = 4,$ Dividing by 4 on both sides

$4^0 = 1,$ Dividing by 4 on both sides

$4^{-1} = \dfrac{1}{4},$ Dividing by 4 on both sides

$4^{-2} = \dfrac{1}{4 \cdot 4} = \dfrac{1}{4^2}.$ Dividing by 4 on both sides

EXAMPLE 4   Express using positive exponents and then simplify.

a) 3^{-2} **b)** $5x^{-4}y^3$ **c)** $\dfrac{1}{5^{-2}}$

SOLUTION

a) $3^{-2} = \dfrac{1}{3^2} = \dfrac{1}{9}$

TECHNOLOGY CONNECTION
1.6

Throughout this text, we have included Technology Connections *to highlight situations in which calculators (primarily graphing calculators) and computers equipped with graphing software can be used to enrich the learning experience. Most* Technology Connections *present discussions and exercises or activities in a generic form—consult a user's manual or an instructor for specific keystrokes.*

Expressions like 4^7 can be simplified using a calculator. To facilitate this, most calculators are equipped with an exponentiation key. Such a key is

often labeled $\boxed{x^y}$ or $\boxed{\wedge}$. To enter 4^7 on most scientific calculators, we press

$\boxed{4}$ $\boxed{x^y}$ $\boxed{7}$ $\boxed{=}$.

On other calculators, we press

$\boxed{4}$ $\boxed{\wedge}$ $\boxed{7}$ $\boxed{\text{ENTER}}$

On the most basic of calculators, there is no choice but to actually multiply 7 factors of 4.

1. List keystrokes that could be used to simplify 2^{-5} on a scientific or graphing calculator.

2. How could 2^{-5} be simplified on a calculator lacking an exponentiation key?

b) $5x^{-4}y^3 = 5\left(\dfrac{1}{x^4}\right)y^3 = \dfrac{5y^3}{x^4}$

c) $\dfrac{1}{5^{-2}} = \dfrac{1}{\dfrac{1}{5^2}} = 1 \cdot \dfrac{5^2}{1} = 25$

Example 4(c) reveals that when a factor of the numerator or the denominator is raised to any power, the factor can be moved to the other side of the fraction bar provided the sign of the exponent is changed. Thus, for example,

$$\frac{a^{-2}b^3}{c^{-4}} = \frac{c^4}{a^2 b^{-3}}.$$

The product and quotient rules apply for all integer exponents.

EXAMPLE 5   Simplify: **(a)** $7^{-3} \cdot 7^8$; **(b)** $\dfrac{b^{-5}}{b^{-4}}$.

SOLUTION

a) $7^{-3} \cdot 7^8 = 7^{-3+8}$ Adding exponents

$\qquad = 7^5$

$Check: 7^{-3} \cdot 7^8 = \dfrac{1}{7^3} \cdot 7^8$

$\qquad = \dfrac{7^8}{7^3} = 7^5.$

b) $\dfrac{b^{-5}}{b^{-4}} = b^{-5-(-4)} = b^{-1}$ Subtracting exponents

$\qquad = \dfrac{1}{b}$ Writing the answer without a negative exponent

Example 5(b) can also be simplified as follows:

$$\frac{b^{-5}}{b^{-4}} = \frac{b^4}{b^5} = b^{4-5} = b^{-1} = \frac{1}{b}.$$

Raising Powers to Powers

Next, consider an expression like $(3^4)^2$:

$(3^4)^2 = (3^4)(3^4)$ We are raising 3^4 to the second power.

$\qquad = (3 \cdot 3 \cdot 3 \cdot 3)(3 \cdot 3 \cdot 3 \cdot 3)$

$\qquad = 3 \cdot 3 \cdot 3 \cdot 3 \cdot 3 \cdot 3 \cdot 3 \cdot 3$ Using the associative law

$\qquad = 3^8.$

Note that in this case, we could have multiplied the exponents:

$(3^4)^2 = 3^{4 \cdot 2} = 3^8.$

Likewise, $(y^8)^3 = (y^8)(y^8)(y^8) = y^{24}$. Once again, we get the same result if we multiply the exponents:

$(y^8)^3 = y^{8 \cdot 3} = y^{24}.$

RAISING A POWER TO A POWER: THE POWER RULE

For any real number a and any integers m and n,

$$(a^m)^n = a^{mn}.$$

(To raise a power to a power, multiply the exponents.)

EXAMPLE 6

Simplify: **(a)** $(3^5)^4$; **(b)** $(y^{-5})^7$; **(c)** $(a^{-3})^{-7}$.

SOLUTION

a) $(3^5)^4 = 3^{5 \cdot 4} = 3^{20}$

b) $(y^{-5})^7 = y^{-5 \cdot 7} = y^{-35}$

c) $(a^{-3})^{-7} = a^{(-3)(-7)} = a^{21}$

Raising a Product or a Quotient to a Power

When an expression inside parentheses is raised to a power, the inside expression is the base. Let's compare $2a^3$ and $(2a)^3$.

$$2a^3 = 2 \cdot a \cdot a \cdot a; \qquad\qquad (2a)^3 = (2a)(2a)(2a)$$
$$= (2 \cdot 2 \cdot 2)(a \cdot a \cdot a)$$
$$= 2^3 a^3$$
$$= 8a^3$$

We see that $2a^3$ and $(2a)^3$ are *not* equivalent. Note also that to simplify $(2a)^3$ we can raise each factor to the power 3. This leads to the following rule.

RAISING A PRODUCT TO A POWER

For any real numbers a and b and any integer n (provided $ab \neq 0$ when $n \leq 0$),

$$(ab)^n = a^n b^n.$$

(To raise a product to the nth power, raise each factor to the nth power.)

EXAMPLE 7

Simplify: **(a)** $(-2x)^3$; **(b)** $(-2x^3 y^{-1})^{-4}$.

SOLUTION

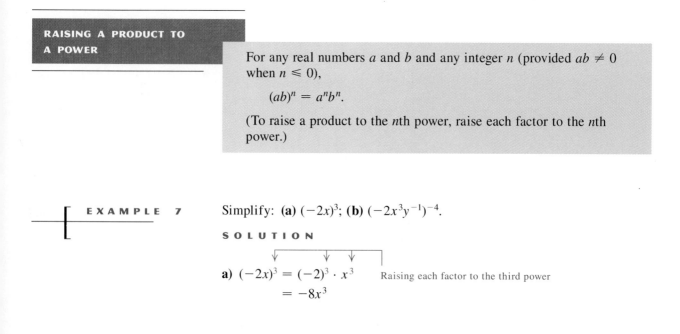

a) $(-2x)^3 = (-2)^3 \cdot x^3$ Raising each factor to the third power
$$= -8x^3$$

b) $(-2x^3y^{-1})^{-4} = (-2)^{-4}(x^3)^{-4}(y^{-1})^{-4}$ Raising each factor to the negative fourth power

$$= \frac{1}{(-2)^4} \cdot x^{-12}y^4$$ Multiplying powers; writing $(-2)^{-4}$ as $\frac{1}{(-2)^4}$

$$= \frac{y^4}{16x^{12}}$$ Note that $x^{-12} = \frac{1}{x^{12}}$.

There is a similar rule for raising a quotient to a power.

RAISING A QUOTIENT TO A POWER

For any integer n, and any real numbers a and b for which a/b, a^n, and b^n exist,

$$\left(\frac{a}{b}\right)^n = \frac{a^n}{b^n}.$$

(To raise a quotient to a power, raise the numerator to the power and divide by the denominator to the power.)

EXAMPLE   8

Simplify: **(a)** $\left(\dfrac{x^2}{3}\right)^4$; **(b)** $\left(\dfrac{y^2z^3}{5}\right)^{-3}$.

SOLUTION

a) $\left(\dfrac{x^2}{3}\right)^4 = \dfrac{(x^2)^4}{3^4} = \dfrac{x^8}{81}$

b) $\left(\dfrac{y^2z^3}{5}\right)^{-3} = \dfrac{(y^2z^3)^{-3}}{5^{-3}}$

$$= \frac{5^3}{(y^2z^3)^3}$$ Moving factors to the other side of the fraction bar and changing each -3 to 3

$$= \frac{125}{y^6z^9}$$

The rule for raising a quotient to a power allows us to derive a useful result for manipulating negative exponents:

$$\left(\frac{a}{b}\right)^{-n} = \frac{a^{-n}}{b^{-n}} = \frac{b^n}{a^n} = \left(\frac{b}{a}\right)^n.$$

Using this result, we can simplify Example 8(b) as follows:

$$\left(\frac{y^2z^3}{5}\right)^{-3} = \left(\frac{5}{y^2z^3}\right)^3$$ Taking the reciprocal of the base and changing the exponent's sign

$$= \frac{5^3}{(y^2z^3)^3} = \frac{125}{y^6z^9}.$$

**DEFINITIONS AND RULES
FOR EXPONENTS**

For any integers m and n (assuming 0 is not raised to a nonpositive power):

Zero as an exponent: $a^0 = 1$

Negative integers as exponents: $a^{-n} = \dfrac{1}{a^n}$

Multiplying with like bases: $a^m \cdot a^n = a^{m+n}$

Dividing with like bases: $\dfrac{a^m}{a^n} = a^{m-n},\ a \neq 0$

Raising a product to a power: $(ab)^n = a^n b^n$

Raising a power to a power: $(a^m)^n = a^{mn}$

Raising a quotient to a power: $\left(\dfrac{a}{b}\right)^n = \dfrac{a^n}{b^n},\ b \neq 0$

EXERCISE SET

1.6

Multiply and simplify. Leave the answer in exponential notation.

1. $7^5 \cdot 7^2$

2. $2^3 \cdot 2^8$

3. $5^6 \cdot 5^3$

4. $6^3 \cdot 6^5$

5. $a^3 \cdot a^0$

6. $x^0 \cdot x^5$

7. $6x^5 \cdot 3x^2$

8. $4a^3 \cdot 2a^7$

9. $(-3m^4)(-7m^9)$

10. $(-2a^5)(7a^4)$

11. $(x^3y^4)(x^7y^6z^0)$

12. $(m^6n^5)(m^4n^7p^0)$

Divide and simplify.

13. $\dfrac{a^9}{a^3}$

14. $\dfrac{x^{12}}{x^3}$

15. $\dfrac{8x^7}{4x^4}$

16. $\dfrac{20a^{20}}{5a^4}$

17. $\dfrac{m^7n^9}{m^2n^5}$

18. $\dfrac{m^{12}n^9}{m^4n^6}$

19. $\dfrac{35x^8y^5}{7x^2y}$

20. $\dfrac{45x^7y^8}{5xy^2}$

21. $\dfrac{-49a^5b^{12}}{7a^2b^2}$

22. $\dfrac{-42a^2b^7}{7ab^4}$

23. $\dfrac{18x^6y^7z^9}{-6x^2yz^3}$

24. $\dfrac{28x^8y^{10}z^{12}}{-7x^4y^2z}$

Simplify.

25. $(-3)^4$

26. $(-2)^6$

27. -3^4

28. -2^6

29. $(-5)^{-2}$

30. $(-4)^{-2}$

31. -5^{-2}

32. -4^{-2}

33. -2^{-5}

34. -5^{-3}

35. -2^{-6}

36. -1^{-8}

Write an equivalent expression without negative exponents.

37. a^{-3}

38. n^{-6}

39. $(5x)^{-3}$

40. $(4xy)^{-5}$

41. x^2y^{-3}

42. $2a^2b^{-5}$

43. x^2y^{-2}

44. $a^2b^{-3}c^4d^{-5}$

45. $\dfrac{y^{-5}}{x^2}$

46. $\dfrac{z^{-4}}{3x^5}$

47. $\dfrac{y^{-5}}{x^{-3}}$

48. $\dfrac{(7x)^{-4}}{y^{-7}}$

49. $\dfrac{x^{-2}y^7}{z^{-4}}$

50. $\dfrac{y^4z^{-3}}{x^{-2}}$

Write an equivalent expression with negative exponents.

51. $\dfrac{1}{3^4}$

52. $\dfrac{1}{9^2}$

53. $\dfrac{1}{(-16)^2}$

54. $\dfrac{1}{(-8)^6}$

55. x^5

56. n^3

57. $6x^2$

58. $-4y^5$

59. $\dfrac{1}{(5y)^3}$

60. $\dfrac{1}{(5x)^5}$

61. $\dfrac{1}{3y^4}$

62. $\dfrac{1}{4b^3}$

Simplify. Should negative exponents appear in the answer, write a second answer using only positive exponents.

63. $8^{-2} \cdot 8^{-4}$

64. $9^{-1} \cdot 9^{-6}$

65. $b^2 \cdot b^{-5}$

66. $a^4 \cdot a^{-3}$

67. $a^{-3} \cdot a^4 \cdot a^2$

68. $x^{-8} \cdot x^5 \cdot x^3$

69. $(14m^2n^3)(-2m^3n^2)$

70. $(6x^5y^{-2})(-3x^2y^3)$

71. $(-2x^{-3})(7x^{-8})$

72. $(6x^{-4}y^3)(-4x^{-8}y^{-2})$

73. $(5a^{-2}b^{-3})(2a^{-4}b)$

74. $(3a^{-5}b^{-7})(2ab^{-2})$

75. $\dfrac{10^{-3}}{10^6}$

76. $\dfrac{12^{-4}}{12^8}$

77. $\dfrac{2^{-7}}{2^{-5}}$

78. $\dfrac{9^{-4}}{9^{-6}}$

79. $\dfrac{y^4}{y^{-5}}$

80. $\dfrac{a^3}{a^{-2}}$

81. $\dfrac{24a^5b^3}{-8a^4b}$

82. $\dfrac{9a^2}{3ab^3}$

83. $\dfrac{14a^4b^{-3}}{-8a^8b^{-5}}$

84. $\dfrac{-24x^6y^7}{18x^{-3}y^9}$

85. $\dfrac{-5x^{-2}y^4z^7}{30x^{-5}y^6z^{-3}}$

86. $\dfrac{9a^6b^{-4}c^7}{27a^{-4}b^5c^9}$

87. $(x^4)^3$

88. $(a^3)^2$

89. $(9^3)^{-4}$

90. $(8^4)^{-3}$

91. $(7^{-8})^{-5}$

92. $(6^{-4})^{-3}$

93. $(6xy)^2$

94. $(5ab)^3$

95. $(a^3b)^4$

96. $(x^3y)^5$

97. $5(x^2y^2)^3$

98. $7(a^3b^4)^2$

99. $(7x^3y^{-4})^{-2}$

100. $(3x^2y^{-5})^{-2}$

101. $(-8x^{-4}y^5z^2)^{-4}$

102. $(-6a^{-2}b^3c)^{-2}$

103. $\dfrac{(3x^3y^4)^3}{6xy^3}$

104. $\dfrac{(5a^3b)^2}{10a^2b}$

105. $\left(\dfrac{-4x^4y^{-2}}{5x^{-1}y^4}\right)^{-4}$

106. $\left(\dfrac{2x^3y^{-2}}{3y^{-3}}\right)^3$

107. $\left(\dfrac{3a^{-2}b^5}{9a^{-4}b^0}\right)^{-2}$

108. $\left(\dfrac{30x^5y^{-7}}{6x^{-2}y^{-6}}\right)^0$

109. $\left(\dfrac{4a^3b^{-9}}{2a^{-2}b^5}\right)^0$

110. $\left(\dfrac{5x^0y^{-7}}{2x^{-2}y^4}\right)^{-2}$

SKILL MAINTENANCE

111. Evaluate $x + (2xy - z)^2$ for $x = 3$, $y = -4$, and $z = -20$.

112. Evaluate $a - (bc - 3a)^2$ for $a = -4$, $b = 5$, and $c = -2$.

113. Find three consecutive odd integers whose sum is 183.

114. Find three consecutive even integers whose sum is 114.

SYNTHESIS

115. ◆ Is 5^{-9} greater or less than 4^{-9}? Why?

116. ◆ Explain why $(-1)^n = 1$ for any even number n.

117. ◆ Explain why $(-17)^{-8}$ is positive.

118. ◆ Explain in your own words why a^0 is defined to be 1. Assume that $a \neq 0$.

Simplify. Assume that all variables represent nonzero integers.

119. $\dfrac{9a^{x-2}}{3a^{2x+2}}$

120. $\dfrac{-12x^{a+1}}{4x^{2-a}}$

121. $[7y(7-8)^{-2} - 8y(8-7)^{-2}]^{(-2)^2}$

122. $\{[(8^{-a})^{-2}]^b\}^{-c} \cdot [(8^0)^a]^c$

123. $(3^{a+2})^a$

124. $(12^{3-a})^{2b}$

125. $\dfrac{4x^{2a+3}y^{2b-1}}{2x^{a+1}y^{b+1}}$

126. $\dfrac{25x^{a+b}y^{b-a}}{-5x^{a-b}y^{b+a}}$

127. $\dfrac{(2^{-2})^a \cdot (2^b)^{-a}}{(2^{-2})^{-b}(2^b)^{-2a}}$

128. $\dfrac{-28x^{b+5}y^{4+c}}{7x^{b-5}y^{c-4}}$

129. $\dfrac{3^{q+3} - 3^2(3^q)}{3(3^{q+4})}$

130. $\left[\left(\dfrac{a^{-2c}}{b^{7c}}\right)^{-3}\left(\dfrac{a^{4c}}{b^{-3c}}\right)^2\right]^{-a}$

1.7   Scientific Notation

Conversions   •   Significant Digits and Rounding   •
Scientific Notation in Problem Solving

There is a variety of symbolism, or *notation,* for numbers. You are already familiar with fractional notation, decimal notation, and percent notation. We now study **scientific notation,** so named because of its usefulness in work with the very large and very small numbers that occur in science.

The following are examples of scientific notation:

7.2×10^5 means   720,000;

3.4×10^{-6} means   0.0000034;

4.89×10^{-3} means   0.00489.

SCIENTIFIC NOTATION

> *Scientific notation* for a number is an expression of the type $N \times 10^n$, where N is in decimal notation, $1 \le N < 10$, and n is an integer.

Conversions

Note that $10^b/10^b = 10^b \cdot 10^{-b} = 1$. To convert to scientific notation, we can multiply by 1, writing 1 in the form $10^b/10^b$ or $10^b \cdot 10^{-b}$.

EXAMPLE 1   *Population Projections.*   It has been estimated that in the year 2025, the world population will be 8,504,000,000 (*Source: The Universal Almanac*). Write scientific notation for this number.

SOLUTION   To write 8,504,000,000 as 8.504×10^n for some integer n, we must move the decimal point 9 places to the left. This is accomplished by dividing—and then multiplying—by 10^9:

$$8,504,000,000 = \frac{8,504,000,000}{10^9} \cdot 10^9 \qquad \text{Multiplying by 1: } \frac{10^9}{10^9} = 1$$

$$= 8.504 \times 10^9. \qquad \text{This is scientific notation.}$$

EXAMPLE 2   Write scientific notation for the mass of a grain of sand:

0.0648 gram (g).

SOLUTION   To write 0.0648 as 6.48×10^n for some integer n, we must move the decimal 2 places to the right. To do this, we multiply—and then divide— by 10^2:

$$0.0648 = \frac{0.0648 \cdot 10^2}{10^2} \qquad \text{Multiplying by 1: } \frac{10^2}{10^2} = 1$$

$$= \frac{6.48}{10^2}$$

$$= 6.48 \times 10^{-2} \text{ g.} \qquad \text{Writing scientific notation}$$

Try to make conversions to scientific notation mentally as often as possible. In doing so, remember that negative powers of 10 are used for small numbers and positive powers of 10 are used for large numbers.

EXAMPLE 3

Convert mentally to scientific notation: **(a)** 82,500,000; **(b)** 0.0000091.

SOLUTION

a) $82,500,000 = 8.25 \times 10^7$ *Check*: Multiplying 8.25 by 10^7 moves the decimal point 7 places to the right.

b) $0.0000091 = 9.1 \times 10^{-6}$ *Check*: Multiplying 9.1 by 10^{-6} moves the decimal point 6 places to the left.

EXAMPLE 4

Convert mentally to decimal notation: **(a)** 4.371×10^7; **(b)** 1.73×10^{-5}.

SOLUTION

a) $4.371 \times 10^7 = 43,710,000$ Moving the decimal point 7 places to the right

b) $1.73 \times 10^{-5} = 0.0000173$ Moving the decimal point 5 places to the left

Significant Digits and Rounding

In the world of science, it is important to know just how accurate a measurement is. Clearly the measurement 5.12 cm is more precise than the measurement 5.1 cm. We say that the number 5.12 has three **significant digits** whereas 5.1 has only two significant digits. When two or more measurements are added, subtracted, multiplied, or divided, the result is only as accurate as the *least* precise measurement used in the computation. Thus scientists have agreed on the following conventions.

1. The sum or difference of two numbers should be rounded off so that it has the same number of significant digits to the right of the decimal as the number in the calculation with the fewest significant digits to the right of the decimal.

   For example,

   $$135.4 \text{ cm} + 50.28 \text{ cm} = 185.68 \text{ cm}$$

   1 digit   2 digits

   should be rounded off to

   1 digit

   185.7 cm.

2. The product or quotient of two numbers should be rounded off so that it contains the same number of significant digits as the number in the calculation with the fewest significant digits.

For example,

$$\underset{\text{2 digits}}{\underline{2.1 \text{ cm}}} \times \underset{\text{3 digits}}{\underline{6.45 \text{ cm}}} = 13.545 \text{ cm}^2$$

should be rounded off to

2 digits
$$\overset{\frown}{14} \text{ cm}^2.$$

For future work in this text with measurements and scientific notation, results will be rounded off according to the above conventions.

EXAMPLE 5

Multiply and write scientific notation for the answer:

$$(7.2 \times 10^5)(4.3 \times 10^9).$$

SOLUTION We have

$$(7.2 \times 10^5)(4.3 \times 10^9) = (7.2 \times 4.3)(10^5 \times 10^9)$$

Using the commutative and associative laws

$$= 30.96 \times 10^{14}$$

Adding exponents

$$= 31 \times 10^{14}.$$

Rounding to 2 significant digits

To find scientific notation for the result, we convert 31 to scientific notation and simplify:

$$31 \times 10^{14} = (3.1 \times 10^1) \times 10^{14} = 3.1 \times 10^{15}.$$

TECHNOLOGY CONNECTION

1.7

Many calculators allow expressions to be entered using scientific notation. To do so, a key normally labeled EE or EXP is used. Often this is a secondary function and a key labeled SHIFT or 2nd must be pressed first. To check Example 5, we press 7.2 EE 5 × 4.3 EE 9. When we then press ENTER or = , the result 3.096E15 or 3.096 15   appears. We must interpret this result as 3.096×10^{15}. Most calculators do not follow conventions for rounding significant digits.

EXAMPLE 6

Divide and write scientific notation for the answer:

$$\frac{3.48 \times 10^{-7}}{4.64 \times 10^6}.$$

SOLUTION

$$\frac{3.48 \times 10^{-7}}{4.64 \times 10^6} = \frac{3.48}{4.64} \times \frac{10^{-7}}{10^6}$$

Separating factors. Our answer must have 3 significant digits.

$$= 0.75 \times 10^{-13}$$

Subtracting exponents; simplifying

$$= (7.5 \times 10^{-1}) \times 10^{-13}$$

Converting 0.75 to scientific notation

$$= 7.50 \times 10^{-14}$$

Adding exponents. We write 7.50 to indicate 3 significant digits.

Scientific Notation in Problem Solving

Scientific notation can be useful in problem solving.

EXAMPLE 7

Astronomy.   Alpha Centauri is the star—apart from the sun—closest to Earth. Its distance from Earth is about 2.4×10^{13} mi. How many light years is it from Earth to Alpha Centauri?

S O L U T I O N

1. Familiarize.   From an astronomy text, we learn that light travels about 5.88×10^{12} mi in one year. Thus 1 light year = 5.88×10^{12} mi. We will let y represent the number of light years from Earth to Alpha Centauri. Let's guess that the answer is 3 light years. Then the distance in miles would be

$$(5.88 \times 10^{12}) \cdot 3 = 17.64 \times 10^{12} = 1.764 \times 10^{13}.$$

1 light year = 5.88×10^{12} mi

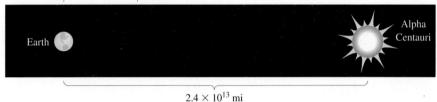

2.4×10^{13} mi

Although our guess is not correct, it does tell us that the distance is more than 3 light years. We are also better able to translate to an equation.

2. Translate.   Note that the distance to Alpha Centauri is y light years, or $(5.88 \times 10^{12})y$ mi. We also are told that the distance is 2.4×10^{13} mi. Since the quantities $(5.88 \times 10^{12})y$ and 2.4×10^{13} both represent the number of miles to Alpha Centauri, we form the equation

$$(5.88 \times 10^{12})y = 2.4 \times 10^{13}.$$

3. Carry out.   We solve the equation:

$$(5.88 \times 10^{12})y = 2.4 \times 10^{13}$$

$$\frac{1}{5.88 \times 10^{12}}(5.88 \times 10^{12})y = \frac{1}{5.88 \times 10^{12}} \times 2.4 \times 10^{13} \qquad \text{Multiplying by } 1/(5.88 \times 10^{12}) \text{ on both sides}$$

$$y = \frac{2.4 \times 10^{13}}{5.88 \times 10^{12}} \qquad \text{Simplifying}$$

$$= \frac{2.4}{5.88} \times \frac{10^{13}}{10^{12}} \qquad \text{Factoring. Our answer must have 2 significant digits (because of the 2.4).}$$

$$\approx 0.41 \times 10 = 4.1.$$

4. Check.   Since light travels 5.88×10^{12} mi in one year, in 4.1 yr it will travel $4.1 \times 5.88 \times 10^{12} = 2.4108 \times 10^{13}$ mi, which is approximately the distance from Earth to Alpha Centauri.

5. State. The distance from Earth to Alpha Centauri is about 4.1 light years.

EXAMPLE 8

Telecommunications. A fiber-optic wire will be used for 350 km of transmission line. The wire has a diameter of 1.2 cm. What is the volume of wire needed for the line?

SOLUTION

1. Familiarize. Drawing a picture, we see that we have a cylinder (a very *long* one). Its length is 350 km and the base has a diameter of 1.2 cm.

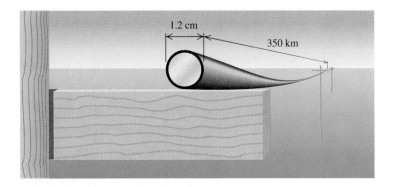

Recall that the formula for the volume of a cylinder is

$$V = \pi r^2 h,$$

where r is the radius and h is the height (in this case, the length of the wire).

2. Translate. We will use the volume formula, but it is important to make the units consistent. Let's express everything in meters:

Length: 350 km = 350,000 m,   or 3.5×10^5 m;
Diameter: 1.2 cm = 0.012 m,   or 1.2×10^{-2} m.

The radius, which we will need in the formula, is half the diameter:

Radius: 0.6×10^{-2} m,   or 6×10^{-3} m.

We now substitute into the above formula:

$$V = \pi (6 \times 10^{-3})^2 (3.5 \times 10^5).$$

3. Carry out. We do the calculation, using 3.14 for π:

$$
\begin{aligned}
V &= 3.14 \times (6 \times 10^{-3})^2 (3.5 \times 10^5) \\
&= 3.14 \times 6^2 \times 10^{-6} \times 3.5 \times 10^5 \\
&= (3.14 \times 6^2 \times 3.5) \times (10^{-6} \times 10^5) \\
&= 395.64 \times 10^{-1} \\
&\approx 4.0 \times 10^1, \text{ or } 40. \quad .
\end{aligned}
$$

Rounding to 2 significant digits. The decimal point emphasizes that we didn't round to the nearest ten.

4. **Check.** About all we can do here is recheck the translation and calculations.

5. **State.** The volume of the wire is about 40. m^3 (cubic meters).

EXERCISE SET

1.7

Convert to scientific notation.

1. 47,000,000,000

2. 2,600,000,000,000

3. 863,000,000,000,000,000

4. 957,000,000,000,000,000

5. 0.000000016

6. 0.000000263

7. 0.00000000007

8. 0.00000000009

9. 407,000,000,000

10. 3,090,000,000,000

11. 0.000000603

12. 0.00000000802

13. 492,700,000,000

14. 953,400,000,000

Convert to decimal notation.

15. 4×10^{-4}

16. 5×10^{-5}

17. 6.73×10^8

18. 9.24×10^7

19. 8.923×10^{-10}

20. 7.034×10^{-2}

21. 9.03×10^{10}

22. 1.01×10^{12}

23. 4.037×10^{-8}

24. 3.007×10^{-9}

25. 8.007×10^{12}

26. 9.001×10^{10}

Simplify and write scientific notation for the answer. Use the correct number of significant digits.

27. $(2.3 \times 10^6)(4.2 \times 10^{-11})$

28. $(6.5 \times 10^3)(5.2 \times 10^{-8})$

29. $(2.34 \times 10^{-8})(5.7 \times 10^{-4})$

30. $(3.26 \times 10^{-6})(8.2 \times 10^{-6})$

31. $(3.2 \times 10^6)(2.6 \times 10^4)$

32. $(3.11 \times 10^3)(1.01 \times 10^{13})$

33. $(3.01 \times 10^{-5})(6.5 \times 10^7)$

34. $(4.08 \times 10^{-10})(7.7 \times 10^5)$

35. $(5.01 \times 10^{-7})(3.02 \times 10^{-6})$

36. $(7.04 \times 10^{-9})(9.01 \times 10^{-7})$

37. $\dfrac{5.1 \times 10^6}{3.4 \times 10^3}$

38. $\dfrac{8.5 \times 10^8}{3.4 \times 10^5}$

39. $\dfrac{7.5 \times 10^{-9}}{2.5 \times 10^{-4}}$

40. $\dfrac{4.0 \times 10^{-6}}{8.0 \times 10^{-3}}$

41. $\dfrac{3.2 \times 10^{-7}}{8.0 \times 10^8}$

42. $\dfrac{12.6 \times 10^8}{4.2 \times 10^{-3}}$

43. $\dfrac{9.36 \times 10^{-11}}{3.12 \times 10^{11}}$

44. $\dfrac{2.42 \times 10^5}{1.21 \times 10^{-5}}$

45. $\dfrac{6.12 \times 10^{19}}{3.06 \times 10^{-7}}$

46. $\dfrac{4.7 \times 10^{-9}}{2.35 \times 10^7}$

47. $\dfrac{780,000,000 \times 0.00071}{0.000005}$

48. $\dfrac{830,000,000 \times 0.12}{3,100,000}$

49. $5.9 \times 10^{23} + 2.4 \times 10^{23}$

50. $1.8 \times 10^{-34} + 5.4 \times 10^{-34}$

Solve.

51. *Astronomy.* Venus has a nearly circular orbit of the sun. If the average distance from the sun to Venus is 1.08×10^8 km, how far does Venus travel in one orbit?

52. *Office supplies.* A ream of copier paper weighs 2.25 kg. How much does a sheet of copier paper weigh?

53. *Printing and engraving.* A ton of five-dollar bills is worth $4,540,000. How many pounds does a five-dollar bill weigh?

54. *Astronomy.*   The average distance of the earth from the sun is about 9.3×10^7 mi. About how far does the earth travel in a yearly orbit about the sun? (Assume a circular orbit.)

55. *Astronomy.*   The brightest star in the night sky, Sirius, is about 4.704×10^{13} mi from Earth. How many light years is it from Earth to Sirius?

56. *Astronomy.*   The diameter of the Milky Way galaxy is approximately 5.88×10^{17} mi. How many light years is it from one end of the galaxy to the other?

Named in tribute to Anders Ångström, a Swedish physicist who measured light waves, 1 Å (read "one Angstrom") equals 10^{-10} meters. One parsec is about 3.26 light years, and one light year equals 9.46×10^{15} meters.

57. How many Angstroms are in one parsec?

58. How many kilometers are in one parsec?

For Exercises 59 and 60, approximate the average distance from the earth to the sun by 1.50×10^{11} meters.

59. Determine the volume of a cylindrical sunbeam that is 3 Å in diameter.

60. Determine the volume of a cylindrical sunbeam that is 5 Å in diameter.

61. *Biology.*   An average of 4.55×10^{11} bacteria live in each pound of U.S. mud.* There are 60 drops in one teaspoon and 6 teaspoons in an ounce. How many bacteria live in a drop of U.S. mud?

62. *Astronomy.*   If a star 5.9×10^{14} mi from Earth were to explode today, its light would not reach us for 100 years. How far does light travel in 13 weeks?

63. *Astronomy.*   The diameter of Jupiter is about 1.43×10^5 km. A day on Jupiter lasts about 10 hr. At what speed is Jupiter's equator spinning?

64. *Finance.*   A *mil* is one thousandth of a dollar. The taxation rate in a certain school district is 5.0 mils for every dollar of assessed valuation. The assessed valuation for the district is 13.4 million dollars. How much tax revenue will be raised?

SKILL MAINTENANCE

65. Subtract: $-\frac{5}{6} - \left(-\frac{3}{4}\right)$.

66. Multiply: $(-7.2)(-4.3)$.

67. Simplify: $-2(x - 3) - 3(4 - x)$.

68. Solve: $4(3x - 7) + 9 = 2$.

SYNTHESIS

69. ◈   Although 1 m = 100 cm, 1 m$^2 \neq$ 100 cm^2. Why?

70. ◈   Write a problem for which the solution is "The area is 3.8×10^{13} m^2."

71. ◈   When a calculator indicates that $5^{17} = 7.629394531 \times 10^{11}$, an approximation is being made. How can you tell? (*Hint:* What should the ones digit be?)

72. ◈   A criminal claims to be carrying $5 million in twenty-dollar bills in a briefcase. Is this possible? Why or why not? (*Hint:* See Exercise 53.)

73. Compare $8 \cdot 10^{-90}$ and $9 \cdot 10^{-91}$. Which is the larger value? How much larger? Write scientific notation for the difference.

74. Write the reciprocal of 8.00×10^{-23} in scientific notation.

75. ▦   Evaluate: $(4096)^{0.05}(4096)^{0.2}$.

76. What is the ones digit in 513^{128}?

77. A grain of sand is placed on the first square of a chessboard, two grains on the second square, four grains on the third, eight on the fourth, and so on. Use scientific notation to approximate the number of grains of sand required for the last square. (*Hint:* Use the fact that $2^{10} \approx 10^3$.)

*Harper's Magazine, April 1996, p. 13.

COLLABORATIVE
C•O•R•N•E•R

Focus: Problem solving, scientific notation, and unit conversion

Time: 15–25 minutes

Group size: 3

ACTIVITY

Given that the earth's average distance from the sun is 1.5×10^{11} meters, determine the earth's orbital speed around the sun in miles per hour. Assume a circular orbit and use the following guidelines.

1. Two students should attempt to solve this problem while the third group member silently observes and writes notes describing the efforts of the other two.
2. After 10–15 minutes, all observers should share their observations with the class as a whole, answering these three questions:

   a) What successful strategies were used?
   b) What unsuccessful strategies were used?
   c) What recommendations do the observers have to make for students working in pairs to solve a problem?

SUMMARY AND REVIEW
1

KEY TERMS

IMPORTANT PROPERTIES AND FORMULAS

Area of a rectangle:	$A = lw$
Area of a square:	$A = s^2$
Area of a parallelogram:	$A = bh$
Area of a trapezoid:	$A = \dfrac{h}{2}(b_1 + b_2)$
Area of a triangle:	$A = \frac{1}{2}bh$
Area of a circle:	$A = \pi r^2$
Circumference of a circle:	$C = \pi d$
Volume of a cube:	$V = s^3$
Volume of a right circular cylinder:	$V = \pi r^2 h$
Perimeter of a square:	$P = 4s$
Distance traveled:	$d = rt$
Simple interest:	$I = Prt$

Addition of Two Real Numbers

1. *Positive numbers*: Add the numbers. The result is positive.
2. *Negative numbers*: Add absolute values. Make the answer negative.
3. *A negative and a positive number*: If the numbers have the same absolute value, the answer is 0. Otherwise, subtract the smaller absolute value from the larger one:
   a) If the positive number is further from 0, make the answer positive.
   b) If the negative number is further from 0, make the answer negative.
4. *One number is zero*: The sum is the other number.

Multiplication of Two Real Numbers

1. To multiply two numbers with *unlike signs*, multiply their absolute values. The answer is *negative*.
2. To multiply two numbers with the *same sign*, multiply their absolute values. The answer is *positive*.

Division of Two Real Numbers

1. To divide two numbers with *unlike signs*, divide their absolute values. The answer is *negative*.
2. To divide two numbers with the *same sign*, divide their absolute values. The answer is *positive*.

The law of opposites: $a + (-a) = 0$

The law of reciprocals: $a \cdot \dfrac{1}{a} = 1,\, a \neq 0$

Absolute value: $|x| = \begin{cases} x, & \text{if } x \geq 0, \\ -x, & \text{if } x < 0 \end{cases}$

For any number a and any nonzero number b,

$$\frac{-a}{b} = \frac{a}{-b} = -\frac{a}{b}.$$

Rules for Order of Operations

1. Calculate within grouping symbols before calculating outside.
2. Simplify all exponential expressions.
3. Do all multiplication and division, in order, from left to right.
4. Do all addition and subtraction, in order, from left to right.

Commutative laws:	$a + b = b + a,$ $ab = ba$
Associative laws:	$a + (b + c) = (a + b) + c,$ $a(bc) = (ab)c$
Distributive law:	$a(b + c) = ab + ac$

The addition principle for equations:

$a = b$ is equivalent to $a + c = b + c$.

The multiplication principle for equations:

For $c \neq 0$, $a = b$ is equivalent to $a \cdot c = b \cdot c$.

Five Steps for Problem Solving
with Algebra

1. *Familiarize* yourself with the problem situation.
2. *Translate* to mathematical language.
3. *Carry out* some mathematical manipulation.
4. *Check* your possible answer in the original problem.
5. *State* the answer clearly.

To solve a formula for a given letter, identify the letter, and:

1. Multiply on both sides to clear fractions or decimals, if that is needed.
2. Combine like terms on each side where convenient.
3. Get all terms with the letter being solved for on one side of the equation and all other terms on the other side, using the addition principle.
4. Combine like terms again, if necessary. This may require factoring.

5. Solve for the letter in question, using the multiplication principle.

Definitions and Rules for Exponents

For any integers m and n (assuming 0 is not raised to a nonpositive power):

Zero as an exponent: $a^0 = 1$

Negative integers as exponents: $a^{-n} = \dfrac{1}{a^n}$

Multiplying with like bases:
  $a^m \cdot a^n = a^{m+n}$ (Product Rule)

Dividing with like bases:
  $\dfrac{a^m}{a^n} = a^{m-n}$; $a \neq 0$ (Quotient Rule)

Raising a product to a power: $(ab)^n = a^n b^n$

Raising a power to a power:
  $(a^m)^n = a^{mn}$ (Power Rule)

Raising a quotient to a power:
  $\left(\dfrac{a}{b}\right)^n = \dfrac{a^n}{b^n}$; $b \neq 0$

Scientific notation for a number is an expression of the type $N \times 10^n$, where $1 \leq N < 10$, N is in decimal notation, and n is an integer.

REVIEW EXERCISES

The following review exercises are for practice. Answers are at the back of the book. If you need to, restudy the section indicated alongside the answer.

1. Translate to an algebraic expression: Five less than the quotient of two numbers.

2. Evaluate
   $$7x^2 - 5y \div zx$$
   for $x = -2$, $y = 3$, and $z = -5$.

3. Name the set consisting of the first six even natural numbers using both roster notation and set-builder notation.

4. Find the area of a triangular sign that has a base of 90 cm and a height of 70 cm.

Find the absolute value.

5. $|-7.3|$ 6. $|4.09|$ 7. $|0|$

Perform the indicated operation.

8. $-9.4 + (-3.7)$ 9. $\left(-\frac{4}{5}\right) + \left(\frac{1}{7}\right)$
10. $\left(-\frac{1}{3}\right) + \frac{4}{5}$ 11. $-7.9 - 3.6$
12. $-\frac{2}{3} - \left(-\frac{1}{2}\right)$ 13. $12.5 - 17.9$
14. $(-2.1)(-3)$ 15. $\left(-\frac{2}{3}\right)\left(\frac{5}{8}\right)$

16. $(1.2)(-4)$

17. $\dfrac{-15}{-3}$

18. $\dfrac{72.8}{-8}$

19. $-7 \div \dfrac{4}{3}$

20. Find $-a$ if $a = -4.01$.

Use a commutative law to write an equivalent expression.

21. $5 + a$

22. $7y$

23. $5x + y$

Use an associative law to write an equivalent expression.

24. $(4 + a) + b$

25. $(xy)7$

26. Obtain an expression that is equivalent to $7mn + 14m$ by factoring.

27. Combine like terms: $5x^3 - 8x^2 + x^3 + 2$.

28. Simplify: $7x - 4[2x + 3(5 - 4x)]$.

Solve. If the solution set is $\varnothing$ or $\mathbb{R}$, classify the equation as a contradiction or as an identity.

29. $x - 4.9 = 1.7$

30. $\frac{2}{3}a = 9$

31. $-9x + 4(2x - 3) = 5(2x - 3) + 7$

32. $3(x - 4) + 2 = x + 2(x - 5)$

33. $5t - (9 - t) = 4t + 2(3 + t)$

34. Translate to an equation. Do not solve. 13 less than twice a number is 21.

35. A number is 17 less than another number. The sum of the numbers is 115. Find the smaller number.

36. One angle of a triangle measures three times the second angle. The third angle measures twice the second angle. Find the measures of the angles.

37. Solve for m: $P = m/S$.

38. Solve for x: $c = mx - rx$.

39. The volume of a film canister is 28.26 cm³. If the radius of the canister is 1.5 cm, determine the height.

40. Multiply and simplify: $(5a^2b^7)(-2a^3b)$.

41. Divide and simplify: $\dfrac{12x^3y^8}{3x^2y^2}$.

42. Evaluate a^0, a^2, and $-a^2$ for $a = -5.3$.

Simplify. Do not use negative exponents in the answer.

43. $3^{-4} \cdot 3^7$

44. $(5a^2)^3$

45. $(-2a^{-3}b^2)^{-3}$

46. $\left(\dfrac{x^2y^3}{z^4}\right)^{-2}$

47. $\left(\dfrac{2a^{-2}b}{4a^3b^{-3}}\right)^4$

Simplify.

48. $\dfrac{7(5 - 2 \cdot 3) - 3^2}{4^2 - 3^2}$

49. $1 - (2 - 5)^2 + 5 \div 10 \cdot 4^2$

50. Convert 0.000000103 to scientific notation.

51. One *parsec* (a unit that is used in astronomy) is 30,860,000,000,000 km. Write scientific notation for this number.

Simplify and write scientific notation for each answer. Use the correct number of significant digits.

52. $(8.7 \times 10^{-9}) \times (4.3 \times 10^{15})$

53. $\dfrac{1.2 \times 10^{-12}}{6.1 \times 10^{-7}}$

54. A sheet of plastic has a thickness of 0.00015 mm. The sheet is 1.2 m by 79 m. Use scientific notation to find the volume of the sheet.

SYNTHESIS

55. ◆ Describe a method that could be used to write equations that have no solution.

56. ◆ Explain how the distributive law can be used when combining like terms.

57. If the smell of gasoline is detectable at 3 parts per billion, what percent of the air is occupied by the gasoline?

58. Evaluate $a + b(c - a^2)^0 + (abc)^{-1}$ for $a = 2$, $b = -3$, and $c = -4$.

59. What's a better deal: a 13-in. diameter pizza for $8 or a 17-in. diameter pizza for $11? Explain.

60. The surface area of a cube is 486 cm². Find the volume of the cube.

61. Solve for z:
$$m = \frac{x}{y - z}.$$

62. Simplify:
$$\frac{(3^{-2})^a \cdot (3^b)^{-2a}}{(3^{-2})^b \cdot (9^{-b})^{-3a}}.$$

63. Each of Bea's test scores counts three times as much as a quiz score. If after 4 quizzes Bea's average is 82.5, what score does she need on the first test in order to raise her average to 85?

64. Fill in the following blank so as to ensure that the equation is an identity.
$$5x - 7(x + 3) - 4 = 2(7 - x) + \underline{\qquad}$$

65. Fill in the following blank so as to ensure that the equation is a contradiction.

$$20 - 7[3(2x + 4) - 10] = 9 - 2(x - 5) + \underline{\hspace{1cm}}$$

66. Use the commutative law for addition once and the distributive law twice to show that

$$a2 + cb + cd + ad = a(d + 2) + c(b + d).$$

67. Find an irrational number between $\frac{1}{2}$ and $\frac{3}{4}$.

CHAPTER TEST
1

1. Translate to an algebraic expression: Three more than the product of two numbers.

2. Evaluate $a^3 - 5b + b \div ac$ for $a = -2$, $b = 6$, and $c = 3$.

3. A triangular stamp's base measures 3 cm and its height 2.5 cm. Find its area.

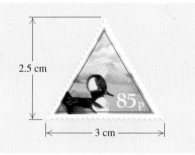

2.5 cm

3 cm

Perform the indicated operation.

4. $-25 + (-16)$

5. $-10.5 + 6.8$

6. $6.21 + (-8.32)$

7. $29.5 - 43.7$

8. $-17.8 - 25.4$

9. $-6.4(5.3)$

10. $-\frac{7}{3} - \left(-\frac{3}{4}\right)$

11. $-\frac{2}{7}\left(-\frac{5}{14}\right)$

12. $\frac{-42.6}{-7.1}$

13. $\frac{2}{5} \div \left(-\frac{3}{10}\right)$

14. Simplify: $5 + (1 - 3)^2 - 7 \div 2^2 \cdot 6$.

15. Use a commutative law to write an expression equivalent to $7x + y$.

Combine like terms.

16. $4y - 15y + 19y$

17. $6a^2b - 5ab^2 + ab^2 - 5a^2b + 2$

18. Simplify: $9x - 3(2x - 5) - 7$.

Solve. If the solution set is $\mathbb{R}$ or $\varnothing$, classify the equation as an identity or a contradiction.

19. $13x - 7 = 41x + 49$

20. $8t - (5 - 2t) = 5(2t - 1)$

21. Solve for P_2: $\dfrac{P_1 V_1}{T_1} = \dfrac{P_2 V_2}{T_2}$.

22. Greg's scores on five tests are 94, 80, 76, 91, and 75. What must Greg score on the sixth test so that his average will be 85?

23. Find three consecutive odd integers such that the sum of 4 times the first, 3 times the second, and 2 times the third is 167.

Simplify. Do not use negative exponents in the answer.

24. $-5(x - 4) - 3(x + 7)$

25. $6b - [7 - 2(9b - 1)]$

26. $(12x^{-4}y^{-7})(-6x^{-6}y)$

27. -3^{-2}

28. $(-6x^2y^{-4})^{-2}$

29. $\left(\dfrac{2x^3y^{-6}}{-4y^{-2}}\right)^2$

30. $(5x^3y)^0$

Simplify and write scientific notation for the answer. Use the correct number of significant digits.

31. $(9.05 \times 10^{-3})(2.22 \times 10^{-5})$

32. $\dfrac{5.6 \times 10^7}{2.8 \times 10^{-3}}$

33. $\dfrac{1.8 \times 10^{-4}}{4.8 \times 10^{-7}}$

Solve.

34. The average distance from the planet Venus to the sun is 6.7×10^7 mi. About how far does Venus travel in one orbit around the sun? (Assume a circular orbit.)

SYNTHESIS

Simplify.

35. $(4x^{3a}y^{b+1})^{2c}$

36. $\dfrac{-27a^{x+1}}{3a^{x-2}}$

37. $\dfrac{(-16x^{x-1}y^{y-2})(2x^{x+1}y^{y+1})}{(-7x^{x+2}y^{y+2})(8x^{x-2}y^{y-1})}$

Graphs, Functions, and Linear Equations

A s a park ranger, I use math to determine the flow of wild and scenic rivers in cubic feet per second and to track the movements of mountain lion, bear, and endangered wildlife populations.

CHRISTINE S. COWLES
Park Ranger
Yosemite National Park, CA

AN APPLICATION

In 1990, the National Park system consisted of about 76.4 million acres. By 1994, the figure was down to 74.9 million acres. Find a linear function that fits this data and then predict the amount of National Park land in 2002.

THIS PROBLEM APPEARS AS
EXERCISE 37 IN SECTION 2.5.

More information on National Parks is available at
http://hepg.awl.com/be/inter_5

raphs are important because they allow us to see relationships. For example, a graph of an equation in two variables helps us see how those two variables are related. Graphs are useful in a wide variety of settings. In this chapter, we emphasize graphs of equations.

A certain kind of relationship between two variables is known as a function. *Functions are very important in mathematics in general, and in problem solving in particular. You will learn what we mean by a function and then begin to use functions to solve problems.*

2.1 Graphs

Points and Ordered Pairs • Quadrants •
Solutions of Equations • Nonlinear Equations

It has often been said that a picture is worth a thousand words. As we turn our attention to the study of graphs, we discover that in mathematics this is quite literally the case. Graphs are a compact means of displaying information and provide a visual approach to problem solving.

Points and Ordered Pairs

On a number line, each point corresponds to a number. On a plane, each point corresponds to a pair of numbers. The idea of using two perpendicular number lines, called **axes**, to identify points in a plane is commonly attributed to the great French mathematician and philosopher René Descartes (1596–1650). Because the variable x is normally represented on the horizontal axis and the variable y is normally represented on the vertical axis, we often refer to the ***x, y* coordinate system**. In honor of Descartes, this representation is also called the **Cartesian coordinate system**.

Note in the figure at right that (2, 3) and (3, 2) are different points. These pairs of numbers are called **ordered pairs** because the order in which the numbers are listed is important. The ordered pair (0, 0) is called the **origin**.

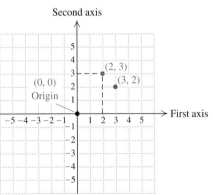

EXAMPLE 1   Plot the points $(-4, 3)$, $(-5, -3)$, $(0, 4)$, and $(2.5, 0)$.

SOLUTION   To plot $(-4, 3)$, we note that the first number, -4, tells us the distance in the first, or horizontal, direction. We go 4 units *left* of the origin. The second number tells us the distance in the second, or vertical, direction. We go 3 units *up*. The point $(-4, 3)$ is then marked, or "plotted."

The points $(-5, -3)$, $(0, 4)$, and $(2.5, 0)$ are also plotted below.

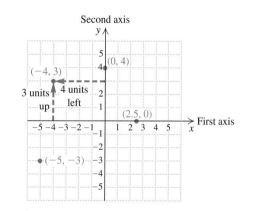

The numbers in an ordered pair are called **coordinates**. In $(-4, 3)$, the *first coordinate* is -4 and the *second coordinate** is 3.

Quadrants

The axes divide the plane into four regions called **quadrants**, as shown here.

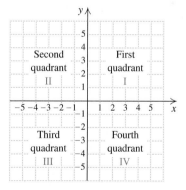

In region I (the *first* quadrant), both coordinates of a point are positive. In region II (the *second* quadrant), the first coordinate is negative and the second coordinate is positive. In the third quadrant, both coordinates are negative, and in the fourth quadrant, the first coordinate is positive while the second coordinate is negative.

Points with one or more 0's as coordinates, such as $(0, -6)$, $(4, 0)$, and $(0, 0)$, are on axes and *not* in quadrants.

---

*The first coordinate is sometimes called the **abscissa** and the second coordinate the **ordinate**.

Solutions of Equations

If an equation has two variables, its solutions are pairs of numbers. When such a solution is written as an ordered pair, the first number listed in the pair usually replaces the variable that occurs first alphabetically.

EXAMPLE 2

Determine whether the pairs $(4, 2)$, $(-1, -4)$, and $(2, 5)$ are solutions of the equation $y = 3x - 1$.

SOLUTION   To determine whether each pair is a solution, we replace x by the first coordinate and y by the second coordinate. When the replacements make the equation true, we say that the ordered pair is a solution.

$$
\begin{array}{c|c}
\multicolumn{2}{c}{y = 3x - 1} \\
\hline
2 & \ ?\ 3(4) - 1 \\
 & 12 - 1 \\
2 & 11
\end{array}
\qquad
\begin{array}{c|c}
\multicolumn{2}{c}{y = 3x - 1} \\
\hline
-4 & \ ?\ 3(-1) - 1 \\
 & -3 - 1 \\
-4 & -4
\end{array}
\qquad
\begin{array}{c|c}
\multicolumn{2}{c}{y = 3x - 1} \\
\hline
5 & \ ?\ 3(2) - 1 \\
 & 6 - 1 \\
5 & 5
\end{array}
$$

Since $2 = 11$ is *false,* the pair $(4, 2)$ is *not* a solution.

Since $-4 = -4$ is *true,* the pair $(-1, -4)$ *is* a solution.

Since $5 = 5$ is *true,* the pair $(2, 5)$ *is* a solution.

In fact, there is an infinite number of solutions of $y = 3x - 1$. Rather than attempt to list all these solutions, we will use a graph as a convenient representation. Thus to *graph* an equation means to make a drawing that represents its solutions.

EXAMPLE 3

Graph the equation $y = x$.

SOLUTION   We label the horizontal axis as the x-axis and the vertical axis as the y-axis.

Next, we find some ordered pairs that are solutions of the equation. In this case, it is easy. Here are a few pairs that satisfy the equation $y = x$:

$$(0, 0), \qquad (1, 1), \qquad (5, 5), \qquad (-1, -1), \qquad (-6, -6).$$

Now we plot these points. We can see that if we were to plot a million solutions, the dots that we drew would merge into a solid line. Observing the pattern, we can draw the line with a ruler. The line is the graph of the equation $y = x$. We label the line $y = x$.

Note that the coordinates of *any* point on the line—for example, $(2.5, 2.5)$—satisfy the equation $y = x$. Note too that the line continues indefinitely in both directions— only part of it is shown.

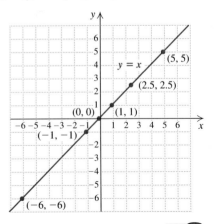

EXAMPLE 4

Graph the equation $y = 2x$.

SOLUTION   We find some ordered pairs that are solutions. This time we list the pairs in a table. To find an ordered pair, we can choose *any* number for x and then determine y. For example, if we choose 3 for x, then $y = 2 \cdot 3 = 6$ (substituting into the equation $y = 2x$). We choose some negative values for x, as well as some positive ones. If a number takes us off the graph paper, we generally do not use it. Next, we plot these points. If we plotted *many* such points, they would appear to make a solid line. We draw the line with a ruler and label it $y = 2x$.

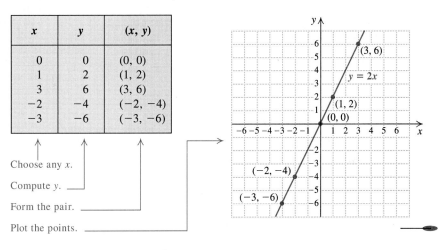

x	y	(x, y)
0	0	(0, 0)
1	2	(1, 2)
3	6	(3, 6)
-2	-4	$(-2, -4)$
-3	-6	$(-3, -6)$

Choose any x.

Compute y.

Form the pair.

Plot the points.

EXAMPLE 5

Graph the equation $y = -\frac{1}{2}x$.

SOLUTION   By choosing even integers for x, we can avoid fractional values when calculating y. For example, if we choose 4 for x, we get $y = \left(-\frac{1}{2}\right)(4)$, or -2. When x is -6, we get $y = \left(-\frac{1}{2}\right)(-6)$, or 3. We find several ordered pairs, plot them, and draw the line.

x	y	(x, y)
4	-2	$(4, -2)$
-6	3	$(-6, 3)$
0	0	(0, 0)
2	-1	$(2, -1)$

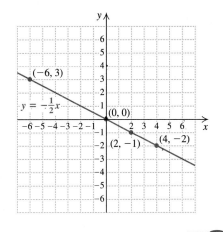

As you can see, the graphs in Examples 3–5 are straight lines. We will refer to any equation whose graph is a straight line as a **linear equation**. Linear equations will be discussed in more detail in Sections 2.3–2.5.

Nonlinear Equations

There are many equations whose graphs are not straight lines. Let's look at some of these **nonlinear equations**.

EXAMPLE 6

Graph: $y = x^2 - 5$.

SOLUTION   We select numbers for x and find the corresponding values for y. For example, if we choose -2 for x, we get $y = (-2)^2 - 5 = 4 - 5 = -1$. The table lists several ordered pairs.

x	y
0	-5
-1	-4
1	-4
-2	-1
2	-1
-3	4
3	4

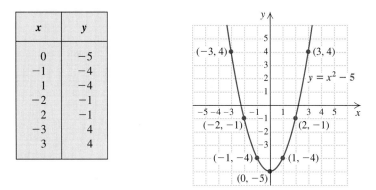

Next, we plot the points. The more points plotted, the clearer the shape of the graph becomes. Since the value of $x^2 - 5$ grows rapidly as x moves away from the origin, the graph rises steeply on either side of the y-axis.   ────●

EXAMPLE 7

Graph: $y = 1/x$.

SOLUTION   We select x-values and find the corresponding y-values. The table lists the ordered pairs $\left(2, \frac{1}{2}\right)$, $\left(-2, -\frac{1}{2}\right)$, $\left(\frac{1}{2}, 2\right)$, and so on.

x	y
3	$\frac{1}{3}$
2	$\frac{1}{2}$
1	1
$\frac{1}{2}$	2
$-\frac{1}{2}$	-2
-1	-1
-2	$-\frac{1}{2}$
-3	$-\frac{1}{3}$

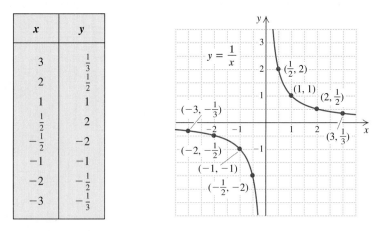

We plot these points, noting that each first coordinate is paired with its reciprocal. Since $1/0$ is undefined, we cannot use 0 as a first coordinate. Thus there are two "branches" to this graph—one on either side of the y-axis. Note that for x-values far to the right or far to the left of 0, the graph approaches, but does not touch, the x-axis.   ────●

EXAMPLE 8   Graph: $y = |x|$.

SOLUTION   We select numbers for x and find the corresponding values for y. For example, if we choose -1 for x, we get $y = |-1| = 1$. Several ordered pairs are listed in the table below.

x	y
-3	3
-2	2
-1	1
0	0
1	1
2	2
3	3

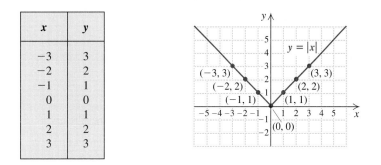

We plot these points, noting that the absolute value of a positive number is the same as the absolute value of its opposite. Thus the x-values 3 and -3 both are paired with the y-value 3. Note that the graph is V-shaped and centered at the origin.

TECHNOLOGY CONNECTION 2.1

All graphing calculators and computers (henceforth *graphers*) utilize a *window*. A window is the rectangular portion of the screen in which a graph appears. Windows are described by four numbers, representing the left and right endpoints of the x-axis and the bottom and top endpoints of the y-axis. A WINDOW key is often used to set these dimensions.

The primary use for graphers is graphing equations. To illustrate, let's graph the equation $y = -4x + 3$. Setting the window so that x and y each range from -10 to 10 results in the graph shown.

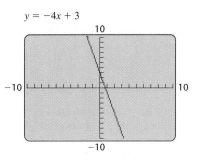

$y = -4x + 3$

Once an equation has been graphed, we can investigate some of its points by using the TRACE feature

that most graphers offer. Usually, a TRACE key is pressed to access this feature. Once it has been pressed, a cursor (often blinking) can be moved along the graph while its x- and y-coordinates appear on the screen.

Many graphers can display a table of ordered pairs for any equation that is entered. By adjusting the TABLE SETUP, we can control the x-values of the ordered pairs through which we scroll.

TABLE MIN = 1.4   ΔTBL = .1

X	Y1	
1.4	-2.6	
1.5	-3	
1.6	-3.4	
1.7	-3.8	
1.8	-4.2	
1.9	-4.6	
2	-5	
X = 1.4		

Graph each of the following equations using a $[-10, 10, -10, 10]$ window. Then use the TRACE feature to find coordinates of at least three points on the graph. If possible, create a table of ordered pairs in which the x-values start at -1 and are 0.1 unit apart.

1. $y = 5x - 3$

2. $y = x^2 - 4x + 3$

3. $y = (x + 4)^2$

4. $y = \sqrt{x + 2}$

5. $y = |x + 2|$

▦ Calculators can shorten our search for ordered pairs, especially when we are uncertain about the shape of a graph. Often, graphs are drawn with the aid of a computer or a graphing calculator. As shown above, determining just a few ordered pairs that solve an equation can be quite time-consuming. With the aid of technology, thousands of ordered pairs can be found in little time (and with no complaining!).

EXERCISE SET

2.1

Plot the points. Label each point with the indicated letter.

1. $A(5, 3)$, $B(2, 4)$, $C(0, 2)$, $D(0, -6)$, $E(3, 0)$, $F(-2, 0)$, $G(1, -3)$, $H(-5, 3)$, $J(-4, 4)$

2. $A(3, 5)$, $B(1, 5)$, $C(0, 4)$, $D(0, -4)$, $E(5, 0)$, $F(-5, 0)$, $G(1, -5)$, $H(-7, 4)$, $J(-5, 5)$

3. $A(3, 0)$, $B(4, 2)$, $C(5, 4)$, $D(6, 6)$, $E(3, -4)$, $F(3, -3)$, $G(3, -2)$, $H(3, -1)$

4. $A(1, 1)$, $B(2, 3)$, $C(3, 5)$, $D(4, 7)$, $E(-2, 1)$, $F(-2, 2)$, $G(-2, 3)$, $H(-2, 4)$, $J(-2, 5)$, $K(-2, 6)$

5. Plot the points $M(2, 3)$, $N(5, -3)$, and $P(-2, -3)$. Draw $\overline{MN}$, $\overline{NP}$, and $\overline{MP}$. ($\overline{MN}$ means the line segment from M to N.) What kind of geometric figure is formed? What is its area?

6. Plot the points $Q(-4, 3)$, $R(5, 3)$, $S(2, -1)$, and $T(-7, -1)$. Draw $\overline{QR}$, $\overline{RS}$, $\overline{ST}$, and $\overline{TQ}$. What kind of figure is formed? What is its area?

Name the quadrant in which each point is located.

7. $(-3, -5)$ **8.** $(2, 17)$ **9.** $(-6, 1)$

10. $(4, -8)$ **11.** $\left(3, \frac{1}{2}\right)$ **12.** $(-1, -8)$

13. $(7, -0.2)$ **14.** $(-4, 31)$

Determine if each ordered pair is a solution of the given equation.

15. $(1, -1)$; $y = 2x - 3$ **16.** $(2, 5)$; $y = 4x - 3$

17. $(3, 4)$; $3s + t = 4$ **18.** $(2, 3)$; $2p + q = 5$

19. $(3, 5)$; $4x - y = 7$ **20.** $(2, 7)$; $5x - y = 3$

21. $\left(0, \frac{3}{5}\right)$; $2a + 5b = 3$ **22.** $\left(0, \frac{3}{2}\right)$; $3f + 4g = 6$

23. $(2, -1)$; $4r + 3s = 5$ **24.** $(2, -4)$; $5w + 2z = 2$

25. $(3, 2)$; $3x - 2y = -4$ **26.** $(1, 2)$; $2x - 5y = -6$

27. $(-1, 3)$; $y = 3x^2$ **28.** $(2, 4)$; $2r^2 - s = 5$

29. $(2, 3)$; $5s^2 - t = 7$ **30.** $(2, 3)$; $y = x^3 - 5$

Graph.

31. $y = -2x$ **32.** $y = -\frac{1}{2}x$

33. $y = x + 3$ **34.** $y = x - 2$

35. $y = 3x - 2$ **36.** $y = -4x + 1$

37. $y = -2x + 3$ **38.** $y = -3x + 1$

39. $y = \frac{2}{3}x + 1$ **40.** $y = \frac{1}{3}x + 2$

41. $y = -\frac{3}{2}x + 1$ **42.** $y = -\frac{2}{3}x - 2$

43. $y = \frac{3}{4}x + 1$ **44.** $y = x^2$

45. $y = -x^2$ **46.** $y = x^2 + 2$

47. $y = x^2 - 2$ **48.** $x = y^2 + 2$

49. $y = |x| + 2$ **50.** $y = -|x|$

51. $y = 3 - x^2$ **52.** $y = x^3 - 2$

53. $y = -\dfrac{1}{x}$ **54.** $y = \dfrac{3}{x}$

SKILL MAINTENANCE

55. *Landscaping.* Grass seed is being spread on a triangular traffic island. If the grass seed can cover an area of 200 ft² and the island's base is 16 ft long, how tall a triangle can the seed fill?

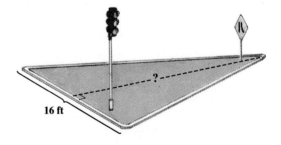

16 ft

56. *Interest rate.* What rate of interest is required in order for a principal of $320 to earn $17.60 in half a year?

57. Subtract: $-3.9 - (-2.5)$.

58. Simplify: $\dfrac{-3 - 7}{2 - 4}$.

SYNTHESIS

59. ◈ Using the equation $y = |x|$, explain why it is "dangerous" to draw a graph after plotting just two points.

60. ◈ Without making a drawing, how can you tell that the graph of $y = x - 30$ passes through three quadrants?

61. ◈ At what point will the line passing through $(a, -1)$ and $(a, 5)$ intersect the line that passes through $(-3, b)$ and $(2, b)$? Why?

62. ◈ Graph $y = 6x$, $y = 3x$, $y = \frac{1}{2}x$, $y = -6x$, $y = -3x$, and $y = -\frac{1}{2}x$ using the same set of axes, and compare the slants of the lines. Describe the pattern that relates the slant of the line to the multiplier of x.

63. ◈ Using the same set of axes, graph $y = 2x$, $y = 2x - 3$, and $y = 2x + 3$. Describe the pattern relating each line to the number that is added to $2x$.

64. Which of the following equations have $\left(-\frac{1}{3}, \frac{1}{4}\right)$ as a solution?

 a) $-\frac{3}{2}x - 3y = -\frac{1}{4}$ b) $8y - 15x = \frac{7}{2}$
 c) $0.16y = -0.09x + 0.1$
 d) $2(-y + 2) - \frac{1}{4}(3x - 1) = 4$

▦ *Graph each equation after plotting at least 10 points.*

65. $y = x^3 + 3x^2 + 3x + 1$; use x-values from -3 to 1

66. $y = 1/(x - 2)$; use values of x from -1 to 5

67. $y = \sqrt{x}$; use values of x from 0 to 10

68. $y = 1/x^2$; use values of x from -3 to 3

69. If $(2, -3)$ and $(-5, 4)$ are the endpoints of a diagonal of a square, what are the coordinates of the other two vertices? What is the area of the square?

70. If $(-10, -2)$, $(-3, 4)$, and $(6, 4)$ are the coordinates of three consecutive vertices of a parallelogram, what are the coordinates of the fourth vertex?

71. Match each sentence with the most appropriate graph found at the top of the next column.

 a) Carpooling to work, Terry spent 10 min on local streets, then 20 min cruising on the freeway, and then 5 min on local streets to his office.
 b) For her commute to work, Sharon drove 10 min to the train station, rode the express for 20 min, and then walked for 5 min to her office.
 c) For his commute to school, Roger walked 10 min to the bus stop, rode the express for 20 min, and then walked 5 min to his class.
 d) Coming home from school, Kristy waited 10 min for the school bus, rode the bus for 20 min, and then walked 5 min to her house.

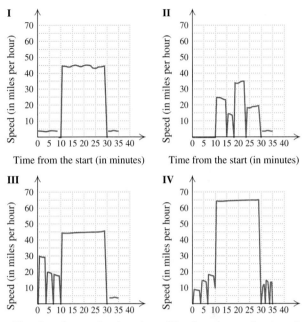

72. Match each sentence with the most appropriate graph.

 a) Roberta worked part time until September, full time until December, and overtime until Christmas.
 b) Clyde worked full time until September, half time until December, and full time until Christmas.
 c) Clarissa worked overtime until September, full time until December, and overtime until Christmas.
 d) Doug worked part time until September, half time until December, and full time until Christmas.

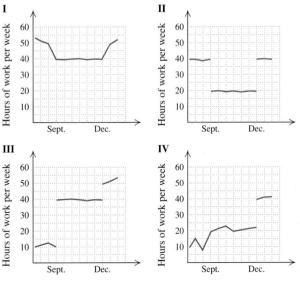

Note: Throughout this text, the icon is used to indicate exercises designed for graphers (graphing calculators or computers).

 In Exercises 73 and 74, use a grapher to draw the graph of each equation. For each equation, select a window that shows the curvature of the graph. Then, if possible, create a table of ordered pairs in which x-values extend, by tenths, from 0 to 0.6.

73. a) $y = -12.4x + 7.8$
    b) $y = -3.5x^2 + 6x - 8$
    c) $y = (x - 3.4)^3 + 5.6$

74. a) $y = 2.3x^4 + 3.4x^2 + 1.2x - 4$
    b) $y = 12.3x - 3.5$
    c) $y = 3(x + 2.3)^2 + 2.3$

2.2 Functions

Functions and Graphs   •   Function Notation and Equations   •   Applications

---

We now develop the idea of a *function*—one of the most important concepts in mathematics. In much the same way that the ordered pairs of Section 2.1 formed correspondences between first coordinates and second coordinates, a function is a special kind of correspondence from one set to another. For example:

To each person in a class   there corresponds   his or her mother.

To each item in a store   there corresponds   its price.

To each real number   there corresponds   the cube of that number.

In each example, the first set is called the **domain**. The second set is called the **range**. For any member of the domain, there is *just one* member of the range to which it corresponds. This kind of correspondence is called a **function**.

EXAMPLE 1   Determine whether each correspondence is a function.

a) $-3 \longrightarrow 5$
    $1 \longrightarrow 2$
    4

b) San Francisco $\longrightarrow$ Giants
    New York $\longrightarrow$ Mets
    Atlanta $\longrightarrow$ Falcons

SOLUTION

a) The correspondence *is* a function because each member of the domain corresponds to *just one* member of the range.

b) The correspondence *is not* a function because a member of the domain (New York) corresponds to more than one member of the range.

FUNCTION

A *function* is a correspondence between a first set, called the *domain*, and a second set, called the *range*, such that each member of the domain corresponds to *exactly one* member of the range.

EXAMPLE 2   Determine whether each correspondence is a function.

Domain	*Correspondence*	*Range*
a) A family	Each person's weight	A set of positive numbers
b) $\{-2, 0, 1, 2\}$	Each number's square	$\{0, 1, 4\}$
c) The set of all states	Each state's members of the U.S. Senate	A set of U.S. senators

SOLUTION

a) The correspondence *is* a function, because each person has *only one* weight.

b) The correspondence *is* a function, because every number has *only one* square.

c) The correspondence *is not* a function, because each state has *two* U.S. senators.

Functions and Graphs

The functions in Examples 1(a) and 2(b) can be expressed as sets of ordered pairs. Example 1(a) can be written $\{(-3, 5), (1, 2), (4, 2)\}$ and Example 2(b) can be written $\{(-2, 4), (0, 0), (1, 1), (2, 4)\}$. We can graph these functions as follows.

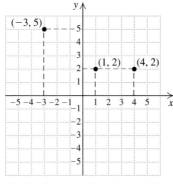

The function $\{(-3, 5), (1, 2), (4, 2)\}$
Domain is $\{-3, 1, 4\}$
Range is $\{5, 2\}$

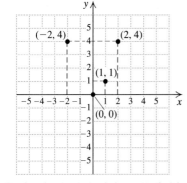

The function $\{(-2, 4), (0, 0), (1, 1), (2, 4)\}$
Domain is $\{-2, 0, 1, 2\}$
Range is $\{4, 0, 1\}$

When a function is given as a set of ordered pairs, the domain is simply the set of all first coordinates and the range is the set of all second coordinates.

EXAMPLE 3

Find the domain and the range of the function f shown here.

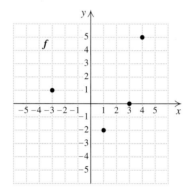

SOLUTION   Here f can be written $\{(-3, 1), (1, -2), (3, 0), (4, 5)\}$. The domain is the set of all first coordinates, $\{-3, 1, 3, 4\}$, and the range is the set of all second coordinates, $\{1, -2, 0, 5\}$. We can also find the domain and the range by directly observing the x- and y-values used in the graph.

EXAMPLE 4

For the function f shown, determine each of the following.

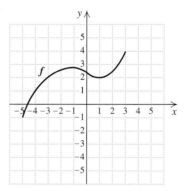

a) What member of the range is paired with 1
b) The domain of f
c) What member of the domain is paired with 1
d) The range of f

SOLUTION

a) To determine what member of the range is paired with 1, we locate 1 on the horizontal axis. Next, we find the point on the graph of f for which 1 is the first coordinate. From that point, we can look to the vertical axis to find the corresponding y-coordinate, 2. The "input" 1 has the "output" 2.

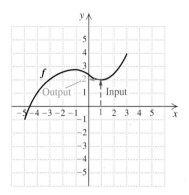

b) The domain of the function is the set of all *x*-values that are in the graph. These extend from −5 to 3 and can be viewed as the curve's shadow, or *projection*, on the *x*-axis. Thus the domain is $\{x|\ -5 \le x \le 3\}$.

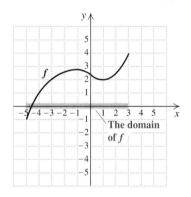

c) To determine what member of the domain is paired with 1, we locate 1 on the vertical axis. (See the graph on the left below.) From there we look left and right to the graph of *f* to find any points for which 1 is the second coordinate. One such point exists, (−4, 1). We observe that −4 is the only element of the domain paired with 1.

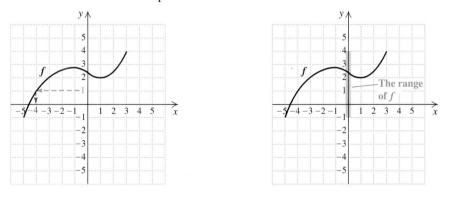

d) The range of the function is the set of all *y*-values that are in the graph. (See the graph on the right above.) These extend from −1 to 4 and can be viewed as the curve's projection on the *y*-axis. Thus the range is $\{y|\ -1 \le y \le 4\}$.

Note that if a graph contains two or more points with the same first coordinate, that graph cannot represent a function (otherwise one member of the domain would correspond to more than one member of the range). This observation is the basis of the *vertical-line test*.

THE VERTICAL-LINE TEST

A graph represents a function if it is not possible to draw a vertical line that intersects the graph more than once.

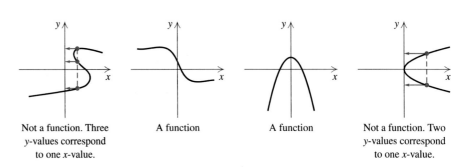

Not a function. Three   A function   A function   Not a function. Two
y-values correspond   *y*-values correspond
to one *x*-value.   to one *x*-value.

Graphs that do not represent functions still do represent *relations*.

RELATION

A *relation* is a correspondence between a first set, called the *domain*, and a second set, called the *range*, such that each member of the domain corresponds to *at least one* member of the range.

Thus, although the correspondences and graphs above are not all functions, they *are* all relations.

Function Notation and Equations

To understand function notation, it helps to imagine a "function machine." Think of putting a member of the domain (an *input*) into the machine. The machine knows the correspondence and produces the appropriate member of the range (the *output*).

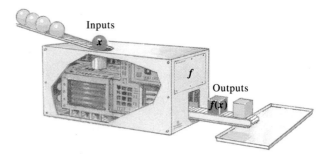

The function pictured has been named *f*. Here *x* represents an arbitrary input, and *f*(*x*)—read "*f* of *x*," "*f* at *x*," or "the value of *f* at *x*"—represents the corresponding output. You should check that in Example 3, *f*(4) is 5, *f*(−3) is 1, and so on. Similarly, in Example 4, *f*(1) = 2. Note that *f*(*x*) *does not mean* *f* times *x*.

Most functions are described by equations. For example, *f*(*x*) = 2*x* + 3 describes the function that takes an input *x*, multiplies it by 2, and then adds 3.

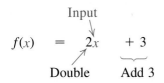

To calculate the output *f*(4), we take the input 4, double it, and add 3 to get 11. That is, we substitute 4 into the formula for *f*(*x*):

$$f(4) = 2 \cdot 4 + 3$$
$$= 11.$$

Sometimes, in place of *f*(*x*) = 2*x* + 3, we write *y* = 2*x* + 3, where it is understood that the value of *y*, the *dependent variable*, is calculated after first choosing a value for *x*, the *independent variable*. To understand why *f*(*x*) notation is so useful, consider two equivalent statements:

a) If *f*(*x*) = 2*x* + 3, then *f*(4) = 11.
b) If *y* = 2*x* + 3, then the value of *y* is 11 when *x* is 4.

The notation used in part (a) is far more concise.

EXAMPLE 5

Find the indicated function value.

a) *f*(5), for *f*(*x*) = 3*x* + 2
b) *g*(−2), for *g*(*r*) = 5*r*² + 3*r*
c) *h*(4), for *h*(*x*) = 7
d) *F*(*a* + 1), for *F*(*x*) = 3*x* + 2

SOLUTION
a) *f*(5) = 3 · 5 + 2 = 17
b) *g*(−2) = 5(−2)² + 3(−2)
$$= 5 \cdot 4 - 6 = 14$$

c) For the function given by *h*(*x*) = 7, all inputs share the same output, 7. Therefore, *h*(4) = 7. The function *h* is an example of a *constant function*.
d) *F*(*a* + 1) = 3(*a* + 1) + 2
$$= 3a + 3 + 2 = 3a + 5$$

Note that whether we write *f*(*x*) = 3*x* + 2, or *f*(*t*) = 3*t* + 2, or *f*(□) = 3□ + 2, we still have *f*(5) = 17. Thus the independent variable can be thought of as a *dummy variable*. The letter chosen for the dummy variable is not as important as the algebraic manipulations to which it is subjected.

When a function is described by an equation, the domain is often unspecified. In such cases, the domain is the set of all numbers for which function values can be calculated.

EXAMPLE 6

For each equation, determine the domain of f.

a) $f(x) = |x|$ **b)** $f(x) = \dfrac{3}{x + 1}$

SOLUTION

a) We ask ourselves, "Is there any number x for which we cannot compute $|x|$?" Since we can find the absolute value of *any* number, the answer is no. Thus the domain of f is $\mathbb{R}$, the set of all real numbers.

b) Is there any number x for which $\dfrac{3}{x + 1}$ cannot be computed? Since $\dfrac{3}{x + 1}$ cannot be computed when $x + 1$ is 0, the answer is yes. To determine what x-value causes the denominator to be 0, we solve an equation:

$x + 1 = 0$ Setting the denominator equal to 0

$\quad x = -1.$ Subtracting 1 on both sides

Thus -1 is *not* in the domain of f, whereas all other real numbers are. The domain of f is $\{x \mid x \text{ is a real number and } x \neq -1\}$.

Applications

Function notation is often used in formulas. For example, to emphasize that the area A of a circle is a function of its radius r, instead of

$A = \pi r^2,$

we can write

$A(r) = \pi r^2.$

When a function is given as a graph in a problem-solving situation, we are often asked to determine certain quantities on the basis of the graph. Often models can be developed and used for calculations. Such a model for Example 7 is developed in Chapter 9. For now, we simply use the graph itself.

EXAMPLE 7

Spread of AIDS.   According to the Federal Centers for Disease Control, there were 30,657 newly reported cases of AIDS in the United States in 1988, 41,639 cases in 1990, 45,839 cases in 1992, 102,605 cases in 1993, 77,561 cases in 1994, and a projected 80,000 cases in 1996.* Estimate the number of newly reported cases for the years 1989 and 1995.

---

*The great increase in the number of cases for 1993 resulted in part from implementation of an expanded definition of AIDS.

SOLUTION

1. and **2. Familiarize** and **Translate**. The given information enables us to plot and connect six points on a graph. We let the horizontal axis represent the year and the vertical axis the number of reported cases. We label the function itself c.

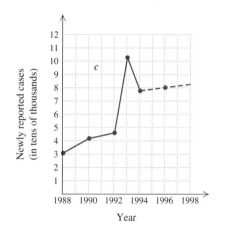

3. **Carry out.**   To estimate the reported number of cases in 1989, we locate the point on the graph that is directly above the year 1989. After doing so, we estimate its second coordinate by moving horizontally from that point to the vertical axis. We see from the graph that $c(1989) \approx 36,000$. Following a similar procedure, we find that $c(1995) \approx 79,000$.

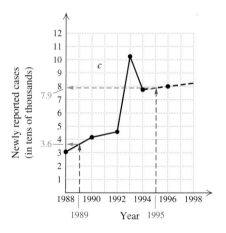

4. **Check.**   A precise check requires consulting an outside information source. Since 36,000 is between 30,657 and 41,639 and 79,000 is between 77,561 and 80,000, our estimates seem plausible.

5. **State.**   There were about 36,000 newly reported cases of AIDS in 1989 and about 79,000 newly reported cases in 1995.

EXERCISE SET

2.2

Determine whether each correspondence is a function.

1. 3 ———→ a
       5 ———→ b
       7 ——→ c
       9 ——→ d
              e

2. 1 ———→ a
       2 ———→ b
       3 ——→ c
       4 ——→ d
       5

3. Girl's Age   Average Daily
       (in months)   Weight Gain (in grams)

       2 —————————————→ 21.8
       9 —————————————→ 11.7
      16 —————————————→ 8.5
      23 —————————————→ 7.0

Source: *American Family Physician*, December 1993,
p. 1435.

4. Boy's Age   Average Daily
       (in months)   Weight Gain (in grams)

       2 —————————————→ 24.3
       9 —————————————→ 11.7
      16 —————————————→ 8.2
      23 —————————————→ 7.0

Source: *American Family Physician*, December 1993,
p. 1435.

5. cat ———————→ dog
       fish ——————→ worm
       dog ———————→ cat
       tiger ——————→ fish
       teacher ———→ student

6. Lake Placid ————→ 1980
       Oslo ——————→ 1976
       Squaw Valley ——→ 1960
       Innsbruck ——————→ 1952
                          1932

*Determine whether each of the following is a function.
Identify any relations that are not functions.*

Domain	Correspondence	Range
7. A family	Each person's eye color	A set of colors
8. A textbook	An even-numbered page in the book	A set of pages
9. A set of avenues	An intersecting road	A set of cross streets
10. A math class	Each person's seat number	A set of numbers
11. A set of numbers	Square each number and then add 4.	A set of numbers
12. A set of shapes	The area of each shape	A set of numbers

For each graph of a function, determine **(a)** $f(1)$; **(b)** *the
domain;* **(c)** *any x-values for which* $f(x) = 2$; *and* **(d)** *the
range.*

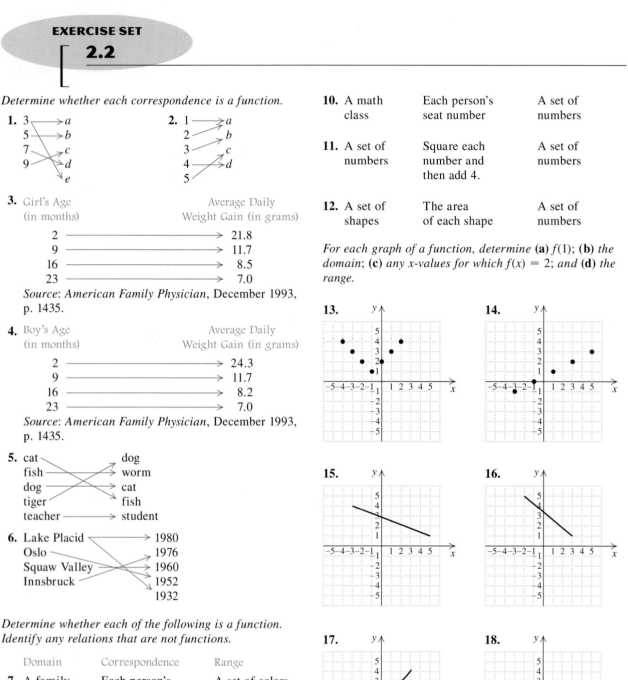

19.
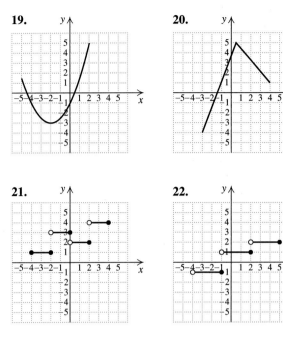

20.

21.

22.

Determine whether each of the following is the graph of a function.

23.

24.

25.
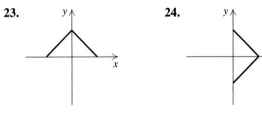

26.

27.
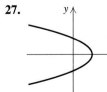

28.
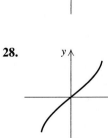

29.
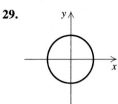

30.
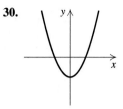

Find the function values.

31. $g(x) = x + 1$

 a) $g(0)$ **b)** $g(-4)$ **c)** $g(-7)$
 d) $g(8)$ **e)** $g(a + 2)$

32. $h(x) = x - 4$

 a) $h(4)$ **b)** $h(8)$ **c)** $h(-3)$
 d) $h(-4)$ **e)** $h(a - 1)$

33. $f(n) = 5n^2 + 4n$

 a) $f(0)$ **b)** $f(-1)$ **c)** $f(3)$
 d) $f(t)$ **e)** $f(2a)$

34. $g(n) = 3n^2 - 2n$

 a) $g(0)$ **b)** $g(-1)$ **c)** $g(3)$
 d) $g(t)$ **e)** $g(2a)$

35. $f(x) = \dfrac{x - 3}{2x - 5}$

 a) $f(0)$ **b)** $f(4)$ **c)** $f(-1)$
 d) $f(3)$ **e)** $f(x + 2)$

36. $s(x) = \dfrac{3x - 4}{2x + 5}$

 a) $s(10)$ **b)** $s(2)$ **c)** $s\left(\tfrac{1}{2}\right)$
 d) $s(-1)$ **e)** $s(x + 3)$

37. Find the domain of f.

 a) $f(x) = \dfrac{2}{x - 3}$ **b)** $f(x) = \dfrac{7}{5 - x}$

 c) $f(x) = 2x + 1$ **d)** $f(x) = x^2 + 3$

 e) $f(x) = \dfrac{3}{2x - 5}$ **f)** $f(x) = |3x - 4|$

38. Find the domain of g.

 a) $g(x) = \dfrac{3}{x - 1}$ **b)** $g(x) = |5 - x|$

 c) $g(x) = \dfrac{9}{x + 3}$ **d)** $g(x) = \dfrac{4}{3x + 4}$

 e) $g(x) = x^3 - 1$ **f)** $g(x) = 7x - 8$

The function A described by $A(s) = s^2 \dfrac{\sqrt{3}}{4}$ gives the area of an equilateral triangle with side s.

s

39. Find the area when a side measures 4 cm.

40. Find the area when a side measures 6 in.

The function V described by $V(r) = 4\pi r^2$ gives the surface area of a sphere with radius r.

41. Find the area when the radius is 3 in.

42. Find the area when the radius is 5 cm.

Chemistry.   *The function F described by*
$$F(C) = \tfrac{9}{5}C + 32$$
gives the Fahrenheit temperature corresponding to the Celsius temperature C.

43. Find the Fahrenheit temperature equivalent to $-10°C$.

44. Find the Fahrenheit temperature equivalent to $5°C$.

Archaeology.   *The function H described by*
$$H(x) = 2.75x + 71.48$$
can be used to predict the height, in centimeters, of a woman whose humerus (the bone from the elbow to the shoulder) is x cm long. Predict the height of a woman whose humerus is the length given.

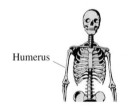

Humerus

45. 32 cm   **46.** 35 cm

Heart attacks and cholesterol.   *For Exercises 47 and 48, use the following graph, which shows the annual heart attack rate per 10,000 men as a function of blood cholesterol level.**

*Copyright 1989, CSPI. Adapted from *Nutrition Action Health-letter* (1875 Connecticut Avenue, N.W., Suite 300, Washington, DC 20009-5728. $24 for 10 issues).

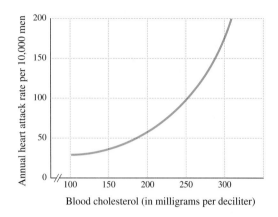

47. Approximate the annual heart attack rate per 10,000 men for those whose blood cholesterol level is 225 mg/dl.

48. Approximate the annual heart attack rate per 10,000 men for those whose blood cholesterol level is 275 mg/dl.

Minivan sales.   *For Exercises 49 and 50, use the following graph, which shows the number of minivans sold as a function of time.*

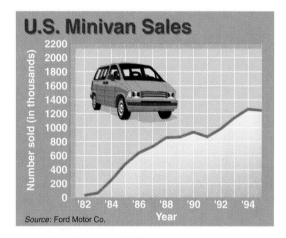

49. Approximate the number of minivans sold in 1992.

50. Approximate the number of minivans sold in 1989.

Blood alcohol level.   *The following table can be used to predict the number of drinks required for a person of a specified weight to be legally intoxicated (blood alcohol level of 0.08 or above) in many states. One 12-oz glass of beer, a 5-oz glass of wine, or a cocktail containing 1 oz of a distilled liquor all count as one drink. Assume that all drinks are consumed within one hour.*

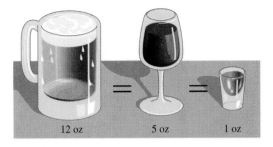

12 oz   5 oz   1 oz

Input, Body Weight (in pounds)	Output, Number of Drinks
100	2.5
160	4
180	4.5
200	5

51. Use the data in the table above to draw a graph and to estimate the number of drinks that a 140-lb person would have to drink to be considered intoxicated.

52. Use the graph from Exercise 51 to estimate the number of drinks a 120-lb person would have to drink to be considered intoxicated.

Electronics.   *An older video cassette recorder has a revolution counter and a booklet with a table relating the counter reading and the time for which a tape has run.*

Counter Reading	Time for Tape (in hours)
000	0
300	1
500	2
675	3
800	4

53. Use the data in the table above to draw a graph of the time that a tape has run as a function of the counter reading and then estimate the time elapsed when the counter has reached 600.

54. Use the graph from Exercise 53 to estimate the time elapsed when the counter has reached 200.

Population growth.   *The town of Falconburg recorded the following dates and populations.*

Input, Year	Output, Population (in tens of thousands)
1989	5.8
1991	6
1993	7
1995	10

55. Use the data in the table above to draw a graph of the population as a function of time. Then estimate what the population was in 1992.

56. Use the graph in Exercise 55 to predict Falconburg's population in the year 1997.

57. *Retailing.*   Shoreside Gifts is experiencing constant growth. They recorded a total of $250,000 in sales in 1992 and $285,000 in 1997. Use a graph that displays the store's total sales as a function of time to predict total sales for the year 2001.

58. Use the graph in Exercise 57 to estimate what the total sales were in 1995.

59. Find three consecutive even integers such that the sum of the first, two times the second, and three times the third is 124.

60. The changes in the salary of a vice president of a corporation for three consecutive years are, respectively, a 10% increase, a 15% increase, and a 5% decrease. What is the percent of total change for those three years?

61. The surface area of a rectangular solid of length l, width w, and height h is given by $S = 2lh + 2lw + 2wh$. Solve for l.

62. Solve the formula in Exercise 61 for w.

63. Simplify: $3(x^2 - 5x) + 2(x - 7)$.

64. Simplify: $(-3a^2b^5)(4a^4b^8)$.

S Y N T H E S I S

Researchers at Yale University have suggested that the following graphs* may represent three different aspects of love.

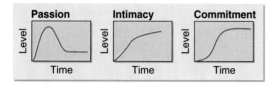

65. ◆ In what unit would you measure time if the horizontal length of each graph were ten units? Why?

66. ◆ Do you agree with the researchers that these graphs should be shaped as they are? Why or why not?

67. ◆ For the function given by $n(z) = ab + wz$, what is the independent variable? How can you tell?

68. ◆ Explain in your own words why every function is a relation, but not every relation is a function.

For Exercises 69 and 70, let $f(x) = 3x^2 - 1$ and $g(x) = 2x + 5$.

69. Find $f(g(-4))$ and $g(f(-4))$.

70. Find $f(g(-1))$ and $g(f(-1))$.

Pregnancy. *For Exercises 71–74, use the following graph of a woman's "stress test." This graph shows the size of a pregnant woman's contractions as a function of time.*

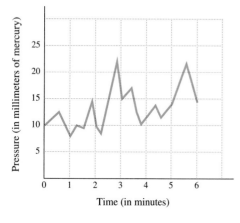

71. How large is the largest contraction that occurred during the test?

72. At what time during the test did the largest contraction occur?

73. ◆ On the basis of the information provided, how large a contraction would you expect 60 seconds after the end of the test? Why?

74. What is the frequency of the largest contraction?

75. The *greatest integer function* $f(x) = [\![x]\!]$ is defined as follows: $[\![x]\!]$ is the greatest integer that is less than or equal to x. For example, if $x = 3.74$, then $[\![x]\!] = 3$; and if $x = -0.98$, then $[\![x]\!] = -1$. Graph the greatest integer function for $-5 \leq x \leq 5$. (The notation $f(x) = \text{INT}[x]$ is used in many graphing calculators and computer programs.)

76. Suppose that a function g is such that $g(-1) = -7$ and $g(3) = 8$. Find a formula for g if $g(x)$ is of the form $g(x) = mx + b$, where m and b are constants.

77. *Energy expenditure.* On the basis of the information given below, what burns more energy: walking $4\frac{1}{2}$ mph for two hours or bicycling 14 mph for one hour?

APPROXIMATE ENERGY EXPENDITURE BY A 150-POUND PERSON IN VARIOUS ACTIVITIES

Activity	Calories per Hour
Walking, $2\frac{1}{2}$ mph	210
Bicycling, $5\frac{1}{2}$ mph	210
Walking, $3\frac{3}{4}$ mph	300
Bicycling, 13 mph	660

Source: Based on material prepared by Robert E. Johnson, M.D., Ph.D., and colleagues, University of Illinois.

---

*From "A Triangular Theory of Love," by R. J. Sternberg, 1986, *Psychological Review*, **93**(2), 119–135. Copyright 1986 by the American Psychological Association, Inc. Reprinted by permission.

COLLABORATIVE
C•O•R•N•E•R

Focus: Functions

Time: 15–20 minutes

Group size: 3–4

The California Department of Motor Vehicles calculates automobile registration fees (VLF) according to the schedule shown below.

ACTIVITY

1. Determine the original sale price of the oldest vehicle owned by a member of your group. Approximate and/or refer to the age of a family member's vehicle, if necessary. Be sure to note the year in which the car was purchased.
2. Use the schedule below to calculate the vehicle license fee (VLF) for the vehicle in part (1) above for each year from the year of purchase to the present.

To speed your work, each group member can find the fee for a few different years.

3. Graph the results from part (2). On the x-axis, plot years beginning with the year of purchase, and on the y-axis, plot $V(x)$, the VLF as a function of year.
4. What is the lowest VLF that the owner of this car will ever have to pay, according to this schedule? Compare your group's answer with other groups' answers.
5. Does your group feel that California's method for calculating registration fees is fair? Why or why not? How could it be improved?
6. Try, as a group, to find an algebraic form for the function $y = V(x)$.
7. *Optional out-of-class extension*: Create a program for a grapher that accepts two inputs (initial value of the vehicle and year of purchase) and produces $V(x)$ as the output.

DMV **VEHICLE LICENSE FEE INFORMATION**

A Public Service Agency

The 2% **Vehicle License Fee (VLF)** is in lieu of a personal property tax on vehicles. Most VLF revenue is returned to City and County Local Governments (see reverse side). The license fee charged is based upon the sale price or vehicle value when initially registered in California. The vehicle value is adjusted for any subsequent sale or transfer, that occurred 8/19/91 or later, excluding sales or transfers between specified relatives.

The VLF is calculated by rounding the sale price to the nearest **odd** hundred dollar. That amount is reduced by a percentage utilizing an eleven year schedule (shown to the right), and 2% of that amount is the fee charged. See the accompanying example for a vehicle purchased last year for $9,199. This would be the second registration year following that purchase.

WHERE DO YOUR DMV FEES GO? SEE REVERSE SIDE.

DMV77 8(REV.8/95)   95 30123

PERCENTAGE SCHEDULE
Rev. & Tax. Code Sec. 10753.2
(Trailer coaches have a different schedule)

1st Year	100%	7th Year	40%
2nd Year	90%	8th Year	30%
3rd Year	80%	9th Year	25%
4th Year	70%	10th year	20%
5th Year	60%	11th Year	
6th Year	50%	onward	15%

VLF CALCULATION EXAMPLE

Purchase Price:	$9,199
Rounded to:	$9,100
Times the Percentage:	90%
Equals Fee Basis of:	$8,190
Times 2% Equals:	$163.80
Rounded to:	$164

2.3 Linear Functions: Graphs and Models

Slope–Intercept Form of an Equation   •   Functions and Graphs
as Models

Different functions have different graphs. In this section, we examine functions with graphs that are straight lines. Such functions and their graphs are called *linear* and can be written in the form $f(x) = mx + b$.

Slope–Intercept Form of an Equation

Examples 3–5 in Section 2.1 showed that for any number m, the graph of $y = mx$ is a straight line passing through the origin. What will happen if we add a number b on the right side to get the equation $y = mx + b$?

EXAMPLE 1   Graph $y = 2x$ and $y = 2x + 3$, using the same set of axes.

SOLUTION   We first make a table of solutions of both equations.

x	y $y = 2x$	y $y = 2x + 3$
0	0	3
1	2	5
−1	−2	1
2	4	7
−2	−4	−1

We then plot these points. Drawing a blue line for $y = 2x + 3$ and a red line for $y = 2x$, we observe that the graph of $y = 2x + 3$ is simply the graph of $y = 2x$ shifted, or *translated*, 3 units up. The lines are parallel.

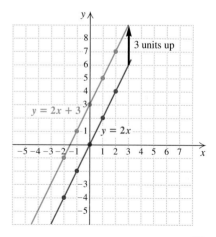

E X A M P L E   2

Graph $f(x) = \frac{1}{3}x$ and $g(x) = \frac{1}{3}x - 2$, using the same set of axes.

S O L U T I O N   We first make a table of solutions of both equations. By choosing multiples of 3, we can avoid fractions.

	$f(x)$	$g(x)$
x	$f(x) = \frac{1}{3}x$	$g(x) = \frac{1}{3}x - 2$
0	0	-2
3	1	-1
-3	-1	-3
6	2	0

We then plot these points. Drawing a blue line for $g(x) = \frac{1}{3}x - 2$ and a red line for $f(x) = \frac{1}{3}x$, we see that the graph of $g(x) = \frac{1}{3}x - 2$ is simply the graph of $f(x) = \frac{1}{3}x$ shifted, or translated, 2 units down.

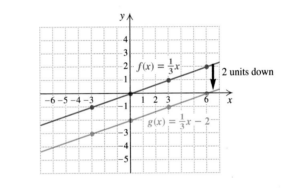

Note that the graph of $y = 2x + 3$ passed through the point $(0, 3)$ and the graph of $g(x) = \frac{1}{3}x - 2$ passed through the point $(0, -2)$. In general, the graph of $y = mx + b$ is a line parallel to $y = mx$, passing through the point $(0, b)$. The point $(0, b)$ is called the **y-intercept**. Often, to save time, we refer to the number b as the y-intercept.

E X A M P L E   3

For each equation, find the y-intercept.

a) $y = -5x + 4$ **b)** $f(x) = 5.3x - 12$

S O L U T I O N

a) The y-intercept is $(0, 4)$, or simply 4.
b) The y-intercept is $(0, -12)$, or simply -12.

In examining the graphs in Examples 1 and 2, note that the slant of the red lines seems to match the slant of the blue lines. This leads us to suspect that it is the number m, in the equation $y = mx + b$, that is responsible for the slant of the line. The following definition enables us to visualize this slant, or *slope*, as a geometric ratio.

SLOPE

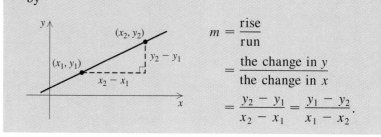

The *slope* of the line passing through (x_1, y_1) and (x_2, y_2) is given by

$$m = \frac{\text{rise}}{\text{run}}$$

$$= \frac{\text{the change in } y}{\text{the change in } x}$$

$$= \frac{y_2 - y_1}{x_2 - x_1} = \frac{y_1 - y_2}{x_1 - x_2}.$$

In the definition above, (x_1, y_1) and (x_2, y_2)—read "x sub-one, y sub-one and x sub-two, y sub-two"—represent two different points on a line. It does not matter which point is considered (x_1, y_1) and which is considered (x_2, y_2) so long as coordinates are subtracted in the same order in both the numerator and the denominator.

The letter m is traditionally used for slope. This usage has its roots in the French verb *monter*, to climb.

EXAMPLE 4

Find the slope of the lines drawn in Examples 1 and 2.

SOLUTION   To find the slope of a line, we can use the coordinates of any two points on that line. We use $(1, 5)$ and $(2, 7)$ to find the slope of the blue line in Example 1:

$$\text{Slope} = \frac{\text{rise}}{\text{run}} = \frac{\text{change in } y}{\text{change in } x} = \frac{y_2 - y_1}{x_2 - x_1} = \frac{7 - 5}{2 - 1} = 2.$$

To find the slope of the red line in Example 1, we use $(-2, -4)$ and $(1, 2)$:

$$\text{Slope} = \frac{\text{rise}}{\text{run}} = \frac{\text{change in } y}{\text{change in } x} = \frac{2 - (-4)}{1 - (-2)} = \frac{6}{3} = 2.$$

We can use $(3, -1)$ and $(6, 0)$ to find the slope of the blue line in Example 2:

$$\text{Slope} = \frac{\text{rise}}{\text{run}} = \frac{\text{change in } y}{\text{change in } x} = \frac{0 - (-1)}{6 - 3} = \frac{1}{3}.$$

You can confirm that the red line in Example 2 also has a slope of $\frac{1}{3}$.

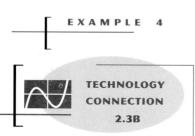

TECHNOLOGY CONNECTION

2.3B

To use a grapher to examine the effect of m when graphing $y = mx + b$, begin with the graph of $y_1 = x + 1$. Then on the same set of axes, graph the lines $y_2 = 2x + 1$ and $y_3 = 3x + 1$. What do you think the graph of $y = 4x + 1$ would look like? Try drawing lines like $y = 2.5x + 1$ and $y = \frac{3}{4}x + 1$ and describe how a positive multiplier of x affects the graph.

To see the effect of a negative value for m, graph the equations $y_1 = x + 1$ and $y_2 = -x + 1$ on the same set of axes. Then graph $y_3 = 2x + 1$ and $y_4 = -2x + 1$ on those same axes. How does a negative multiplier of x affect the graph?

In Example 4, we found that the lines given by $y = 2x + 3$, $y = 2x$, $g(x) = \frac{1}{3}x - 2$, and $f(x) = \frac{1}{3}x$ have slopes 2, 2, $\frac{1}{3}$, and $\frac{1}{3}$, respectively. This supports (but does not prove) the following:

The slope of any line written in the form $y = mx + b$ is m.

A proof of this result is outlined in Exercise 81 on p. 99.

EXAMPLE 5

Determine the slope of the line given by $y = \frac{2}{3}x + 4$, and graph the line.

SOLUTION   Here $m = \frac{2}{3}$, so the slope is $\frac{2}{3}$. This means that from *any* point on the graph, we can locate a second point by simply going *up* 2 units and *to the right* 3 units. Where do we start? Since the y-intercept, $(0, 4)$, is known to be on the graph, we calculate that $(0 + 3, 4 + 2)$, or $(3, 6)$, is also on the graph. Knowing two points, we can draw the graph.

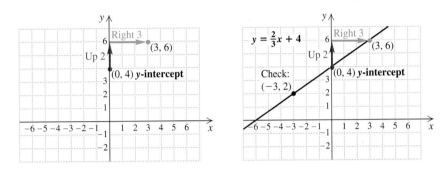

Important: To check the graph, we use some other value for x, say -3, and determine y (in this case, 2). We plot that point and see that it *is* on the line. Were it not, we would know that some error had been made.

When the slope of a line is negative, the line slants downward from left to right.

EXAMPLE 6

Graph: $f(x) = -\frac{1}{2}x + 5$.

SOLUTION   The y-intercept is $(0, 5)$. The slope is $-\frac{1}{2}$, or $\frac{-1}{2}$. From the y-intercept, we go *down* 1 unit and *to the right* 2 units. That gives us the point $(2, 4)$. We can now draw the graph.

As a new type of check, we re-name the slope and find another point:

$$-\frac{1}{2} = \frac{1}{-2}.$$

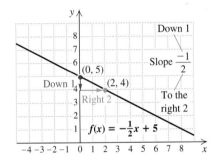

Thus we can go *up* 1 unit and then *to the left* 2 units. This gives the point $(-2, 6)$. Since $(-2, 6)$ is on the line, we have a check.

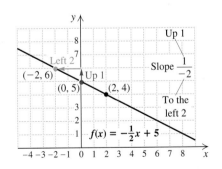

Because an equation of the form $y = mx + b$ makes use of the slope and the y-intercept, it is called the *slope–intercept* form of a linear equation.

THE SLOPE–INTERCEPT EQUATION

Any equation $y = mx + b$ has a graph that is a straight line. It goes through the *y-intercept* $(0, b)$ and has slope m. Any equation of the form $y = mx + b$ is said to be a *slope–intercept equation*.

Note that any graph of $y = mx + b$ will pass the vertical-line test and thus represents a function.

EXAMPLE 7   Find a linear function whose graph has slope $-\frac{2}{3}$ and y-intercept $(0, 4)$.

SOLUTION   We use the slope–intercept form, $f(x) = mx + b$:

$$f(x) = -\frac{2}{3}x + 4. \qquad \text{Substituting } -\frac{2}{3} \text{ for } m \text{ and 4 for } b$$

EXAMPLE 8   Determine the slope and the y-intercept for the graph of $5x - 4y = 8$.

SOLUTION   We convert to a slope–intercept equation:

$$5x - 4y = 8$$
$$-4y = -5x + 8 \qquad \text{Adding } -5x$$
$$y = -\frac{1}{4}(-5x + 8) \qquad \text{Multiplying by } -\frac{1}{4}$$
$$y = \frac{5}{4}x - 2. \qquad \text{Using the distributive law}$$

Because we have an equation of the form $y = mx + b$, we know that the slope is $\frac{5}{4}$ and the y-intercept is $(0, -2)$.

Functions and Graphs as Models

We have seen that slope is the ratio of vertical change to horizontal change. Thus a slope of 3/2 can be taken to mean that the y-values increase 3 units for every 2-unit increase in x-values. Equivalently, we can say that y-values increase $\frac{3}{2}$ units per unit increase in x. In this way, slope is regarded as the *rate of change* of y with respect to x. This viewpoint is useful when formulating models.

EXAMPLE 9   *Cost Projections.* Cleartone Communications charges $50 for a cellular phone and $40 per month for calls made under its economy plan. Formulate and graph a mathematical model for cost. Then use the model to determine the total cost for $3\frac{1}{2}$ months of service.

SOLUTION

1. Familiarize.   The problem describes a situation in which a monthly fee is charged after an initial purchase has been made. After 1 month of ser-

vice, the total cost will be \$50 + \$40 = \$90. After 2 months, the total cost will be \$50 + \$40 · 2 = \$130. This can be generalized in a model if we let $C(t)$ represent the total cost, in dollars, for t months of service.

2. **Translate.** The total cost consists of the \$50 phone purchase plus an additional \$40 for each month of service. Thus,

$$C(t) = 50 + 40t,$$

where $t \geq 0$ (since there cannot be a negative number of months).

3. **Carry out.** Before graphing, we rewrite the model in slope–intercept form: $C(t) = 40t + 50$. We see that the vertical intercept is $(0, 50)$ and the slope—or rate—is \$40 per month.

   We plot $(0, 50)$ and, from there, count *up* \$40 and *to the right* 1 month. This takes us to $(1, 90)$. We then draw a line passing through both points.

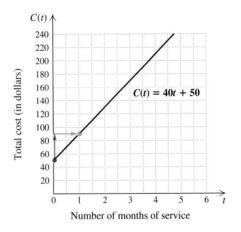

To find the total cost for $3\frac{1}{2}$ months, we determine $C(3.5)$:

$$C(3.5) = 40 \cdot 3.5 + 50 \qquad \text{Replacing } t \text{ by 3.5}$$
$$= 140 + 50$$
$$= 190.$$

4. **Check.** As a check, note that $(2, 130)$ and $(3.5, 190)$ are on the graph (see the following page), as predicted in steps (1) and (3) above. Another way to check is to find the graph's rate of change. This can be done using

any pair of points on the line—we select (2, 130) and (3.5, 190):

$$\text{Rate of change} = \frac{\$190 - \$130}{3.5 \text{ mos} - 2 \text{ mos}}$$

$$= \frac{\$60}{1.5 \text{ mos}} = \$40 \text{ per month.}$$

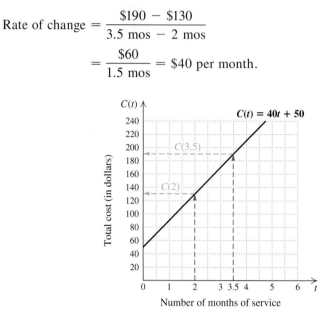

Since this is the rate stated in the problem, our graph and model check.

5. State.   The model $C(t) = 40t + 50$ can be used to determine the total cost, in dollars, of t months of cellular phone service under Cleartone's economy plan. The total cost for $3\frac{1}{2}$ months of service is $190.

EXAMPLE 10

Salvage Value.   Tyline Electric uses the function $S(t) = -700t + 3500$ to determine the *salvage value* $S(t)$, in dollars, of a photocopier t years after its purchase.

a) What do the numbers -700 and 3500 signify?
b) How long will it take the copier to *depreciate* completely?
c) What is the domain of S?

SOLUTION   Drawing, or at least visualizing, a graph can be useful here.

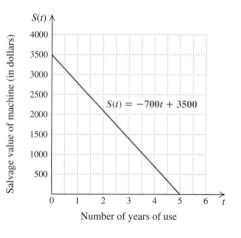

a) At time $t = 0$, we have $S(0) = -700 \cdot 0 + 3500 = 3500$. Thus the number 3500 signifies the original cost of the copier, in dollars.

This function is written in slope–intercept form. Since the output is measured in dollars and the input in years, the number -700 signifies that the value of the copier is declining at a rate of $700 per year.

b) The copier will have depreciated completely when its value drops to 0. To learn when this occurs, we determine when $S(t) = 0$:

$$S(t) = 0$$
$$-700t + 3500 = 0$$
$$-700t = -3500$$
$$t = 5.$$

The copier will have depreciated completely in 5 yr.

c) The number of years of service cannot be negative, nor can the salvage value be negative. In part (b) we found that after 5 yr the salvage value will have dropped to 0. Thus the domain of S is $\{x \mid 0 \le x \le 5\}$. The graph above serves as a visual check of this result.

An equation need not be provided in order for us to determine a rate of change from a graph. In the next example, we examine the rate at which a runner's distance changes with respect to time—the *speed* at which the runner moves.

EXAMPLE 11

Stephanie runs 10 km during each workout. For the first 7 km, her pace is twice as fast as it is for the last 3 km. Which of the following graphs best describes Stephanie's workout?

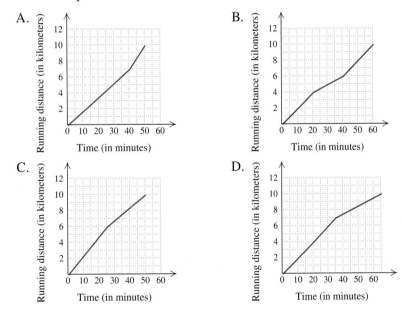

SOLUTION   The slope in graph A increases as we move to the right. This would indicate that Stephanie ran faster for the *last* part of her workout. Thus graph A is not the correct one.

The slopes in graph B indicate that Stephanie slowed down in the middle of her run and then resumed her original speed. Thus graph B does not correctly model the situation either.

According to graph C, Stephanie slowed down not at the 7-km mark, but at the 6-km mark. Thus graph C is also incorrect.

Graph D indicates that Stephanie ran the first 7 km in 35 min, a rate of 0.2 km/min. It also indicates that she ran the final 3 km in 30 min, a rate of 0.1 km/min. This means that Stephanie's rate was twice as fast for the first 7 km, so graph D provides a correct description of her workout.   ⬤━━

EXERCISE SET

2.3

Determine the slope and the y-intercept.

1. $y = 4x + 5$

2. $y = 5x + 3$

3. $f(x) = -2x - 6$

4. $g(x) = -5x + 7$

5. $y = -\frac{3}{8}x - 0.2$

6. $y = \frac{15}{7}x + 2.2$

7. $g(x) = 0.5x - 9$

8. $f(x) = -3.1x + 5$

9. $y = 43x + 197$

10. $y = -52x + 700$

Find a linear function whose graph has the given slope and y-intercept.

11. Slope $\frac{2}{3}$, y-intercept $(0, -7)$

12. Slope $-\frac{3}{4}$, y-intercept $(0, 5)$

13. Slope -4, y-intercept $(0, 2)$

14. Slope 2, y-intercept $(0, -1)$

15. Slope $-\frac{7}{9}$, y-intercept $(0, 3)$

16. Slope $-\frac{4}{11}$, y-intercept $(0, 9)$

17. Slope 5, y-intercept $\left(0, \frac{1}{2}\right)$

18. Slope 6, y-intercept $\left(0, \frac{2}{3}\right)$

For each pair of points, find the slope of the line containing them.

19. $(6, 9)$ and $(4, 5)$

20. $(8, 7)$ and $(2, -1)$

21. $(3, 8)$ and $(9, -4)$

22. $(17, -12)$ and $(-9, -15)$

23. $(-16.3, 12.4)$ and $(-5.2, 8.7)$

24. $(14.4, -7.8)$ and $(-12.5, -17.6)$

For each graph, find the rate of change. Remember to use appropriate units. See Examples 4 and 9.

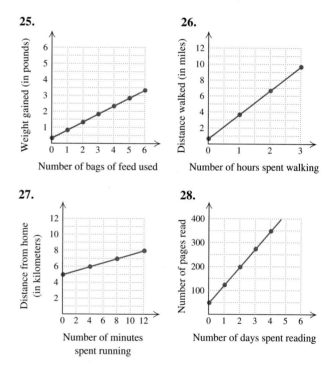

25.

Weight gained (in pounds) / Number of bags of feed used

26.

Distance walked (in miles) / Number of hours spent walking

27.

Distance from home (in kilometers) / Number of minutes spent running

28.

Number of pages read / Number of days spent reading

29.

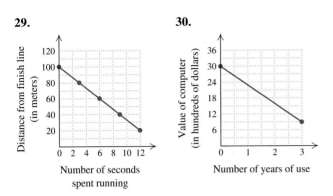

Distance from finish line (in meters) vs. Number of seconds spent running

30.

Value of computer (in hundreds of dollars) vs. Number of years of use

31. *Skiing rate.* A cross-country skier reaches the 3-km mark of a race in 15 min and the 12-km mark 45 min later. Find the speed of the skier.

32. *Running rate.* An Olympic marathoner passes the 5-km point of a race after 30 min and reaches the 25-km point 2 hr later. Find the speed of the marathoner.

33. *Rate of sugar production.* At the beginning of a production run, 4.5 tons of sugar had already been refined. Six hours later, the total amount of refined sugar reached 8.1 tons. Calculate the rate of production.

34. *Work rate.* As a painter begins work, one fourth of a house has already been painted. Eight hours later, the house is two-thirds done. Calculate the painter's work rate.

35. *Rate of descent.* A plane descends to sea level from 12,000 ft after being airborne for $1\frac{1}{2}$ hr. The entire flight time is 2 hr and 10 min. Determine the plane's average rate of descent.

36. *Growth in overseas travel.* In 1988, the number of U.S. visitors overseas was about 11.6 million. In 1995, the number grew to 18.7 million (*Source: Statistical Abstract of the United States*). Determine the rate at which the number of U.S. visitors overseas was growing.

Determine the slope and the y-intercept. Then draw a graph. Be sure to check as in Example 5 or Example 6.

37. $y = \frac{5}{2}x + 1$ **38.** $y = \frac{2}{5}x + 4$

39. $f(x) = -\frac{5}{2}x + 4$ **40.** $f(x) = -\frac{2}{5}x + 3$

41. $2x - y = 5$ **42.** $2x + y = 4$

43. $f(x) = \frac{1}{3}x + 6$ **44.** $f(x) = -3x + 6$

45. $7y + 2x = 7$ **46.** $4y + 20 = x$

47. $f(x) = -0.25x + 2$ **48.** $f(x) = 1.5x - 3$

49. $4x - 5y = 10$ **50.** $5x + 4y = 4$

51. $f(x) = \frac{5}{4}x - 2$ **52.** $f(x) = \frac{4}{3}x + 2$

53. $12 - 4f(x) = 3x$ **54.** $15 + 5f(x) = -2x$

55. *Cricket chirps per minute.* The number of cricket chirps per minute is given by $N(t) = 7.2t - 32$, where t is the temperature in degrees Celsius.
   a) Graph the equation $N(t) = 7.2t - 32$.
   b) Use the graph to predict the number of cricket chirps per minute when it is 5°C.

56. *Cost of a telephone call.* The cost, in dollars, of a long-distance telephone call is given by $C(m) = 0.80m + 1$, where m is the length of the call in minutes.
   a) Graph $C(m) = 0.80m + 1$.
   b) Use the graph to approximate the cost of a 4-min call.
   c) Use the graph to determine how long a phone call can be made for $5.00.

57. *Cable television charges.* Clear County Cable Television charges a $25 installation fee and $20 per month for basic service. Formulate and graph a model that can be used to determine the total cost, $C(t)$, of t months of basic cable TV service. Find the total cost of 6 mos of service.

58. *Deluxe cable service.* Twin Cities Cable Television charges a $35 installation fee and $30 per month for deluxe service. Formulate and graph a model that can be used to determine the total cost, $C(t)$, for t months of deluxe cable TV service. Find the total cost for 8 mos of service.

59. *Cellular phone charges.* The Cellular Connection charges $60 for a cellular phone and $40 per month under its economy plan. Formulate and graph a model that can be used to determine the total cost, $C(t)$, of operating a Cellular Connection phone for t months. Find the total cost for $5\frac{1}{2}$ mos of service and the domain of C.

60. *Telephone charges.*   Tubular Calling charges $50 for a telephone and $25 per month under its economy plan. Formulate and graph a model that can be used to determine the total cost, $C(t)$, of operating a Tubular Calling phone for t months. Find the total cost for $4\frac{1}{2}$ mos of service and the domain of C.

61. *Value of a fax machine.*   FaxMax bought a multifunction fax machine for $750. The machine depreciates at a rate of $25 per month. Find and graph a function F that can be used to determine the value of the fax machine t months after purchase. Then determine the domain of F.

62. *Value of a computer.*   SendUp Graphics bought a computer for $3800. The computer depreciates at a rate of $50 per month. Find and graph a function C that can be used to determine the value of the computer t months after purchase. Then find the domain of C.

In Exercises 63–68, each model is of the form $f(x) = mx + b$. In each case, determine what m and b signify.

63. *Cost of a taxi ride.*   The cost, in dollars, of a taxi ride in Pelham is given by $C(d) = 0.75d + 2$, where d is the number of miles traveled.

64. *Cost of a movie ticket.*   The average price $P(t)$, in dollars, of a movie ticket can be estimated by the function

$$P(t) = 0.1522t + 4.29,$$

where t is the number of years since 1990.

65. *Sales of cotton goods.*   The function given by $f(t) = 2.6t + 17.8$ can be used to estimate the yearly sales of cotton goods, in billions of dollars, t years after 1975.

66. *Cost of renting a truck.*   The cost, in dollars, of a one-day truck rental is given by $C(d) = 0.3d + 20$, where d is the number of miles driven.

67. *Life expectancy of American women.*   The life expectancy of American women t years after 1950 is given by $A(t) = \frac{3}{20}t + 72$.

68. *Natural gas demand.*   The demand, in quadrillions of joules, for natural gas is approximated by $D(t) = \frac{1}{5}t + 20$, where t is the number of years after 1960.

69. Match each sentence with the most appropriate graph.

   a) The rate at which fluids were given intravenously was doubled after 3 hr.

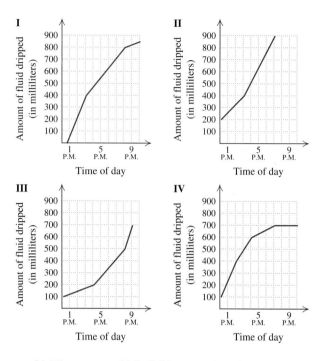

   b) The rate at which fluids were given intravenously was gradually reduced to 0.
   c) The rate at which fluids were given intravenously remained constant for 5 hr.
   d) The rate at which fluids were given intravenously was gradually increased.

70. Match each sentence on the following page with the most appropriate graph below.

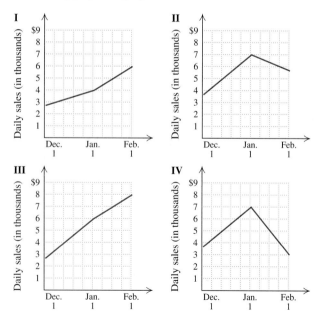

a) After January 1, daily sales continued to rise, but at a slower rate.

b) After January 1, sales decreased faster than they ever grew.

c) The rate of growth in daily sales doubled after January 1.

d) After January 1, daily sales decreased at half the rate that they grew in December.

SKILL MAINTENANCE

71. Simplify: $9\{2x - 3[5x + 2(-3x + y^0 - 2)]\}$.

72. Solve: $3[7x - 2(4 + 5x)] = 5(x + 2) - 7$.

73. Simplify: $(5x^3y^4)^2$.

74. Write in scientific notation: 53,000,000,000.

SYNTHESIS

75. ◈ Belly Up, Inc., is losing $1.5 million per year while Spinning Wheels, Inc., is losing $170 an hour. Which firm would you rather own and why?

76. ◈ Rosie Picshure claims that her firm's profits continue to go up, but the rate of increase is going down.

a) Sketch a graph that might represent her firm's profits as a function of time.

b) Explain why the graph can go up while the rate of increase goes down.

77. ◈ Devise a problem similar to Exercises 31–36 for a classmate to solve.

78. ◈ Is it true that the rate of change for a linear function is constant? Why or why not?

In Exercises 79 and 80, assume that r, p, and s are constants and that x and y are variables. Determine the slope and the y-intercept.

79. $rx + py = s$ **80.** $rx + py = s - ry$

81. Let (x_1, y_1) and (x_2, y_2) be two distinct points on the graph of $y = mx + b$. Use the fact that both pairs are solutions of the equation to prove that m is the slope of the line given by $y = mx + b$. (*Hint*: Use the slope formula.)

Given that $f(x) = mx + b$, classify each of the following as true or false.

82. $f(c + d) = f(c) + f(d)$ **83.** $f(cd) = f(c)f(d)$

84. $f(kx) = kf(x)$

85. $f(c - d) = f(c) - f(d)$

86. Find k so that the line containing $(-3, k)$ and $(4, 8)$ is parallel to the line containing $(5, 3)$ and $(1, -6)$.

87. *Cost of a speeding ticket.* The following penalty schedule is used to determine the cost of a speeding ticket in certain states.

Use this schedule to graph the cost of a speeding ticket as a function of the number of miles per hour over the limit that a driver is going.

88. Find the slope of the line that contains the pair of points.

a) $(5b, -6c)$, $(b, -c)$

b) (b, d), $(b, d + e)$

c) $(c + f, a + d)$, $(c - f, -a - d)$

89. Match each sentence on the following page with the most appropriate graph below.

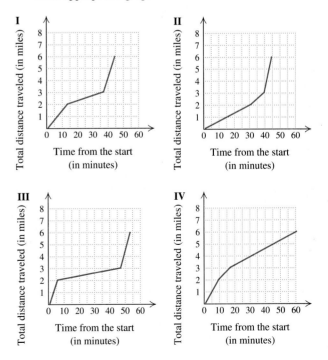

a) Annie drove 2 mi to a lake, swam 1 mi, and then drove 3 mi to a store.

b) During a preseason workout, Rico biked 2 mi, ran for 1 mi, and then walked 3 mi.

c) James bicycled 2 mi to a park, hiked 1 mi over the notch, and then took a 3-mi bus ride back to the park.

d) After hiking 2 mi, Marcy ran for 1 mi before catching a bus for the 3-mi ride into town.

90. 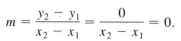 Graph the equations

$$y_1 = 1.4x + 2, \qquad y_2 = 0.6x + 2,$$
$$y_3 = 1.4x + 5, \quad \text{and} \quad y_4 = 0.6x + 5$$

using a grapher. If possible, use the SIMULTANEOUS mode so that you cannot tell which equation is being graphed first. Then decide which line corresponds to each equation.

91. ◈ A student makes a mistake when using a grapher to draw $4x + 5y = 12$ and the following screen appears.

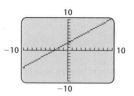

Use algebra to show that a mistake has been made. What do you think the mistake was?

92. ◈ A student makes a mistake when using a grapher to draw $5x - 2y = 3$ and the following screen appears.

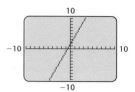

Use algebra to show that a mistake has been made. What do you think the mistake was?

2.4 Another Look at Linear Graphs

Zero Slope and Lines with Undefined Slope •
Graphing Using Intercepts • Solving Equations Graphically •
Recognizing Linear Equations

In Section 2.3, we graphed linear equations using slopes and y-intercepts. We now graph lines that have slope 0 or that have an undefined slope. We also graph lines using both x- and y-intercepts and learn how to use graphs to solve certain equations.

Zero Slope and Lines with Undefined Slope

If two different points have the same second coordinate, what is the slope of the line joining them? In this case, we have $y_2 = y_1$, so

$$m = \frac{y_2 - y_1}{x_2 - x_1} = \frac{0}{x_2 - x_1} = 0.$$

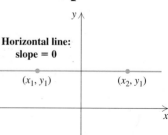

**SLOPE OF A
HORIZONTAL LINE**

Every horizontal line has a slope of 0.

EXAMPLE 1

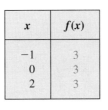

Graph: $f(x) = 3$.

SOLUTION   Recall from Section 2.2 that a function of this type is called a *constant function*. Writing slope–intercept form,

$$f(x) = 0 \cdot x + 3,$$

we see that the *y*-intercept is (0, 3) and the slope is 0. Thus we can graph *f* by plotting the point (0, 3) and, from there, determining a slope of 0. Because $0 = 0/2$ (any nonzero number could be used in place of 2), we can draw the graph by going up 0 units and to the right 2 units. As a check, we also find some ordered pairs. Note that for any choice of *x*-value, $f(x)$ must be 3.

x	$f(x)$
-1	3
0	3
2	3

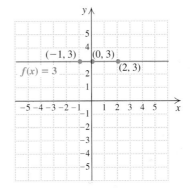

We see from Example 1 that the graph of any constant function of the form $f(x) = b$ or $y = b$ is a horizontal line that crosses the *y*-axis at (0, *b*).

Suppose that two different points are on a vertical line. They then have the same first coordinate. In this case, we have $x_2 = x_1$, so

$$m = \frac{y_2 - y_1}{x_2 - x_1} = \frac{y_2 - y_1}{0}.$$

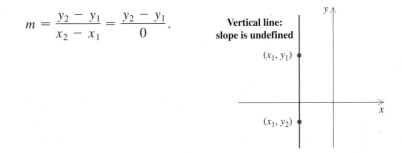

**Vertical line:
slope is undefined**

Since we cannot divide by 0, this is undefined.

SLOPE OF A VERTICAL LINE

The slope of a vertical line is undefined.

EXAMPLE 2   Graph: $x = -2$.

SOLUTION   With y missing, no matter which value of y is chosen, x must be -2. Thus the pairs $(-2, 3)$, $(-2, 0)$, and $(-2, -4)$ all satisfy the equation. The graph is a line parallel to the y-axis. Note that since y is missing, this equation cannot be written in slope–intercept form.

x	y
-2	3
-2	0
-2	-4

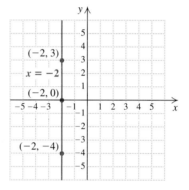

Example 2 shows us that the graph of any equation of the form $x = a$ is a vertical line that crosses the x-axis at $(a, 0)$.

EXAMPLE 3   Find the slope of each given line. If the slope is undefined, state this.

a) $3y + 2 = 8$ **b)** $2x = 10$

SOLUTION

a) We solve for y:

$$3y + 2 = 8$$
$$3y = 6$$
$$y = 2. \quad \text{The graph of } y = 2 \text{ is a horizontal line.}$$

Since $3y + 2 = 8$ is equivalent to $y = 2$, the slope of the line $3y + 2 = 8$ is 0.

b) When y does not appear, we solve for x:

$$2x = 10$$
$$x = 5. \quad \text{The graph of } x = 5 \text{ is a vertical line.}$$

Since $2x = 10$ is equivalent to $x = 5$, the slope of the line $2x = 10$ is undefined.

Graphing Using Intercepts

Any line that is not horizontal or vertical will cross both the x- and y-axes. The points at which the axes are crossed are called the *x-intercept* and the *y-intercept*. Any time these intercepts are not $(0, 0)$, they can be used to graph an equation. Note that to find the y-intercept, we replace x in the equation of the line with 0 and solve for y. To find the x-intercept, we replace y with 0 and solve for x.

TO DETERMINE INTERCEPTS:

The x-intercept is $(a, 0)$. To find a, let $y = 0$ and solve the original equation for x.

The y-intercept is $(0, b)$. To find b, let $x = 0$ and solve the original equation for y.

EXAMPLE 4

Graph the equation $3x + 2y = 12$ by using intercepts.

SOLUTION   *To find the y-intercept, we let $x = 0$ and solve for y:*

$$3 \cdot 0 + 2y = 12$$
$$2y = 12$$
$$y = 6.$$

The y-intercept is $(0, 6)$.

To find the x-intercept, we let $y = 0$ and solve for x:

$$3x + 2 \cdot 0 = 12$$
$$3x = 12$$
$$x = 4.$$

The x-intercept is $(4, 0)$.

We plot the two intercepts and draw the line. A third point could be calculated and used as a check.

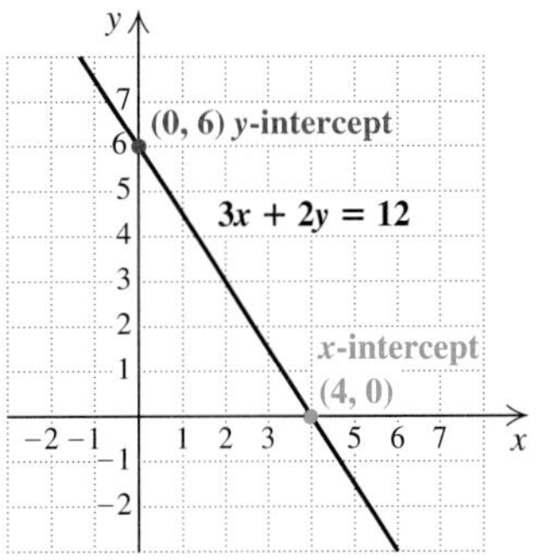

EXAMPLE 5

Graph $f(x) = 2x + 5$ by using intercepts.

SOLUTION   Because the function is in slope–intercept form, we know that the y-intercept is $(0, 5)$. To find the x-intercept, we replace $f(x)$ by 0 and solve for x:

$$0 = 2x + 5$$
$$-5 = 2x$$
$$-\tfrac{5}{2} = x.$$

The x-intercept is $\left(-\tfrac{5}{2}, 0\right)$.

We plot both intercepts and draw the line. As a check, note that the slope of the line is 2, as expected.

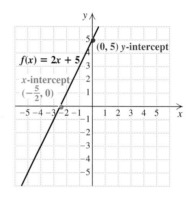

Solving Equations Graphically

Note in Example 5 that the x-intercept, $-\tfrac{5}{2}$, is the solution of $2x + 5 = 0$. Visually, $-\tfrac{5}{2}$ is the x-coordinate of the point at which the graphs of $f(x) = 2x + 5$ and $h(x) = 0$ intersect. Similarly, we can solve $2x + 5 = -3$ by finding the x-coordinate of the point at which the graphs of $f(x) = 2x + 5$ and $g(x) = -3$ intersect. Careful inspection suggests that -4 is that x-value. To check, note that $f(-4) = 2(-4) + 5 = -3$.

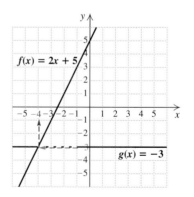

EXAMPLE 6

Solve graphically: $\frac{1}{2}x + 3 = 2$.

SOLUTION   To find the x-value for which $\frac{1}{2}x + 3$ will equal 2, we graph $f(x) = \frac{1}{2}x + 3$ and $g(x) = 2$ on the same set of axes. Since the intersection appears to be $(-2, 2)$, the solution is apparently -2.

Check:

$$
\begin{array}{c|c}
\frac{1}{2}x + 3 = 2 \\
\hline
\frac{1}{2}(-2) + 3 \;?\; 2 \\
-1 + 3 & 2 \\
2 & 2 \quad \text{TRUE}
\end{array}
$$

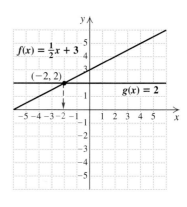

The solution is -2.

EXAMPLE 7

Solve graphically: $-\frac{3}{4}x + 6 = 2x - 1$.

SOLUTION   We graph $f(x) = -\frac{3}{4}x + 6$ and $g(x) = 2x - 1$ on the same set of axes. It appears that the lines intersect at $(2.5, 4)$.

Check:

$$
\begin{array}{c|c}
-\frac{3}{4}x + 6 = 2x - 1 \\
\hline
-\frac{3}{4}(2.5) + 6 \;?\; 2(2.5) - 1 \\
-1.875 + 6 & 5 - 1 \\
4.125 & 4 \qquad \text{FALSE}
\end{array}
$$

Our check shows that 2.5 is *not* the solution, although it may not be off by much. To find the exact solution, we need either a more precise way of determining the point of intersection (see the Technology Connection at the top of the following page) or an algebraic approach:

$$
\begin{aligned}
-\tfrac{3}{4}x + 6 &= 2x - 1 \\
-\tfrac{3}{4}x + 7 &= 2x \qquad && \text{Adding 1 on both sides} \\
7 &= \tfrac{11}{4}x \qquad && \text{Adding } \tfrac{3}{4}x \text{ on both sides} \\
\tfrac{28}{11} &= x. \qquad && \text{Multiplying by } \tfrac{4}{11} \text{ on both sides}
\end{aligned}
$$

The solution is $\frac{28}{11}$, or about 2.55. A check of this answer is left to the student.

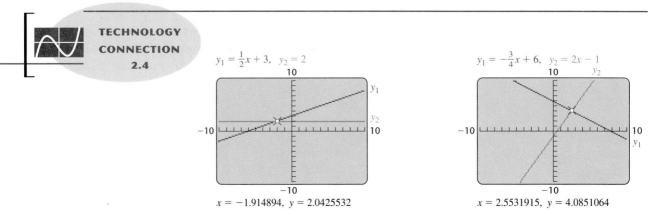

To solve Examples 6 and 7 with a grapher, we can use the TRACE and ZOOM features to magnify the area near the point of intersection. By doing this (more than once, for Example 7), we can find the coordinates of the intersection to the desired degree of accuracy. In the figures shown, we have set a ZOOM "factor" of 5 for both X and Y.

On some graphers, an INTERSECT feature can be used to find the intersection more directly.

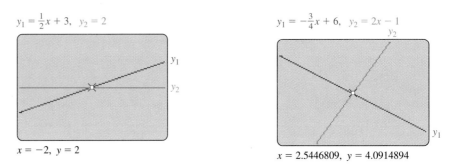

Recognizing Linear Equations

Consider an equation of the form $Ax + By = C$, where A, B, and C are real numbers. Suppose that A and B are nonzero and solve for y:

$$Ax + By = C$$
$$By = -Ax + C \qquad \text{Adding } -Ax \text{ on both sides}$$
$$y = -\frac{A}{B}x + \frac{C}{B}. \qquad \text{Dividing by } B \text{ on both sides}$$

Since the last equation is a slope–intercept equation, we see that $Ax + By = C$ is a linear equation when $A \neq 0$ and $B \neq 0$.

Suppose next that A or B (but not both) is 0. If A is 0, then $By = C$ and $y = C/B$. If B is 0, then $Ax = C$ and $x = C/A$. In the first case, the graph is horizontal; in the second case, the line is vertical. Thus, $Ax + By = C$ is a linear equation when A or B (but not both) is 0. We have now justified the following result.

THE STANDARD FORM OF A LINEAR EQUATION

Any equation of the form $Ax + By = C$, where A, B, and C are real numbers and A and B are not both 0, is linear.

Any equation of the form $Ax + By = C$ is said to be a linear equation in *standard form*.

EXAMPLE 8

Determine whether the equation $y = x^2 - 5$ is linear.

SOLUTION   We attempt to put the equation in standard form:

$$y = x^2 - 5$$
$$-x^2 + y = -5. \qquad \text{Adding } -x^2 \text{ on both sides}$$

This last equation is not linear because it has an x^2-term. We graphed this equation as Example 6 in Section 2.1.

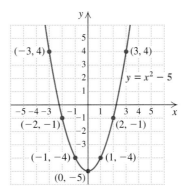

Only linear equations have graphs that are straight lines. Also, only linear graphs have a constant slope. Were you to try to calculate the slope between several pairs of points in Example 8, you would find that the slopes vary.

EXERCISE SET

2.4

For each equation, find the slope. If the slope is undefined, state this.

1. $5x - 6 = 15$
2. $5y - 12 = 3x$
3. $3x = 12 + y$
4. $-12 = 4x - 7$
5. $5y = 6$
6. $19 = -6y$
7. $5x - 7y = 30$
8. $2x - 3y = 18$
9. $12 - 4x = 9 + x$
10. $15 + 7x = 3x - 5$

11. $2y - 4 = 35 + x$
12. $2x - 17 + y = 0$
13. $3y + x = 3y + 2$
14. $x - 4y = 12 - 4y$
15. $4y + 8x = 6$
16. $5y + 6x = -3$
17. $y - 6 = 14$
18. $3y - 5 = 8$
19. $3y - 2x = 5 + 9y - 2x$
20. $17y + 4x + 3 = 7 + 4x$
21. $7x - 3y = -2x + 1$
22. $9x - 4y = 3x + 5$

Graph.

23. $y = 4$

24. $x = -1$

25. $x = 2$

26. $y = 5$

27. $4 \cdot f(x) = 20$

28. $6 \cdot g(x) = 12$

29. $3x = -15$

30. $2x = 10$

31. $4 \cdot g(x) + 3x = 12 + 3x$

32. $6x - 4y + 12 = -4y$

33. $7 - 3x = 4 + 2x$

34. $3 - f(x) = 2$

Find the intercepts. Then graph by using the intercepts and a third point as a check.

35. $x - 2 = y$

36. $x - 4 = y$

37. $3x - 1 = y$

38. $3x - 4 = y$

39. $5x - 4y = 20$

40. $3x + 5y = 15$

41. $f(x) = -2 - 2x$

42. $g(x) = -5 - 5x$

43. $5y = -15 + 3x$

44. $7x = 3y - 21$

45. $g(x) = 2x - 9$

46. $f(x) = 3x - 8$

47. $1.4y - 3.5x = -9.8$

48. $3.6x - 2.1y = 22.68$

49. $5x + 2y = 7$

50. $3x - 4y = 10$

Solve each equation graphically. Then check your answer by solving the same equation algebraically.

51. $x - 3 = 4$

52. $x + 4 = 6$

53. $2x + 1 = 7$

54. $3x - 5 = 1$

55. $\frac{1}{3}x - 2 = 1$

56. $\frac{1}{2}x + 3 = -1$

57. $x + 3 = 5 - x$

58. $x - 7 = 3x - 3$

59. $5 - \frac{1}{2}x = x - 4$

60. $3 - x = \frac{1}{2}x - 3$

61. $2x - 1 = -x + 3$

62. $-3x + 4 = 3x - 4$

Determine whether each equation is linear. Find the slope of any nonvertical lines.

63. $5x - 3y = 15$

64. $3x + 5y + 15 = 0$

65. $16 + 4y = 0$

66. $3x - 12 = 0$

67. $3g(x) = 6x^2$

68. $2x + 4f(x) = 8$

69. $3y = 7xy - 5$

70. $5x - 4xy = 12$

71. $6y - \dfrac{4}{x} = 0$

72. $\dfrac{3y}{4x} = 2x$

73. $\dfrac{f(x)}{x} = x^2$

74. $\dfrac{g(x)}{2} = 3 + x$

SKILL MAINTENANCE

75. We use the formula

$$f = \frac{F(c - v_0)}{c - v_s}$$

when studying the physics of sound. Solve for F.

76. The fare for a taxi ride from Johnson Street to Elm Street is \$5.20. If the rate of the taxi is \$1.00 for the first $\frac{1}{2}$ mi and 30¢ for each additional $\frac{1}{4}$ mi, how far is it from Johnson Street to Elm Street?

Factor.

77. $9x - 15y$

78. $12a + 21ab$

SYNTHESIS

79. ◆ Wind friction, or *air resistance*, increases with speed. Here are some measurements made in a wind tunnel. Plot the data and explain why a linear function does or does not give an approximate fit.

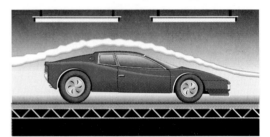

Velocity (in kilometers per hour)	Force of Resistance (in newtons)
10	3
21	4.2
34	6.2
40	7.1
45	15.1
52	29.0

80. ◆ Wind chill is a measure of how cold the wind makes you feel. Below are some measurements of wind chill for a 15-mph breeze. How can you tell from the data that a linear function will give an approximate fit?

Temperature	15-mph Wind Chill
30°	9°
25°	2°
20°	−5°
15°	−11°
10°	−18°
5°	−25°
0°	−31°

Source: National Oceanic & Atmospheric Administration, as reported in the *Burlington Free Press*, 17 January 1992.

81. ◆ When an equation is solved graphically, why is it important that the solution be checked algebraically?

82. ◆ Under what condition(s) will the x- and y-intercepts of a line coincide? What would the equation for such a line look like?

83. Give an equation, in standard form, for the line whose x-intercept is 5 and whose y-intercept is -4.

84. Find the x-intercept of $y = mx + b$, assuming that $m \neq 0$.

In Exercises 85–88, assume that r, p, and s are nonzero constants and that x and y are variables. Determine whether each equation is linear.

85. $rx + 3y = p - s$

86. $py = sx - ry + 2$

87. $r^2x = py + 5$

88. $\dfrac{x}{r} - py = 17$

89. Suppose that two linear equations have the same y-intercept but that equation A has an x-intercept that is half the x-intercept of equation B. How do the slopes compare?

Consider the linear equation

$$ax + 3y = 5x - by + 8.$$

90. Find a and b if the graph is a horizontal line passing through $(0, 2)$.

91. Find a and b if the graph is a vertical line passing through $(4, 0)$.

◢◣ *Solve graphically and then check by solving algebraically.*

92. $5x + 3 = 7 - 2x$

93. $4x - 1 = 3 - 2x$

94. $3x - 2 = 5x - 9$

95. $8 - 7x = -2x - 5$

2.5 Other Equations of Lines

Point–Slope Equations • Parallel and Perpendicular Lines • Which Form to Use?

Specifying a line's slope and one point through which the line passes enables us to draw the line. In this section, we study how this same information can be used to produce an *equation* of the line.

Point–Slope Equations

Suppose that a line of slope m passes through the point (x_1, y_1). For any other point (x, y) to lie on this line, we must have

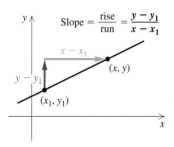

$$\frac{y - y_1}{x - x_1} = m.$$

Note that when x and y are replaced with x_1 and y_1, we have $\frac{0}{0} = m$, a false equation. To avoid this difficulty, we multiply by $x - x_1$ on both sides and simplify:

$$(x - x_1)\frac{y - y_1}{x - x_1} = m(x - x_1)$$

$$y - y_1 = m(x - x_1). \qquad \text{Removing a factor equal to 1: } \frac{x - x_1}{x - x_1} = 1$$

This is the *point–slope* form of a linear equation.

POINT–SLOPE EQUATION

The *point–slope equation* of a line with slope m, passing through (x_1, y_1), is

$$y - y_1 = m(x - x_1).$$

E X A M P L E   1   Find and graph an equation of the line passing through $(3, 4)$ with slope $-\frac{1}{2}$.

S O L U T I O N   We substitute in the equation $y - y_1 = m(x - x_1)$:

$$y - y_1 = m(x - x_1)$$
$$y - 4 = -\frac{1}{2}(x - 3). \qquad \text{Substituting}$$

To graph this point–slope equation, we count off a slope of $-\frac{1}{2}$, starting at $(3, 4)$. Then we draw the line.

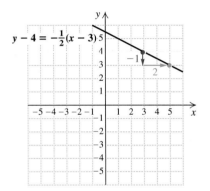

E X A M P L E   2   Find a linear function that has a graph passing through the points $(-1, -5)$ and $(3, -2)$.

S O L U T I O N   We first determine the slope of the line and then use the point–slope equation. Note that

$$m = \frac{-5 - (-2)}{-1 - 3} = \frac{-3}{-4} = \frac{3}{4}.$$

Since the line passes through $(3, -2)$, we have

$$y - (-2) = \frac{3}{4}(x - 3) \qquad \text{Substituting into the point–slope equation}$$
$$y + 2 = \frac{3}{4}x - \frac{9}{4}. \qquad \text{Using the distributive law}$$

Before using function notation, we isolate y:

$$y = \frac{3}{4}x - \frac{17}{4} \qquad \text{Subtracting 2 on both sides: } -\frac{9}{4} - \frac{8}{4} = -\frac{17}{4}$$
$$f(x) = \frac{3}{4}x - \frac{17}{4}. \qquad \text{Using function notation}$$

You can check that using $(-1, -5)$ instead of $(3, -2)$ in $y - y_1 = m(x - x_1)$ will yield the same expression for $f(x)$.

EXAMPLE   3

Medical Insurance.   According to the U.S. Bureau of the Census, the number of people living in the United States with no health insurance grew from 33.4 million in 1989 to 39.7 million in 1993 (*Source*: *Statistical Abstract of the United States*, 1995). Assuming constant growth since 1987, how many people living in the United States in 2001 will lack health insurance?

SOLUTION

1. **Familiarize.**   Constant growth indicates a constant rate of change, so a linear relationship can be assumed. We form the pairs (2, 33.4) and (6, 39.7) as the data points, choosing suitable scales on the two axes. The jagged "break" on the vertical axis is used to avoid including a large portion of unused grid. We let n represent the number of people who are uninsured, in millions, and t the number of years since 1987.

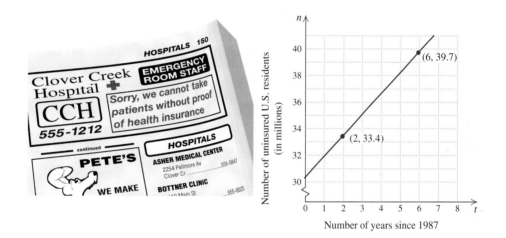

2. **Translate.**   To find an equation relating n and t, we first find the slope of the line. This corresponds to the *growth rate*:

$$m = \frac{39.7 \text{ million people} - 33.4 \text{ million people}}{6 \text{ years} - 2 \text{ years}}$$

$$= \frac{6.3 \text{ million people}}{4 \text{ years}}$$

$$= 1.575 \text{ million people per year.}$$

Next we use the point–slope equation and solve for n:

$n - 33.4 = 1.575(t - 2)$ Writing point–slope form

$n - 33.4 = 1.575t - 3.15$ Using the distributive law

$\qquad\quad n = 1.575t + 30.25.$ Adding 33.4 on both sides

3. **Carry out.**   Using function notation, we have

$n(t) = 1.575t + 30.25.$

To predict the number of people who will be uninsured in 2001, we find

$$n(14) = 1.575 \cdot 14 + 30.25 \qquad \text{2001 is 14 years from 1987.}$$
$$= 52.3.$$

4. **Check.**   To check, we can repeat our calculations. We could also extend the graph to see that (14, 52.3) is on the line.

5. **State.**   Assuming constant growth, there will be 52.3 million uninsured residents of the United States in 2001. ━━━●

Parallel and Perpendicular Lines

If two lines are vertical, they are parallel. How can we tell whether nonvertical lines are parallel? The answer is simple: We look at their slopes (see Examples 1 and 2 in Section 2.3).

SLOPE AND PARALLEL LINES	Two nonvertical lines are parallel if they have the same slope.

EXAMPLE   4

Determine whether the line passing through the points (1, 7) and (4, −2) is parallel to the line given by $f(x) = -3x + 4.2$.

SOLUTION   The slope of the line passing through (1, 7) and (4, −2) is given by

$$m = \frac{7 - (-2)}{1 - 4} = \frac{9}{-3} = -3.$$

Since the graph of $f(x) = -3x + 4.2$ also has a slope of −3, the lines are parallel. ━━━●

If one line is vertical and another is horizontal, they are perpendicular. There are other instances in which two lines are perpendicular.

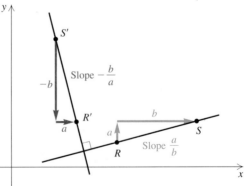

Consider a line $\overleftrightarrow{RS}$, as shown in the graph, with slope a/b. Then think of rotating the figure 90° to get a line $\overleftrightarrow{R'S'}$ perpendicular to $\overleftrightarrow{RS}$. For the new line, the rise and the run are interchanged, but the rise is now negative. Thus the slope of the new line is $-b/a$. Let's multiply the slopes:

$$\frac{a}{b}\left(-\frac{b}{a}\right) = -1.$$

This can help us determine which lines are perpendicular.

SLOPE AND PERPENDICULAR LINES

Two lines are perpendicular if the product of their slopes is -1. (If one line has slope m, the slope of a line perpendicular to it is $-1/m$. That is, we take the reciprocal and change the sign.) Lines are also perpendicular if one is vertical and the other is horizontal.

EXAMPLE 5   Consider the line given by the equation $8y = 7x - 24$.

a) Find an equation for a parallel line passing through $(-1, 2)$.
b) Find an equation for a perpendicular line passing through $(-1, 2)$.

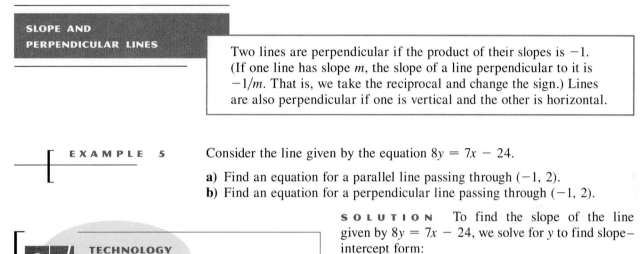

TECHNOLOGY CONNECTION 2.5

To check that $y = \frac{7}{8}x - 3$ and $y = -\frac{8}{7}x + \frac{6}{7}$ are perpendicular, we can use a grapher. To do so, we use the ZSQUARE option of the ZOOM menu to create a "squared" window. This corrects distortion that would otherwise result from differing scales on the axes.

1. Use a grapher to check that
$$y = \tfrac{3}{4}x + 2 \quad \text{and} \quad y = -\tfrac{4}{3}x - 1$$
are perpendicular.
2. Use a grapher to check that
$$y = -\tfrac{2}{5}x - 4 \quad \text{and} \quad y = \tfrac{5}{2}x + 3$$
are perpendicular.
3. To see that this type of check is not foolproof, graph
$$y = \tfrac{31}{40}x + 2 \quad \text{and} \quad y = -\tfrac{40}{30}x - 1.$$
Are the lines perpendicular? Why or why not?

SOLUTION   To find the slope of the line given by $8y = 7x - 24$, we solve for y to find slope–intercept form:

$$8y = 7x - 24$$
$$y = \tfrac{7}{8}x - 3. \qquad \text{Multiplying by } \tfrac{1}{8} \text{ on both sides}$$

The slope is $\frac{7}{8}$.

a) The slope of any parallel line will be $\frac{7}{8}$. The point–slope equation yields

$$y - 2 = \tfrac{7}{8}[x - (-1)] \qquad \text{Substituting } \tfrac{7}{8} \text{ for the slope and } (-1, 2) \text{ for the point}$$
$$y = \tfrac{7}{8}x + \tfrac{23}{8}.$$

b) The slope of a perpendicular line is given by the opposite of the reciprocal of $\frac{7}{8}$, or $-\frac{8}{7}$. The point–slope equation yields

$$y - 2 = -\tfrac{8}{7}[x - (-1)] \qquad \text{Substituting } -\tfrac{8}{7} \text{ for the slope and } (-1, 2) \text{ for the point}$$
$$y = -\tfrac{8}{7}x + \tfrac{6}{7}.$$

Which Form to Use?

We have now studied the slope–intercept, point–slope, and standard forms of a linear equation. Depending on what information we are given and what information we are seeking, one form may be more useful than the others. A referenced summary is given below.

Slope–intercept form, $y = mx + b$	• Useful when an equation is needed and the slope and y-intercept are given. See Example 7 on p. 92. • Useful when a line's slope and y-intercept are needed. See Example 8 on p. 92. • Useful when solving equations graphically. See Example 6 on p. 105. • Commonly used for linear functions.
Standard form, $Ax + By = C$	• Allows for easy calculation of intercepts. See Example 4 on p. 103. • Will prove useful in future work. See Sections 3.1–3.3.
Point–slope form, $y - y_1 = m(x - x_1)$	• Useful when an equation is needed and the slope and a point on the line are given. See Example 1 on p. 110. • Useful when a linear function is needed and two points on its graph are given. See Example 2 on p. 110. • Will prove useful in future work with curves and tangents in calculus.

EXERCISE SET
2.5

Find an equation in point–slope form of the line having the specified slope and containing the point indicated. Then graph the line.

1. $m = 4$, $(2, 3)$ **2.** $m = 5$, $(7, 4)$
3. $m = -2$, $(4, 7)$ **4.** $m = -3$, $(7, 3)$
5. $m = 3$, $(-2, -4)$ **6.** $m = 1$, $(-5, -7)$
7. $m = -2$, $(8, 0)$ **8.** $m = -3$, $(-2, 0)$
9. $m = \frac{2}{5}$, $(-3, 8)$ **10.** $m = \frac{3}{4}$, $(1, -5)$

For each point–slope equation listed, state the slope and a point on the graph.

11. $y - 3 = \frac{2}{7}(x - 1)$ **12.** $y - 4 = 9(x - 2)$
13. $y + 2 = -5(x - 7)$ **14.** $y - 1 = -\frac{2}{9}(x + 5)$
15. $y - 1 = -\frac{5}{3}(x + 2)$ **16.** $y + 7 = -4(x - 9)$

Find an equation of the line having the specified slope and containing the indicated point. Write your final answer as a linear function in slope–intercept form.

17. $m = 5$, $(2, -3)$ **18.** $m = -4$, $(-1, 5)$
19. $m = -\frac{2}{3}$, $(4, -7)$ **20.** $m = -\frac{1}{5}$, $(-2, 1)$
21. $m = -0.6$, $(-3, -4)$ **22.** $m = 2.3$, $(4, -5)$

Find a function for the line containing each pair of points.

23. $(1, 4)$ and $(5, 6)$ **24.** $(2, 6)$ and $(4, 1)$

25. $(2.5, -3)$ and $(6.5, 3)$ **26.** $(2, -1.3)$ and $(7, 1.7)$
27. $(6, 1)$ and $(0, -2)$ **28.** $(-3, 0)$ and $(-1, 5)$
29. $(-2, -3)$ and $(-4, -6)$
30. $(-4, -7)$ and $(-2, -1)$

In Exercises 31–40, assume that a constant rate of change exists for each model formed.

31. *Records in the 400-meter run.*   In 1930, the record for the 400-m run was 46.8 sec. In 1970, it was 43.8 sec. Let R represent the record in the 400-m run and t the number of years since 1930.

a) Find a linear function $R(t)$ that fits the data.
b) Use the function of part (a) to predict the record in 1999; in 2002.
c) When will the record be 40 sec?

32. *Records in the 1500-meter run.*   In 1930, the record for the 1500-m run was 3.85 min. In 1950, it was 3.70 min. Let R represent the record in the 1500-m run and t the number of years since 1930.

a) Find a linear function $R(t)$ that fits the data.
b) Use the function of part (a) to predict the record in 1998; in 2002.
c) When will the record be 3.3 min?

33. *PAC contributions.*   In 1986, Political Action Committees (PACs) contributed $132.7 million to congressional candidates. In 1992, the figure rose to $178.6 million (*Source*: Congressional Research Service and Federal Election Commission). Let *A* represent the amount of PAC contributions and *t* the number of years since 1986.

a) Find a linear function $A(t)$ that fits the data.
b) Use the function of part (a) to predict the amount of PAC contributions in 2002.

PAC Contributions to Congressional Candidates

$132.7 million   1986
$178.6 million   1992

Congressional Research Service & Federal Election Commission

34. *Consumer demand.*   Suppose that 6.5 million lb of coffee are sold when the price is $8 per pound, and 4.0 million lb are sold at $9 per pound.

a) Find a linear function that expresses the amount of coffee sold as a function of the price per pound.
b) Use the function of part (a) to predict how much consumers would be willing to buy at a price of $6 per pound.

35. *Recycling.*   In 1990, Americans recycled 32.9 million tons of garbage. In 1993, the figure grew to 45.0 million tons (*Source: Statistical Abstract of the United States*, 1995). Let *N* represent the number of tons recycled and *t* the number of years since 1990.

a) Find a linear function $N(t)$ that fits the data.
b) Use the function of part (a) to predict the amount recycled in 2001.

36. *Seller's supply.*   Suppose that suppliers are willing to sell 5.0 million lb of coffee at a price of $8 per pound and 7.0 million lb at $9 per pound.

a) Find a linear function that expresses the amount suppliers are willing to sell as a function of the price per pound.
b) Use the function of part (a) to predict how much suppliers would be willing to sell at a price of $6 per pound.

37. *National Park Land.*   In 1990, the National Park system consisted of about 76.4 million acres. By 1994, the figure was down to 74.9 million acres (*Source: Statistical Abstract of the United States,*

1995). Let *A* represent the amount of land in the National Park system, in millions of acres, *t* years after 1990.

a) Find a linear function $A(t)$ that fits the data.
b) Use the function of part (a) to predict the amount of land in the National Park system in 2002.

38. *Pressure at sea depth.*   The pressure 100 ft beneath the ocean's surface is approximately 4 atm (atmospheres), whereas at a depth of 200 ft, the pressure is about 7 atm.

a) Find a linear function that expresses pressure as a function of depth.
b) Use the function of part (a) to determine the pressure at a depth of 690 ft.

39. *Life expectancy of females in the United States.*   In 1990, the life expectancy of females was 78.8 yr. In 1995, it was 79.7 yr (*Source: Statistical Abstract of the United States*, 1995). Let *E* represent life expectancy and *t* the number of years since 1990.

a) Find a linear function $E(t)$ that fits the data.
b) Use the function of part (a) to predict the life expectancy of females in 2004.

40. *Life expectancy of males in the United States.*   In 1990, the life expectancy of males was 71.8 yr. In 1995, it was 72.8 yr. Let *E* represent life expectancy and *t* the number of years since 1990.

a) Find a linear function $E(t)$ that fits the data.
b) Use the function of part (a) to predict the life expectancy of males in 2004.

Without graphing, tell whether the graphs of each pair of equations are parallel.

41. $x + 6 = y,$
$y - x = -2$

42. $2x - 7 = y,$
$y - 2x = 8$

43. $y + 3 = 5x,$
$3x - y = -2$

44. $y + 8 = -6x,$
$-2x + y = 5$

45. $f(x) = 3x + 9,$
$2y = 6x - 2$

46. $f(x) = -7x - 9,$
$-3y = 21x + 7$

Write an equation of the line containing the specified point and parallel to the indicated line.

47. $(3, 7),\ x + 2y = 6$

48. $(0, 3),\ 3x - y = 7$

49. $(2, -1),\ 5x - 7y = 8$

50. $(-4, -5),\ 2x + y = -3$

51. $(-6, 2),\ 3x - 9y = 2$

52. $(-7, 0),\ 5x + 2y = 6$

53. $(-3, -2),\ 3x + 2y = -7$

54. $(-4, 3),\ 6x - 5y = 4$

Without graphing, tell whether the graphs of each pair of equations are perpendicular.

55. $f(x) = 4x - 5,$
$4y = 8 - x$

56. $2x - 5y = -3,$
$2x + 5y = 4$

57. $x + 2y = 5,$
$2x + 4y = 8$

58. $y = -x + 7,$
$f(x) = x + 3$

Write an equation of the line containing the specified point and perpendicular to the indicated line.

59. $(2, 5),\ 2x + y = -3$

60. $(4, 0),\ x - 3y = 0$

61. $(3, -2),\ 3x + 4y = 5$

62. $(-3, -5),\ 5x - 2y = 4$

63. $(0, 9),\ 2x + 5y = 7$

64. $(-3, -4),\ -3x + 6y = 2$

65. $(-4, -7),\ 3x - 5y = 6$

66. $(-4, 5),\ 7x - 2y = 1$

SKILL MAINTENANCE

67. The price of a radio, including 5% sales tax, is $36.75. Find the price of the radio before the tax was added.

68. A basketball team increases its score by 7 points in each of three consecutive games. If the team scored a total of 228 points in all three games, what was its score in the first game?

69. 15% of what number is 12.4?

70. Subtract: $-\frac{7}{4} - \left(-\frac{2}{3}\right)$.

SYNTHESIS

71. ◈ The total number of reported cases of AIDS in the United States grew from 372 in 1981 to 100,000 in 1989 and 200,000 in 1992. Can a linear function be used to predict the number of cases in 2004? Why or why not?

72. ◈ The information in Exercises 34 and 36 indicates that when suppliers charge $8 per pound for coffee, the supply will not meet the demand. How could suppliers determine a price for which their supply will exactly meet the demand?

73. ◈ Explain why the negative slope in Exercise 31 is "good," whereas the negative slope in Exercise 37 is "bad."

74. ◈ A firm offers its entering employees a starting salary with a guaranteed 7% increase each year. Can a linear function be used to express the yearly salary as a function of the number of years an employee has worked? Why or why not?

For Exercises 75–78, assume that a linear equation models each situation.

75. *Depreciation of a computer.* Gina's computer cost $2500 new. The computer's value has dropped to $2150 after 5 mos. What will the computer be worth after 8 mos?

76. *Cellular phone charges.* Rick's cellular phone cost him $70 to purchase and activate. After 4 mos, his total cost for the phone is $190. Predict Rick's total cost for the phone after 9 mos.

77. *Operating expenses.* The total cost for operating Ming's Wings was $7500 after 4 mos and $9250 after 7 mos. Predict the total cost after 10 mos.

78. *Depreciation of a printer.* After 6 mos of use, the value of Pearl's printer had dropped to $900. After 8 mos, the value had gone down to $750. How much did the printer cost when new?

79. *Temperature conversion.* Water freezes at 32° Fahrenheit and at 0° Celsius. Water boils at 212°F and at 100°C. What Celsius temperature corresponds to a room temperature of 70°F?

80. For a linear function f, $f(-1) = 3$ and $f(2) = 4$.
 a) Find an equation for f.
 b) Find $f(3)$.
 c) Find a such that $f(a) = 100$.

81. For a linear function g, $g(3) = -5$ and $g(7) = -1$.

   a) Find an equation for g.

   b) Find $g(-2)$.

   c) Find a such that $g(a) = 75$.

82. Find the value of k so that the graph of $5y - kx = 7$ and the line containing the points $(7, -3)$ and $(-2, 5)$ are parallel.

83. Find the value of k so that the graph of $7y - kx = 9$ and the line containing the points $(2, -1)$ and $(-4, 5)$ are perpendicular.

84. ⊡ Use a grapher with a *squared* window to check your answers to Exercises 41–66.

85. ◈ ⊡ When several data points are available and they appear to be nearly collinear, a procedure known as *linear regression* can be used to find an equation for the line that most closely fits the data.

   a) Use a grapher with a LINEAR REGRESSION option and the table that follows to find a linear function that predicts a woman's life expectancy as a function of the year in which she was born. Compare this with the answer to Exercise 39.

   b) Predict the life expectancy in 2004 and compare your answer with the answer to Exercise 39. Which answer seems more reliable? Why?

LIFE EXPECTANCY OF WOMEN

Year, x	Life expectancy, y (in years)
1920	54.6
1930	61.6
1940	65.2
1950	71.1
1960	73.1
1970	74.7
1980	77.5
1990	78.8

Source: *Statistical Abstract of the United States* and *The World Almanac 1996.*

COLLABORATIVE
C ◆ O ◆ R ◆ N ◆ E ◆ R

Focus: Linear functions and models

Time: 15 minutes to 1 week

Group size: 3

The table at right shows how the political views of first-year college students have changed.

ACTIVITY

1. Each group member should select one of the three categories and find a linear function that can predict the percentage of college students who fall into that category t years after 1970. Then use the function to predict the percentage of students in that category at present.

2. After the entire group has confirmed all three calculations, survey a random sample of first-year students at your college. Survey results may not match your predictions very closely. Why?

How would you describe your political views?

Political Categories	First-year College Students 1970	First-year College Students 1990
Liberal or Far Left	37%	26%
Middle of the Road	45%	54%
Conservative or Far Right	18%	20%

Source: "The American Freshman: Twenty-Five Year Trends" (UCLA, 1992–1993)

2.6 The Algebra of Functions

The Sum, Difference, Product, or Quotient of Two Functions •
Domains and Graphs

We now return to the idea of a function as a machine and examine four ways in which functions can be combined.

The Sum, Difference, Product, or Quotient of Two Functions

Suppose that a is in the domain of two functions, f and g. The input a is paired with $f(a)$ by f and with $g(a)$ by g. The outputs can then be added to get $f(a) + g(a)$.

EXAMPLE 1

Let $f(x) = x + 4$ and $g(x) = x^2 + 1$. Find $f(2) + g(2)$.

SOLUTION   We visualize two function machines. Because 2 is in the domain of each function, we can compute $f(2)$ and $g(2)$.

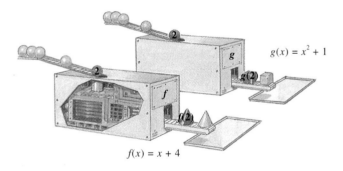

Since

$$f(2) = 2 + 4 = 6 \quad \text{and} \quad g(2) = 2^2 + 1 = 5,$$

we have

$$f(2) + g(2) = 6 + 5 = 11.$$

In Example 1, suppose that we were to write $f(x) + g(x)$ as $(x + 4) + (x^2 + 1)$, or $f(x) + g(x) = x^2 + x + 5$. This could then be regarded as a "new" function. The notation $(f + g)(x)$ is generally used to denote a function formed in this manner. Similar notations exist for subtraction, multiplication, and division of functions.

THE ALGEBRA OF FUNCTIONS

If f and g are functions and x is in the domain of both functions, then:

1. $(f + g)(x) = f(x) + g(x)$;
2. $(f - g)(x) = f(x) - g(x)$;
3. $(f \cdot g)(x) = f(x) \cdot g(x)$;
4. $(f / g)(x) = f(x)/g(x)$, provided $g(x) \neq 0$.

EXAMPLE   2

For $f(x) = x^2 - x$ and $g(x) = x + 2$, find the following.

a) $(f + g)(3)$

b) $(f - g)(x)$ and $(f - g)(-1)$

c) $(f / g)(x)$ and $(f / g)(-4)$

d) $(f \cdot g)(3)$

SOLUTION

a) Since $f(3) = 3^2 - 3 = 6$ and $g(3) = 3 + 2 = 5$, we have

$$(f + g)(3) = f(3) + g(3)$$
$$= 6 + 5 \qquad \text{Substituting}$$
$$= 11.$$

Alternatively, we could first find $(f + g)(x)$:

$$(f + g)(x) = f(x) + g(x)$$
$$= x^2 - x + x + 2$$
$$= x^2 + 2. \qquad \text{Combining like terms}$$

Thus,

$$(f + g)(3) = 3^2 + 2 = 11.$$

b) We have

$$(f - g)(x) = f(x) - g(x)$$
$$= x^2 - x - (x + 2) \qquad \text{Substituting}$$
$$= x^2 - 2x - 2. \qquad \text{Removing parentheses and combining like terms}$$

Thus,

$$(f - g)(-1) = (-1)^2 - 2(-1) - 2 \qquad \text{Using } (f - g)(x) \text{ is faster than using } f(x) - g(x).$$
$$= 1. \qquad \text{Simplifying}$$

c) We have

$$(f / g)(x) = f(x)/g(x)$$
$$= \frac{x^2 - x}{x + 2}.$$

Thus,

$$(f/g)(-4) = \frac{(-4)^2 - (-4)}{-4 + 2} \qquad \text{Substituting}$$

$$= \frac{20}{-2}$$

$$= -10.$$

d) Using our work in part (a), we have

$$(f \cdot g)(3) = f(3) \cdot g(3)$$

$$= 6 \cdot 5$$

$$= 30.$$

It is also possible to compute $(f \cdot g)(3)$ by first multiplying $x^2 - x$ and $x + 2$ using methods we will discuss in Chapter 5.

Domains and Graphs

Although applications involving products and quotients of functions rarely appear in newspapers, situations involving sums or differences of functions often do appear in print. Consider the following graph, reprinted from the *New York Times*.*

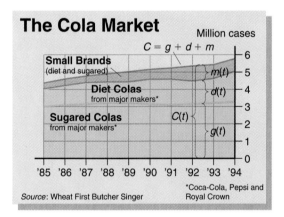

In this graph, the total number of cases of cola sold, $C(t)$, is regarded as a function of the year. The number of cases of sugared cola sold is denoted by $g(t)$, the number of cases of diet cola sold by $d(t)$, and the number of cases of "small brands" cola sold by $m(t)$. Although separate graphs for g, d, and m

---

New York Times, 3/23/95, page D1, article by Glenn Collins.

have not been drawn, we can see that

$$C(t) = g(t) + d(t) + m(t).$$

In the next graph, the functions *are* graphed separately before being added. Here all braces extend to the horizontal axis.

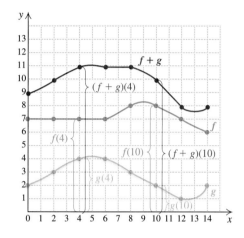

To find $(f + g)(a)$, $(f - g)(a)$, $(f \cdot g)(a)$, or $(f / g)(a)$, we must first be able to find $f(a)$ and $g(a)$. Thus we need to ensure that a is in the domain of both f and g.

EXAMPLE   3   Let

$$f(x) = \frac{5}{x} \quad \text{and} \quad g(x) = \frac{2x - 6}{x + 1}.$$

Find the domain of $f + g$, the domain of $f - g$, and the domain of $f \cdot g$.

SOLUTION   Note that because division by 0 is undefined, we have

Domain of f = $\{x \mid x$ is a real number and $x \neq 0\}$

and

Domain of g = $\{x \mid x$ is a real number and $x \neq -1\}$.

In order to find $f(a) + g(a)$, $f(a) - g(a)$, or $f(a) \cdot g(a)$, we must know that a is in *both* of the above domains. Thus,

Domain of $f + g$ = Domain of $f - g$ = Domain of $f \cdot g$

$= \{x \mid x$ is a real number and $x \neq 0$ and $x \neq -1\}$.

Suppose in Example 3 that we want to find $(f / g)(3)$. Finding $f(3)$ and $g(3)$ poses no problem:

$$f(3) = \frac{5}{3} \quad \text{and} \quad g(3) = \frac{2 \cdot 3 - 6}{3 + 1} = 0;$$

but then

$$(f/g)(3) = f(3)/g(3)$$
$$= \tfrac{5}{3} / \; 0 \; .$$

Thus, although 3 is in the domain of both f and g, it is not in the domain of f/g.

DETERMINING THE DOMAIN

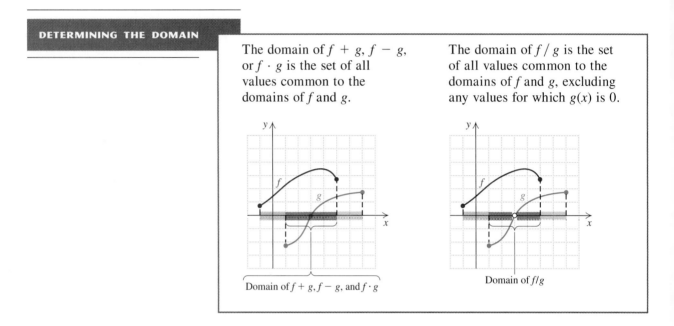

The domain of $f + g$, $f - g$, or $f \cdot g$ is the set of all values common to the domains of f and g.

The domain of f/g is the set of all values common to the domains of f and g, excluding any values for which $g(x)$ is 0.

Domain of $f + g$, $f - g$, and $f \cdot g$

Domain of f/g

EXAMPLE 4

Given $f(x) = 1/x$ and $g(x) = 2x - 7$, find the domains of $f + g$, $f - g$, $f \cdot g$, and f/g.

SOLUTION The domain of f is $\{x \mid x \neq 0\}$ or $\{x \mid x$ is a real number and $x \neq 0\}$. The domain of g is $\mathbb{R}$. Thus the domains of $f + g$, $f - g$, and $f \cdot g$ are the set of all elements common to both the domain of f and the domain of g. We have

the domain of $f + g =$ the domain of $f - g =$ the domain of $f \cdot g$
$$= \{x \mid x \text{ is a real number and } x \neq 0\}.$$

The domain of f/g is $\{x \mid x$ is a real number and $x \neq 0\}$, with the additional restriction that $g(x) \neq 0$. To determine what x-values would make $g(x) = 0$, we solve:

$$2x - 7 = 0 \qquad \text{Replacing } g(x) \text{ with } 2x - 7$$
$$2x = 7$$
$$x = \tfrac{7}{2}.$$

Since $g(x) = 0$ for $x = \tfrac{7}{2}$,

the domain of $f/g = \left\{x \mid x \text{ is a real number and } x \neq 0 \text{ and } x \neq \tfrac{7}{2}\right\}.$

TECHNOLOGY CONNECTION

2.6

A partial check of Example 4 can be performed using a grapher's TABLE feature—assuming the grapher is so equipped. To do so, we set up the table so the TABLE MINIMUM is 0 and the increment of change ($\triangle$Tbl) is 0.7. Next, we let $y_1 = 1/x$ and $y_2 = 2x - 7$. Using the Y-VARS key to write $y_3 = y_1 + y_2$ and $y_4 = y_1/y_2$, we can create the table of values shown here. Note that when x is 3.5, a value for y_3 can be found, but y_4 is undefined. When setting up these functions, you may wish to enter y_1 and y_2 without "selecting" either. Otherwise, the table's columns must be scrolled to display y_3 and y_4.

X	Y₃	Y₄
0	ERROR	ERROR
.7	−4.171	−.2551
1.4	−3.486	−.1701
2.1	−2.324	−.1701
2.8	−1.043	−.2551
3.5	.28571	ERROR
4.2	1.6381	.17007
X = 0		

Use a similar approach to partially check Example 3.

EXERCISE SET

2.6

Let $f(x) = -3x + 1$ and $g(x) = x^2 + 2$. Find the following.

1. $f(2) + g(2)$
2. $f(-1) + g(-1)$
3. $f(5) - g(5)$
4. $f(4) - g(4)$
5. $f(-1) \cdot g(-1)$
6. $f(-2) \cdot g(-2)$
7. $f(-4)/g(-4)$
8. $f(3)/g(3)$
9. $g(1) - f(1)$
10. $g(2)/f(2)$
11. $g(0)/f(0)$
12. $g(6) - f(6)$

Let $F(x) = x^2 - 3$ and $G(x) = 4 - x$. Find the following.

13. $(F + G)(x)$
14. $(F + G)(a)$
15. $(F + G)(-4)$
16. $(F + G)(-5)$
17. $(F - G)(3)$
18. $(F - G)(2)$
19. $(F \cdot G)(-3)$
20. $(F \cdot G)(-4)$
21. $(F/G)(0)$
22. $(F/G)(1)$
23. $(F/G)(-2)$
24. $(F/G)(-1)$

The accompanying graph indicates the number of new cases of AIDS diagnosed each month in the United States. For Exercises 25–28, let $T(t)$ represent the total number of new cases per month, $F(t)$ the number of new cases per month among people born in 1960 or later, $G(t)$ the number of new cases per month among people born before 1960, and t the number of years since January 1986.*

25. Estimate $F(7)$ and interpret its meaning.
26. Estimate $G(5)$ and interpret its meaning.

*Source: The Nation's Health, Official Newspaper of the American Public Health Association, 1/96, p. 24.

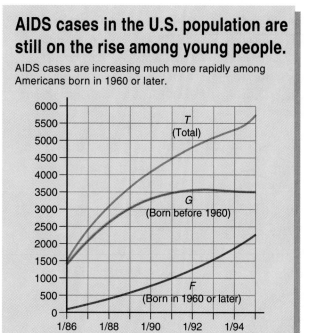

AIDS cases in the U.S. population are still on the rise among young people.

AIDS cases are increasing much more rapidly among Americans born in 1960 or later.

Source: National Cancer Institute

27. Estimate $T(6)$, $G(6)$, and $F(6)$. Show that $(G + F)(6) = T(6)$.

28. Estimate $T(4)$, $G(4)$, and $F(4)$. Show that $(T - F)(4) = G(4)$.

The graph below indicates how the three major airports servicing New York City have been utilized.

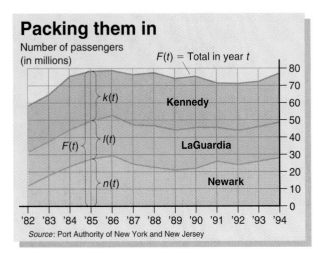

Packing them in

Number of passengers
(in millions)

Source: Port Authority of New York and New Jersey

29. Estimate $(n + l)(\text{'}92)$. What does it represent?

30. Estimate $(k + l)(\text{'}94)$. What does it represent?

31. Estimate $(k - l)(\text{'}92)$. What does it represent?

32. Estimate $(k - n)(\text{'}89)$. What does it represent?

33. Estimate $(n + l + k)(\text{'}93)$. What does it represent?

34. Estimate $(n + l + k)(\text{'}83)$. What does it represent?

For each pair of functions f and g, determine the domain of the sum, difference, and product of the two functions.

35. $f(x) = x^2$,
  $g(x) = 3x - 4$

36. $f(x) = 5x - 1$,
  $g(x) = 2x^2$

37. $f(x) = \dfrac{1}{x - 2}$,
  $g(x) = 4x^3$

38. $f(x) = 3x^2$,
  $g(x) = \dfrac{1}{x - 4}$

39. $f(x) = \dfrac{2}{x}$,
  $g(x) = x^2 - 4$

40. $f(x) = x^3 + 1$,
  $g(x) = \dfrac{5}{x}$

41. $f(x) = 4x + \dfrac{2}{x - 1}$,
  $g(x) = 3x^3$

42. $f(x) = 9 - x^2$,
  $g(x) = \dfrac{3}{x - 5} + 2x$

43. $f(x) = \dfrac{3}{x - 2}$,
  $g(x) = \dfrac{5}{4 - x}$

44. $f(x) = \dfrac{5}{x - 3}$,
  $g(x) = \dfrac{1}{x - 2}$

For each pair of functions f and g, determine the domain of f/g.

45. $f(x) = x^4$,
  $g(x) = x - 3$

46. $f(x) = 2x^3$,
  $g(x) = 5 - x$

47. $f(x) = 3x - 2$,
  $g(x) = 2x - 8$

48. $f(x) = 5 + x$,
  $g(x) = 6 - 2x$

49. $f(x) = \dfrac{3}{x - 4}$,
  $g(x) = 5 - x$

50. $f(x) = \dfrac{1}{2 - x}$,
  $g(x) = 7 - x$

51. $f(x) = \dfrac{2x}{x + 1}$,
  $g(x) = 2x + 5$

52. $f(x) = \dfrac{7x}{x - 2}$,
  $g(x) = 3x + 7$

For Exercises 53–58, consider the functions F and G as shown.

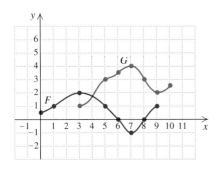

53. Determine $(F + G)(5)$ and $(F + G)(7)$.

54. Determine $(F \cdot G)(6)$ and $(F \cdot G)(9)$.

55. Find the domains of F, G, $F + G$, and F/G.

56. Find the domains of $F - G$, $F \cdot G$, and G/F.

57. Graph $F + G$.

58. Graph $G - F$.

S K I L L   M A I N T E N A N C E

59. Solve: $5x - 7 = 0$.

60. Evaluate $3x - y^2$ for $x = -4$ and $y = 5$.

61. A student's average after 4 tests is 78.5. What score is needed on the fifth test in order to raise the average to 80?

62. Write scientific notation for 0.00000631.

SYNTHESIS

In the graph that follows, W(t) represents the number of gallons of whole milk, L(t) the number of gallons of low-fat milk, and S(t) the number of gallons of skim milk consumed by the average American in a year.

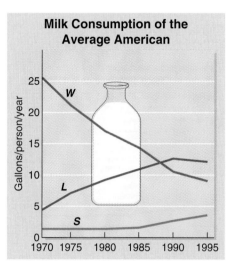

Milk Consumption of the Average American

63. ◈ Explain in words what $(W - S)(t)$ represents and what it would mean to have $(W - S)(t) < 0$.

64. ◈ Consider $(W + L + S)(t)$ and explain why you feel that total milk consumption per person has or has not changed over the years 1970–1995.

65. Find the domain of m/n, if
$$m(x) = 3x \text{ for } -1 < x < 5$$
and
$$n(x) = 2x - 3.$$

66. Find the domains of $f + g$, $f - g$, $f \cdot g$, and f/g, if
$$f = \{(-2, 1), (-1, 2), (0, 3), (1, 4), (2, 5)\}$$
and
$$g = \{(-4, 4), (-3, 3), (-2, 4), (-1, 0), (0, 5), (1, 6)\}.$$

67. For f and g as defined in Exercise 66, find $(f + g)(-2)$, $(f \cdot g)(0)$, and $(f/g)(1)$.

68. Find the domain of F/G, if
$$F(x) = \frac{1}{x - 4} \quad \text{and} \quad G(x) = \frac{x^2 - 4}{x - 3}.$$

69. Find the domain of f/g, if
$$f(x) = \frac{3x}{2x + 5} \quad \text{and} \quad g(x) = \frac{x^4 - 1}{3x + 9}.$$

70. Write equations for two functions f and g such that the domain of $f + g$ is
$$\{x \mid x \text{ is a real number and } x \neq -2 \text{ and } x \neq 5\}.$$

71. Sketch the graph of two functions f and g such that the domain of f/g is
$$\{x \mid -2 \leq x \leq 3 \text{ and } x \neq 1\}.$$

72. ◈ If $f(x) = c$, where c is some nonzero constant, describe how the graphs of $y = g(x)$ and $y = (f + g)(x)$ will differ.

73. ◈ Refer to the graph accompanying Exercises 25–28. Compute
$$\frac{F(7) - F(4)}{7 - 4} \quad \text{and} \quad \frac{G(7) - G(4)}{7 - 4}$$
and explain how these figures support the claim that the number of AIDS cases is increasing most rapidly among Americans born in 1960 or later.

74. ◩ Using the window $[-5, 5, -1, 9]$, graph $y_1 = 5$, $y_2 = x + 2$, and $y_3 = \sqrt{x}$. Then predict what shape the graphs of $y_1 + y_2$, $y_1 + y_3$, and $y_2 + y_3$ will take. Use a grapher to check each prediction.

75. ◩ Let $y_1 = 2.5x + 1.5$, $y_2 = x - 3$, and $y_3 = y_1/y_2$. Depending on whether the CONNECTED or DOT mode is used, the graph of y_3 appears as follows.

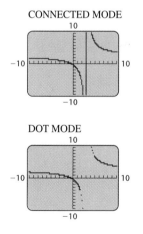

CONNECTED MODE

DOT MODE

Use algebra to determine which graph more accurately represents y_3.

76. ◩ Use the TABLE feature on a grapher to check your answers to Exercises 37, 43, 45, and 51. (See Technology Connection 2.6 on p. 123.)

COLLABORATIVE
C ◆ O ◆ R ◆ N ◆ E ◆ R

Focus: The algebra of functions

Time: 10–15 minutes

Group size: 2–3

The graph and data at right chart the average retirement age $R(x)$ and life expectancy $E(x)$ of U.S. citizens in year x.

ACTIVITY

1. Working as a team, perform the appropriate calculations and then graph $E - R$.
2. What does $(E - R)(x)$ represent? In what fields of study or business might the function $E - R$ prove useful?
3. Should E and R really be calculated separately for men and women? Why or why not?

4. What advice would you give to someone considering early retirement?

Year:	1955	1965	1975	1985	1995e	2005e
Average Retirement Age:	67.3	64.9	63.2	62.8	62.7	61.5
Average Life Expectancy:	73.9	75.5	77.2	78.5	79.1	80.1

e = estimated

If You're Age 50, Consider This...

Source: Bureau of Labor Statistics, courtesy of the Insurance Advisory Board

---

SUMMARY AND REVIEW
2

KEY TERMS

Axes, p. 66
x, y-coordinate system, p. 66
Cartesian coordinate system, p. 66
Ordered pair, p. 66
Origin, p. 66
Coordinates, p. 67
Quadrant, p. 67
Graph, p. 68
Linear equation, p. 69
Nonlinear equation, p. 70

Domain, p. 74
Range, p. 74
Function, p. 74
Projection, p. 77
Relation, p. 78
Input, p. 78
Output, p. 78
Dependent variable, p. 79
Independent variable, p. 79
Constant function, p. 79
Dummy variable, p. 79

Linear function, p. 88
y-intercept, p. 89
Slope, p. 89
Rise, p. 90
Run, p. 90
Rate of change, p. 92
Salvage value, p. 94
Depreciate, p. 94
Zero slope, p. 101
Undefined slope, p. 102
x-intercept, p. 103

IMPORTANT PROPERTIES AND FORMULAS

The Vertical-Line Test

A graph represents a function if it is not possible to draw a vertical line that intersects the graph more than once.

_____

The *x*-intercept is $(a, 0)$. To find a, let $y = 0$ and solve the original equation for x.

The *y*-intercept is $(0, b)$. To find b, let $x = 0$ and solve the original equation for y.

_____

$$\text{Slope} = m = \frac{\text{rise}}{\text{run}} = \frac{\text{change in } y}{\text{change in } x} = \frac{y_2 - y_1}{x_2 - x_1}$$

_____

Every horizontal line has a slope of 0.
The slope of a vertical line is undefined.

_____

The slope–intercept equation of a line is
$$y = mx + b.$$
The point–slope equation of a line is
$$y - y_1 = m(x - x_1).$$
The standard form of a linear equation is
$$Ax + By = C.$$

Parallel lines: The slopes are equal.
Perpendicular lines: The product of the slopes is -1.

_____

The Algebra of Functions

1. $(f + g)(x) = f(x) + g(x)$
2. $(f - g)(x) = f(x) - g(x)$
3. $(f \cdot g)(x) = f(x) \cdot g(x)$
4. $(f/g)(x) = f(x)/g(x)$, provided $g(x) \neq 0$

REVIEW EXERCISES

Determine whether the ordered pair is a solution.

1. $(3, 7)$, $4p - q = 5$

2. $(-2, 4)$, $x - 2y = 12$

3. $\left(0, -\frac{1}{2}\right)$, $3a - 4b = 2$

4. $(8, -2)$, $3c + 2d = 28$

Graph.

5. $y = -3x + 2$

6. $y = -x^2 + 1$

7. $8x + 32 = 0$

8. $y - 2 = 4$

9. For the following graph of f, determine **(a)** $f(2)$; **(b)** the domain of f; **(c)** any *x*-values for which $f(x) = 2$; and **(d)** the range of f.

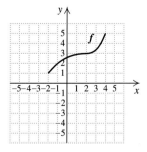

10. The function $C(t) = 645t + 9800$ can be used to estimate the average cost of tuition at a state university t years after 1997. (Figures reflect prices paid by both in-state and out-of-state students.) Predict the average cost of tuition at a state university in 2010.

Find the slope and the y-intercept.

11. $g(x) = -4x - 9$

12. $-6y + 2x = 7$

Find the slope of each line. If the slope is undefined, state this.

13. Containing the points $(4, 5)$ and $(-3, 1)$

14. Containing $(-16.4, 2.8)$ and $(-16.4, 3.5)$

15. Find the rate of change for the graph below. Use appropriate units.

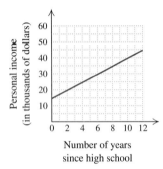

Number of years
since high school

16. Find a linear function whose graph has slope $\frac{2}{7}$ and y-intercept $(0, -6)$.

17. Graph using intercepts: $-2x + 4y = 7$.

Determine whether each of these is a linear equation.

18. $2x - 7 = 0$ **19.** $3x - 8f(x) = 7$

20. $2a + 7b^2 = 3$ **21.** $2p - \dfrac{7}{q} = 1$

22. Find an equation in point–slope form of the line with slope -2 and containing $(-3, 4)$.

23. Use function notation to write an equation for the line containing $(2, 5)$ and $(-4, -3)$.

Determine whether each pair of lines is parallel, perpendicular, or neither.

24. $y + 5 = -x$,
    $x - y = 2$

25. $3x - 5 = 7y$,
    $7y - 3x = 7$

26. In 1955, the U.S. minimum wage was $0.75, and in 1995, it was $4.25. Let W represent the minimum wage, in dollars, t years after 1955.
   a) Find a linear function $W(t)$ that fits the data.
   b) Use the function of part (a) to predict the minimum wage in 2005.

Find an equation of the line.

27. Containing the point $(2, -5)$ and parallel to the line $3x - 5y = 9$

28. Containing the point $(2, -5)$ and perpendicular to the line $3x - 5y = 9$

Let $g(x) = 3x - 6$ and $h(x) = x^2 + 1$. Find the following.

29. $g(0)$

30. $h(-5)$

31. $(g \cdot h)(4)$

32. $(g - h)(-2)$

33. $(g/h)(-1)$

34. $g(a + b)$

35. The domains of $g + h$ and $g \cdot h$

36. The domain of h/g

SKILL MAINTENANCE

37. Simplify: $-\frac{2}{3} - \left(-\frac{4}{5}\right)$.

38. Simplify: $(5a^3b)^2$.

39. Solve: $3(x - 2) + x = 5(x + 4)$.

40. Use scientific notation to write the number of feet in 1000 miles.

SYNTHESIS

41. ◈ Explain why every function is a relation, but not every relation is a function.

42. ◈ Explain why the slope of a vertical line is undefined whereas the slope of a horizontal line is 0.

43. Find the y-intercept of the function given by
$$f(x) + 3 = 0.17x^2 + (5 - 2x)^x - 7.$$

44. Determine the value of a such that the lines
$$3x - 4y = 12 \quad \text{and} \quad ax + 6y = -9$$
are parallel.

45. Homespun Jellies charges $2.49 for each jar of preserves. Shipping charges are $3.75 for handling, plus $0.60 per jar. Find a linear function for determining the cost of shipping x jars of preserves.

46. Match each sentence with the most appropriate graph.

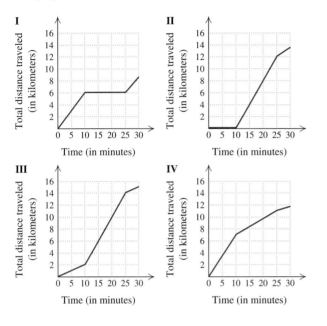

I   II

III   IV

a) Joni walks for 10 min to the train station, rides the train for 15 min, and then walks 5 min to the office.
b) During a workout, Phil bikes for 10 min, runs for 15 min, and then walks for 5 min.
c) Sam pilots his motorboat for 10 min to the middle of the lake, fishes for 15 min, and then motors for another 5 min to another spot.
d) Patti waits 10 min for her train, rides the train for 15 min, and then runs for 5 min to her job.

CHAPTER TEST
2

Determine whether the ordered pair is a solution.

1. $(0, -5)$, $x + 4y = -20$

2. $(1, -4)$, $-2p + 5q = 18$

Graph.

3. $y = -5x + 4$

4. $y = -2x^2 + 3$

5. $f(x) = 5$

6. $3 - x = 9$

7. For the following graph of f, determine **(a)** $f(-2)$; **(b)** the domain of f; **(c)** any x-value for which $f(x) = \frac{1}{2}$; and **(d)** the range of f.

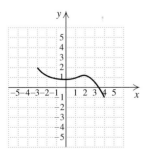

8. The function $S(t) = 1.2t + 21.4$ can be used to estimate the total U.S. sales of books, in billions of dollars, t years after 1992. Predict the total U.S. sales of books in 2002.

Find the slope and the y-intercept.

9. $f(x) = -\frac{3}{5}x + 12$

10. $-5y - 2x = 7$

Find the slope of the line containing the following points. If the slope is undefined, state so.

11. $(-2, -2)$ and $(6, 3)$

12. $(-3.1, 5.2)$ and $(-4.4, 5.2)$

13. Find a linear function whose graph has slope -5 and y-intercept $(0, -1)$.

14. Graph using intercepts: $-2x + 5y = 12$.

15. Which of these are linear equations?

   a) $8x - 7 = 0$
   b) $4b - 9a^2 = 2$
   c) $2x - 5y = 3$

16. Find an equation in point–slope form of the line with slope 4 and containing $(-2, -4)$.

17. Use function notation to write an equation for the line containing $(3, -1)$ and $(4, -2)$.

Determine without graphing whether each pair of lines is parallel, perpendicular, or neither.

18. $4y + 2 = 3x$,
$-3x + 4y = -12$

19. $y = -2x + 5$,
$2y - x = 6$

Find an equation of the line.

20. Containing $(-3, 2)$ and parallel to the line $2x - 5y = 8$

21. Containing $(-3, 2)$ and perpendicular to the line $2x - 5y = 8$

22. Find the following, given that $g(x) = -3x - 4$ and $h(x) = x^2 + 1$.

a) $h(-2)$
b) $(g \cdot h)(3)$
c) The domain of h/g

23. If you rent a van for one day and drive it 250 mi, the cost is $100. If you drive it 300 mi, the cost is $115. Let $C(m)$ represent the cost, in dollars, of driving m miles.

a) Find a linear function that fits the data.
b) Use the function to find how much it will cost to rent the van for one day and drive it 500 mi.

SKILL MAINTENANCE

24. Simplify: $-\frac{4}{5} - \frac{2}{15}$.

25. Simplify: $(a^4b^5)(a^2b^7)$.

26. Solve: $5x - 2(x + 3) = 4x - 9$.

27. Write scientific notation for 0.0000067.

SYNTHESIS

28. The function $f(t) = 5 + 15t$ can be used to determine a bicycle racer's location, in miles from the starting line, measured t hours after passing the 5-mi mark.

a) How far from the start will the racer be 1 hr and 40 min after passing the 5-mi mark?
b) Assuming a constant rate, how fast is the racer traveling?

29. The graph of the function $f(x) = mx + b$ contains the points $(r, 3)$ and $(7, s)$. Express s in terms of r if the graph is parallel to the line $3x - 2y = 7$.

30. Given that $f(x) = 5x^2 + 1$ and $g(x) = 4x - 3$, find an expression for $h(x)$ so that the domain of $f/g/h$ is $\left\{x \mid x \text{ is a real number and } x \neq \frac{3}{4} \text{ and } x \neq \frac{2}{7}\right\}$. Answers may vary.

Systems of Equations and Problem Solving

L ike any businessperson, I need math for pricing, markup, and discount. I use algebra for mixing stains, and trigonometry and geometry in floor layout, design, and measurement.

DONALD WILLIAMS
Hardwood Floorman
Calistoga, CA

AN APPLICATION

Williams' Custom Flooring has 0.5 gal of stain that is 20% brown and 80% neutral. A customer orders 1.5 gal of stain that is 60% brown and 40% neutral. How much pure brown stain and how much neutral stain should be added to the original 0.5 gal in order to make up the order?

THIS PROBLEM APPEARS AS EXERCISE 61 IN SECTION 3.3.

More information on flooring is available at
http://hepg.awl.com/be/inter_5

The most difficult part of problem solving is almost always translating the problem situation to mathematical language. Once a problem has been translated, the rest is usually straightforward. In this chapter, we study systems of equations *and how to solve them using graphing, substitution, and elimination. Systems of equations often provide the easiest way to translate a problem to mathematical language.*

Systems of equations have extensive application to many fields, such as psychology, sociology, business, education, engineering, and science. This chapter also includes a brief introduction to matrices, *which can also be used to solve systems of equations.*

3.1   Systems of Equations in Two Variables

Translating   •   Identifying Solutions   •   Solving Systems Graphically

Translating

Problems involving more than one unknown quantity are often solved most easily by translating to two or more equations.

EXAMPLE 1

Real Estate.   Translate the following problem situation to mathematical language, using two equations.

In 1996, the Simon Property Group and the DeBartolo Realty Corporation merged to form the largest real estate company in the United States, owning 183 shopping centers in 32 states. Prior to merging, Simon owned twice as many properties as DeBartolo. How many properties did each company own before the merger?

DeBartolo shareholders would receive 0.68 share of Simon common stock for each share of DeBartolo common stock. Simon also would agree to repay $1.5 billion in DeBartolo debt. At Tuesday's closing price of $ 23.625 a share for common stock, the transaction is valued at roughly $3 billion.

Executives say the proposed company, Simon DeBartolo Group, would be the largest real estate company in the United States, worth $7.5 billion. **Not included in the deal:** DeBartolo's ownership stake in the San Francisco 49ers, or the Indiana Pacers, owned separately by the Simon

SOLUTION

1. **Familiarize.** We have already seen problems in which we need to look up certain formulas or the meaning of certain words. Here we need only observe that the words shopping centers and properties are being used interchangeably.

   Often problems contain information that has no bearing on the problem being solved. In this case, the number 32 is irrelevant to the question being asked. Instead we focus on the number of properties owned and the phrase "twice as many." Rather than guess and check, let's proceed to the next step, using x for the number of properties originally owned by Simon and y for the number of properties originally owned by DeBartolo.

2. **Translate.** There are two statements to translate. First we look at the total number of properties involved:

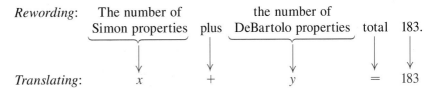

The second statement compares the number of properties that each company held before merging:

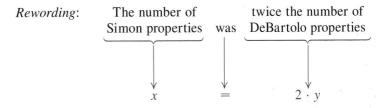

We have now translated the problem to a pair, or **system, of equations**:

$$x + y = 183,$$
$$x = 2y.$$

Problems like Example 1 *can* be solved using one variable; however, as problems become complicated, you will find that using more than one variable (and more than one equation) is often the preferable approach.

EXAMPLE 2

Retail Sales. In one day, Glovers, Inc., sold 20 pairs of gloves. Fleece gloves sold for $24.95 a pair and Gore-Tex® gloves for $37.50. Receipts totaled $687.25. Write a system of equations that could be used to find how many of each kind were sold.

SOLUTION

1. **Familiarize.** To familiarize ourselves with this problem, let's guess that Glovers sold 12 pairs of fleece gloves and 11 pairs of Gore-Tex gloves. Does this guess check? Since a total of 20 pairs of gloves was sold, and since $12 + 11 \neq 20$, our guess cannot be right.

As a second guess, suppose that 12 pairs of fleece gloves and 8 pairs of Gore-Tex gloves were sold. Now the total sold is 20, so our guess is right in that respect. How much money would have been taken in? Since fleece gloves sold for $24.95 and Gore-Tex gloves for $37.50, the total received would have been

$$\underbrace{\text{Money from}}_{\text{fleece gloves}} \quad \text{plus} \quad \underbrace{\text{Money from}}_{\text{Gore-Tex gloves}}$$

$$12(\$24.95) \quad + \quad 8(\$37.50) \quad = \$299.40 + \$300.00$$
$$= \$599.40.$$

Although the total number of pairs is correct, our guess is incorrect because the problem states that the total amount received was $687.25. Since $599.40 is less than $687.25, more of the expensive gloves were sold than we had guessed. We could now adjust our guess accordingly. Instead, let's work toward an algebraic approach that avoids guessing.

2. **Translate.** We let f = the number of pairs of fleece gloves sold and g = the number of pairs of Gore-Tex gloves sold. The information can be organized in a table, which will help with the translating.

Kind of Glove	Fleece	Gore-Tex	Total
Number Sold	f	g	20
Price	$24.95	$37.50	
Amount Taken In	24.95f	37.50g	687.25

$\longrightarrow f + g = 20$

$\longrightarrow 24.95f + 37.50g = 687.25$

The first row of the table and the first sentence of the problem indicate that a total of 20 pairs of gloves was sold:

$$f + g = 20.$$

Since each pair of fleece gloves cost $24.95 and f pairs were sold, 24.95f represents the amount taken in from the sale of fleece gloves. Similarly, 37.50g represents the amount taken in from the sale of g pairs of Gore-Tex gloves. From the third row of the table and the third sentence of the problem, we get the second equation:

$$24.95f + 37.50g = 687.25.$$

Multiplying both sides by 100, we can clear the decimals. This gives the following system of equations as the translation:

$$f + g = 20,$$
$$2495f + 3750g = 68{,}725.$$

Identifying Solutions

A *solution* of a system of equations in two variables is an ordered pair of numbers that makes *both* equations true.

EXAMPLE 3   Determine whether $(-4, 7)$ is a solution of the system

$$x + y = 3,$$
$$5x - y = -27.$$

SOLUTION   We use alphabetical order of the variables. Thus we replace x by -4 and y by 7:

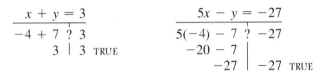

$$\frac{x + y = 3}{-4 + 7 \ ? \ 3}$$
$$3 \mid 3 \quad \text{TRUE}$$

$$\frac{5x - y = -27}{5(-4) - 7 \ ? \ -27}$$
$$-20 - 7$$
$$-27 \mid -27 \quad \text{TRUE}$$

The pair $(-4, 7)$ makes both equations true, so it is a solution of the system. We sometimes describe such a solution by saying, in this case, that $x = -4$ and $y = 7$. Set notation can also be used to list the solution set $\{(-4, 7)\}$.

Solving Systems Graphically

Recall that the graph of an equation is a drawing that represents its solution set. If we graph the equations in Example 3, we find that $(-4, 7)$ is the only point common to both lines. Thus one way to solve a system of two equations is to graph both equations and identify any points of intersection. The coordinates of each point of intersection represent a solution of that system.

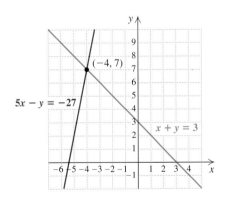

EXAMPLE 4   Solve each system graphically.

a) $y - x = 1,$
      $y + x = 3$

b) $y = -3x + 5,$
      $y = -3x - 2$

c) $3y - 2x = 6,$
      $-12y + 8x = -24$

SOLUTION

a) We graph each equation using any method studied in Chapter 2. All ordered pairs from line L_1 are solutions of the first equation. All ordered pairs from line L_2 are solutions of the second equation. The point of intersection has coordinates that make *both* equations true. Apparently, (1, 2) is the solution. Graphing is not perfectly accurate, so solving by graphing may yield approximate answers. Our check below shows that (1, 2) is indeed the solution.

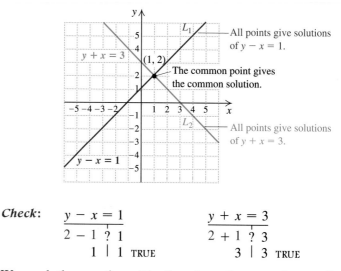

Check:

$$y - x = 1$$
$$2 - 1 \; ? \; 1$$
$$1 \; | \; 1 \quad \text{TRUE}$$

$$y + x = 3$$
$$2 + 1 \; ? \; 3$$
$$3 \; | \; 3 \quad \text{TRUE}$$

b) We graph the equations. The lines have the same slope, -3, and different y-intercepts, so they are parallel. There is no point at which they cross, so the system has no solution.

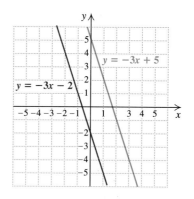

c) We graph the equations and find that the same line is drawn twice. Thus any solution of one of the equations is a solution of the other. Each equation has an infinite number of solutions, some of which are listed on the graph. Each of these is also a solution of the other equation. We check one solution, (0, 2), which is the y-intercept of each equation.

Check:

$$3y - 2x = 6$$
$$\overline{3(2) - 2(0) \;?\; 6}$$
$$6 - 0 \;\Big|$$
$$6 \;\Big|\; 6 \;\text{TRUE}$$

$$-12y + 8x = -24$$
$$\overline{-12(2) + 8(0) \;?\; -24}$$
$$-24 + 0 \;\Big|$$
$$-24 \;\Big|\; -24 \;\text{TRUE}$$

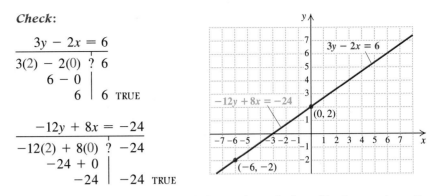

You can check that $(-6, -2)$ is another solution of both equations. In fact, any pair that is a solution of one equation is a solution of the other equation as well. Thus the solution set is $\{(x, y)\,|\,3y - 2x = 6\}$ or, in words, "the set of all pairs (x, y) for which $3y - 2x = 6$." Since the two equations are equivalent, we could have used $-12y + 8x = -24$ in place of $3y - 2x = 6$.

Example 4 illustrates that when we graph a system of two linear equations in two variables, one of the following three outcomes will occur.

1. The lines have one point in common, and that point is the only solution of the system (see part (a)).
2. The lines are parallel, with no point in common, and the system has no solution (see part (b)). This system is called **inconsistent**.

TECHNOLOGY CONNECTION 3.1

A grapher with TRACE and ZOOM features can be used, much as we did on page 106, to solve systems of equations. On many graphers, an INTERSECT option allows you to find the coordinates of the intersection directly. Graphers are especially useful when equations contain fractions or decimals or when the coordinates of the intersection are not integers. To illustrate, consider the following system:

$$3.45x + 4.21y = 8.39,$$
$$7.12x - 5.43y = 6.18.$$

Most graphers require equations to have y alone on one side. After solving for y in each equation, we can obtain the graph shown. By zooming and tracing repeatedly, or by using INTERSECT, we see that, to the nearest hundredth, the coordinates of the intersection are $(1.47, 0.79)$.

Use a grapher to solve each of the following systems. Make sure that all x- and y-coordinates are correct to the nearest hundredth.

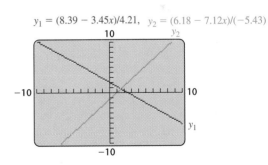

$y_1 = (8.39 - 3.45x)/4.21,\quad y_2 = (6.18 - 7.12x)/(-5.43)$

1. $y = -5.43x + 10.89,$
   $y = 6.29x - 7.04$

2. $y = 123.52x + 89.32,$
   $y = -89.22x + 33.76$

3. $2.18x + 7.81y = 13.78,$
   $5.79x - 3.45y = 8.94$

4. $-9.25x - 12.94y = -3.88,$
   $21.83x + 16.33y = 13.69$

3. The lines coincide, sharing the same graph. Since every solution of one equation is a solution of the other, the system has an infinite number of solutions (see part (c)). The equations are said to be **dependent**.

Any system of equations that has at least one solution is said to be **consistent**. The systems of Examples 4(a) and 4(c) are both consistent.

   Graphing is helpful when solving systems because it allows us to "see" the solution. It can also be used on systems of nonlinear equations, and in many applications, it provides a satisfactory answer. However, graphing has the disadvantage of often yielding inexact answers when fractional or decimal solutions are involved. In Section 3.2, we will develop two algebraic methods of solving systems. Both methods produce exact answers.

EXERCISE SET
3.1

Translate each problem situation to a system of equations. Do not attempt to solve, but save for later use.

1. The difference between two numbers is 11. Twice the smaller plus 3 times the larger is 123. What are the numbers?

2. The sum of two numbers is −42. The first number minus the second number is 52. What are the numbers?

3. *Retail sales.*   Paint Town sold 45 paintbrushes, one kind at $8.50 each and another at $9.75 each. In all, $398.75 was taken in for the brushes. How many of each kind were sold?

4. *Retail sales.*   Mountainside Fleece sold 40 neckwarmers. Solid color neckwarmers sold for $9.90 each and print ones sold for $12.75 each. In all, $421.65 was taken in for the neckwarmers. How many of each type were sold?

5. *Geometry.*   Two angles are supplementary.* One angle is 3° less than twice the other. Find the measures of the angles.

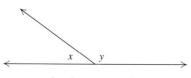
Supplementary angles

6. *Geometry.*   Two angles are complementary.† The sum of the measures of the first angle and half the second angle is 64°. Find the measures of the angles.

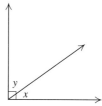
Complementary angles

7. *Basketball scoring.*   Amma scored 18 times during one basketball game. She scored a total of 30 points, two for each field goal and one for each free throw. How many field goals did she make? How many free throws?

8. *Fundraising.*   The St. Mark's Community Barbecue served 250 dinners. A child's plate cost $3.50 and an adult's plate cost $7.00. A total of $1347.50 was collected. How many of each type of plate was served?

9. *Sales of pharmaceuticals.*   The Diabetic Express recently charged $15.75 for a vial of Humulin insulin and $12.95 for a vial of Novolin insulin. If a total of $959.35 was collected for 65 vials of insulin, how many vials of each type were sold?

---

*The sum of the measures of two supplementary angles is 180°.

†The sum of the measures of two complementary angles is 90°.

10. *Court dimensions.* The perimeter of a standard basketball court is 288 ft. The length is 44 ft longer than the width. Find the dimensions.

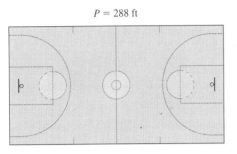

$$P = 288 \text{ ft}$$

11. *Court dimensions.* The perimeter of a standard tennis court used for doubles is 228 ft. The width is 42 ft less than the length. Find the dimensions.

12. *Basketball scoring.* The Central College Cougars made 40 field goals in a recent basketball game, some 2-pointers and the rest 3-pointers. Altogether the 40 baskets counted for 89 points. How many of each type of field goal was made?

13. *Lumber production.* Snookers Lumber can convert logs into either lumber or plywood. In a given day, the mill turns out twice as many units of plywood as lumber. It makes a profit of $25 on a unit of lumber and $40 on a unit of plywood. How many units of each type must be produced and sold in order to make a profit of $10,920?

14. *Video rentals.* J. P.'s Video rents general interest films for $3.00 each and children's films for $1.50 each. In one day, a total of $213 was taken in from the rental of 77 videos. How many of each type of video was rented?

15. *Hockey rankings.* Hockey teams receive 2 points for a win and 1 point for a tie. The Wildcats once won a championship with 60 points. They won 9 more games than they tied. How many wins and how many ties did the Wildcats have?

16. *Radio airplay.* Omar must play 12 commercials during his 1-hr radio show. Each commercial is either 30 sec or 60 sec long. If the total commercial time during that hour is 10 min, how many commercials of each type does Omar play?

17. *Nontoxic floor wax.* A nontoxic floor wax can be made from lemon juice and food-grade linseed oil. The amount of oil should be twice the amount of lemon juice. How much of each ingredient is needed to make 32 oz of floor wax? (The mix should be spread with a rag and buffed when dry.)

18. *Airplane seating.* An airplane has a total of 152 seats. The number of coach-class seats is 5 more than 6 times the number of first-class seats. How many of each type of seat are there on the plane?

Determine whether the ordered pair is a solution of the given system of equations. Remember to use alphabetical order of variables.

19. $(1, 2)$; $\begin{aligned} 4x - \ y &= 2, \\ 10x - 3y &= 4 \end{aligned}$

20. $(-1, -2)$; $\begin{aligned} 2x + y &= -4, \\ x - y &= 1 \end{aligned}$

21. $(2, 5)$; $\begin{aligned} y &= 3x - 1, \\ 2x + y &= 4 \end{aligned}$

22. $(-1, -2)$; $\begin{aligned} x + 3y &= -7, \\ 3x - 2y &= 12 \end{aligned}$

23. $(1, 5)$; $\begin{aligned} x + y &= 6, \\ y &= 2x + 3 \end{aligned}$

24. $(5, 2)$; $\begin{aligned} a + b &= 7, \\ 2a - 8 &= b \end{aligned}$

25. $(3, 1)$; $\begin{aligned} 3x + 4y &= 13, \\ 5x - 4y &= 11 \end{aligned}$

26. $(4, -2)$; $\begin{aligned} -3x - 2y &= -8, \\ y &= 2x - 5 \end{aligned}$

Solve each system graphically. Be sure to check your solution. If a system has an infinite number of solutions, use set-builder notation to write the solution set. If a system has no solution, state this.

27. $\begin{aligned} x - y &= 3, \\ x + y &= 5 \end{aligned}$ **28.** $\begin{aligned} x + y &= 4, \\ x - y &= 2 \end{aligned}$ **29.** $\begin{aligned} 3x + \ y &= 5, \\ x - 2y &= 4 \end{aligned}$

30. $\begin{aligned} 2x - y &= 4, \\ 5x - y &= 13 \end{aligned}$ **31.** $\begin{aligned} 4y &= x + 8, \\ 3x - 2y &= 6 \end{aligned}$ **32.** $\begin{aligned} 4x - \ y &= 9, \\ x - 3y &= 16 \end{aligned}$

33. $\begin{aligned} x &= y - 1, \\ 2x &= 3y \end{aligned}$ **34.** $\begin{aligned} a &= 1 + b, \\ b &= 5 - 2a \end{aligned}$ **35.** $\begin{aligned} 2u + v &= 3, \\ 2u &= v + 7 \end{aligned}$

36. $\begin{aligned} x &= 4, \\ y &= -5 \end{aligned}$ **37.** $\begin{aligned} x &= -3, \\ y &= 2 \end{aligned}$ **38.** $\begin{aligned} 2a + b &= 4, \\ b &= 4a + 1 \end{aligned}$

39. $\begin{aligned} 2b + a &= 11, \\ a - b &= 5 \end{aligned}$ **40.** $\begin{aligned} y &= -\tfrac{1}{3}x - 1, \\ 4x - 3y &= 18 \end{aligned}$

41. $\begin{aligned} y &= -\tfrac{1}{4}x + 1, \\ 2y &= x - 4 \end{aligned}$ **42.** $\begin{aligned} 6x - 2y &= 2, \\ 9x - 3y &= 1 \end{aligned}$

43. $\begin{aligned} y - x &= 5, \\ 2x - 2y &= 10 \end{aligned}$ **44.** $\begin{aligned} y &= -x - 1, \\ 4x - 3y &= 24 \end{aligned}$

45. $\begin{aligned} y &= 3 - x, \\ 2x + 2y &= 6 \end{aligned}$ **46.** $\begin{aligned} 2x - 3y &= 6, \\ 3y - 2x &= -6 \end{aligned}$

47. For the systems in the odd-numbered exercises 27–45, which are consistent?

48. For the systems in the even-numbered exercises 28–46, which are consistent?

49. For the systems in the odd-numbered exercises 27–45, which contain dependent equations?

50. For the systems in the even-numbered exercises 28–46, which contain dependent equations?

SKILL MAINTENANCE

Solve.

51. $3x + 4 = x - 2$ **52.** $\frac{3}{5}x + 2 = \frac{2}{5}x - 5$

53. $4x - 5x = 8x - 9 + 11x$

54. Solve $Q = \frac{1}{4}(a - b)$ for b.

Factor.

55. $3x - 21$ **56.** $2a + ab - ac$

SYNTHESIS

57. ◈ Write a problem for a classmate to solve that can be translated into a system of two equations. Devise the problem so that the solution is "Shelly gave 9 haircuts and 5 shampoos."

58. ◈ Write a problem for a classmate to solve that requires writing a system of two equations. Devise the problem so that the solution is "The Lakers made 6 three-point baskets and 31 two-point baskets."

Arms sales.   For Exercises 59–62, consider the following graph.*

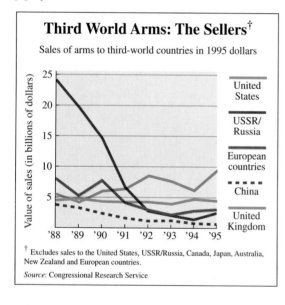

Third World Arms: The Sellers[†]

Sales of arms to third-world countries in 1995 dollars

[†] Excludes sales to the United States, USSR/Russia, Canada, Japan, Australia, New Zealand and European countries.

Source: Congressional Research Service

---

*Based on information provided by the Congressional Research Service, 1996.

59. ◈ Does the graph above support the statement, "The United States has always sold twice as many arms to third-world countries as China has"? Why or why not?

60. In what years did U.S. arms sales to third-world countries exceed USSR/Russia arms sales to third-world countries?

61. What was the last year in which European arms sales to third-world countries exceeded U.S. arms sales to third-world countries?

62. Determine the most recent year in which USSR/Russia arms sales to third-world countries exceeded the combined sales of the United States, Europe, and China.

63. Write a system of equations for which:
   a) (5, 1) is a solution,
   b) there is no solution, and
   c) there is an infinite number of solutions.

64. A system of linear equations has $(1, -1)$ and $(-2, 3)$ as solutions. Determine:
   a) a third point that is a solution, and
   b) how many solutions there are.

65. The solution of the following system is $(4, -5)$. Find A and B.
$$Ax - 6y = 13,$$
$$x - By = -8.$$

Translate to a system of equations. Do not solve.

66. *Ages.*   Burl is twice as old as his son. Ten years ago, Burl was 3 times as old as his son. How old are they now?

67. *Work experience.*   Lou and Juanita are mathematics professors at a state university. Together, they have 46 years of service. Two years ago, Lou had taught 2.5 times as many years as Juanita. How long has each taught at the university?

68. *Design.*   A piece of posterboard has a perimeter of 156 in. If you cut 6 in. off the width, the length becomes 4 times the width. What are the dimensions of the original piece of posterboard?

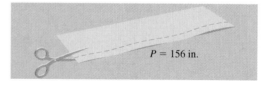

$P = 156$ in.

69. *Nontoxic scouring powder.* A nontoxic scouring powder is made up of 4 parts baking soda and 1 part vinegar. How much of each ingredient is needed for a 16-oz mixture?

Solve graphically.

70. $y = |x|,$
   $x + 4y = 15$

71. $x - y = 0,$
   $y = x^2$

In Exercises 72–75, use a grapher to solve each system of linear equations for x and y. Round all coordinates to the nearest hundredth.

72. $y = 8.23x + 2.11,$
   $y = -9.11x - 4.66$

73. $y = -3.44x - 7.72,$
   $y = 4.19x - 8.22$

74. $14.12x + 7.32y = 2.98,$
   $21.88x - 6.45y = -7.22$

75. $5.22x - 8.21y = -10.21,$
   $-12.67x + 10.34y = 12.84$

COLLABORATIVE
C ◆ O ◆ R ◆ N ◆ E ◆ R

Focus: Systems of linear equations

Time: 20 minutes

Group size: 3

The box score at right, from a basketball game between the Boston Celtics and the Detroit Pistons, contains information on how many field goals and free throws each player attempted and made. For example, the line "Hill 9–17 7–9 25" means that Detroit's Grant Hill made 9 field goals out of 17 attempts and 7 free throws out of 9 attempts, for a total of 25 points. (Each free throw is worth 1 point and each field goal is worth either 2 or 3 points, depending on how far from the basket it was shot.)

ACTIVITY

1. Work as a group to develop a system of two equations in two unknowns that can be used to determine how many 2-pointers and how many 3-pointers were made by Boston's David Wesley.
2. Each group member should solve the system from part (1) in a different way: one person by graphing, one person by making a table and methodically checking all combinations of 2- and 3-pointers, and one person by guesswork. Compare answers when this has been completed.
3. Determine, as a group, how many 2- and 3-pointers the Detroit Pistons made as a team.

■ **Pistons 99, Celtics 89:** In Auburn Hills, Grant Hill had 25 points and 11 rebounds, leading the Detroit Pistons over the Boston Celtics.
   Hill also had eight assists. Otis Thorpe added 16 points and Joe Dumars had 15.
   David Wesley led the Celtics with 25.

BOSTON (89)
Fox 6-9 3-4 17, Williams 5-9 4-4 14, Radja 4-10 2-2 10, Minor 0-3 0-0 0, Wesley 9-13 3-5 25, Walker 5-14 6-8 16, Barros 1-5 0-0 3, Day 1-6 1-2 4, Conlon 0-0 0-0 0, Szabo 0-0 0-0 0. Totals 31-69 19-25 89.

Detroit (99)
Hill 9-17 7-9 25, Long 3-5 2-2 8, Thorpe 7-8 2-3 16, Hunter 2-14 2-2 7, Dumars 5-10 3-3 15, Mills 5-11 0-0 13, Curry 3-3 0-0 7, Ratliff 2-3 0-0 4, Green 1-1 2-2 4, Mahorn 0-0 0-2 0. Totals 37-72 18-23 99.
Boston 26 16 21 26—89
Detroit 27 29 17 26—99

3.2 Solving by Substitution or Elimination

The Substitution Method   •   The Elimination Method   •
Comparing Methods

The Substitution Method

One algebraic (nongraphical) method for solving systems of equations, the *substitution method,* relies on having one variable isolated.

EXAMPLE 1

Solve the system

$$x + y = 4, \quad (1)$$
$$x = y + 1. \quad (2)$$

SOLUTION   Equation (2) says that x and $y + 1$ name the same number. Thus we can substitute $y + 1$ for x in equation (1):

$$x + y = 4 \qquad \text{Equation (1)}$$
$$(y + 1) + y = 4. \qquad \text{Substituting } y + 1 \text{ for } x$$

We solve this last equation, using methods learned earlier:

$$(y + 1) + y = 4$$
$$2y + 1 = 4 \qquad \text{Removing parentheses and combining like terms}$$
$$2y = 3 \qquad \text{Subtracting 1 on both sides}$$
$$y = \frac{3}{2}. \qquad \text{Dividing by 2}$$

We now return to the original pair of equations and substitute $\frac{3}{2}$ for y in either equation so that we can solve for x. For this problem, calculations are slightly easier if we use equation (2):

$$x = y + 1 \qquad \text{Equation (2)}$$
$$= \frac{3}{2} + 1 \qquad \text{Substituting } \frac{3}{2} \text{ for } y$$
$$= \frac{3}{2} + \frac{2}{2} = \frac{5}{2}.$$

We obtain the ordered pair $\left(\frac{5}{2}, \frac{3}{2}\right)$. A check ensures that it is a solution:

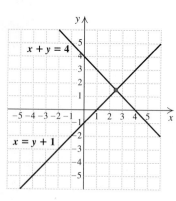

Check:

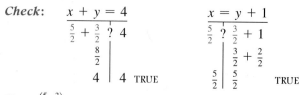

Since $\left(\frac{5}{2}, \frac{3}{2}\right)$ checks, it is the solution.

The exact solution to Example 1 would have been difficult to find graphically because it involves fractions. Despite this, the graph shown does serve as a check and provides a visualization of the problem.

If neither equation in a system has a variable alone on one side, we first isolate a variable in one equation and then substitute.

EXAMPLE 2   Solve the system

$$2x + y = 6, \quad (1)$$
$$3x + 4y = 4. \quad (2)$$

SOLUTION   First we select an equation and solve for one variable. To isolate y, we can add $-2x$ on both sides of equation (1) to get

$$y = 6 - 2x. \quad (3)$$

Then we substitute $6 - 2x$ for y in equation (2) and solve for x:

$$3x + 4(6 - 2x) = 4 \qquad \text{Substituting } 6 - 2x \text{ for } y \text{ in equation (2). Use parentheses!}$$
$$3x + 24 - 8x = 4 \qquad \text{Distributing to remove parentheses}$$
$$3x - 8x = 4 - 24$$
$$-5x = -20$$
$$x = 4.$$

Next, we substitute 4 for x in either equation (1), (2), or (3). It is easiest to use equation (3) since it has already been solved for y:

$$y = 6 - 2x$$
$$= 6 - 2(4)$$
$$= 6 - 8 = -2.$$

The pair $(4, -2)$ appears to be the solution.

Check:

$$
\begin{array}{c|c}
2x + y = 6 & \\
\hline
2(4) + (-2) \;?\; 6 & \\
8 - 2 \;\Big|\; & \\
6 \;\Big|\; 6 \;\text{TRUE} &
\end{array}
\qquad
\begin{array}{c|c}
3x + 4y = 4 & \\
\hline
3(4) + 4(-2) \;?\; 4 & \\
12 - 8 \;\Big|\; & \\
4 \;\Big|\; 4 \;\text{TRUE} &
\end{array}
$$

Since $(4, -2)$ checks, it is the solution.

Some systems have no solution, as we saw graphically in Section 3.1. How do we recognize such systems if we are solving by an algebraic method?

EXAMPLE 3   Solve the system

$$y = -3x + 5,$$
$$y = -3x - 2.$$

SOLUTION   We solved this system graphically in Example 4(b) of Section 3.1, and found that the lines are parallel and the system has no solution. Let's now try to solve the system by substitution. We substitute $-3x - 2$

for y in the first equation:

$$-3x - 2 = -3x + 5 \qquad \text{Substituting } -3x - 2 \text{ for } y$$

$$-2 = 5. \qquad \text{Adding } 3x \text{ on both sides; } -2 = 5 \text{ is a contradiction.}$$

When we add $3x$ to get the x-terms on one side, the x-terms drop out and we end up with a contradiction. When solving algebraically yields a false equation, the system has no solution.   ━━━●

The Elimination Method

The *elimination method* for solving systems of equations makes use of the *addition principle*: If $a = b$, then $a + c = b + c$. Consider the following system:

$$2x - 3y = 0, \qquad (1)$$
$$-4x + 3y = -1. \qquad (2)$$

The key to the advantage of the elimination method for solving this system involves the $-3y$ in one equation and the $3y$ in the other. These terms are opposites. If we add all terms on the left side of the equations, $-3y$ and $3y$ add to 0, and in effect, the variable y is "eliminated."

To use the addition principle for equations, note that according to equation (2), $-4x + 3y$ and -1 are the same number. Thus we can use a vertical form and add $-4x + 3y$ on the left side of equation (1) and -1 on the right side:

$$\begin{array}{ll} 2x - 3y = 0 & (1) \\ \underline{-4x + 3y = -1} & (2) \\ -2x + 0y = -1. & \text{Adding} \end{array}$$

This eliminates the variable y, which is why this is called the elimination method. We now have an equation with just one variable, x, for which we solve:

$$-2x = -1$$
$$x = \tfrac{1}{2}.$$

Next, we substitute $\tfrac{1}{2}$ for x in equation (1) and solve for y:

$$2 \cdot \tfrac{1}{2} - 3y = 0 \qquad \text{Substituting; we also could have used equation (2).}$$
$$1 - 3y = 0$$
$$-3y = -1, \text{ so } y = \tfrac{1}{3}.$$

Check:

$$\begin{array}{c|c} 2x - 3y = 0 & \\ \hline 2\left(\tfrac{1}{2}\right) - 3\left(\tfrac{1}{3}\right) \ ? \ 0 & \\ 1 - 1 & \\ 0 & 0 \ \text{TRUE} \end{array} \qquad \begin{array}{c|c} -4x + 3y = -1 & \\ \hline -4\left(\tfrac{1}{2}\right) + 3\left(\tfrac{1}{3}\right) \ ? \ -1 & \\ -2 + 1 & \\ -1 & -1 \ \text{TRUE} \end{array}$$

Since $\left(\tfrac{1}{2}, \tfrac{1}{3}\right)$ checks, it is the solution.

To eliminate a variable, we must sometimes multiply before adding.

EXAMPLE 4   Solve the system

$$5x + 4y = 22, \qquad (1)$$
$$-3x + 8y = 18. \qquad (2)$$

SOLUTION   If we add, we will not eliminate a variable. However, if the $4y$ in equation (1) were $-8y$, we would. Thus we multiply by -2 on both sides of the first equation:

$$
\begin{array}{ll}
-10x - 8y = -44 & \text{Multiplying by } -2 \text{ on both sides of equation (1)} \\
\underline{-3x + 8y = 18} & \\
-13x + 0 = -26 & \text{Adding} \\
x = 2. & \text{Solving for } x
\end{array}
$$

Then

$$
\begin{array}{ll}
-3 \cdot 2 + 8y = 18 & \text{Substituting 2 for } x \text{ in equation (2)} \\
-6 + 8y = 18 & \\
\left. \begin{array}{l} 8y = 24 \\ y = 3. \end{array} \right\} & \text{Solving for } y
\end{array}
$$

We obtain $(2, 3)$, or $x = 2$, $y = 3$. This checks, so it is the solution.

Sometimes we must multiply twice in order to make two terms become opposites.

EXAMPLE 5   Solve the system

$$2x + 3y = 17, \qquad (1)$$
$$5x + 7y = 29. \qquad (2)$$

SOLUTION   We multiply so that the x-terms are eliminated.

$$
\begin{array}{lll}
2x + 3y = 17, & \xrightarrow{\text{Multiplying by 5}} & 10x + 15y = 85 \\
5x + 7y = 29 & \xrightarrow{\text{Multiplying by } -2} & \underline{-10x - 14y = -58} \\
& & 0 + y = 27 \qquad \text{Adding} \\
& & y = 27.
\end{array}
$$

Next, we substitute to find x:

$$
\begin{array}{ll}
2x + 3 \cdot 27 = 17 & \text{Substituting 27 for } y \text{ in equation (1)} \\
2x + 81 = 17 & \\
2x = -64 & \\
x = -32. & \text{Solving for } x
\end{array}
$$

Check:

$$
\begin{array}{c|c}
2x + 3y = 17 & 5x + 7y = 29 \\
\hline
2(-32) + 3(27) \; ? \; 17 & 5(-32) + 7(27) \; ? \; 29 \\
-64 + 81 & -160 + 189 \\
17 \;\big|\; 17 \quad \text{TRUE} & 29 \;\big|\; 29 \quad \text{TRUE}
\end{array}
$$

We obtain $(-32, 27)$, or $x = -32$, $y = 27$, as the solution.

EXAMPLE 6   Solve the system

$$3y - 2x = 6, \qquad (1)$$
$$-12y + 8x = -24. \qquad (2)$$

SOLUTION   We graphed this system in Example 4(c) of Section 3.1, and found that the lines coincide and the system has an infinite number of solutions. Suppose we were to solve this system using the elimination method:

$$\begin{array}{l} 12y - 8x = 24 \qquad \text{\small Multiplying equation (1) by 4 on both sides} \\ \underline{-12y + 8x = -24} \\ 0 = 0. \qquad \text{\small We obtain an identity; } 0 = 0 \text{ is always true.} \end{array}$$

Note that both variables have been eliminated and what remains is an identity. Any pair that is a solution of equation (1) is also a solution of equation (2). The equations are dependent and the solution set is infinite: $\{(x, y) \mid 3y - 2x = 6\}$.

SPECIAL CASES

When solving a system of two linear equations in two variables:

1. If an identity is obtained, such as $0 = 0$, then the system has an infinite number of solutions. The equations are dependent and, since a solution exists, the system is consistent.*

2. If a contradiction is obtained, such as $0 = 7$, then the system has no solution. The system is inconsistent.

Should decimals or fractions appear, it often helps to *clear* before solving.

EXAMPLE 7   Solve the system

$$0.2x + 0.3y = 1.7,$$
$$\tfrac{1}{7}x + \tfrac{1}{5}y = \tfrac{29}{35}.$$

SOLUTION   We have

$$0.2x + 0.3y = 1.7, \xrightarrow{\text{\small Multiplying by 10}} 2x + 3y = 17$$
$$\tfrac{1}{7}x + \tfrac{1}{5}y = \tfrac{29}{35} \xrightarrow{\text{\small Multiplying by 35}} 5x + 7y = 29.$$

We multiplied by 10 to clear the decimals. Multiplication by 35, the least common denominator, clears the fractions. The problem is now identical to Example 5. The solution is $(-32, 27)$, or $x = -32$, $y = 27$.

Comparing Methods

The following table is a summary that compares the graphical, substitution, and elimination methods for solving systems of equations.

---

*Consistent systems and dependent equations are discussed in greater detail in Section 3.4.

Method	Strengths	Weaknesses
Graphical	Can "see" solutions. Works with any system that can be graphed.	Inexact when solutions involve numbers that are not integers. Solution may not appear on the part of the graph drawn.
Substitution	Yields exact solutions. Easy to use when a variable is alone on one side.	Introduces extensive computations with fractions when solving more complicated systems. Cannot "see" solutions quickly.
Elimination	Yields exact solutions. Easy to use when fractions or decimals appear in the system. The preferred method for systems of 3 or more equations in 3 or more variables (see Section 3.4).	Cannot "see" solutions quickly.

When deciding which method to use, consider this table and directions from your instructor. The situation is analogous to having a piece of wood to cut and three different types of saws available. Although all three saws can cut the wood, the "best" choice depends on the particular piece of wood, the type of cut being made, and your level of skill with each saw.

EXERCISE SET
3.2

For Exercises 1–42, if a system has an infinite number of solutions, use set-builder notation to write the solution set. If a system has no solution, state this.

Solve using the substitution method.

1. $3x + 5y = 3,$
   $x = 8 - 4y$

2. $2x - 3y = 13,$
   $y = 5 - 4x$

3. $3x - 6 = y,$
   $9x - 2y = 3$

4. $x = 3y - 3,$
   $x + 2y = 9$

5. $4x + y = 1,$
   $x - 2y = 16$

6. $5m + n = 8,$
   $3m - 4n = 14$

7. $-3b + a = 7,$
   $5a + 6b = 14$

8. $4x + 12y = 4,$
   $-5x + y = 11$

9. $5p + 7q = 1,$
   $4p - 2q = 16$

10. $3x - y = 1,$
   $2x + 2y = 2$

11. $5x + 3y = 4,$
   $x - 4y = 3$

12. $3x - y = 7,$
   $2x + 2y = 5$

13. $y - 2x = 1,$
   $2x - 3 = y$

14. $x + 2y = 6,$
   $x = 4 - 2y$

Solve using the elimination method.

15. $x + 3y = 7,$
   $-x + 4y = 7$

16. $x + y = 9,$
   $2x - y = -3$

17. $2x + y = 6,$
   $x - y = 3$

18. $x - 2y = 6,$
   $-x + 3y = -4$

(handwritten: $12x = 6$, $x = \frac{1}{2}$)

19. $6x - 3y = 18,$
$6x + 3y = -12$

20. $9x + 3y = -3,$
$2x - 3y = -8$

21. $3x + 2y = 3,$
$9x - 8y = -2$

22. $5x + 3y = 19,$
$2x - 5y = 11$

23. $5x - 7y = -16,$
$2x + 8y = 26$

24. $5r - 3s = 24,$
$3r + 5s = 28$

25. $0.7x - 0.3y = 0.5,$
$-0.4x + 0.7y = 1.3$

26. $0.3x - 0.2y = 4,$
$0.2x + 0.3y = 1$

27. $6x + 7y = 9,$
$8x + 9y = 11$

28. $8x + 9y = 15,$
$9x + 6y = 21$

29. $\frac{1}{3}x + \frac{1}{5}y = 7,$
$\frac{1}{6}x - \frac{2}{5}y = -4$

30. $\frac{2}{5}x + \frac{1}{2}y = 2,$
$\frac{1}{2}x - \frac{1}{6}y = 3$

31. $6x + 10y = 14,$
$3x + 5y = 7$

32. $12x - 6y = -15,$
$-4x + 2y = 5$

Solve using any appropriate method.

33. $a - 2b = 16,$
$b + 3 = 3a$

34. $5x - 9y = 7,$
$7y - 3x = -5$

35. $10x + y = 306,$
$10y + x = 90$

36. $3(a - b) = 15,$
$4a = b + 1$

37. $3y = x - 2,$
$x = 2 + 3y$

38. $x + 2y = 6,$
$x = 4 - 2y$

39. $2x - 7y = 9,$
$7y - 2x = -5$

40. $4x - 7y = 6,$
$4x - 7y = 2$

41. $0.05x + 0.25y = 22,$
$0.15x + 0.05y = 24$

42. $1.3x - 0.2y = 12,$
$0.4x + 17y = 89$

SKILL MAINTENANCE

43. What simple interest rate is charged if a principal of $320 earns $17.60 in $\frac{1}{2}$ year?

44. Simplify: $-9(y + 7) - 6(y - 4)$.

Find each solution set.

45. $3x - 14 = x + 2(x - 7)$

46. $x + 2(3x + 5) = 7x - 3$

SYNTHESIS

47. ◈ Write a system of linear equations that would be most easily solved using the substitution method. Explain why substitution would be easier to use than the elimination method.

48. ◈ Write a system of linear equations that would be most easily solved using the elimination method. Explain why elimination would be easier to use than the substitution method.

49. ◈ Can Exercises 32 and 40 be solved mentally (with no writing)? If so, how? If not, why not?

50. ◈ Outline a procedure that can be used to write an inconsistent system of equations.

51. If $(1, 2)$ and $(-3, 4)$ are two solutions of $f(x) = mx + b$, find m and b.

52. If $(0, -3)$ and $\left(-\frac{3}{2}, 6\right)$ are two solutions of $px - qy = -1$, find p and q.

53. Determine a and b for which $(-4, -3)$ will be a solution of the system
$$ax + by = -26,$$
$$bx - ay = 7.$$

54. Solve for x and y in terms of a and b:
$$5x + 2y = a,$$
$$x - y = b.$$

Solve.

55. $\dfrac{x + y}{2} - \dfrac{x - y}{5} = 1,$
$\dfrac{x - y}{2} + \dfrac{x + y}{6} = -2$

56. 🖩 $3.5x - 2.1y = 106.2,$
$4.1x + 16.7y = -106.28$

Each of the following is a system of nonlinear equations. However, each is reducible to linear, since an appropriate substitution (say, u for 1/x and v for 1/y) yields a linear system. Make such a substitution, solve for the new variables, and then solve for the original variables.

57. $\dfrac{2}{x} + \dfrac{1}{y} = 0,$
$\dfrac{5}{x} + \dfrac{2}{y} = -5$

58. $\dfrac{1}{x} - \dfrac{3}{y} = 2,$
$\dfrac{6}{x} + \dfrac{5}{y} = -34$

59. ◈ ▨ A student solving the system
$$17x + 19y = 102,$$
$$136x + 152y = 826$$
graphs both equations on a grapher and gets the following screen.

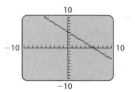

The student then (incorrectly) concludes that the equations are dependent and the solution set is infinite. How can algebra be used to convince the student that a mistake has been made?

COLLABORATIVE
C ◆ O ◆ R ◆ N ◆ E ◆ R

Focus: Systems of linear equations

Time: 20–30 minutes

Group size: 2–4

For aerobic exercise, a person's "target heart rate" T, in beats per minute, is given by $T = \frac{3}{4}(220 - a)$, where a is the person's age, in years.

ACTIVITY

1. Working independently, each group member should calculate the target heart rate for someone whose age is known only by that group member. That target heart rate should then be shared with the other group members.

2. Each group member should then calculate the age associated with each heart rate in two ways: (1) by using the formula above to check a series of guesses; and (2) by substituting into the formula above and solving for a.

3. Express the formula above in the form $y = mx + b$, where x represents age. What does a negative value for m signify?

4. Formulate a system of equations that could be used to determine the age of a person whose target heart rate is twice her or his age. Then have one group member solve the system using the substitution method while the other group members solve the system graphically. Check by comparing your answers.

3.3   Solving Applications: Systems of Two Equations

Total-Value and Mixture Problems   •   Motion Problems

You are in a much better position to solve problems now that systems of equations can be used. Using systems often makes the translating step easier.

EXAMPLE 1

Real Estate.   In 1996, the Simon Property Group and the DeBartolo Realty Corporation merged to form the largest real estate company in the United States, owning 183 shopping centers in 32 states. Prior to merging, Simon owned twice as many properties as DeBartolo. How many properties did each company own before the merger?

SOLUTION   The *Familiarize* and *Translate* steps have been done in Example 1 of Section 3.1. The resulting system of equations is

$$x + y = 183,$$
$$x = 2y,$$

where $x =$ the number of properties originally owned by Simon and $y =$ the number of properties originally owned by DeBartolo.

3. **Carry out.** We solve the system of equations. Since one equation already has a variable isolated, let's use the substitution method:

$$
\begin{aligned}
x + y &= 183 \\
2y + y &= 183 \qquad \text{Substituting } 2y \text{ for } x \\
3y &= 183 \qquad \text{Combining like terms} \\
y &= 61.
\end{aligned}
$$

We return to the second equation and substitute 61 for y and compute x:

$$x = 2y = 2 \cdot 61 = 122.$$

Apparently, Simon owned 122 properties and DeBartolo 61.

4. **Check.** The sum of 122 and 61 is 183, so the total number of properties is correct. Since 122 is twice 61, the numbers check.

5. **State.** Prior to merging, Simon owned 122 properties and DeBartolo owned 61.

Total-Value and Mixture Problems

EXAMPLE 2

Retail Sales. In one day, Glovers, Inc., sold 20 pairs of gloves. Fleece gloves sold for $24.95 a pair and Gore-Tex gloves for $37.50. The company took in $687.25. How many of each kind were sold?

SOLUTION The *Familiarize* and *Translate* steps were done in Example 2 of Section 3.1.

3. **Carry out.** We are to solve the system of equations

$$
\begin{aligned}
f + g &= 20, & (1) \\
2495f + 3750g &= 68{,}725 & (2)
\end{aligned}
$$

where $f =$ the number of pairs of fleece gloves sold and $g =$ the number of pairs of Gore-Tex gloves sold. What method shall we use? We could graph, but the large numbers would make that cumbersome. Since no variable appears alone and the equations are in the form $Ax + By = C$, let's use the elimination method. We eliminate f by multiplying equation (1) by -2495 and adding it to equation (2):

$$
\begin{aligned}
-2495f - 2495g &= -49{,}900 \qquad \text{Multiplying equation (1) by } -2495 \\
\underline{2495f + 3750g} &= \underline{68{,}725} \\
1255g &= 18{,}825 \qquad \text{Adding} \\
g &= 15. \qquad \text{Solving for } g
\end{aligned}
$$

To find f, we substitute 15 for g in equation (1) and solve for f:

$$
\begin{aligned}
f + g &= 20 \qquad \text{Equation (1)} \\
f + 15 &= 20 \qquad \text{Substituting 15 for } g \\
f &= 5. \qquad \text{Solving for } f
\end{aligned}
$$

We obtain (5, 15), or $f = 5$, $g = 15$.

4. **Check.** We check in the original problem. Recall that f is the number of pairs of fleece gloves and g the number of pairs of Gore-Tex gloves:

Number of gloves: $f + g = 5 + 15 = 20$

Money from fleece gloves: $\$24.95f = 24.95 \times 5 = \$124.75$

Money from Gore-Tex gloves: $\$37.50g = 37.50 \times 15 = \underline{\$562.50}$

$$\text{Total} = \$687.25$$

The numbers check.

5. **State.** Glovers sold 5 pairs of fleece gloves and 15 pairs of Gore-Tex gloves.

Example 2 involved two types of items (gloves), the quantity of each type sold, and the total value of the items. We refer to this type of problem as a *total-value problem*.

EXAMPLE 3

Blending Teas. Tara's Tea Terrace sells loose Black tea for 95¢ an ounce and Lapsang Souchong for \$1.43 an ounce. Tara wants to make a 1-lb mixture of the two types, called Imperial Blend, that sells for \$1.10 an ounce. How much tea of each type should Tara use?

SOLUTION

1. **Familiarize.** This problem is similar to Example 2. Rather than pairs of fleece or Gore-Tex gloves, we have ounces of Black tea and ounces of Lapsang Souchong. Rather than two different prices per pair, we have two different prices per ounce. Finally, rather than knowing the total value of the gloves, we know the weight and the price per ounce of the Imperial Blend. It is important to note that we can find the total value of the blend by multiplying 16 ounces (1 lb) times \$1.10 per ounce. Although we could make and check a guess, we proceed to let $b =$ the number of ounces of Black tea and $l =$ the number of ounces of Lapsang Souchong.

2. **Translate.** Since a 16-oz batch is being made, we must have

$$b + l = 16.$$

To find a second equation, note that the total value of the 16-oz blend must match the combined value of the separate ingredients:

Rewording: The value of the Black tea   plus   the value of the Lapsang Souchong   is   the value of the Imperial Blend.

Translating: $b \cdot 95$ $+$ $l \cdot 143$ $=$ $16 \cdot 110$

These equations can also be obtained from a table.

	Black Tea	Lapsang Souchong	Imperial Blend	
Number of Ounces	b	l	16	$\longrightarrow b + l = 16$
Price per Ounce	95¢	143¢	110¢	
Value of Tea	$95b$	$143l$	$16 \cdot 110$, or 1760	$\longrightarrow 95b + 143l = 1760$

We have translated to a system of equations:

$$b + \quad l = 16, \qquad (1)$$
$$95b + 143l = 1760. \qquad (2)$$

3. **Carry out.** We can solve using substitution. When equation (1) is solved for b, we have $b = 16 - l$. Substituting $16 - l$ for b in equation (2), we find l:

$$95(16 - l) + 143l = 1760 \qquad \text{Substituting}$$
$$1520 - 95l + 143l = 1760 \qquad \text{Using the distributive law}$$
$$48l = 240 \qquad \text{Combining like terms; subtracting 1520 on both sides}$$
$$l = 5. \qquad \text{Dividing by 48 on both sides}$$

We have $l = 5$ and, from equation (1) above, $b + l = 16$. Thus, $b = 11$.

4. **Check.** If 11 oz of Black tea and 5 oz of Lapsang Souchong are combined, a 16-oz, or 1-lb, blend will result. The value of 11 oz of Black tea is 11(0.95), or $10.45. The value of 5 oz of Lapsang Souchong is 5(1.43), or $7.15, so the combined value of the blend is $10.45 + $7.15 = $17.60. A 16-oz batch, priced at $1.10 an ounce, would also be worth $17.60, so our answer checks.

5. **State.** The Imperial Blend should be made by combining 11 oz of Black tea with 5 oz of Lapsang Souchong.   ━●

EXAMPLE 4   *Student Loans.*   Enid's student loans totaled $9600. Part was a Perkins loan made at 5% interest and the rest was a Federal Education Loan made at 8% interest. After one year, Enid's loans accumulated $633 in interest. What was the original amount of each loan?

SOLUTION

1. **Familiarize.** We begin with a guess. If $7000 was borrowed at 5% and $2600 was borrowed at 8%, the two loans would total $9600. The interest would then be 0.05($7000), or $350, and 0.08($2600), or $208, for a total of $558 in interest. Our guess was wrong, but checking the guess familiarized us with the problem.

2. **Translate.** Let p = the amount of the Perkins loan and f = the amount of the Federal Education Loan. We then organize a table in which each column comes from the formula for simple interest:

   Principal · Rate · Time = Interest.

	Perkins Loan	Federal Loan	Total
Principal	p	f	$9600
Rate of Interest	5%	8%	
Time	1 yr	1 yr	
Interest	$0.05p$	$0.08f$	$633

$\longrightarrow p + f = 9600$

$\longrightarrow 0.05p + 0.08f = \$633$

The total amount borrowed is found in the first row of the table:

$$p + f = 9600.$$

A second equation, representing the accumulated interest, can be found in the last row:

$$0.05p + 0.08f = 633, \quad \text{or} \quad 5p + 8f = 63,300. \qquad \text{Clearing decimals}$$

3. **Carry out.** The resulting system can be solved by elimination or substitution:

$$p + \ f = 9600,$$
$$5p + 8f = 63,300.$$

   We find that $p = 4500$ and $f = 5100$.

4. **Check.** The total amount borrowed is $4500 + $5100, or $9600. The interest on $4500 at 5% for 1 yr is 0.05($4500), or $225. The interest on $5100 at 8% for 1 yr is 0.08($5100), or $408. The total amount of interest is $225 + $408, or $633, so the numbers check.

5. **State.** The Perkins loan was for $4500 and the Federal Education Loan was for $5100.

Before proceeding to Example 5, briefly scan Examples 2–4 for similarities. Note that in each case, one of the equations in the system is a simple sum while the other equation represents a sum of products. Example 5 continues this pattern with what is commonly called a *mixture problem.*

When solving a problem, see if it is patterned or modeled after a problem that you have already solved.

EXAMPLE 5

Mixing Fertilizers. Yardbird Gardening, Inc., carries two brands of fertilizer containing nitrogen and water. "Gently Green" is 5% nitrogen and "Sun Saver" is 15% nitrogen. Yardbird Gardening needs to combine the two types of solutions to make 100 L of a solution that is 12% nitrogen. How much of each brand should be used?

SOLUTION

1. **Familiarize.** We sketch a picture and make a guess to gain familiarity with the problem.

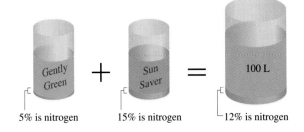

Suppose that 40 L of Gently Green and 60 L of Sun Saver are mixed. The resulting mixture will be the right size, 100 L, but will it be the right strength? To find out, note that 40 L of Gently Green would contribute $0.05(40) = 2$ L of nitrogen to the mixture while 60 L of Sun Saver would contribute $0.15(60) = 9$ L of nitrogen to the mixture. Altogether, 40 L of Gently Green and 60 L of Sun Saver would make 100 L of a mixture that has $2 + 9 = 11$ L of nitrogen. Since this would mean that the final mixture is only 11% nitrogen, our guess of 40 L and 60 L is incorrect. Still, the process of checking our guess has familiarized us with the problem.

2. **Translate.** Let $g =$ the number of liters of Gently Green and $s =$ the number of liters of Sun Saver. The information can be organized in a table.

	Gently Green	Sun Saver	Mixture	
Number of Liters	g	s	100	$\longrightarrow$ $g + s = 100$
Percent of Nitrogen	5%	15%	12%	
Amount of Nitrogen	$0.05g$	$0.15s$	0.12 × 100, or 12 liters	$\longrightarrow$ $0.05g + 0.15s = 12$

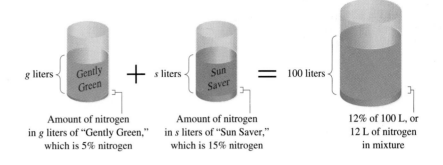

If we add g and s in the first row, we get one equation. It represents the total amount of mixture: $g + s = 100$.

If we add the amounts of nitrogen listed in the third row, we get a second equation. This equation represents the amount of nitrogen in the mixture: $0.05g + 0.15s = 12$.

After clearing decimals, we have translated the problem to the system

$$
\begin{aligned}
g + \quad s &= \quad 100, \quad (1) \\
5g + 15s &= 1200. \quad (2)
\end{aligned}
$$

3. **Carry out.**   We use the elimination method to solve the system:

$$
\begin{array}{rl}
-5g - \quad 5s = -500 & \text{Multiplying equation (1) by } -5 \text{ on both sides} \\
\underline{5g + 15s = \quad 1200} & \\
10s = \quad 700 & \text{Adding} \\
s = \quad 70; & \text{Solving for } s \\
g + 70 = 100 & \text{Substituting into equation (1)} \\
g = \quad 30. & \text{Solving for } g
\end{array}
$$

4. **Check.**   Remember, g is the number of liters of Gently Green and s is the number of liters of Sun Saver.

Total amount of mixture: $g + s = 30 + 70 = 100$

Total amount of nitrogen: 5% of 30 + 15% of 70 = 1.5 + 10.5 = 12

Percentage of nitrogen in mixture: $\dfrac{\text{Total amount of nitrogen}}{\text{Total amount of mixture}} = \dfrac{12}{100} = 12\%$

The numbers do check in the original problem.

5. **State.**   Yardbird Gardening should mix 30 L of Gently Green with 70 L of Sun Saver.

Motion Problems

When a problem deals with distance, speed (rate), and time, recall the following.

**DISTANCE, RATE, AND
TIME EQUATIONS**

If r represents rate, t represents time, and d represents distance, then:

$$d = rt, \qquad r = \frac{d}{t}, \quad \text{and} \quad t = \frac{d}{r}.$$

Be sure to remember at least one of these equations. The others can be obtained by using algebraic manipulations as needed.

EXAMPLE 6

Train Travel. A freight train leaves Ames traveling east at a speed of 60 km/h. Two hours later, a passenger train leaves Ames traveling in the same direction on a parallel track at 90 km/h. At what point will the passenger train catch up to the freight train?

SOLUTION

1. **Familiarize.** Let's make a guess—say, 180 km—and check to see if it is correct. The freight train, traveling 60 km/h, would reach a point 180 km from Ames in $\frac{180}{60} = 3$ hr. The passenger train, traveling 90 km/h, would cover 180 km in $\frac{180}{90} = 2$ hr. Since 3 hr is *not* two hours more than 2 hr, our guess of 180 km is incorrect. Although our guess is wrong, we see that the time that the trains are running and the point at which they meet are both unknown. Let $t =$ the number of hours that the freight train is running before they meet and $d =$ the distance at which the trains meet. Since the freight train has a 2-hr head start, the passenger train runs for $t - 2$ hours before catching up to the freight train.

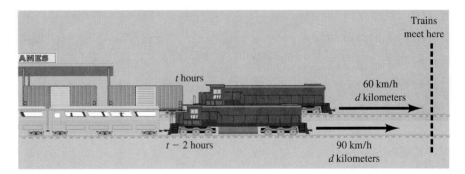

2. **Translate.** We can organize the information in a chart. Each row is determined by the formula *Distance = Rate · Time*.

	Distance	Rate	Time	
Freight Train	d	60	t	⟶ $d = 60t$
Passenger Train	d	90	$t - 2$	⟶ $d = 90(t - 2)$

Using *Distance* = *Rate* · *Time* twice, we get two equations:

$$d = 60t, \qquad (1)$$
$$d = 90(t - 2). \qquad (2)$$

3. Carry out.   We solve the system using substitution:

$$60t = 90(t - 2) \qquad \text{Substituting } 60t \text{ for } d \text{ in equation (2)}$$
$$60t = 90t - 180$$
$$-30t = -180$$
$$t = 6.$$

The time for the freight train is 6 hr, which means that the time for the passenger train is 6 − 2, or 4 hr. Remember that it is distance, not time, that the problem asked for. Thus for $t = 6$, we have $d = 60 \cdot 6 = 360$ km.

4. Check.   At 60 km/h, the freight train will travel 60 · 6, or 360 km, in 6 hr. At 90 km/h, the passenger train will travel 90 · (6 − 2) = 360 km in 4 hr. The numbers check.

5. State.   The trains will meet at a point 360 km east of Ames.   ━━●

EXAMPLE 7

Marine Travel.   A coast guard patrol boat travels 4 hr downstream with a 6-mph current. Returning, against the current, takes 5 hr. Find the speed of the boat in still water.

SOLUTION

1. Familiarize.   We imagine the situation and make a sketch. Note that the current *speeds up* the boat traveling downstream and *slows down* the boat traveling upstream. Since the distances traveled each way must be the same, we can easily check a guess of the boat's speed in still water. Suppose the speed of the boat in still water is 40 mph. The boat would then go 40 + 6 = 46 mph downstream and 40 − 6 = 34 mph upstream. The boat would travel 46 · 4 = 184 mi downstream and 34 · 5 = 170 mi upstream. Since 184 ≠ 170, our guess of 40 mph is incorrect. Rather than guess again, let's have $r =$ the speed, in miles per hour, of the boat in still water. Then $r + 6 =$ the boat's speed downstream, and $r - 6 =$ the boat's speed upstream. We also let $d =$ the distance traveled, in miles.

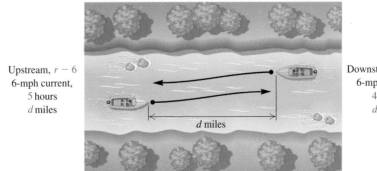

Upstream, $r - 6$
6-mph current,
5 hours
d miles

d miles

Downstream, $r + 6$
6-mph current,
4 hours
d miles

2. **Translate.** The information can be organized in a chart. The distances traveled are the same, so we use *Distance = Rate* (or *Speed*) · *Time*. Each row of the chart gives an equation.

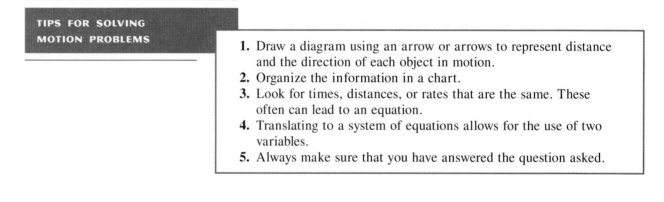

	Distance	Rate	Time	
Downstream	d	$r + 6$	4	$\longrightarrow d = (r + 6)4$
Upstream	d	$r - 6$	5	$\longrightarrow d = (r - 6)5$

The two equations constitute a system:

$$d = (r + 6)4, \quad (1)$$
$$d = (r - 6)5. \quad (2)$$

3. **Carry out.** We solve the system using substitution:

$(r - 6)5 = (r + 6)4$ Substituting $(r - 6)5$ for d in equation (1)

$5r - 30 = 4r + 24$ Using the distributive law

$r = 54.$ Solving for r

4. **Check.** When $r = 54$, the speed downstream is $54 + 6 = 60$ mph, and the speed upstream is $54 - 6 = 48$ mph. The distance downstream, $60 \cdot 4 = 240$ mi, matches the distance upstream, $48 \cdot 5 = 240$ mi, so we have a check.

5. **State.** The speed of the boat in still water is 54 mph.

TIPS FOR SOLVING MOTION PROBLEMS

1. Draw a diagram using an arrow or arrows to represent distance and the direction of each object in motion.
2. Organize the information in a chart.
3. Look for times, distances, or rates that are the same. These often can lead to an equation.
4. Translating to a system of equations allows for the use of two variables.
5. Always make sure that you have answered the question asked.

EXERCISE SET

3.3

1.–18. *For Exercises 1–18, solve Exercises 1–18 from Exercise Set 3.1.*

19. *Inventory.* The Everton College store paid $1728 for an order of 45 calculators. The store paid $9 for each scientific calculator. The others, all graphing calculators, cost the store $58 each. How many of each type of calculator was ordered?

20. *Sales of food.* High Flyin' Wings charges $12 for a bucket of chicken wings and $7 for a chicken dinner. After filling 28 orders for buckets and dinners, High Flyin' Wings had collected $281. How many buckets and how many dinners did they sell?

21. *Blending coffees.* The Coffee Counter charges $9.00 per pound for Kenyan French Roast coffee and $8.00 per pound for Sumatran coffee. How much of each type should be used to make a 20-lb blend that sells for $8.40 per pound?

22. *Mixed nuts.* The Nutty Professor sells cashews for $6.75 per pound and Brazil nuts for $5.00 per pound. How much of each type should be used to make a 50-lb mixture that sells for $5.70 per pound?

23. *Blending granola.* Deep Thought Granola is 25% nuts and dried fruit. Oat Dream Granola is 10% nuts and dried fruit. How much of Deep Thought and how much of Oat Dream should be mixed to form a 20-lb batch of granola that is 19% nuts and dried fruit?

24. *Catering.* Casella's Catering is planning a wedding reception. The bride and groom would like to serve a nut mixture containing 25% peanuts. Casella has available mixtures that are either 40% or 10% peanuts. How much of each type should be mixed to get a 10-lb mixture that is 25% peanuts?

25. *Ink remover.* Etch Clean Graphics uses one cleanser that is 25% acid and a second that is 50% acid. How many liters of each should be mixed to get 10 L of a solution that is 40% acid?

26. *Livestock feed.* Soybean meal is 16% protein and corn meal is 9% protein. How many pounds of each should be mixed to get a 350-lb mixture that is 12% protein?

27. *Student loans.* Lomasi's two student loans totaled $12,000. One of her loans was at 6% simple interest and the other at 9%. After one year, Lomasi owed $855 in interest. What was the amount of each loan?

28. *Investments.* An executive nearing retirement made two investments totaling $15,000. In one year, these investments yielded $1432 in simple interest. Part of the money was invested at 9% and the rest at 10%. How much was invested at each rate?

29. *Food science.* The following bar graph shows the milk fat percentages in three dairy products. How many pounds each of whole milk and cream should be mixed to form 200 lb of milk for cream cheese?

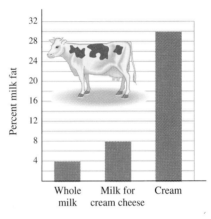

30. *Automotive maintenance.* "Arctic Antifreeze" is 18% alcohol and "Frost No-More" is 10% alcohol. How many liters of each should be mixed to get 20 L of a mixture that is 15% alcohol?

31. *Architecture.* The rectangular ground floor of the John Hancock building has a perimeter of 860 ft. The length is 100 ft more than the width. Find the length and the width.

x

$x + 100$

32. *Real estate.* The perimeter of a lot is 190 m. The width is one fourth of the length. Find the dimensions.

33. *Teller work.* Ashford goes to a bank and gets change for a $50 bill consisting of all $5 bills and $1 bills. There are 22 bills in all. How many of each kind are there?

34. *Making change.* Cecilia makes a $9.25 purchase at the bookstore with a $20 bill. The store has no bills and gives her the change in quarters and fifty-cent pieces. There are 30 coins in all. How many of each kind are there?

35. *Train travel.* A train leaves Danville Junction and travels north at a speed of 75 km/h. Two hours later, a second train leaves on a parallel track and travels north at 125 km/h. How far from the station will they meet?

36. *Car travel.* Two cars leave Denver traveling in opposite directions. One car travels at a speed of 80 km/h and the other at 96 km/h. In how many hours will they be 528 km apart?

37. *Canoeing.* Alvin paddled for 4 hr with a 6-km/h current to reach a campsite. The return trip against the same current took 10 hr. Find the speed of Alvin's canoe in still water.

38. *Boating.* Mia's motorboat took 3 hr to make a trip downstream with a 6-mph current. The return trip against the same current took 5 hr. Find the speed of the boat in still water.

39. ▦ *Point of no return.* A plane flying the 3458-mi trip from New York City to London has a 50-mph tailwind. The flight's *point of no return* is the point at which the flight time required to return to New York is the same as the time required to continue to London. If the speed of the plane in still air is 360 mph, how far is New York from the point of no return?

40. ▦ *Point of no return.* A plane is flying the 2553-mi trip from Los Angeles to Honolulu into a 60-mph headwind. If the speed of the plane in still air is 310 mph, how far from Los Angeles is the plane's point of no return? (See Exercise 39.)

SKILL MAINTENANCE

41. Simplify: $-3(x - 7) - 2[x - (4 + 3x)]$.

42. Solve: $3(x - 5) = 7(x - 6)$.

43. Write an equation of the line containing $(2, -5)$ with slope $-\frac{3}{4}$.

44. Graph: $y = -\frac{1}{2}x + 5$.

45. Find a linear function whose graph is perpendicular to the line $2x + 3y = 7$ with a y-intercept of $(0, -5)$.

46. If $f(x) = 3x^2 - 7x$, find $f(-1)$.

SYNTHESIS

47. ◆ List three or four study tips of your own for someone beginning this exercise set.

48. ◆ Suppose that in Example 3 you are asked only for the amount of Black tea needed for the Imperial Blend. Would the method of solving the problem change? Why or why not?

49. ◆ In what ways are Examples 3 and 4 alike? In what sense are the systems of equations similar?

50. ◆ Write a problem similar to Example 2 for a classmate to solve. Design the problem so that the solution is "The florist sold 14 hanging plants and 9 flats of petunias."

51.–54. *For Exercises 51–54, solve Exercises 66–69 from Exercise Set 3.1.*

55. *Automotive maintenance.* The radiator in Michelle's car contains 16 L of antifreeze and water. This mixture is 30% antifreeze. How much of this mixture should she drain and replace with pure antifreeze so that there will be a mixture of 50% antifreeze?

56. *Exercise.* Natalie jogs and walks to school each day. She averages 4 km/h walking and 8 km/h jogging. From home to school is 6 km and Natalie makes the trip in 1 hr. How far does she jog in a trip?

57. *Book sales.* A limited edition of a book published by a historical society was offered for sale to members. The cost was one book for $12 or two books for $20. The society sold 880 books, for a total of $9840. How many members ordered two books?

58. The tens digit of a two-digit positive integer is 2 more than 3 times the units digit. If the digits are interchanged, the new number is 13 less than half the given number. Find the given integer. (*Hint:* Let x = the tens-place digit and y = the units-place digit; then $10x + y$ is the number.)

59. *Train travel.* A train leaves Union Station for Central Station, 216 km away, at 9 A.M. One hour later, a train leaves Central Station for Union Station. They meet at noon. If the second train had started at 9 A.M. and the first train at 10:30 A.M., they would still have met at noon. Find the speed of each train.

60. *Fuel economy.*   Ellen Jordan's station wagon gets 18 miles per gallon (mpg) in city driving and 24 mpg in highway driving. The car is driven 465 mi on 23 gal of gasoline. How many miles were driven in the city and how many were driven on the highway?

61. *Wood stains.*   Williams' Custom Flooring has 0.5 gal of stain that is 20% brown and 80% neutral. A customer orders 1.5 gal of a stain that is 60% brown and 40% neutral. How much pure brown stain and how much neutral stain should be added to the original 0.5 gal in order to make up the order?*

62. *Gender.*   Phil and Phyllis are siblings. Phyllis has twice as many brothers as she has sisters. Phil has the same number of brothers as sisters. How many girls and how many boys are in the family?

63.   ⬚   See Exercise 61 above. Let x = the amount of pure brown stain added to the original 0.5 gal. Find a function $P(x)$ that can be used to determine the percentage of brown stain in the 1.5-gal mixture. On a grapher, draw the graph of P and use ZOOM and TRACE or the TABLE feature to confirm the answer to Exercise 61.

---

*This problem was suggested by Chris Burditt of Yountville, California.

3.4   Systems of Equations in Three Variables

Identifying Solutions   •   Solving Systems in Three Variables   •   Dependency, Inconsistency, and Geometric Considerations

Some problems translate easily to two equations. Others more naturally call for a translation to three or more equations. In this section, we learn how to solve systems of three linear equations. Later, we will use such systems in problem-solving situations.

Identifying Solutions

A **linear equation in three variables** is an equation equivalent to one in the form $Ax + By + Cz = D$, where A, B, C, and D are real numbers. We refer to the form $Ax + By + Cz = D$ as *standard form* for a linear equation in three variables.

A solution of a system of three equations in three variables is an ordered triple (p, q, r) that makes *all three* equations true.

EXAMPLE 1   Determine whether $\left(\frac{3}{2}, -4, 3\right)$ is a solution of the system

$$4x - 2y - 3z = 5,$$
$$-8x - y + z = -5,$$
$$2x + y + 2z = 5.$$

SOLUTION   We substitute $\left(\frac{3}{2}, -4, 3\right)$ into the three equations, using alphabetical order:

$$\frac{4x - 2y - 3z = 5}{4 \cdot \frac{3}{2} - 2(-4) - 3 \cdot 3 \ ? \ 5}$$
$$6 + 8 - 9 \ \Big| $$
$$5 \ \Big| \ 5 \ \text{TRUE}$$

$$\frac{-8x - y + z = -5}{-8 \cdot \frac{3}{2} - (-4) + 3 \ ? \ -5}$$
$$-12 + 4 + 3 \ \Big| $$
$$-5 \ \Big| \ -5 \ \text{TRUE}$$

$$\frac{2x + y + 2z = 5}{2 \cdot \frac{3}{2} + (-4) + 2 \cdot 3 \ ? \ 5}$$
$$3 - 4 + 6 \ \Big| $$
$$5 \ \Big| \ 5 \ \text{TRUE}$$

The triple makes all three equations true, so it is a solution.   ⎯⎯●

Solving Systems in Three Variables

Graphical methods for solving linear equations in three variables are problematic, because a three-dimensional coordinate system is required and the graph of a linear equation in three variables is a plane. The substitution method *can* be used but becomes very cumbersome unless one or more of the equations has only two variables. Fortunately, the elimination method allows us to manipulate a system of three equations in three variables so that a simpler system of two equations in two variables is formed. Once that simpler system has been solved, we can substitute into one of the three original equations and solve for the third variable.

EXAMPLE 2   Solve the following system of equations:

$$x + \ y + \ z = 4, \qquad (1)$$
$$x - 2y - \ z = 1, \qquad (2)$$
$$2x - \ y - 2z = -1. \qquad (3)$$

SOLUTION   We select *any* two of the three equations and work to get one equation in two variables. Let's add equations (1) and (2):

$$x + \ y + z = 4 \qquad (1)$$
$$\underline{x - 2y - z = 1} \qquad (2)$$
$$2x - \ y \qquad = 5. \qquad (4) \qquad \text{Adding to eliminate } z$$

Next, we select a different pair of equations and eliminate the *same variable* we did above. Let's use equations (1) and (3) to again eliminate z. Be careful here! A common error is to eliminate a different variable in this step.

$$\begin{array}{l} x + y + \ z = \ 4, \\ 2x - y - 2z = -1 \end{array} \xrightarrow[\text{by 2}]{\text{Multiplying equation (1)}} \begin{array}{l} 2x + 2y + 2z = \ 8 \\ \underline{2x - \ y - 2z = -1} \\ 4x + \ y \qquad = \ 7 \qquad (5) \end{array}$$

Now we solve the resulting system of equations (4) and (5). That solution will give us two of the numbers in the solution of the original system.

$$2x - y = 5 \quad \text{(4)}$$

$$\underline{4x + y = 7} \quad \text{(5)}$$

$$6x = 12 \quad \text{Adding}$$

$$x = 2$$

Note that we now have two equations in two variables. Had we eliminated different variables above, this would not be the case.

We can use either equation (4) or (5) to find y. We choose equation (5):

$$4x + y = 7 \quad \text{(5)}$$

$$4 \cdot 2 + y = 7 \quad \text{Substituting 2 for } x \text{ in equation (5)}$$

$$8 + y = 7$$

$$y = -1.$$

We now have $x = 2$ and $y = -1$. To find the value for z, we use any of the original three equations and substitute to find the third number, z. Let's use equation (1) and substitute our two numbers in it:

$$x + y + z = 4 \quad \text{(1)}$$

$$2 + (-1) + z = 4 \quad \text{Substituting 2 for } x \text{ and } -1 \text{ for } y$$

$$1 + z = 4$$

$$z = 3.$$

We have obtained the triple $(2, -1, 3)$. It should check in *all three* equations:

$$\frac{x + y + z = 4}{2 + (-1) + 3 \; ? \; 4}$$
$$4 \mid 4 \quad \text{TRUE}$$

$$\frac{x - 2y - z = 1}{2 - 2(-1) - 3 \; ? \; 1}$$
$$1 \mid 1 \quad \text{TRUE}$$

$$\frac{2x - y - 2z = -1}{2 \cdot 2 - (-1) - 2 \cdot 3 \; ? \; -1}$$
$$-1 \mid -1 \quad \text{TRUE}$$

The solution is $(2, -1, 3)$.

SOLVING SYSTEMS OF THREE LINEAR EQUATIONS

To use the elimination method to solve systems of three linear equations:

1. Write all equations in the standard form $Ax + By + Cz = D$.
2. Clear any decimals or fractions.
3. Choose a variable to eliminate. Then select two of the three equations and work to get one equation in two variables.
4. Next use a different pair of equations and eliminate the same variable that you did in step (3).
5. Solve the system of equations that resulted from steps (3) and (4).
6. Substitute the solution from step (5) into one of the original three equations and solve for the third variable. Then check.

E X A M P L E   3   Solve the system

$$4x - 2y - 3z = 5, \qquad (1)$$
$$-8x - y + z = -5, \qquad (2)$$
$$2x + y + 2z = 5. \qquad (3)$$

S O L U T I O N

1., 2. The equations are already in standard form with no fractions or decimals.

3. Next, select a variable to eliminate. We decide on y because the y-terms are opposites of each other in equations (2) and (3). We add:

$$-8x - y + z = -5 \qquad (2)$$
$$\underline{2x + y + 2z = 5} \qquad (3)$$
$$-6x + 3z = 0. \qquad (4) \qquad \text{Adding}$$

4. We use another pair of equations to create a second equation in x and z. That is, we eliminate the same variable, y, as in step (3). We use equations (1) and (3):

$$4x - 2y - 3z = 5, \qquad\qquad\qquad 4x - 2y - 3z = 5$$
$$2x + y + 2z = 5 \xrightarrow[\text{by 2}]{\text{Multiplying equation (3)}} \underline{4x + 2y + 4z = 10}$$
$$8x + z = 15. \qquad (5)$$

5. Now we solve the resulting system of equations (4) and (5). That allows us to find two parts of the ordered triple.

$$-6x + 3z = 0, \qquad\qquad\qquad -6x + 3z = 0$$
$$8x + z = 15 \xrightarrow[\text{by } -3]{\text{Multiplying equation (5)}} \underline{-24x - 3z = -45}$$
$$-30x = -45$$
$$x = \frac{-45}{-30} = \frac{3}{2}$$

We use equation (5) to find z:

$$8x + z = 15$$
$$8 \cdot \tfrac{3}{2} + z = 15 \qquad \text{Substituting } \tfrac{3}{2} \text{ for } x$$
$$12 + z = 15$$
$$z = 3.$$

6. Finally, we use any of the original equations and substitute to find the third number, y. We choose equation (3):

$$2x + y + 2z = 5 \qquad (3)$$
$$2 \cdot \tfrac{3}{2} + y + 2 \cdot 3 = 5 \qquad \text{Substituting } \tfrac{3}{2} \text{ for } x \text{ and 3 for } z$$
$$3 + y + 6 = 5$$
$$y + 9 = 5$$
$$y = -4.$$

The solution is $\left(\tfrac{3}{2}, -4, 3\right)$. The check was performed as Example 1.

Sometimes, certain variables are missing at the outset.

EXAMPLE 4

Solve the system

$$x + y + z = 180, \quad (1)$$
$$x \quad - z = -70, \quad (2)$$
$$2y - z = 0. \quad (3)$$

SOLUTION

1., 2. The equations appear in standard form with no fractions or decimals.

3., 4. Note that there is no y in equation (2). Thus, at the outset, we already have y eliminated from one equation. We need another equation with y eliminated. Let's use equations (1) and (3):

$$
\begin{array}{l}
x + y + z = 180, \\
2y - z = 0
\end{array}
\xrightarrow[\text{(1) by } -2]{\text{Multiplying equation}}
\begin{array}{r}
-2x - 2y - 2z = -360 \\
2y - z = 0 \\
\hline
-2x \quad - 3z = -360. \quad (4)
\end{array}
$$

5., 6. Now we solve the resulting system of equations (2) and (4):

$$
\begin{array}{r}
x - z = -70, \\
-2x - 3z = -360
\end{array}
\xrightarrow[\text{(2) by } 2]{\text{Multiplying equation}}
\begin{array}{r}
2x - 2z = -140 \\
-2x - 3z = -360 \\
\hline
-5z = -500 \\
z = 100.
\end{array}
$$

Continuing as in Examples 2 and 3, we get the solution $(30, 50, 100)$. The check is left to the student.

Dependency, Inconsistency, and Geometric Considerations

Each equation in Examples 2, 3, and 4 has a graph that is a plane in three dimensions. The solutions are points common to the planes of each system. Since three planes can have an infinite number of points in common or no points at all in common, we need to generalize the concept of *consistency*.

One solution: planes intersecting in exactly one point. System is consistent.

The planes intersect along a common line. An infinite number of points are common to the three planes. System is consistent.

Three parallel planes. There is no common point of intersection. System is inconsistent.

Planes intersect two at a time, but there is no point common to all three. System is inconsistent.

CONSISTENCY

A system of equations that has at least one solution is said to be **consistent**.

A system of equations that has no solution is said to be **inconsistent**.

EXAMPLE 5   Solve:

$$y + 3z = 4, \qquad (1)$$
$$-x - y + 2z = 0, \qquad (2)$$
$$x + 2y + z = 1. \qquad (3)$$

SOLUTION   The variable x is missing in equation (1). By adding equations (2) and (3), we can find a second equation in which x is missing:

$$-x - y + 2z = 0 \qquad (2)$$
$$\underline{x + 2y + z = 1} \qquad (3)$$
$$y + 3z = 1. \qquad (4) \qquad \text{Adding}$$

Equations (1) and (4) form a system in y and z. We solve as before:

$$\begin{aligned} y + 3z &= 4, \\ y + 3z &= 1 \end{aligned} \quad \xrightarrow[\text{(1) by } -1]{\text{Multiplying equation}} \quad \begin{aligned} -y - 3z &= -4 \\ \underline{y + 3z} &= \underline{1} \end{aligned}$$

$$\text{This is a contradiction.} \longrightarrow 0 = -3. \qquad \text{Adding}$$

Since we end up with a *false* equation, or contradiction, we know that the system has no solution. It is *inconsistent*.

The notion of *dependency* can also be extended.

DEPENDENCY

If a system of n linear equations is equivalent to a system of fewer than n of them, we say that the equations are *dependent*. If such is not the case, we call the equations *independent*.

EXAMPLE 6   Solve:

$$2x + y + z = 3, \qquad (1)$$
$$x - 2y - z = 1, \qquad (2)$$
$$3x + 4y + 3z = 5. \qquad (3)$$

SOLUTION   Using equations (1) and (2), we add to eliminate z:

$$2x + y + z = 3$$
$$\underline{x - 2y - z = 1}$$
$$3x - y \qquad = 4. \qquad (4)$$

Next, we use equations (2) and (3) to eliminate z again:

$$x - 2y - z = 1, \xrightarrow[\text{(2) by 3}]{\text{Multiplying equation}} \begin{array}{r} 3x - 6y - 3z = 3 \\ 3x + 4y + 3z = 5 \\ \hline 6x - 2y \qquad = 8. \end{array} \quad (5)$$

$$3x + 4y + 3z = 5$$

We now try to solve the resulting system of equations (4) and (5):

$$3x - y = 4, \xrightarrow[\text{(4) by } -2]{\text{Multiplying equation}} \begin{array}{r} -6x + 2y = -8 \\ 6x - 2y = 8 \\ \hline 0 = 0. \end{array} \quad (6)$$

$$6x - 2y = 8$$

Equation (6), which is an identity, indicates that equations (1), (2), and (3) are *dependent*. This means that the original system of three equations is equivalent to a system of two equations. One way to see this is to observe that two times equation (1), minus equation (2), is equation (3). Thus removing equation (3) from the system does not affect the solution of the system. In writing an answer to this problem, we simply state that "the equations are dependent."

Recall that when dependent equations appeared in Section 3.1, the solution sets were always infinite in size and were written in set-builder notation. There, all systems of dependent equations were *consistent*. This is not always the case for systems of three or more equations. The following figures illustrate some possibilities geometrically.

The planes intersect along a common line. The equations are dependent and the system is consistent. There is an infinite number of solutions.

The planes coincide. The equations are dependent and the system is consistent. There is an infinite number of solutions.

Two planes coincide. The third plane is parallel. The equations are dependent and the system is inconsistent. There is no solution.

EXERCISE SET

3.4

1. Determine whether $(2, -1, -2)$ is a solution of the system

$$\begin{aligned} x + y - 2z &= 5, \\ 2x - y - z &= 7, \\ -x - 2y + 3z &= 6. \end{aligned}$$

2. Determine whether $(1, -2, 3)$ is a solution of the system

$$\begin{aligned} x + y + z &= 2, \\ x - 2y - z &= 2, \\ 3x + 2y + z &= 2. \end{aligned}$$

Solve each system. If a system's equations are dependent or if there is no solution, state this.

3. $2x - y + z = 10,$
$4x + 2y - 3z = 10,$
$x - 3y + 2z = 8$

4. $x + y + z = 6,$
$2x - y + 3z = 9,$
$-x + 2y + 2z = 9$

5. $x - y + z = 6,$
$2x + 3y + 2z = 2,$
$3x + 5y + 4z = 4$

6. $2x - y - 3z = -1,$
$2x - y + z = -9,$
$x + 2y - 4z = 17$

7. $6x - 4y + 5z = 31,$
$5x + 2y + 2z = 13,$
$x + y + z = 2$

8. $2x - 3y + z = 5,$
$x + 3y + 8z = 22,$
$3x - y + 2z = 12$

9. $x + y + z = 0,$
$2x + 3y + 2z = -3,$
$-x + 2y - 3z = -1$

10. $3a - 2b + 7c = 13,$
$a + 8b - 6c = -47,$
$7a - 9b - 9c = -3$

11. $2x + y - 3z = -4,$
$4x - 2y + z = 9,$
$3x + 5y - 2z = 5$

12. $2x + 3y + z = 17,$
$x - 3y + 2z = -8,$
$5x - 2y + 3z = 5$

13. $2x + y + 2z = 11,$
$3x + 2y + 2z = 8,$
$x + 4y + 3z = 0$

14. $2x + y + z = -2,$
$2x - y + 3z = 6,$
$3x - 5y + 4z = 7$

15. $-2x + 8y + 2z = 4,$
$x + 6y + 3z = 4,$
$3x - 2y + z = 0$

16. $x - y + z = 4,$
$5x + 2y - 3z = 2,$
$4x + 3y - 4z = -2$

17. $a + 2b + c = 1,$
$7a + 3b - c = -2,$
$a + 5b + 3c = 2$

18. $4x - y - z = 4,$
$2x + y + z = -1,$
$6x - 3y - 2z = 3$

19. $5x + 3y + \frac{1}{2}z = \frac{7}{2},$
$0.5x - 0.9y - 0.2z = 0.3,$
$3x - 2.4y + 0.4z = -1$

20. $r + \frac{3}{2}s + 6t = 2,$
$2r - 3s + 3t = 0.5,$
$r + s + t = 1$

21. $3p + 2r = 11,$
$q - 7r = 4,$
$p - 6q = 1$

22. $4a + 9b = 8,$
$8a + 6c = -1,$
$6b + 6c = -1$

23. $x + y + z = 105,$
$10y - z = 11,$
$2x - 3y = 7$

24. $x + y + z = 57,$
$-2x + y = 3,$
$x - z = 6$

25. $2a - 3b = 2,$
$7a + 4c = \frac{3}{4},$
$2c - 3b = 1$

26. $a - 3c = 6,$
$b + 2c = 2,$
$7a - 3b - 5c = 14$

27. $x + y + z = 180,$
$y = 2 + 3x,$
$z = 80 + x$

28. $l + m = 7,$
$3m + 2n = 9,$
$4l + n = 5$

29. $x + y = 0,$
$x + z = 1,$
$2x + y + z = 2$

30. $x + z = 0,$
$x + y + 2z = 3,$
$y + z = 2$

31. $y + z = 1,$
$x + y + z = 1,$
$x + 2y + 2z = 2$

32. $x + y + z = 1,$
$-x + 2y + z = 2,$
$2x - y = -1$

SKILL MAINTENANCE

33. If $f(x) = 2x + 7$, find $f(a + 1)$.

34. If $g(x) = 1/(x - 7)$, find the domain of g.

35. Solve $K = \frac{1}{2}t(a - b)$ for b.

36. Solve $K = \frac{1}{2}t(a - b)$ for a.

SYNTHESIS

37. ◆ Explain in your own words what it means for the equations of a system to be dependent.

38. ◆ Describe a method for writing an inconsistent system of three equations in three variables.

39. ◆ Is it possible for a system of three equations to have exactly two ordered triples in its solution set? Why or why not?

40. ◆ Suggest a procedure that could be used to solve a system of four equations in four variables.

Solve.

41. $\dfrac{x + 2}{3} - \dfrac{y + 4}{2} + \dfrac{z + 1}{6} = 0,$
$\dfrac{x - 4}{3} + \dfrac{y + 1}{4} - \dfrac{z - 2}{2} = -1,$
$\dfrac{x + 1}{2} + \dfrac{y}{2} + \dfrac{z - 1}{4} = \dfrac{3}{4}$

42. $w + x + y + z = 2,$
$w + 2x + 2y + 4z = 1,$
$w - x + y + z = 6,$
$w - 3x - y + z = 2$

43. $w + x - y + z = 0,$
$w - 2x - 2y - z = -5,$
$w - 3x - y + z = 4,$
$2w - x - y + 3z = 7$

For Exercises 44 and 45, let u represent $1/x$, v represent $1/y$, and w represent $1/z$. Solve for u, v, and w, and then solve for x, y, and z.

44. $\dfrac{2}{x} - \dfrac{1}{y} - \dfrac{3}{z} = -1,$
$\dfrac{2}{x} - \dfrac{1}{y} + \dfrac{1}{z} = -9,$
$\dfrac{1}{x} + \dfrac{2}{y} - \dfrac{4}{z} = 17$

45. $\dfrac{2}{x} + \dfrac{2}{y} - \dfrac{3}{z} = 3,$
$\dfrac{1}{x} - \dfrac{2}{y} - \dfrac{3}{z} = 9,$
$\dfrac{7}{x} - \dfrac{2}{y} + \dfrac{9}{z} = -39$

Determine k so that each system is dependent.

46. $x - 3y + 2z = 1,$
$2x + y - z = 3,$
$9x - 6y + 3z = k$

47. $5x - 6y + kz = -5,$
$x + 3y - 2z = 2,$
$2x - y + 4z = -1$

In each case, three solutions of an equation in x, y, and z are given. Find the equation.

48. $Ax + By + Cz = 12$;
$$\left(1, \tfrac{3}{4}, 3\right), \left(\tfrac{4}{3}, 1, 2\right), \text{ and } (2, 1, 1)$$

49. $z = b - mx - ny$;
$$(1, 1, 2), (3, 2, -6), \text{ and } \left(\tfrac{3}{2}, 1, 1\right)$$

COLLABORATIVE

C ◆ O ◆ R ◆ N ◆ E ◆ R

Focus: Systems of three linear equations

Time: 10–15 minutes

Group size: 3

Consider the six steps outlined on page 163 along with the following system:

$$2x + 4y = 3 - 5z,$$
$$0.3x = 0.2y + 0.7z + 1.4,$$
$$0.04x + 0.03y = 0.07 + 0.04z.$$

ACTIVITY

1. Working independently, each group member should solve the system above. One person should begin by eliminating x, one should first eliminate y, and one

should first eliminate z. Write neatly so that others can follow your steps.

2. Once each group member has solved the system, compare your answers. If the answers do not check, exchange notebooks and check each other's work. If a mistake is detected, allow the person who made the mistake to make the repair.

3. Decide as a group which of the three approaches above (if any) ranks as easiest and which (if any) ranks as most difficult. Then compare your rankings with the other groups in the class.

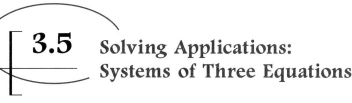

3.5 Solving Applications: Systems of Three Equations

Applications of Three Equations in Three Unknowns

Solving systems of three or more equations is important in many applications. Such systems arise in the natural and social sciences, business, and engineering. In mathematics, purely numerical applications also arise.

EXAMPLE 1

The sum of three numbers is 4. The first number minus twice the second, minus the third is 1. Twice the first number minus the second, minus twice the third is −1. Find the numbers.

S O L U T I O N

1. **Familiarize.**   There are three statements involving the same three numbers. Let's label these numbers x, y, and z.

2. **Translate.**   We can translate directly as follows.

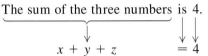

The sum of the three numbers is 4.

$$x + y + z \qquad = 4$$

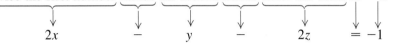

The first number minus twice the second minus the third is 1.

$$x \qquad - \qquad 2y \qquad - \qquad z = 1$$

Twice the first number minus the second minus twice the third is -1.

$$2x \qquad - \qquad y \qquad - \qquad 2z \qquad = -1$$

We now have a system of three equations:

$$x +\ y +\ z = 4,$$
$$x - 2y -\ z = 1,$$
$$2x -\ y - 2z = -1.$$

3. **Carry out.**   We need to solve the system of equations. Note that we found the solution, $(2, -1, 3)$, in Example 2 of Section 3.4.

4. **Check.**   The first statement of the problem says that the sum of the three numbers is 4. That checks. The second statement says that the first number minus twice the second, minus the third is 1: $2 - 2(-1) - 3 = 1$. That checks. The check of the third statement is left to the student.

5. **State.**   The three numbers are 2, -1, and 3.   ———●

E X A M P L E   2

Architecture.   In a triangular cross section of a roof, the largest angle is 70° greater than the smallest angle. The largest angle is twice as large as the remaining angle. Find the measure of each angle.

S O L U T I O N

1. **Familiarize.**   The first thing we do is make a drawing, or a sketch.

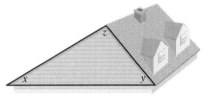

Since we don't know the size of any angle, we use x, y, and z for the measures of the angles. Recall that the measures of the angles in any triangle add up to 180°.

2. **Translate.**   This geometric fact about triangles gives us one equation:

$$x + y + z = 180.$$

Two of the statements can be translated almost directly.

The largest angle is 70° greater than the smallest angle.

$$z \qquad = \qquad x + 70$$

The largest angle is twice as large as the remaining angle.

$$z \qquad = \qquad 2y$$

We now have a system of three equations:

$$
\begin{array}{lll}
x + y + z = 180, & \qquad & x + y + z = 180, \\
x + 70 = z, & \text{or} & x - z = -70, \qquad \text{Rewriting in} \\
2y = z; & & 2y - z = 0. \qquad \text{standard form}
\end{array}
$$

3. **Carry out.**   The system was solved in Example 4 of Section 3.4. The solution is (30, 50, 100).

4. **Check.**   The sum of the numbers is 180, so that checks. The measure of the largest angle, 100°, is 70° greater than the measure of the smallest angle, 30°, so that checks. The measure of the largest angle is also twice the measure of the remaining angle, 50°. Thus we have a check.

5. **State.**   The angles in the triangle measure 30°, 50°, and 100°. ——●

EXAMPLE   3

Cholesterol Levels.   Americans have become very conscious of their cholesterol levels. Recent studies indicate that a child's intake of cholesterol should be no more than 300 mg per day. By eating 1 egg, 1 cupcake, and 1 slice of pizza, a child consumes 302 mg of cholesterol. If the child eats 2 cupcakes and 3 slices of pizza, he or she takes in 65 mg of cholesterol. By eating 2 eggs and 1 cupcake, a child consumes 567 mg of cholesterol. How much cholesterol is in each item?

S O L U T I O N

1. **Familiarize.**   After we have read the problem a few times, it becomes clear that an egg contains considerably more cholesterol than the other foods. Let's guess that one egg contains 200 mg of cholesterol and one cupcake contains 50 mg. Because of the third sentence in the problem, it would follow that a slice of pizza contains 52 mg of cholesterol since $200 + 50 + 52 = 302$.

To see if our guess satisfies the other statements in the problem, we find the amount of cholesterol that 2 cupcakes and 3 slices of pizza would contain: $2 \cdot 50 + 3 \cdot 52 = 256$. Since this does not match the 65 mg listed in the fourth sentence of the problem, our guess was incorrect. Rather than guess again, we examine how we checked our guess and let e, c, and s = the number of milligrams of cholesterol in an egg, a cupcake, and a slice of pizza, respectively.

2. **Translate.** By rewording some of the sentences in the problem, we can translate it into three equations.

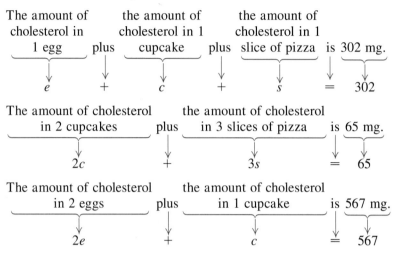

The amount of cholesterol in 1 egg plus the amount of cholesterol in 1 cupcake plus the amount of cholesterol in 1 slice of pizza is 302 mg.

$$e + c + s = 302$$

The amount of cholesterol in 2 cupcakes plus the amount of cholesterol in 3 slices of pizza is 65 mg.

$$2c + 3s = 65$$

The amount of cholesterol in 2 eggs plus the amount of cholesterol in 1 cupcake is 567 mg.

$$2e + c = 567$$

We now have a system of three equations:

$$e + c + s = 302,$$
$$2c + 3s = 65,$$
$$2e + c = 567.$$

3. **Carry out.** We solve and get $e = 274$, $c = 19$, $s = 9$, or (274, 19, 9).

4. **Check.** The sum of 274, 19, and 9 is 302 so the total cholesterol in 1 egg, 1 cupcake, and 1 slice of pizza checks. Two cupcakes and three slices of pizza would contain $2 \cdot 19 + 3 \cdot 9 = 65$ mg, while two eggs and one cupcake would contain $2 \cdot 274 + 19 = 567$ mg of cholesterol. The answer checks.

5. **State.** An egg contains 274 mg of cholesterol, a cupcake contains 19 mg of cholesterol, and a slice of pizza contains 9 mg of cholesterol.

**EXERCISE SET
3.5**

Solve.

1. The sum of three numbers is 57. The second is 3 more than the first. The third is 6 more than the first. Find the numbers.

2. The sum of three numbers is 5. The first number minus the second plus the third is 1. The first minus the third is 3 more than the second. Find the numbers.

3. The sum of three numbers is 26. Twice the first minus the second is 2 less than the third. The third is the second minus three times the first. Find the numbers.

4. The sum of three numbers is 105. The third is 11 less than 10 times the second. Twice the first is 7 more than 3 times the second. Find the numbers.

5. *Geometry.* In triangle *ABC*, the measure of angle *B* is three times that of angle *A*. The measure of angle *C* is 20° more than that of angle *A*. Find the angle measures.

6. *Geometry.* In triangle *ABC*, the measure of angle *B*

is twice the measure of angle *A*. The measure of angle *C* is 80° more than that of angle *A*. Find the angle measures.

7. *Automobile pricing.*   A recent basic model of a particular automobile had a price of $12,685. The basic model with the added features of automatic transmission and power door locks was $14,070. The basic model with air conditioning (AC) and power door locks was $13,580. The basic model with AC and automatic transmission was $13,925. What was the individual cost of each of the three options?

8. *Telemarketing.*   Sven, Tillie, and Isaiah can process 740 telephone orders per day. Sven and Tillie together can process 470 orders, while Tillie and Isaiah together can process 520 orders per day. How many orders can each person process alone?

9. *Lens production.*   When Sight-Rite's three polishing machines, A, B, and C, are all working, 5700 lenses can be polished in one week. When only A and B are working, 3400 lenses can be polished in one week. When only B and C are working, 4200 lenses can be polished in one week. How many lenses can be polished in a week by each machine?

10. *Welding rates.*   Elrod, Dot, and Wendy can weld 74 linear feet per hour when working together. Elrod and Dot together can weld 44 linear feet per hour, while Elrod and Wendy can weld 50 linear feet per hour. How many linear feet per hour can each weld alone?

11. *Restaurant management.*   Kyle works at Dunkin® Donuts, where a 10-oz cup of coffee costs 95¢, a 14-oz cup costs $1.15, and a 20-oz cup costs $1.50. During one busy period, Kyle served 34 cups of coffee, emptying five 96-oz pots while collecting a total of $39.60. How many cups of each size did Kyle fill?

10 oz   14 oz   20 oz
$0.95 $1.15 $1.50

12. *Restaurant management.*   McDonald's® recently sold small soft drinks for 89¢, medium soft drinks for 99¢, and large soft drinks for $1.19. During a lunch-time rush, Chris sold 55 soft drinks for a total of $54.95. The number of small and large drinks,

combined, was 5 fewer than the number of medium drinks. How many drinks of each size were sold?

small   medium   large
$0.89 $0.99 $1.19

13. *Investments.*   A business class divided an imaginary investment of $80,000 among three mutual funds. The first fund grew by 10%, the second by 6%, and the third by 15%. Total earnings were $8850. The earnings from the first fund were $750 more than the earnings from the third. How much was invested in each fund?

14. *Advertising.*   In a recent year, companies spent a total of $84.8 billion on newspaper, television, and radio ads. The total amount spent on television and radio ads was only $2.6 billion more than the amount spent on newspaper ads alone. The amount spent on newspaper ads was $5.1 billion more than what was spent on television ads. How much was spent on each form of advertising? (*Hint*: Let the variables represent numbers of billions of dollars.)

15. *Nutrition.*   A dietician in a hospital prepares meals under the guidance of a physician. Suppose that for a particular patient a physician prescribes a meal to have 800 calories, 55 g of protein, and 220 mg of vitamin C. The dietician prepares a meal of roast beef, baked potatoes, and broccoli according to the data in the following table.

	Calories	Protein (in grams)	Vitamin C (in milligrams)
Roast Beef, 3 oz	300	20	0
Baked Potato	100	5	20
Broccoli, 156 g	50	5	100

How many servings of each food are needed in order to satisfy the doctor's orders?

16. *Nutrition.*   Repeat Exercise 15 but replace the broccoli with asparagus, for which one 180-g serving contains 50 calories, 5 g of protein, and 44 mg of vitamin C. Which meal would you prefer eating?

17. *Obstetrics.*   In the United States, the highest incidence of fraternal twin births occurs among Asian-Americans, then African-Americans, and then Caucasians. Out of every 15,400 births, the total number of fraternal twin births for all three is 739, where there are 185 more for Asian-Americans than African-Americans and 231 more for Asian-Americans than Caucasians. How many births of fraternal twins are there for each group out of every 15,400 births?

18. *Crying rate.*   The sum of the average number of times a man, a woman, and a one-year-old child cry each month is 71.7. A one-year old cries 46.4 more times than a man. The average number of times a one-year-old cries per month is 28.3 more than the average number of times combined that a man and a woman cry. What is the average number of times per month that each cries?

19. *Basketball scoring.*   The New York Knicks recently scored a total of 92 points on a combination of 2-point field goals, 3-point field goals, and 1-point foul shots. Altogether, the Knicks made 50 baskets and 19 more 2-pointers than foul shots. How many shots of each kind were made?

20. *History.*   Find the year in which the first U.S. transcontinental railroad was completed. The following are some facts about the number. The sum of the digits in the year is 24. The ones digit is 1 more than the hundreds digit. Both the tens and the ones digits are multiples of 3.

SKILL MAINTENANCE

21. Solve for x: $3(5 - x) + 7 = 5(x + 3) - 9$.

22. Compute: $(5 - 3^2 \div 2) \cdot (-4)^2$.

23. Simplify: $\dfrac{(a^2 b^3)^5}{a^7 b^{16}}$.

24. Give a slope–intercept equation for a line with slope $-\frac{3}{5}$ and y-intercept $(0, -7)$.

25. If $g(x) = (x - 5)/(x + 7)$, find the domain of g.

26. If $f(x) = 3x^2 - 7x$, find $f(-4)$.

SYNTHESIS

27. ◆ Write a problem for a classmate to solve. Design the problem so that it translates to a system of three equations in three variables.

28. ◆ Exercise 10 can be solved mentally after a careful reading of the problem. How is this possible?

29. Find a three-digit positive integer such that the sum of all three digits is 14, the tens digit is 2 more than the ones digit, and if the digits are reversed, the number is unchanged.

30. *Ages.*   Tammy's age is the sum of the ages of Carmen and Dennis. Carmen's age is 2 more than the sum of the ages of Dennis and Mark. Dennis's age is 4 times Mark's age. The sum of all four ages is 42. How old is Tammy?

31. *Ticket revenue.*   A concert audience of 100 people consists of adults, students, and children. The ticket prices are $10 for adults, $3 for students, and 50¢ for children. The total amount of money taken in is $100. How many adults, students, and children are in attendance? Does there seem to be some information missing? Do some more careful reasoning.

32. *Sharing raffle tickets.*   Hal gives Tom as many raffle tickets as Tom has and Gary as many as Gary has. In like manner, Tom then gives Hal and Gary as many tickets as each then has. Similarly, Gary gives Hal and Tom as many tickets as each then has. If each finally has 40 tickets, with how many tickets does Tom begin?

33. Find the sum of the angle measures at the tips of the star in this figure.

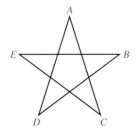

3.6   Elimination Using Matrices

Matrices and Systems • Row-Equivalent Operations

In solving systems of equations, we perform computations with the constants. The variables play no important role until the end. Thus we can simplify writing a system by omitting the variables. For example, the system

$$3x + 4y = 5,$$
$$x - 2y = 1$$

simplifies to

$$\begin{array}{ccc} 3 & 4 & 5 \\ 1 & -2 & 1 \end{array}$$

if we ignore the variables, the operation of addition, and the equals signs.

Matrices and Systems

In the example above, we have written a rectangular array of numbers. Such an array is called a **matrix** (plural, **matrices**). We ordinarily write brackets around matrices. The following are matrices:

$$\begin{bmatrix} -3 & 1 \\ 0 & 5 \end{bmatrix}, \quad \begin{bmatrix} 2 & 0 & -1 & 3 \\ -5 & 2 & 7 & -1 \\ 4 & 5 & 3 & 0 \end{bmatrix}, \quad \begin{bmatrix} 2 & 3 \\ 7 & 15 \\ -2 & 23 \\ 4 & 1 \end{bmatrix}$$

The individual numbers are called *elements* or *entries.*

The **rows** of a matrix are horizontal, and the **columns** are vertical.

$$\begin{bmatrix} 5 & -2 & 2 \\ 1 & 0 & 1 \\ 0 & 1 & 2 \end{bmatrix}$$

← row 1
← row 2
← row 3

↑ column 1 ↑ column 2 ↑ column 3

Let us now use matrices to solve systems of linear equations.

EXAMPLE 1

Solve the system

$$5x - 4y = -1,$$
$$-2x + 3y = 2.$$

SOLUTION   We write a matrix using only coefficients and constants, listing x-coefficients in the first column and y-coefficients in the second. A dashed line separates the coefficients from the constants:

$$\begin{bmatrix} 5 & -4 & \vdots & -1 \\ -2 & 3 & \vdots & 2 \end{bmatrix}.$$

Our goal is to transform

$$\begin{bmatrix} 5 & -4 & | & -1 \\ -2 & 3 & | & 2 \end{bmatrix} \quad \text{into the form} \quad \begin{bmatrix} a & b & | & c \\ 0 & d & | & e \end{bmatrix}.$$

The variables can then be reinserted to form equations from which we can complete the solution.

We do calculations that are similar to those that we would do if we wrote the entire equations. The first step, if possible, is to multiply and/or interchange the rows so that each number in the first column below the first number is a multiple of that number. In this case, we do so by multiplying Row 2 by 5. This corresponds to multiplying the second equation by 5 on both sides.

$$\begin{bmatrix} 5 & -4 & | & -1 \\ -10 & 15 & | & 10 \end{bmatrix} \qquad \text{New Row 2} = 5(\text{Row 2})$$

Next, we multiply the first row by 2 and add the result to the second row. This corresponds to multiplying the first equation by 2 and adding the result to the second equation in order to eliminate a variable. Write out these computations as necessary—we perform them mentally.

$$\begin{bmatrix} 5 & -4 & | & -1 \\ 0 & 7 & | & 8 \end{bmatrix} \qquad \text{New Row 2} = 2(\text{Row 1}) + (\text{Row 2})$$

If we now reinsert the variables, we have

$$5x - 4y = -1, \qquad (1)$$
$$7y = 8. \qquad (2)$$

We can now proceed as before, solving equation (2) for y:

$$7y = 8 \qquad (2)$$
$$y = \tfrac{8}{7}.$$

Next, we substitute $\frac{8}{7}$ for y back in equation (1). This is called *back-substitution*.

$$5x - 4y = -1 \qquad (1)$$
$$5x - 4 \cdot \tfrac{8}{7} = -1 \qquad \text{Substituting } \tfrac{8}{7} \text{ for } y \text{ in equation (1)}$$
$$x = \tfrac{5}{7} \qquad \text{Solving for } x$$

The solution is $\left(\frac{5}{7}, \frac{8}{7}\right)$. A check is left to the student.

EXAMPLE 2

Solve the system

$$2x - y + 4z = -3,$$
$$x \qquad - 4z = 5,$$
$$6x - y + 2z = 10.$$

SOLUTION We first write a matrix, using only the constants. Where there are missing terms, we must write 0's:

$$\begin{bmatrix} 2 & -1 & 4 & | & -3 \\ 1 & 0 & -4 & | & 5 \\ 6 & -1 & 2 & | & 10 \end{bmatrix}$$
(P1)
(P2)
(P3)

(P1), (P2), and (P3) designate the equations that are in the first, second, and third position, respectively.

Our goal is to transform the matrix to one of the form

$$\begin{bmatrix} a & b & c & | & d \\ 0 & e & f & | & g \\ 0 & 0 & h & | & i \end{bmatrix}.$$

A matrix of this form can be rewritten as a system of equations that is equivalent to the original system, and from which a solution can be easily found.

The first step, if possible, is to interchange the rows so that each number in the first column below the first number is a multiple of that number. In this case, we do so by interchanging Rows 1 and 2:

$$\begin{bmatrix} 1 & 0 & -4 & | & 5 \\ 2 & -1 & 4 & | & -3 \\ 6 & -1 & 2 & | & 10 \end{bmatrix}$$

This corresponds to interchanging the first two equations.

Next, we multiply the first row by -2 and add it to the second row:

$$\begin{bmatrix} 1 & 0 & -4 & | & 5 \\ 0 & -1 & 12 & | & -13 \\ 6 & -1 & 2 & | & 10 \end{bmatrix}.$$

This corresponds to multiplying new equation (P1) by -2 and adding it to new equation (P2). We perform the calculations mentally.

Now we multiply the first row by -6 and add it to the third row:

$$\begin{bmatrix} 1 & 0 & -4 & | & 5 \\ 0 & -1 & 12 & | & -13 \\ 0 & -1 & 26 & | & -20 \end{bmatrix}.$$

This corresponds to multiplying equation (P1) by -6 and adding it to equation (P3).

Next, we multiply Row 2 by -1 and add it to the third row:

$$\begin{bmatrix} 1 & 0 & -4 & | & 5 \\ 0 & -1 & 12 & | & -13 \\ 0 & 0 & 14 & | & -7 \end{bmatrix}.$$

This corresponds to multiplying equation (P2) by -1 and adding it to equation (P3).

Reinserting the variables gives us

$$\begin{aligned} x \quad\quad - 4z &= 5, & \text{(P1)} \\ - y + 12z &= -13, & \text{(P2)} \\ 14z &= -7. & \text{(P3)} \end{aligned}$$

We now solve (P3) for z and get $z = -\frac{1}{2}$. Next, we back-substitute $-\frac{1}{2}$ for z in (P2) and solve for y: $-y + 12\left(-\frac{1}{2}\right) = -13$, so $y = 7$. Since there is no y-term in (P1), we need only substitute $-\frac{1}{2}$ for z to solve for x: $x - 4\left(-\frac{1}{2}\right) = 5$, so $x = 3$. The solution is $\left(3, 7, -\frac{1}{2}\right)$. The check is left to the student.

The operations used in the preceding example correspond to those used to produce equivalent systems of equations. We call the matrices **row-equivalent** and the operations that produce them **row-equivalent operations**.

Row-Equivalent Operations

ROW-EQUIVALENT OPERATIONS

Each of the following row-equivalent operations produces an equivalent matrix:

a) Interchanging any two rows.
b) Multiplying all elements of a row by the same nonzero number.
c) Multiplying all elements of a row by a nonzero number and adding the result to another row.

The best overall method for solving systems of equations is by row-equivalent matrices; even computers are programmed to use them. Matrices are part of a branch of mathematics known as linear algebra. They are also studied in many courses in finite mathematics.

EXERCISE SET

3.6

Solve using matrices.

1. $5x - 3y = 13,$
$4x + y = 7$

2. $3x - 3y = 11,$
$9x - 2y = 5$

3. $x + 4y = 8,$
$3x + 5y = 3$

4. $x + 4y = 5,$
$-3x + 2y = 13$

5. $6x - 2y = 4,$
$7x + y = 13$

6. $3x + 4y = 7,$
$-5x + 2y = 10$

7. $4x - y - 3z = 1,$
$8x + y - z = 5,$
$2x + y + 2z = 5$

8. $3x + 2y + 2z = 3,$
$x + 2y - z = 5,$
$2x - 4y + z = 0$

9. $p - 2q - 3r = 3,$
$2p - q - 2r = 4,$
$4p + 5q + 6r = 4$

10. $x + 2y - 3z = 9,$
$2x - y + 2z = -8,$
$3x - y - 4z = 3$

11. $3p + 2r = 11,$
$q - 7r = 4,$
$p - 6q = 1$

12. $4a + 9b = 8,$
$8a + 6c = -1,$
$6b + 6c = -1$

13. $2x + 2y - 2z - 2w = -10,$
$w + y + z + x = -5,$
$x - y + 4z + 3w = -2,$
$w - 2y + 2z + 3x = -6$

14. $-w - 3y + z + 2x = -8,$
$x + y - z - w = -4,$
$w + y + z + x = 22,$
$x - y - z - w = -14$

Solve using matrices.

15. *Coin value.*   A collection of 34 coins consists of dimes and nickels. The total value is $1.90. How many dimes and how many nickels are there?

16. *Coin value.*   A collection of 43 coins consists of dimes and quarters. The total value is $7.60. How many dimes and how many quarters are there?

17. *Mixed granola.*   Grace sells two kinds of granola. One is worth $4.05 per pound and the other is worth $2.70 per pound. She wants to blend the two granolas to get a 15-lb mixture worth $3.15 per pound. How much of each kind of granola should be used?

18. *Trail mix.*   Phil mixes nuts worth $1.60 per pound with oats worth $1.40 per pound to get 20 lb of trail mix worth $1.54 per pound. How many pounds of nuts and how many pounds of oats should be used?

19. *Investments.* Elena receives $212 per year in simple interest from three investments totaling $2500. Part is invested at 7%, part at 8%, and part at 9%. There is $1100 more invested at 9% than at 8%. Find the amount invested at each rate.

20. *Investments.* Miguel receives $306 per year in simple interest from three investments totaling $3200. Part is invested at 8%, part at 9%, and part at 10%. There is $1900 more invested at 10% than at 9%. Find the amount invested at each rate.

SKILL MAINTENANCE

Solve.

21. $0.1x - 12 = 3.6x - 2.34 - 4.9x$

22. $180 = 2x - 11x$

23. $4(9 - x) - 6(8 - 3x) = 5(3x + 4)$

24. Solve $5b = c - ab$ for b.

SYNTHESIS

25. ◈ Explain how you can recognize a dependent system when solving with matrices.

26. ◈ Explain how you can recognize an inconsistent system when solving with matrices.

27. The sum of the digits in a four-digit number is 10. Twice the sum of the thousands digit and the tens digit is 1 less than the sum of the other two digits. The tens digit is twice the thousands digit. The ones digit equals the sum of the thousands digit and the hundreds digit. Find the four-digit number.

28. Solve for x and y:

$$ax + by = c,$$
$$dx + ey = f.$$

3.7 Determinants and Cramer's Rule

Determinants of 2 × 2 Matrices • Cramer's Rule: 2 × 2 Systems •
Cramer's Rule: 3 × 3 Systems

Determinants of 2 × 2 Matrices

When a matrix has m rows and n columns, it is called an "m by n" matrix. Thus its *dimensions* are denoted by $m \times n$. If a matrix has the same number of rows and columns, it is called a **square matrix**. Associated with every square matrix is a number called its **determinant**, defined as follows for 2 × 2 matrices.

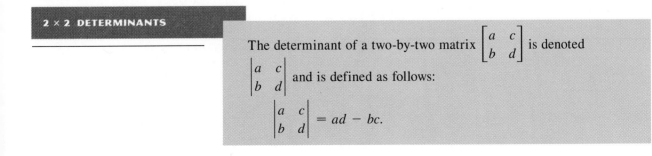

2 × 2 DETERMINANTS

The determinant of a two-by-two matrix $\begin{bmatrix} a & c \\ b & d \end{bmatrix}$ is denoted $\begin{vmatrix} a & c \\ b & d \end{vmatrix}$ and is defined as follows:

$$\begin{vmatrix} a & c \\ b & d \end{vmatrix} = ad - bc.$$

EXAMPLE 1

Evaluate: $\begin{vmatrix} 2 & -5 \\ 6 & 7 \end{vmatrix}$.

SOLUTION   We multiply and subtract as follows:

$$\begin{vmatrix} 2 & -5 \\ 6 & 7 \end{vmatrix} = 2 \cdot 7 - 6 \cdot (-5) = 14 + 30 = 44.$$

Cramer's Rule: 2 × 2 Systems

One of the many uses for determinants is in solving systems of linear equations in which the number of variables is the same as the number of equations and the constants are not all 0. Let's consider a system of two equations:

$$a_1 x + b_1 y = c_1,$$
$$a_2 x + b_2 y = c_2.$$

If we use the elimination method, a series of steps can show that

$$x = \frac{c_1 b_2 - c_2 b_1}{a_1 b_2 - a_2 b_1} \quad \text{and} \quad y = \frac{a_1 c_2 - a_2 c_1}{a_1 b_2 - a_2 b_1}.$$

Determinants can be used in these expressions for x and y.

**CRAMER'S RULE:
2 × 2 SYSTEMS**

The solution of the system

$$a_1 x + b_1 y = c_1,$$
$$a_2 x + b_2 y = c_2,$$

if it is unique, is given by

$$x = \frac{\begin{vmatrix} c_1 & b_1 \\ c_2 & b_2 \end{vmatrix}}{\begin{vmatrix} a_1 & b_1 \\ a_2 & b_2 \end{vmatrix}}, \quad y = \frac{\begin{vmatrix} a_1 & c_1 \\ a_2 & c_2 \end{vmatrix}}{\begin{vmatrix} a_1 & b_1 \\ a_2 & b_2 \end{vmatrix}}.$$

The equations above make sense only if the determinant in the denominator is not 0. If the denominator *is* 0, then one of two things happens.

1. If the denominator is 0 and the other two determinants in the numerators are also 0, then the equations in the system are dependent.
2. If the denominator is 0 and at least one of the other determinants in the numerators is not 0, then the system is inconsistent.

To use Cramer's rule, we find the determinants and compute x and y as shown above. Note that both denominators contain a_1, a_2, b_1, and b_2 in the same position as in the original equations. In the numerator of x, c_1 and c_2 replace a_1 and a_2. In the numerator of y, c_1 and c_2 replace b_1 and b_2.

EXAMPLE 2   Solve using Cramer's rule:

$$2x + 5y = 7,$$
$$5x - 2y = -3.$$

SOLUTION   We have

$$x = \frac{\begin{vmatrix} 7 & 5 \\ -3 & -2 \end{vmatrix}}{\begin{vmatrix} 2 & 5 \\ 5 & -2 \end{vmatrix}} \qquad \text{Using Cramer's rule}$$

$$= \frac{7(-2) - (-3)5}{2(-2) - 5 \cdot 5} = -\frac{1}{29}$$

and

$$y = \frac{\begin{vmatrix} 2 & 7 \\ 5 & -3 \end{vmatrix}}{\begin{vmatrix} 2 & 5 \\ 5 & -2 \end{vmatrix}} \qquad \text{Using Cramer's rule}$$

$$= \frac{2(-3) - 5 \cdot 7}{-29} = \frac{41}{29}. \qquad \text{The denominator is the same as in the expression for } x.$$

The solution is $\left(-\frac{1}{29}, \frac{41}{29}\right)$. The check is left to the student.

Cramer's Rule: 3 × 3 Systems

A similar method has been developed for solving systems of three equations, but before stating the rule, we must extend our terminology.

3 × 3 DETERMINANTS

The determinant of a three-by-three matrix is defined as follows:

Subtract.   Add.

$$\begin{vmatrix} a_1 & b_1 & c_1 \\ a_2 & b_2 & c_2 \\ a_3 & b_3 & c_3 \end{vmatrix} = a_1 \begin{vmatrix} b_2 & c_2 \\ b_3 & c_3 \end{vmatrix} \overset{\downarrow}{-} a_2 \begin{vmatrix} b_1 & c_1 \\ b_3 & c_3 \end{vmatrix} \overset{\downarrow}{+} a_3 \begin{vmatrix} b_1 & c_1 \\ b_2 & c_2 \end{vmatrix}$$

Note that the a's come from the first column. Note too that the 2 × 2 determinants above can be obtained by crossing out the row and the column in which the a occurs. This is shown at the top of p. 182.

For a_1: $\begin{vmatrix} a_1 & b_1 & c_1 \\ a_2 & b_2 & c_2 \\ a_3 & b_3 & c_3 \end{vmatrix}$ For a_2: $\begin{vmatrix} a_1 & b_1 & c_1 \\ a_2 & b_2 & c_2 \\ a_3 & b_3 & c_3 \end{vmatrix}$

For a_3: $\begin{vmatrix} a_1 & b_1 & c_1 \\ a_2 & b_2 & c_2 \\ a_3 & b_3 & c_3 \end{vmatrix}$

EXAMPLE 3   Evaluate:

$$\begin{vmatrix} -1 & 0 & 1 \\ -5 & 1 & -1 \\ 4 & 8 & 1 \end{vmatrix}.$$

SOLUTION   We have

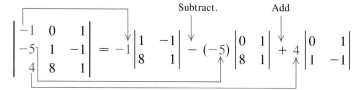

$$= -1(1 + 8) + 5(0 - 8) + 4(0 - 1)$$

Evaluating the three determinants

$$= -9 - 40 - 4 = -53.$$

CRAMER'S RULE:
3 × 3 SYSTEMS

The solution of the system

$$a_1x + b_1y + c_1z = d_1,$$
$$a_2x + b_2y + c_2z = d_2,$$
$$a_3x + b_3y + c_3z = d_3$$

is found by considering the following determinants:

$$D = \begin{vmatrix} a_1 & b_1 & c_1 \\ a_2 & b_2 & c_2 \\ a_3 & b_3 & c_3 \end{vmatrix}, \qquad D_x = \begin{vmatrix} d_1 & b_1 & c_1 \\ d_2 & b_2 & c_2 \\ d_3 & b_3 & c_3 \end{vmatrix},$$

In D_x, the d's replace the a's.

In D_y, the d's replace the b's.

$$D_y = \begin{vmatrix} a_1 & d_1 & c_1 \\ a_2 & d_2 & c_2 \\ a_3 & d_3 & c_3 \end{vmatrix}, \qquad D_z = \begin{vmatrix} a_1 & b_1 & d_1 \\ a_2 & b_2 & d_2 \\ a_3 & b_3 & d_3 \end{vmatrix}.$$

In D_z, the d's replace the c's.

If a unique solution exists, it is given by

$$x = \frac{D_x}{D}, \qquad y = \frac{D_y}{D}, \qquad z = \frac{D_z}{D}.$$

EXAMPLE 4   Solve using Cramer's rule:

$$x - 3y + 7z = 13,$$
$$x + y + z = 1,$$
$$x - 2y + 3z = 4.$$

SOLUTION   We compute D, D_x, D_y, and D_z:

$$D = \begin{vmatrix} 1 & -3 & 7 \\ 1 & 1 & 1 \\ 1 & -2 & 3 \end{vmatrix} = -10; \qquad D_x = \begin{vmatrix} 13 & -3 & 7 \\ 1 & 1 & 1 \\ 4 & -2 & 3 \end{vmatrix} = 20;$$

$$D_y = \begin{vmatrix} 1 & 13 & 7 \\ 1 & 1 & 1 \\ 1 & 4 & 3 \end{vmatrix} = -6; \qquad D_z = \begin{vmatrix} 1 & -3 & 13 \\ 1 & 1 & 1 \\ 1 & -2 & 4 \end{vmatrix} = -24.$$

Then

$$x = \frac{D_x}{D} = \frac{20}{-10} = -2;$$

$$y = \frac{D_y}{D} = \frac{-6}{-10} = \frac{3}{5};$$

$$z = \frac{D_z}{D} = \frac{-24}{-10} = \frac{12}{5}.$$

The solution is $\left(-2, \frac{3}{5}, \frac{12}{5}\right)$. The check is left to the student.

In Example 4, we need not have evaluated D_z. Once x and y were found, we could have substituted them into one of the equations to find z.

To use Cramer's rule, we divide by D, provided $D \neq 0$. If $D = 0$ and at least one of the other determinants is not 0, then the system is inconsistent. If *all* the determinants are 0, then the equations in the system are dependent.

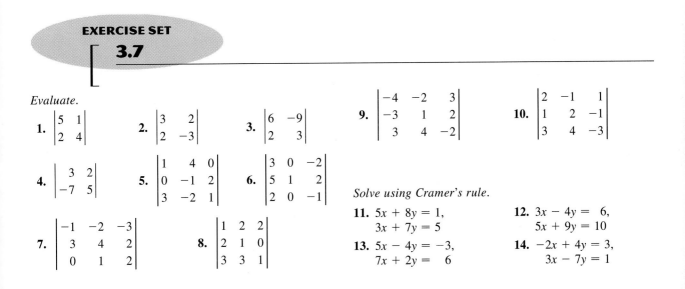

EXERCISE SET

3.7

Evaluate.

1. $\begin{vmatrix} 5 & 1 \\ 2 & 4 \end{vmatrix}$

2. $\begin{vmatrix} 3 & 2 \\ 2 & -3 \end{vmatrix}$

3. $\begin{vmatrix} 6 & -9 \\ 2 & 3 \end{vmatrix}$

4. $\begin{vmatrix} 3 & 2 \\ -7 & 5 \end{vmatrix}$

5. $\begin{vmatrix} 1 & 4 & 0 \\ 0 & -1 & 2 \\ 3 & -2 & 1 \end{vmatrix}$

6. $\begin{vmatrix} 3 & 0 & -2 \\ 5 & 1 & 2 \\ 2 & 0 & -1 \end{vmatrix}$

7. $\begin{vmatrix} -1 & -2 & -3 \\ 3 & 4 & 2 \\ 0 & 1 & 2 \end{vmatrix}$

8. $\begin{vmatrix} 1 & 2 & 2 \\ 2 & 1 & 0 \\ 3 & 3 & 1 \end{vmatrix}$

9. $\begin{vmatrix} -4 & -2 & 3 \\ -3 & 1 & 2 \\ 3 & 4 & -2 \end{vmatrix}$

10. $\begin{vmatrix} 2 & -1 & 1 \\ 1 & 2 & -1 \\ 3 & 4 & -3 \end{vmatrix}$

Solve using Cramer's rule.

11. $5x + 8y = 1,$
   $3x + 7y = 5$

12. $3x - 4y = 6,$
   $5x + 9y = 10$

13. $5x - 4y = -3,$
   $7x + 2y = 6$

14. $-2x + 4y = 3,$
   $3x - 7y = 1$

15. $3x - y + 2z = 1,$
$x - y + 2z = 3,$
$-2x + 3y + z = 1$

16. $3x + 2y - z = 4,$
$3x - 2y + z = 5,$
$4x - 5y - z = -1$

17. $2x - 3y + 5z = 27,$
$x + 2y - z = -4,$
$5x - y + 4z = 27$

18. $x - y + 2z = -3,$
$x + 2y + 3z = 4,$
$2x + y + z = -3$

19. $r - 2s + 3t = 6,$
$2r - s - t = -3,$
$r + s + t = 6$

20. $a - 3c = 6,$
$b + 2c = 2,$
$7a - 3b - 5c = 14$

SKILL MAINTENANCE

Solve.

21. $0.5x - 2.34 + 2.4x = 7.8x - 9$

22. $5x + 7x = -144$

23. A piece of wire 32.8 ft long is to be cut into two pieces, and those pieces are each to be bent to make a square. The length of a side of one square is to be 2.2 ft greater than the length of a side of the other. How should the wire be cut?

SYNTHESIS

Solve.

24. $\begin{vmatrix} y & -2 \\ 4 & 3 \end{vmatrix} = 44$

25. $\begin{vmatrix} 2 & x & -1 \\ -1 & 3 & 2 \\ -2 & 1 & 1 \end{vmatrix} = -12$

26. $\begin{vmatrix} m+1 & -2 \\ m-2 & 1 \end{vmatrix} = 27$

27. Show that an equation of the line through (x_1, y_1) and (x_2, y_2) can be written

$$\begin{vmatrix} x & y & 1 \\ x_1 & y_1 & 1 \\ x_2 & y_2 & 1 \end{vmatrix} = 0.$$

28. ◆ Cramer's rule states that whenever the equations $a_1x + b_1y = c_1$ and $a_2x + b_2y = c_2$ are dependent, we have

$$\begin{vmatrix} a_1 & b_1 \\ a_2 & b_2 \end{vmatrix} = 0.$$

Explain why this occurs.

3.8 Business and Economic Applications

Break-Even Analysis   •   Supply and Demand

Break-Even Analysis

When a company manufactures x units of a product, it invests money. This is **total cost** and can be thought of as a function C, where $C(x)$ is the total cost of producing x units. When the company sells x units of the product, it takes in money. This is **total revenue** and can be thought of as a function R, where $R(x)$ is the total revenue from the sale of x units. **Total profit** is the money taken in less the money spent, or total revenue minus total cost. Total profit from the production and sale of x units is a function P given by

Profit = Revenue − Cost,   or $P(x) = R(x) - C(x).$

If $R(x)$ is greater than $C(x)$, the company makes money. If $C(x)$ is greater than $R(x)$, the company has a loss. When $R(x) = C(x)$, the company breaks even.

There are two kinds of costs. First, there are costs like rent, insurance, machinery, and so on. These costs, which must be paid whether a product is produced or not, are called *fixed costs*. When a product is being produced, there are costs for labor, materials, marketing, and so on. These are called *variable costs*, because they vary according to the amount being produced. The sum of the fixed cost and the variable cost gives the *total cost* of producing a product.

EXAMPLE 1

Manufacturing Radios. Ergs, Inc., is planning to make a new kind of radio. Fixed costs will be $90,000, and it will cost $15 to produce each radio (variable costs). Each radio sells for $26.

a) Find the total cost $C(x)$ of producing x radios.
b) Find the total revenue $R(x)$ from the sale of x radios.
c) Find the total profit $P(x)$ from the production and sale of x radios.
d) What profit or loss will the company realize from the production and sale of 3000 radios? of 14,000 radios?
e) Graph the total-cost, total-revenue, and total-profit functions using the same set of axes. Determine the break-even point.

SOLUTION

a) Total cost is given by

$$C(x) = \text{(Fixed costs) plus (Variable costs)},$$
$$\text{or} \quad C(x) = \quad 90{,}000 \quad + \quad 15x,$$

where x is the number of radios produced.

b) Total revenue is given by

$$R(x) = 26x.$$ $26 times the number of radios sold. We assume that every radio produced is sold.

c) Total profit is given by

$$P(x) = R(x) - C(x)$$
$$= 26x - (90{,}000 + 15x)$$
$$= 11x - 90{,}000.$$

d) Profits will be

$$P(3000) = 11 \cdot 3000 - 90{,}000 = -\$57{,}000$$

when 3000 radios are produced and sold, and

$$P(14,000) = 11 \cdot 14,000 - 90,000 = \$64,000$$

when 14,000 radios are produced and sold. Thus the company loses money if only 3000 radios are sold, but makes money if 14,000 are sold.

e) The graphs of each of the three functions are shown below:

$$R(x) = 26x, \qquad (1)$$
$$C(x) = 90,000 + 15x, \qquad (2)$$
$$P(x) = 11x - 90,000. \qquad (3)$$

$R(x)$, $C(x)$, and $P(x)$ are all in dollars.

Equation (1) has a graph that goes through the origin and has a slope of 26. Equation (2) has an intercept on the $-axis of 90,000 and has a slope of 15. Equation (3) has an intercept on the $-axis of $-90,000$ and has a slope of 11. It is shown by the dashed line. The red dashed line shows a "negative" profit, which is a loss. (That is what is known as "being in the red.") The black dashed line shows a "positive" profit, or gain. (That is what is known as "being in the black.")

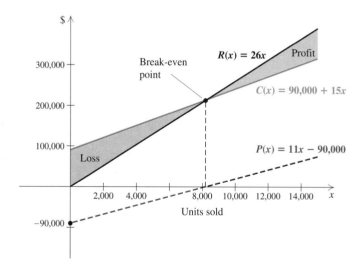

Profits occur where the revenue is greater than the cost. Losses occur where the revenue is less than the cost. The **break-even point** occurs where the graphs of R and C cross. Thus to find the break-even point, we solve a system:

$$R(x) = 26x,$$
$$C(x) = 90,000 + 15x.$$

Since both revenue and cost are in *dollars* and they are equal at the break-even point, the system can be rewritten as

$$d = 26x, \qquad (1)$$
$$d = 90,000 + 15x \qquad (2)$$

and solved using substitution:

$26x = 90,000 + 15x$ Substituting $26x$ for d in equation (2)

$11x = 90,000$

$x \approx 8181.8.$

The firm will break even if it produces and sells about 8182 radios (8181 will yield a tiny loss and 8182 a tiny gain), and takes in a total of $R(8182) = 26 \cdot 8182 = \$212,732$ in revenue. Note that the x-coordinate of the break-even point can also be found by solving $P(x) = 0$.

Supply and Demand

As the price of coffee varies, the amount sold varies. The table and graph below both show that consumer *demand* goes down as the price goes up. As price goes down, demand goes up.

DEMAND FUNCTION, D

Price, p, per Kilogram	Quantity, $D(p)$ (in millions of kilograms)
$ 8.00	25
9.00	20
10.00	15
11.00	10
12.00	5

As the price of coffee varies, the amount available varies. The table and graph below both show that sellers will supply less as the price goes down, but will supply more as the price goes up.

SUPPLY FUNCTION, S

Price, p, per Kilogram	Quantity, $S(p)$ (in millions of kilograms)
$ 9.00	5
9.50	10
10.00	15
10.50	20
11.00	25

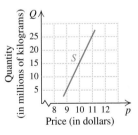

Let's look at the above graphs together. We see that as price increases, demand decreases. As price increases, supply increases. The point of intersection is called the **equilibrium point**. At that price, the amount that the seller will supply is the same amount that the consumer will buy. The situation is analogous to a buyer and a seller negotiating the price of an item. The equilibrium point is the price and quantity that they finally agree on.

Any ordered pair of coordinates from the graph is (price, quantity), because the horizontal axis is the price axis and the vertical axis is the quantity axis. If D is a demand function and S is a supply function, then the equilibrium point is where demand equals supply:

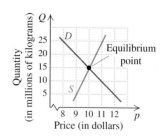

$$D(p) = S(p).$$

EXAMPLE 2

Find the equilibrium point for the demand and supply functions given:

$$D(p) = 1000 - 60p, \qquad (1)$$
$$S(p) = 200 + 4p. \qquad (2)$$

SOLUTION   Since both demand and supply are *quantities* and they are equal at the equilibrium point, we rewrite the system as

$$q = 1000 - 60p, \qquad (1)$$
$$q = 200 + 4p. \qquad (2)$$

We substitute $200 + 4p$ for q in equation (1) and solve:

$$200 + 4p = 1000 - 60p$$
$$200 + 64p = 1000 \qquad \text{Adding } 60p \text{ on both sides}$$
$$64p = 800 \qquad \text{Adding } -200 \text{ on both sides}$$
$$p = \frac{800}{64} = 12.5.$$

Thus the equilibrium price is $12.50 per unit.
    To find the equilibrium quantity, we substitute $12.50 into either $D(p)$ or $S(p)$. We use $S(p)$:

$$S(12.5) = 200 + 4(12.5) = 200 + 50 = 250.$$

Thus the equilibrium quantity is 250 units, and the equilibrium point is ($12.50, 250).

EXERCISE SET

3.8

For each of the following pairs of total cost and total-revenue functions, find **(a)** *the total-profit function and* **(b)** *the break-even point.*

1. $C(x) = 25x + 270,000$;
$R(x) = 70x$

2. $C(x) = 45x + 300,000$;
$R(x) = 65x$

3. $C(x) = 10x + 120,000$;
$R(x) = 60x$

4. $C(x) = 30x + 49,500$;
$R(x) = 85x$

5. $C(x) = 20x + 10,000$;
$R(x) = 100x$

6. $C(x) = 40x + 22,500$;
$R(x) = 85x$

7. $C(x) = 22x + 16,000$;
$R(x) = 40x$

8. $C(x) = 15x + 75,000;$
$R(x) = 55x$

9. $C(x) = 50x + 195,000;$
$R(x) = 125x$

10. $C(x) = 34x + 928,000;$
$R(x) = 128x$

Find the equilibrium point for each of the following pairs of demand and supply functions.

11. $D(p) = 1000 - 10p,$
$S(p) = 230 + p$

12. $D(p) = 2000 - 60p,$
$S(p) = 460 + 94p$

13. $D(p) = 760 - 13p,$
$S(p) = 430 + 2p$

14. $D(p) = 800 - 43p,$
$S(p) = 210 + 16p$

15. $D(p) = 7500 - 25p,$
$S(p) = 6000 + 5p$

16. $D(p) = 8800 - 30p,$
$S(p) = 7000 + 15p$

17. $D(p) = 1600 - 53p,$
$S(p) = 320 + 75p$

18. $D(p) = 5500 - 40p,$
$S(p) = 1000 + 85p$

Solve.

19. *Manufacturing lamps.*   City Lights, Inc., is planning to manufacture a new type of lamp. For the first year, the fixed costs for setting up production are $22,500. The variable costs for producing each lamp are estimated to be $40. The revenue from each lamp is to be $85. Find the following.

a) The total cost $C(x)$ of producing x lamps
b) The total revenue $R(x)$ from the sale of x lamps
c) The total profit $P(x)$ from the production and sale of x lamps
d) The profit or loss from the production and sale of 3000 lamps; of 400 lamps
e) The break-even point

20. *Computer manufacturing.*   Sky View Electronics is planning to introduce a new line of computers. For the first year, the fixed costs for setting up production are $125,100. The variable costs for producing each computer are $750. The revenue from each computer is $1050. Find the following.

a) The total cost $C(x)$ of producing x computers
b) The total revenue $R(x)$ from the sale of x computers
c) The total profit $P(x)$ from the production and sale of x computers
d) The profit or loss from the production and sale of 400 computers; of 700 computers
e) The break-even point

21. *Manufacturing caps.*   Martina's Custom Printing is planning on adding painter's caps to its product line. For the first year, the fixed costs for setting up production are $16,404. The variable costs for producing a dozen caps are $6.00. The revenue on each dozen caps will be $18.00. Find the following.

a) The total cost $C(x)$ of producing x dozen caps
b) The total revenue $R(x)$ from the sale of x dozen caps
c) The total profit $P(x)$ from the production and sale of x dozen caps
d) The profit or loss from the production and sale of 3000 dozen caps; of 1000 dozen caps
e) The break-even point

22. *Sport coat production.*   Sarducci's is planning a new line of sport coats. For the first year, the fixed costs for setting up production are $10,000. The variable costs for producing each coat are $20. The revenue from each coat is to be $100. Find the following.

a) The total cost $C(x)$ of producing x coats
b) The total revenue $R(x)$ from the sale of x coats
c) The total profit $P(x)$ from the production and sale of x coats
d) The profit or loss from the production and sale of 2000 coats; of 50 coats
e) The break-even point

SKILL MAINTENANCE

23. Graph: $y - 3 = \frac{2}{5}(x - 1)$.

24. When two consecutive integers are added and then doubled, the result is 3 less than 5 times the smaller number. Find both numbers.

25. Solve: $9x = 5x - \{3(2x - 7) - 4\}$.

26. Solve $v - st = rw$ for t.

SYNTHESIS

27. ◈ Variable costs and fixed costs are often compared to the slope and the y-intercept, respectively, of an equation for a line. Explain why you feel this analogy is or is not valid.

28. ◈ In this section, we examined supply and demand functions for coffee. Does it seem realistic to you for the graph of D to have a constant slope? Why or why not?

29. *Loudspeaker production.*   Fidelity Speakers, Inc., has fixed costs of $15,400 and variable costs of $100 for each pair of speakers produced. If the speakers sell for $250 a pair, how many pairs of speakers must be produced (and sold) in order to have enough profit to cover the fixed costs of two new facilities? Assume that all fixed costs are identical.

30. *Yo-yo production.*   Bing Boing Hobbies is willing to produce 100 yo-yo's at $2.00 each and 500 yo-yo's at

$8.00 each. Research indicates that the public will buy 500 yo-yo's at $1.00 each and 100 yo-yo's at $9.00 each. Find the equilibrium point.

📈 *Use a grapher to solve.*

31. *Dog food production.* Puppy Love, Inc., will soon begin producing a new line of puppy food. The marketing department predicts that the demand function will be $D(p) = -14.97p + 987.35$ and the supply function will be $S(p) = 98.55p - 5.13$.

a) To the nearest cent, what price per unit should be charged in order to have equilibrium between supply and demand?

b) The production of the puppy food involves $87,985 in fixed costs and $5.15 per unit in variable costs. If the price per unit is the value you found in part (a), how many units must be sold in order to break even?

32. *Computer production.* The Number Cruncher Computer Corporation is planning a new line of computers, each of which will sell for $970. The fixed costs in setting up production are $1,235,580 and the variable costs for each computer are $697.

a) What is the break-even point? (Round to the nearest whole number.)

b) The marketing department at Number Cruncher is not sure that $970 is the best price: Their demand function for the new computers is given by $D(p) = -304.5p + 374,580$ and their supply function is given by $S(p) = 788.7p - 576,504$. To the nearest dollar, what price p would result in equilibrium between supply and demand?

SUMMARY AND REVIEW
3

KEY TERMS

IMPORTANT PROPERTIES AND FORMULAS

When solving a system of two linear equations in two variables:

1. If an identity is obtained, such as $0 = 0$, then the system has an infinite number of solutions. The equations are dependent and, since a solution exists, the system is consistent.

2. If a contradiction is obtained, such as $0 = 7$, then the system has no solution. The system is inconsistent.

To use the elimination method to solve systems of three linear equations:

1. Write all equations in the standard form $Ax + By + Cz = D$.
2. Clear any decimals or fractions.
3. Choose a variable to eliminate. Then select two of the three equations and work to get one equation in two variables.
4. Next use a different pair of equations and eliminate the same variable that you did in step (3).
5. Solve the system of equations that resulted from steps (3) and (4).
6. Substitute the solution from step (5) into one of the original three equations and solve for the third variable. Then check.

Row-Equivalent Operations

Each of the following row-equivalent operations produces an equivalent matrix:

a) Interchanging any two rows.
b) Multiplying each element of a row by the same nonzero number.
c) Multiplying each element of a row by a nonzero number and adding the result to another row.

Determinant of a 2 × 2 Matrix

$$\begin{vmatrix} a & c \\ b & d \end{vmatrix} = ad - bc$$

Determinant of a 3 × 3 Matrix

$$\begin{vmatrix} a_1 & b_1 & c_1 \\ a_2 & b_2 & c_2 \\ a_3 & b_3 & c_3 \end{vmatrix} = a_1 \begin{vmatrix} b_2 & c_2 \\ b_3 & c_3 \end{vmatrix} - a_2 \begin{vmatrix} b_1 & c_1 \\ b_3 & c_3 \end{vmatrix} \\ + a_3 \begin{vmatrix} b_1 & c_1 \\ b_2 & c_2 \end{vmatrix}$$

Cramer's Rule: 2 × 2 Systems

The solution of the system

$$a_1 x + b_1 y = c_1,$$
$$a_2 x + b_2 y = c_2$$

if it is unique, is given by

$$x = \frac{\begin{vmatrix} c_1 & b_1 \\ c_2 & b_2 \end{vmatrix}}{\begin{vmatrix} a_1 & b_1 \\ a_2 & b_2 \end{vmatrix}}, \qquad y = \frac{\begin{vmatrix} a_1 & c_1 \\ a_2 & c_2 \end{vmatrix}}{\begin{vmatrix} a_1 & b_1 \\ a_2 & b_2 \end{vmatrix}}.$$

Cramer's Rule: 3 × 3 Systems

The solution of the system

$$a_1 x + b_1 y + c_1 z = d_1,$$
$$a_2 x + b_2 y + c_2 z = d_2,$$
$$a_3 x + b_3 y + c_3 z = d_3$$

is found by considering the following determinants:

$$D = \begin{vmatrix} a_1 & b_1 & c_1 \\ a_2 & b_2 & c_2 \\ a_3 & b_3 & c_3 \end{vmatrix}, \qquad D_x = \begin{vmatrix} d_1 & b_1 & c_1 \\ d_2 & b_2 & c_2 \\ d_3 & b_3 & c_3 \end{vmatrix},$$

$$D_y = \begin{vmatrix} a_1 & d_1 & c_1 \\ a_2 & d_2 & c_2 \\ a_3 & d_3 & c_3 \end{vmatrix}, \qquad D_z = \begin{vmatrix} a_1 & b_1 & d_1 \\ a_2 & b_2 & d_2 \\ a_3 & b_3 & d_3 \end{vmatrix}.$$

If a unique solution exists, it is given by

$$x = \frac{D_x}{D}, \qquad y = \frac{D_y}{D}, \qquad z = \frac{D_z}{D}.$$

REVIEW EXERCISES

For Exercises 1–9, if a system has an infinite number of solutions, use set-builder notation to write the solution set. If a system has no solution, state this.

Solve graphically.

1. $3x + 2y = -4,$
$y = 3x + 7$

2. $2x + 3y = 12,$
$4x - y = 10$

Solve using the substitution method.

3. $x - 3y = -2,$
$7y - 4x = 6$

4. $y = x + 2,$
$y - x = 8$

5. $9x - 6y = 2,$
$x = 4y + 5$

Solve using the elimination method.

6. $8x - 2y = 10,$
$-4y - 3x = -17$

7. $4x - 7y = 18,$
$9x + 14y = 40$

8. $3x - 5y = -4,$
$5x - 3y = 4$

9. $1.5x - 3 = -2y,$
$3x + 4y = 6$

Solve.

10. Sean has $37 to spend. He can spend all the money on two compact discs and a cassette, or he can buy one CD and two cassettes and have $5.00 left over. What is the price of a CD? What is the price of a cassette?

11. A train leaves Watsonville at noon traveling north at a speed of 44 mph. One hour later, another train, going 55 mph, travels north on a parallel track. How many hours will the second train travel before it overtakes the first train?

12. "Orange-Thirst" is 15% orange juice and "Quencho" is 5% orange juice. How many liters of each should be mixed together in order to get 10 L of a mixture that is 10% orange juice?

Solve. If a system's equations are dependent or if there is no solution, state this.

13. $x + 4y + 3z = 2,$
$2x + y + z = 10,$
$-x + y + 2z = 8$

14. $4x + 2y - 6z = 34,$
$2x + y + 3z = 3,$
$6x + 3y - 3z = 37$

15. $2x - 5y - 2z = -4,$
$7x + 2y - 5z = -6,$
$-2x + 3y + 2z = 4$

16. $-5x + 5y = -6,$
$2x - 2y = 4$

17. $3x + y \quad\;\;\; = 2,$
$x + 3y + z = 0,$
$x + \quad\;\;\; z = 2$

Solve.

18. In triangle ABC, the measure of angle A is 4 times the measure of angle C, and the measure of angle B is 45° more than the measure of angle C. What are the measures of the angles of the triangle?

19. Find the three-digit number in which the sum of the digits is 11, the tens digit is 3 less than the sum of the hundreds and ones digits, and the ones digit is 5 less than the hundreds digit.

20. Lynn has $159 in her purse, consisting of $20, $5, and $1 bills. The number of $20 bills is the same as the total number of $1 and $5 bills. If she has 14 bills in her purse, how many of each denomination does she have?

Solve using matrices. Show your work.

21. $3x + 4y = -13,$
$5x + 6y = 8$

22. $3x - y + z = -1,$
$2x + 3y + z = 4,$
$5x + 4y + 2z = 5$

Evaluate.

23. $\begin{vmatrix} -2 & 4 \\ -3 & 5 \end{vmatrix}$

24. $\begin{vmatrix} 2 & 3 & 0 \\ 1 & 4 & -2 \\ 2 & -1 & 5 \end{vmatrix}$

Solve using Cramer's rule. Show your work.

25. $2x + 3y = 6,$
$x - 4y = 14$

26. $2x + y + z = -2,$
$2x - y + 3z = 6,$
$3x - 5y + 4z = 7$

27. Find the equilibrium point for the demand and supply functions

$$S(p) = 60 + 7p$$

and

$$D(p) = 120 - 13p.$$

28. Kregel Furniture is planning to produce a new type of bed. For the first year, the fixed costs for setting up production are $35,000. The variable costs for producing each bed are $175. The revenue from each bed is $300. Find the following.

   a) The total cost $C(x)$ of producing x beds
   b) The total revenue $R(x)$ from the sale of x beds
   c) The total profit $P(x)$ from the production and sale of x beds

d) The profit or loss from the production and sale of 1200 beds; of 200 beds

e) The break-even point

SKILL MAINTENANCE

29. Solve: $4x - 5x + 8 = -9x + 2x$.

30. Solve $Q = at - 4t$ for t.

31. Write an equation of the line containing $(3, 5)$ and perpendicular to the line $3x - 7y = 4$.

32. The Hammondton Hoopsters increased its score by 7 points in each of three consecutive games. In those three games, the team scored a total of 228 points. What was their score in each game?

SYNTHESIS

33. ◈ How would you go about solving a problem that involved four variables?

34. ◈ Explain how a system of equations can be both dependent and inconsistent.

35. Solve graphically:

$$y = x + 2,$$
$$y = x^2 + 2.$$

36. The graph of $f(x) = ax^2 + bx + c$ contains the points $(-2, 3)$, $(1, 1)$, and $(0, 3)$. Find a, b, and c and give a formula for the function.

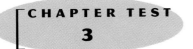

CHAPTER TEST

3

Solve, if possible, using the substitution method.

1. $x + 3y = -8,$
$4x - 3y = 23$

2. $2x + 4y = -6,$
$y = 3x - 9$

Solve, if possible, using the elimination method.

3. $4x - 6y = 3,$
$6x - 4y = -3$

4. $4y + 2x = 18,$
$3x + 6y = 26$

5. The perimeter of a rectangle is 96. The length of the rectangle is 6 less than twice the width. Find the dimensions of the rectangle.

6. Between her home mortgage (loan), car loan, and credit card bill (loan), Rema is $75,300 in debt. Rema's credit card bill accumulates 1.5% interest, her car loan 1% interest, and her mortgage 0.6% interest each month. After one month, her total accumulated interest is $460.50. The interest on Rema's credit card bill was $4.50 more than the interest on her car loan. Find the amount of each loan.

Solve. If a system's equations are dependent or if there is no solution, state this.

7. $-3x + y - 2z = 8,$
$-x + 2y - z = 5,$
$2x + y + z = -3$

8. $6x + 2y - 4z = 15,$
$-3x - 4y + 2z = -6,$
$4x - 6y + 3z = 8$

9. $2x + 2y = 0,$
$4x + 4z = 4,$
$2x + y + z = 2$

10. $3x + 3z = 0,$
$2x + 2y = 2,$
$3y + 3z = 3$

Solve using matrices.

11. $7x - 8y = 10,$
$9x + 5y = -2$

12. $x + 3y - 3z = 12,$
$3x - y + 4z = 0,$
$-x + 2y - z = 1$

Evaluate.

13. $\begin{vmatrix} 4 & -2 \\ 3 & 7 \end{vmatrix}$

14. $\begin{vmatrix} 3 & 4 & 2 \\ 2 & -5 & 4 \\ 4 & 5 & -3 \end{vmatrix}$

15. Solve using Cramer's rule:

$$8x - 3y = 5,$$
$$2x + 6y = 3.$$

16. An electrician, a carpenter, and a plumber are hired to work on a house. The electrician earns $21 per hour, the carpenter $19.50 per hour, and the plumber $24 per hour. The first day on the job, they worked a total of 21.5 hr and earned a total of $469.50. If the plumber worked 2 more hours than the carpenter did, how many hours did the electrician work?

17. Find the equilibrium point for the demand and supply functions

$$D(p) = 79 - 8p \quad \text{and} \quad S(p) = 37 + 6p.$$

18. Sweet Spot Manufacturing is planning a new type of tennis racket. For the first year, the fixed costs for setting up production are $40,000. The variable costs for producing each racket are $30. The sales department predicts that 1500 rackets can be sold during the first year. The revenue from each racket is $80. Find the following.

a) The total cost $C(x)$ of producing x tennis rackets
b) The total revenue $R(x)$ from the sale of x tennis rackets
c) The total profit $P(x)$
d) The profit or loss if the expected sales of 1500 rackets occurs
e) The break-even point

SKILL MAINTENANCE

19. The price of a radio, including 5% sales tax, is $36.75. Find the price of the radio before the tax was added.

20. Solve $P = 4a - 3b$ for a.

21. Solve: $-3x - 5 + 6x = 8x - 14$.

22. Find an equation of the line passing through the points $(5, -1)$ and $(-3, 4)$.

SYNTHESIS

23. The graph of the function $f(x) = mx + b$ contains the points $(-1, 3)$ and $(-2, -4)$. Find m and b.

24. At a county fair, an adult's ticket sold for $5.50, a senior citizen's ticket for $4.00, and a child's ticket for $1.50. On opening day, the number of adults' and senior citizens' tickets sold was 30 more than the number of children's tickets sold. The number of adults' tickets sold was 6 more than 4 times the number of senior citizens' tickets sold. Total receipts from the ticket sales were $11,219.50. How many of each type of ticket were sold?

CUMULATIVE REVIEW
1-3

Solve.

1. $-14.3 + 29.17 = x$

2. $x + 9.4 = -12.6$

3. $3.9(-11) = x$

4. $-2.4x = -48$

5. $4x + 7 = -14$

6. $-3 + 5x = 2x + 15$

7. $3n - (4n - 2) = 7$

8. $6y - 5(3y - 4) = 10$

9. $14 + 2c = -3(c + 4) - 6$

10. $5x - [4 - 2(6x - 1)] = 12$

Simplify. Do not leave negative exponents in your answers.

11. $x^4 \cdot x^{-6} \cdot x^{13}$

12. $(4x^{-3}y^2)(-10x^4y^{-7})$

13. $(6x^2y^3)^2(-2x^0y^4)^3$

14. $\dfrac{y^4}{y^{-6}}$

15. $\dfrac{-10a^7b^{-11}}{25a^{-4}b^{22}}$

16. $\left(\dfrac{3x^4y^{-2}}{4x^{-5}}\right)^4$

17. $(1.95 \times 10^{-3})(5.73 \times 10^8)$

18. $\dfrac{2.42 \times 10^5}{6.05 \times 10^{-2}}$

19. Solve $A = \frac{1}{2}h(b + t)$ for b.

20. Determine whether $(-3, 4)$ is a solution of $5a - 2b = -23$.

Graph.

21. $y = -2x + 3$

22. $y = x^2 - 1$

23. $4x + 16 = 0$

24. $-3x + 2y = 6$

25. Find the slope and the y-intercept of the line with equation $-4y + 9x = 12$.

26. Find the slope, if it exists, of the line containing the points $(2, 7)$ and $(-1, 3)$.

27. Find an equation of the line with slope -3 and containing the point $(2, -11)$.

28. Find an equation of the line containing the points $(-6, 3)$ and $(4, 2)$.

29. Determine whether the lines are parallel or perpendicular:

$$2x = 4y + 7,$$
$$x - 2y = 5.$$

30. Find an equation of the line containing the point $(2, 1)$ and perpendicular to the line $x - 2y = 5$.

31. For the graph of f shown, determine the domain, the range, $f(-3)$, and any value of x for which $f(x) = 5$.

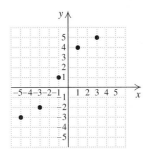

32. Determine the domain of the function given by

$$f(x) = \frac{7}{2x - 1}.$$

Given $g(x) = 4x - 3$ and $h(x) = -2x^2 + 1$, find the following function values.

33. $h(4)$

34. $-g(0)$

35. $(g \cdot h)(-1)$

36. $g(a) - h(2a)$

Solve.

37. $3x + y = 4,$
$\quad 6x - y = 5$

38. $4x + 4y = 4,$
$\quad 5x - 3y = -19$

39. $\quad 6x - 10y = -22,$
$\quad -11x - 15y = 27$

40. $x + y + z = -5,$
$\quad 2x + 3y - 2z = 8,$
$\quad x - y + 4z = -21$

41. $\quad 2x + 5y - 3z = -11,$
$\quad -5x + 3y - 2z = -7,$
$\quad 3x - 2y + 5z = 12$

Evaluate.

42. $\begin{vmatrix} 2 & -3 \\ 4 & 1 \end{vmatrix}$

43. $\begin{vmatrix} 1 & 0 & 1 \\ -1 & 2 & 1 \\ 2 & 1 & 3 \end{vmatrix}$

44. The sum of two numbers is 26. Three times the smaller plus twice the larger is 60. Find the numbers.

45. "Soakem" is 34% salt and the rest water. "Rinsem" is 61% salt and the rest water. How many ounces of each would be needed to obtain 120 oz of a mixture that is 50% salt?

46. Find three consecutive odd numbers such that the sum of 4 times the first number and 5 times the third number is 47.

47. Belinda's scores on four tests are 83, 92, 100, and 85. What must the score be on the fifth test so that the average will be 90?

48. The perimeter of a rectangle is 32 cm. If 5 times the width equals 3 times the length, what are the dimensions of the rectangle?

49. There are 4 more nickels than dimes in a piggy bank. The total amount of money in the bank is $2.45. How many of each type of coin are in the bank?

50. One month Ladi and Bo spent $680 for electricity, rent, and telephone. The electric bill was $\frac{1}{4}$ of the rent and the rent was $400 more than the phone bill. How much was the electric bill?

51. A hockey team played 64 games one season. It won 15 more games than it tied and lost 10 more games than it won. How many games did it win? lose? tie?

52. Reggie, Jenna, and Achmed are counting calories. For lunch one day, Reggie ate two cookies and a banana, for a total of 260 calories. Jenna had a cup of yogurt and a banana, for a total of 245 calories. Achmed ate a cookie, a cup of yogurt, and two bananas, for a total of 415 calories. How many calories are in each item?

SYNTHESIS

53. Simplify: $(6x^{a+2}y^{b+2})(-2x^{a-2}y^{y+1})$.

54. An automotive dealer discovers that when $1000 is spent on radio advertising, weekly sales increase by $101,000. When $1250 is spent on radio advertising, weekly sales increase by $126,000. Assuming that sales increase according to a linear equation, by what amount would sales increase when $1500 is spent on radio advertising?

55. Given that $f(x) = mx + b$ and that $f(5) = -3$ when $f(-4) = 2$, find m and b.

Inequalities and Problem Solving

AN APPLICATION

A $3.00 toll is charged to cross the bridge from Sanibel Island to mainland Florida. A six-month pass, which costs $15.00, reduces the toll to $0.50. A one-year pass, costing $60, allows for free crossings. How many crossings per month does it take, on average for the six-month pass to be the more economical choice?

THIS PROBLEM APPEARS AS EXERCISE 83 IN SECTION 4.2.

More information on Sanibel Island is available at
http://hepg.awl.com/be/inter_5

*I*n the travel business, we use math quite a lot. We must be able to tell clients how many miles and for how long they will be traveling. We also use math to calculate exchange rates, taxes, per-person room rates, and commissions, as well to teach clients how to use military time.

SANDRA DAVIS
Travel Agent
Philadelphia, PA

*I*nequalities are mathematical sentences containing symbols such as < (is less than). Principles similar to those used for solving equations enable us to solve inequalities and the problems that translate to inequalities. In this chapter, we develop procedures for solving a variety of inequalities and systems of inequalities.

4.1   Inequalities and Applications

Solving Inequalities   •   Interval Notation   •   The Addition Principle   •
The Multiplication Principle   •   Using the Principles Together   •
Problem Solving

Solving Inequalities

We can extend our equation-solving skills to the solving of inequalities. An **inequality** is any sentence containing $<$, $>$, $\leq$, $\geq$, or $\neq$ (see Section 1.2)—for example,

$$-2 < a, \qquad x > 4, \qquad x + 3 \leq 6, \qquad 6 - 7y \geq 10y - 4, \text{ and } 5x \neq 10.$$

Any replacement for the variable that makes an inequality true is called a **solution**. The set of all solutions is called the **solution set**. When all solutions of an inequality are found, we say that we have **solved** the inequality.

EXAMPLE 1

Determine whether the given number is a solution of the inequality.

a) $x + 3 < 6$;   5   **b)** $2x - 3 > -5$;   1

SOLUTION

a) We substitute to get $5 + 3 < 6$, or $8 < 6$, a false sentence. Thus, 5 is not a solution.
b) We substitute to get $2 \cdot 1 - 3 > -5$, or $-1 > -5$, a true sentence. Thus, 1 is a solution.

The *graph* of an inequality is a drawing that represents its solutions. An inequality in one variable can be graphed on a number line. Inequalities in two variables can be graphed on a coordinate plane, and are considered later in this chapter.

EXAMPLE 2

Graph $x < 4$ on a number line.

SOLUTION   The solutions are all real numbers less than 4, so we shade all numbers less than 4. Since 4 is not a solution, we use an open dot at 4.

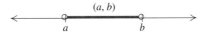

We can write the solution set using *set-builder notation* (see Section 1.1):

$$\{x\mid x < 4\}.$$

This is read

"The set of all x such that x is less than 4."

Interval Notation

Another way to write solutions of an inequality in one variable is to use **interval notation.** Interval notation uses parentheses, ( ), and brackets, [ ].

If a and b are real numbers such that $a < b$, we define the **open interval (a, b)** as the set of all numbers x for which $a < x < b$. Thus,

$$(a, b) = \{x\mid a < x < b\}.$$

Its graph excludes the endpoints:

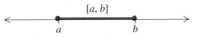

CAUTION! Do not confuse the *interval* (a, b) with the *ordered pair* (a, b). The context in which the notation appears usually makes the meaning clear.

The **closed interval $[a, b]$** is defined as the set of all numbers x for which $a \le x \le b$. Thus,

$$[a, b] = \{x\mid a \le x \le b\}.$$

Its graph includes the endpoints*:

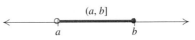

There are two kinds of **half-open intervals,** defined as follows:

1. $(a, b] = \{x\mid a < x \le b\}$. This is open on the left. Its graph is as follows:

---

*Some books use the representations ⊢────→ and ⊢────⊣ instead of, respectively,

○────○ and ●────● .

2. $[a, b) = \{x \mid a \leq x < b\}$. This is open on the right. Its graph is as follows:

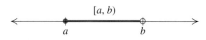

We use the symbols ∞ and $-\infty$ to represent positive and negative infinity, respectively. Thus the notation (a, ∞) represents the set of all real numbers greater than a, and $(-\infty, a)$ represents the set of all real numbers less than a.

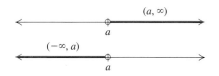

The notations $[a, \infty)$ and $(-\infty, a]$ are used when we want to include the endpoint a.

EXAMPLE 3   Graph $y \geq -2$ on a number line and write the solution set using both set-builder and interval notations.

SOLUTION   Using set-builder notation, we write the solution set as $\{y \mid y \geq -2\}$.

Using interval notation, we write the solution set as $[-2, \infty)$.

To graph the solution, we shade all numbers to the right of -2 and use a solid dot to indicate that -2 is also a solution.

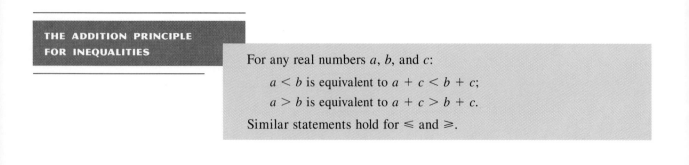

The Addition Principle

Two inequalities are equivalent if they have the same solution set. For example, the inequalities $x > 4$ and $4 < x$ are equivalent. Just as the addition principle for equations produces equivalent equations, the addition principle for inequalities produces equivalent inequalities.

THE ADDITION PRINCIPLE FOR INEQUALITIES

For any real numbers a, b, and c:

$a < b$ is equivalent to $a + c < b + c$;

$a > b$ is equivalent to $a + c > b + c$.

Similar statements hold for $\leq$ and $\geq$.

As with equations, we try to get the variable alone on one side in order to determine solutions easily.

EXAMPLE   4

Solve and graph: **(a)** $x + 5 > 1$; **(b)** $4x - 1 \geqslant 5x - 2$.

SOLUTION

a)

$$x + 5 > 1$$
$$x + 5 + (-5) > 1 + (-5) \qquad \text{Using the addition principle to add } -5 \text{ on both sides}$$
$$x > -4$$

When an inequality— like this last one— has an infinite number of solutions, we cannot possibly check them all. Instead, we can perform a partial check by substituting one member of the solution set (here we use -1) into the original inequality:

$$\frac{x + 5 > 1}{-1 + 5 \; ? \; 1}$$
$$\qquad\quad 4 \; | \; 1 \quad \text{TRUE}$$

Since $4 > 1$ is true, we have our check. The solution set is $\{x \mid x > -4\}$, or $(-4, \infty)$. The graph is as follows:

b)

$$4x - 1 \geqslant 5x - 2$$
$$4x - 1 + 2 \geqslant 5x - 2 + 2 \qquad \text{Adding 2 on both sides}$$
$$4x + 1 \geqslant 5x \qquad \text{Simplifying}$$
$$4x + 1 - 4x \geqslant 5x - 4x \qquad \text{Adding } -4x \text{ on both sides}$$
$$1 \geqslant x \qquad \text{Simplifying}$$

We know that $1 \geqslant x$ has the same meaning as $x \leqslant 1$. You can check that any number less than or equal to 1 is a solution. The solution set is $\{x \mid 1 \geqslant x\}$ or, more commonly, $\{x \mid x \leqslant 1\}$. Using interval notation, we write that the solution set is $(-\infty, 1]$. The graph is as follows:

The Multiplication Principle

The multiplication principle for inequalities differs from the multiplication principle for equations.

Consider this true inequality:

$$4 < 9.$$

If we multiply both sides of $4 < 9$ by 2, we get another true inequality:

$$4 \cdot 2 < 9 \cdot 2, \quad \text{or} \quad 8 < 18.$$

If we multiply both sides of $4 < 9$ by -2, we get a false inequality:

$$\text{FALSE} \longrightarrow 4(-2) < 9(-2), \quad \text{or} \quad -8 < -18. \longleftarrow \text{FALSE}$$

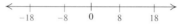

This is because negation reverses relative position on the number line. However, if the inequality symbol is reversed, we get a true inequality:

$$-8 > -18. \longleftarrow \text{TRUE}$$

The $<$ symbol has been reversed!

THE MULTIPLICATION PRINCIPLE FOR INEQUALITIES

For any real numbers a and b, and for any *positive* number c,

$a < b$ is equivalent to $ac < bc$;

$a > b$ is equivalent to $ac > bc$.

For any real numbers a and b, and for any *negative* number c,

$a < b$ is equivalent to $ac > bc$;

$a > b$ is equivalent to $ac < bc$.

Similar statements hold for $\leq$ and $\geq$.

Since division by c is the same as multiplication by $1/c$, there is no need for a separate division principle.

CAUTION!   Remember that to multiply or divide both sides of an inequality by a negative number, we must reverse the inequality symbol.

EXAMPLE 5   Solve and graph: **(a)** $3y < \frac{3}{4}$; **(b)** $-5x \geq -80$.

SOLUTION

a) $3y < \frac{3}{4}$

The symbol stays the same.

$$\frac{1}{3} \cdot 3y < \frac{1}{3} \cdot \frac{3}{4} \qquad \text{Multiplying by } \frac{1}{3} \text{ on both sides}$$

$$y < \frac{1}{4}$$

Any number less than $\frac{1}{4}$ is a solution. The solution set is $\left\{y \mid y < \frac{1}{4}\right\}$, or $\left(-\infty, \frac{1}{4}\right)$. The graph is as follows:

b) $-5x \geqslant -80$

                    The symbol must be reversed.

$-\frac{1}{5} \cdot (-5x) \leqslant -\frac{1}{5} \cdot (-80)$ Multiplying by $-\frac{1}{5}$ or dividing by -5 on both sides

          $x \leqslant 16$

The solution set is $\{x \mid x \leqslant 16\}$, or $(-\infty, 16]$. The graph is as follows:

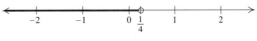

Using the Principles Together

We use the addition and multiplication principles together in solving inequalities in much the same way as in solving equations.

EXAMPLE 6   Solve: **(a)** $16 - 7y \geqslant 10y - 4$; **(b)** $-3(x + 8) - 5x > 4x - 9$.

SOLUTION

a) $16 - 7y \geqslant 10y - 4$

$-16 + 16 - 7y \geqslant -16 + 10y - 4$ Adding -16 on both sides

          $-7y \geqslant 10y - 20$

$-10y + (-7y) \geqslant -10y + 10y - 20$ Adding $-10y$ on both sides

          $-17y \geqslant -20$

                    The symbol must be reversed.

$-\frac{1}{17} \cdot (-17y) \leqslant -\frac{1}{17} \cdot (-20)$ Multiplying by $-\frac{1}{17}$, or dividing by -17, on both sides

          $y \leqslant \frac{20}{17}$

The solution set is $\left\{y \mid y \leqslant \frac{20}{17}\right\}$, or $\left(-\infty, \frac{20}{17}\right]$.

b) $-3(x + 8) - 5x > 4x - 9$

     $-3x - 24 - 5x > 4x - 9$ Using the distributive law

          $-24 - 8x > 4x - 9$

$-24 - 8x + 8x > 4x - 9 + 8x$ Adding $8x$ on both sides

          $-24 > 12x - 9$

     $-24 + 9 > 12x - 9 + 9$ Adding 9 on both sides

          $-15 > 12x$

                    The symbol stays the same.

          $-\frac{5}{4} > x$ Multiplying by $\frac{1}{12}$ and simplifying

The solution set is $\left\{x \mid -\frac{5}{4} > x\right\}$, or $\left\{x \mid x < -\frac{5}{4}\right\}$, or $\left(-\infty, -\frac{5}{4}\right)$.

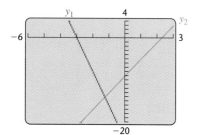

To check Example 6(b) with a grapher, let $y_1 = -3(x + 8) - 5x$ and $y_2 = 4x - 9$. Graph y_1 and y_2 in the window $[-6, 3, -20, 4]$ and identify those x-values for which $y_1 > y_2$.

can then show that $y_1 > y_2$ for x-values to the left of the point of intersection, or for the interval $(-\infty, -1.25)$.

On many graphers, the interval on the number line that is the solution can be found directly. This is accomplished by using the Y-VARS and TEST keys to enter $y_3 = y_1 > y_2$. The solution set is then displayed as an interval (shown by a horizontal line 1 unit above the x-axis).

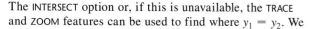

The INTERSECT option or, if this is unavailable, the TRACE and ZOOM features can be used to find where $y_1 = y_2$. We

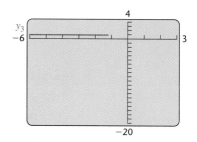

Problem Solving

Many problem-solving situations translate to inequalities. In addition to "is less than" and "is more than," other phrases are commonly used.

Phrase	Translation
a "is at most" 17	$a \leq 17$
a "is at least" 5	$a \geq 5$
a "can't exceed" 12	$a \leq 12$
a "is a better buy than" b	$a < b$

E X A M P L E   7

Records in the Men's 200-m Dash.   Michael Johnson set a world record of 19.32 sec in the men's 200-m dash in the 1996 Olympics. The function

$$R(t) = -0.0513t + 19.32$$

can be used to predict the world record in the men's 200-m dash t years after 1996. Determine (in terms of an inequality) those years for which the world record will be less than 19.0 sec.

S O L U T I O N

1. **Familiarize.**   We already have a formula. To become more familiar with it, we might make a substitution for t. Suppose we want to know the record after 20 years, in the year 2016. We substitute 20 for t:

$$R(20) = -0.0513(20) + 19.32 = 18.294 \text{ sec.}$$

We see that by 2016, the record will be less than 19.0 sec. To predict the exact year in which the 19.0-sec mark will be broken, we could make other guesses that are less than 20. Instead, we proceed to the next step.

2. **Translate.**   The record $R(t)$ is to be *less than* 19.0 sec. Thus we have

$R(t) < 19.0.$

We replace $R(t)$ with $-0.0513t + 19.32$ to find the times t that solve the inequality:

$-0.0513t + 19.32 < 19.0.$

3. **Carry out.**   We solve the inequality:

$-0.0513t + 19.32 < 19.0$

$-0.0513t < -0.32$ Adding -19.32 on both sides

$t > 6.24.$ Dividing by -0.0513 and rounding

4. **Check.**   A partial check is to substitute a value for t greater than 6.24. We did that in the *Familiarize* step.

5. **State.**   The record will be less than 19.0 sec for races occurring more than 6.24 years after 1996, or approximately $\{t|\ t > 2002\}$. ————●

EXAMPLE 8

On a new job, Rose can be paid in one of two ways:

Plan A: A salary of $600 per month, plus a commission of 4% of sales;

Plan B: A salary of $800 per month, plus a commission of 6% of sales in excess of $10,000.

For what amount of monthly sales is plan A better than plan B, if we assume that sales are always more than $10,000?

SOLUTION

1. **Familiarize.**   Listing the given information in a table will be helpful.

Plan A: Monthly Income	Plan B: Monthly Income
$600 salary 4% of sales *Total*: $600 + 4% of sales	$800 salary 6% of sales over $10,000 *Total*: $800 + 6% of sales over $10,000

Next, suppose that Rose sold a certain amount—say, $12,000— in one month. Which plan would be better? Under plan A, she would earn $600 plus 4% of $12,000, or

$600 + 0.04(12,000) = \$1080.$

Since with plan B commissions are paid only on sales in excess of

$10,000, Rose would earn $800 plus 6% of ($12,000 − $10,000), or

$$800 + 0.06(12,000 − 10,000) = \$920.$$

This shows that for monthly sales of $12,000, plan A is better. Similar calculations will show that for sales of $30,000 a month, plan B is better. To determine *all* values for which plan A earns more money, we must solve an inequality that is based on the calculations above.

2. **Translate.**   We let S represent the amount of monthly sales. Examining the calculations in the *Familiarize* step, we see that monthly income from plan A is $600 + 0.04S$ and from plan B is $800 + 0.06(S − 10,000)$. We want to find all values of S for which

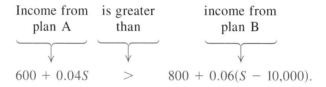

Income from plan A	is greater than	income from plan B

$$600 + 0.04S \quad > \quad 800 + 0.06(S − 10,000).$$

3. **Carry out.**   We solve the inequality:

$$600 + 0.04S > 800 + 0.06(S − 10,000)$$
$$600 + 0.04S > 800 + 0.06S − 600 \qquad \text{Using the distributive law}$$
$$600 + 0.04S > 200 + 0.06S \qquad \text{Combining like terms}$$
$$400 > 0.02S \qquad \text{Subtracting both 200 and 0.04S on both sides}$$
$$20,000 > S, \text{ or } S < 20,000. \qquad \text{Dividing by 0.02 on both sides}$$

4. **Check.**   For $S = 20,000$, the income from plan A is

$$600 + 4\% \cdot 20,000, \text{ or } \$1400.$$

The income from plan B is

$$800 + 6\% \cdot (20,000 − 10,000), \text{ or } \$1400.$$

This confirms that for sales totaling $20,000, Rose's pay is the same under either plan.

In the *Familiarize* step, we saw that for sales of $12,000, plan A pays more. Since $12,000 < 20,000$, this is a partial check. Since we cannot check all possible values of S, we will stop here.

5. **State.**   For monthly sales of less than $20,000, plan A is better.

EXERCISE SET

4.1

Determine whether the given numbers are solutions of the inequality.

1. $x − 2 \geqslant 6$; $−4, 0, 4, 8$

2. $3x + 5 \leqslant −10$; $−5, −10, 0, 27$

3. $t − 8 > 2t − 3$; $0, −8, −9, −3$

4. $5y − 7 < 5 − y$; $2, −3, 0, 3$

Graph each inequality, and write the solution set using both set-builder and interval notation.

5. $y < 5$

6. $x > 4$

7. $x \geqslant −4$

8. $t \leqslant 6$

9. $t > −2$

10. $y < −3$

11. $x \leqslant −5$

12. $x \geqslant −6$

Solve. Then graph.

13. $x + 8 > 3$

14. $x + 5 > 2$

15. $a + 7 \leq -13$

16. $a + 9 \leq -12$

17. $x - 9 \leq 10$

18. $t + 14 \geq 9$

19. $y - 9 > -18$

20. $y - 8 > -14$

21. $y - 18 \leq -4$

22. $x - 11 \leq -2$

23. $9t < -81$

24. $8x \geq 24$

25. $0.5x < 25$

26. $0.3x < -18$

27. $-8y \leq 3.2$

28. $-9x \geq -8.1$

29. $-\frac{5}{6}y \leq -\frac{3}{4}$

30. $-\frac{3}{4}x \geq -\frac{5}{8}$

31. $5y + 13 > 28$

32. $2x + 7 < 19$

33. $-9x + 3x \geq -24$

34. $5y + 2y \leq -21$

35. Let $f(x) = 8x - 9$ and $g(x) = 3x - 11$. Find all values of x for which $f(x) < g(x)$.

36. Let $f(x) = 2x - 7$ and $g(x) = 5x - 9$. Find all values of x for which $f(x) < g(x)$.

37. Let $f(x) = 0.4x + 5$ and $g(x) = 1.2x - 4$. Find all values of x for which $g(x) \geq f(x)$.

38. Let $f(x) = \frac{3}{8} + 2x$ and $g(x) = 3x - \frac{1}{8}$. Find all values of x for which $g(x) \geq f(x)$.

Solve.

39. $4(3y - 2) \geq 9(2y + 5)$

40. $4m + 5 \geq 14(m - 2)$

41. $3(2 - 5x) + 2x < 2(4 + 2x)$

42. $2(0.5 - 3y) + y > (4y - 0.2)8$

43. $5[3m - (m + 4)] > -2(m - 4)$

44. $[8x - 3(3x + 2)] - 5 \geq 3(x + 4) - 2x$

45. $19 - (2x + 3) \leq 2(x + 3) + x$

46. $13 - (2c + 2) \geq 2(c + 2) + 3c$

47. $\frac{1}{4}(8y + 4) - 17 < -\frac{1}{2}(4y - 8)$

48. $\frac{1}{3}(6x + 24) - 20 > -\frac{1}{4}(12x - 72)$

49. $2[4 - 2(3 - x)] - 1 \geq 4[2(4x - 3) + 7] - 25$

50. $5[3(7 - t) - 4(8 + 2t)] - 20 \leq -6[2(6 + 3t) - 4]$

Solve.

51. *Truck rentals.* Metro Concerts can rent a truck for either $55 with unlimited mileage or $29 plus 40¢ per mile. For what mileages would the unlimited mileage plan save money?

52. *Truck rentals.* Campus Entertainment rents a truck for $45 plus 20¢ per mile. A budget of $75 has been set for the rental. For what mileages will they not exceed the budget?

53. *Insurance claims.* After a serious automobile accident, most insurance companies will replace the damaged car with a new one if repair costs exceed 80% of the NAPA, or "blue-book," value of the car. Miguel's car recently sustained $9200 worth of damage but was not replaced. What was the blue-book value of his car?

54. *Phone rates.* A long-distance telephone call using Down East Calling costs 20 cents for the first minute and 16 cents for each additional minute. The same call, placed on Long Call Systems, costs 19 cents for the first minute and 18 cents for each additional minute. For what length phone calls is Down East Calling less expensive?

55. *Moving costs.* Musclebound Movers charges $85 plus $40 an hour to move households across town. Champion Moving charges $60 an hour for cross-town moves. For what lengths of time is Champion more expensive?

56. *Wages.* Toni can be paid in one of two ways:

Plan A: A salary of $400 per month, plus a commission of 8% of gross sales;

Plan B: A salary of $610 per month, plus a commission of 5% of gross sales.

For what amount of gross sales should Toni select plan A?

57. *Checking-account rates.* The Hudson Bank offers two checking-account plans. Their Anywhere plan charges 20¢ per check whereas their Acu-checking plan costs $2 per month plus 12¢ per check. For what numbers of checks per month will the Acu-checking plan cost less?

58. *Wages.* Branford can be paid for his masonry work in one of two ways:

Plan A: $300 plus $9.00 per hour;

Plan B: Straight $12.50 per hour.

Suppose that the job takes n hours. For what values of n is plan B better for Branford?

59. *Insurance benefits.* Bayside Insurance offers two plans. Under plan A, Giselle would pay the first $50 of her medical bills and 20% of all bills after that. Under plan B, Giselle would pay the first $250 of bills, but only 10% of the rest. For what amount of medical bills will plan B save Giselle money? (Assume that her bills will exceed $250.)

60. *Wedding costs.* The Arnold Inn offers two plans for wedding parties. Under plan A, the inn charges $30 for each person in attendance. Under plan B, the inn charges $1300 plus $20 for each person in excess of the first 25 who attend. For what size parties will plan B cost less? (Assume that more than 25 guests will attend.)

61. *Investing.* Lillian is about to invest $20,000, part at 6% and the rest at 8%. What is the most that she can invest at 6% and still be guaranteed at least $1500 in interest per year?

62. *Temperature conversion.* The function

$$C(F) = \frac{5}{9}(F - 32)$$

can be used to find the Celsius temperature $C(F)$ that corresponds to $F°$ Fahrenheit.

a) Gold is solid at Celsius temperatures less than 1063° C. Find the Fahrenheit temperatures for which gold is solid.

b) Silver is solid at Celsius temperatures less than 960.8° C. Find the Fahrenheit temperatures for which silver is solid.

63. *Manufacturing.* Ergs, Inc., is planning to make a new kind of radio. Fixed costs will be $90,000, and variable costs will be $15 for the production of each radio. The total-cost function for x radios is

$$C(x) = 90,000 + 15x.$$

The company makes $26 in revenue for each radio sold. The total-revenue function for x radios is

$$R(x) = 26x.$$

(See Section 3.8.)

a) When $R(x) < C(x)$, the company loses money. Find the values of x for which the company loses money.

b) When $R(x) > C(x)$, the company makes a profit. Find the values of x for which the company makes a profit.

64. *Publishing.* The demand and supply functions for a locally produced poetry book are approximated by

$$D(p) = 2000 - 60p \quad \text{and}$$
$$S(p) = 460 + 94p,$$

where p is the price in dollars (see Section 3.8).

a) Find those values of p for which demand exceeds supply.

b) Find those values of p for which demand is less than supply.

SKILL MAINTENANCE

65. Graph: $f(x) = 2x - 1$.

66. Graph: $5y - 10 = 2x$.

67. Solve: $-3x + 5 = 11$.

68. Solve: $-2(4 - x) = 7x$.

69. Simplify: $|-16|$.

70. Simplify: $-|-4|$.

SYNTHESIS

71. ◈ Explain in your own words why the inequality symbol must be reversed when both sides of an inequality are multiplied by a negative number.

72. ◈ A Presto photocopier costs $510 and an Exact Image photocopier costs $590. Write a problem that involves the cost of the copiers, the cost per page of photocopies, and the number of copies for which the Presto machine is the more expensive machine to own.

73. ◈ Explain how the addition principle can be used to avoid ever needing to multiply or divide on both sides of an inequality by a negative number.

Solve. Assume that a, b, c, d, and m are positive constants.

74. $3ax + 2x \geqslant 5ax - 4$; assume $a > 1$

75. $6by - 4y \leqslant 7by + 10$

76. $a(by - 2) \geqslant b(2y + 5)$; assume $a > 2$

77. $c(6x - 4) < d(3 + 2x)$; assume $3c > d$

78. $c(2 - 5x) + dx > m(4 + 2x)$; assume $5c + 2m < d$

79. $a(3 - 4x) + cx < d(5x + 2)$; assume $c > 4a + 5d$

Determine whether the statement is true or false. If false, give an example that shows this.

80. For any real numbers a, b, c, and d, if $a < b$ and $c < d$, then $a - c < b - d$.

81. For all real numbers x and y, if $x < y$, then $x^2 < y^2$.

82. ◈ Are the inequalities

$$x < 3 \quad \text{and} \quad x + \frac{1}{x} < 3 + \frac{1}{x}$$

equivalent? Why or why not?

83. ◈ Are the inequalities

$$x < 3 \quad \text{and} \quad 0 \cdot x < 0 \cdot 3$$

equivalent? Why or why not?

Solve.

84. $x + 5 \leqslant 5 + x$

85. $x + 8 < 3 + x$

86. $x^2 > 0$

87. Assume that the graphs of $y_1 = -\frac{1}{2}x + 5$, $y_2 = x - 1$, and $y_3 = 2x - 3$ are as shown below. Solve each inequality, referring only to the figure.

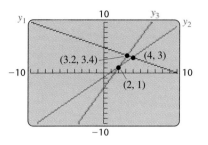

a) $-\frac{1}{2}x + 5 > x - 1$

b) $x - 1 \le 2x - 3$

c) $2x - 3 \ge -\frac{1}{2}x + 5$

88. Using an approach similar to that in Technology Connection 4.1, use a grapher to check your answers to Exercises 13, 37, 53, and 59.

COLLABORATIVE

C ◆ O ◆ R ◆ N ◆ E ◆ R

Focus: Inequalities and problem solving

Time: 10–15 minutes

Group size: 2

In the United States, the amount of solid waste being recycled is slowly catching up to the amount being generated. In 1991, each person generated, on average, 4.3 lb of solid waste every day, of which 0.8 lb was recycled. In 1994, each person generated, on average, 4.4 lb of solid waste, of which 1.0 lb was recycled. (*Source*: Characterization of Municipal Solid Waste in the United States: 1995 Update, Executive Summary, United States Environmental Protection Agency, March 1996.)

ACTIVITY

Assume that the amount of solid waste being generated and the amount being recycled are both increasing linearly. One group member should find a linear function w for which $w(t)$ represents the number of pounds of waste generated per person per day t years after 1991. The other group member should find a linear function r for which $r(t)$ represents the number of pounds recycled per person per day t years after 1991. Finally, working together, the group should determine those years for which the amount recycled will meet or exceed the amount generated.

4.2   Intersections, Unions, and Compound Inequalities

Intersections of Sets and Conjunctions of Sentences   •   Unions of Sets and Disjunctions of Sentences   •   More on Domains of Functions

---

We now consider **compound inequalities**—that is, sentences formed by two or more inequalities, joined by the word *and* or the word *or*.

Intersections of Sets and Conjunctions of Sentences

The **intersection** of two sets A and B is the set of all members that are common to both A and B. We denote the intersection of sets A and B as

$A \cap B.$

The intersection of two sets is often pictured as shown here.

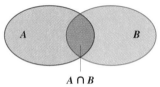

$A \cap B$

EXAMPLE 1

Find the intersection: $\{1, 2, 3, 4, 5\} \cap \{-2, -1, 0, 1, 2, 3\}$.

SOLUTION   The numbers 1, 2, and 3 are common to both sets, so the intersection is $\{1, 2, 3\}$.

When two or more sentences are joined by the word *and* to make a compound sentence, the new sentence is called a **conjunction** of the sentences. The following is a conjunction of inequalities:

$-2 < x \quad and \quad x < 1.$

For a conjunction to be true, each individual sentence must be true. *The solution set of a conjunction is the intersection of the solution sets of the individual sentences.* Consider the conjunction

$-2 < x \quad and \quad x < 1.$

The graphs of each separate sentence are shown below, and the intersection is the last graph. We use both set-builder and interval notations.

$\{x \mid -2 < x\}$ ←—|—|—|—|—|—⊕—|—|—|—|—|—|—|—|→ $(-2, \infty)$
                      −7 −6 −5 −4 −3 −2 −1 0 1 2 3 4 5 6 7

$\{x \mid x < 1\}$ ←—————————————⊕—|—|—|—|—|—|→ $(-\infty, 1)$
                      −7 −6 −5 −4 −3 −2 −1 0 1 2 3 4 5 6 7

$\{x \mid -2 < x\} \cap \{x \mid x < 1\}$ ←—|—|—|—|—|—⊕—————⊕—|—|—|—|—|→ $(-2, 1)$
$= \{x \mid -2 < x \ and \ x < 1\}$ −7 −6 −5 −4 −3 −2 −1 0 1 2 3 4 5 6 7

Because there are numbers that are both greater than -2 and less than 1, the conjunction $-2 < x \ and \ x < 1$ can be abbreviated by $-2 < x < 1$. Thus the interval $(-2, 1)$ can be represented as $\{x \mid -2 < x < 1\}$, the set of all numbers that are *simultaneously* greater than -2 *and* less than 1. Note that, in general, for $a < b$,

$a < x \quad and \quad x < b \quad$ **can be abbreviated** $\quad a < x < b;$

and $\quad b > x \quad and \quad x > a \quad$ **can be abbreviated** $\quad b > x > a.$

EXAMPLE 2

Solve and graph: $-1 \leq 2x + 5 < 13$.

SOLUTION   This inequality is an abbreviation for the conjunction

$-1 \leq 2x + 5 \quad and \quad 2x + 5 < 13.$

The word *and* corresponds to set *intersection*. To solve the conjunction, we solve each of the two inequalities separately and then find the intersection of the solution sets:

$$-1 \leq 2x + 5 \quad and \quad 2x + 5 < 13$$

$$-6 \leq 2x \qquad and \qquad 2x < 8 \qquad \text{Subtracting 5 on both sides of each inequality}$$

$$-3 \leq x \qquad and \qquad x < 4. \qquad \text{Dividing by 2 on both sides of each inequality}$$

We now abbreviate the answer:

$$-3 \leq x < 4.$$

The solution set is $\{x \mid -3 \leq x < 4\}$, or, in interval notation, $[-3, 4)$. The graph is the intersection of the two separate solution sets.

$\{x \mid -3 \leq x\}$
$[-3, \infty)$

$\{x \mid x < 4\}$
$(-\infty, 4)$

$\{x \mid -3 \leq x\} \cap \{x \mid x < 4\}$
$= \{x \mid -3 \leq x < 4\}$
$[-3, 4)$

The steps in Example 2 are sometimes combined as follows:

$$-1 \leq 2x + 5 < 13$$
$$-1 - 5 \leq 2x + 5 - 5 < 13 - 5$$
$$-6 \leq 2x < 8$$
$$-3 \leq x < 4.$$

Such an approach saves some writing and will prove useful in Section 4.3.

EXAMPLE 3

Solve and graph: $2x - 5 \geq -3$ *and* $5x + 2 \geq 17$.

SOLUTION   We first solve each inequality separately:

$$2x - 5 \geq -3 \quad and \quad 5x + 2 \geq 17$$
$$2x \geq 2 \qquad and \qquad 5x \geq 15$$
$$x \geq 1 \qquad and \qquad x \geq 3.$$

Next, we find the intersection of the two separate solution sets.

$\{x \mid x \geq 1\}$
$[1, \infty)$

$\{x \mid x \geq 3\}$
$[3, \infty)$

$\{x \mid x \geq 1\} \cap \{x \mid x \geq 3\}$
$= \{x \mid x \geq 3\}$
$[3, \infty)$

The numbers common to both sets are those that are greater than or equal to 3. Thus the solution set is $\{x|\ x \geq 3\}$, or, in interval notation, $[3, \infty)$. You should check that any number in $[3, \infty)$ satisfies the conjunction whereas numbers outside $[3, \infty)$ do not.

INTERSECTION

> The word "and" corresponds to "intersection" and to the symbol "$\cap$". Any solution of a conjunction must make each part of the conjunction true.

Sometimes there is no way to solve both parts of a conjunction at once.

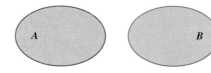

When $A \cap B = \varnothing$, A and B are said to be *disjoint*.

$$A \cap B = \varnothing.$$

EXAMPLE 4   Solve and graph: $2x - 3 > 1\ and\ 3x - 1 < 2$.

SOLUTION   We solve each inequality separately:

$$
\begin{aligned}
2x - 3 > 1 \quad &and \quad 3x - 1 < 2 \\
2x > 4 \quad &and \quad 3x < 3 \\
x > 2 \quad &and \quad x < 1.
\end{aligned}
$$

The solution set is the intersection of the individual inequalities.

$\{x|\ x > 2\}$ 〈——————○————————→ $(2, \infty)$
  $-7\ -6\ -5\ -4\ -3\ -2\ -1\ \ 0\ \ 1\ \ 2\ \ 3\ \ 4\ \ 5\ \ 6\ \ 7$

$\{x|\ x < 1\}$ 〈——————————○————————→ $(-\infty, 1)$
  $-7\ -6\ -5\ -4\ -3\ -2\ -1\ \ 0\ \ 1\ \ 2\ \ 3\ \ 4\ \ 5\ \ 6\ \ 7$

$\{x|\ x > 2\} \cap \{x|\ x < 1\}$ 〈——————————————————→ $\varnothing$
$= \{x|\ x > 2\ and\ x < 1\} = \varnothing$ $-7\ -6\ -5\ -4\ -3\ -2\ -1\ \ 0\ \ 1\ \ 2\ \ 3\ \ 4\ \ 5\ \ 6\ \ 7$

Since no number is both greater than 2 and less than 1, the solution set is the empty set, $\varnothing$.

Unions of Sets and Disjunctions of Sentences

The **union** of two sets A and B is the collection of elements belonging to A and/or B. We denote the union of A and B by

$A \cup B$.

The union of two sets is often pictured as shown at left.

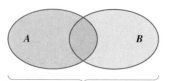

$A \cup B$

EXAMPLE 5

Find the union: $\{2, 3, 4\} \cup \{3, 5, 7\}$.

SOLUTION   The numbers in either or both sets are 2, 3, 4, 5, and 7, so the union is $\{2, 3, 4, 5, 7\}$.

When two or more sentences are joined by the word *or* to make a compound sentence, the new sentence is called a **disjunction** of the sentences. Here are three examples:

$x < -3 \quad or \quad x > 3;$

y is an odd number $\quad or \quad$ y is a prime number;

$x < 0 \quad or \quad x = 0 \quad or \quad x > 0.$

For a disjunction to be true, at least one of the individual sentences must be true. *The solution set of a disjunction is the union of the individual solution sets.* Consider the disjunction

$x < -3 \quad or \quad x > 3.$

The graphs of each separate sentence are shown below, and the union is the last graph. Again, we use both set-builder and interval notations.

$\{x \mid x < -3\}$ $(-\infty, -3)$

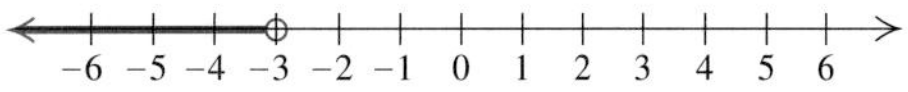

$\{x \mid x > 3\}$ $(3, \infty)$

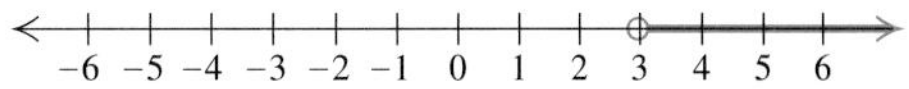

$\{x \mid x < -3\} \cup \{x \mid x > 3\}$ $(-\infty, -3) \cup (3, \infty)$
$= \{x \mid x < -3 \ or \ x > 3\}$

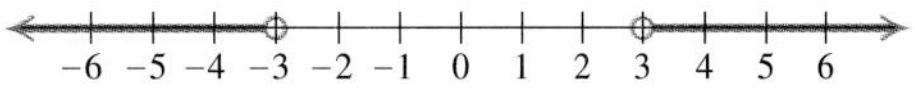

Answers to disjunctions can rarely be written concisely. The solution set of $x < -3 \ or \ x > 3$ is simply written $\{x \mid x < -3 \ or \ x > 3\}$, or $(-\infty, -3) \cup (3, \infty)$.

UNION

The word "or" corresponds to "union" and to the symbol "$\cup$". For a number to be a solution of the disjunction, it must be in *at least one* of the solution sets.

EXAMPLE 6

Solve and graph: $7 + 2x < -1 \ or \ 13 - 5x \le 3$.

SOLUTION   We solve each inequality separately, retaining the word *or*:

$7 + 2x < -1 \quad or \quad 13 - 5x \le 3$

$2x < -8 \quad or \quad -5x \le -10$

Reverse the symbol.

$x < -4 \quad or \quad x \ge 2.$

To find the solution set of the disjunction, we consider the individual graphs. We graph $x < -4$ and then $x \geq 2$. Then we take the union of the graphs.

$\{x \mid x < -4\}$ $(-\infty, -4)$

$\{x \mid x \geq 2\}$ $[2, \infty)$

$\{x \mid x < -4 \text{ or } x \geq 2\}$ $(-\infty, -4) \cup [2, \infty)$

The solution set is $\{x \mid x < -4 \ \text{or} \ x \geq 2\}$, or, in interval notation, $(-\infty, -4) \cup [2, \infty)$.

> **CAUTION!**   A compound inequality like
>
> $$x < -4 \quad \text{or} \quad x \geq 2,$$
>
> as in Example 6, *cannot* be expressed as $2 \leq x < -4$ because to do so would be to say that x is *simultaneously* less than -4 and greater than or equal to 2. No number is both less than -4 *and* greater than 2, but many are less than -4 *or* greater than 2.

EXAMPLE 7

Solve: $-2x - 5 < -2 \ \text{or} \ x - 3 < -10$.

SOLUTION   We solve the individual inequalities separately, retaining the word *or*:

$$-2x - 5 < -2 \quad \text{or} \quad x - 3 < -10$$
$$-2x < 3 \qquad \text{or} \qquad x < -7$$

Reverse the symbol.

$$x > -\tfrac{3}{2} \quad \text{or} \qquad x < -7.$$

Keep the word "or."

The solution set is $\left\{x \mid x < -7 \text{ or } x > -\tfrac{3}{2}\right\}$. In interval notation, the solution is $(-\infty, -7) \cup \left(-\tfrac{3}{2}, \infty\right)$.

EXAMPLE 8

Solve: $3x - 11 < 4 \ \text{or} \ 4x + 9 \geq 1$.

SOLUTION   We solve the individual inequalities separately, retaining the word *or*:

$$3x - 11 < 4 \quad \text{or} \quad 4x + 9 \geq 1$$
$$3x < 15 \quad \text{or} \qquad 4x \geq -8$$
$$x < 5 \quad \text{or} \qquad x \geq -2.$$

Keep the word "or."

To find the solution set, we first look at the individual graphs.

$\{x \mid x < 5\}$ 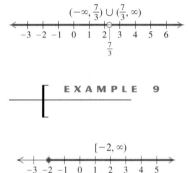 $(-\infty, 5)$

$\{x \mid x \geq -2\}$ $[-2, \infty)$

$\begin{aligned} &\{x \mid x < 5\} \cup \{x \mid x \geq -2\} \\ &= \{x \mid x < 5 \text{ or } x \geq -2\} \end{aligned}$ $(-\infty, \infty) = \mathbb{R}$

Since *all* numbers are less than 5 or greater than or equal to -2, the two sets fill the entire number line. Thus the solution set is $\mathbb{R}$, the set of all real numbers.

More on Domains of Functions

In Section 2.2, we saw that if $g(x) = (5x - 2)/(3x - 7)$, then the domain of $g = \left\{x \mid x \text{ is a real number and } x \neq \frac{7}{3}\right\}$. We can now represent such a set using interval notation:

$$\left\{x \mid x \text{ is a real number and } x \neq \tfrac{7}{3}\right\} = \left(-\infty, \tfrac{7}{3}\right) \cup \left(\tfrac{7}{3}, \infty\right).$$

$\left(-\infty, \frac{7}{3}\right) \cup \left(\frac{7}{3}, \infty\right)$

EXAMPLE 9   Find the domain of f if $f(x) = \sqrt{x + 2}$.

SOLUTION   The expression $\sqrt{x + 2}$ is not a real number when $x + 2$ is negative. Thus, if $f(x) = \sqrt{x + 2}$, the domain of f is the set of all x-values for which $x + 2 \geq 0$. Since $x + 2 \geq 0$ is equivalent to $x \geq -2$, we have

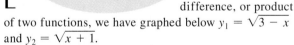
$[-2, \infty)$

$$\text{Domain of } f = \{x \mid x \geq -2\} = [-2, \infty).$$

TECHNOLOGY CONNECTION
4.2

To visualize the domain of a sum, difference, or product of two functions, we have graphed below $y_1 = \sqrt{3 - x}$ and $y_2 = \sqrt{x + 1}$.

1. Determine algebraically the domains for y_1 and y_2 and trace each curve to verify the x-values used.
2. Determine the domains of $y_1 + y_2$, $y_1 - y_2$, $y_2 - y_1$, and $y_1 \cdot y_2$ algebraically and then check by graphing each sum, difference, or product of y_1 and y_2.

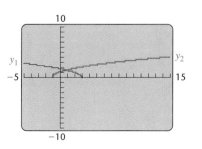

EXERCISE SET

4.2

Find each indicated intersection or union.

1. $\{9, 10, 11\} \cap \{9, 11, 13\}$

2. $\{2, 4, 8\} \cup \{8, 9, 10\}$

3. $\{1, 5, 10, 15\} \cup \{5, 15, 20\}$

4. $\{2, 5, 9, 11\} \cap \{5, 8, 12\}$

5. $\{a, b, c, d\} \cap \{b, f, g\}$

6. $\{a, b, c\} \cup \{a, c\}$

7. $\{r, s, t\} \cup \{r, u, t, s, v\}$

8. $\{m, n, o, p\} \cap \{m, o, p\}$

9. $\{2, 5, 7, 9\} \cap \{5, 7\}$

10. $\{1, 5, 9\} \cup \{4, 6, 8\}$

11. $\{3, 5, 7\} \cup \varnothing$

12. $\{3, 5, 7\} \cap \varnothing$

Graph and write interval notation.

13. $2 < x < 7$

14. $0 \le y \le 4$

15. $-6 \le y \le -2$

16. $-9 \le x < -5$

17. $x < -2 \ or \ x > 1$

18. $x < -2 \ or \ x > 3$

19. $x \le -1 \ or \ x > 4$

20. $x \le -5 \ or \ x > 2$

21. $-3 \le -x < 5$

22. $x > -7 \ and \ x < -2$

23. $x > -2 \ and \ x < 4$

24. $3 > -x \ge -1$

25. $5 > a \ or \ a > 7$

26. $t \ge 2 \ or \ -3 > t$

27. $x \ge 5 \ or \ -x \ge 4$

28. $-x < 3 \ or \ x < -6$

29. $5 > x \ and \ x \ge -6$

30. $6 > -x \ge 0$

31. $x < 7 \ and \ x \ge 3$

32. $x \ge -3 \ and \ x < 3$

33. $t < 2 \ or \ t < 5$

34. $t > 4 \ or \ t > -1$

35. $x > -1 \ or \ x \le 3$

36. $4 > x \ or \ x \ge -3$

37. $x \ge 5 \ and \ x > 7$

38. $x \le -4 \ and \ x < 1$

Solve and graph each solution set.

39. $-3 < t + 2 < 7$

40. $-2 < t + 1 \le 5$

41. $2 < x + 3 \ and \ x + 1 \le 5$

42. $-1 < x + 2 \ and \ x - 4 < 3$

43. $-5 \le 2a - 1 \ and \ 3a + 1 < 7$

44. $-4 \le 3n + 2 \ and \ 2n - 3 \le 5$

45. $x + 7 \le -2 \ or \ x + 7 \ge 5$

46. $x + 5 < -3 \ or \ x + 5 \ge 4$

47. $2 \le f(x) \le 8$, where $f(x) = 3x - 1$

48. $10 \ge g(x) \ge -2$, where $g(x) = 3x - 5$

49. $-18 \le f(x) < 0$, where $f(x) = -2x - 7$

50. $4 > g(t) \ge 2$, where $g(t) = -3t - 8$

51. $f(x) \le 2 \ or \ f(x) \ge 8$, where $f(x) = 3x - 1$

52. $g(x) \le -2 \ or \ g(x) \ge 10$, where $g(x) = 3x - 5$

53. $f(x) < -1 \ or \ f(x) > 1$, where $f(x) = 2x - 7$

54. $g(x) < -7 \ or \ g(x) > 7$, where $g(x) = 3x + 5$

55. $6 > 2a - 1 \ or \ -4 \le -3a + 2$

56. $3a - 7 > -10 \ or \ 5a + 2 \le 22$

57. $a + 4 < -1 \ and \ 3a - 5 < 7$

58. $1 - a < -2 \ and \ 2a + 1 > 9$

59. $3x + 2 < 2 \ or \ 4 - 2x < 14$

60. $2x - 1 > 5 \ or \ 3 - 2x \ge 7$

61. $2t - 7 \le 5 \ or \ 5 - 2t > 3$

62. $5 - 3a \le 8 \ or \ 2a + 1 > 7$

For $f(x)$ as given, use interval notation to write the domain of f.

63. $f(x) = \dfrac{7}{x - 5}$

64. $f(x) = \dfrac{2}{x + 3}$

65. $f(x) = \sqrt{x + 4}$

66. $f(x) = \sqrt{x + 7}$

67. $f(x) = \dfrac{x + 3}{2x - 5}$

68. $f(x) = \dfrac{x - 1}{3x + 4}$

69. $f(x) = \sqrt{12 - 3x}$

70. $f(x) = \sqrt{8 - 4x}$

SKILL MAINTENANCE

Solve.

71. $2x - 3y = 7,$
$3x + 2y = -10$

72. $3x - 9(x + 4) = 20(3x + 7)$

73. $5(2x + 3) = 3(x - 4)$

Graph.

74. $3x - 4y = -12$

75. $f(x) = 5$

76. $x = -2$

SYNTHESIS

77. ◆ Describe the circumstances under which $A \cap B = A$.

78. ◆ Describe the circumstances under which $[a, b] \cup [c, d] = [a, d]$.

79. ◆ Explain why the conjunction $3 < x$ and $x < 5$ can be rewritten as $3 < x < 5$, but the disjunction $3 < x$ or $x < 5$ cannot be rewritten as $3 < x < 5$.

80. ◆ Explain how the use of the word *or* in a compound inequality differs from the use of the word *or* in everyday English. (*Hint*: Consider the expression and/or.)

81. Use the accompanying graph of $f(x) = 2x - 5$ to solve $-7 < 2x - 5 < 7$.

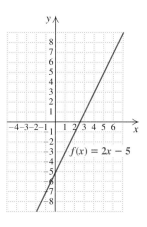

82. Use the accompanying graph of $g(x) = 4 - x$ to solve $4 - x < -2$ or $4 - x > 7$.

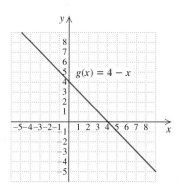

83. *Minimizing tolls.* A $3.00 toll is charged to cross the bridge from Sanibel Island to mainland Florida.

A six-month pass, costing $15.00, reduces the toll to $0.50. A one-year pass, costing $60, allows for free crossings. How many crossings per month does it take, on average, for the six-month pass to be the more economical choice?

84. *Converting dress sizes.* The function
$$f(x) = 2(x + 10)$$
can be used to convert dress sizes x in the United States to dress sizes $f(x)$ in Italy. For what dress sizes in the United States will dress sizes in Italy be between 32 and 46?

85. *Pressure at sea depth.* The function
$$P(d) = 1 + \frac{d}{33}$$
gives the pressure, in atmospheres (atm), at a depth of d feet in the sea. For what depths d is the pressure at least 1 atm and at most 7 atm?

86. *Temperatures of liquids.* The formula
$$C = \frac{5}{9}(F - 32)$$
can be used to convert Fahrenheit temperatures F to Celsius temperatures C.

a) Gold is liquid for Celsius temperatures C such that $1063° \le C < 2660°$. Find a comparable inequality for Fahrenheit temperatures.

b) Silver is liquid for Celsius temperatures C such that $960.8° \le C < 2180°$. Find a comparable inequality for Fahrenheit temperatures.

87. *Solid waste generation.* The function
$$w(t) = 0.05t + 4.3$$
can be used to estimate the number of pounds of solid waste, $w(t)$, produced daily, on average, by each person in the United States, t years after 1991. For what years will waste production range from 5.0 to 5.25 lb per person per day?

88. *Records in the women's 100-m dash.* Florence Griffith Joyner set a world record of 10.49 sec in the women's 100-m dash in 1988. The function

$$R(t) = -0.0433t + 10.49$$

can be used to predict the world record in the women's 100-m dash t years after 1988. Predict (in terms of an inequality) those years for which the world record was between 11.5 and 10.8 sec. (Measure from the middle of 1988.)

Solve and graph.

89. $4a - 2 \le a + 1 \le 3a + 4$

90. $4m - 8 > 6m + 5$ *or* $5m - 8 < -2$

91. $x - 10 < 5x + 6 \le x + 10$

92. $3x < 4 - 5x < 5 + 3x$

Determine whether each sentence is true or false for all real numbers a, b, and c.

93. If $-b < -a$, then $a < b$.

94. If $a \le c$ and $c \le b$, then $b > a$.

95. If $a < c$ and $b < c$, then $a < b$.

96. If $-a < c$ and $-c > b$, then $a > b$.

For $f(x)$ as given, use interval notation to write the domain of f.

97. $f(x) = \dfrac{\sqrt{5 + 2x}}{x - 1}$

98. $f(x) = \dfrac{\sqrt{3 - 4x}}{x + 7}$

99.  Let $y_1 = -1$, $y_2 = 2x + 5$, and $y_3 = 13$. Then use the graphs of y_1, y_2, and y_3 to check the solution to Example 2.

100. Let $y_1 = -2x - 5$, $y_2 = -2$, $y_3 = x - 3$, and $y_4 = -10$. Then use the graphs of y_1, y_2, y_3, and y_4 to check the solution to Example 7.

101. Use a grapher to check your answers to Exercises 33–36 and Exercises 53–56.

102. On many graphers, the TEST key provides access to inequality symbols, while the LOGIC option of that same key accesses the conjunction *and* and the disjunction *or*. Thus compound inequalities like Exercise 23 can be checked by forming expressions like $y_3 = y_1$ *and* y_2 (where $y_1 = x > -2$ and $y_2 = x < 4$). As in Technology Connection 4.1, the interval(s) in the solution set are shown as a horizontal line 1 unit above the x-axis. (Be careful to "deselect" y_1 and y_2 so that only y_3 is drawn.) Use this approach to check your answers to Exercises 29 and 60.

COLLABORATIVE

C ◆ O ◆ R ◆ N ◆ E ◆ R

Focus: Compound inequalities and solution sets

Time: 20–30 minutes

Group size: 2–3

At present (1997), the U.S. Postal Service charges 23 cents per ounce plus an additional 9-cent delivery fee (1 oz or less costs 32 cents; more than 1 oz, but not more than 2 oz, costs 55 cents; and so on). Rapid Delivery charges $1.05 per pound plus an additional $2.50 delivery fee (up to 16 oz costs $3.55, more than 16 oz, but less than 32 oz, costs $4.60; and so on). Let x be the weight, in ounces, of an item being mailed.*

---

*Based on an article by Michael Contino in *Mathematics Teacher*, May 1995. Email: mike_contino@cams.edu

ACTIVITY

Working together, group members should graph the function p, where $p(x)$ represents the cost, in dollars, of mailing x ounces at a post office. Using the same set of axes, the group should also graph the function r, where $r(x)$ represents the cost, in dollars, of mailing x ounces with Rapid Delivery. Finally, determine those weights for which the Postal Service is less expensive. Express your answer using both set-builder and interval notation.

4.3 Absolute-Value Equations and Inequalities

Equations with Absolute Value   •   Inequalities with Absolute Value

Equations with Absolute Value

Recall from Section 1.2 the definition of absolute value.

ABSOLUTE VALUE

The absolute value of x, denoted $|x|$, is defined as

$$|x| = \begin{cases} x, & \text{if } x \geq 0, \\ -x, & \text{if } x < 0. \end{cases}$$

(When x is nonnegative, the absolute value of x is x. When x is negative, the absolute value of x is the opposite of x.)

Since distance is always nonnegative, we can think of a number's absolute value as its distance from zero on a number line.

EXAMPLE 1   Find the solution set: **(a)** $|x| = 4$; **(b)** $|x| = 0$; **(c)** $|x| = -7$.

SOLUTION

a) We interpret $|x| = 4$ to mean that the number x is 4 units from zero on a number line. There are two such numbers, 4 and -4. Thus the solution set is $\{-4, 4\}$.

A second way to visualize this problem is to graph $f(x) = |x|$ (see Section 2.1). We also graph $g(x) = 4$. The x-values of the points of intersection are the solutions of $|x| = 4$.

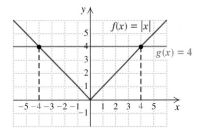

b) We interpret $|x| = 0$ to mean that x is 0 units from zero on a number line. The only number that satisfies this is 0 itself. Thus the solution set is $\{0\}$.

c) Since distance is always nonnegative, it doesn't make sense to talk about a number that is -7 units from zero. Remember that the absolute value of a number is never negative. Thus $|x| = -7$ has no solution; the solution set is $\varnothing$.

Example 1 leads us to the following principle for solving equations.

THE ABSOLUTE-VALUE PRINCIPLE FOR EQUATIONS

For any positive number p and any algebraic expression X:

a) The solutions of $|X| = p$ are those numbers that satisfy $X = -p \ or \ X = p$.

b) The equation $|X| = 0$ is equivalent to the equation $X = 0$.

c) The equation $|X| = -p$ has no solution.

E X A M P L E   2   Find the solution set: **(a)** $|2x + 5| = 13$; **(b)** $|4 - 7x| = -8$.

S O L U T I O N

a) We use the absolute-value principle, replacing X by $2x + 5$ and p by 13:

$$|X| = p$$
$$|2x + 5| = 13$$
$$2x + 5 = -13 \quad or \quad 2x + 5 = 13$$
$$2x = -18 \quad or \quad 2x = 8$$
$$x = -9 \quad or \quad x = 4.$$

Check: For -9:

$$\frac{|2x + 5| = 13}{|2(-9) + 5| \ ? \ 13}$$
$$|-18 + 5|$$
$$|-13|$$
$$13 \ \Big| \ 13 \ \text{TRUE}$$

For 4:

$$\frac{|2x + 5| = 13}{|2 \cdot 4 + 5| \ ? \ 13}$$
$$|8 + 5|$$
$$|13|$$
$$13 \ \Big| \ 13 \ \text{TRUE}$$

The number $2x + 5$ is 13 units from zero if x is replaced by -9 or 4. The solution set is $\{-9, 4\}$.

TECHNOLOGY CONNECTION

4.3A

To check Example 2(a), we let $y_1 = \text{ABS}(2x + 5)$ and $y_2 = 13$. Using the window $[-12, 8, -2, 18]$, we use INTERSECT or TRACE and ZOOM to find the points of intersection, $(-9, 13)$ and $(4, 13)$. The x-coordinates, -9 and 4, are the solutions.

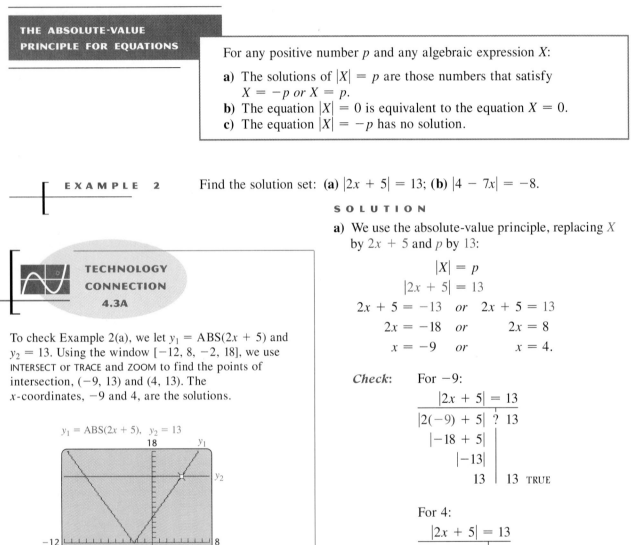

$y_1 = \text{ABS}(2x + 5), \ y_2 = 13$

Intersection
$X = 4, \ Y = 13$

1. Use a grapher to show that Example 2(b) has no solution.

b) The absolute-value principle reminds us that absolute value is always nonnegative. The equation $|4 - 7x| = -8$ has no solution. The solution set is $\varnothing$.

The absolute-value principle can be used together with the addition and multiplication principles to solve many types of equations with absolute value.

EXAMPLE 3

Given that $f(x) = 2|x + 3| + 1$, find all x for which $f(x) = 15$.

SOLUTION Since we are looking for $f(x) = 15$, we substitute:

$$
\begin{aligned}
f(x) &= 15 \\
2|x + 3| + 1 &= 15 & &\text{Replacing } f(x) \text{ by } 2|x + 3| + 1 \\
\left.\begin{aligned}
2|x + 3| &= 14 \\
|x + 3| &= 7
\end{aligned}\right\} & & \begin{aligned}&\text{Adding } -1 \text{ and multiplying by } \tfrac{1}{2} \text{ on} \\ &\text{both sides to isolate } |x + 3|\end{aligned} \\
x + 3 = -7 \quad &or \quad x + 3 = 7 & & \begin{aligned}&\text{Replacing } X \text{ by } x + 3 \text{ and } p \text{ by } 7 \text{ in} \\ &\text{the absolute-value principle}\end{aligned} \\
x = -10 \quad &or \quad x = 4.
\end{aligned}
$$

We leave it to the student to check that $f(-10) = f(4) = 15$. The solution set is $\{-10, 4\}$.

EXAMPLE 4

Solve: $|x - 2| = 3$.

SOLUTION Because this equation is of the form $|a - b| = c$, it can be solved in two different ways.

Method 1. This approach is helpful in calculus. The expressions $|a - b|$ and $|b - a|$ can be used to represent the *distance between a and b* on the number line. For example, the distance between 7 and 8 is given by $|8 - 7|$ or $|7 - 8|$. From this viewpoint, the equation $|x - 2| = 3$ states that the distance between x and 2 is 3 units. We draw a number line and locate all numbers that are 3 units from 2.

The solutions of $|x - 2| = 3$ are -1 and 5.

Method 2. We interpret the equation as stating that the number $x - 2$ is 3 units from zero. Using the absolute-value principle, we replace X by $x - 2$ and p by 3:

$$
\begin{aligned}
|X| &= p \\
|x - 2| &= 3 \\
x - 2 = -3 \quad &or \quad x - 2 = 3 & &\text{Using the absolute-value principle} \\
x = -1 \quad &or \quad x = 5.
\end{aligned}
$$

The check consists of observing that both methods give the same solutions. The solution set is $\{-1, 5\}$.

Sometimes an equation has two absolute-value expressions. Consider $|a| = |b|$. This means that a and b are the same distance from zero.

If a and b are the same distance from zero, then either they are the same number or they are opposites.

EXAMPLE 5

Solve: $|2x - 3| = |x + 5|$.

SOLUTION   Either $2x - 3 = x + 5$ (they are the same number) or $2x - 3 = -(x + 5)$ (they are opposites). We solve each equation separately:

$$2x - 3 = x + 5 \quad or \quad 2x - 3 = -(x + 5)$$
$$x - 3 = 5 \qquad or \quad 2x - 3 = -x - 5$$
$$x = 8 \qquad or \quad 3x - 3 = -5$$
$$3x = -2$$
$$x = -\tfrac{2}{3}.$$

We leave the check to the student. The solutions are 8 and $-\tfrac{2}{3}$. The solution set is $\left\{-\tfrac{2}{3}, 8\right\}$.

Inequalities with Absolute Value

Our methods for solving equations with absolute value can be adapted for solving inequalities.

EXAMPLE 6

Solve $|x| < 4$. Then graph.

SOLUTION   The solutions of $|x| < 4$ are all numbers whose *distance from zero is less than* 4. By substituting or by looking at the number line, we can see that numbers like $-3, -2, -1, -\tfrac{1}{2}, -\tfrac{1}{4}, 0, \tfrac{1}{4}, \tfrac{1}{2}, 1, 2$, and 3 are all solutions. In fact, the solutions are all the numbers between -4 and 4. The solution set is $\{x|\ -4 < x < 4\}$. In interval notation, the solution is $(-4, 4)$. The graph is as follows:

$$\overset{\longleftarrow\ \longrightarrow}{\underset{-7\ -6\ -5\ -4\ -3\ -2\ -1\ \ 0\ \ 1\ \ 2\ \ 3\ \ 4\ \ 5\ \ 6\ \ 7}{}}$$
$$|x| < 4$$

We can also visualize this by graphing $f(x) = |x|$ and $g(x) = 4$, as in Example 1. The solution set consists of all x-values for which $(x, f(x))$ is below the horizontal line $g(x) = 4$. These x-values comprise the interval $(-4, 4)$.

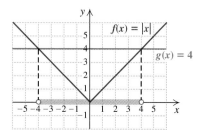

EXAMPLE 7

Solve $|x| \geqslant 4$. Then graph.

SOLUTION   The solutions of $|x| \geqslant 4$ are all numbers whose *distance from zero is greater than or equal to 4*—in other words, those numbers x such that $x \leqslant -4$ or $4 \leqslant x$. The solution set is $\{x \mid x \leqslant -4 \text{ or } x \geqslant 4\}$. In interval notation, the solution is $(-\infty, -4] \cup [4, \infty)$. We can check mentally with numbers like -4.1, -5, 4.1, and 5. The graph is as follows:

$$|x| \geqslant 4$$

Like Examples 1 and 6, this can be visualized by graphing $f(x) = |x|$ and $g(x) = 4$. The solution set consists of all x-values for which $(x, f(x))$ is on or above the horizontal line $g(x) = 4$. These x-values comprise $(-\infty, -4] \cup [4, \infty)$.

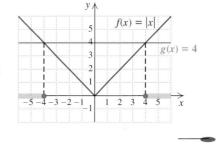

Examples 1, 6, and 7 illustrate three types of problems in which absolute-value symbols appear. The following is a general principle for solving such problems.

PRINCIPLES FOR SOLVING ABSOLUTE-VALUE PROBLEMS

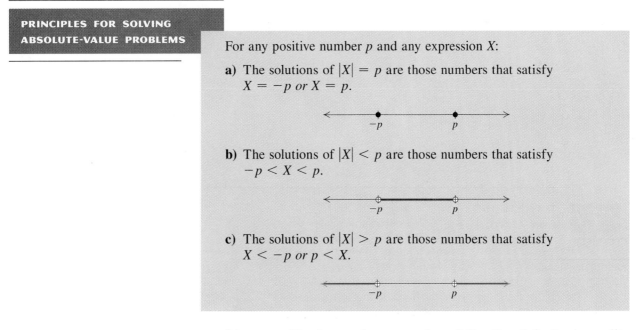

For any positive number p and any expression X:

a) The solutions of $|X| = p$ are those numbers that satisfy
$X = -p \text{ or } X = p.$

b) The solutions of $|X| < p$ are those numbers that satisfy
$-p < X < p.$

c) The solutions of $|X| > p$ are those numbers that satisfy
$X < -p \text{ or } p < X.$

Of course, if p is negative, any value of X will satisfy the inequality $|X| > p$ since absolute value is never negative. By the same reasoning, $|X| < p$

has no solution when p is not positive. Thus the inequality $|2x - 7| > -3$ is true for any real number x, and the inequality $|2x - 7| < -3$ has no solution.

Note that an inequality of the form $|X| < p$ corresponds to a *con*junction, whereas an inequality of the form $|X| > p$ corresponds to a *dis*junction.

EXAMPLE 8

Solve $|3x - 2| < 4$. Then graph.

SOLUTION We use part (b) of the principles listed above. In this case, X is $3x - 2$ and p is 4:

$$|X| < p$$
$$|3x - 2| < 4 \qquad \text{Replacing } X \text{ by } 3x - 2 \text{ and } p \text{ by } 4$$
$$-4 < 3x - 2 < 4 \qquad \text{The number } 3x - 2 \text{ must be within 4 units of zero.}$$
$$-2 < \quad 3x \quad < 6 \qquad \text{Adding 2}$$
$$-\tfrac{2}{3} < \quad x \quad < 2. \qquad \text{Multiplying by } \tfrac{1}{3}$$

The solution set is $\{x \mid -\tfrac{2}{3} < x < 2\}$. In interval notation, the solution is $\left(-\tfrac{2}{3}, 2\right)$. The graph is as follows:

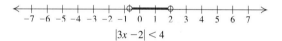

$$|3x - 2| < 4$$

EXAMPLE 9

Given that $f(x) = |4x + 2|$, find all x for which $f(x) \geq 6$.

SOLUTION We have

$$f(x) \geq 6,$$
$$\text{or} \quad |4x + 2| \geq 6. \qquad \text{Substituting}$$

To solve, we use part (c) of the principles listed above. In this case, X is

---

TECHNOLOGY CONNECTION

4.3B

To solve an inequality like $|4x + 2| \geq 6$ with a grapher, simply graph the equations $y_1 = \text{ABS}(4x + 2)$ and $y_2 = 6$. Then use INTERSECT or TRACE and ZOOM to find all points on the graph of $y_1 = |4x + 2|$ that are *on or above* the line $y = 6$. The x-values of those points solve the inequality. How can the same graph be used to solve the inequality $|4x + 2| < 6$ or the equation $|4x + 2| = 6$? Try using this procedure to solve Example 8 on a grapher.

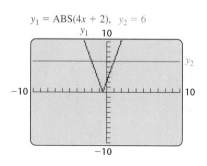

$y_1 = \text{ABS}(4x + 2), \quad y_2 = 6$

$4x + 2$ and p is 6:

$$|X| \geq p$$

$$|4x + 2| \geq 6 \qquad \text{Replacing } X \text{ by } 4x + 2 \text{ and } p \text{ by } 6$$

$$4x + 2 \leq -6 \quad or \quad 6 \leq 4x + 2 \qquad \text{The number } 4x + 2 \text{ must be at least 6 units from zero.}$$

$$4x \leq -8 \quad or \quad 4 \leq 4x \qquad \text{Adding } -2$$

$$x \leq -2 \quad or \quad 1 \leq x. \qquad \text{Multiplying by } \tfrac{1}{4}$$

The solution set is $\{x \mid x \leq -2 \text{ or } x \geq 1\}$. In interval notation, the solution is $(-\infty, -2] \cup [1, \infty)$. The graph is as follows:

$|4x + 2| \geq 6$

EXERCISE SET
4.3

Solve.

1. $|x| = 7$

2. $|x| = 9$

3. $|x| = -5$

4. $|x| = -3$

5. $|y| = 8.6$

6. $|p| = 0$

7. $|m| = 0$

8. $|t| = 5.5$

9. $|5x + 2| = 3$

10. $|2x - 3| = 4$

11. $|7x - 2| = -9$

12. $|3x - 10| = -8$

13. $|2y| - 5 = 13$

14. $|5x| - 3 = 37$

15. $7|z| + 2 = 16$

16. $5|q| - 2 = 9$

17. $\left| \dfrac{4 - 5x}{6} \right| = 7$

18. $\left| \dfrac{2x - 1}{3} \right| = 5$

19. $|t - 7| + 3 = 4$

20. $|m + 5| + 9 = 16$

21. $3|2x - 5| - 7 = -1$

22. $5 - 2|3x - 4| = -5$

23. Let $f(x) = |3x - 4|$. Find all x for which $f(x) = 8$.

24. Let $f(x) = |2x - 7|$. Find all x for which $f(x) = 10$.

25. Let $f(x) = |x| - 2$. Find all x for which $f(x) = 6.3$.

26. Let $f(x) = |x| + 7$. Find all x for which $f(x) = 18$.

27. Let $f(x) = \left| \dfrac{3x - 2}{5} \right|$. Find all x for which $f(x) = 2$.

28. Let $f(x) = \left| \dfrac{1 - 2x}{3} \right|$. Find all x for which $f(x) = 1$.

Each pair of numbers represents two points on a number line. Find the distance between the points.

29. 25, 14

30. 32, 17

31. -9, 24

32. -18, -37

33. -8, -42

34. -9, -36

Solve.

35. $|x + 4| = |2x - 7|$

36. $|3x + 5| = |x - 6|$

37. $|x - 9| = |x + 6|$

38. $|x + 4| = |x - 3|$

39. $|5t + 7| = |4t + 3|$

40. $|3a - 1| = |2a + 4|$

41. $|n - 3| = |3 - n|$

42. $|y - 2| = |2 - y|$

43. $|7 - a| = |a + 5|$

44. $|6 - t| = |t + 7|$

45. $\left| \tfrac{1}{2}x - 5 \right| = \left| \tfrac{1}{4}x + 3 \right|$

46. $\left| 2 - \tfrac{2}{3}x \right| = \left| 4 + \tfrac{7}{8}x \right|$

Solve and graph.

47. $|a| \leq 6$

48. $|x| < 2$

49. $|x| > 7$

50. $|a| \geq 3$

51. $|t| > 0$

52. $|t| \geq 1.7$

53. $|x - 3| < 5$

54. $|x - 1| < 3$

55. $|x + 2| \leq 5$

56. $|x + 4| \leq 1$

57. $|x - 3| + 2 > 7$

58. $|x - 4| + 5 > 1$

59. $|2y - 7| > -1$

60. $|3y - 4| > 8$

61. $|3a - 4| + 2 \geq 7$

62. $|2a - 5| + 1 \geq 8$

63. $|y - 3| < 12$

64. $|p - 2| < 3$

65. $9 - |x + 4| \leqslant 5$

66. $12 - |x - 5| \leqslant 9$

67. $|4 - 3y| > 8$

68. $|7 - 2y| < -6$

69. $|3 - 4x| < -5$

70. $7 + |4a - 5| \leqslant 26$

71. $\left| \dfrac{2 - 5x}{4} \right| \geqslant \dfrac{2}{3}$

72. $\left| \dfrac{1 + 3x}{5} \right| > \dfrac{7}{8}$

73. $|m + 5| + 9 \leqslant 16$

74. $|t - 7| + 3 \geqslant 4$

75. $25 - 2|a + 3| > 19$

76. $30 - 4|a + 2| > 12$

77. Let $f(x) = |2x - 3|$. Find all x for which $f(x) \leqslant 4$.

78. Let $f(x) = |5x + 2|$. Find all x for which $f(x) \leqslant 3$.

79. Let $f(x) = 2 + |3x - 4|$. Find all x for which $f(x) \geqslant 13$.

80. Let $f(x) = |2 - 9x|$. Find all x for which $f(x) \geqslant 25$.

81. Let $f(x) = 7 + |2x - 1|$. Find all x for which $f(x) < 16$.

82. Let $f(x) = 5 + |3x + 2|$. Find all x for which $f(x) < 19$.

SKILL MAINTENANCE

83. *Dinner prices.* The Danville Volunteer Fire Department served 250 dinners. A child's dinner cost $3 and an adult's dinner cost $8. The total amount of money collected was $1410. How many of each type of dinner was served?

84. *Agriculture.* The perimeter of a rectangular cornfield is 628 m. The length of the field is 6 m greater than the width. Find the area of the field.

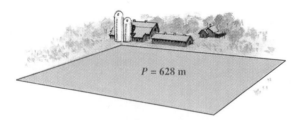

$P = 628$ m

85. Evaluate $2x^2 + 3y$ for $x = 5$ and $y = 6$.

86. Evaluate $7a + 9b$ for $a = 8$ and $b = -3$.

SYNTHESIS

87. ◆ Explain why the inequality $|x + 7| < 1$ can be interpreted as "the distance between x and -7 is less than 1."

88. ◆ Explain why the inequality $|x + 5| \geqslant 2$ can be interpreted as "the number x is at least 2 units from -5."

89. ◆ Explain in your own words why -7 is not a solution of $|x| < 5$.

90. ◆ Explain in your own words why $[6, \infty)$ is only part of the solution of $|x| \geqslant 6$.

91. From the definition of absolute value, $|x| = x$ only when $x \geqslant 0$. Thus, $|3x + 6| = 3x + 6$ only when $3x + 6 \geqslant 0$, which means that $x \geqslant -2$. Solve $|2x - 5| = 2x - 5$ using this same reasoning.

Solve.

92. $|x + 3| = x + 3$

93. $|7x - 2| = x + 4$

94. $|x + 2| > x$

95. $2 \leqslant |x - 1| \leqslant 5$

Find an equivalent inequality with absolute value.

96. $-3 < x < 3$

97. $-5 \leqslant y \leqslant 5$

98. $x \leqslant -6 \ or \ 6 \leqslant x$

99. $x < -4 \ or \ 4 < x$

100. $x < -8 \ or \ 2 < x$

101. $-5 < x < 1$

102. x is less than 2 units from 7.

103. x is more than 1 unit from 5.

104. *Motion of a spring.* A weighted spring is bouncing up and down so that its distance d above the ground satisfies the inequality $|d - 6 \text{ ft}| \leqslant \frac{1}{2}$ ft (see the figure below). Find all possible distances d.

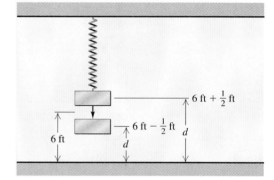

105. Use the accompanying graph of $f(x) = |2x - 6|$ to solve $|2x - 6| \leqslant 4$.

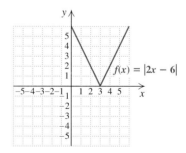

106. ◈ Describe a procedure that could be used to solve any equation of the form $g(x) < c$ graphically (see Example 6).

107. ◈ Is it possible for an equation in x of the form $|ax + b| = c$ to have exactly one solution? Why or why not?

108. 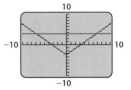 Use a grapher to check the solutions to Examples 3 and 5.

109. Use a grapher to check your answers to Exercises 1, 9, 15, 41, 53, 63, 71, and 95.

110. ◈ 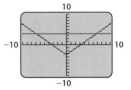 Isabel is using the following graph to solve $|x - 3| < 4$.

How can you tell that a mistake has been made?

4.4   Inequalities in Two Variables

Graphs of Linear Inequalities   •   Systems of Linear Inequalities

In Section 4.1, we graphed inequalities in one variable on a number line. Now we graph inequalities in two variables on a plane.

Graphs of Linear Inequalities

When the equals sign in a linear equation is replaced by an inequality sign, a **linear inequality** is formed. Solutions of linear inequalities are ordered pairs.

EXAMPLE 1   Determine whether $(-3, 2)$ and $(6, -7)$ are solutions of the inequality $5x - 4y > 13$.

SOLUTION   Below, on the left, we replace x by -3 and y by 2. On the right, we replace x by 6 and y by -7.

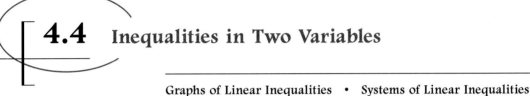

$$\begin{array}{c|c}
5x - 4y > 13 \\ \hline
5(-3) - 4 \cdot 2 \ ? \ 13 \\
-15 - 8 \\
-23 \ | \ 13 \quad \text{FALSE}
\end{array}$$

$$\begin{array}{c|c}
5x - 4y > 13 \\ \hline
5(6) - 4(-7) \ ? \ 13 \\
30 + 28 \\
58 \ | \ 13 \quad \text{TRUE}
\end{array}$$

Since $-23 > 13$ is false, $(-3, 2)$ is not a solution.

Since $58 > 13$ is true, $(6, -7)$ is a solution.

The graph of a linear equation is a straight line. The graph of a linear inequality is a half-plane, bordered by the graph of the *related equation*. To find an inequality's related equation, we simply replace the inequality sign with an equals sign.

EXAMPLE 2 Graph: $y \leq x$.

SOLUTION We first graph the related equation $y = x$. Every solution of $y = x$ is an ordered pair, like $(3, 3)$, in which both coordinates are the same. The graph of $y = x$ is shown on the left below. Every point on the line is a solution of $y \leq x$, so the line is part of the graph of $y \leq x$.

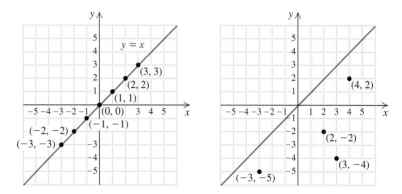

Notice that in the graph on the right each ordered pair on the half-plane below $y = x$ contains a y-coordinate that is less than the x-coordinate. All these pairs represent solutions of $y \leq x$. We check one pair, $(4, 2)$, as follows:

$$\frac{y \leq x}{2 \mid 4} \quad \text{TRUE}$$

It turns out that *any* point on the same side of $y = x$ as $(4, 2)$ is also a solution. Thus, if one point in a half-plane is a solution, then *all* points in that half-plane are solutions. We complete the drawing of the solution set by shading the half-plane below $y = x$. The complete solution set consists of the shaded half-plane and the line. Note too that for any inequality of the form $y \leq mx + b$ or $y < mx + b$, we shade *below* the graph of $y = mx + b$.

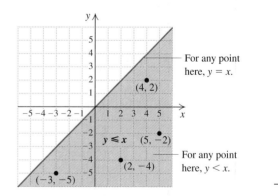

EXAMPLE 3 Graph: $8x + 3y > 24$.

SOLUTION First we sketch the line $8x + 3y = 24$. Since the inequality sign is $>$, points on this line are not part of the graph of

$8x + 3y > 24$, so the line is drawn dashed. Points representing solutions of $8x + 3y > 24$ are in either the half-plane above the line or the half-plane below the line. To determine which, we select a point that is not on the line and determine whether it is a solution of $8x + 3y > 24$. We try $(-3, 4)$ as a test point:

$$
\begin{array}{c|c}
\multicolumn{2}{c}{8x + 3y > 24} \\
\hline
8(-3) + 3 \cdot 4 & ?\ 24 \\
-24 + 12 & \\
-12 & 24 \quad \text{FALSE}
\end{array}
$$

Since $-12 > 24$ is *false,* $(-3, 4)$ is not a solution. Thus no point in the half-plane containing $(-3, 4)$ is a solution. The points in the other half-plane *are* solutions, so we shade that half-plane and obtain the following graph.

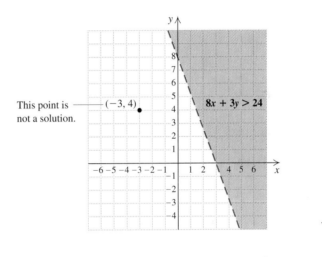

This point is not a solution. $(-3, 4)$

$8x + 3y > 24$

STEPS FOR GRAPHING
LINEAR INEQUALITIES

1. Replace the inequality sign with an equals sign and graph this related equation. If the inequality symbol is $<$ or $>$, draw the line dashed. If the inequality symbol is $\leq$ or $\geq$, draw the line solid.
2. The graph consists of a half-plane on one side or the other of the line, and, if the line is solid, the line as well. To determine which half-plane to shade, test a point not on the line. If that point is a solution of the inequality, shade the half-plane containing the point. If it is not, shade the other half-plane.

EXAMPLE 4

Graph: $6x - 2y < 12$.

SOLUTION   We first graph the related equation, $6x - 2y = 12$, as a dashed line. This line passes through the points $(2, 0)$, $(0, -6)$, and $(3, 3)$, and serves as the boundary of the solution set of the inequality. To determine which half-plane to shade, we test a point *not* on the line. The pair $(0, 0)$ is easy to substitute:

$$
\begin{array}{r}
6x - 2y < 12 \\
\hline
6 \cdot 0 - 2 \cdot 0 \;?\; 12 \\
0 - 0 \;\big|\; \\
0 \;\big|\; 12 \;\text{TRUE}
\end{array}
$$

Since the inequality $0 < 12$ is *true,* the point $(0, 0)$ is a solution, as is every other point in the half-plane containing $(0, 0)$. The graph is shown below.

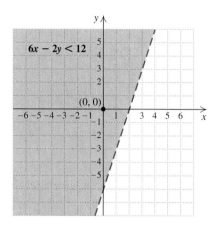

EXAMPLE 5

Graph $x > -3$ on a plane.

SOLUTION   There is a missing variable in this inequality. If we graph the inequality on a line, its graph is as follows:

$$
\overset{\longleftarrow \;\;\;|\;\;\;|\;\;\;|\;\;\;|\;\;\;\oplus\;\;\;|\;\;\;|\;\;\;|\;\;\;|\;\;\;|\;\;\;|\;\;\;|\;\;\;\longrightarrow}{-7\;\;-6\;\;-5\;\;-4\;\;-3\;\;-2\;\;-1\;\;\;0\;\;\;1\;\;\;2\;\;\;3\;\;\;4\;\;\;5\;\;\;6\;\;\;7}
$$

However, we can also write this inequality as $x + 0y > -3$ and graph it on a plane. We can use the same technique as in the examples above. First we graph the related equation $x = -3$ in the plane, using a dashed line. Then we test some point, say, $(2, 5)$:

$$
\begin{array}{r}
x + 0y > -3 \\
\hline
2 + 0 \cdot 5 \;?\; -3 \\
2 \;\big|\; -3 \;\text{TRUE}
\end{array}
$$

Since $(2, 5)$ is a solution, all points in the half-plane containing $(2, 5)$ are solutions. We shade that half-plane. Another approach is to simply note that the solutions of $x > -3$ are all pairs with first coordinates greater than -3.

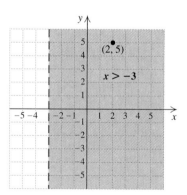

> **EXAMPLE 6**   Graph $y \leq 4$ on a plane.

SOLUTION   One approach is to graph $y = 4$, using a solid line, and then use $(2, -3)$ as a test point:

$$
\begin{array}{c}
0x + y \leq 4 \\
\hline
0 \cdot 2 + (-3) \; ? \; 4 \\
-3 \; | \; 4 \; \text{TRUE}
\end{array}
$$

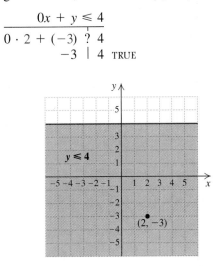

We see that $(2, -3)$ is a solution, so all points in the half-plane containing $(2, -3)$ are solutions. Another approach is to simply note that the solutions of $y \leq 4$ are all pairs with y-coordinates less than or equal to 4.

TECHNOLOGY CONNECTION 4.4

On many graphers, an inequality like $y < 1.2x + 3.49$ can be drawn using the SHADE option of the DRAW key. Often this feature will only shade regions *between* two curves. Thus you may need to enter a y-value, like -100 or 100, that you know will bound the shaded region out of sight of the viewing window. To shade $y < 1.2x + 3.49$, enter $y_1 = 1.2x + 3.49$. Then select SHADE and enter Shade$(-100, y_1)$. You should see a graph similar to this:

$y < 1.2x + 3.49$

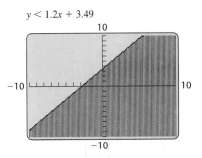

Use a grapher to draw the graphs of each of the following inequalities. Note that for most graphers you must first solve for y if it is not already isolated. You may also need to determine in advance what half-plane is shaded.

1. $y > x + 3.5$
2. $7y \leq 2x + 5$
3. $8x - 2y < 11$
4. $11x + 13y + 4 \geq 0$

Systems of Linear Inequalities

To graph a system of equations, we graph the individual equations and then find the intersection of the individual graphs. We do the same thing for a system of inequalities, that is, we graph each inequality and find the intersection of the individual graphs.

> **EXAMPLE 7**   Graph the system
>
> $$x + y \leq 4,$$
> $$x - y < 4.$$

SOLUTION   To graph $x + y \leq 4$, we graph $x + y = 4$ using a solid line. Since the test point $(0, 0)$ *is* a solution, we shade all points below the line red. The arrows near the ends of the line are another way of indicating the half-plane containing solutions.

Next, we graph $x - y < 4$. We graph $x - y = 4$ using a dashed line and consider $(0, 0)$ as a test point. Again, $(0, 0)$ is a solution, so we shade that side of the line blue. The solution set of the system is the region that is shaded purple (both red and blue) and part of the line $x + y = 4$.

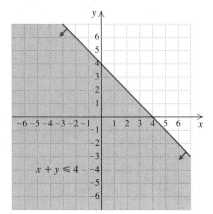 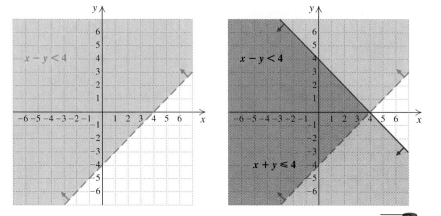

EXAMPLE 8

Graph: $-2 < x \leq 3$.

SOLUTION   This is a system of inequalities:

$$-2 < x,$$
$$x \leq 3.$$

We graph the equation $-2 = x$, and see that the graph of the first inequality is the half-plane to the right of the line $-2 = x$. It is shaded red.

We graph the second inequality, starting with the line $x = 3$, and find that its graph is the line and also the half-plane to its left. It is shaded blue.

The solution set of the system is the region that is the intersection of the individual graphs. Since it is shaded both blue and red, it appears to be purple. All points in this region have x-coordinates that are greater than -2 but do not exceed 3.

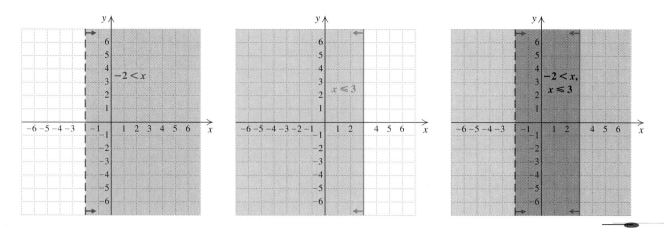

A system of inequalities may have a graph that consists of a polygon and its interior. In Section 4.5, we will have use for the vertices of such a graph.

EXAMPLE 9   Graph the system of inequalities. Find the coordinates of any vertices formed.

$$6x - 2y \leq 12, \qquad (1)$$
$$y - 3 \leq 0, \qquad (2)$$
$$x + y \geq 0. \qquad (3)$$

SOLUTION   We graph the lines

$$6x - 2y = 12,$$
$$y - 3 = 0,$$
and $$x + y = 0$$

using solid lines. The regions for each inequality are indicated by the arrows near the ends of the lines. We note where the regions overlap and shade the region of solutions purple.

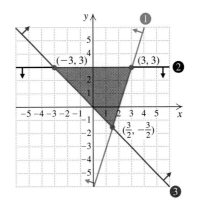

To find the vertices, we solve three different systems of equations. The system of related equations from inequalities (1) and (2) is

$$6x - 2y = 12,$$
$$y - 3 = 0.$$

Solving, we obtain the vertex (3, 3).

The system of related equations from inequalities (1) and (3) is

$$6x - 2y = 12,$$
$$x + y = 0.$$

Solving, we obtain the vertex $\left(\frac{3}{2}, -\frac{3}{2}\right)$.

The system of related equations from inequalities (2) and (3) is

$$y - 3 = 0,$$
$$x + y = 0.$$

Solving, we obtain the vertex (−3, 3).

EXERCISE SET

4.4

Determine whether each ordered pair is a solution of the given inequality.

1. $(-4, 2)$; $2x + 3y < -1$

2. $(3, -6)$; $4x + 2y \le -2$

3. $(8, 14)$; $2y - 3x \ge 9$

4. $(7, 20)$; $3x - y > -1$

Graph on a plane.

5. $y < \frac{1}{2}x$

6. $y > 2x$

7. $y \le x - 4$

8. $y < x + 3$

9. $y \ge x + 4$

10. $y > x - 2$

11. $x - y \ge 5$

12. $x + y < 4$

13. $2x + 3y < 6$

14. $3x + 4y \le 12$

15. $2y - x \le 4$

16. $2y - 3x > 6$

17. $2x - 2y \ge 8 + 2y$

18. $3x - 2 \le 5x + y$

19. $y \ge 3$

20. $x < -5$

21. $x \le 6$

22. $y > -3$

23. $-2 < y < 5$

24. $-4 < y < -1$

25. $-4 \le x \le 4$

26. $-3 \le y \le 4$

27. $0 \le y \le 3$

28. $0 \le x \le 6$

Graph each system.

29. $y > x,$
$y < -x + 5$

30. $y < x,$
$y > -x + 1$

31. $y \ge x,$
$y \le -x + 2$

32. $y \ge x,$
$y \le -x + 4$

33. $y \le -2,$
$x \ge -1$

34. $y \ge -3,$
$x \ge 1$

35. $x > -2,$
$y < -2x + 3$

36. $x < 3,$
$y > -3x + 2$

37. $y \le 4,$
$y \ge -x + 2$

38. $y \ge -2,$
$y \ge x + 3$

39. $x + y \le 3,$
$x - y \le 4$

40. $x + y < 1,$
$x - y < 2$

41. $y + 3x > 0,$
$y + 3x < 2$

42. $y - 2x \ge 1,$
$y - 2x \le 3$

Graph each system of inequalities. Find the coordinates of any vertices formed.

43. $y \le 2x - 1,$
$y \ge -2x + 1,$
$x \le 3$

44. $2y - x \le 2,$
$y - 3x \ge -4,$
$y \ge -1$

45. $x + 2y \le 12,$
$2x + y \le 12,$
$x \ge 0,$
$y \ge 0$

46. $x - y \le 2,$
$x + 2y \ge 8,$
$y \le 4$

47. $8x + 5y \le 40,$
$x + 2y \le 8,$
$x \ge 0,$
$y \ge 0$

48. $4y - 3x \ge -12,$
$4y + 3x \ge -36,$
$y \le 0,$
$x \le 0$

49. $y - x \ge 1,$
$y - x \le 3,$
$2 \le x \le 5$

50. $3x + 4y \ge 12,$
$5x + 6y \le 30,$
$1 \le x \le 3$

SKILL MAINTENANCE

51. One side of a square is 5 less than a side of an equilateral triangle. If the perimeter of the square is the same as the perimeter of the triangle, what is the length of a side of the square? of a side of the triangle?

Solve.

52. $4y - 3x = 8,$
$2x + 5y = -1$

53. $5(3x - 4) = -2(x + 5)$

54. $4(3x + 4) = 2 - x$

SYNTHESIS

55. ◈ Do all systems of linear inequalities have solutions? Why or why not?

56. ◈ Explain how a system of linear inequalities could have a solution set containing exactly one pair.

57. ◈ Is $(0, 0)$ always the best test point to use when graphing inequalities? Why or why not?

58. ◈ When graphing linear inequalities, Ron makes a habit of always shading above the line when the symbol $\ge$ is used. Is this wise? Why or why not?

Graph.

59. $x + y > 8,$
$x + y \le -2$

60. $x - 2y \le 0,$
$-2x + y \le 2,$
$x \le 2,$
$y \le 2,$
$x + y \le 4$

61. $x + y \ge 1,$
$-x + y \ge 2,$
$x \le 4,$
$y \ge 0,$
$y \le 4,$
$x \le 2$

62. Write four systems of four inequalities that describe a 2-unit–by–2-unit square that has (0, 0) as one of the vertices.

63. *Luggage size.* Unless an additional fee is paid, most major airlines will not check any luggage that is more than 62 in. long. The U.S. Postal Service will ship a package only if the sum of the package's length and girth (distance around its midsection) does not exceed 108 in. Concert Productions is ordering several 62-in. long trunks that will be both mailed and checked as luggage. Using w and h for width and height (in inches), respectively, write and graph an inequality that represents all acceptable combinations of width and height.

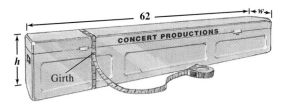

64. *Widths of a basketball floor.* Sizes of basketball floors vary due to building sizes and other constraints such as cost. The length L is to be at

most 94 ft and the width W is to be at most 50 ft. Graph a system of inequalities that describes the possible dimensions of a basketball floor.

65. *Hockey wins and losses.* The Skating Stars figure that they need at least 60 points for the season in order to make the playoffs. A win is worth 2 points and a tie is worth 1 point. Graph a system of inequalities that describes the situation. (*Hint*: Let w = the number of wins and t = the number of ties.)

66. *Elevators.* Many elevators have a capacity of 1 metric ton (1000 kg). Suppose that c children, each weighing 35 kg, and a adults, each 75 kg, are on an elevator. Graph a system of inequalities that indicates when the elevator is overloaded.

Exercises 67 and 68 can be worked only with graphers that have a SHADE *feature.*

67. Use a grapher to graph each inequality.

  a) $3x + 6y > 2$ **b)** $x - 5y \leq 10$
  c) $13x - 25y + 10 \leq 0$ **d)** $2x + 5y > 0$

68. Use a grapher to check your answers to Exercises 29–42. Then use TRACE or INTERSECT to determine any point(s) of intersection.

COLLABORATIVE
C ◆ O ◆ R ◆ N ◆ E ◆ R

Focus: Linear inequalities

Time: 15–25 minutes

Group size: 3

Under a proposed "Rule of 85," full-time faculty in the California State Teachers Retirement System (kindergarten through community college) who are a years old with y years of service would have the option of retirement if $a + y \geq 85$.

ACTIVITY

1. Decide, as a group, how old you think someone would have to be in order to teach full-time. Express this age range as an inequality involving a.

2. Decide, as a group, the number of years someone could teach full-time before retiring. Express this answer as a compound inequality involving y.

3. Using the Rule of 85 and the answers to parts (1) and (2) above, write a system of inequalities. Then, using a scale of 5 yr per square, graph the system. To facilitate comparisons with graphs from other groups, plot a on the horizontal axis and y on the vertical axis.

4. Compare the graphs from all groups. Try to reach consensus on the graph that most clearly illustrates what the status would be of someone who would have the option of retirement under the Rule of 85.

5. If your instructor is agreeable to the idea, attempt to represent him or her with a point on your graph.

4.5 Applications Using Linear Programming

Objective Functions and Constraints   •   Linear Programming

There are many problems in real life in which we want to find a greatest value (a maximum) or a least value (a minimum). For example, most businesses would like to know how to make the *most* profit and how to make their expenses the *least* possible. Some such problems can be solved using systems of inequalities.

Objective Functions and Constraints

Often a quantity we wish to maximize depends on two or more other quantities. For instance, a gardener's profits P might depend on the number of shrubs s and the number of trees t that are planted. If the gardener makes a $5 profit from each shrub and a $9 profit from each tree, the total profit is given by the **objective function**

$$P = 5s + 9t.$$

Thus the gardener might be tempted to simply plant lots of trees since they yield the greater profit. This would be a good idea were it not for the fact that the number of trees and shrubs planted—and thus the total profit—is subject to the demands, or **constraints,** of the situation. For example, the gardener might be required to plant at least 3 shrubs. Thus the objective function would be subject to the *constraint*

$$s \geq 3.$$

He or she might also be required to plant no more than 10 plants. This would subject the objective function to a *second* constraint:

$$s + t \leq 10.$$

Finally, the gardener might be told to spend no more than $350 on the plants. If the shrubs cost $20 each and the trees cost $50 each, the objective function is subject to a *third* constraint:

$$\underbrace{\text{The cost of the shrubs}}_{20s} \text{ plus } \underbrace{\text{the cost of the trees}}_{50t} \underbrace{\text{cannot exceed}}_{\leq} \underbrace{\$350.}_{350}$$

In short, the gardener wishes to maximize the objective function

$$P = 5s + 9t$$

subject to the constraints

$$s + t \leq 10,$$
$$s \geq 3,$$
$$20s + 50t \leq 350,$$
$$\left. \begin{array}{l} s \geq 0, \\ t \geq 0. \end{array} \right\} \longleftarrow \text{Because the number of trees and shrubs cannot be negative}$$

These constraints form a system of linear inequalities that can be graphed.

Linear Programming

The gardener's problem is "How many shrubs and trees should be planted, subject to the constraints listed, in order to maximize profit?" To solve such a problem, we use an important result from a branch of mathematics known as **linear programming.**

THE CORNER PRINCIPLE

Suppose that an objective function $F = ax + by + c$ depends on x and y. Suppose also that F is subject to constraints on x and y, which form a system of linear inequalities. If F has a minimum or a maximum value, it can then be found as follows:

1. Graph the system of inequalities and find the vertices.
2. Find the value of the objective function at each vertex. The largest and the smallest of those values are the maximum and the minimum of the function, respectively.

This result was proven during World War II, when linear programming was developed to help with shipping troops and supplies to Europe.

EXAMPLE 1

Solve the gardener's problem discussed above.

SOLUTION   We are asked to maximize $P = 5s + 9t$, subject to the constraints

$$s + t \leq 10,$$
$$s \geq 3,$$
$$20s + 50t \leq 350,$$
$$s \geq 0,$$
$$t \geq 0.$$

We graph the system (see the following page), using the techniques of Section 4.4. The portion of the graph that is shaded represents all pairs that satisfy the constraints. It is sometimes called the *feasible region.*

According to the corner principle, P is maximized at one of the vertices of the shaded region. To determine the coordinates of the vertices, we solve

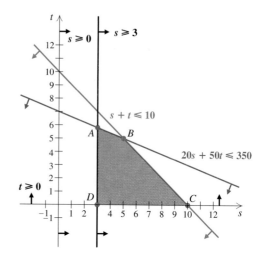

the following systems:

$$20s + 50t = 350,$$
$$s = 3;$$

$$s + t = 10,$$
$$20s + 50t = 350;$$

$$s + t = 10,$$
$$t = 0;$$

$$t = 0,$$
$$s = 3.$$

The solutions of the systems are (3, 5.8), (5, 5), (10, 0), and (3, 0), respectively. We now find the value of P at each of these points:

Vertex (s, t)	Profit $P = 5s + 9t$	
A (3, 5.8)	5(3) + 9(5.8) = 67.2	
B (5, 5)	5(5) + 9(5) = 70	←———Maximum
C (10, 0)	5(10) + 9(0) = 50	
D (3, 0)	5(3) + 9(0) = 15	←———Minimum

The largest value of P occurs at (5, 5). Thus profit is maximized at $70 if the gardener plants 5 shrubs and 5 trees. Incidentally, we have also shown that profit is minimized at $15 if 3 shrubs and 0 trees are planted.   ——●

EXAMPLE 2   ***Test Scores.*** Tani is taking a test in which multiple-choice questions are worth 10 points each and short-answer questions are worth 15 points each. It takes her 3 min to answer each multiple-choice question and 6 min for each short-answer question. The total time allowed is 60 min, and no more than 16 questions can be answered. Assuming that all her answers are correct, how many items of each type should Tani answer in order to get the best score?

SOLUTION

1. Familiarize. Tabulating information will help us to see the picture.

Type	Number of Points for Each	Time Required for Each	Number Answered
Multiple-choice	10	3 min	x
Short-answer	15	6 min	y
Total time: 60 min			
Total number of items: 16 or fewer			

Note that we use x to represent the number of multiple-choice questions and y to represent the number of short-answer questions that are answered.

2. Translate. In this case, it helps to extend the table.

Type	Number of Points for Each	Time Required for Each	Number Answered	Total Time for Type	Total Points for Type
Multiple-choice	10	3 min	x	$3x$	$10x$
Short-answer	15	6 min	y	$6y$	$15y$
Total			$x + y \leq 16$	$3x + 6y \leq 60$	$10x + 15y$

Because no more than 16 items may be answered

Because the time cannot be more than 60 min

This is the total score on the test.

Suppose that the total score on the test is T. We write T as the objective function in terms of x and y:

$$T = 10x + 15y.$$

We wish to maximize T subject to these facts (constraints) about x and y:

$$x + y \leq 16,$$
$$3x + 6y \leq 60,$$
$$x \geq 0,$$
$$y \geq 0.$$

Because the number of items answered cannot be negative

3. **Carry out.** The mathematical manipulation consists of graphing the system and evaluating T at each vertex. The graph is as follows:

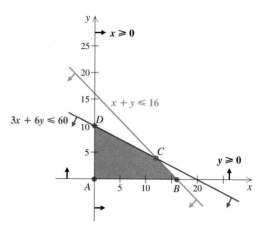

We can find the coordinates of each vertex by solving a system of two linear equations. The coordinates of point A are obviously $(0, 0)$. To find the coordinates of point C, we solve the system

$$x + y = 16, \qquad (1)$$
$$3x + 6y = 60, \qquad (2)$$

as follows:

$$-3x - 3y = -48 \qquad \text{Multiplying both sides of equation (1) by } -3$$
$$\underline{3x + 6y = 60}$$
$$3y = 12 \qquad \text{Adding}$$
$$y = 4.$$

Then we find that $x = 12$. Thus the coordinates of vertex C are $(12, 4)$.

Continuing to find the coordinates of the vertices and computing the test score for each ordered pair, we obtain the following:

Vertex (x, y)	Score $T = 10x + 15y$
A $(0, 0)$	0
B $(16, 0)$	160
C $(12, 4)$	180
D $(0, 10)$	150

The greatest score in the table is 180, obtained when 12 multiple-choice and 4 short-answer questions are answered.

4. **Check.** We can check that $T \le 180$ for any other pair in the shaded region. This is left to the student.

5. **State.** In order to maximize her score, Tani should answer 12 multiple-choice questions and 4 short-answer questions.

Find the maximum and the minimum values of each objective function and the values of x and y at which they occur.

1. $F = 4x + 28y$,
 subject to
 $5x + 3y \leq 34$,
 $3x + 5y \leq 30$,
 $x \geq 0$,
 $y \geq 0$

2. $G = 14x + 16y$,
 subject to
 $3x + 2y \leq 12$,
 $2y - x \leq 4$,
 $x \geq 0$,
 $y \geq 0$

3. $P = 16x - 2y + 40$,
 subject to
 $6x + 8y \leq 48$,
 $0 \leq y \leq 4$,
 $0 \leq x \leq 7$

4. $Q = 24x - 3y + 52$,
 subject to
 $5x + 4y \leq 20$,
 $0 \leq y \leq 4$,
 $0 \leq x \leq 3$

5. $F = 2y - 3x$,
 subject to
 $y \leq 2x + 1$,
 $y \geq -2x + 3$,
 $x \leq 3$

6. $G = 5x + 2y + 4$,
 subject to
 $y \leq 2x + 1$,
 $y \geq -x + 3$,
 $x \leq 5$

Solve.

7. *Lunch time profits.* Elrod's lunch cart sells burritos and chili. To stay in business, Elrod must sell at least 10 orders of chili and 30 burritos each day. Because of limited space, no more than 40 orders of chili or 70 burritos can be made. The total number of orders cannot exceed 90. If profit is $1.65 per chili order and $1.05 per burrito, how many of each item should Elrod sell in order to maximize profit?

8. *Milling.* Johnson Lumber can convert logs into either lumber or plywood. In a given week, the mill can turn out 400 units of production, of which 100 units of lumber and 150 units of plywood are required by regular customers. The profit on a unit of lumber is $20 and on a unit of plywood is $30. How many units of each type should the mill produce in order to maximize profit?

9. *Cycle production.* Yawaka manufactures motorcycles and bicycles. To stay in business, the number of bicycles made cannot exceed 3 times the number of motorcycles made. Yawaka lacks the facilities to produce more than 60 motorcycles or more than 120 bicycles. The total production of motorcycles and bicycles cannot exceed 160. The profit on a motorcycle is $1340 and on a bicycle is $200. Find the number of each that should be manufactured in order to maximize profit.

10. *Gas mileage.* Roschelle owns a car and a moped. She has at most 12 gal of gasoline to be used between the car and the moped. The car's tank holds at most 10 gal and the moped's 3 gal. The mileage for the car is 20 mpg and for the moped is 100 mpg. How many gallons of gasoline should each vehicle use if Roschelle wants to travel as far as possible? What is the maximum number of miles?

11. *Test scores.* Edy is about to take a test that contains short-answer questions worth 4 points each and word problems worth 7 points each. Edy must do at least 5 short-answer questions, but time restricts doing more than 10. She must do at least 3 word problems, but time restricts doing more than 10. Edy can do no more than 18 questions in total. How many of each type of question must Edy do in order to maximize her score? What is this maximum score?

12. *Test scores.* Phil is about to take a test that contains matching questions worth 10 points each and essay questions worth 25 points each. He must do at least 3 matching questions, but time restricts doing more than 12. Phil must do at least 4 essays, but time restricts doing more than 15. If no more than 20 questions can be answered, how many of each type should Phil do in order to maximize his score? What is this maximum score?

13. *Farming.* Thano's farm consists of 240 acres of cropland. Thano wishes to plant this acreage in corn or oats. Profit per acre in corn production is $400 and in oats it is $300. Furthermore, the total number of hours of labor available during the production period is 320. Each acre of land in corn production uses 2 hr of labor during the production period, while production of oats requires 1 hr per acre. Determine how the land should be divided between corn and oats in order to maximize profit.

14. *Investing.* Rosa is planning to invest up to $40,000 in corporate or municipal bonds, or both. She must invest from $6000 to $22,000 in corporate bonds, and she does not want to invest more than $30,000 in municipal bonds. The interest on corporate bonds is 8% and on municipal bonds is $7\frac{1}{2}\%$. This is simple interest for one year. How much should Rosa invest in each type of bond in order to earn the most interest? What is the maximum income?

15. *Investing.* Jamaal is planning to invest up to $22,000 in City Bank or State Bank, or both. He wants to invest at least $2000 but no more than $14,000 in City Bank. State Bank does not insure more than a $15,000 investment, so he will invest no more than that in State Bank. The interest in City Bank is 6% and in State Bank is $6\frac{1}{2}\%$. This is simple interest for one year. How much should he invest in each bank in order to earn the most interest? What is the maximum income?

16. *Coffee blending.* The Coffee Peddler has 1440 lb of Sumatran coffee and 700 lb of Kona coffee. A batch of Hawaiian Blend requires 8 lb of Kona and 12 lb of Sumatran, and yields a profit of $90. A batch of Classic Blend requires 4 lb of Kona and 16 lb of Sumatran, and yields a $55 profit. How many batches of each kind should be made in order to maximize profit? What is the maximum profit? (*Hint:* Organize the information in a table.)

17. *Textile production.* It takes Cosmic Stitching 2 hr of cutting and 4 hr of sewing to make a knit suit. To make a worsted suit, it takes 4 hr of cutting and 2 hr of sewing. At most 20 hr per day are available for cutting and at most 16 hr per day are available for sewing. The profit on a knit suit is $34 and on a worsted suit is $31. How many of each kind of suit should be made in order to maximize profit?

18. *Biscuit production.* The Hockeypuck Biscuit Factory makes two types of biscuits, Biscuit Jumbos and Mitimite Biscuits. The oven can cook at most 200 biscuits per hour. Jumbos each require 2 oz of flour, Mitimites require 1 oz of flour, and there is at most 300 oz of flour available. The income from Jumbos is $1.00 and from Mitimites is $0.80. How many of each type of biscuit should be made in order to maximize income? What is the maximum income?

SYNTHESIS

19. ◈ Write a linear programming problem for a classmate to solve. Devise the problem so that profit must be maximized subject to at least two (nontrivial) constraints.

20. *Airplane production.* Alpha Tours has two types of airplanes, P-1 and P-2, and contracts requiring accommodations for a minimum of 2000 first-class, 1500 tourist-class, and 2400 economy-class passengers. Airplane P-1 costs $30 per mile to operate and can accommodate 40 first-class, 40 tourist-class, and 120 economy-class passengers, whereas airplane P-2 costs $25 per mile to operate and can accommodate 80 first-class, 30 tourist-class, and 40 economy-class passengers. How many of each type of airplane should be used in order to minimize the operating cost?

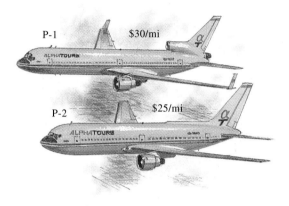

21. *Airplane production.* A new airplane P-3 becomes available, having an operating cost of $37.50 per mile and accommodating 40 first-class, 40 tourist-class, and 80 economy-class passengers. If airplane P-1 of Exercise 20 were replaced by airplane P-3, how many P-2's and how many P-3's would be needed in order to minimize the operating cost?

22. *Furniture production.* Guy's Home Furnishings produces chairs and sofas. The chairs require 20 ft of wood, 1 lb of foam rubber, and 2 sq yd of material. The sofas require 100 ft of wood, 50 lb of foam rubber, and 20 sq yd of material. Guy has in stock 1900 ft of wood, 500 lb of foam rubber, and 240 sq yd of material. The chairs can be sold for $80 each and the sofas for $1200 each. How many of each should be produced in order to maximize income?

SUMMARY AND REVIEW
4

KEY TERMS

Inequality, p. 198
Solution, p. 198
Solution set, p. 198
Set-builder notation, p. 199
Interval notation, p. 199
Open interval, p. 199
Closed interval, p. 199

Half-open interval, p. 199
Compound inequality, p. 209
Intersection, p. 210
Conjunction, p. 210
Union, p. 212
Disjunction, p. 213
Absolute value, p. 219

Linear inequality, p. 227
Related equation, p. 227
Objective function, p. 236
Constraint, p. 236
Linear programming, p. 237
Feasible region, p. 237

IMPORTANT PROPERTIES AND FORMULAS

The Addition Principle for Inequalities

For any real numbers a, b, and c:

$a < b$ is equivalent to $a + c < b + c$;

$a > b$ is equivalent to $a + c > b + c$.

Similar statements hold for $\leq$ and $\geq$.

The Multiplication Principle for Inequalities

For any real numbers a and b, and for any *positive* number c,

$a < b$ is equivalent to $ac < bc$;

$a > b$ is equivalent to $ac > bc$.

For any real numbers a and b, and for any *negative* number c,

$a < b$ is equivalent to $ac > bc$;

$a > b$ is equivalent to $ac < bc$.

Similar statements hold for $\leq$ and $\geq$.

Set intersection:

$$A \cap B = \{x \mid x \text{ is in } A \text{ and } x \text{ is in } B\}$$

Set union:

$$A \cup B = \{x \mid x \text{ is in } A \text{ or in } B, \text{ or both}\}$$

For any real numbers a and b with $a < b$, $a < x$ and $x < b$ can be abbreviated $a < x < b$.

Intersection corresponds to "and"; union corresponds to "or."

$|x| = x$ if $x \geq 0$; $|x| = -x$ if $x < 0$.

The Absolute-Value Principles for Equations and Inequalities

For any positive number p and any algebraic expression X:

a) The solutions of $|X| = p$ are those numbers that satisfy $X = -p$ or $X = p$.
b) The solutions of $|X| < p$ are those numbers that satisfy $-p < X < p$.
c) The solutions of $|X| > p$ are those numbers that satisfy $X < -p$ or $p < X$.

If $|X| = 0$, then $X = 0$. If p is negative, then $|X| = p$ has no solution.

The Corner Principle

Suppose that an objective function $F = ax + by + c$ depends on x and y. Suppose also that F is subject to constraints on x and y, which form a system of linear inequalities. If F has a minimum or a maximum value, it can then be found as follows:

1. Graph the system of inequalities and find the vertices.
2. Find the value of the objective function at each vertex. The largest and the smallest of those values are the maximum and the minimum of the function, respectively.

REVIEW EXERCISES

Graph each inequality and write the solution set using both set-builder and interval notation.

1. $x \leq -4$

2. $a + 7 \leq -14$

3. $y - 5 \geq -12$

4. $4y > -15$

5. $-0.3y < 9$

6. $-6x - 5 < 13$

7. $-\frac{1}{2}x - \frac{1}{4} > \frac{1}{2} - \frac{1}{4}x$

8. $0.3y - 7 < 2.6y + 15$

9. $-2(x - 5) \geq 6(x + 7) - 12$

10. Let $f(x) = 3x - 5$ and $g(x) = 11 - x$. Find all values of x for which $f(x) \leq g(x)$.

Solve.

11. Jessica can choose between two summer jobs. She can work as a checker in a discount store for $5.40 an hour, or she can mow lawns for $9.00 an hour. In order to mow lawns, she must buy a $450 lawn-mower. How many hours of labor will it take Jessica to make more money mowing lawns?

12. Jerry is going to invest $4500, part at 6% and the rest at 7%. What is the most he can invest at 6% and still be guaranteed $300 in interest each year?

13. Find the intersection:

$\{1, 2, 5, 6, 9\} \cap \{1, 3, 5, 9\}$.

14. Find the union:

$\{1, 2, 5, 6, 9\} \cup \{1, 3, 5, 9\}$.

Graph and write interval notation.

15. $x \leq 3$ *and* $x > -5$

16. $x \leq 3$ *or* $x > -5$

Solve and graph each solution set.

17. $-4 < x + 3 \leq 5$

18. $-15 < -4x - 5 < 0$

19. $3x < -9$ *or* $-5x < -5$

20. $2x + 5 < -17$ *or* $-4x + 10 \leq 34$

21. $2x + 7 \leq -5$ *or* $x + 7 \geq 15$

22. $f(x) < -5$ *or* $f(x) > 5$, where $f(x) = 3 - 5x$

23. Write the domain of f using interval notation if $f(x) = \sqrt{x - 3}$.

Solve.

24. $|x| = 6$

25. $|x| \geq 3.5$

26. $|x - 2| = 7$

27. $|2x + 5| < 12$

28. $|3x - 4| \geq 15$

29. $|2x + 5| = |x - 9|$

30. $|5x + 6| = -8$

31. $\left| \dfrac{x + 4}{8} \right| \leq 1$

32. $2|x - 5| - 7 > 3$

33. Let $f(x) = |3x - 5|$. Find all x for which $f(x) < 0$.

34. Graph $x - 2y \geq 6$ on a plane.

Graph each system of inequalities. Find the coordinates of any vertices formed.

35. $x + 3y > -1$,
     $x + 3y < 4$

36. $x - 3y \leq 3$,
     $x + 3y \geq 9$,
     $y \leq 6$

37. Find the maximum and the minimum values of

$F = 3x + y + 4$

subject to

$y \leq 2x + 1$,

$x \leq 7$,

$y \geq 3$.

38. Simply Hardware Computer Company has two manufacturing plants. The West plant cannot produce more than 60 computers a month, while the East plant cannot produce more than 120 computers a month. The Computer Outpost sells at least 160

Simply Hardware computers each month. It costs $40 to ship a computer to The Computer Outpost from the West plant and $25 to ship from the East plant. How many computers should be shipped from each plant in order to minimize cost?

SKILL MAINTENANCE

Solve.

39. $5x - 4y = -10,$
$4x + 2y = 5$

40. $3(x + 4) = 2(x - 5)$

41. Graph: $f(x) = -2x - 6.$

42. The perimeter of a rectangular field is 786 ft. The length is 9 ft longer than the width. Find the area of the field.

SYNTHESIS

43. ◈ Explain in your own words why $|x| = p$ has two solutions when p is positive and no solution when p is negative.

44. ◈ Explain why the graph of the solution of a system of linear inequalities is the intersection, not the union, of the individual graphs.

45. Solve: $|2x + 5| \leq |x + 3|$.

46. Classify as true or false: If $x < 3$, then $x^2 < 9$. If false, give an example showing why.

47. Just-For-Fun manufactures marbles with a 1.1-cm diameter and a ± 0.03-cm manufacturing tolerance, or allowable variation in diameter. Write the tolerance as an inequality with absolute value.

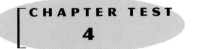

CHAPTER TEST 4

Graph each inequality and write the solution set using both set-builder and interval notation.

1. $x - 2 < 12$

2. $-0.6y < 30$

3. $-4y - 3 \geq 5$

4. $3a - 5 \leq -2a + 6$

5. $4(5 - x) < 2x + 5$

6. $-8(2x + 3) + 6(4 - 5x) \geq 2(1 - 7x) - 4(4 + 6x)$

7. Let $f(x) = -5x - 1$ and $g(x) = -9x + 3$. Find all values of x for which $f(x) > g(x)$.

8. Lia can rent a van for either $40 per day with unlimited mileage or $30 per day with 100 free miles and an extra charge of 15¢ for each mile over 100. For what numbers of miles traveled would the unlimited mileage plan save Lia money?

9. A refrigeration repair company charges $40 for the first half-hour of work and $30 for each additional hour. Blue Mountain Camp has budgeted $100 to repair its walk-in cooler. For what lengths of a service call will the budget not be exceeded?

10. Find the intersection:
$\{1, 3, 5, 7, 9\} \cap \{3, 5, 11, 13\}$.

11. Find the union:
$\{1, 3, 5, 7, 9\} \cup \{3, 5, 11, 13\}$.

12. Write the domain of f using interval notation if $f(x) = \sqrt{7 - x}$.

Solve and graph each solution set.

13. $-3 < x - 2 < 4$

14. $-11 \leq -5x - 2 < 0$

15. $3x - 2 < 7 \ or \ x - 2 > 4$

16. $-3x > 12 \ or \ 4x > -10$

17. $-\frac{1}{3} \leq \frac{1}{6}x - 1 < \frac{1}{4}$

18. $|x| = 9$

19. $|x| > 3$

20. $|4x - 1| < 4.5$

21. $|-5x - 3| \geq 10$

22. $|2 - 5x| = -10$

23. $g(x) < -3 \ or \ g(x) > 3$, where $g(x) = 4 - 2x$

24. Let $f(x) = |x + 10|$ and $g(x) = |x - 12|$. Find all values of x for which $f(x) = g(x)$.

Graph the system of inequalities. Find the coordinates of any vertices formed.

25. $x + y \geq 3,$
$x - y \geq 5$

26. $2y - x \geq -7,$
$2y + 3x \leq 15,$
$y \leq 0,$
$x \leq 0$

27. Find the maximum and the minimum values of

$$F = 5x + 3y$$

subject to

$$x + y \leq 15,$$
$$1 \leq x \leq 6,$$
$$0 \leq y \leq 12.$$

28. LaKenya wants to invest $60,000 in mutual funds and municipal bonds. She does not want to invest more than 50%, or less than 20%, of her money in mutual funds. The minimum investment for municipal bonds is $10,000, and they are guaranteed only up to $40,000, so she will not invest more than $40,000 in municipal bonds. The mutual funds should produce a return of 10%, and the municipal bonds a return of 12%. How much should she invest in each in order to maximize her income? What is her maximum income?

SKILL MAINTENANCE

Solve.

29. $4(x - 2) - 3(2x + 7) = 4$

30. $3x + 5y = 8,$
$7x + 9y = -11$

31. Graph: $2x - y = -6$.

32. A disc jockey must play 15 commercials during 1 hr of a radio show. Each commercial is either 30 sec or 60 sec long. The total commercial time during that hour is 10 min. How many of each type of commercial were played?

SYNTHESIS

Solve. Write the solution set using interval notation.

33. $|2x - 5| \leq 7$ *and* $|x - 2| \geq 2$

34. $7x < 8 - 3x < 6 + 7x$

Polynomials and Polynomial Functions

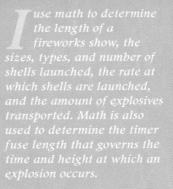

AN APPLICATION

Fireworks are typically launched from a mortar with an upward velocity (initial speed) of about 64 ft/sec. The height $h(t)$, in feet, of a "weeping willow" display, t seconds after having been launched from an 80-ft-high rooftop, is given by

$$h(t) = -16t^2 + 64t + 80.$$

After how long will the cardboard shell from the fireworks reach the ground?

THIS PROBLEM APPEARS AS EXAMPLE 5 IN SECTION 5.8.

More information on pyrotechnics is available at
http://hepg.awl.com/be/inter_5

I use math to determine the length of a fireworks show, the sizes, types, and number of shells launched, the rate at which shells are launched, and the amount of explosives transported. Math is also used to determine the timer fuse length that governs the time and height at which an explosion occurs.

PHIL RAMSEY
Pyrotechnician
Frankfort, IN

> *A polynomial is a type of algebraic expression that contains one or more terms. We define polynomials more specifically in this chapter and learn how to manipulate them so that we can use polynomials and polynomial functions in problem solving.*

5.1   Introduction to Polynomials and Polynomial Functions

Algebraic Expressions and Polynomials   •   Polynomial Functions   •
Adding Polynomials   •   Opposites and Subtraction

In this section, we introduce a type of algebraic expression known as a *polynomial.* After developing some vocabulary, we study addition and subtraction of polynomials, and evaluate *polynomial functions.*

Algebraic Expressions and Polynomials

In Chapter 1, we introduced algebraic expressions like

$$5x^2, \quad \frac{3}{x^2 + 5}, \quad 9a^3b^4, \quad 3x^{-2}, \quad 6x^2 + 3x + 1, \quad -9, \quad \text{and} \quad 5 - 2x.$$

Of the expressions listed, $5x^2$, $9a^3b^4$, $3x^{-2}$, and -9 are examples of *terms.* A **term** is simply a number or a product of a number and a variable or variables raised to a power. The **degree of a term** with whole-number exponents is the sum of the exponents of the variables, if there are variables. Thus the term $5x^2$ has degree 2 and the term $9a^3b^4$ has degree 7. Nonzero constant terms have degree 0, and the term 0 is said to have no degree.

A number like 5 in the term $5x^2$ is called the **coefficient** of that term. Thus the coefficient of $9a^3b^4$ is 9 and the coefficient of $-2x$ is -2. The coefficient of a constant term is just that constant.

When all variables in a term are raised to whole-number powers, the term is a **monomial.** Of the expressions above, $5x^2$, $9a^3b^4$, and -9 are monomials.

A **polynomial** is a monomial or a sum of monomials. Of the expressions listed, $5x^2$, $9a^3b^4$, $6x^2 + 3x + 1$, -9, and $5 - 2x$ are polynomials. In fact, with the exception of $9a^3b^4$, these are all polynomials *in one variable.* The expression $9a^3b^4$ is a *polynomial in two variables.* Note that $5 - 2x$ is the sum of 5 and $-2x$. Thus, 5 and $-2x$ are the terms in the polynomial $5 - 2x$.

The **leading term** of a polynomial is the term of highest degree. Its coefficient is called the **leading coefficient.** The **degree of a polynomial** is the same as the degree of its leading term.

E X A M P L E   1

For each polynomial given, find the degree of each term, the degree of the polynomial, the leading term, and the leading coefficient.

a) $2x^3 + 8x^2 - 17x - 3$
b) $6x^2 + 8x^2y^3 - 17xy - 24xy^2z^4 + 2y + 3$

S O L U T I O N

	(a)				**(b)**					
Term	$2x^3$	$8x^2$	$-17x$	-3	$6x^2$	$8x^2y^3$	$-17xy$	$-24xy^2z^4$	$2y$	3
Degree	3	2	1	0	2	5	2	7	1	0
Degree of Polynomial	3				7					
Leading Term	$2x^3$				$-24xy^2z^4$					
Leading Coefficient	2				-24					

A polynomial of degree 0 or 1 is called **linear**. A polynomial in one variable is said to be **quadratic** if it is of degree 2 and **cubic** if it is of degree 3. The following are some names for certain kinds of polynomials.

Type	Definition	Examples
Monomial	A polynomial of one term	$4, -3p, 5x^2, -7a^2b^3, 0, xyz$
Binomial	A polynomial of two terms	$2x + 7, a - 3b, 5x^2 + 7y^3$
Trinomial	A polynomial of three terms	$x^2 - 7x + 12, 4a^2 + 2ab + b^2$

We generally arrange polynomials in one variable so that the exponents *decrease* from left to right. This is called **descending order.** Some polynomials may be written with exponents *increasing* from left to right, which is **ascending order.** Generally, if an exercise is written in one kind of order, the answer is written in that same order.

E X A M P L E   2

Arrange in ascending order: $12 + 2x^3 - 7x + x^2$.

S O L U T I O N

$$12 + 2x^3 - 7x + x^2 = 12 - 7x + x^2 + 2x^3$$

Polynomials in several variables can be arranged with respect to the powers of one of the variables.

EXAMPLE 3   Arrange in descending powers of x: $y^4 + 2 - 5x^2 + 3x^3y + 7xy^2$.

SOLUTION

$y^4 + 2 - 5x^2 + 3x^3y + 7xy^2 = 3x^3y - 5x^2 + 7xy^2 + y^4 + 2$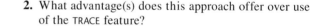

Polynomial Functions

A polynomial function is one like

$$P(x) = 5x^7 + 3x^5 - 4x^2 - 5,$$

where the expression used to describe the function is a polynomial. To evaluate a polynomial function, we substitute a number, just as we did in Chapter 2. In this text, we concentrate on polynomial functions in one variable.

EXAMPLE 4   For the polynomial function $P(x) = -x^2 + 4x - 1$, find the following: **(a)** $P(2)$; **(b)** $P(10)$; **(c)** $P(-10)$.

SOLUTION

a) $P(2) = -2^2 + 4(2) - 1$ We square the input before taking its opposite.
$ = -4 + 8 - 1 = 3$

b) $P(10) = -10^2 + 4(10) - 1$
$ = -100 + 40 - 1 = -61$

c) $P(-10) = -(-10)^2 + 4(-10) - 1$
$ = -100 - 40 - 1 = -141$

TECHNOLOGY CONNECTION 5.1A

A visual way to evaluate a polynomial function is to enter and graph the function as y_1. If a CALC menu exists, selecting VALUE allows us to enter the value of x in which we are interested. The corresponding y-value and point on the graph then appear.

1. Use this approach to check Examples 4(b) and 4(c).
2. What advantage(s) does this approach offer over use of the TRACE feature?

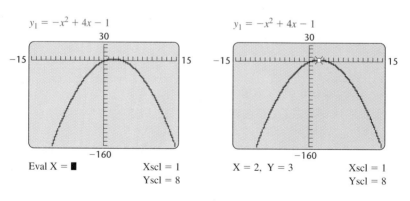

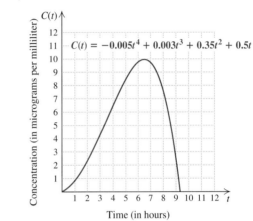

EXAMPLE 5

Veterinary Medicine. Gentamicin is an antibiotic frequently used by veterinarians. The concentration, in micrograms per milliliter (mcg/ml), of Gentamicin in a horse's bloodstream t hours after injection can be approximated by the polynomial function

$$C(t) = -0.005t^4 + 0.003t^3 + 0.35t^2 + 0.5t.$$

a) What is the concentration 2 hr after injection?

b) Use the graph at right to estimate $C(4)$.

SOLUTION

a) We evaluate the function for $t = 2$:

$$C(2) = -0.005(2)^4 + 0.003(2)^3 + 0.35(2)^2 + 0.5(2)$$
$$= -0.005(16) + 0.003(8) + 0.35(4) + 0.5(2)$$
$$= -0.08 + 0.024 + 1.4 + 1$$
$$= -0.08 + 2.424 = 2.344.$$

We carry out the calculation using the rules for order of operations.

On many graphers, once y_1 has been entered, a TBL SET feature can be set so that the grapher ASKS for a value for the independent variable. When choices for x are entered, the corresponding values for y_1 appear.

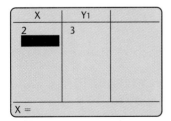

3. Check Examples 4(b) and 4(c) using a table.

It is also possible to find values of y_1 by selecting y_1 from a function menu (sometimes labeled Y-VARS) and then enclosing the desired x-value in parentheses.

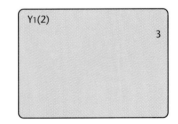

Recalling the last entry, often by pressing 2nd ENTRY ◄ ◄, allows for changing inputs without repeatedly accessing the function menu.

4. Check Examples 4(b) and 4(c) by finding $y_1(10)$ and $y_1(-10)$.

The concentration after 2 hr is 2.344 mcg/ml.

b) To estimate $C(4)$, the concentration after 4 hr, we locate 4 on the horizontal axis. From there we move vertically to the graph of the function and then horizontally to the $C(t)$-axis. This locates a value of about 6.5. Thus,

$$C(4) \approx 6.5.$$

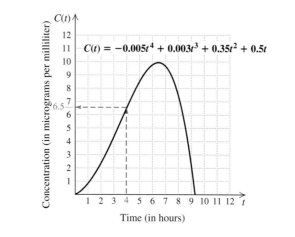

Adding Polynomials

Recall from Section 1.3 that when two terms have the same variable(s) raised to the same power(s), they are **similar**, or **like**, **terms** and can be "combined" or "collected."

EXAMPLE 6 Combine like terms.

a) $3x^2 - 4y + 2x^2$
b) $9x^3 + 5x - 4x^2 - 2x^3 + 5x^2$
c) $3x^2y + 5xy^2 - 3x^2y - xy^2$

SOLUTION

a) $3x^2 - 4y + 2x^2 = 3x^2 + 2x^2 - 4y$ Rearranging terms using the commutative law for addition

$\qquad\qquad\qquad\quad = (3 + 2)x^2 - 4y$ Using the distributive law

$\qquad\qquad\qquad\quad = 5x^2 - 4y$

b) $9x^3 + 5x - 4x^2 - 2x^3 + 5x^2 = 7x^3 + x^2 + 5x$ We usually perform the middle steps mentally and write just the answer.

c) $3x^2y + 5xy^2 - 3x^2y - xy^2 = 4xy^2$

The sum of two polynomials can be found by writing a plus sign between them and then combining like terms.

EXAMPLE 7 Add: $(-3x^3 + 2x - 4) + (4x^3 + 3x^2 + 2)$.

SOLUTION

$$(-3x^3 + 2x - 4) + (4x^3 + 3x^2 + 2) = x^3 + 3x^2 + 2x - 2$$

Using columns is sometimes helpful. To do so, we write the polynomials one under the other, listing like terms under one another and leaving spaces for missing terms.

EXAMPLE 8

Add: $4ax^2 + 4bx - 5$ and $-6ax^2 + 8$.

SOLUTION

$$
\begin{array}{r}
4ax^2 + 4bx - 5 \\
-6ax^2 \qquad\ + 8 \\
\hline
-2ax^2 + 4bx + 3
\end{array}
$$

EXAMPLE 9

Add: $13x^3y + 3x^2y - 5y$ and $x^3y + 4x^2y - 3xy$.

SOLUTION

$(13x^3y + 3x^2y - 5y) + (x^3y + 4x^2y - 3xy) = 14x^3y + 7x^2y - 3xy - 5y$

Opposites and Subtraction

If the sum of two polynomials is 0, the polynomials are *opposites,* or *additive inverses,* of each other. For example,

$$(3x^2 - 5x + 2) + (-3x^2 + 5x - 2) = 0,$$

so the opposite of $(3x^2 - 5x + 2)$ is $(-3x^2 + 5x - 2)$. We can say the same thing using algebraic symbolism, as follows:

The opposite of $(3x^2 - 5x + 2)$ is $(-3x^2 + 5x - 2)$.

$$- \qquad (3x^2 - 5x + 2) = \quad -3x^2 + 5x - 2$$

To form the opposite of a polynomial, we can think of distributing the "$-$" sign, or multiplying each term of the polynomial by -1, and removing the parentheses. The effect is to change the sign of each term in the polynomial.

THE OPPOSITE OF A POLYNOMIAL

The *opposite* of a polynomial P can be written as $-P$ or, equivalently, by replacing each term with its opposite.

EXAMPLE 10

Write two equivalent expressions for the opposite of

$$7xy^2 - 6xy - 4y + 3.$$

SOLUTION

a) $-(7xy^2 - 6xy - 4y + 3)$ Writing the opposite of P as $-P$

b) $-7xy^2 + 6xy + 4y - 3$ Multiplying each term by -1 and removing parentheses

To subtract a polynomial, we add its opposite.

EXAMPLE 11

Subtract: $(-5x^2 + 4) - (2x^2 + 3x - 1)$.

SOLUTION

$$(-5x^2 + 4) - (2x^2 + 3x - 1)$$
$$= (-5x^2 + 4) + (-2x^2 - 3x + 1) \quad \text{Adding the opposite of the}$$
$$\text{polynomial being subtracted}$$
$$= -7x^2 - 3x + 5$$

With practice, you will find that you can skip some steps, by mentally taking the opposite of each term and then combining like terms. Eventually, all you will write is the answer.

To use columns for subtraction, we mentally change the signs of the terms being subtracted.

EXAMPLE 12

Subtract:

$$(4x^2y - 6x^3y^2 + x^2y^2) - (4x^2y + x^3y^2 + 3x^2y^3 - 8x^2y^2).$$

SOLUTION

Write: (Subtract)

$$\begin{array}{l} 4x^2y - 6x^3y^2 \qquad\quad + x^2y^2 \\ -(4x^2y + x^3y^2 + 3x^2y^3 - 8x^2y^2) \end{array}$$

Think: (Add)

$$\begin{array}{l} 4x^2y - 6x^3y^2 \qquad\quad + x^2y^2 \\ -4x^2y - x^3y^2 - 3x^2y^3 + 8x^2y^2 \\ \hline \quad -7x^3y^2 - 3x^2y^3 + 9x^2y^2 \end{array}$$

Take the opposite of each term mentally and add.

TECHNOLOGY CONNECTION

5.1B

One way to check problems like Example 11 is to note that if the subtraction is correct, then

$$(-5x^2 + 4) - (2x^2 + 3x - 1) = -7x^2 - 3x + 5$$

is an identity. Thus the graphs of

$$y_1 = (-5x^2 + 4) - (2x^2 + 3x - 1)$$

and

$$y_2 = -7x^2 - 3x + 5$$

should coincide.

1. Graph y_1 and y_2 as described above, using the window $[-10, 10, -160, 30]$ with Xscl $= 1$ and Yscl $= 8$. Do the graphs coincide? How can you tell?

Example 11 can also be checked by graphing $y_3 = y_2 - y_1$ (but not graphing y_1 or y_2), where the Y-VARS key is used and y_2 and y_1 are as described above. If, as expected, the graphs of y_2 and y_1 are the same, then the graph of y_3 will be the x-axis.

2. Perform the just-described check. What advantage(s) does this check have over the check used in (1) above?

EXERCISE SET

5.1

Determine the degree of each term and the degree of the polynomial.

1. $-7x^5 - x^3 + x^2 + 3x - 9$
2. $t^3 - 5t^2 + t + 1$
3. $y^3 + 2y^7 + x^2y^4 - 8$
4. $u^2 + 3v^5 - u^3v^4 - 7$
5. $a^5 + 4a^2b^4 + 6ab + 4a - 3$
6. $8p^6 + 2p^4t^4 - 7p^3t + 5p^2 - 14$

Arrange in descending order. Then find the leading term and the leading coefficient.

7. $15 - 4y^3 + 7y - 6y^2$
8. $2 - 5y + 6y^2 + 11y^3 - 18y^4$
9. $5x^2 + 3x^7 - x + 12$
10. $9 - 3x - 10x^4 + 7x^2$
11. $a + 5a^3 - a^7 - 19a^2 + 8a^5$
12. $a^3 - 7 + 11a^4 + a^9 - 5a^2$

Arrange in ascending powers of x.

13. $7x - 9 + 3x^4 - 5x^2$
14. $-3x^4 + 2x^3 - x + 9$
15. $-9x^3y + 3xy^3 + x^2y^2 + 2x^4$
16. $5x^2y^2 - 9xy + 8x^3y^2 - 5x^4$
17. $4ax - 7ab + 4x^6 - 7ax^2$
18. $5xy^8 - 3ax^5 + 4ax^3 - 12a + 5x^5$

Find the specified function values.

19. Find $P(4)$ and $P(0)$: $P(x) = 3x^2 - 2x + 5$.
20. Find $Q(3)$ and $Q(-1)$: $Q(x) = -4x^3 + 7x^2 - 1$.
21. Find $P(-2)$ and $P\left(\frac{1}{3}\right)$: $P(y) = 8y^3 - 12y - 5$.
22. Find $Q(-3)$ and $Q(0)$:
    $Q(y) = 9y^3 + 8y^2 - 4y - 9$.

Evaluate each polynomial for $x = 4$.

23. $-7x + 5$
24. $4x - 13$
25. $x^3 - 5x^2 + x$
26. $7 - x + 3x^2$

Evaluate each polynomial function for $x = -1$.

27. $f(x) = -5x^3 + 3x^2 - 4x - 3$
28. $g(x) = -4x^3 + 2x^2 + 5x - 7$

Electing officers. *For a club consisting of n people, the number of ways in which a president, vice president, and treasurer can be elected is given by the polynomial function given by*

$$p(n) = n^3 - 3n^2 + 2n.$$

29. The Stage Right drama club has 12 members. In how many ways can a president, vice president, and treasurer be elected?
30. The Southside Rugby Club has 20 members. In how many ways can they elect a president, vice president, and treasurer?

Falling distance. *The distance $s(t)$, in feet, traveled by a body falling freely from rest in t seconds is approximated by the function given by*

$$s(t) = 16t^2.$$

31. A paintbrush falls from a scaffold and takes 3 sec to hit the ground. How high is the scaffold?

$s(t) = 16t^2$

32. A stone is dropped from a cliff and takes 8 sec to hit the ground. How high is the cliff?

Total revenue. *An electronics firm is marketing a new kind of stereo. The firm determines that when it sells x stereos, its total revenue is*

$$R(x) = 280x - 0.4x^2 \text{ dollars.}$$

33. What is the total revenue from the sale of 75 stereos?
34. What is the total revenue from the sale of 100 stereos?

Total cost. *The electronics firm determines that the total cost, in dollars, of producing x stereos is given by*

$$C(x) = 5000 + 0.6x^2.$$

35. What is the total cost of producing 75 stereos?

36. What is the total cost of producing 100 stereos?

Daily accidents. *The number of daily accidents (the average number of accidents per day) involving drivers of age a is approximated by the polynomial function*

$$P(a) = 0.4a^2 - 40a + 1039.$$

37. Find the number of daily accidents involving a 20-year-old driver.

38. Find the number of daily accidents involving a 25-year-old driver.

Medicine. *Ibuprofen is a medication used to relieve pain. The polynomial function*

$$M(t) = 0.5t^4 + 3.45t^3 - 96.65t^2 + 347.7t, \quad 0 \leqslant t \leqslant 6$$

can be used to estimate the number of milligrams of ibuprofen in the bloodstream t hours after 400 mg of the medication has been swallowed. Use the following graph for Exercises 39–42.

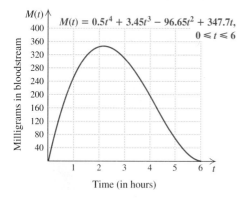

39. Estimate the number of milligrams of ibuprofen in the bloodstream 2 hr after 400 mg has been swallowed.

40. Estimate the number of milligrams of ibuprofen in the bloodstream 4 hr after 400 mg has been swallowed.

41. Approximate $M(5)$.

42. Approximate $M(3)$.

Surface area of a right circular cylinder. *The surface area of a right circular cylinder is given by the polynomial*

$$2\pi rh + 2\pi r^2,$$

where h = the height, r = the radius of the base, and h and r are given in the same units.

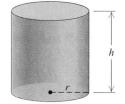

43. 🔢 A 16-oz beverage can has height 6.3 in. and radius 1.2 in. Find the surface area of the can. (Use a calculator with a $\boxed{\pi}$ key or use 3.141592654 for π.)

44. 🔢 A 12-oz beverage can has height 4.7 in. and radius 1.2 in. Find the surface area of the can. (Use a calculator with a $\boxed{\pi}$ key or use 3.141592654 for π.)

Combine like terms.

45. $4a + 7 - 4 + 2a^3 - 6a + 3$

46. $8x + 12 - 8 - 7x + 5x^2 + 10$

47. $3a^2b + 4b^2 - 9a^2b - 6b^2$

48. $5x^2y^2 + 4x^3 - 8x^2y^2 - 12x^3$

49. $8x^2 - 3xy + 12y^2 + x^2 - y^2 + 5xy + 4y^2$

50. $a^2 - 2ab + b^2 + 9a^2 + 5ab - 4b^2 + a^2$

Add.

51. $(5a + 6b - 3c) + (4a - 2b + 2c)$

52. $(6x - 5y + 3z) + (9x + 12y - 8z)$

53. $(a^2 - 3b^2 + 4c^2) + (-5a^2 + 2b^2 - c^2)$

54. $(x^2 - 5y^2 - 9z^2) + (-6x^2 + 9y^2 - 2z^2)$

55. $(x^2 + 2x - 3xy - 7) + (-3x^2 - x + 2xy + 6)$

56. $(3a^2 - 2b + ab + 6) + (-a^2 + 5b - 5ab - 2)$

57. $(7x^2y - 3xy^2 + 4xy) + (-2x^2y - xy^2 + xy)$

58. $(7ab - 3ac + 5bc) + (13ab - 15ac - 8bc)$

59. $(2r^2 + 12r - 11) + (6r^2 - 2r + 4) + (r^2 - r - 2)$

60. $(5x^2 + 19x - 23) + (-7x^2 - 11x + 12) + (-x^2 - 9x + 8)$

61. $\left(\frac{1}{8}xy - \frac{3}{5}x^3y^2 + 4.3y^3\right) + \left(-\frac{1}{3}xy - \frac{3}{4}x^3y^2 - 2.9y^3\right)$

62. $\left(\frac{2}{3}xy + \frac{5}{6}xy^2 + 5.1x^2y\right) + \left(-\frac{4}{5}xy + \frac{3}{4}xy^2 - 3.4x^2y\right)$

Write two equivalent expressions for the opposite, or additive inverse, of each polynomial.

63. $5x^3 - 7x^2 + 3x - 6$

64. $-8y^4 - 18y^3 + 4y - 9$

65. $-12y^5 + 4ay^4 - 7by^2$

66. $7ax^3y^2 - 8by^4 - 7abx - 12ay$

Subtract.

67. $(8x - 4) - (-5x + 2)$

68. $(9y + 3) - (-4y - 2)$

69. $(-3x^2 + 2x + 9) - (x^2 + 5x - 4)$

70. $(-9y^2 + 4y + 8) - (4y^2 + 2y - 3)$

71. $(5a - 2b + c) - (3a + 2b - 2c)$

72. $(8x - 4y + z) - (4x + 6y - 3z)$

73. $(3x^2 - 2x - x^3) - (5x^2 - 8x - x^3)$

74. $(8y^2 - 3y - 4y^3) - (3y^2 - 9y - 7y^3)$

75. $(5a^2 + 4ab - 3b^2) - (9a^2 - 4ab + 2b^2)$

76. $(9y^2 - 14yz - 8z^2) - (12y^2 - 8yz + 4z^2)$

77. $(6ab - 4a^2b + 6ab^2) - (3ab^2 - 10ab - 12a^2b)$

78. $(10xy - 4x^2y^2 - 3y^3) - (-9x^2y^2 + 4y^3 - 7xy)$

79. $\left(\frac{5}{8}x^4 - \frac{1}{4}x^2 - \frac{1}{2}\right) - \left(-\frac{3}{8}x^4 + \frac{3}{4}x^2 + \frac{1}{2}\right)$

80. $\left(\frac{5}{6}y^4 - \frac{1}{2}y^2 - 7.8y + \frac{1}{3}\right) - \left(-\frac{3}{8}y^4 + \frac{3}{4}y^2 + 3.4y - \frac{1}{5}\right)$

Total profit. *Total profit is defined as total revenue minus total cost. In Exercises 81 and 82, $R(x)$ and $C(x)$ are the revenue and cost from the sale of x stereos.*

81. If $R(x) = 280x - 0.4x^2$ and $C(x) = 5000 + 0.6x^2$, find the profit from the sale of 70 stereos.

82. If $R(x) = 280x - 0.7x^2$ and $C(x) = 8000 + 0.5x^2$, find the profit from the sale of 100 stereos.

SKILL MAINTENANCE

83. Graph: $f(x) = \frac{2}{3}x - 1$.

84. Solve: $|2x - 3| = 7$.

85. Multiply: $3(y - 2)$.

86. Simplify: $3(x - 4) - 5(x + 16)$.

SYNTHESIS

87. ◈ Is the sum of two binomials always a binomial? Why or why not?

88. ◈ Ani claims that she can add any two polynomials but finds subtraction difficult. What advice would you offer her?

89. ◈ Write a problem in which revenue and cost functions are given and a profit function, $P(x)$, is required. Devise the problem so that $P(0) < 0$ and $P(100) > 0$.

90. ◈ Write a problem in which revenue and cost

functions are given and a profit function, $P(x)$, is required. Devise the problem so that $P(10) < 0$ and $P(50) > 0$.

For $P(x)$ and $Q(x)$ as given, find the following.

$$P(x) = 13x^5 - 22x^4 - 36x^3 + 40x^2 - 16x + 75,$$

$$Q(x) = 42x^5 - 37x^4 + 50x^3 - 28x^2 + 34x + 100$$

91. $2[P(x)] + Q(x)$

92. $3[P(x)] - Q(x)$

93. $2[Q(x)] - 3[P(x)]$

94. $4[P(x)] + 3[Q(x)]$

95. *Volume of a display.* The number of spheres in a triangular pyramid with x layers is given by the function

$$N(x) = \frac{1}{6}x^3 + \frac{1}{2}x^2 + \frac{1}{3}x.$$

The volume of a sphere of radius r is given by the function

$$V(r) = \frac{4}{3}\pi r^3,$$

where π can be approximated as 3.14.

Chocolate Heaven has a window display of truffles piled in a triangular pyramid formation 5 layers deep. If the diameter of each truffle is 3 cm, find the volume of chocolate in the display.

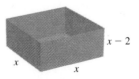

96. Find a polynomial function that gives the outside surface area of a box like this one, with an open top and dimensions as shown.

97. Develop a formula for the surface area of a right circular cylinder in which h = the height, in *centimeters*, and r = the radius, in *meters*. (See Exercises 43 and 44.)

Perform the indicated operation. Assume that the exponents are natural numbers.

98. $(2x^{2a} + 4x^a + 3) + (6x^{2a} + 3x^a + 4)$

99. $(3x^{6a} - 5x^{5a} + 4x^{3a} + 8) -$
$(2x^{6a} + 4x^{4a} + 3x^{3a} + 2x^{2a})$

100. $(2x^{5b} + 4x^{4b} + 3x^{3b} + 8) -$
$(x^{5b} + 2x^{3b} + 6x^{2b} + 9x^b + 8)$

101.  Use a grapher to check your answers to Exercises 45, 59, and 73.

102. ⌷ Use a grapher to check your answers to Exercises 20, 33, and 34.

103. ⌷ A student who is trying to graph

$p(x) = 0.05x^4 - x^2 + 5$ gets the following screen.

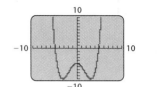

How can the student tell at a glance that a mistake has been made?

COLLABORATIVE
C ◆ O ◆ R ◆ N ◆ E ◆ R

Focus: Polynomial functions

Time: 20 minutes

Group size: 5

ACTIVITY

1. All group members should shake hands with each other. Without "double counting," how many handshakes occurred?
2. Complete the table in the next column.
3. Join another group to determine the number of handshakes for a group of size 10.

Group Size	Number of Handshakes
1	
2	
3	
4	
5	

4. Try to find a function of the form $H(n) = an^2 + bn$, for which $H(n)$ is the number of different handshakes that are possible in a group of n people. Make sure that $H(n)$ produces all of the values in the table above. (*Hint:* Look for a pattern in the table.)
5. Find H(class size).

5.2   Multiplication of Polynomials

Multiplying Monomials   •   Multiplying Monomials and Binomials   •
Multiplying Any Two Polynomials   •   The Product of Two
Binomials: FOIL   •   Squares of Binomials   •   Products of Sums
and Differences   •   Function Notation

Just like numbers, polynomials can be multiplied. The product of two polynomials $P(x)$ and $Q(x)$ is a polynomial $R(x)$ that gives the same value as $P(x) \cdot Q(x)$ for any replacement of x.

Multiplying Monomials

To multiply monomials, we first multiply their coefficients. Then we multiply the variables using the rules for exponents and the commutative and associative laws. With practice, we can work mentally, writing only the answer.

EXAMPLE 1

Multiply and simplify: **(a)** $(-8x^4y^7)(5x^3y^2)$; **(b)** $(-2x^2yz^5)(-6x^5y^{10}z^2)$.

SOLUTION

a) $(-8x^4y^7)(5x^3y^2) = -8 \cdot 5 \cdot x^4 \cdot x^3 \cdot y^7 \cdot y^2$ Using the associative and commutative laws

$$= -40x^{4+3}y^{7+2}$$ Multiplying coefficients; adding exponents

$$= -40x^7y^9$$

b) $(-2x^2yz^5)(-6x^5y^{10}z^2) = (-2)(-6) \cdot x^2 \cdot x^5 \cdot y \cdot y^{10} \cdot z^5 \cdot z^2$

$$= 12x^7y^{11}z^7$$ Multiplying coefficients; adding exponents

Multiplying Monomials and Binomials

The distributive law is the basis for multiplying polynomials other than monomials. We first multiply a monomial and a binomial.

EXAMPLE 2

Multiply: **(a)** $2x(3x - 5)$; **(b)** $3a^2b(a^2 - b^2)$.

SOLUTION

a) $2x \cdot (3x - 5) = 2x \cdot 3x - 2x \cdot 5$ Using the distributive law

$$= 6x^2 - 10x$$ Multiplying monomials

b) $3a^2b(a^2 - b^2) = 3a^2b \cdot a^2 - 3a^2b \cdot b^2$ Using the distributive law

$$= 3a^4b - 3a^2b^3$$

The distributive law is also used when multiplying two binomials. In this case, however, we begin by distributing a *binomial* rather than a monomial. With practice, some of the following steps can be combined.

EXAMPLE 3

Multiply: $(y^3 - 5)(2y^3 + 4)$.

SOLUTION

$$(y^3 - 5)(2y^3 + 4) = (y^3 - 5)2y^3 + (y^3 - 5)4$$ "Distributing" the $y^3 - 5$

$$= 2y^3(y^3 - 5) + 4(y^3 - 5)$$ Using the commutative law for multiplication

$$= 2y^3 \cdot y^3 - 2y^3 \cdot 5 + 4 \cdot y^3 - 4 \cdot 5$$ Using the distributive law (twice)

$$= 2y^6 - 10y^3 + 4y^3 - 20$$ Multiplying the monomials

$$= 2y^6 - 6y^3 - 20$$ Combining like terms

Multiplying Any Two Polynomials

Repeated use of the distributive law enables us to multiply any two polynomials.

EXAMPLE 4   Multiply: $(p + 2)(p^4 - 2p^3 + 3)$.

SOLUTION   By the distributive law, we have

$$(p + 2)(p^4 - 2p^3 + 3)$$
$$= (p + 2)(p^4) - (p + 2)(2p^3) + (p + 2)(3)$$
$$= p^4(p + 2) - 2p^3(p + 2) + 3(p + 2)$$
$$= p^4 \cdot p + p^4 \cdot 2 - 2p^3 \cdot p - 2p^3 \cdot 2 + 3 \cdot p + 3 \cdot 2$$
$$= p^5 + 2p^4 - 2p^4 - 4p^3 + 3p + 6$$
$$= p^5 - 4p^3 + 3p + 6. \qquad \text{Combining like terms}$$

THE PRODUCT OF TWO POLYNOMIALS

> The *product* of two polynomials $P(x)$ and $Q(x)$ is found by multiplying each term of $P(x)$ by every term of $Q(x)$ and then combining like terms.

It is also possible to use columns, multiplying each term at the top by every term at the bottom, keeping like terms in columns, and leaving spaces for missing terms. Then we add.

EXAMPLE 5   Multiply: $(5x^3 + x - 4)(-2x^2 + 3x + 6)$.

SOLUTION

$$
\begin{array}{r}
5x^3 + x - 4 \\
-2x^2 + 3x + 6 \\
\hline
30x^3 \phantom{{}+{}} + 6x - 24 \\
15x^4 \phantom{{}+{}} + 3x^2 - 12x \\
-10x^5 \phantom{{}+{}} - 2x^3 + 8x^2 \\
\hline
-10x^5 + 15x^4 + 28x^3 + 11x^2 - 6x - 24
\end{array}
$$

Multiplying by 6
Multiplying by $3x$
Multiplying by $-2x^2$
Adding

EXAMPLE 6   Multiply $4x^4y - 7x^2y + 3y$ by $2y - 3x^2y$.

SOLUTION

$$
\begin{array}{r}
4x^4y - 7x^2y + 3y \\
-3x^2y + 2y \\
\hline
8x^4y^2 - 14x^2y^2 + 6y^2 \\
-12x^6y^2 + 21x^4y^2 - 9x^2y^2 \\
\hline
-12x^6y^2 + 29x^4y^2 - 23x^2y^2 + 6y^2
\end{array}
$$

Writing descending powers of x
Multiplying by $2y$
Multiplying by $-3x^2y$
Adding

The Product of Two Binomials: FOIL

We now consider what are called *special products*. These lead to faster ways to multiply in certain situations.

Let us find a faster special-product rule for the product of two binomials. Consider $(x + 7)(x + 4)$. We multiply each term of $(x + 7)$ by each term of $(x + 4)$.

$$(x + 7)(x + 4) = x \cdot x + x \cdot 4 + 7 \cdot x + 7 \cdot 4.$$

This multiplication illustrates a pattern that occurs whenever two binomials are multiplied:

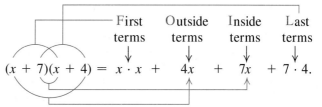

We use the mnemonic device FOIL to remember this method for multiplying.

A visualization of $(x + 7)(x + 4)$ using areas

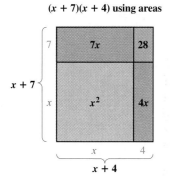

THE FOIL METHOD

To multiply two binomials $A + B$ and $C + D$, multiply the First terms AC, the Outside terms AD, the Inside terms BC, and then the Last terms BD. Then combine like terms, if possible.

$$(A + B)(C + D) = AC + AD + BC + BD$$

1. Multiply First terms: AC.
2. Multiply Outside terms: AD.
3. Multiply Inside terms: BC.
4. Multiply Last terms: BD.
$$\downarrow$$
FOIL

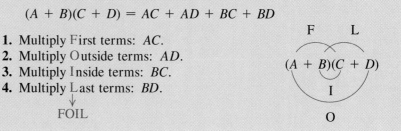

EXAMPLE 7

Multiply.

a) $(x + 5)(x - 8)$
b) $(2x + 3y)(x - 4y)$
c) $(5xy + 2x)(x^2 + 2xy^2)$

SOLUTION

$$\begin{array}{cccc} \text{F} & \text{O} & \text{I} & \text{L} \end{array}$$

a) $(x + 5)(x - 8) = x^2 - 8x + 5x - 40$
$ = x^2 - 3x - 40$ Combining like terms

b) $(2x + 3y)(x - 4y) = 2x^2 - 8xy + 3xy - 12y^2$ Using FOIL

$ = 2x^2 - 5xy - 12y^2$ Combining like terms

c) $(5xy + 2x)(x^2 + 2xy^2) = 5x^3y + 10x^2y^3 + 2x^3 + 4x^2y^2$

There are no like terms to combine.

Squares of Binomials

A visualization of
$(A + B)^2$ using areas

A fast method for squaring binomials can be developed using FOIL:

$$(A + B)^2 = (A + B)(A + B)$$
$$= A^2 + AB + AB + B^2$$
$$= A^2 + 2AB + B^2;$$

$$(A - B)^2 = (A - B)(A - B)$$
$$= A^2 - AB - AB + B^2$$
$$= A^2 - 2AB + B^2.$$

SQUARING A BINOMIAL

$$(A + B)^2 = A^2 + 2AB + B^2;$$
$$(A - B)^2 = A^2 - 2AB + B^2$$

The square of a binomial is the square of the first term, plus twice the product of the two terms, plus the square of the last term.

EXAMPLE 8

Multiply: **(a)** $(y - 5)^2$; **(b)** $(2x + 3y)^2$; **(c)** $\left(\frac{1}{2}x - 3y^4\right)^2$.

SOLUTION

$$(A - B)^2 = A^2 - 2 \cdot A \cdot B + B^2$$

a) $(y - 5)^2 = y^2 - 2 \cdot y \cdot 5 + 5^2$

$ = y^2 - 10y + 25$

It can be helpful to memorize the words of the rules and say them while you are calculating.

b) $(2x + 3y)^2 = (2x)^2 + 2 \cdot 2x \cdot 3y + (3y)^2$

$ = 4x^2 + 12xy + 9y^2$ Raising a product to a power

TECHNOLOGY CONNECTION

5.2A

c) $\left(\frac{1}{2}x - 3y^4\right)^2 = \left(\frac{1}{2}x\right)^2 - 2 \cdot \frac{1}{2}x \cdot 3y^4 + (3y^4)^2$

$\phantom{\left(\frac{1}{2}x - 3y^4\right)^2} = \frac{1}{4}x^2 - 3xy^4 + 9y^8$ Raising a product to a power; multiplying exponents

To verify that
$(x + 3)^2 \neq x^2 + 9$, let
$y_1 = (x + 3)^2$ and $y_2 = x^2 + 9$.
Then compare y_1 and y_2 using a table of values or a graph.

CAUTION! Note that $(3 + 5)^2 \neq 3^2 + 5^2$ (since $64 \neq 9 + 25$). More generally,

$$(A + B)^2 \neq A^2 + B^2 \quad \text{and} \quad (A - B)^2 \neq A^2 - B^2.$$

Products of Sums and Differences

Another pattern emerges when we are multiplying a sum and difference of the same two terms. Note the following:

$$
\begin{array}{cccc}
\text{F} & \text{O} & \text{I} & \text{L} \\
\downarrow & \downarrow & \downarrow & \downarrow
\end{array}
$$

$$(A + B)(A - B) = A^2 - AB + AB - B^2$$
$$= A^2 - B^2. \qquad {\scriptstyle -AB + AB = 0}$$

THE PRODUCT OF A SUM AND A DIFFERENCE

$(A + B)(A - B) = A^2 - B^2$ This is called a *difference of two squares.*

The product of the sum and difference of the same two terms is the square of the first term minus the square of the second term.

EXAMPLE 9   Multiply.

a) $(y + 5)(y - 5)$ **b)** $(2xy^2 + 3x)(2xy^2 - 3x)$

c) $(0.2t - 1.4m)(0.2t + 1.4m)$ **d)** $\left(\frac{2}{3}n - m^3\right)\left(\frac{2}{3}n + m^3\right)$

SOLUTION

$$
\begin{array}{cccccc}
(A & + & B)(A & - & B) & = A^2 - B^2 \\
\downarrow & & \downarrow \quad \downarrow & & \downarrow & \downarrow \quad \downarrow
\end{array}
$$

a) $(y + 5)(y - 5) = y^2 - 5^2$ Replacing A by y and B by 5
$$= y^2 - 25 \qquad {\scriptstyle \text{Try to do problems like this mentally.}}$$

b) $(2xy^2 + 3x)(2xy^2 - 3x) = (2xy^2)^2 - (3x)^2$
$$= 4x^2y^4 - 9x^2 \qquad {\scriptstyle \text{Raising a product to a power}}$$

c) $(0.2t - 1.4m)(0.2t + 1.4m) = (0.2t)^2 - (1.4m)^2$
$$= 0.04t^2 - 1.96m^2$$

d) $\left(\frac{2}{3}n - m^3\right)\left(\frac{2}{3}n + m^3\right) = \left(\frac{2}{3}n\right)^2 - (m^3)^2$
$$= \frac{4}{9}n^2 - m^6$$

EXAMPLE 10   Multiply.

a) $(5y + 4 + 3x)(5y + 4 - 3x)$

b) $(3xy^2 + 4y)(-3xy^2 + 4y)$

c) $(a - 5b)(a + 5b)(a^2 - 25b^2)$

SOLUTION

a) $(5y + 4 + 3x)(5y + 4 - 3x) = (5y + 4)^2 - (3x)^2$
$$= 25y^2 + 40y + 16 - 9x^2$$

<TECHNOLOGY CONNECTION 5.2B>

The check outlined in Technology Connection 5.1B can be used to check multiplication of polynomials in one variable. To check Example 3, using x in place of y, let
$y_1 = (x^3 - 5)(2x^3 + 4)$,
$y_2 = 2x^6 - 6x^3 - 20$, and
$y_3 = y_2 - y_1$. Since the multiplication is correct, $y_2 = y_1$ and $y_3 = 0$. Making use of a SPLIT mode—available on many graphers—you can view both the graph and a table of values for y_3. We do not graph y_1 or y_2.

X	Y3
−5	0
−4	0

X = −5

Had we found $y_3 \neq 0$, we would have known that a mistake had been made.

1. Use the procedure above to check Examples 4 and 5.

Note that $(5y + 4 + 3x)(5y + 4 - 3x)$ can be multiplied using columns, but not as quickly.

b) $(3xy^2 + 4y)(-3xy^2 + 4y) = (4y + 3xy^2)(4y - 3xy^2)$ Rewriting
$$= (4y)^2 - (3xy^2)^2$$
$$= 16y^2 - 9x^2y^4$$

c) $(a - 5b)(a + 5b)(a^2 - 25b^2) = (a^2 - 25b^2)(a^2 - 25b^2)$
$$= (a^2 - 25b^2)^2$$
$$= (a^2)^2 - 2(a^2)(25b^2) + (25b^2)^2$$

Squaring a binomial
$$= a^4 - 50a^2b^2 + 625b^4$$

Function Notation

Let's stop for a moment and look back at what we have done in this section. We have shown, for example, that
$$(x - 2)(x + 2) = x^2 - 4,$$
that is, $x^2 - 4$ and $(x - 2)(x + 2)$ are equivalent expressions.

From the viewpoint of functions, if
$$f(x) = x^2 - 4$$
and
$$g(x) = (x - 2)(x + 2),$$
then for any given input x, the outputs $f(x)$ and $g(x)$ are identical. Thus the graphs of these functions are identical and we say that f and g represent the same function. Functions like these are graphed in detail in Chapter 8.

x	$f(x)$	$g(x)$
3	5	5
2	0	0
1	−3	−3
0	−4	−4
−1	−3	−3
−2	0	0
−3	5	5

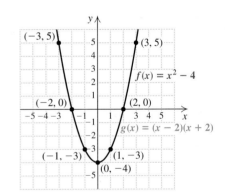

Our work with multiplying can be used when evaluating functions.

<EXAMPLE 11>

Given $f(x) = x^2 - 4x + 5$, find and simplify each of the following.

a) $f(a + 3)$
b) $f(a + h) - f(a)$

SOLUTION

a) To find $f(a + 3)$, we replace x by $a + 3$. Then we simplify:

$$f(a + 3) = (a + 3)^2 - 4(a + 3) + 5$$
$$= a^2 + 6a + 9 - 4a - 12 + 5$$
$$= a^2 + 2a + 2.$$

b) $f(a + h) - f(a) = [(a + h)^2 - 4(a + h) + 5] - [a^2 - 4a + 5]$
$$= a^2 + 2ah + h^2 - 4a - 4h + 5 - a^2 + 4a - 5$$
$$= 2ah + h^2 - 4h$$

EXERCISE SET
5.2

Multiply.

1. $3a^2 \cdot 7a$

2. $-5x^3 \cdot 2x$

3. $5x(-4x^2y)$

4. $-3ab^2(2a^2b^2)$

5. $(2x^3y^2)(-5x^2y^4)$

6. $(7a^2bc^4)(-8ab^3c^2)$

7. $8x(2 - x)$

8. $3a(a^2 - 4a)$

9. $5cd(3c^2d - 5cd^2)$

10. $a^2(2a^2 - 5a^3)$

11. $(2x + 5)(3x - 4)$

12. $(2a + 3b)(4a - b)$

13. $(m + 2n)(m - 3n)$

14. $(m - 5)(m + 5)$

15. $(y + 8x)(2y - 7x)$

16. $(x + y)(x - 2y)$

17. $(a^2 - 2b^2)(a^2 - 3b^2)$

18. $(2m^2 - n^2)(3m^2 - 5n^2)$

19. $(x - 4)(x^2 + 4x + 16)$

20. $(y + 3)(y^2 - 3y + 9)$

21. $(x + y)(x^2 - xy + y^2)$

22. $(a - b)(a^2 + ab + b^2)$

23. $(a^2 + a - 1)(a^2 + 4a - 5)$

24. $(x^2 - 2x + 1)(x^2 + x + 2)$

25. $(4a^2b - 2ab + 3b^2)(ab - 2b + a)$

26. $(2x^2 + y^2 - 2xy)(x^2 - 2y^2 - xy)$

27. $\left(x - \frac{1}{2}\right)\left(x - \frac{1}{4}\right)$

28. $\left(b - \frac{1}{3}\right)\left(b - \frac{1}{3}\right)$

29. $(1.2x - 3y)(2.5x + 5y)$

30. $(40a - 0.24b)(0.3a + 10b)$

31. Let $P(x) = 3x^2 - 5$ and $Q(x) = 4x^2 - 7x + 2$. Find $P(x) \cdot Q(x)$.

32. Let $P(x) = x^2 - x + 1$ and $Q(x) = x^3 + x^2 + x + 2$. Find $P(x) \cdot Q(x)$.

Multiply.

33. $(a + 8)(a + 5)$

34. $(x + 3)(x + 2)$

35. $(y - 4)(y + 3)$

36. $(y - 1)(y + 5)$

37. $(x + 3)^2$

38. $(y - 7)^2$

39. $(x - 2y)^2$

40. $(2s + 3t)^2$

41. $(2x + 9)(x + 2)$

42. $(3b + 2)(2b - 5)$

43. $(10a - 0.12b)^2$

44. $(10p^2 + 2.3q)^2$

45. $(2x - 3y)(2x + y)$

46. $(2a - 3b)(2a - b)$

47. $\left(2a + \frac{1}{3}\right)^2$

48. $\left(3c - \frac{1}{2}\right)^2$

49. $(2x^3 - 3y^2)^2$

50. $(3s^2 + 4t^3)^2$

51. $(a^2b^2 + 1)^2$

52. $(x^2y - xy^2)^2$

53. Let $P(x) = 4x - 1$. Find $P(x) \cdot P(x)$.

54. Let $Q(x) = 3x^2 + 1$. Find $Q(x) \cdot Q(x)$.

Multiply.

55. $(c + 2)(c - 2)$

56. $(x - 3)(x + 3)$

57. $(2a + 1)(2a - 1)$

58. $(3 - 2x)(3 + 2x)$

59. $(3m - 2n)(3m + 2n)$

60. $(3x + 5y)(3x - 5y)$

61. $(x^3 + yz)(x^3 - yz)$

62. $(2a^3 + 5ab)(2a^3 - 5ab)$

63. $(-mn + m^2)(mn + m^2)$

64. $(-3b + a^2)(3b + a^2)$

65. $(x + 1)(x - 1)(x^2 + 1)$

66. $(y - 2)(y + 2)(y^2 + 4)$

67. $(a - b)(a + b)(a^2 - b^2)$

68. $(2x - y)(2x + y)(4x^2 - y^2)$

69. $(a + b + 1)(a + b - 1)$

70. $(m + n + 2)(m + n - 2)$

71. $(2x + 3y + 4)(2x + 3y - 4)$

72. $(3a - 2b + c)(3a - 2b - c)$

73. *Compounding interest.*   Suppose that P dollars is invested in a savings account at interest rate i, compounded annually, for 2 yr. The amount A in the account after 2 yr is given by
$$A = P(1 + i)^2.$$
Find an equivalent expression for A.

74. *Compounding interest.*   Suppose that P dollars is invested in a savings account at interest rate i, compounded semiannually, for 1 yr. The amount A in the account after 1 yr is given by
$$A = P\left(1 + \frac{i}{2}\right)^2.$$
Find an equivalent expression for A.

75. Given $f(x) = 5x + x^2$, find and simplify.
a) $f(t - 1)$
b) $f(a + h) - f(a)$

76. Given $f(x) = 4 + 3x - x^2$, find and simplify.
a) $f(p + 1)$
b) $f(a + h) - f(a)$

SKILL MAINTENANCE

77. *Wages.*   Takako worked a total of 17 days last month at her father's restaurant. She earned $50 a day during the week and $60 a day during the weekend. Last month Takako earned $940. How many weekdays did she work?

78. *Geometry.*   The perimeter of a triangle is 174. The lengths of the three sides are consecutive even numbers. What are the lengths of the sides of the triangle?

79. *Manufacturing.*   In a factory, there are three machines A, B, and C. When all three are running, they produce 222 suitcases per day. If A and B work but C does not, they produce 159 suitcases per day. If B and C work but A does not, they produce 147 suitcases. What is the daily production of each machine?

80. *Value of coins.*   There are 50 dimes in a roll of dimes, 40 nickels in a roll of nickels, and 40 quarters in a roll of quarters. Kacie has 13 rolls of coins, which have a total value of $89. There are three more rolls of dimes than nickels. How many of each type of roll does she have?

SYNTHESIS

81. ◈ Find two binomials whose product is $x^2 - 25$ and explain how you decided on those two binomials.

82. ◈ Find two binomials whose product is $x^2 - 6x + 9$ and explain how you decided on those two binomials.

83. ◈ A student incorrectly claims that since $2x^2 \cdot 2x^2 = 4x^4$, it follows that $5x^5 \cdot 5x^5 = 25x^{25}$. What mistake is the student making?

84. ◈ We have seen that $(a - b)(a + b) = a^2 - b^2$. Explain how this result can be used to develop a fast way of multiplying $95 \cdot 105$.

Multiply. Assume that variables in exponents represent natural numbers.

85. $(ab^{3n})^{2n}$

86. $[(-x^a y^b)^4]^a$

87. $(z^{n^2})^{n^3}(z^{4n^3})^{n^2}$

88. $(a^x b^{2y})\left(\frac{1}{2}a^{3x}b\right)^2$

89. $(a^x b^y)^{w+z}$

90. $y^3 z^n(y^{3n}z^3 - 4yz^{2n})$

91. $[(a + b)(a - b)][5 - (a + b)][5 + (a + b)]$

92. $[x + y + 1][x^2 - x(y + 1) + (y + 1)^2]$

93. $(y - 1)^6(y + 1)^6$

94. $(a - b + c - d)(a + b + c + d)$

95. $\left(\frac{2}{3}x + \frac{1}{3}y + 1\right)\left(\frac{2}{3}x - \frac{1}{3}y - 1\right)$

96. $\left(x - \frac{1}{7}\right)\left(x^2 + \frac{1}{7}x + \frac{1}{49}\right)$

97. $(4x^2 + 2xy + y^2)(4x^2 - 2xy + y^2)$

98. $(x^2 - 7x + 12)(x^2 + 7x + 12)$

99. $(x^a + y^b)(x^a - y^b)(x^{2a} + y^{2b})$

100. $(x - 1)(x^2 + x + 1)(x^3 + 1)$

101. $(x^{a-b})^{a+b}$

102. $(M^{x+y})^{x+y}$

103. Draw rectangles similar to those on p. 261 to show that $(x + 2)(x + 5) = x^2 + 7x + 10$.

104. ⊿ Use a grapher to check your answers to Exercises 23, 35, and 65.

105. ⊿ Use a grapher to determine if each of the following is an identity.
a) $(x - 1)^2 = x^2 - 1$
b) $(x - 2)(x + 3) = x^2 + x - 6$
c) $(x - 1)^3 = x^3 - 3x^2 + 3x - 1$
d) $(x + 1)^4 = x^4 + 1$
e) $(x + 1)^4 = x^4 + 4x^3 + 8x^2 + 4x + 1$

COLLABORATIVE
C ◆ O ◆ R ◆ N ◆ E ◆ R

Focus: Polynomial multiplication

Time: 15–20 minutes

Group size: 2

Consider the following dialogue:

Jinny: Cal, let me do a number trick with you. Think of a number between 1 and 7. I'll have you perform some manipulations to this number, you'll tell me the result, and I'll tell you your number.

Cal: Okay. I've thought of a number.

Jinny: Good. Write it down so I can't see it, double it, and then subtract x from the result.

Cal: Hey, this is algebra!

Jinny: I know. Now square your binomial and subtract x^2.

Cal: How did you know I had an x^2? I *thought* this was rigged!

Jinny: It is. Now, divide by 4 and tell me either your constant term or your x-term. I'll tell you the other term and the number you chose.

Cal: Okay. The constant term is 16.

Jinny: Then the other term is $-4x$ and the number you chose was 4.

Cal: You're right! How did you do it?

ACTIVITY

1. Each group member should follow Jinny's instructions. Then determine how Jinny determined Cal's number and the other term.
2. Suppose that, at the end, Cal told Jinny the x-term. How would Jinny have determined Cal's number and the other term?
3. Would Jinny's "trick" work with *any* real number? Why do you think she specified numbers between 1 and 7?
4. Each group member should create a new number "trick" and perform it on the other group member. Be sure to include a variable so that both members can gain practice with polynomials.

5.3 Common Factors and Factoring by Grouping

Terms with Common Factors • Factoring by Grouping

Factoring is the reverse of multiplication. To **factor** an expression means to write an equivalent expression that is a product. Skill at factoring will assist us when working with polynomial functions and solving polynomial equations later in this chapter.

Terms with Common Factors

When factoring, we look for factors common to every term in an expression and then use the distributive law.

EXAMPLE 1

Factor out a common factor: $4y^2 - 8$.

SOLUTION

$$4y^2 - 8 = 4 \cdot y^2 - 4 \cdot 2 \qquad \text{Noting that 4 is a common factor}$$
$$= 4(y^2 - 2) \qquad \text{Using the distributive law}$$

In some cases, there is more than one common factor. In Example 2(a) below, for instance, 5 is a common factor, x^3 is a common factor, and $5x^3$ is a common factor. If there is more than one common factor, we factor out the *largest common factor,* that is, the factor that has the largest coefficient and the highest degree. In Example 2(a), the largest common factor is $5x^3$.

EXAMPLE 2

Factor out a common factor.

a) $5x^4 + 20x^3$ **b)** $12x^2y - 20x^3y$ **c)** $10p^6q^2 - 4p^5q^3 - 2p^4q^4$

SOLUTION

a) $5x^4 + 20x^3 = 5x^3(x + 4)$ Try to write your answer directly. Multiply mentally to check your answer.

b) $12x^2y - 20x^3y = 4x^2y(3 - 5x)$

    Check: $4x^2y \cdot 3 = 12x^2y$ and $4x^2y(-5x) = -20x^3y$,
        so $4x^2y(3 - 5x) = 12x^2y - 20x^3y$.

c) $10p^6q^2 - 4p^5q^3 - 2p^4q^4 = 2p^4q^2(5p^2 - 2pq - q^2)$

The check is left to the student.

The polynomials in Examples 1 and 2 cannot be factored further unless (in the cases of Examples 1 and 2c) irrational numbers or (in the case of Example 2b) fractions are used. In both examples, we have **factored completely** over the set of integers. The factors used are said to be **prime polynomials** over the set of integers.

When a factor contains more than one term, it is usually desirable for the leading coefficient to be positive. To achieve this may require factoring out a common factor with a negative coefficient.

EXAMPLE 3

Factor out a common factor with a negative coefficient.

a) $-4x - 24$ **b)** $-2x^3 + 6x^2 - 10x$

SOLUTION

a) $-4x - 24 = -4(x + 6)$

b) $-2x^3 + 6x^2 - 10x = -2x(x^2 - 3x + 5)$

EXAMPLE 4

Height of a Thrown Object. Suppose that a baseball is thrown upward with an initial velocity of 64 ft/sec. Its height in feet, $h(t)$, after t seconds is given by

$$h(t) = -16t^2 + 64t.$$

Find an equivalent expression for $h(t)$ by factoring out a common factor.

To check Example 4 with a table, let $y_1 = -16x^2 + 64x$ and $y_2 = -16x(x - 4)$. Then compare values of y_1 and y_2.

ΔTBL = 1

X	Y1	Y2
0	0	0
1	48	48
2	64	64
3	48	48
4	0	0
5	-80	-80
6	-192	-192

X = 0

1. How can $y_3 = y_2 - y_1$ and a table be used as a check?

$h(t) = -16t^2 + 64t$

SOLUTION   We factor out $-16t$ as follows:

$$h(t) = -16t^2 + 64t = -16t(t - 4).$$

Note that we can obtain function values using either expression for $h(t)$, since factoring forms equivalent expressions. For example,

$$h(1) = -16 \cdot 1^2 + 64 \cdot 1 = 48$$

and $h(1) = -16 \cdot 1(1 - 4) = 48.$ Using the factorization

In Example 4, we could have evaluated $-16t^2 + 64t$ and $-16t(t - 4)$ using any value for t. The results should always match. Thus a quick partial check of any factorization is to evaluate the factorization and the original polynomial for one or two convenient replacements. The check in Example 4 becomes foolproof if three replacements are used. In general, an nth-degree factorization is correct if it checks for $n + 1$ different replacements.

Factoring by Grouping

The largest common factor is sometimes a binomial.

EXAMPLE 5   Factor: $(a - b)(x + 5) + (a - b)(x - y^2)$.

SOLUTION   Here the largest common factor is the binomial $a - b$:

$$(a - b)(x + 5) + (a - b)(x - y^2) = (a - b)[(x + 5) + (x - y^2)]$$
$$= (a - b)[2x + 5 - y^2].$$

Often, in order to identify a common binomial factor, we must regroup into two groups of two terms each.

EXAMPLE 6   Factor: **(a)** $y^3 + 3y^2 + 4y + 12$; **(b)** $4x^3 - 15 + 20x^2 - 3x$.

SOLUTION

a) $y^3 + 3y^2 + 4y + 12 = (y^3 + 3y^2) + (4y + 12)$ Grouping

$\qquad\qquad\qquad\qquad\quad = y^2(y + 3) + 4(y + 3)$ Factoring out common factors

$\qquad\qquad\qquad\qquad\quad = (y + 3)(y^2 + 4)$ Factoring out $y + 3$

b) When we try grouping $4x^3 - 15 + 20x^2 - 3x$ as

$$(4x^3 - 15) + (20x^2 - 3x),$$

we are unable to factor $4x^3 - 15$. When this happens, we can rearrange the polynomial and try a different grouping:

$$4x^3 - 15 + 20x^2 - 3x = 4x^3 + 20x^2 - 3x - 15 \qquad \text{Using the commutative law to rearrange the terms}$$

$$= 4x^2(x + 5) - 3(x + 5)$$
$$= (x + 5)(4x^2 - 3).$$

In Section 1.3, we saw that the expressions $b - a$ and $-(a - b)$ or $-1(a - b)$ are equivalent. Remembering this can help if we wish to reverse a subtraction.

EXAMPLE 7

Factor: $ax - bx + by - ay$.

SOLUTION   We have

$$ax - bx + by - ay = (ax - bx) + (by - ay) \qquad \text{Grouping}$$

$$= x(a - b) + y(b - a) \qquad \text{Factoring each binomial}$$

$$= x(a - b) + y(-1)(a - b) \qquad \text{Factoring out } -1 \text{ to reverse } b - a$$

$$= x(a - b) - y(a - b) \qquad \text{Simplifying}$$

$$= (a - b)(x - y). \qquad \text{Factoring out } a - b$$

We can always check our factoring by multiplying:

Check: $(a - b)(x - y) = ax - ay - bx + by = ax - bx + by - ay.$

Some polynomials with four terms, like $x^3 + x^2 + 3x - 3$, are prime. Not only is there no common monomial factor, but no matter how we group terms, there is no common binomial factor:

$$x^3 + x^2 + 3x - 3 = x^2(x + 1) + 3(x - 1); \qquad \text{No common factor}$$
$$x^3 + 3x + x^2 - 3 = x(x^2 + 3) + (x^2 - 3); \qquad \text{No common factor}$$
$$x^3 - 3 + x^2 + 3x = (x^3 - 3) + x(x + 3). \qquad \text{No common factor}$$

EXERCISE SET
5.3

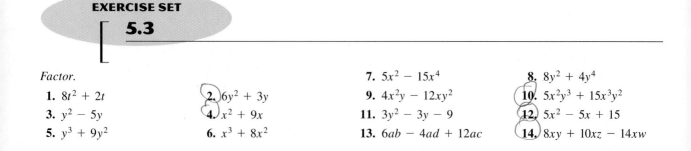

Factor.

1. $8t^2 + 2t$

2. $6y^2 + 3y$

3. $y^2 - 5y$

4. $x^2 + 9x$

5. $y^3 + 9y^2$

6. $x^3 + 8x^2$

7. $5x^2 - 15x^4$

8. $8y^2 + 4y^4$

9. $4x^2y - 12xy^2$

10. $5x^2y^3 + 15x^3y^2$

11. $3y^2 - 3y - 9$

12. $5x^2 - 5x + 15$

13. $6ab - 4ad + 12ac$

14. $8xy + 10xz - 14xw$

15. $9x^3y^6z^2 - 12x^4y^4z^4 + 15x^2y^5z^3$

16. $14a^4b^3c^5 + 21a^3b^5c^4 - 35a^4b^4c^3$

Factor out a factor with a negative coefficient.

17. $-5x + 15$

18. $-5x - 40$

19. $-6y - 72$

20. $-8t + 72$

21. $-2x^2 + 4x - 12$

22. $-2x^2 + 12x + 40$

23. $-3y^2 + 24x$

24. $-7x^2 - 56y$

25. $-3y^3 + 12y^2 - 15y$

26. $-4m^4 - 32m^3 + 64m$

27. $-x^2 + 5x - 9$

28. $-p^3 - 4p^2 + 11$

29. $-a^4 + 2a^3 - 13a$

30. $-m^3 - m^2 + m - 2$

31. *Height of a rocket.* A model rocket is launched upward with an initial velocity of 96 ft/sec. Its height in feet, $h(t)$, after t seconds is given by

$$h(t) = -16t^2 + 96t.$$

a) Find an equivalent expression for $h(t)$ by factoring out a common factor with a negative coefficient.

b) Check your factoring by evaluating both expressions for $h(t)$ at $t = 2$.

32. *Height of a baseball.* A baseball is popped up with an upward velocity of 72 ft/sec. Its height in feet, $h(t)$, after t seconds is given by

$$h(t) = -16t^2 + 72t.$$

a) Find an equivalent expression for $h(t)$ by factoring out a common factor with a negative coefficient.

b) Perform a partial check of part (a) by evaluating both expressions for $h(t)$ at $t = 2$.

33. *Counting spheres in a pile.* The number N of spheres in a triangular pile like the one shown here is a polynomial function given by

$$N(x) = \frac{1}{6}x^3 + \frac{1}{2}x^2 + \frac{1}{3}x,$$

where x is the number of layers and $N(x)$ is the number of spheres. Find an equivalent expression for $N(x)$ by factoring out a common factor.

34. *Number of games in a league.* If there are n teams in a league and each team plays every other team once, we can find the total number of games played by using the polynomial function $f(n) = \frac{1}{2}n^2 - \frac{1}{2}n$.

Find an equivalent expression for $f(n)$ by factoring out a common factor.

35. *Surface area of a silo.* A silo is a structure that is shaped like a right circular cylinder with a half sphere on top. The surface area of a silo of height h and radius r (including the area of the base) is given by the polynomial $2\pi rh + \pi r^2$. Find an equivalent expression by factoring out a common factor.

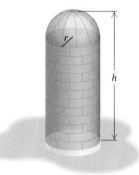

36. *Number of diagonals.* The number of diagonals of a polygon having n sides is given by the polynomial function

$$P(n) = \frac{1}{2}n^2 - \frac{3}{2}n.$$

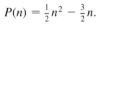

Find an equivalent expression for $P(n)$ by factoring out a common factor.

37. *Total revenue.* Household Sound is marketing a new kind of stereo. The firm determines that when it sells x stereos, the total revenue R is given by the polynomial function

$$R(x) = 280x - 0.4x^2 \text{ dollars.}$$

Find an equivalent expression for $R(x)$ by factoring out $0.4x$.

38. *Total cost.* Household Sound determines that the total cost C of producing x stereos is given by the polynomial function

$$C(x) = 0.18x + 0.6x^2.$$

Find an equivalent expression for $C(x)$ by factoring out $0.6x$.

Factor.

39. $a(b - 2) + c(b - 2)$ **40.** $a(x^2 - 3) - 2(x^2 - 3)$

41. $(x + 7)(x - 1) + (x + 7)(x - 2)$

42. $(a + 5)(a - 2) + (a + 5)(a + 1)$

43. $a^2(x - y) + 5(y - x)$

44. $3x^2(x - 6) + 2(6 - x)$

45. $ac + ad + bc + bd$ **46.** $xy + xz + wy + wz$

47. $b^3 - b^2 + 2b - 2$ **48.** $y^3 - y^2 + 3y - 3$

49. $a^3 - 3a^2 + 6 - 2a$ **50.** $t^3 + 6t^2 - 2t - 12$

51. $24x^3 - 36x^2 + 72x$ **52.** $12a^4 - 21a^3 - 9a^2$

53. $x^6 - x^5 - x^3 + x^4$ **54.** $y^4 - y^3 - y + y^2$

55. $2y^4 + 6y^2 + 5y^2 + 15$ **56.** $2xy - x^2y - 6 + 3x$

SKILL MAINTENANCE

Let $f(x) = 3x - 5$.

57. Draw the graph of f.   **58.** Find $f(2a + 1)$.

59. *Mixing rice.* Countryside Rice is 90% white rice and 10% wild rice. Mystic Rice is 50% wild rice. How much of each type should be used to create a 25-lb batch of rice that is 35% wild rice?

60. *Geometry.* The first angle of a triangle is 4 times the second angle. The third angle is 75°. What are the measures of the first two angles?

SYNTHESIS

61. ◆ Under what conditions would it be easier to evaluate a polynomial *after* it has been factored?

62. ◆ Explain in your own words why $-(a - b) = b - a$.

63. ◆ Is it true that if a polynomial's coefficients and exponents are all prime numbers, then the polynomial itself is prime? Why or why not?

64. ◆ Following Example 4, we stated that checking the factorization of a second-degree polynomial by making a single replacement is only a *partial* check. Write an *incorrect* factorization and explain how evaluating both the polynomial and the factorization might not catch the mistake.

Complete each of the following.

65. $x^5y^4 + \underline{\hspace{1cm}} = x^3y(\underline{\hspace{1cm}} + xy^5)$

66. $a^3b^7 - \underline{\hspace{1cm}} = \underline{\hspace{1cm}} (ab^4 - c^2)$

Factor.

67. $rx^2 - rx + 5r + sx^2 - sx + 5s$

68. $3a^2 + 6a + 30 + 7a^2b + 14ab + 70b$

69. $5x^2 - x^2y + 10x - 2xy + 15xz - 3xyz$

70. $a^4x^4 + a^4x^2 + 5a^4 + a^2x^4 + a^2x^2 + 5a^2 + 5x^4 + 5x^2 + 25$
(*Hint*: Use three groups of three.)

Factor. Assume that all exponents are natural numbers.

71. $2x^{3a} + 8x^a + 4x^{2a}$ **72.** $3a^{n+1} + 6a^n - 15a^{n+2}$

73. $4x^{a+b} + 7x^{a-b}$

74. $7y^{2a+b} - 5y^{a+b} + 3y^{a+2b}$

75. ▱ Use the TABLE feature of a grapher to check your answers to Exercises 17, 29, and 53.

76. ▱ Use a grapher to show that
$$(x^2 - 3x + 2)^4 = x^8 + 81x^4 + 16$$
is *not* an identity.

5.4   Factoring Trinomials

Factoring Trinomials of the Type $x^2 + bx + c$ •
Factoring Trinomials of the Type $ax^2 + bx + c$, $a \neq 1$

Our study of the factoring of trinomials begins with trinomials of the type $x^2 + bx + c$. We then move on to the type $ax^2 + bx + c$, where $a \neq 1$.

Factoring Trinomials of the Type $x^2 + bx + c$

When trying to factor trinomials of the type $x^2 + bx + c$, we can use a trial-and-error procedure.

Constant Term Positive

Recall the FOIL method of multiplying two binomials:

$$(x + 3)(x + 5) = x^2 + \underbrace{5x + 3x}_{} + 15$$

$$= x^2 + \quad 8x \quad + 15.$$

The product is a trinomial in which the leading coefficient is 1. To factor $x^2 + 8x + 15$, we think of FOIL: The first term, x^2, is the product of the First terms of two binomial factors, so the first term in each binomial will be x. The challenge is to find two numbers p and q such that

$$x^2 + 8x + 15 = (x + p)(x + q).$$

Note that the Outer and Inner products, qx and px, can be written as $(p + q)x$. The Last product, pq, will be a constant. Thus the numbers p and q must be selected so that their product is 15 and their sum is 8. In this case, we know from above that these numbers are 3 and 5. The factorization is

$$(x + 3)(x + 5), \quad \text{or} \quad (x + 5)(x + 3). \qquad \text{Using a commutative law}$$

In general,

$$(x + p)(x + q) = x^2 + (p + q)x + pq.$$

To factor $x^2 + (p + q)x + pq$, we use FOIL in reverse:

$$x^2 + (p + q)x + pq = (x + p)(x + q).$$

EXAMPLE 1

Factor: $x^2 + 9x + 8$.

SOLUTION We think of FOIL in reverse. The first term of each factor is x. We are looking for numbers p and q such that

$$x^2 + 9x + 8 = (x + p)(x + q) = x^2 + (p + q)x + pq.$$

Thus we look for factors of 8 whose sum is 9.

Pair of Factors	Sum of Factors
2, 4	6
1, 8	9

← The numbers we need are 1 and 8.

The factorization is thus $(x + 1)(x + 8)$. We can check by multiplying to see if the product is the original trinomial.

When factoring trinomials with a leading coefficient of 1, it suffices to consider all pairs of factors along with their sums, as we did above. At times, however, you may be tempted to form factors without calculating any sums. It is essential that you check any attempt made in this manner! For example, if we attempt the factorization

$$x^2 + 9x + 8 \stackrel{?}{=} (x + 2)(x + 4),$$

a check reveals that $(x + 2)(x + 4) = x^2 + 6x + 8 \neq x^2 + 9x + 8$. This type of trial-and-error procedure becomes easier to use with time. As you gain experience, you will find that many trials can be performed mentally.

When the constant term of a trinomial is positive, the constant terms in the binomial factors both have the same sign. This ensures a positive product. The sign used is that of the trinomial's middle term.

EXAMPLE 2   Factor: $y^2 - 9y + 20$.

SOLUTION   Since the constant term is positive and the coefficient of the middle term is negative, we look for a factorization of 20 in which both factors are negative. Their sum must be -9.

Pair of Factors	Sum of Factors	
$-1, -20$	-21	
$-2, -10$	-12	
$-4, \ -5$	$-9 \leftarrow$	—— The numbers we need are -4 and -5.

The factorization is $(y - 4)(y - 5)$.

Constant Term Negative

When the constant term of a trinomial is negative, we look for one negative factor and one positive factor. The sum of the factors must still be the coefficient of the middle term.

EXAMPLE 3   Factor: $x^3 - x^2 - 30x$.

SOLUTION   *Always* look first for a common factor! This time there is one, x. We factor it out:

$$x^3 - x^2 - 30x = x(x^2 - x - 30).$$

Now we consider $x^2 - x - 30$. We need a factorization of -30 in which one factor is positive, the other factor is negative, and the sum of the factors is -1. Since the sum is to be negative, the negative factor must be further from 0 than the positive factor is. Thus we need only consider pairs of factors in which the negative term has the larger absolute value.

Pair of Factors	Sum of Factors	
$1, -30$	-29	
$3, -10$	-7	
$5, \ -6$	$-1 \leftarrow$	—— The numbers we want are 5 and -6.

The factorization of $x^2 - x - 30$ is $(x + 5)(x - 6)$. *Don't forget to include the factor that was factored out earlier*! In this case, the factorization of the original trinomial is $x(x + 5)(x - 6)$.

EXAMPLE 4

Factor: $2x^2 + 34x - 220$.

SOLUTION *Always* look first for a common factor! This time we can factor out 2:

$$2x^2 + 34x - 220 = 2(x^2 + 17x - 110).$$

We next look for a factorization of -110 in which one factor is positive, the other factor is negative, and the sum of the factors is 17. Since the sum is to be positive, the positive factor must be further from 0 than the negative factor is. Thus we examine only pairs of factors in which the positive term has the larger absolute value.

Pair of Factors	Sum of Factors
$-1, \ 110$	109
$-2, \ \ 55$	53
$-5, \ \ 22$	$17 \leftarrow$

The numbers we need are -5 and 22.

The factorization of $x^2 + 17x - 110$ is $(x - 5)(x + 22)$. The factorization of the original trinomial, $2x^2 + 34x - 220$, is $2(x - 5)(x + 22)$.

Some polynomials are not factorable using integers.

EXAMPLE 5

Factor: $x^2 - x - 7$.

SOLUTION There are no factors of -7 whose sum is -1. This trinomial is *not* factorable into binomials with integer coefficients. The polynomial is *prime*.

TECHNOLOGY CONNECTION
5.4

The method described in Technology Connection 5.1B can be used to check Example 4: Let
$y_1 = 2x^2 + 34x - 220$,
$y_2 = 2(x - 5)(x + 22)$, and
$y_3 = y_2 - y_1$.

1. How should the graphs of y_1 and y_2 compare?
2. What should the graph of y_3 look like?
3. Check Example 3 with a grapher.
4. Show that $(2x + 5)(x - 3)$ is *not* a factorization of $2x^2 + x - 15$.

TIPS FOR FACTORING
$x^2 + bx + c$

1. If necessary, rewrite the trinomial in descending order. Search for factors of c that add up to b. Remember the following:

   - If c is positive, the signs of the factors are the same as the sign of b.
   - If c is negative, one factor is positive and the other is negative.
   - If the sum of the two factors is the opposite of b, changing the signs of both factors will give the desired factors whose sum is b.

2. Check the result by multiplying the binomials.

These tips still apply when a trinomial has more than one variable.

EXAMPLE 6

Factor: $x^2 - 2xy - 48y^2$.

SOLUTION   We look for numbers p and q such that

$$x^2 - 2xy - 48y^2 = (x + py)(x + qy).$$

Our thinking is much the same as if we were factoring $x^2 - 2x - 48$. We look for factors of -48 whose sum is -2. Those factors are 6 and -8. Thus,

$$x^2 - 2xy - 48y^2 = (x + 6y)(x - 8y).$$

We leave the check to the student.

Sometimes a trinomial like $x^6 + 2x^3 - 15$ can be factored if we first think of it as $(x^3)^2 + 2x^3 - 15$. To do this, make a substitution (perhaps just mentally) in which $u = x^3$. The trinomial then becomes

$$u^2 + 2u - 15.$$

We try factoring this trinomial and if a factorization is found, we replace all occurrences of u by x^3. Since

$$u^2 + 2u - 15 = (u - 3)(u + 5),$$

the original polynomial can be factored as

$$x^6 + 2x^3 - 15 = (x^3 - 3)(x^3 + 5).$$

Factoring Trinomials of the Type $ax^2 + bx + c$, $a \neq 1$

Now we look at trinomials in which the leading coefficient is not 1. We consider two methods. Use what works best for you or what your instructor chooses for you.

Method 1: The FOIL Method

We first consider the **FOIL method** for factoring trinomials of the type

$$ax^2 + bx + c, \quad \text{where } a \neq 1.$$

Consider the following multiplication.

$$
\begin{array}{c}
\quad\quad\quad \text{F} \quad\quad \text{O} \quad\ \text{I} \quad\ \text{L} \\
(3x + 2)(4x + 5) = 12x^2 + 15x + 8x + 10 \\
= 12x^2 + \quad 23x \quad + 10
\end{array}
$$

To factor $12x^2 + 23x + 10$, we must reverse what we just did. We look for two binomials whose product is this trinomial. The product of the First terms must be $12x^2$. The product of the Outside terms plus the product of the Inside terms must be $23x$. The product of the Last terms must be 10. We know from the preceding discussion that the answer is

$$(3x + 2)(4x + 5).$$

In general, however, finding such an answer involves trial and error. We use the following method.

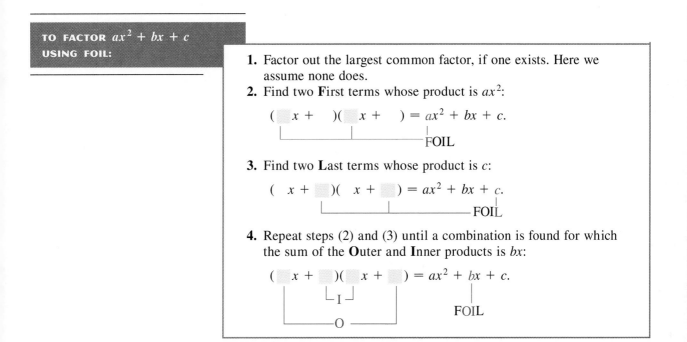

TO FACTOR $ax^2 + bx + c$
USING FOIL:

1. Factor out the largest common factor, if one exists. Here we assume none does.
2. Find two **F**irst terms whose product is ax^2:

$$(\quad x + \quad)(\quad x + \quad) = ax^2 + bx + c.$$
FOIL

3. Find two **L**ast terms whose product is c:

$$(\quad x + \quad)(\quad x + \quad) = ax^2 + bx + c.$$
FOIL

4. Repeat steps (2) and (3) until a combination is found for which the sum of the **O**uter and **I**nner products is bx:

$$(\quad x + \quad)(\quad x + \quad) = ax^2 + bx + c.$$
I
O
FOIL

EXAMPLE 7

Factor: $3x^2 + 10x - 8$.

SOLUTION

1. First, observe that there is no common factor (other than 1 or -1).
2. Next, factor the first term, $3x^2$. The only possibility for factors is $3x \cdot x$. Thus, if a factorization exists, it must be of the form

   $(3x + \quad)(x + \quad)$.

   We need to find the right numbers for the blanks.
3. Note that the constant term, -8, can be factored as $(-8)(1)$, $8(-1)$, $(-2)4$, and $2(-4)$, as well as $(1)(-8)$, $(-1)8$, $4(-2)$, and $(-4)2$.
4. Find a pair of factors for which the sum of the products (the "outside" and "inside" parts of FOIL) is the middle term, $10x$. Each possibility should be checked by multiplying:

   $(3x - 8)(x + 1) = 3x^2 - 5x - 8$.

   This gives a middle term with a negative coefficient. Since a positive coefficient is needed, a second possibility must be tried:

   $(3x + 8)(x - 1) = 3x^2 + 5x - 8$.

   Note that changing the signs of the two constant terms changes only the sign of the middle term. We try again:

$$(3x - 2)(x + 4) = 3x^2 + 10x - 8. \qquad \text{\small This is what we wanted.}$$

Thus the desired factorization is $(3x - 2)(x + 4)$. ————•

EXAMPLE 8

Factor: $6x^6 - 19x^5 + 10x^4$.

SOLUTION

1. First, factor out the common factor x^4:

$$x^4(6x^2 - 19x + 10).$$

2. Note that $6x^2 = 6x \cdot x$ and $6x^2 = 3x \cdot 2x$. Thus, $6x^2 - 19x + 10$ may factor into

$$(3x + \quad)(2x + \quad) \quad \text{or} \quad (6x + \quad)(x + \quad).$$

3. We factor the last term, 10. The possibilities are $10 \cdot 1$, $(-10)(-1)$, $5 \cdot 2$, and $(-5)(-2)$, as well as $1 \cdot 10$, $(-1)(-10)$, $2 \cdot 5$, and $(-2)(-5)$.

4. There are 8 possibilities for *each* factorization in step (2). We need factors for which the sum of the products (the "outer" and "inner" parts of FOIL) is the middle term, $-19x$. Since the x-coefficient is negative, we consider pairs of negative factors. Each possible factorization must be checked by multiplying:

$$(3x - 10)(2x - 1) = 6x^2 - 23x + 10.$$

We try again:

$$(3x - 5)(2x - 2) = 6x^2 - 16x + 10.$$

Actually this last attempt could have been rejected by simply noting that $2x - 2$ has a common factor, 2. Since the *largest* common factor was removed in step (1), no other common factors can exist. We try again, reversing the -5 and -2:

$$(3x - 2)(2x - 5) = 6x^2 - 19x + 10. \qquad \text{\small This is what we wanted.}$$

The factorization of $6x^2 - 19x + 10$ is $(3x - 2)(2x - 5)$. But do not forget the common factor! We must include it to get the complete factorization of the original trinomial:

$$6x^6 - 19x^5 + 10x^4 = x^4(3x - 2)(2x - 5).$$ ————•

TIPS FOR FACTORING WITH FOIL

1. If the largest common factor has been factored out of the original trinomial, then no binomial factor can have a common factor (other than 1 or -1).
2. If a and c are both positive, then the signs in the factors will be the same as the sign of b.
3. When a possible factoring produces the opposite of the desired middle term, reverse the signs of the constants in the factors.
4. Be systematic about your trials. Keep track of those possibilities that you have tried and those that you have not.

Keep in mind that this method of factoring involves trial and error. With practice, you will find yourself making fewer and better guesses.

Method 2: The Grouping Method

The second method for factoring trinomials of the type $ax^2 + bx + c, a \neq 1$, is known as the *grouping method*. It involves not only trial and error and FOIL but also factoring by grouping. We know that

$$x^2 + 7x + 10 = x^2 + 2x + 5x + 10$$
$$= x(x + 2) + 5(x + 2)$$
$$= (x + 2)(x + 5),$$

but what if the leading coefficient is not 1? Consider $6x^2 + 23x + 20$. The method is similar to what we just did with $x^2 + 7x + 10$, but we need two more steps.* First, multiply the leading coefficient, 6, and the constant, 20, to get 120. Then find a factorization of 120 in which the sum of the factors is the coefficient of the middle term: 23. The middle term is then split into a sum or difference using these factors.

(1) Multiply 6 and 20: $6 \cdot 20 = 120$.

$6x^2 + 23x + 20$

(2) Factor 120: $120 = 8 \cdot 15$, and $8 + 15 = 23$.

(3) Split the middle term: $23x = 8x + 15x$.

(4) Factor by grouping.

We factor by grouping as follows:

$$6x^2 + 23x + 20 = 6x^2 + 8x + 15x + 20$$
$$= 2x(3x + 4) + 5(3x + 4)$$ Factoring by
$$= (3x + 4)(2x + 5).$$ grouping

TO FACTOR $ax^2 + bx + c$ USING GROUPING:

1. Make sure that any common factors have been factored out.
2. Multiply the leading coefficient a and the constant c.
3. Try to factor the product ac so that the sum of the factors is b. That is, find integers p and q so that $pq = ac$ and $p + q = b$.
4. Split the middle term. That is, write bx as $px + qx$.
5. Factor by grouping.

EXAMPLE 9

Factor: $3x^2 + 10x - 8$.

SOLUTION

1. First, look for a common factor. There is none (other than 1 or -1).
2. Multiply the leading coefficient and the constant, 3 and -8:

$$3(-8) = -24.$$

---

*The rationale behind these steps is outlined in Exercise 103.

3. Try to factor -24 so that the sum of the factors is 10:

$$-24 = 12(-2) \quad \text{and} \quad 12 + (-2) = 10.$$

4. Split $10x$ using the results of step (3):

$$10x = 12x - 2x.$$

5. Finally, factor by grouping:

$$\begin{aligned} 3x^2 + 10x - 8 &= 3x^2 + 12x - 2x - 8 \\ &= 3x(x + 4) - 2(x + 4) \\ &= (x + 4)(3x - 2). \end{aligned}$$

$\left.\begin{aligned} \\ \\ \end{aligned}\right\}$ Factoring by grouping

EXERCISE SET
5.4

Factor.

1. $x^2 + 8x + 12$

2. $x^2 + 6x + 5$

3. $t^2 - 8t + 15$

4. $y^2 - 12y + 27$

5. $x^2 - 27 - 6x$

6. $t^2 - 15 - 2t$

7. $2n^2 - 20n + 50$

8. $2a^2 - 16a + 32$

9. $a^3 + a^2 - 72a$

10. $x^3 + 3x^2 - 54x$

11. $14x + x^2 + 45$

12. $12y + y^2 + 32$

13. $y^2 + 2y - 63$

14. $p^2 + 3p - 40$

15. $t^2 - 14t + 45$

16. $a^2 - 11a + 28$

17. $3x + x^2 - 10$

18. $x + x^2 - 6$

19. $3x^2 + 15x + 18$

20. $5y^2 + 40y + 35$

21. $56 + x - x^2$

22. $32 + 4y - y^2$

23. $32y + 4y^2 - y^3$

24. $56x + x^2 - x^3$

25. $x^4 + 11x^2 - 80$

26. $y^4 + 5y^2 - 84$

27. $x^2 + 12x + 13$

28. $x^2 - 3x + 7$

29. $p^2 - 5pq - 24q^2$

30. $x^2 + 12xy + 27y^2$

31. $y^2 + 8yz + 16z^2$

32. $x^2 - 14xy + 49y^2$

33. $p^4 + 80p^2 + 79$

34. $x^4 + 50x^2 + 49$

35. $x^8 - 7x^4 + 10$

36. $x^6 + 2x^3 - 63$

37. $6x^2 - 5x - 25$

38. $3x^2 - 16x - 12$

39. $10y^3 - 12y - 7y^2$

40. $6x^3 - 15x - x^2$

41. $24a^2 - 14a + 2$

42. $3a^2 - 10a + 8$

43. $35y^2 + 34y + 8$

44. $9a^2 + 18a + 8$

45. $4t + 10t^2 - 6$

46. $8x + 30x^2 - 6$

47. $8x^2 - 16 - 28x$

48. $18x^2 - 24 - 6x$

49. $a^6 - a^3 - 6$

50. $t^8 - 5t^4 + 6$

51. $14x^4 - 19x^3 - 3x^2$

52. $70x^4 - 68x^3 + 16x^2$

53. $12a^2 - 4a - 16$

54. $12a^2 - 14a - 20$

55. $9x^2 + 15x + 4$

56. $6y^2 - y - 2$

57. $8 - 6z - 9z^2$

58. $3 + 35a - 12a^2$

59. $-8t^2 - 8t + 30$

60. $-36a^2 + 21a - 3$

61. $18xy^3 + 3xy^2 - 10xy$

62. $3x^3y^2 - 5x^2y^2 - 2xy^2$

63. $24x^2 - 2 - 47x$

64. $15y^2 - 10 - 47y$

65. $63x^3 + 111x^2 + 36x$

66. $50y^3 + 115y^2 + 60y$

67. $24x^4 + 2x^2 - 15$

68. $40y^4 + 4y^2 - 12$

69. $12a^2 - 17ab + 6b^2$

70. $20p^2 - 23pq + 6q^2$

71. $2x^2 + xy - 6y^2$

72. $8m^2 - 6mn - 9n^2$

73. $6x^2 - 29xy + 28y^2$

74. $10p^2 + 7pq - 12q^2$

75. $9x^2 - 30xy + 25y^2$

76. $4p^2 + 12pq + 9q^2$

77. $9x^2y^2 + 5xy - 4$

78. $7a^2b^2 + 13ab + 6$

SKILL MAINTENANCE

79. *Height of a baseball.*   A baseball is thrown upward with an initial velocity of 80 ft/sec from a 224-ft-high cliff. Its height in feet, $h(t)$, after t seconds is given by

$$h(t) = -16t^2 + 80t + 224.$$

What is the height of the ball after 0 sec, 1 sec, 3 sec, 4 sec, and 6 sec?

80. Graph: $f(x) = -\frac{3}{4}x + 2$.

81. If $g(x) = -5x^2 - 7x$, find $g(-3)$.

82. *Height of a rocket.*   A model rocket is launched upward with an initial velocity of 96 ft/sec from a height of 880 ft. Its height in feet, $h(t)$, after t

seconds is given by

$$h(t) = -16t^2 + 96t + 880.$$

What is its height after 0 sec, 1 sec, 3 sec, 8 sec, and 10 sec?

SYNTHESIS

83. ◈ Explain how to conclude that $x^2 + 5x + 200$ is a prime polynomial without performing any trials.

84. ◈ Explain how to conclude that $x^2 - 59x + 6$ is a prime polynomial without performing any trials.

85. ◈ Describe in your own words an approach that can be used to factor any "nonprime" trinomial of the form $ax^2 + bx + c$.

86. ◈ Suppose $(rx + p)(sx - q) = ax^2 - bx + c$ is true. Explain how this can be used to factor $ax^2 + bx + c$.

Factor. Assume that variables in exponents represent positive integers.

87. $2a^4b^6 - 3a^2b^3 - 20$

88. $5x^8y^6 + 35x^4y^3 + 60$

89. $x^2 - \frac{4}{25} + \frac{3}{5}x$

90. $y^2 - \frac{8}{49} + \frac{2}{7}y$

91. $y^2 + 0.4y - 0.05$

92. $4x^{2a} - 4x^a - 3$

93. $x^{2a} + 5x^a - 24$

94. $x^2 + ax + bx + ab$

95. $bdx^2 + adx + bcx + ac$

96. $2ar^2 + 4asr + as^2 - asr$

97. $a^2p^{2a} + a^2p^a - 2a^2$

98. $(x + 3)^2 - 2(x + 3) - 35$

99. $6(x - 7)^2 + 13(x - 7) - 5$

100. Find all integers m for which $x^2 + mx + 75$ can be factored.

101. Find all integers q for which $x^2 + qx - 32$ can be factored.

102. One of the factors of $x^2 - 345x - 7300$ is $x + 20$. Find the other factor.

103. To better understand factoring $ax^2 + bx + c$ by grouping, suppose that $ax^2 + bx + c = (mx + r)(nx + s)$. Show that if $P = ms$ and $Q = rn$, then $P + Q = b$ and $PQ = ac$.

104. ▱ Use the TABLE feature to check your answers to Exercises 11, 65, and 89.

105. ▱ Let $y_1 = 3x^2 + 10x - 8$, $y_2 = (x + 4)(3x - 2)$, and $y_3 = y_2 - y_1$ to check Example 9 graphically.

106. ◈ Explain how the following graph of

$$y = x^2 + 3x - 2 - (x - 2)(x + 1)$$

can be used to show that

$$x^2 + 3x - 2 \neq (x - 2)(x + 1).$$

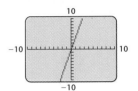

5.5   Factoring Perfect-Square Trinomials and Differences of Squares

Perfect-Square Trinomials   •   Differences of Squares   •
More Factoring by Grouping

We now introduce a faster way to factor trinomials that are squares of binomials. A faster way to factor differences of squares is also developed.

Perfect-Square Trinomials

Consider the trinomial

$$x^2 + 6x + 9.$$

To factor it, we can proceed as in Section 5.4 and look for factors of 9 that add to 6. These factors are 3 and 3 and the factorization is

$$x^2 + 6x + 9 = (x + 3)(x + 3) = (x + 3)^2.$$

Note that the result is the square of a binomial. Because of this, we call $x^2 + 6x + 9$ a **perfect-square trinomial.** Although trial and error can be used to factor any perfect-square trinomial, a faster procedure can be developed. In order to do so, we must first recognize when a trinomial is a perfect square.

TO RECOGNIZE A PERFECT-SQUARE TRINOMIAL:

- Two of the terms must be squares, such as A^2 and B^2.
- There must be no minus sign before A^2 or B^2.
- The remaining term is twice the product of A and B, $2AB$, or its opposite, $-2AB$.

EXAMPLE 1

Determine whether each polynomial is a perfect-square trinomial.

a) $x^2 + 10x + 25$
b) $4x + 16 + 3x^2$
c) $100y^2 + 81 - 180y$

SOLUTION

a) • Two of the terms in $x^2 + 10x + 25$ are squares: x^2 and 25.
 • There is no minus sign before either x^2 or 25.
 • The remaining term, $10x$, is twice the product of the square roots, x and 5.

Thus the trinomial is a perfect square.

b) In $4x + 16 + 3x^2$, only one term, 16, is a square ($3x^2$ is not a square because 3 is not a perfect square integer and $4x$ is not a square because x is not a square).

Thus the trinomial is not a perfect square.

c) It can help to first write the polynomial in descending order:

$$100y^2 - 180y + 81.$$

- Two of the terms, $100y^2$ and 81, are squares.
- There is no minus sign before either $100y^2$ or 81.
- If the product of the square roots, $10y$ and 9, is doubled, we get the opposite of the remaining term: $2(10y)(9) = 180y$ (the opposite of $-180y$).

Thus, $100y^2 + 81 - 180y$ is a perfect-square trinomial.

To factor a perfect-square trinomial, we use the same equations that we used for squaring a binomial. The pattern learned in Section 5.2 is reversed.

**FACTORING A
PERFECT-SQUARE TRINOMIAL**

$$A^2 + 2AB + B^2 = (A + B)^2;$$
$$A^2 - 2AB + B^2 = (A - B)^2$$

EXAMPLE 2   Factor.

a) $x^2 - 10x + 25$
b) $16y^2 + 49 + 56y$
c) $-20xy + 4y^2 + 25x^2$

SOLUTION

a) $x^2 - 10x + 25 = (x - 5)^2$ We find the square terms and write the
quantities that were squared with a minus sign
Note the sign!   between them.

b) $16y^2 + 49 + 56y = 16y^2 + 56y + 49$ Using a commutative law

$$= (4y + 7)^2$$ We find the square terms and
write the quantities that were
squared with a plus sign
between them.

c) $-20xy + 4y^2 + 25x^2 = 4y^2 - 20xy + 25x^2$ Writing descending
order with respect to y

$$= (2y - 5x)^2$$

This square can also be expressed as

$$25x^2 - 20xy + 4y^2 = (5x - 2y)^2.$$

As always, any factorization can be checked by multiplying:

$$(5x - 2y)^2 = (5x - 2y)(5x - 2y) = 25x^2 - 20xy + 4y^2.$$

When factoring, always look first for a factor common to all the terms.

EXAMPLE 3   Factor: **(a)** $2x^2 - 12xy + 18y^2$; **(b)** $-4y^2 - 144y^8 + 48y^5$.

SOLUTION

a) We first look for a common factor. This time, there is a common factor, 2.

$$2x^2 - 12xy + 18y^2 = 2(x^2 - 6xy + 9y^2)$$ Factoring out the 2
$$= 2(x - 3y)^2$$ Factoring the perfect-square trinomial

b) $-4y^2 - 144y^8 + 48y^5 = -4y^2(1 + 36y^6 - 12y^3)$ Factoring out the
common factor

$$= -4y^2(36y^6 - 12y^3 + 1)$$ Changing order.
Note that $(y^3)^2 = y^6$.

$$= -4y^2(6y^3 - 1)^2$$ Factoring the perfect-
square trinomial

Differences of Squares

When an expression like $x^2 - 9$ is recognized as a difference of two squares, we can reverse another pattern first seen in Section 5.2.

FACTORING A DIFFERENCE OF TWO SQUARES

$$A^2 - B^2 = (A + B)(A - B)$$

To factor a difference of two squares, write the product of the sum and the difference of the two quantities being squared.

EXAMPLE 4

Factor: **(a)** $x^2 - 9$; **(b)** $25y^6 - 49x^2$; **(c)** $a^2 - 7$.

SOLUTION

a) $x^2 - 9 = x^2 - 3^2 = (x + 3)(x - 3)$

$$A^2 \quad - \quad B^2 \ = \ (\ A \ + \ B\ )(\ A \ - \ B\ )$$
$$\downarrow \qquad \downarrow \qquad \quad \downarrow \quad \ \ \downarrow \quad \downarrow \quad \ \ \downarrow$$

b) $25y^6 - 49x^2 = (5y^3)^2 - (7x)^2 = (5y^3 + 7x)(5y^3 - 7x)$

c) Because $\sqrt{7}$ is irrational, $a^2 - 7$ cannot be factored over the set of rational numbers. It is a prime polynomial.

As always, the first step in factoring is to look for common factors.

EXAMPLE 5

Factor: **(a)** $5 - 5x^2y^6$; **(b)** $16x^4y - 81y$.

SOLUTION

a)
$$\begin{aligned}
5 - 5x^2y^6 &= 5(1 - x^2y^6) && \text{Factoring out the common factor} \\
&= 5[1^2 - (xy^3)^2] && \text{Rewriting } x^2y^6 \text{ as a quantity squared} \\
&= 5(1 + xy^3)(1 - xy^3) && \text{Factoring the difference of squares}
\end{aligned}$$

b)
$$\begin{aligned}
16x^4y - 81y &= y(16x^4 - 81) && \text{Factoring out the common factor} \\
&= y[(4x^2)^2 - 9^2] \\
&= y(4x^2 + 9)(4x^2 - 9) && \text{Factoring the difference of squares} \\
&= y(4x^2 + 9)(2x + 3)(2x - 3) && \text{Factoring } 4x^2 - 9, \text{ which is} \\
& && \textit{also} \text{ a difference of squares}
\end{aligned}$$

In Example 5(b), it is tempting to try to factor $(4x^2 + 9)$. Note that it is a sum of two squares. Apart from possibly removing a common factor, it is impossible to factor a sum of squares using real numbers. Note also in Example 5(b) that $4x^2 - 9$ *could* be factored further. Whenever a factor itself can be factored, try to do so. That way you will be factoring *completely*.

More Factoring by Grouping

Sometimes, when factoring a polynomial with four terms, we may be able to factor further.

EXAMPLE 6

Factor: $x^3 + 3x^2 - 4x - 12$.

SOLUTION

$$\begin{aligned}
x^3 + 3x^2 - 4x - 12 &= x^2(x + 3) - 4(x + 3) & \text{Factoring by grouping} \\
&= (x + 3)(x^2 - 4) & \text{Factoring out } x + 3 \\
&= (x + 3)(x + 2)(x - 2) & \text{Factoring } x^2 - 4
\end{aligned}$$

A difference of squares can have four or more terms. For example, one of the squares may be a trinomial. In this case, a type of grouping can be used.

EXAMPLE 7

Factor: **(a)** $x^2 + 6x + 9 - y^2$; **(b)** $a^2 - b^2 + 8b - 16$.

SOLUTION

a)
$$\begin{aligned}
x^2 + 6x + 9 - y^2 &= (x^2 + 6x + 9) - y^2 & \text{Grouping as a trinomial minus } y^2 \\
& & \text{to show a difference of squares} \\
&= (x + 3)^2 - y^2 \\
&= (x + 3 + y)(x + 3 - y)
\end{aligned}$$

b) Grouping $a^2 - b^2 + 8b - 16$ into two groups of two terms does not yield a common binomial factor, so we look for a trinomial square. In this case, the trinomial square is being subtracted from a^2:

$$\begin{aligned}
a^2 - b^2 + 8b - 16 &= a^2 - (b^2 - 8b + 16) & \text{Factoring out } -1 \text{ and} \\
& & \text{rewriting as subtraction} \\
&= a^2 - (b - 4)^2 & \text{Factoring the perfect-square trinomial} \\
&= (a + (b - 4))(a - (b - 4)) & \text{Factoring a differ-} \\
& & \text{ence of squares} \\
&= (a + b - 4)(a - b + 4) & \text{Removing} \\
& & \text{parentheses}
\end{aligned}$$

EXERCISE SET

5.5

Factor.

1. $x^2 + 8x + 16$

2. $t^2 + 6t + 9$

3. $a^2 - 16a + 64$

4. $a^2 - 14a + 49$

5. $2a^2 + 8a + 8$

6. $4a^2 - 16a + 16$

7. $y^2 + 36 - 12y$

8. $y^2 + 36 + 12y$

9. $24a^2 + a^3 + 144a$

10. $-18y^2 + y^3 + 81y$

11. $32x^2 + 48x + 18$

12. $2x^2 - 40x + 200$

13. $64 + 25y^2 - 80y$

14. $1 - 8d + 16d^2$

15. $a^4 - 10a^2 + 25$

16. $y^4 + 8y^2 + 16$

17. $0.25x^2 + 0.30x + 0.09$

18. $0.04x^2 - 0.28x + 0.49$

19. $p^2 - 2pq + q^2$

20. $m^2 + 2mn + n^2$

21. $25a^2 - 30ab + 9b^2$

22. $49p^2 - 84pq + 36q^2$

23. $t^8 + 2t^4s^4 + s^8$

24. $a^4 + 2a^2b^2 + b^4$

25. $x^2 - 16$

26. $y^2 - 100$

27. $p^2 - 49$

28. $m^2 - 64$

29. $a^2b^2 - 81$

30. $p^2q^2 - 25$

31. $6x^2 - 6y^2$

32. $8x^2 - 8y^2$

33. $7xy^4 - 7xz^4$

34. $25ab^4 - 25az^4$

35. $4a^3 - 49a$

36. $9x^4 - 25x^2$

37. $3x^8 - 3y^8$

38. $9a^4 - a^2b^2$

39. $9a^4 - 25a^2b^4$

40. $16x^6 - 121x^2y^4$

41. $\frac{1}{25} - x^2$

42. $\frac{1}{16} - y^2$

43. $(a + b)^2 - 9$

44. $(p + q)^2 - 25$

45. $x^2 - 6x + 9 - y^2$

46. $a^2 - 8a + 16 - b^2$

47. $m^2 - 2mn + n^2 - 25$

48. $x^2 + 2xy + y^2 - 9$

49. $36 - (x + y)^2$

50. $49 - (a + b)^2$

51. $r^2 - 2r + 1 - 4s^2$

52. $c^2 + 4cd + 4d^2 - 9p^2$

53. $16 - a^2 + 2ab - b^2$

54. $9 - x^2 + 2xy - y^2$

55. $m^3 - 7m^2 - 4m + 28$

56. $x^3 + 8x^2 - x - 8$

57. $a^3 - ab^2 - 2a^2 + 2b^2$

58. $p^2q - 25q + 3p^2 - 75$

SKILL MAINTENANCE

Solve.

59. $x - y + z = 6,$
$2x + y - z = 0,$
$x + 2y + z = 3$

60. $|5 - 7x| \geq 9$

61. $|5 - 7x| \leq 9$

62. $5 - 7x > -9 + 12x$

SYNTHESIS

63. ◈ Describe a procedure that could be used to find a polynomial with four terms that can be factored as a difference of two squares.

64. ◈ Under what conditions can a sum of two squares be factored?

65. ◈ Are the product and power rules for exponents (see Section 1.6) important when factoring differences of squares? Why or why not?

66. ◈ Without finding the entire factorization, determine the number of factors of $x^{256} - 1$. Explain how you arrived at your answer.

Factor. Assume that variables in exponents represent positive integers.

67. $-\frac{3}{4}p^2 + \frac{6}{5}p - \frac{12}{25}$

68. $-\frac{8}{27}r^2 - \frac{10}{9}rs - \frac{1}{6}s^2 + \frac{2}{3}rs$

69. $\frac{1}{36}x^8 + \frac{2}{9}x^4 + \frac{4}{9}$

70. $0.09x^8 + 0.48x^4 + 0.64$

71. $a^2 + 2ab + b^2 - c^2 + 6c - 9$

72. $r^2 - 8r - 25 - s^2 - 10s + 16$

73. $x^{2a} - y^2$

74. $x^{4a} - y^{2b}$

75. $4y^{4a} + 20y^{2a} + 20y^{2a} + 100$

76. $25y^{2a} - (x^{2b} - 2x^b + 1)$

77. $8(a - 3)^2 - 64(a - 3) + 128$

78. $3(x + 1)^2 + 12(x + 1) + 12$

79. $5c^{100} - 80d^{100}$

80. $9x^{2n} - 6x^n + 1$

81. $c^{2w+1} + 2c^{w+1} + c$

82. If $P(x) = x^2$, use factoring to simplify
$$P(a + h) - P(a).$$

83. If $P(x) = x^4$, use factoring to simplify
$$P(a + h) - P(a).$$

84. *Volume of carpeting.* The volume of a carpet that is rolled up can be estimated by the polynomial $\pi R^2 h - \pi r^2 h$.

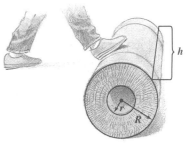

a) Factor the polynomial.

b) Use both the original and the factored forms to find the volume of a roll for which $R = 50$ cm, $r = 10$ cm, and $h = 4$ m. Use 3.14 for π.

85. ⚏ Use a grapher to check your answers to Exercises 1, 35, and 55 graphically by examining $y_1 =$ the original polynomial, $y_2 =$ the factored polynomial, and $y_3 = y_2 - y_1$.

86. ⚏ Check your answers to Exercises 1, 35, and 55 by using tables of values (see Exercise 85).

5.6 Factoring Sums or Differences of Cubes

Formulas for Factoring Sums or Differences of Cubes • Using the Formulas

Formulas for Factoring Sums or Differences of Cubes

We have seen that a difference of two squares can be factored but (unless a common factor exists) a *sum* of two squares cannot be factored. The situation is different with cubes: The difference *or sum* of two cubes can always be factored. To see this, consider the following products:

$$(A + B)(A^2 - AB + B^2) = A(A^2 - AB + B^2) + B(A^2 - AB + B^2)$$
$$= A^3 - A^2B + AB^2 + A^2B - AB^2 + B^3$$
$$= A^3 + B^3 \quad \text{Combining like terms}$$

and

$$(A - B)(A^2 + AB + B^2) = A(A^2 + AB + B^2) - B(A^2 + AB + B^2)$$
$$= A^3 + A^2B + AB^2 - A^2B - AB^2 - B^3$$
$$= A^3 - B^3. \quad \text{Combining like terms}$$

These equations (reversed) allow us to factor a sum or a difference of two cubes.

FACTORING A SUM OR DIFFERENCE OF TWO CUBES

$$A^3 + B^3 = (A + B)(A^2 - AB + B^2);$$
$$A^3 - B^3 = (A - B)(A^2 + AB + B^2)$$

Using the Formulas

When factoring a sum or difference of cubes, it can be helpful to remember that $2^3 = 8$, $3^3 = 27$, $4^3 = 64$, $5^3 = 125$, $6^3 = 216$, and so on.

EXAMPLE 1

Factor: $x^3 - 27$.

SOLUTION We have

$$x^3 - 27 = x^3 - 3^3.$$

In one set of parentheses, we write the first quantity that is cubed, x, a minus sign, and the second quantity that is cubed, 3:

$$(x - 3)(\quad\quad).$$

To get the other factor, we think of $x - 3$ and do the following:

———— Square the first term: x^2.
———— Multiply the terms and then change the sign: $3x$.
———— Square the second term: $(-3)^2$, or 9.

$$(x - 3)(x^2 + 3x + 9).$$

Thus, $x^3 - 27 = (x - 3)(x^2 + 3x + 9)$.

In Example 1, note that $x^2 + 3x + 9$ cannot be factored further. In general, unless A^3 and B^3 share a common factor or A and B are themselves powers, the trinomials $A^2 + AB + B^2$ and $A^2 - AB + B^2$ are prime.

EXAMPLE 2   Factor.

a) $125x^3 + y^3$ **b)** $m^6 + 64$

c) $128y^7 - 250x^6y$ **d)** $r^6 - s^6$

SOLUTION

a) We have

$$125x^3 + y^3 = (5x)^3 + y^3.$$

In one set of parentheses, we write the first quantity that is cubed, $5x$, a plus sign, and the second quantity that is cubed, y:

$$(5x + y)(\quad\quad).$$

To get the other factor, we think of $5x + y$ and do the following:

———— Square the first term: $(5x)^2$, or $25x^2$.
———— Multiply the terms and then change the sign: $-5xy$.
———— Square the second term: y^2.

$$(5x + y)(25x^2 - 5xy + y^2).$$

Thus, $125x^3 + y^3 = (5x + y)(25x^2 - 5xy + y^2)$.

b) We have

$$m^6 + 64 = (m^2)^3 + 4^3. \qquad \text{Rewriting as quantities cubed}$$

Next, we reuse the pattern used in part (a) above:

$$A^3 + B^3 = (A + B)(A^2 - A \cdot B + B^2)$$
$$(m^2)^3 + 4^3 = (m^2 + 4)((m^2)^2 - m^2 \cdot 4 + 4^2)$$
$$= (m^2 + 4)(m^4 - 4m^2 + 16).$$

c) We have

$$128y^7 - 250x^6y = 2y(64y^6 - 125x^6) \qquad \text{Remember: } \textit{Always} \text{ look for a common factor.}$$
$$= 2y[(4y^2)^3 - (5x^2)^3] \qquad \text{Rewriting as quantities cubed}$$

To factor $(4y^2)^3 - (5x^2)^3$, it is essential to remember the pattern used in Example 1:

$$A^3 \;-\; B^3 \;=\; (A\;-\;B)(\;A^2\;+\;A\cdot B\;+\;B^2\,)$$
$$(4y^2)^3 - (5x^2)^3 = (4y^2 - 5x^2)((4y^2)^2 + 4y^2\cdot 5x^2 + (5x^2)^2)$$
$$= (4y^2 - 5x^2)(16y^4 + 20x^2y^2 + 25x^4).$$

Thus,

$$128y^7 - 250x^6y = 2y(4y^2 - 5x^2)(16y^4 + 20x^2y^2 + 25x^4).$$

d) We have

$$r^6 - s^6 = (r^3)^2 - (s^3)^2$$
$$= (r^3 + s^3)(r^3 - s^3). \qquad \text{Factoring a difference of two } squares$$

Next, we factor the sum and difference of two cubes:

$$r^6 - s^6 = (r + s)(r^2 - rs + s^2)(r - s)(r^2 + rs + s^2).$$

In Example 2(d), suppose we first factored $r^6 - s^6$ as a difference of two cubes:

$$(r^2)^3 - (s^2)^3 = (r^2 - s^2)(r^4 + r^2s^2 + s^4)$$
$$= (r + s)(r - s)(r^4 + r^2s^2 + s^4).$$

In this case, we might have missed some factors; $r^4 + r^2s^2 + s^4$ can be factored as $(r^2 - rs + s^2)(r^2 + rs + s^2)$, but we probably would never have suspected that such a factorization exists.

Try to remember the following:

USEFUL FACTORING FACTS

Sum of cubes:	$A^3 + B^3 = (A + B)(A^2 - AB + B^2)$;
Difference of cubes:	$A^3 - B^3 = (A - B)(A^2 + AB + B^2)$;
Difference of squares:	$A^2 - B^2 = (A + B)(A - B)$;
Sum of squares:	$A^2 + B^2$ cannot be factored using real numbers if the largest common factor has been removed.

EXERCISE SET

5.6

Factor.

1. $t^3 - 8$

2. $x^3 + 64$

3. $x^3 + 27$

4. $z^3 - 1$

5. $m^3 - 64$

6. $x^3 - 27$

7. $8a^3 + 1$

8. $27x^3 + 1$

9. $8 - 27b^3$

10. $64 - 125x^3$

11. $8x^3 + 27$

12. $27y^3 + 64$

13. $y^3 - z^3$

14. $x^3 - y^3$

15. $x^3 + \frac{1}{27}$

16. $a^3 + \frac{1}{8}$

17. $2y^3 - 128$

18. $8t^3 - 8$

19. $8a^3 + 1000$

20. $54x^3 + 2$

21. $rs^3 + 64r$

22. $ab^3 + 125a$

23. $2y^3 - 54z^3$

24. $5x^3 - 40z^3$

25. $y^3 + 0.125$

26. $x^3 + 0.001$

27. $125c^6 - 8d^6$

28. $64x^6 - 8t^6$

29. $3z^5 - 3z^2$

30. $2y^4 - 128y$

31. $t^6 + 1$

32. $z^6 - 1$

33. $p^6 - q^6$

34. $t^6 + 64y^6$

35. $a^9 + b^{12}c^{15}$

36. $x^{12} - y^3z^{12}$

SKILL MAINTENANCE

37. The width of a rectangle is 7 ft less than its length. If the width is increased by 2 ft, the perimeter is then 66 ft. What is the area of the original rectangle?

38. If $f(x) = 7 - x^2$, find $f(-3)$.

39. Find the slope and the y-intercept of the line given by $4x - 3y = 8$.

Solve.

40. $|x| = 27$

41. $|5x - 6| \le 39$

42. $|5x - 6| > 39$

SYNTHESIS

43. ◈ Explain how the formula for factoring a *difference* of two cubes can be used to factor $x^3 + 8$.

44. ◈ How could you use factoring to convince someone that $x^3 + y^3 \ne (x + y)^3$?

45. ◈ Is the following statement true or false and why?

If $A^3 + B^3$ has a common factor, then $A + B$ has a common factor.

Factor.

46. $x^{6a} - y^{3b}$

47. $2x^{3a} + 16y^{3b}$

48. $(x + 5)^3 + (y - 5)^3$

49. $\frac{1}{16}x^{3a} + \frac{1}{2}y^{6a}z^{9b}$

50. $5x^3y^6 - \frac{5}{8}$

51. $x^3 - (x + y)^3$

52. $x^{6a} - (x^{2a} + 1)^3$

53. $(x^{2a} - 1)^3 - x^{6a}$

54. $t^4 - 8t^3 - t + 8$

55. If $P(x) = x^3$, use factoring to simplify
$$P(a + h) - P(a).$$

56. If $Q(x) = x^6$, use factoring to simplify
$$Q(a + h) - Q(a).$$

57. Using one set of axes, graph the following.
a) $f(x) = x^3$
b) $g(x) = x^3 - 8$
c) $h(x) = (x - 2)^3$

58. ◈ Explain how the geometric model below can be used to verify the formula for factoring $a^3 - b^3$.

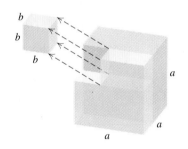

59. 〰 Use a grapher to check Example 1: Let $y_1 = x^3 - 27$, $y_2 = (x - 3)(x^2 + 3x + 9)$, and $y_3 = y_1 - y_2$.

60. 〰 Use the approach of Exercise 59 to check your answers to Exercises 7, 17, and 29.

5.7   Factoring: A General Strategy

Mixed Factoring Problems

Factoring is an important algebraic skill. Once you recognize the kind of expression you have, factoring can be done without too much difficulty.

A STRATEGY FOR FACTORING

A. Always factor out the largest common factor.

B. Look at the number of terms.

Two terms: Try factoring as a difference of squares first. Next, try factoring as a sum or a difference of cubes. Do *not* try to factor a *sum* of squares.

Three terms: Try factoring as a perfect-square trinomial. Next, try trial and error, using the FOIL method or the grouping method.

Four or more terms: Try factoring by grouping and factoring out a common binomial factor. Next, try grouping into a difference of squares, one of which is a trinomial.

C. Always *factor completely*. If a factor with more than one term can itself be factored further, do so.

EXAMPLE 1

Factor: $10a^2x - 40b^2x$.

SOLUTION

A. Look first for a common factor:

$$10a^2x - 40b^2x = 10x(a^2 - 4b^2). \qquad \text{Factoring out the largest common factor}$$

B. The factor $a^2 - 4b^2$ has two terms. It is a difference of squares. We factor it, keeping the common factor:

$$10a^2x - 40b^2x = 10x(a + 2b)(a - 2b).$$

C. Have we factored completely? Yes, because no factor with more than one term can be factored further.

EXAMPLE 2

Factor: $x^6 - 64$.

SOLUTION

A. Look for a common factor. There is none (other than 1 or -1).

B. There are two terms, a difference of squares: $(x^3)^2 - (8)^2$. We factor it:

$$x^6 - 64 = (x^3 + 8)(x^3 - 8).$$

C. One factor is a sum of two cubes, and the other factor is a difference of two cubes. We factor both:

$$x^6 - 64 = (x + 2)(x^2 - 2x + 4)(x - 2)(x^2 + 2x + 4).$$

The factorization is complete because no factor can be factored further.

EXAMPLE 3

Factor: $10x^6 + 40y^2$.

SOLUTION

A. Factor out the largest common factor: $10x^6 + 40y^2 = 10(x^6 + 4y^2)$.

B. In the parentheses, there are two terms, a sum of squares, which cannot be factored.

C. We cannot factor further. ——•

EXAMPLE 4

Factor: $2x^2 + 50a^2 - 20ax$.

SOLUTION

A. Factor out the largest common factor: $2(x^2 + 25a^2 - 10ax)$.

B. Next, we rearrange the trinomial in descending powers of x: $2(x^2 - 10ax + 25a^2)$. The trinomial is a perfect-square trinomial:

$$2x^2 + 50a^2 - 20ax = 2(x - 5a)^2.$$

C. No factor with more than one term can be factored further. ——•

EXAMPLE 5

Factor: $6x^2 - 20x - 16$.

SOLUTION

A. Factor out the largest common factor: $2(3x^2 - 10x - 8)$.

B. The trinomial factor is not a square. We factor using trial and error:

$$6x^2 - 20x - 16 = 2(x - 4)(3x + 2).$$

C. We cannot factor further. ——•

EXAMPLE 6

Factor: $3x + 12 + ax^2 + 4ax$.

SOLUTION

A. There is no common factor (other than 1 or -1).

B. There are four terms. We try grouping to find a common binomial factor:

$$3x + 12 + ax^2 + 4ax = 3(x + 4) + ax(x + 4) \qquad \text{Factoring two grouped binomials}$$

$$= (x + 4)(3 + ax). \qquad \text{Removing the common binomial factor}$$

C. We cannot factor further. ——•

EXAMPLE 7

Factor: $y^2 - 9a^2 + 12y + 36$.

SOLUTION

A. There is no common factor (other than 1 or -1).

B. There are four terms. We try grouping to remove a common binomial factor, but find none. Next, we try grouping as a difference of squares:

$$(y^2 + 12y + 36) - 9a^2 \qquad \text{Grouping}$$

$$= (y + 6)^2 - (3a)^2 \qquad \text{Rewriting as a difference of squares}$$

$$= (y + 6 + 3a)(y + 6 - 3a). \qquad \text{Factoring the difference of squares}$$

C. No factor with more than one term can be factored further. ——•

EXAMPLE 8   Factor: $x^3 - xy^2 + x^2y - y^3$.

SOLUTION

A. There is no common factor (other than 1 or -1).

B. There are four terms. We try grouping to remove a common binomial factor:

$$x^3 - xy^2 + x^2y - y^3$$
$$= x(x^2 - y^2) + y(x^2 - y^2) \qquad \text{Factoring two grouped binomials}$$
$$= (x^2 - y^2)(x + y). \qquad \text{Removing the common binomial factor}$$

C. The factor $x^2 - y^2$ can be factored further:

$$x^3 - xy^2 + x^2y - y^3 = (x + y)(x - y)(x + y), \text{ or } (x + y)^2(x - y).$$

No factor can be factored further, so we have factored completely.

EXERCISE SET
5.7

Factor completely.

1. $5m^4 - 20$

2. $x^2 - 144$

3. $a^2 - 81$

4. $2a^2 - 11a + 12$

5. $8x^2 - 18x - 5$

6. $2xy^2 - 50x$

7. $a^2 + 25 + 10a$

8. $p^2 + 64 + 16p$

9. $3x^2 + 15x - 252$

10. $2y^2 + 10y - 132$

11. $9x^2 - 25y^2$

12. $16a^2 - 81b^2$

13. $m^6 - 1$

14. $64t^6 - 1$

15. $x^2 + 6x - y^2 + 9$

16. $t^2 + 10t - p^2 + 25$

17. $343x^3 + 27y^3$

18. $128a^3 + 250b^3$

19. $8m^3 + m^6 - 20$

20. $-37x^2 + x^4 + 36$

21. $ac + cd - ab - bd$

22. $xw - yw + xz - yz$

23. $4c^2 - 4cd + d^2$

24. $70b^2 - 3ab - a^2$

25. $24 + 9t^2 + 8t + 3t^3$

26. $4a - 14 + 2a^3 - 7a^2$

27. $2x^3 + 6x^2 - 8x - 24$

28. $3x^3 + 6x^2 - 27x - 54$

29. $54a^3 - 16b^3$

30. $54x^3 - 250y^3$

31. $36y^2 - 35 + 12y$

32. $2b - 28a^2b + 10ab$

33. $a^8 - b^8$

34. $2x^4 - 32$

35. $a^3b - 16ab^3$

36. $x^3y - 25xy^3$

37. $(a - 3)(a + 7) + (a - 3)(a - 1)$

38. $x^2(x + 3) - 4(x + 3)$

39. $7a^4 - 14a^3 + 21a^2 - 7a$ **40.** $a^3 - ab^2 + a^2b - b^3$

41. $42ab + 27a^2b^2 + 8$

42. $-23xy + 20x^2y^2 + 6$

43. $p + 64p^4$

44. $125a + 8a^4$

45. $a^2 - b^2 - 6b - 9$

46. $m^2 - n^2 - 8n - 16$

SKILL MAINTENANCE

47. Graph: $f(x) = -\frac{3}{4}x + 6$.

48. *Exam scores.* There are 75 questions on a college entrance examination. Two points are awarded for each correct answer, and one half point is deducted for each incorrect answer. A score of 100 indicates how many correct and how many incorrect answers, assuming that all questions are answered?

49. *Perimeter.* A pentagon with all five sides the same size has the same perimeter as an octagon in which all eight sides are the same size. One side of the pentagon is 2 less than 3 times the length of one side of the octagon. Find the perimeters.

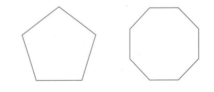

50. Solve: $|x - 7| < 3$.

S Y N T H E S I S

51. ◆ Emily has factored a polynomial as $(a - b)(x - y)$, while Jorge has factored the same polynomial as $(b - a)(y - x)$. Who is correct and why?

52. ◆ In your own words, outline a procedure that can be used to factor any polynomial.

53. ◆ Explain how one could construct a polynomial that is a difference of squares that contains a sum of two cubes and a difference of two cubes as factors.

54. ◆ Explain how one could construct a polynomial with four terms that can be factored by grouping three terms together.

Factor.

55. $60x^2 + 97xy^2 + 30y^4$

56. $28a^3 + 25a^2bc + 3ab^2c^2$

57. $-16 + 17(5 - y^2) - (5 - y^2)^2$

58. $(x - p)^2 - p^2$

59. $a^4 - 50a^2b^2 + 49b^4$

60. $(y - 1)^4 - (y - 1)^2$

61. $27x^{6s} + 64y^{3t}$

62. $x^6 - 2x^5 + x^4 - x^2 + 2x - 1$

63. $4x^2 + 4xy + y^2 - r^2 + 6rs - 9s^2$

64. $(1 - x)^3 - (x - 1)^6$

65. $24t^{2a} - 6$

66. $a^{2w+1} + 2a^{w+1} + a$

67. $\dfrac{x^{27}}{1000} - 1$

68. $a - by^8 + b - ay^8$

69. $3(x + 1)^2 - 9(x + 1) - 12$

70. $3a^2 + 3b^2 - 3c^2 - 3d^2 + 6ab - 6cd$

71. $3(a + 2)^2 + 30(a + 2) + 75$

72. $(m - 1)^3 - (m + 1)^3$

73. If $\left(x + \dfrac{2}{x}\right)^2 = 6$, find $x^3 + \dfrac{8}{x^3}$.

5.8 Applications of Polynomial Equations

The Principle of Zero Products • Polynomial Functions and Graphs • Problem Solving

---

Whenever two polynomials are set equal to each other, we have a **polynomial equation.** Some examples of polynomial equations are

$$4x^3 + x^2 + 5x = 6x - 3,$$
$$x^2 - x = 6,$$

and

$$3y^4 + 2y^2 + 2 = 0.$$

The *degree of a polynomial equation* is the same as the highest degree of any term in the equation. Thus, from top to bottom, the degree of each equation listed above is 3, 2, and 4. A second-degree polynomial equation in one variable is often called a **quadratic equation.** Of the equations listed above, only $x^2 - x = 6$ is a quadratic equation.

Polynomial equations, and quadratic equations in particular, occur frequently in applications, so the ability to solve them is an important skill. One way of solving certain polynomial equations involves factoring.

The Principle of Zero Products

When we multiply two or more numbers, the product is 0 if one of the factors is 0. Conversely, if a product is 0, then at least one of the factors must be 0. This property of 0 gives us a new principle for solving equations.

**THE PRINCIPLE OF
ZERO PRODUCTS**

For any real numbers a and b:

If $ab = 0$, then $a = 0$ or $b = 0$ (or both).

If $a = 0$ or $b = 0$, then $ab = 0$.

To solve an equation using the principle of zero products, we first write it in *standard form*: with 0 on one side of the equation and the leading coefficient positive.

EXAMPLE 1

Solve: $x^2 - x = 6$.

SOLUTION   To apply the principle of zero products, we need 0 on one side of the equation. Thus we add -6 on both sides:

$x^2 - x - 6 = 0.$ Getting 0 on one side

In order to express the polynomial as a product, we factor:

$(x - 3)(x + 2) = 0.$ Factoring

Since $(x - 3)(x + 2)$ is 0, the principle of zero products says that at least one factor is 0. Thus,

$x - 3 = 0$ *or* $x + 2 = 0.$ Using the principle of zero products

Each of these linear equations is then solved separately:

$x = 3$ *or* $x = -2.$

We check as follows:

Check:
$$\frac{x^2 - x = 6}{3^2 - 3 \; ? \; 6}$$
$$9 - 3$$
$$6 \mid 6 \text{ TRUE}$$

$$\frac{x^2 - x = 6}{(-2)^2 - (-2) \; ? \; 6}$$
$$4 + 2$$
$$6 \mid 6 \text{ TRUE}$$

Both 3 and -2 are solutions. The solution set is $\{3, -2\}$.

TO USE THE PRINCIPLE OF ZERO PRODUCTS:

1. Obtain a 0 on one side of the equation using the addition principle.
2. Factor the nonzero side of the equation.
3. Set each factor that is not a constant equal to 0.
4. Solve the resulting equations.

CAUTION! When using the principle of zero products, you must make sure that there is a 0 on one side of the equation. If neither side of the equation is 0, the procedure will not work.

For example, consider $x^2 - x = 6$ in Example 1 as

$$x(x - 1) = 6.$$

Suppose we reasoned as follows, setting factors equal to 6:

$$x = 6 \quad or \quad x - 1 = 6 \longleftarrow \text{This is wrong!}$$
$$x = 7.$$

Neither 6 nor 7 checks, as shown below:

$$\begin{array}{c|c} x^2 - x = 6 \\ \hline 6^2 - 6 \ ? \ 6 \\ 36 - 6 \\ \hspace{1em} 30 \ | \ 6 \ \text{FALSE} \end{array} \qquad \begin{array}{c|c} x^2 - x = 6 \\ \hline 7^2 - 7 \ ? \ 6 \\ 49 - 7 \\ \hspace{1em} 42 \ | \ 6 \ \text{FALSE} \end{array}$$

EXAMPLE 2

Solve.

a) $5b^2 = 10b$ b) $x^2 - 6x + 9 = 0$ c) $3x^3 - 30x = 9x^2$

SOLUTION

a) We have

$$5b^2 = 10b$$
$$5b^2 - 10b = 0 \qquad \text{Getting 0 on one side}$$
$$5b(b - 2) = 0 \qquad \text{Factoring}$$
$$5b = 0 \quad or \quad b - 2 = 0 \qquad \text{Using the principle of zero products}$$
$$b = 0 \quad or \quad \hspace{2em} b = 2. \qquad \text{We leave the checks to the student.}$$

The solutions are 0 and 2. The solution set is $\{0, 2\}$.

b) We have

$$x^2 - 6x + 9 = 0$$
$$(x - 3)(x - 3) = 0 \qquad \text{Factoring}$$
$$x - 3 = 0 \quad or \quad x - 3 = 0 \qquad \text{Using the principle of zero products}$$
$$x = 3 \quad or \quad \hspace{2em} x = 3. \qquad \textit{Check}: 3^2 - 6 \cdot 3 + 9 = 9 - 18 + 9 = 0.$$

There is only one solution, 3. The solution set is {3}.

c) We have

$$3x^3 - 30x = 9x^2$$

$$3x^3 - 9x^2 - 30x = 0 \quad \text{Getting 0 on one side and writing in descending order}$$

$$3x(x^2 - 3x - 10) = 0 \quad \text{Factoring out a common factor}$$

$$3x(x + 2)(x - 5) = 0 \quad \text{Factoring the trinomial}$$

$$3x = 0 \quad or \quad x + 2 = 0 \quad or \quad x - 5 = 0 \quad \text{Using the principle of zero products}$$

$$x = 0 \quad or \quad x = -2 \quad or \quad x = 5. \quad \text{The checks are left to the student.}$$

The solutions are 0, −2, and 5. The solution set is {0, −2, 5}.

EXAMPLE 3

Given that $f(x) = 3x^2 - 4x$, find all values of a for which $f(a) = 4$.

SOLUTION   We want all numbers a for which $f(a) = 4$. Since $f(a) = 3a^2 - 4a$, we must have

$$3a^2 - 4a = 4 \quad \text{Setting } f(a) \text{ equal to 4}$$

$$3a^2 - 4a - 4 = 0 \quad \text{Getting 0 on one side}$$

$$(3a + 2)(a - 2) = 0 \quad \text{Factoring}$$

$$3a + 2 = 0 \quad or \quad a - 2 = 0$$

$$a = -\tfrac{2}{3} \quad or \quad a = 2.$$

Check: $f\left(-\tfrac{2}{3}\right) = 3\left(-\tfrac{2}{3}\right)^2 - 4\left(-\tfrac{2}{3}\right) = 3 \cdot \tfrac{4}{9} + \tfrac{8}{3} = \tfrac{4}{3} + \tfrac{8}{3} = \tfrac{12}{3} = 4.$

$f(2) = 3(2)^2 - 4(2) = 3 \cdot 4 - 8 = 12 - 8 = 4.$

To have $f(a) = 4$, we must have $a = -\tfrac{2}{3}$ or $a = 2$.

EXAMPLE 4

Find the domain of F if $F(x) = \dfrac{x - 2}{x^2 + 2x - 15}$.

SOLUTION   The domain of F is the set of all values for which

$$\frac{x - 2}{x^2 + 2x - 15}$$

is a real number. Since division by 0 is undefined, $F(x)$ cannot be calculated for any x-value for which the denominator, $x^2 + 2x - 15$, is 0. To make sure these values are *excluded,* we solve:

$$x^2 + 2x - 15 = 0 \quad \text{Setting the denominator equal to 0}$$

$$(x - 3)(x + 5) = 0 \quad \text{Factoring}$$

$$x - 3 = 0 \quad or \quad x + 5 = 0$$

$$x = 3 \quad or \quad x = -5. \quad \text{These are the values to } excluded.$$

The domain of F is $\{x \mid x \text{ is a real number and } x \neq -5 \text{ and } x \neq 3\}$.

Polynomial Functions and Graphs

Let us return for a moment to the equation in Example 1, $x^2 - x = 6$. One way to begin solving this equation is to make a graph. We could either graph by hand or use computer graphics or a graphing calculator to draw the graph of the function $f(x) = x^2 - x$ and then check visually for an x-value that is paired with 6. See figure (a) at the top of the following page.

TECHNOLOGY CONNECTION

5.8

We have already used graphers to solve linear equations to any specified degree of accuracy (see p. 106). While some graphers can use a SOLVE, INTERSECT, or ROOT feature, virtually all can zoom and trace. To solve a quadratic equation like $x^2 - 1.83x - 6.64 = 0$ with ZOOM and TRACE, first graph $f(x) = x^2 - 1.83x - 6.64$:

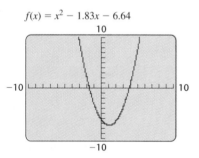

$f(x) = x^2 - 1.83x - 6.64$

The x-coordinate of each x-intercept solves the equation $x^2 - 1.83x - 6.64 = 0$. There appears to be an x-intercept between -2 and -1, and another between 3 and 4. Let's first examine the negative intercept. By adjusting the window or zooming repeatedly, we see that this intercept is very close to -1.82.

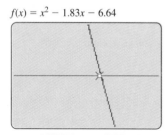

$f(x) = x^2 - 1.83x - 6.64$

$x = -1.821277, \ y = .00998461$

It appears that one zero of the function given by $f(x) = x^2 - 1.83x - 6.64$ is approximately -1.82. Note that $f(-1.82) = 0.003$, which is close enough to 0 to show that -1.82 approximates a zero, or root, of the function. Repeat the process of zooming in on the positive x-intercept to confirm that the other zero is about 3.65.

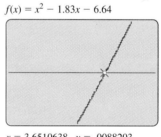

$f(x) = x^2 - 1.83x - 6.64$

$x = 3.6510638, \ y = .0088203$

Use a grapher to find any solutions that exist accurate to two decimal places.

1. $x^2 + 3x - 10 = 0$ (Check by factoring.)
2. $x^2 - 2x - 24 = 0$ (Check by factoring.)
3. $-3.55x^2 + 15.73x - 22.44 = 0$
4. $-0.65x^2 - 8.80x - 26.62 = 0$
5. $x^3 - 11.22x^2 + 15.94x + 37.49 = 0$
6. $2.75x^3 - 18.98x^2 + 13.68x + 72.77 = 0$

For an approach that avoids intercepts, graph a function representing each side of the following equations. The x-values of the intersections will be solutions.

7. $1.2x^2 + 3.25x = 4.3$
8. $-0.9x^2 + 4.13x = 1.87$

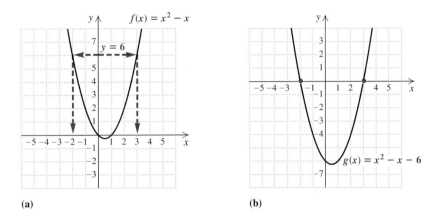

(a)   (b)

Equivalently, we could graph the function $g(x) = x^2 - x - 6$ and look for values of x for which $g(x) = 0$. See figure (b) above. Here you can visualize what we call the *roots,* or *zeros,* of a polynomial function.

It appears from the graph that $f(x) = 6$ and $g(x) = 0$ when $x \approx -2$ or $x \approx 3$. Although making a graph is not the fastest or most precise method of solving a polynomial equation, it gives us a visualization and is useful with problems that are more difficult to solve algebraically.

Problem Solving

Some problems can be translated to quadratic equations, which we can now solve. The problem-solving process is the same as for other kinds of problems.

EXAMPLE 5

Fireworks Displays.   Fireworks are typically launched from a mortar with an upward velocity (initial speed) of about 64 ft/sec. The height $h(t)$, in feet, of a "weeping willow" display, t seconds after having been launched from an 80-ft-high rooftop, is given by

$$h(t) = -16t^2 + 64t + 80.$$

After how long will the cardboard shell from the fireworks reach the ground?

SOLUTION

1. **Familiarize.** We draw a picture and label it, using the information provided (see the figure). If we wanted to, we could evaluate $h(t)$ for a few values of t. Note that t cannot be negative, since it represents time from launch.

2. **Translate.** The relevant function has been provided. Since we are asked to determine how long it will take for the shell to hit the ground, we are interested in the value of t for which $h(t) = 0$:

$$-16t^2 + 64t + 80 = 0.$$

3. Carry out.   We solve by factoring:

$$-16t^2 + 64t + 80 = 0$$
$$\left.\begin{array}{l}-16(t^2 - 4t - 5) = 0 \\ -16(t - 5)(t + 1) = 0\end{array}\right\} \quad \text{Factoring}$$
$$t - 5 = 0 \quad or \quad t + 1 = 0$$
$$t = 5 \quad or \qquad t = -1.$$

The solutions appear to be 5 and -1.

4. Check.   Since t cannot be negative, we need check only 5:

$$h(5) = -16 \cdot 5^2 + 64 \cdot 5 + 80 = 0.$$

The number 5 checks.

5. State.   The cardboard shell will hit the ground about 5 sec after the fireworks display has been launched.   ——●

The following problem involves the **Pythagorean theorem,** which relates the lengths of the sides of a right triangle. A **right triangle** has a 90°, or right, angle, which is denoted by the symbol ⌐ or ⌐. The longest side, opposite the 90° angle, is called the **hypotenuse.** The other sides, called **legs**, form the two sides of the right angle.

THE PYTHAGOREAN THEOREM

The sum of the squares of the legs of a right triangle is equal to the square of the hypotenuse:

$$a^2 + b^2 = c^2.$$

EXAMPLE 6

Carpentry.   In order to build a deck at a right angle to their house, Lucinda and Felipe decide to plant a stake in the ground a precise distance from the back wall of their house. This stake will combine with two marks on the house to form a right triangle. From a course in geometry, Lucinda remembers that there are three consecutive integers that can work as sides of a right triangle. Find the measurements of that triangle.

SOLUTION

1. Familiarize.   Recall that x, $x + 1$, and $x + 2$ can be used to represent three unknown consecutive integers. Since $x + 2$ is the largest number, it must represent the hypotenuse. The legs serve as the sides of the right angle, so one leg must be formed by the marks on the house. We make a drawing in which

$x =$ the distance between the marks on the house,

$x + 1 =$ the length of the other leg,

and

$x + 2 =$ the length of the hypotenuse.

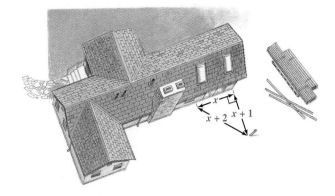

2. **Translate.** Applying the Pythagorean theorem, we translate as follows:

$$a^2 + b^2 = c^2$$
$$x^2 + (x + 1)^2 = (x + 2)^2.$$

3. **Carry out.** We solve the equation as follows:

$$x^2 + (x^2 + 2x + 1) = x^2 + 4x + 4 \qquad \text{Squaring the binomials}$$
$$2x^2 + 2x + 1 = x^2 + 4x + 4 \qquad \text{Combining like terms}$$
$$x^2 - 2x - 3 = 0 \qquad \text{Adding } -x^2 \text{ and } -4x \text{ and } -4 \text{ on both sides}$$
$$(x - 3)(x + 1) = 0 \qquad \text{Factoring}$$
$$x - 3 = 0 \quad or \quad x + 1 = 0 \qquad \text{Using the principle of zero products}$$
$$x = 3 \quad or \qquad x = -1.$$

4. **Check.** The integer -1 cannot be a length of a side because it is negative. When $x = 3$, $x + 1 = 4$, and $x + 2 = 5$. Since $3^2 + 4^2 = 5^2$, the lengths 3, 4, and 5 determine a right triangle. Thus, 3, 4, and 5 check.

5. **State.** Lucinda and Felipe should use a triangle with sides having a ratio of $3:4:5$. Thus, if the marks on the house are 3 ft apart, they should locate the stake at the point in the yard that is precisely 4 ft from one mark and 5 ft from the other mark.   ⟶●

EXAMPLE 7

Display of a Stamp. A valuable stamp is 4 cm wide and 5 cm long. The stamp is to be mounted on a sheet of paper that is $5\frac{1}{2}$ times the area of the stamp. Determine the dimensions of the paper that will ensure a uniform border around the stamp.

SOLUTION

1. **Familiarize.** We make a drawing and label it, using x to represent the width of the border, in centimeters. Since the border extends uniformly around the entire stamp, the length of the sheet of paper must be $5 + 2x$ and the width must be $4 + 2x$.

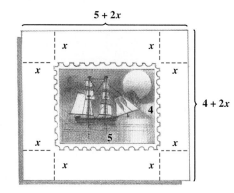

2. **Translate.** We rephrase the information given and translate as follows:

$$\underbrace{\text{Area of sheet}}_{(5 + 2x)(4 + 2x)} \quad \underbrace{\text{is}}_{=} \quad \underbrace{5\tfrac{1}{2} \text{ times}}_{5\tfrac{1}{2} \cdot} \quad \underbrace{\text{area of stamp.}}_{5 \cdot 4}$$

3. **Carry out.** We solve the equation:

$$(5 + 2x)(4 + 2x) = 5\tfrac{1}{2} \cdot 5 \cdot 4$$

$20 + 10x + 8x + 4x^2 = 110$	Multiplying
$4x^2 + 18x - 90 = 0$	Finding standard form
$2x^2 + 9x - 45 = 0$	Multiplying by $\frac{1}{2}$ on both sides
$(2x + 15)(x - 3) = 0$	Factoring
$2x + 15 = 0 \quad or \quad x - 3 = 0$	Principle of zero products
$x = -7\tfrac{1}{2} \quad or \qquad x = 3.$	

4. **Check.** We check 3 in the original problem. (Note that $-7\tfrac{1}{2}$ is not a solution because measurements cannot be negative.) If the border is 3 cm wide, the paper will have a length of $5 + 2 \cdot 3$, or 11 cm and a width of $4 + 2 \cdot 3$, or 10 cm. The area of the paper is thus $11 \cdot 10$, or 110 cm². Since the area of the stamp is 20 cm² and 110 cm² is $5\tfrac{1}{2}$ times 20 cm², the number 3 checks.

5. **State.** The sheet of paper should be 11 cm long and 10 cm wide.

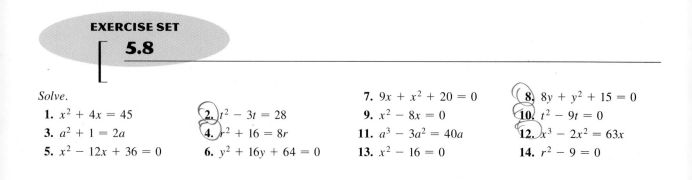

EXERCISE SET
5.8

Solve.

1. $x^2 + 4x = 45$
2. $t^2 - 3t = 28$
3. $a^2 + 1 = 2a$
4. $r^2 + 16 = 8r$
5. $x^2 - 12x + 36 = 0$
6. $y^2 + 16y + 64 = 0$
7. $9x + x^2 + 20 = 0$
8. $8y + y^2 + 15 = 0$
9. $x^2 - 8x = 0$
10. $t^2 - 9t = 0$
11. $a^3 - 3a^2 = 40a$
12. $x^3 - 2x^2 = 63x$
13. $x^2 - 16 = 0$
14. $r^2 - 9 = 0$

15. $(t - 6)(t + 6) = 45$

16. $(a - 4)(a + 4) = 20$

17. $3x^2 - 8x + 4 = 0$

18. $9x^2 - 15x + 4 = 0$

19. $4t^3 + 11t^2 + 6t = 0$

20. $8y^3 + 10y^2 + 3y = 0$

21. $(y - 3)(y + 2) = 14$

22. $(z + 4)(z - 2) = -5$

23. $x(5 + 12x) = 28$

24. $a(1 + 21a) = 10$

25. $a^2 - \frac{1}{64} = 0$

26. $x^2 - \frac{1}{25} = 0$

27. $t^4 - 26t^2 + 25 = 0$

28. $t^4 - 13t^2 + 36 = 0$

29. Let $f(x) = x^2 + 12x + 40$. Find a such that $f(a) = 8$.

30. Let $f(x) = x^2 + 14x + 50$. Find a such that $f(a) = 5$.

31. Let $g(x) = 2x^2 + 5x$. Find a such that $g(a) = 12$.

32. Let $g(x) = 2x^2 - 15x$. Find a such that $g(a) = -7$.

33. Let $h(x) = 12x + x^2$. Find a such that $h(a) = -27$.

34. Let $h(x) = 4x - x^2$. Find a such that $h(a) = -32$.

Find the domain of the function f given by each of the following.

35. $f(x) = \dfrac{3}{x^2 - 4x - 5}$

36. $f(x) = \dfrac{2}{x^2 - 7x + 6}$

37. $f(x) = \dfrac{x}{6x^2 - 54}$

38. $f(x) = \dfrac{2x}{5x^2 - 20}$

39. $f(x) = \dfrac{x - 5}{9x - 18x^2}$

40. $f(x) = \dfrac{1 + x}{3x - 15x^2}$

41. $f(x) = \dfrac{7}{5x^3 - 35x^2 + 50x}$

42. $f(x) = \dfrac{3}{2x^3 - 2x^2 - 12x}$

Solve.

43. The square of a number plus the number is 156. What is the number?

44. The square of a number plus the number is 132. What is the number?

45. A book is 5 cm longer than it is wide. Find the length and the width if the area is 84 cm².

46. An envelope is 4 cm longer than it is wide. The area is 96 cm². Find the length and the width.

47. *Geometry.* If each of the sides of a square is lengthened by 6 cm, the area becomes 144 cm². Find the length of a side of the original square.

48. *Geometry.* If each of the sides of a square is lengthened by 4 m, the area becomes 49 m². Find the length of a side of the original square.

49. *Framing a picture.* A picture frame measures 12 cm by 20 cm, and 84 cm² of picture shows. Find the width of the frame.

50. *Framing a picture.* A picture frame measures 14 cm by 20 cm, and 160 cm² of picture shows. Find the width of the frame.

51. *Landscaping.* A rectangular garden is 30 ft by 40 ft. Part of the garden is removed in order to install a walkway of uniform width around it. The area of the new garden is one-half the area of the old garden. How wide is the walkway?

52. *Landscaping.* A rectangular lawn measures 60 ft by 80 ft. Part of the lawn is torn up to install a sidewalk of uniform width around it. The area of the new lawn is 2400 ft². How wide is the sidewalk?

53. *Tent design.* The triangular entrance to a tent is 2 ft taller than it is wide. The area of the entrance is 12 ft². Find the height and the base.

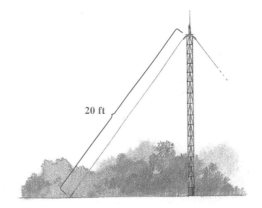

54. Three consecutive even integers are such that the square of the first plus the square of the third is 136. Find the three integers.

55. Three consecutive even integers are such that the square of the third is 76 more than the square of the second. Find the three integers.

56. *Antenna wires.* A wire is stretched from the ground to the top of an antenna tower, as shown. The wire is 20 ft long. The height of the tower is 4 ft greater than the distance d from the tower's base to the bottom of the wire. Find the distance d and the height of the tower.

57. *Parking lot design.*   A rectangular parking lot is 50 ft longer than it is wide. Determine the dimensions of the parking lot if it measures 250 ft diagonally.

58. *Sailing.*   A triangular sail is 9 m taller than it is wide. The area is 56 m². Find the height and the base of the sail.

59. *Ladder location.*   The foot of an extension ladder is 9 ft from a wall. The height that the ladder reaches on the wall and the length of the ladder are consecutive integers. How long is the ladder?

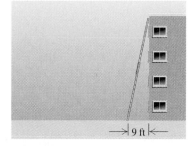

60. *Ladder location.*   The foot of an extension ladder is 10 ft from a wall. The ladder is 2 ft longer than the height that it reaches on the wall. How far up the wall does the ladder reach?

61. *Garden design.*   Ignacio is planning a garden that is 25 m longer than it is wide. The garden will have an area of 7500 m². What will its dimensions be?

62. *Garden design.*   A flower bed is to be 3 m longer than it is wide. The flower bed will have an area of 108 m². What will its dimensions be?

63. *Fireworks.*   Suppose that a bottle rocket is launched upward with an initial velocity of 96 ft/sec from a height of 880 ft. Its height in feet, $h(t)$, after t seconds (see top of next column) is given by

$$h(t) = -16t^2 + 96t + 880.$$

After how long will the rocket reach the ground?

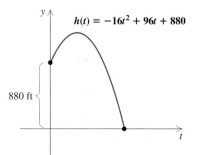

$$h(t) = -16t^2 + 96t + 880$$

880 ft

64. *Safety flares.*   Suppose that a flare is launched upward with an initial velocity of 80 ft/sec from a height of 224 ft. Its height in feet, $h(t)$, after t seconds is given by

$$h(t) = -16t^2 + 80t + 224.$$

After how long will the flare reach the ground?

65. *Camcorder production.*   Suppose that the cost of making x video cameras is $C(x) = \frac{1}{9}x^2 + 2x + 1$, where $C(x)$ is in thousands of dollars. If the revenue from the sale of x video cameras is given by $R(x) = \frac{5}{36}x^2 + 2x$, where $R(x)$ is in thousands of dollars, how many cameras must be sold in order for the firm to break even?

66. *Cabinet making.*   Dovetail Woodworking determines that the revenue R, in thousands of dollars, from the sale of x sets of cabinets is given by $R(x) = 2x^2 + x$. If the cost C, in thousands of dollars, of producing x sets of cabinets is given by $C(x) = x^2 - 2x + 10$, how many sets must be produced and sold in order for the company to break even?

SKILL MAINTENANCE

67. *Driving.*   At noon, two cars start from the same location going in opposite directions at different speeds. After 7 hr, they are 651 mi apart. If one car is traveling 15 mph slower than the other car, what are their respective speeds?

68. *Television sales.*   At the beginning of the month, J.C.'s Appliances had 150 televisions in stock. During the month, they sold 45% of their conventional televisions and 60% of their surround-sound televisions. If they sold a total of 78 televisions, how many of each type did they sell?

Solve.

69. $2x - 14 + 9x > -8x + 16 + 10x$

70. $x + y = 0,$
$z - y = -2,$
$x - z = 6$

SYNTHESIS

71. ◈ Suppose you are given a detailed graph of $y = p(x)$, where $p(x)$ is some polynomial in x. How could the graph be used to help solve the equation $p(x) = 0$?

72. ◈ Can the number of solutions of a quadratic equation exceed two? Why or why not?

73. ◈ Explain how one could write a quadratic equation that has -3 and 5 as solutions.

74. ◈ If the graph of $f(x) = ax^2 + bx + c$ has no x-intercepts, what can you conclude about the equation $ax^2 + bx + c = 0$?

Solve.

75. $(8x + 11)(12x^2 - 5x - 2) = 0$

76. $(x + 1)^3 = (x - 1)^3 + 26$

77. $(x - 2)^3 = x^3 - 2$

78. Use the following graph of $f(x) = x^2 - 2x - 3$ to solve $x^2 - 2x - 3 = 0$ and to solve $x^2 - 2x - 3 < 5$.

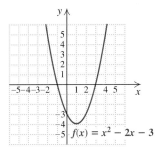

79. Use the following graph of $g(x) = -x^2 - 2x + 3$ to solve $-x^2 - 2x + 3 = 0$ and to solve $-x^2 - 2x + 3 \geqslant -5$.

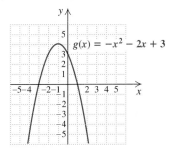

80. Find a polynomial function f for which $f(2) = 0$, $f(-1) = 0$, $f(3) = 0$, and $f(0) = 30$.

81. Find a polynomial function g for which $g(-3) = 0$, $g(1) = 0$, $g(5) = 0$, and $g(0) = 45$.

82. The sum of two numbers is 17, and the sum of their squares is 205. Find the numbers.

83. *Box construction.*   A rectangular piece of tin is twice as long as it is wide. Squares 2 cm on a side are cut out of each corner, and the ends are turned up to make a box whose volume is 480 cm³. What are the dimensions of the piece of tin?

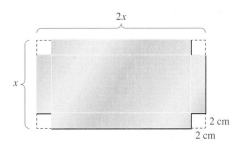

84. *Navigation.*   A tugboat and a freighter leave the same port at the same time at right angles. The freighter travels 7 km/h slower than the tugboat. After 4 hr, they are 68 km apart. Find the speed of each boat.

85. ⚓ *Skydiving.*   During the first 13 sec of a jump, a skydiver falls approximately $11.12t^2$ feet in t seconds. A small heavy object (with less wind resistance) falls about $15.4t^2$ feet in t seconds. Suppose that a skydiver jumps from 30,000 ft, and 1 sec later a camera falls out of the airplane. How long will it take the camera to catch up to the skydiver?

86. ⚓ Use the TABLE feature of a grapher to check that -5 and 3 are not in the domain of F as shown in Example 4.

87. ⚓ Use the TABLE feature of a grapher to check your answers to Exercises 37, 39, and 41.

⚓ *In Exercises 88–92, use a grapher to find any solutions that exist accurate to two decimal places.*

88. $x^2 - 2x - 8 = 0$ (Check by factoring.)

89. $x^2 + 3x - 4 = 0$ (Check by factoring.)

90. $-x^2 + 13.80x = 47.61$

91. $-x^2 + 3.63x + 34.34 = x^2$

92. $x^3 - 3.48x^2 + x = 3.48$

93. ◈ ⚓ Mary Louise is attempting to solve $x^3 + 20x^2 + 4x + 80 = 0$ with a grapher. Unfortunately, when she graphs $y_1 = x^3 + 20x^2 + 4x + 80$ in a standard $[-10, 10, -10, 10]$ window, she sees no graph at all, let alone any x-intercept. Can this problem be solved graphically? If so, how? If not, why not?

SUMMARY AND REVIEW
5

KEY TERMS

IMPORTANT PROPERTIES AND FORMULAS

Factoring Formulas

$A^2 + 2AB + B^2 = (A + B)^2;$

$A^2 - 2AB + B^2 = (A - B)^2;$

$A^2 - B^2 = (A + B)(A - B);$

$A^3 + B^3 = (A + B)(A^2 - AB + B^2);$

$A^3 - B^3 = (A - B)(A^2 + AB + B^2)$

To Factor $ax^2 + bx + c$ Using FOIL

1. Factor out the largest common factor, if one exists. Here we assume none does.
2. Find two **First** terms whose product is ax^2:

$$(\_\_x + \quad)(\_\_x + \quad) = ax^2 + bx + c.$$
FOIL

3. Find two **Last** terms whose product is c:

$$(\ x + \_\_)(\ x + \_\_) = ax^2 + bx + c.$$
FOIL

4. Repeat steps (2) and (3) until a combination is found for which the sum of the **O**uter and **I**nner products is bx:

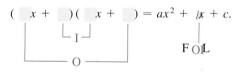

$$(\_\_x + \quad)(\_\_x + \quad) = ax^2 + bx + c.$$
FOIL

To Factor $ax^2 + bx + c$ Using Grouping

1. Make sure that any common factors have been factored out.
2. Multiply the leading coefficient a and the constant c.
3. Try to factor the product ac so that the sum of the factors is b. That is, find integers p and q so that $pq = ac$ and $p + q = b$.
4. Split the middle term. That is, write bx as $px + qx$.
5. Factor by grouping.

To Factor a Polynomial

A. Always factor out the largest common factor.
B. Look at the number of terms.

Two terms: Try factoring as a difference of squares first. Next, try factoring as a sum or a difference of cubes. Do *not* try to factor a *sum* of squares.

Three terms: Try factoring as a perfect-square trinomial. Next, try trial and error, using the FOIL method or the grouping method.

Four or more terms: Try factoring by grouping and factoring out a common binomial factor.

Next, try grouping into a difference of squares, one of which is a trinomial.

C. Always *factor completely*. If a factor with more than one term can itself be factored further, do so.

The Principle of Zero Products

For any real numbers a and b:

If $ab = 0$, then $a = 0$ or $b = 0$ (or both).
If $a = 0$ or $b = 0$, then $ab = 0$.

REVIEW EXERCISES

1. Given the polynomial
$$2xy^6 - 7x^8y^3 + 2x^3 - 3,$$
determine the degree of each term and the degree of the polynomial.

2. Given the polynomial
$$4x - 5x^3 + 2x^2 - 7,$$
arrange in descending order and determine the leading term and the leading coefficient.

3. Arrange in ascending powers of x:
$$3x^6y - 7x^8y^3 + 2x^3 - 3x^2.$$

4. Find $P(0)$ and $P(-1)$:
$$P(x) = x^3 - x^2 + 4x.$$

5. Evaluate the polynomial function for $x = -2$:
$$P(x) = 4 - 2x - x^2.$$

Combine like terms.

6. $4x^2y - 3xy^2 - 5x^2y + xy^2$

7. $3ab - 10 + 5ab^2 - 2ab + 7ab^2 + 14$

Add.

8. $(-6x^3 - 4x^2 + 3x + 1) + (5x^3 + 2x + 6x^2 + 1)$

9. $(3x^4 + 3x^3 - 8x + 9) + (-6x^4 + 4x + 7 + 3x)$

10. $(-9xy^2 - xy + 6x^2y) + (-5x^2y - xy + 4xy^2) + (12x^2y - 3xy^2 + 6xy)$

Subtract.

11. $(3x - 5) - (-6x + 2)$

12. $(4a - b + 3c) - (6a - 7b - 4c)$

13. $(8x^2 - 4xy + y^2) - (2x^2 + 3xy - 2y^2)$

Multiply.

14. $(3x^2y)(-6xy^3)$

15. $(x^4 - 2x^2 + 3)(x^4 + x^2 - 1)$

16. $(4ab + 3c)(2ab - c)$

17. $(2x + 5y)(2x - 5y)$

18. $(2x - 5y)^2$

19. $(x + 3)(2x - 1)$

20. $(x^2 + 4y^3)^2$

21. $(x - 5)(x^2 + 5x + 25)$

22. $\left(x - \frac{1}{3}\right)\left(x - \frac{1}{6}\right)$

Factor.

23. $6x^2 + 5x$

24. $9y^4 - 3y^2$

25. $15x^4 - 18x^3 + 21x^2 - 9x$

26. $a^2 - 12a + 27$

27. $3m^2 + 14m + 8$

28. $25x^2 + 20x + 4$

29. $4y^2 - 16$

30. $5x^2 + x^3 - 14x$

31. $ax + 2bx - ay - 2by$

32. $3y^3 + 6y^2 - 5y - 10$

33. $a^4 - 81$

34. $4x^4 + 4x^2 + 20$

35. $27x^3 - 8$

36. $0.064b^3 - 0.125c^3$

37. $y^5 + y$

38. $2z^8 - 16z^6$

39. $54x^6y - 2y$

40. $36x^2 - 120x + 100$

41. $6t^2 + 17pt + 5p^2$

42. $x^3 + 2x^2 - 9x - 18$

43. $a^2 - 2ab + b^2 - 4t^2$

Solve.

44. $x^2 - 20x = -100$

45. $6b^2 - 13b + 6 = 0$

46. $8y^2 = 14y$

47. $r^2 = 16$

48. Let $f(x) = x^2 - 7x - 40$. Find a such that $f(a) = 4$.

49. Find the domain of the function f given by

$$f(x) = \frac{x - 3}{3x^2 + 19x - 14}.$$

50. The area of a square is 5 more than 4 times the length of a side. What is the length of a side of the square?

51. The sum of the squares of three consecutive odd numbers is 83. Find the numbers.

52. A photograph is 3 in. longer than it is wide. When a 2-in. border is placed around the photograph, the total area of the photograph and the border is 108 in². Find the dimensions of the photograph.

SKILL MAINTENANCE

Solve.

53. $3x + 2y + z = 3,$
$2x - y + 2z = 16,$
$x + y - z = -9$

54. $-19x + 10 + 15x > 2x - 4 - 12x$

55. $|10 - 3x| \le 14$

56. Graph: $f(x) = \frac{2}{3}x + 2$.

57. There are 70 questions on a test. The questions are either multiple-choice, true–false, or fill-in. There are twice as many true–false as fill-in and 5 more multiple-choice than true–false. How many of each type of question are there on the test?

SYNTHESIS

58. ◆ Explain how to find the roots of a polynomial function from its graph.

59. ◆ Explain in your own words why there must be a 0 on one side of an equation before you can use the principle of zero products.

Factor.

60. $128x^6 - 2y^6$

61. $(x - 1)^3 - (x + 1)^3$

Multiply.

62. $[a - (b - 1)][(b - 1)^2 + a(b - 1) + a^2]$

63. $(z^{n^2})^{n^3}(z^{4n^3})^{n^2}$

64. Solve: $64x^3 = x$.

CHAPTER TEST
5

Given the polynomial $3xy^3 - 4x^2y + 5x^5y^4 - 2x^4y$.

1. Determine the degree of the polynomial.

2. Arrange in descending powers of x.

3. Determine the leading term of the polynomial $8a - 2 + a^2 - 4a^3$.

4. Given $P(x) = 2x^3 + 3x^2 - x + 4$, find $P(0)$ and $P(-2)$.

5. Given $P(x) = x^2 - 5x$, find and simplify
$$P(a + h) - P(a).$$

6. Combine like terms:
$$5xy - 2xy^2 - 2xy + 5xy^2.$$

Add.

7. $(-6x^3 + 3x^2 - 4y) + (3x^3 - 2y - 7y^2)$

8. $(5m^3 - 4m^2n - 6mn^2 - 3n^3) +$
$(9mn^2 - 4n^3 + 2m^3 + 6m^2n)$

Subtract.

9. $(9a - 4b) - (3a + 4b)$

10. $(6y^2 - 2y - 5y^3) - (4y^2 - 7y - 6y^3)$

Multiply.

11. $(-4x^2y)(-16xy^2)$

12. $(6a - 5b)(2a + b)$

13. $(x - y)(x^2 - xy - y^2)$

14. $(3m^2 + 4m - 2)(-m^2 - 3m + 5)$

15. $(4y - 9)^2$

16. $(x - 2y)(x + 2y)$

Factor.

17. $15x^2 - 5x^4$

18. $y^3 + 5y^2 - 4y - 20$

19. $p^2 - 12p - 28$

20. $12m^2 + 20m + 3$

21. $9y^2 - 25$

22. $3r^3 - 3$

23. $9x^2 + 25 - 30x$ **24.** $x^8 - y^8$

25. $y^2 + 8y + 16 - 100t^2$ **26.** $20a^2 - 5b^2$

27. $24x^2 - 46x + 10$ **28.** $16a^7b + 54ab^7$

29. $4y^4x + 36yx^2 + 8y^2x^3 - 16xy$

Solve.

30. $x^2 - 18 = 3x$ **31.** $5y^2 = 125$

32. $2x^2 + 21 = -17x$ **33.** $9x^2 + 3x = 0$

34. Let $f(x) = 3x^2 - 15x + 11$. Find a such that $f(a) = 11$.

35. Find the domain of the function f given by
$$f(x) = \frac{3 - x}{x^2 + 2x + 1}.$$

36. A photograph is 3 cm longer than it is wide. Its area is 40 cm². Find its length and its width.

Solve.

37. $|3x + 8| < 10$

38. $-3x + 4 - 5x > 8 - 9x - 3$

39. $2x - y + z = 9,$
$x - y + z = 4,$
$x + 2y - z = 5$

40. Find the slope and the y-intercept of the line given by
$$x + 3y = 8.$$

41. a) Multiply: $(x^2 + x + 1)(x^3 - x^2 + 1)$.
b) Factor: $x^5 + x + 1$.

42. Factor: $6x^{2n} - 7x^n - 20$.

CHAPTER 6

Rational Expressions, Equations, and Functions

Few people realize that there's a close relationship between music and math. The metric intervals between notes are based on math. Math is essential in drumming. When I am creating or teaching new rhythms, each measure and rhythmic pattern must be properly subdivided.

SONNY EMORY
Percussionist
Earth, Wind & Fire

AN APPLICATION

To duplicate a common African polyrhythm, a drummer needs to play sextuplets (6 beats per measure) on a tom-tom while simultaneously playing quarter notes (4 beats per measure) on a bass drum. Into how many equally sized parts must a measure be divided in order to precisely execute this rhythm?

THIS PROBLEM APPEARS AS EXERCISE 65 IN SECTION 6.2.

More information on drums is available at **http://hepg.awl.com/be/inter_5**

A rational expression is an expression that indicates division, as the fractional symbols in arithmetic do. In this chapter, we add, subtract, multiply, and divide rational expressions, and use them in equations and functions. We then use rational expressions to solve problems that we could not have solved before.

6.1 Rational Expressions and Functions: Multiplying and Dividing

Rational Functions • Multiplying •
Simplifying Rational Expressions • Dividing and Simplifying

An expression that consists of a polynomial divided by a nonzero polynomial is called a **rational expression.** The following are examples of rational expressions:

$$\frac{7}{8}, \quad \frac{a}{b}, \quad \frac{8}{y+5}, \quad \frac{x^2 + 7xy - 4}{x^3 - y^3}, \quad \frac{1 + z^3}{1 - z^6}.$$

Rational Functions

Like polynomials, certain rational expressions are used to describe functions. Such functions are called **rational functions**.

EXAMPLE 1

The function given by

$$T(t) = \frac{t^2 + 5t}{2t + 5}$$

gives the time required for two machines, working together, to complete a job that the first machine could do alone in t hours and the other machine could do in $t + 5$ hours. How long will the two machines, working together, require for the job if the first machine alone would take **(a)** 1 hour? **(b)** 5 hours?

SOLUTION

a) $T(1) = \dfrac{1^2 + 5 \cdot 1}{2 \cdot 1 + 5} = \dfrac{1 + 5}{2 + 5} = \dfrac{6}{7}$ hr

b) $T(5) = \dfrac{5^2 + 5 \cdot 5}{2 \cdot 5 + 5} = \dfrac{25 + 25}{10 + 5} = \dfrac{50}{15} = \dfrac{10}{3}$ hr

In Section 5.8, we found the domain of a rational function by determining when the denominator is 0. For Example 1 above, the denominator is 0 when t is $-\frac{5}{2}$, so the domain is $\left(-\infty, -\frac{5}{2}\right) \cup \left(-\frac{5}{2}, \infty\right)$.

Although graphing rational functions is beyond the scope of this course, it is educational to examine a computer-generated graph of the above function. Note that the graph consists of two unconnected "branches." Since $-\frac{5}{2}$ is not in the domain of T, if a vertical line were drawn at $-\frac{5}{2}$, it would never be touched by the graph of T.

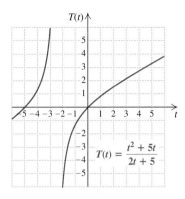

Multiplying

The calculations that are performed with rational expressions resemble those performed in arithmetic.

PRODUCTS OF RATIONAL EXPRESSIONS

To multiply two rational expressions, multiply numerators and multiply denominators:

$$\frac{A}{B} \cdot \frac{C}{D} = \frac{AC}{BD}, \quad \text{where } B \neq 0, D \neq 0.$$

EXAMPLE 2

Multiply: $\dfrac{x+1}{y-3} \cdot \dfrac{x^2}{y+1}$.

SOLUTION

$$\frac{x+1}{y-3} \cdot \frac{x^2}{y+1} = \frac{(x+1)x^2}{(y-3)(y+1)} \qquad \text{Multiplying the numerators and multiplying the denominators}$$

Recall from arithmetic that multiplication by 1 can be used to find equivalent expressions:

$$\frac{3}{5} = \frac{3}{5} \cdot \frac{2}{2} \qquad \text{Multiplying by } \tfrac{2}{2}, \text{ which is 1}$$

$$= \frac{6}{10}.$$

In a similar manner, multiplication by 1 can be used to find equivalent rational expressions in algebra:

$$\frac{x-5}{x+2} = \frac{x-5}{x+2} \cdot \frac{x+3}{x+3}$$

$$= \frac{(x-5)(x+3)}{(x+2)(x+3)}. \qquad \text{Multiplying by } \tfrac{x+3}{x+3}, \text{ which, provided } x \neq -3, \text{ is 1}$$

The expressions

$$\frac{x - 5}{x + 2} \quad \text{and} \quad \frac{(x - 5)(x + 3)}{(x + 2)(x + 3)}$$

are equivalent: So long as x is replaced by a number that can be used in either expression, both expressions will represent the same value.

EXAMPLE 3

Multiply to find an equivalent expression:

$$\frac{4 - x}{y - x} \cdot \frac{-1}{-1}.$$

SOLUTION   We have

$$\frac{4 - x}{y - x} \cdot \frac{-1}{-1} = \frac{-4 + x}{-y + x} = \frac{x - 4}{x - y}.$$

Multiplying by $-1/-1$ is the same as multiplying by 1, so

$$\frac{4 - x}{y - x} \quad \text{is equivalent to} \quad \frac{x - 4}{x - y}.$$

Simplifying Rational Expressions

As in arithmetic, rational expressions are *simplified* by "removing" a factor equal to 1. This reverses the process shown above:

$$\frac{6}{10} = \frac{3 \cdot 2}{5 \cdot 2} = \frac{3}{5} \cdot \frac{2}{2} = \frac{3}{5}. \qquad \text{We "removed" the factor that equals 1: } \frac{2}{2} = 1.$$

Similarly,

$$\frac{(x - 5)(x + 3)}{(x + 2)(x + 3)} = \frac{x - 5}{x + 2} \cdot \frac{x + 3}{x + 3} = \frac{x - 5}{x + 2}. \qquad \text{We "removed" the factor that}$$
$$\text{equals 1: } \frac{x + 3}{x + 3} = 1.$$

EXAMPLE 4

Simplify: $\dfrac{7a^2 + 21a}{14a}.$

SOLUTION   We first factor the numerator and the denominator, looking for the largest factor common to both. Once the greatest common factor is identified, we use it to write 1 and simplify:

$$\frac{7a^2 + 21a}{14a} = \frac{7a(a + 3)}{7 \cdot 2 \cdot a} \qquad \text{Factoring; the greatest common factor is } 7a.$$

$$= \frac{7a}{7a} \cdot \frac{a + 3}{2} \qquad \text{Factoring the rational expression}$$

$$= 1 \cdot \frac{a + 3}{2} \qquad \frac{7a}{7a} = 1$$

$$= \frac{a + 3}{2}. \qquad \text{Removing the factor 1}$$

A rational expression is said to be **simplified** when no factors equal to 1 can be removed. When simplifying, it is important to always make sure that this is the case. Suppose we removed 7/7 instead of $(7a)/(7a)$ in Example 4. We would then have

$$\frac{7a^2 + 21a}{14a} = \frac{7(a^2 + 3a)}{7 \cdot 2a} = \frac{a^2 + 3a}{2a}.$$ Removing a factor equal to 1: $\frac{7}{7} = 1$

Here, since 7 was not the *greatest* common factor, we would need to simplify further:

$$\frac{a^2 + 3a}{2a} = \frac{a(a + 3)}{a \cdot 2} = \frac{a + 3}{2}.$$ Removing another factor equal to 1: $a/a = 1$. The rational expression is now simplified.

EXAMPLE 5

Simplify: **(a)** $\dfrac{x^2 - 4}{2x^2 - 3x - 2}$; **(b)** $\dfrac{9x^2 + 6xy - 3y^2}{12x^2 - 12y^2}$.

SOLUTION

a) $\dfrac{x^2 - 4}{2x^2 - 3x - 2} = \dfrac{(x - 2)(x + 2)}{(2x + 1)(x - 2)}$ Factoring the numerator and the denominator

$$= \frac{x - 2}{x - 2} \cdot \frac{x + 2}{2x + 1}$$ Factoring the rational expression; $\frac{x-2}{x-2} = 1$

$$= \frac{x + 2}{2x + 1}$$ Removing a factor equal to 1

b) $\dfrac{9x^2 + 6xy - 3y^2}{12x^2 - 12y^2} = \dfrac{3(x + y)(3x - y)}{12(x + y)(x - y)}$ Factoring the numerator and the denominator

$$= \frac{3(x + y)}{3(x + y)} \cdot \frac{3x - y}{4(x - y)}$$ Factoring the rational expression

$$= \frac{3x - y}{4(x - y)}$$ Removing a factor equal to 1

For purposes of later work, we usually do not multiply out the numerator and the denominator. ➥

Canceling

Canceling is a shortcut that you may have used for removing a factor equal to 1 when working with fractional notation or rational expressions. With great concern, we mention it here as a possible way to speed up your work. Canceling *can* be done to remove factors equal to 1 in products. It *cannot* be done in sums or when adding expressions together. Our concern is that canceling be done with care and understanding. Example 5(b) might have been done faster as follows:

$$\frac{9x^2 + 6xy - 3y^2}{12x^2 - 12y^2} = \frac{3\cancel{(x + y)}(3x - y)}{\cancel{3} \cdot 4\cancel{(x + y)}(x - y)}$$ When a factor that equals 1 is noted, it is "canceled" as shown.

$$= \frac{3x - y}{4(x - y)}.$$ Removing a factor equal to 1: $\frac{3(x + y)}{3(x + y)} = 1$

TECHNOLOGY CONNECTION

6.1

To check that the simplified form of a rational expression (in one variable) is equivalent to the original expression, we can let y_1 = the original expression, y_2 = the simplified expression, and $y_3 = y_1 - y_2$ (or $y_2 - y_1$). If y_1 and y_2 are indeed equivalent, tables or the TRACE feature can be used to show that y_3 is 0 (unless y_1 or y_2 is undefined).

If you are uninterested in the separate graphs or tables of y_1 and y_2, you can check problems like Example 5(a) by simply entering

$$y_3 = (x^2 - 4)/(2x^2 - 3x - 2)$$
$$\quad - (x + 2)/(2x + 1).$$

1. Use a grapher to check Example 4. Be careful to use parentheses as needed.
2. Use a grapher to show that $(x + 3)/x \neq 3$. (See the Caution! box on p. 316.)

CAUTION!   Canceling is often performed incorrectly:

$$\frac{\cancel{x} + 3}{\cancel{x}} = 3, \qquad \frac{4x + 3}{\cancel{2}} = 2x + 3, \qquad \frac{\cancel{5}}{\cancel{5} + x} = \frac{1}{x}$$

     Wrong!   Wrong!   Wrong!

To check that these are not equivalent, substitute a number for x.

In each of these situations, the expressions canceled are *not* factors that equal 1. Factors are parts of products. For example, in $x \cdot 3$, x and 3 are factors, but in $x + 3$, x and 3 are *not* factors, but terms. If you can't factor, you can't cancel! If in doubt, don't cancel!

After multiplying, we usually simplify, if possible. That is one reason why we leave the numerator and the denominator in factored form. Even so, we might need to factor them further in order to simplify.

EXAMPLE 6

Multiply. Then simplify by removing a factor equal to 1.

a) $\dfrac{x + 2}{x - 3} \cdot \dfrac{x^2 - 4}{x^2 + x - 2}$ **b)** $\dfrac{1 - a^3}{a^2} \cdot \dfrac{a^5}{a^2 - 1}$

SOLUTION

a) $\dfrac{x + 2}{x - 3} \cdot \dfrac{x^2 - 4}{x^2 + x - 2} = \dfrac{(x + 2)(x^2 - 4)}{(x - 3)(x^2 + x - 2)}$ Multiplying the numerators and also the denominators

$= \dfrac{(x + 2)(x - 2)(x + 2)}{(x - 3)(x + 2)(x - 1)}$ Factoring the numerator and the denominator and identifying common factors

$= \dfrac{(x + 2)\cancel{(x + 2)}(x - 2)}{(x - 3)\cancel{(x + 2)}(x - 1)}$ Removing a factor equal to 1: $\dfrac{x + 2}{x + 2} = 1$

$= \dfrac{(x + 2)(x - 2)}{(x - 3)(x - 1)}$ Simplifying

b) $\dfrac{1 - a^3}{a^2} \cdot \dfrac{a^5}{a^2 - 1}$

$= \dfrac{(1 - a^3)a^5}{a^2(a^2 - 1)}$

$= \dfrac{(1 - a)(1 + a + a^2)a^5}{a^2(a - 1)(a + 1)}$ Factoring a difference of two cubes and a difference of two squares

$= \dfrac{-1(a - 1)(1 + a + a^2)a^5}{a^2(a - 1)(a + 1)}$ *Important!* Factoring out -1 reverses the subtraction.

$= \dfrac{\cancel{(a - 1)}\cancel{a^2} \cdot a^3(-1)(1 + a + a^2)}{\cancel{(a - 1)}\cancel{a^2}(a + 1)}$ Rewriting a^5 as $a^2 \cdot a^3$; removing a factor equal to 1: $\dfrac{(a - 1)a^2}{(a - 1)a^2} = 1$

$= \dfrac{-a^3(1 + a + a^2)}{a + 1}$ Simplifying

Dividing and Simplifying

Two expressions are reciprocals of each other if their product is 1. As in arithmetic, to find the reciprocal of a rational expression, we interchange numerator and denominator.

The reciprocal of $\dfrac{x}{x^2 + 3}$ is $\dfrac{x^2 + 3}{x}$.

The reciprocal of $y - 8$ is $\dfrac{1}{y - 8}$.

**QUOTIENTS OF
RATIONAL EXPRESSIONS**

For any rational expressions A/B and C/D, with $B, C, D \neq 0$,

$$\frac{A}{B} \div \frac{C}{D} = \frac{A}{B} \cdot \frac{D}{C}.$$

(To divide two rational expressions, multiply by the reciprocal of the divisor. We often say that we "*invert* and multiply.")

E X A M P L E   7

Divide. Simplify by removing a factor equal to 1 if possible.

a) $\dfrac{x - 2}{x + 1} \div \dfrac{x + 5}{x - 3}$

b) $\dfrac{a^2 - 1}{a - 1} \div \dfrac{a^2 - 2a + 1}{a + 1}$

S O L U T I O N

a) $\dfrac{x - 2}{x + 1} \div \dfrac{x + 5}{x - 3} = \dfrac{x - 2}{x + 1} \cdot \dfrac{x - 3}{x + 5}$ Multiplying by the reciprocal

$\qquad\qquad = \dfrac{(x - 2)(x - 3)}{(x + 1)(x + 5)}$ Multiplying the numerators and the denominators

b) $\dfrac{a^2 - 1}{a - 1} \div \dfrac{a^2 - 2a + 1}{a + 1} = \dfrac{a^2 - 1}{a - 1} \cdot \dfrac{a + 1}{a^2 - 2a + 1}$ Multiplying by the reciprocal

$\qquad\qquad = \dfrac{(a^2 - 1)(a + 1)}{(a - 1)(a^2 - 2a + 1)}$ Multiplying the numerators and the denominators

$\qquad\qquad = \dfrac{(a + 1)(a - 1)(a + 1)}{(a - 1)(a - 1)(a - 1)}$ Factoring the numerator and the denominator

$\qquad\qquad = \dfrac{(a + 1)(a - 1)(a + 1)}{(a - 1)(a - 1)(a - 1)}$ Removing a factor equal to 1:
$\dfrac{a - 1}{a - 1} = 1$

$\qquad\qquad = \dfrac{(a + 1)(a + 1)}{(a - 1)(a - 1)}$ Simplifying

> The procedures covered in this chapter are by their nature rather long. It may help to write out lots of steps as you do the problems. If you have difficulty, consider taking a clean sheet of paper and starting over. Don't squeeze your work into a small amount of space. When using lined paper, consider using two spaces at a time, with the paper's line locating the fraction bar.

EXERCISE SET
6.1

For each rational function, find the function values indicated, provided the value exists.

1. $v(t) = \dfrac{4t^2 - 5t + 2}{t + 3}$; $v(0), v(3), v(7)$

2. $s(x) = \dfrac{5x^2 + 4x - 12}{6 - x}$; $s(4), s(-1), s(3)$

3. $r(y) = \dfrac{3y^3 - 2y}{y - 5}$; $r(0), r(4), r(5)$

4. $f(r) = \dfrac{5r^2 - r^3}{r - 4}$; $f(2), f(5), f(-1)$

5. $g(x) = \dfrac{2x^3 - 9}{x^2 - 4x + 4}$; $g(0), g(2), g(-1)$

6. $r(t) = \dfrac{t^2 - 5t + 4}{t^2 - 9}$; $r(1), r(2), r(-3)$

Multiply to obtain equivalent expressions. Do not simplify. Assume that all denominators are nonzero.

7. $\dfrac{5x}{5x} \cdot \dfrac{x - 3}{x + 2}$

8. $\dfrac{3 - a^2}{a - 7} \cdot \dfrac{-1}{-1}$

9. $\dfrac{t - 2}{t + 3} \cdot \dfrac{-1}{-1}$

10. $\dfrac{x - 4}{x + 5} \cdot \dfrac{x - 5}{x - 5}$

Simplify by removing a factor equal to 1.

11. $\dfrac{15x}{5x^2}$

12. $\dfrac{7a^3}{21a}$

13. $\dfrac{18t^3}{27t^7}$

14. $\dfrac{8y^5}{4y^9}$

15. $\dfrac{2a - 10}{2}$

16. $\dfrac{3a + 12}{3}$

17. $\dfrac{15}{25a - 30}$

18. $\dfrac{21}{6x - 9}$

19. $\dfrac{3x - 12}{3x + 15}$

20. $\dfrac{4y - 20}{4y + 12}$

21. $\dfrac{5x + 20}{x^2 + 4x}$

22. $\dfrac{3x + 21}{x^2 + 7x}$

23. $\dfrac{3a - 1}{2 - 6a}$

24. $\dfrac{6 - 5a}{10a - 12}$

25. $\dfrac{8t - 16}{t^2 - 4}$

26. $\dfrac{t^2 - 9}{5t + 15}$

27. $\dfrac{2t - 1}{1 - 4t^2}$

28. $\dfrac{3a - 2}{4 - 9a^2}$

29. $\dfrac{12 - 6x}{5x - 10}$

30. $\dfrac{21 - 7x}{3x - 9}$

31. $\dfrac{a^2 - 25}{a^2 + 10a + 25}$

32. $\dfrac{a^2 - 16}{a^2 - 8a + 16}$

33. $\dfrac{x^2 + 9x + 8}{x^2 - 3x - 4}$

34. $\dfrac{t^2 - 8t - 9}{t^2 + 5t + 4}$

35. $\dfrac{16 - t^2}{t^2 - 8t + 16}$

36. $\dfrac{25 - p^2}{p^2 + 10p + 25}$

Multiply and simplify.

37. $\dfrac{5a^3}{3b} \cdot \dfrac{7b^3}{10a^7}$

38. $\dfrac{25a}{9b^8} \cdot \dfrac{3b^5}{5a^2}$

39. $\dfrac{3x - 6}{5x} \cdot \dfrac{x^3}{5x - 10}$

40. $\dfrac{5t^3}{4t - 8} \cdot \dfrac{6t - 12}{10t}$

41. $\dfrac{y^2 - 16}{2y + 6} \cdot \dfrac{y + 3}{y - 4}$

42. $\dfrac{m^2 - n^2}{4m + 4n} \cdot \dfrac{m + n}{m - n}$

43. $\dfrac{x^2 - 16}{x^2} \cdot \dfrac{x^2 - 4x}{x^2 - x - 12}$

44. $\dfrac{y^2 + 10y + 25}{y^2 - 9} \cdot \dfrac{y^2 + 3y}{y + 5}$

45. $\dfrac{7a - 14}{4 - a^2} \cdot \dfrac{5a^2 + 6a + 1}{35a + 7}$

46. $\dfrac{a^2 - 1}{2 - 5a} \cdot \dfrac{15a - 6}{a^2 + 5a - 6}$

47. $\dfrac{6 - 2t}{t^2 + 4t + 4} \cdot \dfrac{t^3 + 2t^2}{t^8 - 9t^6}$

48. $\dfrac{x^2 - 6x + 9}{12 - 4x} \cdot \dfrac{x^6 - 9x^4}{x^3 - 3x^2}$

49. $\dfrac{x^2 - 2x - 35}{2x^3 - 3x^2} \cdot \dfrac{4x^3 - 9x}{7x - 49}$

50. $\dfrac{y^2 - 10y + 9}{y^2 - 1} \cdot \dfrac{y + 4}{y^2 - 5y - 36}$

51. $\dfrac{c^3 + 8}{c^5 - 4c^3} \cdot \dfrac{c^6 - 4c^5 + 4c^4}{c^2 - 2c + 4}$

52. $\dfrac{x^3 - 27}{x^4 - 9x^2} \cdot \dfrac{x^5 - 6x^4 + 9x^3}{x^2 + 3x + 9}$

53. $\dfrac{a^3 - b^3}{3a^2 + 9ab + 6b^2} \cdot \dfrac{a^2 + 2ab + b^2}{a^2 - b^2}$

54. $\dfrac{x^3 + y^3}{x^2 + 2xy - 3y^2} \cdot \dfrac{x^2 - y^2}{3x^2 + 6xy + 3y^2}$

55. $\dfrac{4x^2 - 9y^2}{8x^3 - 27y^3} \cdot \dfrac{4x^2 + 6xy + 9y^2}{4x^2 + 12xy + 9y^2}$

56. $\dfrac{3x^2 - 3y^2}{27x^3 - 8y^3} \cdot \dfrac{6x^2 + 5xy - 6y^2}{6x^2 + 12xy + 6y^2}$

Divide and simplify.

57. $\dfrac{9x^5}{8y^2} \div \dfrac{3x}{16y^9}$

58. $\dfrac{16a^7}{3b^5} \div \dfrac{8a^3}{6b}$

59. $\dfrac{6x + 12}{x^8} \div \dfrac{x + 2}{x^3}$

60. $\dfrac{3y + 15}{y^7} \div \dfrac{y + 5}{y^2}$

61. $\dfrac{x^2 - 4}{x^3} \div \dfrac{x^5 - 2x^4}{x + 4}$

62. $\dfrac{y^2 - 9}{y^2} \div \dfrac{y^5 + 3y^4}{y + 2}$

63. $\dfrac{25x^2 - 4}{x^2 - 9} \div \dfrac{2 - 5x}{x + 3}$

64. $\dfrac{4a^2 - 1}{a^2 - 4} \div \dfrac{2a - 1}{2 - a}$

65. $\dfrac{5y - 5x}{15y^3} \div \dfrac{x^2 - y^2}{3x + 3y}$

66. $\dfrac{x^2 - y^2}{4x + 4y} \div \dfrac{3y - 3x}{12x^2}$

67. $\dfrac{x^2 - 16}{x^2 - 10x + 25} \div \dfrac{3x - 12}{x^2 - 3x - 10}$

68. $\dfrac{y^2 - 36}{y^2 - 8y + 16} \div \dfrac{3y - 18}{y^2 - y - 12}$

69. $\dfrac{y^3 + 3y}{y^2 - 9} \div \dfrac{y^2 + 5y - 14}{y^2 + 4y - 21}$

70. $\dfrac{a^3 + 4a}{a^2 - 16} \div \dfrac{a^2 + 8a + 15}{a^2 + a - 20}$

71. $\dfrac{x^3 - 64}{x^3 + 64} \div \dfrac{x^2 - 16}{x^2 - 4x + 16}$

72. $\dfrac{8y^3 - 27}{64y^3 - 1} \div \dfrac{4y^2 - 9}{16y^2 + 4y + 1}$

73. $\dfrac{8a^3 + b^3}{2a^2 + 3ab + b^2} \div \dfrac{8a^2 - 4ab + 2b^2}{4a^2 + 4ab + b^2}$

74. $\dfrac{x^3 + 8y^3}{2x^2 + 5xy + 2y^2} \div \dfrac{x^3 - 2x^2y + 4xy^2}{8x^2 - 2y^2}$

SKILL MAINTENANCE

75. Solve by substitution:
$$3x + y = 13,$$
$$x = y + 1.$$

76. Evaluate:
$$\begin{vmatrix} 3 & -2 \\ 4 & 7 \end{vmatrix}.$$

77. Solve: $\frac{2}{3}(3x - 4) = 8$.

78. A concert committee needs to take in $4000 from ticket sales in order to break even. If a total of 400 tickets is to be sold at full price and 200 tickets sold at half price, how should the tickets be priced?

SYNTHESIS

79. ◆ Is it possible to understand how to simplify rational expressions without first understanding how to multiply rational expressions? Why or why not?

80. ◆ Nancy *incorrectly* simplifies $\dfrac{x + 2}{x}$ as
$$\dfrac{x + 2}{x} = \dfrac{\cancel{x} + 2}{\cancel{x}} = 1 + 2 = 3.$$

She insists this is correct because it checked when x was replaced by 1. Explain her misconception.

81. ◆ Tony *incorrectly* argues that since
$$\dfrac{a^2 - 4}{a - 2} = \dfrac{a^2}{a} + \dfrac{-4}{-2} = a + 2,$$
it follows that
$$\dfrac{x^2 + 9}{x + 1} = \dfrac{x^2}{x} + \dfrac{9}{1} = x + 9.$$
Explain his misconception.

82. Let
$$g(x) = \dfrac{2x + 3}{4x - 1}.$$
Determine each of the following.
a) $g(x + h)$
b) $g(2x - 2) \cdot g(x)$
c) $g\left(\frac{1}{2}x + 1\right) \cdot g(x)$

83. Graph the function given by
$$f(x) = \dfrac{x^2 - 9}{x - 3}.$$
(*Hint*: Determine the domain of f and simplify.)

Perform the indicated operations and simplify.

84. $\left[\dfrac{r^2 - 4s^2}{r + 2s} \div (r + 2s)\right] \cdot \dfrac{2s}{r - 2s}$

85. $\left[\dfrac{d^2 - d}{d^2 - 6d + 8} \cdot \dfrac{d - 2}{d^2 + 5d}\right] \div \dfrac{5d}{d^2 - 9d + 20}$

Simplify.

86. $\dfrac{x(x + 1) - 2(x + 3)}{(x + 1)(x + 2)(x + 3)}$

87. $\dfrac{10x - 20 - 5x^2}{x^3 + 8}$

88. $\dfrac{m^2 - t^2}{m^2 + t^2 + m + t + 2mt}$

89. $\dfrac{a^3 - 2a^2 + 2a - 4}{a^3 - 2a^2 - 3a + 6}$

90. $\dfrac{x^3 + x^2 - y^3 - y^2}{x^2 - 2xy + y^2}$

91. $\dfrac{u^6 + v^6 + 2u^3v^3}{u^3 - v^3 + u^2v - uv^2}$

92. $\dfrac{x^5 - x^3 + x^2 - 1 - (x^3 - 1)(x + 1)^2}{(x^2 - 1)^2}$

93. Let

$$f(x) = \dfrac{4}{x^2 - 1} \quad \text{and} \quad g(x) = \dfrac{4x^2 + 8x + 4}{x^3 - 1}.$$

Find each of the following.

a) $(f \cdot g)(x)$
b) $(f / g)(x)$
c) $(g / f)(x)$

94. 〰 Use a grapher to check Example 6. Use the method described on p. 315.

95. 〰 Use a grapher to check your answers to Exercises 27, 45, and 71. Use the method described on p. 315.

96. 〰 Use a grapher to show that

$$\dfrac{x^2 - 16}{x + 2} \neq x - 8.$$

97. ◈ 〰 To check Example 4, Kara lets

$$y_1 = \dfrac{7x^2 + 21x}{14x} \quad \text{and} \quad y_2 = \dfrac{x + 3}{2}.$$

Since the graphs of y_1 and y_2 appear to be identical, Kara believes that the domains of the functions described by y_1 and y_2 are the same, $\mathbb{R}$. How could you convince Kara otherwise?

6.2 Rational Expressions and Functions: Adding and Subtracting

---

When Denominators Are the Same   •   When Denominators Are Different

---

Rational expressions are added in much the same way as the fractions of arithmetic.

When Denominators Are the Same

ADDITION WITH LIKE DENOMINATORS

To add or subtract when denominators are the same, add or subtract the numerators and keep the same denominator.

$$\dfrac{A}{C} + \dfrac{B}{C} = \dfrac{A + B}{C} \quad \text{and} \quad \dfrac{A}{C} - \dfrac{B}{C} = \dfrac{A - B}{C}, \quad \text{where } C \neq 0.$$

EXAMPLE 1

Add: $\dfrac{3 + x}{x} + \dfrac{4}{x}$.

SOLUTION

$$\frac{3 + x}{x} + \frac{4}{x} = \frac{7 + x}{x} \qquad \text{This expression cannot be simplified further because } x \text{ is not a factor of } 7 + x.$$

EXAMPLE 2

Add: $\dfrac{4x^2 - 5xy}{x^2 - y^2} + \dfrac{2xy - y^2}{x^2 - y^2}$.

SOLUTION

$$\frac{4x^2 - 5xy}{x^2 - y^2} + \frac{2xy - y^2}{x^2 - y^2} = \frac{4x^2 - 3xy - y^2}{x^2 - y^2} \qquad \begin{array}{l}\text{Adding the numerators and} \\ \text{combining like terms. The} \\ \text{denominator is unchanged.}\end{array}$$

$$= \frac{(x - y)(4x + y)}{(x - y)(x + y)} \qquad \begin{array}{l}\text{Factoring the numerator and} \\ \text{the denominator and looking} \\ \text{for common factors}\end{array}$$

$$= \frac{\cancel{(x - y)}(4x + y)}{\cancel{(x - y)}(x + y)} \qquad \begin{array}{l}\text{Removing a factor equal to 1:} \\ \dfrac{x - y}{x - y} = 1\end{array}$$

$$= \frac{4x + y}{x + y} \qquad \text{Simplifying}$$

Recall from Chapter 1 that a fraction bar is a grouping symbol. Thus, when a numerator containing a polynomial is subtracted, care must be taken to subtract, or change the sign of, *each* term in that polynomial.

EXAMPLE 3

If

$$f(x) = \frac{4x + 5}{x + 3} - \frac{x - 2}{x + 3},$$

find a simplified form of $f(x)$.

TECHNOLOGY CONNECTION

6.2

To check Example 3 with a grapher, compare the graphs of

$$y_1 = \frac{4x + 5}{x + 3} - \frac{x - 2}{x + 3} \quad \text{and} \quad y_2 = \frac{3x + 7}{x + 3}$$

on the same set of axes. Since the equations are equivalent, one curve (it has two branches) should appear. Equivalently, you can show that $y_3 = y_2 - y_1$ is 0 for all x not equal to -3. The TABLE feature can assist in either type of check.

SOLUTION

$$f(x) = \frac{4x + 5}{x + 3} - \frac{x - 2}{x + 3}$$

$$= \frac{4x + 5 - (x - 2)}{x + 3} \qquad \begin{array}{l}\text{The parentheses} \\ \text{remind us to subtract} \\ \textit{both} \text{ terms.}\end{array}$$

$$= \frac{4x + 5 - x + 2}{x + 3}$$

$$= \frac{3x + 7}{x + 3}$$

When Denominators Are Different

In order to add rational expressions such as

$$\frac{7}{12xy^2} + \frac{8}{15x^3y} \quad \text{or} \quad \frac{x}{x^2 - y^2} + \frac{y}{x^2 - 4xy + 3y^2},$$

we must first find common denominators. As in arithmetic, our work is easier when we use the *least common multiple* (LCM) of the denominators.

LEAST COMMON MULTIPLE

To find the least common multiple (LCM) of two or more expressions, find the prime factorization of each expression and form a product that uses each factor the greatest number of times that it occurs in any one prime factorization.

EXAMPLE 4

Find the least common multiple of each pair of polynomials.

a) $21x$ and $3x^2$

b) $x^2 + x - 12$ and $x^2 - 16$

SOLUTION

a) We write the prime factorizations of $21x$ and $3x^2$:

$$21x = 3 \cdot 7 \cdot x \quad \text{and} \quad 3x^2 = 3 \cdot x \cdot x.$$

The factors 3, 7, and x must appear in the LCM if $21x$ is to be a factor of the LCM. Note that $3x^2$, the other polynomial, is not a factor of $3 \cdot 7 \cdot x$. This is because the prime factors of $3x^2$—namely, $3, x$, and x—do not all appear in $3 \cdot 7 \cdot x$. By including a second factor of x, we form a product that contains both $21x$ and $3x^2$ as factors:

$$\text{LCM} = \underbrace{3 \cdot 7 \cdot x}_{} \cdot x$$

21x is a factor.

$3x^2$ is a factor.

Note that each factor (3, 7, and x) is used the greatest number of times that it occurs as a factor of either $21x$ or $3x^2$. The LCM is $3 \cdot 7 \cdot x \cdot x$, or $21x^2$.

b) We factor both expressions:

$$x^2 + x - 12 = (x - 3)(x + 4),$$
$$x^2 - 16 = (x + 4)(x - 4).$$

The LCM must contain each polynomial as a factor. By multiplying the factors of $x^2 + x - 12$ by $x - 4$, we form a product that contains both $x^2 + x - 12$ and $x^2 - 16$ as factors:

$$\text{LCM} = (x - 3)(x + 4)(x - 4)$$

$x^2 + x - 12$ is a factor.

$x^2 - 16$ is a factor.

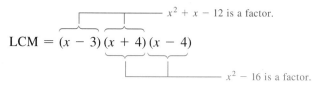

Before adding or subtracting rational expressions with unlike denominators, we determine the *least common denominator,* or LCD, by finding the LCM of the denominators. Then we multiply each rational expression by a form of 1, as needed, to form an equivalent expression that has the LCD.

EXAMPLE 5

Add: $\dfrac{2}{21x} + \dfrac{5}{3x^2}$.

SOLUTION   In Example 4(a), we found that the LCD is $3 \cdot 7 \cdot x \cdot x$, or $21x^2$. We now multiply each rational expression by 1, using expressions for 1 that give us the LCD in each expression. In this case, we use x/x and $7/7$:

$$\frac{2}{21x} \cdot \frac{x}{x} + \frac{5}{3x^2} \cdot \frac{7}{7} = \frac{2x}{21x^2} + \frac{35}{21x^2} \qquad \text{We now have a common denominator.}$$

$$= \frac{2x + 35}{21x^2}. \qquad \text{This expression cannot be simplified.}$$

Note that multiplying by x/x in the first expression gave us the LCD, $21x^2$. Similarly, multiplying by $7/7$ in the second expression also gave us the denominator $21x^2$.

EXAMPLE 6

Add: $\dfrac{x^2}{x^2 + 2xy + y^2} + \dfrac{2x - 2y}{x^2 - y^2}$.

SOLUTION   Before looking for an LCD, we factor each denominator:

$$\frac{x^2}{x^2 + 2xy + y^2} + \frac{2x - 2y}{x^2 - y^2} = \frac{x^2}{(x + y)(x + y)} + \frac{2x - 2y}{(x + y)(x - y)}.$$

Although we need not *always* factor the numerators, we do so if it enables us to simplify. In this case, the rightmost rational expression can be simplified:

$$\frac{x^2}{(x + y)(x + y)} + \frac{2(x - y)}{(x + y)(x - y)} = \frac{x^2}{(x + y)(x + y)} + \frac{2}{x + y}.$$

Factoring and removing a factor equal to 1: $\dfrac{x - y}{x - y} = 1$

Note that the LCM of $(x + y)(x + y)$ and $(x + y)$ is $(x + y)(x + y)$. To get the LCD in the second expression, we multiply by 1, using $(x + y)/(x + y)$. Then we add and, if possible, simplify.

$$\frac{x^2}{(x + y)(x + y)} + \frac{2}{x + y} = \frac{x^2}{(x + y)(x + y)} + \frac{2}{x + y} \cdot \frac{x + y}{x + y}$$

$$= \frac{x^2}{(x + y)(x + y)} + \frac{2x + 2y}{(x + y)(x + y)} \qquad \text{We now have the LCD.}$$

$$= \frac{x^2 + 2x + 2y}{(x + y)(x + y)} \qquad \text{Since the numerator cannot be factored, we cannot simplify further.}$$

EXAMPLE 7   Subtract: $\dfrac{2y + 1}{y^2 - 7y + 6} - \dfrac{y + 3}{y^2 - 5y - 6}$.

SOLUTION

$$\dfrac{2y + 1}{y^2 - 7y + 6} - \dfrac{y + 3}{y^2 - 5y - 6} = \dfrac{2y + 1}{(y - 6)(y - 1)} - \dfrac{y + 3}{(y - 6)(y + 1)}$$

The LCD is
$(y - 6)(y - 1)(y + 1)$.

$$= \dfrac{2y + 1}{(y - 6)(y - 1)} \cdot \dfrac{y + 1}{y + 1} - \dfrac{y + 3}{(y - 6)(y + 1)} \cdot \dfrac{y - 1}{y - 1}$$

Multiplying by 1 to get the LCD in each expression

$$= \dfrac{(2y + 1)(y + 1) - (y + 3)(y - 1)}{(y - 6)(y - 1)(y + 1)}$$

$$= \dfrac{2y^2 + 3y + 1 - (y^2 + 2y - 3)}{(y - 6)(y - 1)(y + 1)}$$

The parentheses are important.

$$= \dfrac{2y^2 + 3y + 1 - y^2 - 2y + 3}{(y - 6)(y - 1)(y + 1)}$$

$$= \dfrac{y^2 + y + 4}{(y - 6)(y - 1)(y + 1)}$$

We leave the denominator in factored form.

EXAMPLE 8   Add: $\dfrac{3}{8a} + \dfrac{1}{-8a}$.

SOLUTION

$$\dfrac{3}{8a} + \dfrac{1}{-8a} = \dfrac{3}{8a} + \dfrac{-1}{-1} \cdot \dfrac{1}{-8a}$$

When denominators are opposites, we multiply one rational expression by $-1/-1$ to get the LCD.

$$= \dfrac{3}{8a} + \dfrac{-1}{8a} = \dfrac{2}{8a}$$

$$= \dfrac{2 \cdot 1}{2 \cdot 4a} = \dfrac{1}{4a}$$

Simplifying by removing a factor equal to 1: $\dfrac{2}{2} = 1$

EXAMPLE 9   Subtract: $\dfrac{5x}{x - 2y} - \dfrac{3y - 7}{2y - x}$.

SOLUTION

$$\dfrac{5x}{x - 2y} - \dfrac{3y - 7}{2y - x} = \dfrac{5x}{x - 2y} - \dfrac{-1}{-1} \cdot \dfrac{3y - 7}{2y - x}$$

Note that $x - 2y$ and $2y - x$ are opposites.

$$= \dfrac{5x}{x - 2y} - \dfrac{7 - 3y}{x - 2y}$$

Performing the multiplication. *Note*: $-1(2y - x) = -2y + x$
$= x - 2y$.

$$= \dfrac{5x - (7 - 3y)}{x - 2y}$$

Subtracting; the parentheses are important.

$$= \dfrac{5x - 7 + 3y}{x - 2y}$$

In Example 9, you may have noticed that when $3y - 7$ is multiplied by -1 and subtracted, the result is $-7 + 3y$, which is equivalent to the original $3y - 7$. Thus, instead of multiplying the numerator by -1 and then subtracting, we could have simply *added* $3y - 7$ to $5x$, as in the following:

$$\frac{5x}{x - 2y} - \frac{3y - 7}{2y - x} = \frac{5x}{x - 2y} + (-1) \cdot \frac{3y - 7}{2y - x} \qquad \text{Rewriting subtraction as addition}$$

$$= \frac{5x}{x - 2y} + \frac{1}{-1} \cdot \frac{3y - 7}{2y - x} \qquad \text{Writing } -1 \text{ as } \frac{1}{-1}$$

$$= \frac{5x}{x - 2y} + \frac{3y - 7}{x - 2y} \qquad \text{The opposite of } 2y - x \text{ is } x - 2y.$$

$$= \frac{5x + 3y - 7}{x - 2y}. \qquad \text{This checks with the answer to Example 9.}$$

EXAMPLE 10   Perform the indicated operations and simplify:

$$\frac{2x}{x^2 - 4} + \frac{5}{2 - x} - \frac{1}{2 + x}.$$

SOLUTION   We have

$$\frac{2x}{x^2 - 4} + \frac{5}{2 - x} - \frac{1}{2 + x}$$

$$= \frac{2x}{(x - 2)(x + 2)} + \frac{5}{2 - x} - \frac{1}{2 + x} \qquad \text{Factoring}$$

$$= \frac{2x}{(x - 2)(x + 2)} + \frac{-1}{-1} \cdot \frac{5}{(2 - x)} - \frac{1}{x + 2} \qquad \text{Multiplying by } \frac{-1}{-1} \text{ since } 2 - x \text{ is the opposite of } x - 2$$

$$= \frac{2x}{(x - 2)(x + 2)} + \frac{-5}{x - 2} - \frac{1}{x + 2} \qquad \text{The LCD is } (x - 2)(x + 2).$$

$$= \frac{2x}{(x - 2)(x + 2)} + \frac{-5}{x - 2} \cdot \frac{x + 2}{x + 2} - \frac{1}{x + 2} \cdot \frac{x - 2}{x - 2} \qquad \text{Multiplying by 1 to get the LCD}$$

$$= \frac{2x - 5(x + 2) - (x - 2)}{(x - 2)(x + 2)} = \frac{2x - 5x - 10 - x + 2}{(x - 2)(x + 2)}$$

$$= \frac{-4x - 8}{(x - 2)(x + 2)} = \frac{-4(x + 2)}{(x - 2)(x + 2)}$$

$$= \frac{-4(x + 2)}{(x - 2)(x + 2)} \qquad \text{Removing a factor equal to 1: } \frac{x + 2}{x + 2} = 1$$

$$= \frac{-4}{x - 2}, \text{ or } -\frac{4}{x - 2}.$$

Another correct answer is $\dfrac{4}{2 - x}$. It is found by writing $-\dfrac{4}{x - 2}$ as $\dfrac{4}{-(x - 2)}$ and removing parentheses.

Our work in Example 10 indicates that if

$$f(x) = \frac{2x}{x^2 - 4} + \frac{5}{2 - x} - \frac{1}{2 + x}$$

and

$$g(x) = \frac{-4}{x - 2},$$

then, for $x \neq -2$ and $x \neq 2$, we have $f = g$. Note that whereas the domain of f includes all real numbers except -2 or 2, the domain of g excludes only 2.

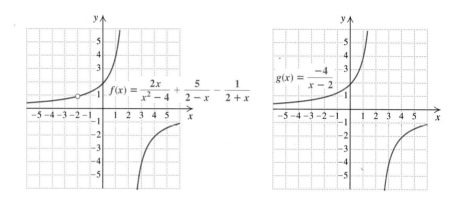

Whenever rational expressions are simplified, a quick partial check is to evaluate both the original and the simplified expressions for a convenient choice of x. For instance, to check Example 10, if $x = 1$, we have

$$f(1) = \frac{2 \cdot 1}{1^2 - 4} + \frac{5}{2 - 1} - \frac{1}{2 + 1}$$

$$= \frac{2}{-3} + \frac{5}{1} - \frac{1}{3}$$

$$= 5 - \frac{3}{3} = 4$$

and

$$g(1) = \frac{-4}{1 - 2}$$

$$= \frac{-4}{-1} = 4.$$

Since both functions include the pair $(1, 4)$, our algebra was *probably* correct. Although this is only a partial check (on rare occasions, an incorrect answer might "check"), because it is so easy to perform, it is nonetheless very useful. Further evaluation provides a more definitive check.

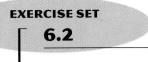

EXERCISE SET

6.2

Perform the indicated operations. Simplify when possible.

1. $\dfrac{4}{3a} + \dfrac{8}{3a}$

2. $\dfrac{3}{2y} + \dfrac{5}{2y}$

3. $\dfrac{3}{4a^2 b} - \dfrac{7}{4a^2 b}$

4. $\dfrac{5}{3m^2 n^2} - \dfrac{4}{3m^2 n^2}$

5. $\dfrac{a - 5b}{a + b} + \dfrac{a + 7b}{a + b}$

6. $\dfrac{x - 3y}{x + y} + \dfrac{x + 5y}{x + y}$

7. $\dfrac{4y + 2}{y - 2} - \dfrac{y - 3}{y - 2}$

8. $\dfrac{3t + 2}{t - 4} - \dfrac{t - 2}{t - 4}$

9. $\dfrac{3x - 4}{x^2 - 5x + 4} + \dfrac{3 - 2x}{x^2 - 5x + 4}$

10. $\dfrac{5x - 4}{x^2 - 6x - 7} + \dfrac{5 - 4x}{x^2 - 6x - 7}$

11. $\dfrac{3a - 2}{a^2 - 25} - \dfrac{4a - 7}{a^2 - 25}$

12. $\dfrac{2a - 5}{a^2 - 9} - \dfrac{3a - 8}{a^2 - 9}$

13. $\dfrac{a^2}{a - b} + \dfrac{b^2}{b - a}$

14. $\dfrac{s^2}{r - s} + \dfrac{r^2}{s - r}$

15. $\dfrac{3}{x} - \dfrac{8}{-x}$

16. $\dfrac{2}{a} - \dfrac{5}{-a}$

17. $\dfrac{x - 7}{x^2 - 16} - \dfrac{x - 1}{16 - x^2}$

18. $\dfrac{y - 4}{y^2 - 25} - \dfrac{9 - 2y}{25 - y^2}$

19. $\dfrac{t^2 + 3}{t^4 - 16} + \dfrac{7}{16 - t^4}$

20. $\dfrac{y^2 - 5}{y^4 - 81} + \dfrac{4}{81 - y^4}$

21. $\dfrac{m - 3n}{m^3 - n^3} - \dfrac{2n}{n^3 - m^3}$

22. $\dfrac{r - 6s}{r^3 - s^3} - \dfrac{5s}{s^3 - r^3}$

23. $\dfrac{a + 2}{a - 4} + \dfrac{a - 2}{a + 3}$

24. $\dfrac{a + 3}{a - 5} + \dfrac{a - 2}{a + 4}$

25. $2 + \dfrac{x - 3}{x + 1}$

26. $3 + \dfrac{y + 2}{y - 5}$

27. $\dfrac{4xy}{x^2 - y^2} + \dfrac{x - y}{x + y}$

28. $\dfrac{5ab}{a^2 - b^2} + \dfrac{a + b}{a - b}$

29. $\dfrac{8}{2x^2 - 7x + 5} + \dfrac{3x + 2}{2x^2 - x - 10}$

30. $\dfrac{7}{3y^2 + y - 4} + \dfrac{9y + 2}{3y^2 - 2y - 8}$

31. $\dfrac{4}{x + 1} + \dfrac{x + 2}{x^2 - 1} + \dfrac{3}{x - 1}$

32. $\dfrac{-2}{y + 2} + \dfrac{5}{y - 2} + \dfrac{y + 3}{y^2 - 4}$

33. $\dfrac{x + 6}{5x + 10} - \dfrac{x - 2}{4x + 8}$

34. $\dfrac{a + 3}{5a + 25} - \dfrac{a - 1}{3a + 15}$

35. $\dfrac{5ab}{a^2 - b^2} - \dfrac{a - b}{a + b}$

36. $\dfrac{6xy}{x^2 - y^2} - \dfrac{x + y}{x - y}$

37. $\dfrac{x}{x^2 + 9x + 20} - \dfrac{4}{x^2 + 7x + 12}$

38. $\dfrac{x}{x^2 + 11x + 30} - \dfrac{5}{x^2 + 9x + 20}$

39. $\dfrac{3y}{y^2 - 7y + 10} - \dfrac{2y}{y^2 - 8y + 15}$

40. $\dfrac{5x}{x^2 - 6x + 8} - \dfrac{3x}{x^2 - x - 12}$

41. $\dfrac{2x + 1}{x - y} + \dfrac{5x^2 - 5xy}{x^2 - 2xy + y^2}$

42. $\dfrac{2 - 3a}{a - b} + \dfrac{3a^2 + 3ab}{a^2 - b^2}$

43. $\dfrac{3y + 2}{y^2 + 5y - 24} + \dfrac{7}{y^2 + 4y - 32}$

44. $\dfrac{3x + 2}{x^2 - 7x + 10} + \dfrac{2x}{x^2 - 8x + 15}$

45. $\dfrac{a - 3}{a^2 - 16} - \dfrac{3a - 2}{a^2 + 2a - 24}$

46. $\dfrac{t + 4}{t^2 - 9} - \dfrac{3t - 1}{t^2 + 2t - 3}$

47. $\dfrac{2}{a^2 - 5a + 4} + \dfrac{-2}{a^2 - 4}$

48. $\dfrac{3}{a^2 - 7a + 6} + \dfrac{-3}{a^2 - 9}$

49. $3 + \dfrac{t}{t + 2} - \dfrac{2}{t^2 - 4}$

50. $2 + \dfrac{t}{t - 3} - \dfrac{3}{t^2 - 9}$

51. $\dfrac{2}{y + 3} - \dfrac{y}{y - 1} + \dfrac{y^2 + 2}{y^2 + 2y - 3}$

52. $\dfrac{1}{x+1} - \dfrac{x}{x-2} + \dfrac{x^2+2}{x^2-x-2}$

53. $\dfrac{5y}{1-2y} - \dfrac{2y}{2y+1} + \dfrac{3}{4y^2-1}$

54. $\dfrac{4x}{x^2-1} + \dfrac{3x}{1-x} - \dfrac{4}{x-1}$

55. $\dfrac{2}{x^2-5x+6} - \dfrac{4}{x^2-2x-3} + \dfrac{2}{x^2+4x+3}$

56. $\dfrac{1}{t^2+5t+6} - \dfrac{2}{t^2+3t+2} + \dfrac{1}{t^2-3t-4}$

SKILL MAINTENANCE

57. Simplify. Use only positive exponents in your answer.

$$\dfrac{15x^{-7}y^{12}z^4}{35x^{-2}y^6z^{-3}}$$

58. Find an equation for the line that passes through the point $(-2, 3)$ and is perpendicular to the line $f(x) = -\frac{4}{5}x + 7$.

59. *Value of coins.* There are 50 dimes in a roll of dimes, 40 nickels in a roll of nickels, and 40 quarters in a roll of quarters. Robert has a total of 12 rolls of coins with a total value of $70.00. If he has 3 more rolls of nickels than dimes, how many of each roll of coins does he have?

60. *Audiotapes.* Anna wants to buy tapes for her work at the campus radio station. She needs some 30-min tapes and some 60-min tapes. If she buys 12 tapes with a total recording time of 10 hr, how many tapes of each length did she buy?

SYNTHESIS

61. ◆ When two rational expressions are added or subtracted, should the numerator of the sum or difference be factored? Why or why not?

62. ◆ Janine found that the sum of two rational expressions was $(3 - x)/(x - 5)$. The answer given at the back of the book is $(x - 3)/(5 - x)$. Is Janine's answer incorrect? Why or why not?

63. ◆ Many students make the mistake of always multiplying denominators when looking for a common denominator. Use Example 7 to explain why this approach can yield results that are more difficult to simplify.

64. ◆ Is the sum of two rational expressions always a rational expression? Why or why not?

65. *Music.* To duplicate a common African polyrhythm, a drummer needs to play sextuplets (6 beats per measure) on a tom-tom while simultaneously playing quarter notes (4 beats per measure) on a bass drum. Into how many equally sized parts must a measure be divided, in order to precisely execute this rhythm?

66. *Astronomy.* The earth, Jupiter, Saturn, and Uranus all revolve around the sun. The earth takes 1 yr, Jupiter 12 yr, Saturn 30 yr, and Uranus 84 yr. How frequently do these four planets line up with each other?

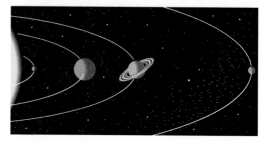

Find the LCM.

67. $x^8 - x^4, \quad x^5 - x^2, \quad x^5 - x^3, \quad x^5 + x^2$

68. $2a^3 + 2a^2b + 2ab^2, \quad a^6 - b^6,$
$2b^2 + ab - 3a^2, \quad 2a^2b + 4ab^2 + 2b^3$

69. The LCM of two expressions is $8a^4b^7$. One of the expressions is $2a^3b^7$. List all the possibilities for the other expression.

70. Determine the domain and the range of the function graphed below.

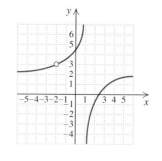

If

$$f(x) = \dfrac{x^3}{x^2-4} \quad and \quad g(x) = \dfrac{x^2}{x^2+3x-10},$$

find each of the following.

71. $(f + g)(x)$ **72.** $(f - g)(x)$

73. $(f \cdot g)(x)$ **74.** $(f / g)(x)$

Perform the indicated operations and simplify.

75. $5(x - 3)^{-1} + 4(x + 3)^{-1} - 2(x + 3)^{-2}$

76. $4(y - 1)(2y - 5)^{-1} + 5(2y + 3)(5 - 2y)^{-1} +$ $(y - 4)(2y - 5)^{-1}$

77. $\dfrac{x + 4}{6x^2 - 20x} \cdot \left(\dfrac{x}{x^2 - x - 20} + \dfrac{2}{x + 4} \right)$

78. $\dfrac{x^2 - 7x + 12}{x^2 - x - 29/3} \cdot \left(\dfrac{3x + 2}{x^2 + 5x - 24} + \dfrac{7}{x^2 + 4x - 32} \right)$

79. $\dfrac{8t^5}{2t^2 - 10t + 12} \div \left(\dfrac{2t}{t^2 - 8t + 15} - \dfrac{3t}{t^2 - 7t + 10} \right)$

80. $\dfrac{9t^3}{3t^3 - 12t^2 + 9t} \div \left(\dfrac{t + 4}{t^2 - 9} - \dfrac{3t - 1}{t^2 + 2t - 3} \right)$

81. Use a grapher to check your answers to Exercises 9, 25, and 49. Use the method discussed on p. 321.

82. Let

$$f(x) = 2 + \frac{x - 3}{x + 1}.$$

Use algebra, together with a grapher, to determine the domain and the range of f.

6.3   Complex Rational Expressions

Multiplying by 1 • Dividing Two Rational Expressions

A **complex rational expression** is a rational expression that contains rational expressions within its numerator and/or its denominator. Here are some examples:

$$\frac{x + \dfrac{5}{x}}{4x}, \quad \frac{\dfrac{x - y}{x + y}}{\dfrac{2x - y}{3x + y}}, \quad \frac{\dfrac{2}{3}}{\dfrac{4}{5}}, \quad \frac{\dfrac{3x}{5} - \dfrac{2}{x}}{\dfrac{4x}{3} + \dfrac{7}{6x}}.$$

The rational expressions within each complex rational expression are red.

Two methods are used to simplify complex rational expressions.

Method 1: Multiplying by 1

One method of simplifying a complex rational expression is to multiply the entire expression by 1. To write 1, we use the LCD of the expressions within the complex rational expression.

EXAMPLE 1   Simplify:

$$\frac{\dfrac{1}{a^3b} + \dfrac{1}{b}}{\dfrac{1}{a^2b^2} - \dfrac{1}{b^2}}.$$

SOLUTION   The denominators within the complex rational expression are a^3b, b, a^2b^2, and b^2. Thus the LCD is a^3b^2. We multiply by 1, using $(a^3b^2)/(a^3b^2)$:

$$\frac{\dfrac{1}{a^3b} + \dfrac{1}{b}}{\dfrac{1}{a^2b^2} - \dfrac{1}{b^2}} = \frac{\dfrac{1}{a^3b} + \dfrac{1}{b}}{\dfrac{1}{a^2b^2} - \dfrac{1}{b^2}} \cdot \frac{a^3b^2}{a^3b^2}$$

Multiplying by 1, using the LCD

$$= \frac{\left(\dfrac{1}{a^3b} + \dfrac{1}{b}\right)a^3b^2}{\left(\dfrac{1}{a^2b^2} - \dfrac{1}{b^2}\right)a^3b^2}$$

Multiplying the numerator and the denominator. Remember to use parentheses.

$$= \frac{\dfrac{1}{a^3b} \cdot a^3b^2 + \dfrac{1}{b} \cdot a^3b^2}{\dfrac{1}{a^2b^2} \cdot a^3b^2 - \dfrac{1}{b^2} \cdot a^3b^2}$$

Using the distributive law to carry out the multiplications

$$= \frac{\dfrac{a^3b}{a^3b} \cdot b + \dfrac{b}{b} \cdot a^3b}{\dfrac{a^2b^2}{a^2b^2} \cdot a - \dfrac{b^2}{b^2} \cdot a^3}$$

Removing factors that equal 1; study this carefully.

$$= \frac{b + a^3b}{a - a^3}$$

Simplifying

$$= \frac{b(1 + a^3)}{a(1 - a^2)}$$

Factoring

$$= \frac{b(1 + a)(1 - a + a^2)}{a(1 + a)(1 - a)}$$

Factoring further and identifying a factor that equals 1

$$= \frac{b(1 - a + a^2)}{a(1 - a)}.$$

Simplifying

USING MULTIPLICATION BY 1 TO SIMPLIFY A COMPLEX RATIONAL EXPRESSION

1. Find the LCD of all expressions *within* the complex rational expression.
2. Multiply the complex rational expression by 1, using the LCD to form the expression for 1.
3. Distribute and simplify so that the numerator and the denominator of the complex rational expression are simply polynomials.
4. Factor and simplify, if possible.

Note that using the LCD to form 1 clears the numerator and the denominator of the complex rational expression of all rational expressions.

EXAMPLE 2

Simplify:

$$\frac{\dfrac{3}{2x-2}-\dfrac{1}{x+1}}{\dfrac{1}{x-1}+\dfrac{x}{x^2-1}}.$$

SOLUTION   In this case, to find the LCD, we have to factor first:

$$\frac{\dfrac{3}{2x-2}-\dfrac{1}{x+1}}{\dfrac{1}{x-1}+\dfrac{x}{x^2-1}}=\frac{\dfrac{3}{2(x-1)}-\dfrac{1}{x+1}}{\dfrac{1}{x-1}+\dfrac{x}{(x-1)(x+1)}}$$

The LCD is $2(x-1)(x+1)$.

$$=\frac{\dfrac{3}{2(x-1)}-\dfrac{1}{x+1}}{\dfrac{1}{x-1}+\dfrac{x}{(x-1)(x+1)}}\cdot\frac{2(x-1)(x+1)}{2(x-1)(x+1)}$$

Multiplying by 1, using the LCD

$$=\frac{\dfrac{3}{2(x-1)}\cdot 2(x-1)(x+1)-\dfrac{1}{x+1}\cdot 2(x-1)(x+1)}{\dfrac{1}{x-1}\cdot 2(x-1)(x+1)+\dfrac{x}{(x-1)(x+1)}\cdot 2(x-1)(x+1)}$$

Using the distributive law

$$=\frac{\dfrac{2(x-1)}{2(x-1)}\cdot 3(x+1)-\dfrac{x+1}{x+1}\cdot 2(x-1)}{\dfrac{x-1}{x-1}\cdot 2(x+1)+\dfrac{(x-1)(x+1)}{(x-1)(x+1)}\cdot 2x}$$

Removing factors that equal 1

$$=\frac{3(x+1)-2(x-1)}{2(x+1)+2x}$$

Simplifying

$$=\frac{3x+3-2x+2}{2x+2+2x}$$

Using the distributive law

$$=\frac{x+5}{4x+2}.$$

Method 2: Dividing Two Rational Expressions

Another method for simplifying complex rational expressions involves first adding or subtracting, as necessary, to get one rational expression in the numerator and one rational expression in the denominator. The problem is thereby simplified to one involving the division of two rational expressions.

EXAMPLE 3

Simplify:

$$\frac{\dfrac{3}{x} - \dfrac{2}{x^2}}{\dfrac{3}{x-2} + \dfrac{1}{x^2}}.$$

SOLUTION

$$\frac{\dfrac{3}{x} - \dfrac{2}{x^2}}{\dfrac{3}{x-2} + \dfrac{1}{x^2}} = \frac{\dfrac{3}{x} \cdot \dfrac{x}{x} - \dfrac{2}{x^2}}{\dfrac{3}{x-2} \cdot \dfrac{x^2}{x^2} + \dfrac{1}{x^2} \cdot \dfrac{x-2}{x-2}}$$

Multiplying $3/x$ by 1 to get x^2 as a common denominator

Multiplying by 1, twice, to get $x^2(x-2)$ as a common denominator

$$= \frac{\dfrac{3x}{x^2} - \dfrac{2}{x^2}}{\dfrac{3x^2}{(x-2)x^2} + \dfrac{x-2}{x^2(x-2)}}$$

There is now a common denominator in the numerator and a common denominator in the denominator of the complex rational expression.

$$= \frac{\dfrac{3x - 2}{x^2}}{\dfrac{3x^2 + x - 2}{(x-2)x^2}}$$

Subtracting in the numerator and adding in the denominator. We now have one rational expression divided by another rational expression.

$$= \frac{3x - 2}{x^2} \div \frac{3x^2 + x - 2}{(x-2)x^2}$$

Rewriting with a division symbol

$$= \frac{3x - 2}{x^2} \cdot \frac{(x-2)x^2}{3x^2 + x - 2}$$

To divide, multiply by the reciprocal of the divisor.

$$= \frac{(3x - 2)(x-2)x^2}{x^2(3x - 2)(x+1)}$$

Factoring and removing a factor equal to 1: $\dfrac{x^2(3x-2)}{x^2(3x-2)} = 1$

$$= \frac{x-2}{x+1}$$

USING DIVISION TO SIMPLIFY A COMPLEX RATIONAL EXPRESSION

1. Add or subtract, as necessary, to get one rational expression in the numerator.
2. Add or subtract, as necessary, to get one rational expression in the denominator.
3. Perform the indicated division (invert the divisor and multiply).
4. Simplify, if possible, by removing any factors that equal 1.

EXAMPLE   4

Simplify:

$$\frac{1 + \dfrac{2}{x}}{1 - \dfrac{4}{x^2}}.$$

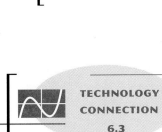

TECHNOLOGY
CONNECTION
6.3

To check Example 4 with a grapher, we can show that the graphs of

$$y_1 = \frac{1 + \dfrac{2}{x}}{1 - \dfrac{4}{x^2}}$$

and

$$y_2 = \frac{x}{x - 2}$$

coincide, or we can show that (except for $x = -2$ or 0) their tables of values coincide. We could also check by showing that (except for $x = -2$ or 0 or 2) $y_2 - y_1 = 0$.

1. Use a grapher to check Example 3. What values, if any, can x *not* equal?

S O L U T I O N   We have

$$\frac{1 + \dfrac{2}{x}}{1 - \dfrac{4}{x^2}} = \frac{\left.\begin{array}{c}\dfrac{x}{x} + \dfrac{2}{x}\end{array}\right\}}{\left.\begin{array}{c}\dfrac{x^2}{x^2} - \dfrac{4}{x^2}\end{array}\right\}} \quad \begin{array}{l}\text{Finding a common denominator} \\[2em] \text{Finding a common denominator}\end{array}$$

$$= \frac{\dfrac{x + 2}{x}}{\dfrac{x^2 - 4}{x^2}} \quad \begin{array}{l}\text{Adding in the numerator} \\[1.5em] \text{Subtracting in the denominator}\end{array}$$

$$= \frac{x + 2}{x} \cdot \frac{x^2}{x^2 - 4} \quad \text{Multiplying by the reciprocal of the divisor}$$

$$= \frac{(x + 2) \cdot x^2}{x(x + 2)(x - 2)} \quad \text{Factoring}$$

$$= \frac{(x + 2)x \cdot x}{x(x + 2)(x - 2)} \quad \text{Removing a factor equal to 1: } \frac{(x + 2)x}{(x + 2)x} = 1$$

$$= \frac{x}{x - 2}. \quad \text{Simplifying}$$

As a quick partial check, we select a convenient value for x—say, 1:

$$\frac{1 + \dfrac{2}{1}}{1 - \dfrac{4}{1^2}} = \frac{1 + 2}{1 - 4} = \frac{3}{-3} = -1 \quad \text{We evaluated the original expression for } x = 1.$$

and

$$\frac{1}{1 - 2} = \frac{1}{-1} = -1. \quad \text{We evaluated the simplified result for } x = 1.$$

Since both expressions yield the same result, our simplification is probably correct. More evaluation would provide a more definitive check.

If negative exponents occur, we first find an equivalent expression using positive exponents and then proceed as in the examples above.

EXAMPLE   5

Simplify:

$$\frac{a^{-1} + b^{-1}}{a^{-3} + b^{-3}}.$$

SOLUTION

$$\frac{a^{-1} + b^{-1}}{a^{-3} + b^{-3}} = \frac{\dfrac{1}{a} + \dfrac{1}{b}}{\dfrac{1}{a^3} + \dfrac{1}{b^3}}$$ Rewriting with positive exponents. We continue, using Method 2.

$$= \frac{\dfrac{1}{a} \cdot \dfrac{b}{b} + \dfrac{1}{b} \cdot \dfrac{a}{a}}{\dfrac{1}{a^3} \cdot \dfrac{b^3}{b^3} + \dfrac{1}{b^3} \cdot \dfrac{a^3}{a^3}}$$ Finding a common denominator

Finding a common denominator

$$= \frac{\dfrac{b}{ab} + \dfrac{a}{ab}}{\dfrac{b^3}{a^3b^3} + \dfrac{a^3}{a^3b^3}}$$

$$= \frac{\dfrac{b + a}{ab}}{\dfrac{b^3 + a^3}{a^3b^3}}$$ Adding in the numerator

Adding in the denominator

$$= \frac{b + a}{ab} \cdot \frac{a^3b^3}{b^3 + a^3}$$ Multiplying by the reciprocal of the divisor

$$= \frac{(b + a) \cdot ab \cdot a^2b^2}{ab(b + a)(b^2 - ab + a^2)}$$ Factoring and looking for common factors

$$= \frac{\cancel{(b + a)} \cdot \cancel{ab} \cdot a^2b^2}{\cancel{ab}\cancel{(b + a)}(b^2 - ab + a^2)}$$ Removing a factor equal to 1: $\dfrac{(b + a)ab}{(b + a)ab} = 1$

$$= \frac{a^2b^2}{b^2 - ab + a^2}$$

It is difficult to say which method is best to use. For expressions like

$$\frac{\dfrac{3x + 1}{x - 5}}{\dfrac{2 - x}{x + 3}} \quad \text{or} \quad \frac{\dfrac{3}{x} - \dfrac{2}{x}}{\dfrac{1}{x + 1} + \dfrac{5}{x + 1}},$$

the second method is probably easier to use since it is little or no work to write the expression as a quotient of two rational expressions.

On the other hand, expressions like

$$\frac{\dfrac{3}{a^2b} - \dfrac{4}{bc^3}}{\dfrac{1}{b^3c} + \dfrac{2}{ac^4}} \quad \text{or} \quad \frac{\dfrac{5}{a^2 - b^2} + \dfrac{2}{a^2 + 2ab + b^2}}{\dfrac{1}{a - b} + \dfrac{4}{a + b}}$$

require fewer steps if we use the first method. Either method works for any complex rational expression.

EXERCISE SET
6.3

Simplify. If possible, use a second method or evaluation as a check.

1. $\dfrac{5 + \dfrac{1}{a}}{\dfrac{1}{a} - 2}$

2. $\dfrac{\dfrac{1}{y} + 7}{\dfrac{1}{y} - 5}$

3. $\dfrac{x - x^{-1}}{x + x^{-1}}$

4. $\dfrac{y + y^{-1}}{y - y^{-1}}$

5. $\dfrac{\dfrac{3}{x} + \dfrac{4}{y}}{\dfrac{4}{x} - \dfrac{3}{y}}$

6. $\dfrac{\dfrac{5}{z} + \dfrac{2}{y}}{\dfrac{4}{z} - \dfrac{1}{y}}$

7. $\dfrac{\dfrac{x^2 - y^2}{xy}}{\dfrac{x - y}{y}}$

8. $\dfrac{\dfrac{a^2 - b^2}{ab}}{\dfrac{a - b}{b}}$

9. $\dfrac{\dfrac{3x}{y} - x}{2y - \dfrac{y}{x}}$

10. $\dfrac{1 - \dfrac{2}{3x}}{x - \dfrac{4}{9x}}$

11. $\dfrac{a^{-1} + b^{-1}}{\dfrac{a^2 - b^2}{ab}}$

12. $\dfrac{x^{-1} + y^{-1}}{\dfrac{x^2 - y^2}{xy}}$

13. $\dfrac{\dfrac{1}{x + h} - \dfrac{1}{x}}{h}$

14. $\dfrac{\dfrac{1}{a - h} - \dfrac{1}{a}}{h}$

15. $\dfrac{\dfrac{a^2 - 4}{a^2 + 3a + 2}}{\dfrac{a^2 - 5a - 6}{a^2 - 6a - 7}}$

16. $\dfrac{\dfrac{x^2 - x - 12}{x^2 - 2x - 15}}{\dfrac{x^2 + 8x + 12}{x^2 - 5x - 14}}$

17. $\dfrac{\dfrac{2}{y - 3} + \dfrac{1}{y + 1}}{\dfrac{3}{y + 1} + \dfrac{4}{y - 3}}$

18. $\dfrac{\dfrac{1}{x - 2} + \dfrac{3}{x - 1}}{\dfrac{2}{x - 1} + \dfrac{5}{x - 2}}$

19. $\dfrac{a(a + 3)^{-1} - 2(a - 1)^{-1}}{a(a + 3)^{-1} - (a - 1)^{-1}}$

20. $\dfrac{a(a + 2)^{-1} - 3(a - 3)^{-1}}{a(a + 2)^{-1} - (a - 3)^{-1}}$

21. $\dfrac{\dfrac{x}{x^2 + 3x - 4} - \dfrac{1}{x^2 + 3x - 4}}{\dfrac{x}{x^2 + 6x + 8} + \dfrac{3}{x^2 + 6x + 8}}$

22. $\dfrac{\dfrac{x}{x^2 + 5x - 6} + \dfrac{6}{x^2 + 5x - 6}}{\dfrac{x}{x^2 - 5x + 4} - \dfrac{2}{x^2 - 5x + 4}}$

23. $\dfrac{\dfrac{3}{a^2 - 9} + \dfrac{2}{a + 3}}{\dfrac{4}{a^2 - 9} + \dfrac{1}{a + 3}}$

24. $\dfrac{\dfrac{2}{a^2 - 1} + \dfrac{1}{a + 1}}{\dfrac{3}{a^2 - 1} + \dfrac{2}{a - 1}}$

25. $\dfrac{\dfrac{4}{x^2 - 1} - \dfrac{3}{x + 1}}{\dfrac{5}{x^2 - 1} - \dfrac{2}{x - 1}}$

26. $\dfrac{\dfrac{5}{x^2 - 4} - \dfrac{3}{x - 2}}{\dfrac{4}{x^2 - 4} - \dfrac{2}{x + 2}}$

27. $\dfrac{\dfrac{y}{y^2 - 1} + \dfrac{3}{1 - y^2}}{\dfrac{y^2}{y^2 - 1} + \dfrac{9}{1 - y^2}}$

28. $\dfrac{\dfrac{y}{y^2 - 4} + \dfrac{5}{4 - y^2}}{\dfrac{y^2}{y^2 - 4} + \dfrac{25}{4 - y^2}}$

29. $\dfrac{\dfrac{y^2}{y^2 - 9} - \dfrac{y}{y + 3}}{\dfrac{y}{y^2 - 9} - \dfrac{1}{y - 3}}$

30. $\dfrac{\dfrac{y^2}{y^2 - 25} - \dfrac{y}{y - 5}}{\dfrac{y}{y^2 - 25} - \dfrac{1}{y + 5}}$

31. $\dfrac{\dfrac{a}{a + 3} + \dfrac{4}{5a}}{\dfrac{a}{2a + 6} + \dfrac{3}{a}}$

32. $\dfrac{\dfrac{a}{a + 2} + \dfrac{5}{a}}{\dfrac{a}{2a + 4} + \dfrac{1}{3a}}$

33. $\dfrac{\dfrac{1}{x^2 - 3x + 2} + \dfrac{1}{x^2 - 4}}{\dfrac{1}{x^2 + 4x + 4} + \dfrac{1}{x^2 - 4}}$

34. $\dfrac{\dfrac{1}{x^2 + 3x + 2} + \dfrac{1}{x^2 - 1}}{\dfrac{1}{x^2 - 1} + \dfrac{1}{x^2 - 4x + 3}}$

35. $\dfrac{\dfrac{3}{a^2 - 4a + 3} + \dfrac{3}{a^2 - 5a + 6}}{\dfrac{3}{a^2 - 3a + 2} + \dfrac{3}{a^2 + 3a - 10}}$

36. $\dfrac{\dfrac{1}{a^2 + 7a + 10} - \dfrac{2}{a^2 - 7a + 12}}{\dfrac{2}{a^2 - a - 6} - \dfrac{1}{a^2 + a - 20}}$

37. $\dfrac{\dfrac{y}{y^2 - 4} - \dfrac{2y}{y^2 + y - 6}}{\dfrac{2y}{y^2 - 4} - \dfrac{y}{y^2 + 5y + 6}}$

38. $\dfrac{\dfrac{y}{y^2 - 1} - \dfrac{3y}{y^2 + 5y + 4}}{\dfrac{3y}{y^2 - 1} - \dfrac{y}{y^2 - 4y + 3}}$

39. $\dfrac{\dfrac{3}{x^2 + 2x - 3} - \dfrac{1}{x^2 - 3x - 10}}{\dfrac{3}{x^2 - 6x + 5} - \dfrac{1}{x^2 + 5x + 6}}$

40. $\dfrac{\dfrac{1}{a^2 + 7a + 12} + \dfrac{1}{a^2 + a - 6}}{\dfrac{1}{a^2 + 2a - 8} + \dfrac{1}{a^2 + 5a + 4}}$

SKILL MAINTENANCE

41. If $f(x) = x^2 - 3$, find $f(-5)$.

42. Solve for y: $\dfrac{a}{x + y} = b$.

43. Graph: $f(x) = -3x + 7$.

44. Solve: $|2x - 5| = 7$.

45. *Earnings.*   Antonio received $28 in tips on Monday, $22 in tips on Tuesday, and $36 in tips on Wednesday. How much will Antonio need to receive in tips on Thursday if his average for the four days is to be $30?

46. *Framing.*   Glenn has two rectangular frames. The first frame is 3 cm shorter, and 4 cm narrower, than the second frame. If the perimeter of the second frame is 1 cm less than twice the perimeter of the first, what is the perimeter of each frame?

SYNTHESIS

47. ◈ To simplify a complex rational expression in which the sum of two fractions is divided by the difference of the same two fractions, which method is easiest? Why?

48. ◈ In arithmetic, we are taught that

$$\frac{a}{b} \div \frac{c}{d} = \frac{a}{b} \cdot \frac{d}{c}$$

(to divide by a fraction, we invert and multiply). Use Method 1 to explain *why* we do this.

Simplify.

49. $\dfrac{5x^{-2} + 10x^{-1}y^{-1} + 5y^{-2}}{3x^{-2} - 3y^{-2}}$

50. $(a^2 - ab + b^2)^{-1}(a^2b^{-1} + b^2a^{-1}) \times$
$(a^{-2} - b^{-2})(a^{-2} + 2a^{-1}b^{-1} + b^{-2})^{-1}$

51. *Astronomy.*   When two galaxies are moving in opposite directions at velocities v_1 and v_2, an observer in one of the galaxies would see the other galaxy receding at speed

$$\frac{v_1 + v_2}{1 + \dfrac{v_1 v_2}{c^2}},$$

where c is the speed of light. Determine the observed speed if v_1 and v_2 are both one-fourth the speed of light.

In Exercises 52 and 53, find the reciprocal of the expression shown. Then simplify, if possible.

52. $1 + \dfrac{1}{1 + \dfrac{1}{1 + \dfrac{1}{x}}}$

53. $x^2 + x + 1 + \dfrac{1}{x} + \dfrac{1}{x^2}$

54. For $f(x) = \dfrac{1}{1 + x}$, find $f(f(a))$.

55. For $g(x) = \dfrac{x + 1}{x - 2}$, find $g(g(a))$.

Find and simplify

$$\frac{f(x + h) - f(x)}{h}$$

for each rational function f in Exercises 56–59.

56. $f(x) = \dfrac{3}{x^2}$

57. $f(x) = \dfrac{5}{x}$

58. $f(x) = \dfrac{1}{1 - x}$

59. $f(x) = \dfrac{x}{1 + x}$

60. If

$$F(x) = \frac{3 + \dfrac{1}{x}}{2 - \dfrac{8}{x^2}},$$

find the domain of *F*.

61. If

$$G(x) = \frac{x - \dfrac{1}{x^2 - 1}}{\dfrac{1}{9} - \dfrac{1}{x^2 - 16}},$$

find the domain of *G*.

62. Let

$$f(x) = \left[\frac{\dfrac{x + 3}{x - 3} + 1}{\dfrac{x + 3}{x - 3} - 1} \right]^4.$$

Find a simplified form of $f(x)$ and specify the domain of *f*.

63. Use a grapher to check your answers to Exercises 3, 17, 31, and 61.

64. Use a grapher to check your answers to Exercises 1, 10, 35, and 60.

65. ◆ Use algebra to determine the domain of the function given by

$$f(x) = \frac{\dfrac{1}{x - 2}}{\dfrac{x}{x - 2} - \dfrac{5}{x - 2}}.$$

Then explain how a grapher could be used to check your answer.

◆ **COLLABORATIVE**
C ◆ O ◆ R ◆ N ◆ E ◆ R

Focus: Complex rational expressions

Time: 15–20 minutes

Group size: 3–4

ACTIVITY

Consider the steps in Examples 2 and 3 for simplifying a complex rational expression by each of the two methods. Then, work as a group to simplify

$$\frac{\dfrac{5}{x + 1} - \dfrac{1}{x}}{\dfrac{2}{x^2} + \dfrac{4}{x}}$$

subject to the following conditions:

1. The group should decide which method will more easily simplify this expression.
2. Using the method selected in part (1), one group member should perform the first step in the simplification and then pass the problem on to another member of the group. That person then performs the next step and passes the problem on to another group member. This process continues until, eventually, the simplification is complete.
3. Check your work by repeating this "pass-around" procedure with the method not chosen in step (1) above.
4. What method *was* easier? Why? Compare your responses with those of other groups.

6.4   Rational Equations

Solving Rational Equations   •   Rational Equations and Graphs

Solving Rational Equations

In Sections 6.1–6.3, we learned how to *simplify expressions*. We now learn to *solve* a new type of *equation*. A **rational equation** is an equation that contains one or more rational expressions. Here are some examples:

$$\frac{2}{3} - \frac{5}{6} = \frac{1}{t}, \qquad \frac{a-1}{a-5} = \frac{4}{a^2-25}, \qquad x^3 + \frac{6}{x} = 5.$$

As you will see in Section 6.5, equations of this type occur frequently in applications. To solve rational equations, recall that one way to *clear fractions* is to multiply by the LCD.

TO SOLVE A RATIONAL EQUATION:	Multiply on both sides by the LCD. This is called *clearing fractions* and produces an equation similar to those we have already solved.

Recall that division by 0 is undefined. Note, too, that variables usually appear in at least one denominator of a rational equation. Thus certain numbers can often be ruled out as possible solutions before we ever attempt to solve a given rational equation.

EXAMPLE 1   Solve: $\dfrac{x+4}{3x} + \dfrac{x+8}{5x} = 2$.

SOLUTION   Because the left side of this equation is undefined when x is 0, we state at the outset that $x \neq 0$. Next, we multiply both sides of the equation by the LCD, $3 \cdot 5 \cdot x$:

$$3 \cdot 5 \cdot x\left(\frac{x+4}{3x} + \frac{x+8}{5x}\right) = 3 \cdot 5 \cdot x \cdot 2 \qquad \text{Multiplying by the LCD to clear fractions}$$

$$3 \cdot 5 \cdot x \cdot \frac{x+4}{3x} + 3 \cdot 5 \cdot x \cdot \frac{x+8}{5x} = 3 \cdot 5 \cdot x \cdot 2 \qquad \text{Using the distributive law}$$

$$\frac{3 \cdot 5 \cdot x \cdot (x+4)}{3x} + \frac{3 \cdot 5 \cdot x \cdot (x+8)}{5x} = 30x \qquad \text{Locating factors equal to 1}$$

$$5(x+4) + 3(x+8) = 30x. \qquad \text{Removing factors equal to 1: } \frac{3x}{3x} = 1; \frac{5x}{5x} = 1$$

Then

$$5x + 20 + 3x + 24 = 30x$$ Using the distributive law
$$8x + 44 = 30x$$
$$44 = 22x$$
$$2 = x.$$ This should check since $x \neq 0$.

Check:

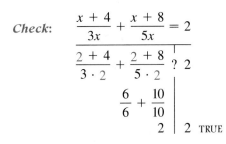

The number 2 is the solution.   ————•

Note that when we clear fractions, all denominators "disappear." This leaves an equation without rational expressions, which we know how to solve.

EXAMPLE 2   Solve: $\dfrac{x - 1}{x - 5} = \dfrac{4}{x - 5}$.

S O L U T I O N   To ensure that neither denominator is 0, we state at the outset the restriction that $x \neq 5$. Then we proceed as before, multiplying by the LCD, $x - 5$, on both sides:

$$(x - 5) \cdot \frac{x - 1}{x - 5} = (x - 5) \cdot \frac{4}{x - 5}$$
$$x - 1 = 4$$
$$x = 5.$$ But recall that $x \neq 5$.

In this case, it is important to remember that, because of our restriction above, 5 cannot be a solution. A check confirms the necessity of that restriction.

Check:
$$\frac{x - 1}{x - 5} = \frac{4}{x - 5}$$
$$\frac{5 - 1}{5 - 5} \; ? \; \frac{4}{5 - 5}$$
$$\frac{4}{0} \; \bigg| \; \frac{4}{0}$$ Division by 0 is undefined.

This equation has no solution.   ————•

To help see why 5 is not a solution to Example 2, consider the fact that the multiplication principle for equations requires that we multiply by a *nonzero* number on both sides if we are to form an equivalent equation. When both sides of an equation are multiplied by an expression containing variables, it is possible that certain replacements will make that expression equal to 0. Thus it is safe to say that *if* a solution of

$$\frac{x-1}{x-5} = \frac{4}{x-5}$$

exists, then it is also a solution of $x - 1 = 4$. We *cannot* conclude that every solution of $x - 1 = 4$ is a solution of the original equation.

EXAMPLE 3   Solve: $\dfrac{x^2}{x-3} = \dfrac{9}{x-3}$.

SOLUTION   Note that $x \neq 3$. Since the LCD is $x - 3$, we multiply by $x - 3$ on both sides:

$$(x-3) \cdot \frac{x^2}{x-3} = (x-3) \cdot \frac{9}{x-3}$$

$$x^2 = 9 \qquad \text{Simplifying}$$

$$x^2 - 9 = 0 \qquad \text{Getting 0 on one side}$$

$$(x-3)(x+3) = 0 \qquad \text{Factoring}$$

$$x = 3 \quad or \quad x = -3. \qquad \text{Using the principle of zero products}$$

Although 3 is a solution of $x^2 = 9$, it must be rejected as a solution of the rational equation. You should perform a check to confirm that -3 *is* a solution despite the fact that 3 is not.

EXAMPLE 4   Solve: $\dfrac{2}{x+5} + \dfrac{1}{x-5} = \dfrac{16}{x^2-25}$.

SOLUTION   To find all restrictions and to assist in finding the LCD, we factor:

$$\frac{2}{x+5} + \frac{1}{x-5} = \frac{16}{(x+5)(x-5)}. \qquad \text{Factoring } x^2 - 25$$

Note that $x \neq -5$ and $x \neq 5$. We multiply by the LCD, $(x+5)(x-5)$, and then use the distributive law:

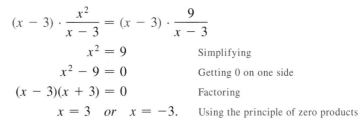

$$2(x-5) + (x+5) = 16$$

$$2x - 10 + x + 5 = 16$$

$$3x - 5 = 16$$

$$3x = 21$$

$$x = 7.$$

The check is left to the student. The solution is 7.

EXAMPLE 5

Let $f(x) = x + \dfrac{6}{x}$. Find all values of a for which $f(a) = 5$.

SOLUTION   Since $f(a) = a + \dfrac{6}{a}$, the problem asks that we find all values of a for which

$$a + \frac{6}{a} = 5.$$

To solve for a, we first note that $a \neq 0$. Next, we multiply by the LCD, a, on both sides:

$$a\left(a + \frac{6}{a}\right) = 5 \cdot a \qquad \text{Multiplying on both sides by } a; \text{ parentheses are important.}$$

$$a \cdot a + a \cdot \frac{6}{a} = 5a \qquad \text{Using the distributive law}$$

$$a^2 + 6 = 5a \qquad \text{Simplifying}$$

$$a^2 - 5a + 6 = 0 \qquad \text{Getting 0 on one side}$$

$$(a - 3)(a - 2) = 0 \qquad \text{Factoring}$$

$$a = 3 \quad or \quad a = 2. \qquad \text{Using the principle of zero products}$$

Check: $f(3) = 3 + \dfrac{6}{3} = 3 + 2 = 5;$

$$f(2) = 2 + \frac{6}{2} = 2 + 3 = 5.$$

The solutions are 2 and 3. For $a = 2$ or $a = 3$, we have $f(a) = 5$. ⟶

TECHNOLOGY CONNECTION

6.4

There are several ways in which Example 5 can be checked. One way is to confirm that the graphs of $y_1 = x + 6/x$ and $y_2 = 5$ intersect at $x = 2$ and $x = 3$. You can also use a table to check that $y_1 = y_2$ when x is 2 and again when x is 3.

Another way of checking is to use a graph and/or a table to show that $y_3 = y_2 - y_1$ is 0 when x is 2 or 3.

Use a grapher to check Examples 1–3.

Rational Equations and Graphs

One way to visualize the solution to Example 5 is to make a graph. This can be done by graphing

$$f(x) = x + \frac{6}{x}$$

with a computer, with a calculator, or by hand. We then inspect the graph for any x-values that are paired with 5. (Note that no y-value is paired with 0, since 0 is not in the domain of f.) It appears from the graph that $f(x) = 5$ when $x \approx 2$ or $x \approx 3$. Although making a graph is not the fastest or most precise method of solving a rational equation, it provides visualization and is useful when problems are too difficult to solve algebraically.

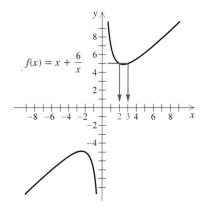

EXERCISE SET
6.4

Solve.

1. $\dfrac{1}{3} + \dfrac{4}{5} = \dfrac{x}{9}$

2. $\dfrac{7}{8} + \dfrac{2}{5} = \dfrac{x}{20}$

3. $\dfrac{x}{3} - \dfrac{x}{4} = 12$

4. $\dfrac{y}{5} - \dfrac{y}{3} = 15$

5. $\dfrac{5}{8} - \dfrac{1}{a} = \dfrac{2}{5}$

6. $\dfrac{1}{3} - \dfrac{1}{x} = \dfrac{5}{6}$

7. $\dfrac{2}{3} - \dfrac{1}{5} = \dfrac{7}{3x}$

8. $\dfrac{1}{2} - \dfrac{2}{7} = \dfrac{3}{2x}$

9. $\dfrac{2}{6} + \dfrac{1}{2x} = \dfrac{1}{3}$

10. $\dfrac{12}{15} - \dfrac{1}{3x} = \dfrac{4}{5}$

11. $y + \dfrac{4}{y} = -5$

12. $\dfrac{4}{3y} - \dfrac{3}{y} = \dfrac{10}{3}$

13. $\dfrac{y-1}{y-3} = \dfrac{2}{y-3}$

14. $\dfrac{x-2}{x-4} = \dfrac{2}{x-4}$

15. $\dfrac{3}{x-2} = \dfrac{5}{x+4}$

16. $\dfrac{5}{4t} = \dfrac{7}{5t-2}$

17. $\dfrac{x^2-1}{x+2} = \dfrac{3}{x+2}$

18. $\dfrac{x^2+4}{x-1} = \dfrac{5}{x-1}$

19. $\dfrac{4}{a-7} = \dfrac{-2a}{a+3}$

20. $\dfrac{6}{a+1} = \dfrac{a}{a-1}$

21. $\dfrac{50}{t-2} - \dfrac{16}{t} = \dfrac{30}{t}$

22. $\dfrac{60}{t-5} - \dfrac{18}{t} = \dfrac{40}{t}$

23. $\dfrac{3}{x} + \dfrac{x}{x+2} = \dfrac{4}{x^2+2x}$

24. $\dfrac{x}{x+1} + \dfrac{5}{x} = \dfrac{1}{x^2+x}$

In Exercises 25–30, a rational function f is given. Find all values of a for which f(a) is the indicated value.

25. $f(x) = 2x - \dfrac{15}{x}$; $f(a) = 1$

26. $f(x) = 2x - \dfrac{6}{x}$; $f(a) = 1$

27. $f(x) = \dfrac{x-5}{x+1}$; $f(a) = \dfrac{3}{5}$

28. $f(x) = \dfrac{x-3}{x+2}$; $f(a) = \dfrac{1}{5}$

29. $f(x) = \dfrac{12}{x} - \dfrac{12}{2x}$; $f(a) = 8$

30. $f(x) = \dfrac{6}{x} - \dfrac{6}{2x}$; $f(a) = 5$

Solve.

31. $\dfrac{5}{x+2} - \dfrac{3}{x-2} = \dfrac{2x}{4-x^2}$

32. $\dfrac{y+3}{y+2} - \dfrac{y}{y^2-4} = \dfrac{y}{y-2}$

33. $\dfrac{2}{a+4} + \dfrac{2a-1}{a^2+2a-8} = \dfrac{1}{a-2}$

34. $\dfrac{3}{x^2-6x+9} + \dfrac{x-2}{3x-9} = \dfrac{x}{2x-6}$

35. $\dfrac{2}{x+3} - \dfrac{3x+5}{x^2+4x+3} = \dfrac{5}{x+1}$

36. $\dfrac{3-2y}{y+1} - \dfrac{10}{y^2-1} = \dfrac{2y+3}{1-y}$

37. $\dfrac{x-1}{x^2-2x-3} + \dfrac{x+2}{x^2-9} = \dfrac{2x+5}{x^2+4x+3}$

38. $\dfrac{2x+1}{x^2-3x-10} + \dfrac{x-1}{x^2-4} = \dfrac{3x-1}{x^2-7x+10}$

39. $\dfrac{3}{x^2-x-12} + \dfrac{1}{x^2+x-6} = \dfrac{4}{x^2+3x-10}$

40. $\dfrac{3}{x^2-2x-3} - \dfrac{1}{x^2-1} = \dfrac{2}{x^2-8x+7}$

SKILL MAINTENANCE

41. Factor completely: $81x^4 - y^4$.

42. Determine whether each of the following systems is consistent or inconsistent.

  a) $2x - 3y = 4,$
     $4x - 6y = 7$
  b) $x + 3y = 2,$
     $2x - 3y = 1$

43. Solve: $|x - 2| > 3$.

44. *Test questions.* There are 70 questions on a test. The questions are either multiple-choice, true–false, or fill-in. There are twice as many true–false as fill-in and 5 fewer multiple-choice than true–false. How many of each type of question are there on the test?

45. Find two consecutive even numbers whose product is 288.

46. Simplify: $(a^2 b^3)^5 / (a^{-4} b^2)^3$.

SYNTHESIS

47. ◈ Is the following statement true or false: "For any real numbers a, b, and c, if $ac = bc$, then $a = b$"? Explain why you answered as you did.

48. ◈ When checking a possible solution of a rational equation, is it sufficient to check that the "solution" does not make any denominator equal to 0? Why or why not?

49. ◈ Explain how one can easily produce rational equations for which no solution exists. (*Hint*: Examine Example 2.)

50. ◈ Below are unlabeled graphs of $f(x) = x + 2$ and $g(x) = (x^2 - 4)/(x - 2)$. How could you determine which graph represents f and which graph represents g?

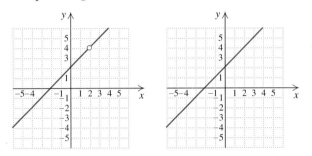

For each pair of functions f and g, find all values of a for which $f(a) = g(a)$.

51. $f(x) = \dfrac{2 + \dfrac{x}{2}}{2 - \dfrac{x}{4}}, \quad g(x) = \dfrac{2}{\dfrac{x}{4} - 2}$

52. $f(x) = \dfrac{x - \dfrac{3}{2}}{x + \dfrac{2}{3}}, \quad g(x) = \dfrac{x + \dfrac{1}{2}}{x - \dfrac{2}{3}}$

53. $f(x) = \dfrac{1}{1 + x} + \dfrac{x}{1 - x}, \quad g(x) = \dfrac{1}{1 - x} - \dfrac{x}{1 + x}$

54. $f(x) = \dfrac{x + 3}{x + 2} - \dfrac{x + 4}{x + 3}, \quad g(x) = \dfrac{x + 5}{x + 4} - \dfrac{x + 6}{x + 5}$

55. ▦ $f(x) = \dfrac{0.793}{x} + 18.15, \quad g(x) = \dfrac{6.034}{x} - 43.17$

56. ▦ $f(x) = \dfrac{2.315}{x} - \dfrac{12.6}{17.4}, \quad g(x) = \dfrac{6.71}{x} + 0.763$

Recall that identities are true for any possible replacement of the variable(s). Determine whether each of the following equations is an identity.

57. $\dfrac{x^2 + 6x - 16}{x - 2} = x + 8, \ x \neq 2$

58. $\dfrac{x^3 + 8}{x^2 - 4} = \dfrac{x^2 - 2x + 4}{x - 2}, \ x \neq -2, x \neq 2$

59. ▧ Use a grapher to check your answers to Exercises 3, 27, and 33.

60. ▧ Use a grapher to check your answers to Exercises 4, 18, and 28.

61. ▧ Use a grapher with a TABLE feature to show that 2 is not in the domain of f, if $f(x) = (x^2 - 4)/(x - 2)$. (See Exercise 50.)

62. ◈ ▧ Can Exercise 61 be answered on a grapher using only graphs? Why or why not?

6.5   Solving Applications Using Rational Equations

Problems Involving Work • Problems Involving Motion

Now that we are able to solve rational equations, we can solve certain problems that we could not have handled before. The five problem-solving steps remain the same.

Problems Involving Work

Lon can mow a lawn in 4 hr. Penny can mow the same lawn in 5 hr. How long would it take both of them, working together, to mow the lawn?

S O L U T I O N

1. **Familiarize.** We familiarize ourselves with the problem by considering two *incorrect* ways of translating the problem to mathematical language.

   a) One *incorrect* way to translate the problem is to add the two times:

   $$4 \text{ hr} + 5 \text{ hr} = 9 \text{ hr}.$$

   Now think about this. Lon can do the job *alone* in 4 hr. If Lon and Penny work together, whatever time it takes them must be *less* than 4 hr.

   b) Another *incorrect* approach is to assume that each person mows half the lawn. Were this the case,

   Lon would mow $\frac{1}{2}$ the lawn in $\frac{1}{2}(4 \text{ hr})$, or 2 hr

   and

   Penny would mow $\frac{1}{2}$ the lawn in $\frac{1}{2}(5 \text{ hr})$, or $2\frac{1}{2}$ hr.

   But time would be wasted since Lon would finish $\frac{1}{2}$ hr before Penny. Were Lon to help Penny after completing his half, the entire job would take between 2 and $2\frac{1}{2}$ hr. This information provides a partial check on any answer we get—the answer should be between 2 and $2\frac{1}{2}$ hr.

Let's consider how much of the job each person completes in 1 hr, 2 hr, 3 hr, and so on. Since Lon takes 4 hr to mow the entire lawn, in 1 hr he mows $\frac{1}{4}$ of the lawn. Since Penny takes 5 hr to mow the entire lawn, in 1 hr she mows $\frac{1}{5}$ of the lawn. Together, Lon and Penny mow

$\frac{1}{4} + \frac{1}{5}$ of the lawn in 1 hr.

In 2 hr, Lon mows $\frac{1}{4} \cdot 2$ of the lawn and Penny mows $\frac{1}{5} \cdot 2$ of the lawn. Together, they mow

$\frac{1}{4} \cdot 2 + \frac{1}{5} \cdot 2$ of the lawn in 2 hr.

Continuing this pattern, we can form a table like the following one.

Time	Fraction of the Lawn Mowed		
	By Lon	**By Penny**	**Together**
1 hr	$\frac{1}{4}$	$\frac{1}{5}$	$\frac{1}{4} + \frac{1}{5}$, or $\frac{9}{20}$
2 hr	$\frac{1}{4} \cdot 2$	$\frac{1}{5} \cdot 2$	$\frac{1}{4} \cdot 2 + \frac{1}{5} \cdot 2$, or $\frac{9}{10}$
3 hr	$\frac{1}{4} \cdot 3$	$\frac{1}{5} \cdot 3$	$\frac{1}{4} \cdot 3 + \frac{1}{5} \cdot 3$, or $1\frac{7}{20}$
t hr	$\frac{1}{4} \cdot t$	$\frac{1}{5} \cdot t$	$\frac{1}{4} \cdot t + \frac{1}{5} \cdot t$

From the table, we note that what is needed is the number of hours t required for Lon and Penny to mow exactly one lawn.

2. **Translate.** From the table, we see that t must be some number for which

$$\frac{1}{4} \cdot t + \frac{1}{5} \cdot t = 1,$$

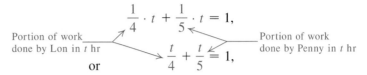

Portion of work done by Lon in t hr

Portion of work done by Penny in t hr

or

$$\frac{t}{4} + \frac{t}{5} = 1,$$

where 1 represents the idea that one entire job is completed in t hours.

3. **Carry out.** We solve the equation:

$$\frac{t}{4} + \frac{t}{5} = 1$$

$$20\left(\frac{t}{4} + \frac{t}{5}\right) = 20 \cdot 1 \qquad \text{Multiplying by the LCD}$$

$$20 \cdot \frac{t}{4} + 20 \cdot \frac{t}{5} = 20 \qquad \text{Using the distributive law}$$

$$5t + 4t = 20$$

$$9t = 20$$

$$t = \frac{20}{9}, \text{ or } 2\frac{2}{9} \text{ hr.}$$

4. **Check.** In $\frac{20}{9}$ hr, Lon mows $\frac{1}{4} \cdot \frac{20}{9}$, or $\frac{5}{9}$, of the lawn and Penny mows $\frac{1}{5} \cdot \frac{20}{9}$, or $\frac{4}{9}$, of the lawn. Together, they mow $\frac{5}{9} + \frac{4}{9}$, or 1 lawn. The fact that our solution is between 2 and $2\frac{1}{2}$ hr (see step 1 above) is also a check.

5. **State.** It will take $2\frac{2}{9}$ hr for Lon and Penny, working together, to mow the lawn.

EXAMPLE 2

It takes Red 9 hr more than Hannah to construct a stone wall. Working together, they can build the wall in 20 hr. How long would it take each, working alone, to build the wall?

SOLUTION

1. **Familiarize.** Unlike Example 1, this problem does not provide us with the times required by the individuals to do the job alone. Let's have

h = the amount of time it would take Hannah working alone and $h + 9$ = the amount of time it would take Red working alone.

2. **Translate.** Using the same reasoning as in Example 1, we see that Hannah can build $\dfrac{1}{h}$ of a wall in 1 hr and Red can build $\dfrac{1}{h + 9}$ of a wall in 1 hr. In 20 hr, Hannah builds $\left(\dfrac{1}{h}\right)20$, or $\dfrac{20}{h}$ of the wall and Red builds $\left(\dfrac{1}{h + 9}\right)20$, or $\dfrac{20}{h + 9}$ of the wall. Since Hannah and Red complete 1 entire wall in 20 hr, we have

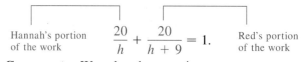

Hannah's portion of the work $\dfrac{20}{h} + \dfrac{20}{h + 9} = 1.$ Red's portion of the work

3. **Carry out.** We solve the equation:

$$\frac{20}{h} + \frac{20}{h + 9} = 1$$

$$h(h + 9)\left(\frac{20}{h} + \frac{20}{h + 9}\right) = h(h + 9)1 \qquad \text{Multiplying by the LCD}$$

$$(h + 9)20 + h \cdot 20 = h(h + 9) \qquad \text{Distributing and simplifying}$$

$$40h + 180 = h^2 + 9h$$

$$0 = h^2 - 31h - 180 \qquad \text{Getting 0 on one side}$$

$$0 = (h - 36)(h + 5) \qquad \text{Factoring}$$

$$h - 36 = 0 \quad or \quad h + 5 = 0 \qquad \text{Principle of zero products}$$

$$h = 36 \quad or \qquad h = -5.$$

4. **Check.** Since negative time has no meaning in the problem, -5 is not a solution to the original problem. The number 36 checks since, if Hannah takes 36 hr alone and Red takes $36 + 9 = 45$ hr alone, in 20 hr they would have completed

$$\frac{20}{36} + \frac{20}{45} = \frac{5}{9} + \frac{4}{9} = 1 \text{ wall.}$$

5. **State.** It would take Hannah 36 hr to build the wall alone, and Red 45 hr.

The equations used in Examples 1 and 2 can be generalized, as follows.

**MODELING WORK
PROBLEMS**

If

a = the time needed for A to complete the work alone,

b = the time needed for B to complete the work alone, and

t = the time needed for A and B to complete the work together,

then

$$\frac{t}{a} + \frac{t}{b} = 1.$$

Problems Involving Motion

Problems dealing with distance, rate (or speed), and time are called **motion problems**. To translate them, we use either the basic motion formula, $d = rt$, or the formulas $r = d/t$ or $t = d/r$, which can be derived from $d = rt$.

EXAMPLE 3

A racer is bicycling 15 km/h faster than a person on a mountain bike. In the time it takes the racer to travel 80 km, the person on the mountain bike has gone 50 km. Find the speed of each bicyclist.

SOLUTION

1. **Familiarize.** Let's guess that the person on the mountain bike is going 10 km/h. The racer would then be traveling 10 + 15, or 25 km/h. At 25 km/h, the racer will travel 80 km in $\frac{80}{25} = 3.2$ hr. Going 10 km/h, the mountain bike will cover 50 km in $\frac{50}{10} = 5$ hr. Since 3.2 ≠ 5, our guess was wrong, but we can see that if r = the rate, in kilometers per hour, of the slower bike, then the rate of the racer = $r + 15$.

   Drawing a sketch and constructing a table can be helpful.

	Distance	Speed	Time
Mountain Bike	50	r	t
Racing Bike	80	$r + 15$	t

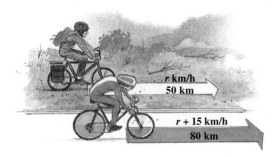

2. **Translate.** By looking at how we checked our guess, we see that in the **Time** column of the table, the t's can be replaced, using the formula *Time = Distance/Rate*, as follows.

	Distance	Speed	Time
Mountain Bike	50	r	$50/r$
Racing Bike	80	$r + 15$	$80/(r + 15)$

Since we are told that the times must be the same, we can write an equation:

$$\frac{50}{r} = \frac{80}{r + 15}.$$

3. **Carry out.** We solve the equation:

$$\frac{50}{r} = \frac{80}{r + 15}$$

$$r(r + 15)\frac{50}{r} = r(r + 15)\frac{80}{r + 15} \qquad \text{Multiplying by the LCD}$$

$$50r + 750 = 80r \qquad \text{Simplifying}$$

$$750 = 30r$$

$$25 = r.$$

4. **Check.**   If our answer checks, the mountain bike is going 25 km/h and the racing bike is going $25 + 15 = 40$ km/h.

   Traveling 80 km at 40 km/h, the racer is riding for $\frac{80}{40} = 2$ hr. Traveling 50 km at 25 km/h, the person on the mountain bike is riding for $\frac{50}{25} = 2$ hr. Our answer checks since the two times are the same.

5. **State.**   The speed of the racer is 40 km/h, and the speed of the person on the mountain bike is 25 km/h.   ⬤▬

In the following example, although the distance is the same in both directions, the key to the translation lies in an additional piece of given information.

EXAMPLE   4   Sandy's tugboat goes 10 mph in still water. It travels 24 mi upstream and 24 mi back in a total time of 5 hr. What is the speed of the current?

SOLUTION

1. **Familiarize.**   Let's guess that the speed of the current is 4 mph. The tugboat would then be moving $10 - 4 = 6$ mph upstream and $10 + 4 = 14$ mph downstream. The tugboat would require $\frac{24}{6} = 4$ hr to travel 24 mi upstream and $\frac{24}{14} = 1\frac{5}{7}$ hr to travel 24 mi downstream. Since the total time, $4 + 1\frac{5}{7} = 5\frac{5}{7}$ hr, is not the 5 hr mentioned in the problem, we know that our guess is wrong.

   Suppose that the current's speed $= c$ mph. The tugboat would then travel $10 - c$ mph when going upstream and $10 + c$ mph when going downstream.

   A sketch and table can help display the information.

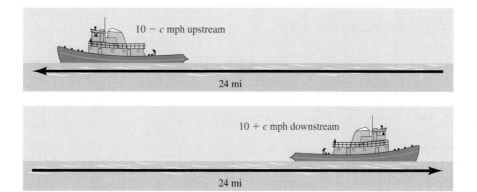

	Distance	Speed	Time
Upstream	24	$10 - c$	t_1
Downstream	24	$10 + c$	t_2

2. **Translate.**   From examining our guess, we see that the time traveled can be represented using the formula *Time = Distance/Rate*:

	Distance	Speed	Time
Upstream	24	$10 - c$	$24/(10 - c)$
Downstream	24	$10 + c$	$24/(10 + c)$

Since the total time upstream and back is 5 hr, we use the last column of the table to form an equation:

$$\frac{24}{10 - c} + \frac{24}{10 + c} = 5.$$

3. **Carry out.**   We solve the equation:

$$\frac{24}{10 - c} + \frac{24}{10 + c} = 5$$

$$(10 - c)(10 + c)\left[\frac{24}{10 - c} + \frac{24}{10 + c}\right] = (10 - c)(10 + c)5 \qquad \text{Multiplying by the LCD}$$

$$24(10 + c) + 24(10 - c) = (100 - c^2)5$$

$$480 = 500 - 5c^2 \qquad \text{Simplifying}$$

$$5c^2 - 20 = 0$$

$$5(c^2 - 4) = 0$$

$$5(c - 2)(c + 2) = 0$$

$$c = 2 \quad or \quad c = -2.$$

4. **Check.**   Since speed cannot be negative in this problem, -2 cannot be a solution. You should confirm that 2 checks in the original problem.

5. **State.**   The speed of the current is 2 mph.

EXERCISE SET

6.5

Solve.

1. The reciprocal of 3, plus the reciprocal of 6, is the reciprocal of what number?

2. The reciprocal of 5, plus the reciprocal of 7, is the reciprocal of what number?

3. The sum of a number and 6 times its reciprocal is −5. Find the number.

4. The sum of a number and 21 times its reciprocal is −10. Find the number.

5. The reciprocal of the product of two consecutive integers is $\frac{1}{42}$. Find the two integers.

6. The reciprocal of the product of two consecutive integers is $\frac{1}{72}$. Find the two integers.

7. *Painting.* Otto can paint a room in 4 hr. Sally can paint the same room in 3 hr. Working together, how long will it take them to paint the room?

8. *Mail order.* Zoe, an experienced shipping clerk, can fill a certain order in 5 hr. Willy, a new clerk, needs 9 hr to complete the same job. Working together, how long will it take them to fill the order?

9. *Filling a pool.* A swimming pool can be filled in 12 hr if water enters through a pipe alone or in 30 hr if water enters through a hose alone. If water is entering through both the pipe and the hose, how long will it take to fill the pool?

10. *Filling a tank.* A tank can be filled in 18 hr by pipe A alone and in 22 hr by pipe B alone. How long will it take to fill the tank if both pipes are working?

11. *Printing.* Pronto Press can print an order of booklets in 4.5 hr. Red Dot Printers can do the same job in 5.5 hr. How long will it take if both presses are used?

12. *Wood cutting.* Damon can clear a lot in 5.5 hr. His partner, Tyron, can complete the same job in 7.5 hr. How long will it take them to clear the lot working together?

13. *Sanding.* Mavis can sand the living room floor in 3 hr. When she works together with Henri, the job takes 2 hr. How long would it take Henri, working by himself, to sand the floor?

14. *Cutting firewood.* Jake can cut and split a cord of firewood in 6 fewer hr than Skyler can. When they work together, it takes them 4 hr. How long would it take each of them to do the job alone?

15. *Painting.* Sara takes 3 hr longer to paint a floor than it takes Kate. When they work together, it takes them 2 hr. How long would each take to do the job alone?

16. *Painting.* Claudia can paint a neighbor's house 4 times as fast as Jan can. The year they worked together it took them 8 days. How long would it take each to paint the house alone?

17. *Newspaper delivery.* Zsuzanna can deliver papers 3 times as fast as Stan can. If they work together, it takes them 1 hr. How long would it take each to deliver the papers alone?

18. *Waxing a car.* Rosita can wax her car in 2 hr. When she works together with Helga, they can wax the car in 45 min. How long would it take Helga, working by herself, to wax the car?

19. *Sorting recyclables.* Together, it takes John and Deb 2 hr 55 min to sort recyclables. Alone, John would require 2 more hr than Deb. How long would it take Deb to do the job alone? (*Hint*: Convert minutes to hours or hours to minutes.)

20. *Paving.* Together, Larry and Mo require 4 hr 48 min to pave a driveway. Alone, Larry would require 4 hr more than Mo. How long would it take Mo to do the job alone? (*Hint*: Convert minutes to hours.)

21. *Copying.* A new photocopier works twice as fast as an old one. When the machines work together, a university can produce all its staff manuals in 15 hr. Find the time it would take each machine, working alone, to complete the same job.

22. *Completing a puzzle.*   Working together, Hans and Gina can complete a jigsaw puzzle in 1.5 hr. Hans takes 4 hr longer than Gina does when working alone. How long would it take Gina alone to complete the puzzle?

23. *Boating.*   The current in the Lazy River moves at a rate of 4 mph. Ken's dinghy motors 6 mi upstream in the same time it takes to motor 12 mi downstream. What is the speed of the dinghy in still water?

24. *Kayaking.*   The speed of the current in Catamount Creek is 3 mph. Ahmad's kayak can travel 4 mi upstream in the same time it takes to travel 10 mi downstream. What is the speed of Ahmad's kayak in still water?

25. *Moving sidewalks.*   The moving sidewalk at O'Hare Airport in Chicago moves 1.8 ft/sec. Walking on the moving sidewalk, Camille travels 105 ft forward in the time it takes to travel 51 ft in the opposite direction. How fast would Camille be walking on a nonmoving sidewalk?

26. *Moving sidewalks.*   Newark Airport's moving sidewalk moves at a speed of 1.7 ft/sec. Walking on the moving sidewalk, Benny can travel 120 ft forward in the same time it takes to travel 52 ft in the opposite direction. How fast would Benny be walking on a nonmoving sidewalk?

27. *Train speed.*   The speed of the A&M freight train is 14 mph less than the speed of the A&M passenger train. The passenger train travels 400 mi in the same time that the freight train travels 330 mi. Find the speed of each train.

28. *Walking.*   Rosanna walks 2 mph slower than Simone. In the time it takes Simone to walk 8 mi, Rosanna walks 5 mi. Find the speed of each person.

29. *Bus travel.*   A local bus travels 7 mph slower than the express. The express travels 90 mi in the time it takes the local to travel 75 mi. Find the speed of each bus.

30. *Train speed.*   The A train goes 12 mph slower than the B train. The A train travels 230 mi in the same time that the B train travels 290 mi. Find the speed of each train.

31. *Boating.*   Suzie has a boat that travels 15 km/h in still water. She motors 140 km downstream in the same time it takes to travel 35 km upstream. What is the speed of the river?

32. *Boating.*   A paddleboat travels 2 km/h in still water.

The boat is paddled 4 km downstream in the same time it takes to go 1 km upstream. What is the speed of the river?

33. *Moped speed.*   Jaime's moped travels 8 km/h faster than Mara's. Jaime travels 69 km in the same time that Mara travels 45 km. Find the speed of each person's moped.

34. *Shipping.*   A barge moves 7 km/h in still water. It travels 45 km upriver and 45 km downriver in a total time of 14 hr. What is the speed of the current?

35. *Swimming.*   Al swims 55 m per minute in still water. He swims 150 m upstream and 150 m downstream in a total time of 5.5 min. What is the speed of the current?

36. *Air travel.*   A plane travels 100 mph in still air. It travels 240 mi into the wind and 240 mi with the wind in a total time of 5 hr. Find the wind speed.

37. *Travel by van.*   Cecilia's van covered 120 mi at a certain speed. Had Cecilia driven 10 mph faster, the trip would have been 2 hr shorter. How fast did Cecilia drive?

38. *Boating.*   Carlos' Boston Whaler cruised 45 mi upstream and 45 mi back in a total of 8 hr. The speed of the river is 3 mph. Find the speed of the boat in still water.

SKILL MAINTENANCE

39. Determine the domain of f if
$$f(x) = \frac{x - 5}{x^2 - 4x - 5}.$$

40. Solve: $|x - 2| = 9$.

41. Combine similar terms:
$$4y - 5xy^2 + 6xy - 3xy^2 - 2y.$$

42. 240 is 16% of what number?

SYNTHESIS

43. ◆ Two steamrollers are paving a parking lot. Working together, will the two steamrollers take less than half as long as the slower steamroller would working alone? Why or why not?

44. ◆ Two fuel lines are filling a freighter with oil. Will the faster fuel line take more or less than twice as long to fill the freighter by itself? Why?

45. ◆ Write a work problem for a classmate to solve. Devise the problem so that the solution is "Liane and Michele will take 4 hr to complete the job, working together."

46. ◈ Write a work problem for a classmate to solve. Devise the problem so that the solution is "Jen takes 5 hr and Pablo takes 6 hr to complete the job alone."

47. *Filling a tub.*   A tub can be filled in 10 min and drained in 8 min. How long will it take to empty a full tub if the water is left on?

48. *Filling a tank.*   A tank can be filled in 9 hr and drained in 11 hr. How long will it take to fill the tank if the drain is left open?

49. *Escalators.*   Together, a 100-cm-wide escalator and a 60-cm-wide escalator can empty a 1575-person auditorium in 14 min (*Source*: *McGraw-Hill Encyclopedia of Science and Technology*). The wider escalator moves twice as many people as the narrower one. How many people per hour does the 60-cm-wide escalator move?

50. *Aviation.*   A Coast Guard plane has enough fuel to fly for 6 hr, and its speed in still air is 240 mph. The plane departs with a 40-mph tailwind and returns to the same airport flying into the same wind. How far can the plane travel under these conditions?

51. *Boating.*   Shoreline Travel operates a 3-hr paddleboat cruise on the Missouri River. If the speed of the boat in still water is 12 mph, how far upriver can the pilot travel against a 5-mph current before it is time to turn around?

52. *Boating.*   The speed of a motor boat in still water is 3 times the speed of a river's current. A trip up the river and back takes 10 hr, and the total distance of the trip is 100 km. Find the speed of the current.

53. *Travel by car.*   Melissa drives to work at 50 mph and arrives 1 min late. She drives to work at 60 mph and arrives 5 min early. How far does Melissa live from work?

54. At what time after 4:00 will the minute hand and the hour hand of a clock first be in the same position?

55. At what time after 10:30 will the hands of a clock first be perpendicular?

Average speed is defined as total distance divided by total time.

56. Lenore drove 200 km. For the first 100 km of the trip, she drove at a speed of 40 km/h. For the second half of the trip, she traveled at a speed of 60 km/h. What was the average speed for the entire trip? (It was *not* 50 km/h.)

57. For the first 50 mi of a 100-mi trip, Chip drove 40 mph. What speed would he have to travel for the last half of the trip so that the average speed for the entire trip would be 45 mph?

COLLABORATIVE
C ✦ O ✦ R ✦ N ✦ E ✦ R

Focus: Testing a mathematical model

Time: 20–30 minutes

Group size: 2–3

Materials: An empty 1-gal plastic jug, a kitchen or laboratory sink, a stopwatch or a watch capable of measuring seconds, an inexpensive pen or pair of scissors or a nail or knife for poking holes in plastic.

Problems like Exercises 47 and 48 can be solved algebraically and then checked at home or in a laboratory.

ACTIVITY

1. Turn the water faucet on full force. While one group member fills the empty jug with water, the other group member(s) should record how many seconds this takes.

2. After carefully poking several holes in the bottom of the jug, record how many seconds it takes the full jug to empty.

3. Using the information found in parts (1) and (2) above, use algebra to predict how long it will take to fill the punctured jug.

4. Test your prediction by again turning the water on full force and timing how long it takes for the pierced jug to be filled.

5. How accurate was your prediction? How might your prediction have been made more accurate?

6.6   Division of Polynomials

Divisor a Monomial   •   Divisor a Polynomial

A rational expression indicates division. Division of polynomials, like division of real numbers, relies on our multiplication and subtraction skills.

Divisor a Monomial

To divide a monomial by a monomial, we can subtract exponents when bases are the same (see Section 1.6). For example,

$$\frac{45x^{10}}{3x^4} = 15x^{10-4} = 15x^6, \qquad \frac{48a^2b^5}{-3ab^2} = \frac{48}{-3}a^{2-1}b^{5-2} = -16ab^3.$$

To divide a polynomial by a monomial, the division is regarded as a sum of quotients of monomials. This uses the fact that since

$$\frac{A}{C} + \frac{B}{C} = \frac{A+B}{C}, \quad \text{we know that} \quad \frac{A+B}{C} = \frac{A}{C} + \frac{B}{C}.$$

EXAMPLE 1

Divide $12x^3 + 8x^2 + x + 4$ by $4x$.

SOLUTION

$$
\begin{aligned}
(12x^3 + 8x^2 + x + 4) \div (4x) &= \frac{12x^3 + 8x^2 + x + 4}{4x} && \text{Writing a fractional expression} \\
&= \frac{12x^3}{4x} + \frac{8x^2}{4x} + \frac{x}{4x} + \frac{4}{4x} && \text{Writing as a sum of quotients} \\
&= 3x^2 + 2x + \frac{1}{4} + \frac{1}{x} && \text{Performing the four indicated divisions}
\end{aligned}
$$

EXAMPLE 2

Divide: $(8x^4y^5 - 3x^3y^4 + 5x^2y^3) \div x^2y^3$.

SOLUTION

$$
\begin{aligned}
\frac{8x^4y^5 - 3x^3y^4 + 5x^2y^3}{x^2y^3} &= \frac{8x^4y^5}{x^2y^3} - \frac{3x^3y^4}{x^2y^3} + \frac{5x^2y^3}{x^2y^3} && \text{Try to perform this step mentally.} \\
&= 8x^2y^2 - 3xy + 5
\end{aligned}
$$

DIVISION BY A MONOMIAL

To divide a polynomial by a monomial, divide each term of the polynomial by the monomial.

Divisor a Polynomial

When the divisor has more than one term, we use a procedure very similar to long division in arithmetic.

EXAMPLE 3

Divide $2x^2 - 7x - 15$ by $x - 5$.

SOLUTION   We have

$$
\begin{array}{r}
2x \\
x - 5 \overline{)2x^2 - 7x - 15} \qquad \text{Divide } 2x^2 \text{ by } x: \ 2x^2/x = 2x. \\
-(2x^2 - 10x) \qquad \text{Multiply } x - 5 \text{ by } 2x. \\
3x \qquad \text{Subtract by mentally changing signs} \\
\text{and adding: } -7x + 10x = 3x.
\end{array}
$$

We now "bring down" the other term in the dividend, -15.

$$
\begin{array}{r}
2x \ + \ 3 \\
x - 5 \overline{)2x^2 - 7x - 15} \\
2x^2 - 10x \\
\hline
3x - 15 \qquad \text{Divide } 3x \text{ by } x: \ 3x/x = 3. \\
-(3x - 15) \qquad \text{Multiply } x - 5 \text{ by } 3. \\
\hline
0 \qquad \text{Subtract.}
\end{array}
$$

The quotient is $2x + 3$.

Check: $(x - 5)(2x + 3) = 2x^2 - 7x - 15$. The answer checks.   ⟶

To understand why we perform long division as we do, note that Example 3 amounts to "filling in" an unknown polynomial:

$$(x - 5)(\ \ ? \ \ ) = 2x^2 - 7x - 15.$$

We see that $2x$ must be in the unknown polynomial if we are to get the first term, $2x^2$, from the multiplication. To see what else is needed, note that

$$(x - 5)(2x \ \ \ \ ) = 2x^2 - 10x \neq 2x^2 - 7x - 15.$$

The $2x^2 - 10x$ can be considered a (poor) approximation of $2x^2 - 7x - 15$. To see how far off the approximation is, we subtract:

$$
\left.
\begin{array}{r}
2x^2 - 7x - 15 \\
-(2x^2 - 10x) \\
\hline
3x - 15
\end{array}
\right\}
\qquad \text{Note where this appeared in the long division above.}
$$

To get the needed terms, $3x - 15$, we add the term 3 in the unknown polynomial. We use 3 because $(x - 5) \cdot 3$ is $3x - 15$:

$$
\begin{aligned}
(x - 5)(2x + 3) &= 2x^2 - 10x + 3x - 15 \\
&= 2x^2 - 7x - 15.
\end{aligned}
$$

If we now subtract from $2x^2 - 7x - 15$, the remainder is 0.

Should a nonzero remainder occur, when do we stop dividing? We continue until the degree of the remainder is less than the degree of the divisor.

EXAMPLE 4

Divide $x^2 + 5x + 8$ by $x + 3$.

SOLUTION   We have

$$
\begin{array}{r}
x \\
x + 3 \overline{) x^2 + 5x + 8} \\
\underline{x^2 + 3x} \\
2x
\end{array}
$$

Divide the first term of the dividend by the first term of the divisor: $x^2/x = x$.

Multiply x above by $x + 3$.

Subtract.

The subtraction we have done is $(x^2 + 5x) - (x^2 + 3x)$. Remember: To subtract, add the opposite (change the sign of every term, then add).

We now "bring down" the next term of the dividend—in this case, 8—and repeat the process:

$$
\begin{array}{r}
x + 2 \\
x + 3 \overline{) x^2 + 5x + 8} \\
\underline{x^2 + 3x} \\
2x + 8 \\
\underline{2x + 6} \\
2
\end{array}
$$

Divide the first term by the first term: $2x/x = 2$.

The 8 has been "brought down."

Multiply 2 by $x + 3$.

Subtract: $(2x + 8) - (2x + 6)$.

The quotient is $x + 2$, with remainder 2. Note that the degree of the remainder is 0 and the degree of the divisor, $x + 3$, is 1. Since $0 < 1$, the process stops.

Check: $(x + 3)(x + 2) + 2 = x^2 + 5x + 6 + 2$ Add the remainder to the product.

$$= x^2 + 5x + 8$$

We can write our answer as $x + 2$, R2, or we can write our answer as

$$\text{Quotient} + \frac{\text{Remainder}}{\text{Divisor}}$$

$$\underbrace{x + 2}_{} + \left(\frac{2}{x + 3} \right),$$

which is how answers for the problems in this section are listed at the back of the book. Answers given in this form can be checked by multiplying:

$$(x + 3)\left[(x + 2) + \frac{2}{x + 3} \right] = (x + 3)(x + 2) + (x + 3)\frac{2}{x + 3}$$

Using the distributive law

$$= x^2 + 5x + 6 + 2$$

$$= x^2 + 5x + 8. \qquad \text{This was the dividend above.}$$

You may have noticed that in each example all polynomials were written in descending order. When this is not the case, we rearrange terms before dividing.

TIPS FOR DIVIDING POLYNOMIALS

1. Arrange polynomials in descending order.
2. If there are missing terms in the dividend, either write them with 0 coefficients or leave space for them.
3. Continue the long division process until the degree of the remainder is less than the degree of the divisor.

EXAMPLE 5

Divide: $(9a^2 + a^3 - 5) \div (a^2 - 1)$.

SOLUTION   We rewrite the problem in descending order:

$$(a^3 + 9a^2 - 5) \div (a^2 - 1).$$

Thus,

$$
\require{enclose}
\begin{array}{r}
a + 9 \\
a^2 - 1 \enclose{longdiv}{a^3 + 9a^2 + 0a - 5} \\
\underline{a^3 - a} \\
9a^2 + a - 5 \\
\underline{9a^2 - 9} \\
a + 4
\end{array}
$$

When there is a missing term, we can write it in, as in this example, or leave space, as in Example 6.

Subtracting: $a^3 + 9a^2 - (a^3 - a) = 9a^2 + a$

The degree of the remainder is less than the degree of the divisor, so we are finished.

The answer is $a + 9 + \dfrac{a + 4}{a^2 - 1}$.

EXAMPLE 6

Let $f(x) = 125x^3 - 8$ and $g(x) = 5x - 2$. If $F(x) = (f/g)(x)$, find an expression for $F(x)$.

SOLUTION   Recall that $(f/g)(x) = f(x)/g(x)$. Thus,

$$F(x) = \frac{125x^3 - 8}{5x - 2}$$

and

$$
\require{enclose}
\begin{array}{r}
25x^2 + 10x + 4 \\
5x - 2 \enclose{longdiv}{125x^3 - 8} \\
\underline{125x^3 - 50x^2} \\
50x^2 \\
\underline{50x^2 - 20x} \\
20x - 8 \\
\underline{20x - 8} \\
0.
\end{array}
$$

Leaving space for the missing terms.

Subtracting:
$125x^3 - (125x^3 - 50x^2) = 50x^2$

Subtracting

Note that, because $F(x) = f(x)/g(x)$, $g(x)$ cannot be 0. Since $g(x)$ is 0 for $x = \frac{2}{5}$ (check this), we have

$$F(x) = 25x^2 + 10x + 4, \quad \text{provided } x \ne \tfrac{2}{5}.$$

EXERCISE SET
6.6

Divide and check.

1. $\dfrac{24x^6 + 18x^5 - 36x^2}{6x^2}$ **2.** $\dfrac{30y^8 - 15y^6 + 40y^4}{5y^4}$

3. $\dfrac{28a^3 + 7a^2 - 3a - 14}{7a}$

4. $\dfrac{-40x^3 + 20x^2 - 3x + 7}{5x}$

5. $(26y^3 - 9y^2 - 8y) \div (2y^2)$

6. $(6a^4 + 9a^2 - 8) \div (2a)$

7. $(18x^7 - 27x^4 - 3x^2) \div (-3x^2)$

8. $(36y^6 - 18y^4 - 12y^2) \div (-6y)$

9. $(a^2b - a^3b^3 - a^5b^5) \div (a^2b)$

10. $(x^3y^2 - x^3y^3 - x^4y^2) \div (x^2y^2)$

11. $(6p^2q^2 - 9p^2q + 12pq^2) \div (-3pq)$

12. $(16y^4z^2 - 8y^6z^4 + 12y^8z^3) \div (4y^4z)$

13. $(x^2 + 10x + 21) \div (x + 3)$

14. $(y^2 - 8y + 16) \div (y - 4)$

15. $(a^2 - 8a - 16) \div (a + 4)$

16. $(y^2 - 10y - 25) \div (y - 5)$

17. $(x^2 - 11x + 23) \div (x - 5)$

18. $(x^2 - 11x + 23) \div (x - 7)$

19. $(y^2 - 25) \div (y + 5)$

20. $(a^2 - 81) \div (a - 9)$

21. $(y^3 - 4y^2 + 3y - 6) \div (y - 2)$

22. $(x^3 - 5x^2 + 4x - 7) \div (x - 3)$

23. $(2x^3 + 3x^2 - x - 3) \div (x + 2)$

24. $(3x^3 - 5x^2 - 3x - 2) \div (x - 2)$

25. $(a^3 - a + 12) \div (a - 4)$

26. $(x^3 - x + 6) \div (x + 2)$

27. $(10y^3 + 6y^2 - 9y + 10) \div (5y - 2)$

28. $(6x^3 - 11x^2 + 11x - 2) \div (2x - 3)$

29. $(2x^4 - x^3 - 5x^2 + x - 6) \div (x^2 + 2)$

30. $(3x^4 + 2x^3 - 11x^2 - 2x + 5) \div (x^2 - 2)$

For Exercises 31–38, $f(x)$ and $g(x)$ are as given. Find a simplified expression for $F(x)$ if $F(x) = (f/g)(x)$. (See Example 6.)

31. $f(x) = 64x^3 - 8, \ g(x) = 4x - 2$

32. $f(x) = 8x^3 + 27, \ g(x) = 2x + 3$

33. $f(x) = 6x^2 - 11x - 10, \ g(x) = 3x + 2$

34. $f(x) = 8x^2 - 22x - 21, \ g(x) = 2x - 7$

35. $f(x) = x^4 - 3x^2 - 54, \ g(x) = x^2 - 9$

36. $f(x) = x^4 - 24x^2 - 25, \ g(x) = x^2 - 25$

37. $f(x) = 2x^5 - 3x^4 - 2x^3 + 8x^2 - 5, \ g(x) = x^2 - 1$

38. $f(x) = 3x^4 - x^3 - 10x^2 + 4x - 8, \ g(x) = x^2 - 4$

SKILL MAINTENANCE

Solve.

39. $x^2 - 5x = 0$ **40.** $25y^2 = 64$

41. Find three consecutive positive integers such that the product of the first and second integers is 26 less than the product of the second and third integers.

42. If $f(x) = 2x^3$, find $f(-3a)$.

Solve.

43. $|2x - 3| = 7$ **44.** $|3x - 1| < 8$

SYNTHESIS

45. ◈ Do addition, subtraction, and multiplication of polynomials always result in a polynomial? Does division? Why or why not?

46. ◈ Explain how you could construct a polynomial of degree 4 that has a remainder of 3 when divided by $x + 1$.

47. ◈ Can the quotient of two sums always be rewritten as a sum of two quotients? Why or why not?

48. ◈ Explain how factoring could be used to solve Example 6.

Divide.

49. $(x^4 - x^3y + x^2y^2 + 2x^2y - 2xy^2 + 2y^3) \div$
    $(x^2 - xy + y^2)$

50. $(4a^3b + 5a^2b^2 + a^4 + 2ab^3) \div (a^2 + 2b^2 + 3ab)$

51. $(a^7 + b^7) \div (a + b)$

52. Find k such that when $x^3 - kx^2 + 3x + 7k$ is divided
    by $x + 2$, the remainder is 0.

53. When $x^2 - 3x + 2k$ is divided by $x + 2$, the
    remainder is 7. Find k.

54. Let

$$f(x) = \frac{3x + 7}{x + 2}.$$

a) Use division to find an expression equivalent to
    $f(x)$. Then graph f.
b) On the same set of axes, sketch both
    $g(x) = 1/(x + 2)$ and $h(x) = 1/x$.

c) How do the graphs of f, g, and h compare?

55. ◈ Jamaladeen incorrectly states that

$$(x^3 + 9x^2 - 6) \div (x^2 - 1) = x + 9 + \frac{x + 4}{x^2 - 1}.$$

Without performing any long division, how could
you show Jamaladeen that his division cannot
possibly be correct?

56. ◈ ⊞ Check Example 3 by setting
    $y_1 = (2x^2 - 7x - 15)/(x - 5)$ and $y_2 = 2x + 3$.
    Then use either the TRACE feature (after selecting the
    ZOOM Z INTEGER option) or the TABLE feature (with
    TblMin = 0 and $\triangle$Tbl = 1) to show that $y_1 \neq y_2$ for
    $x = 5$.

57. ⊞ Use a grapher to check Example 5. Perform the
    check using $y_1 = (9x^2 + x^3 - 5)/(x^2 - 1)$,
    $y_2 = x + 9 + (x + 4)/(x^2 - 1)$, and $y_3 = y_2 - y_1$.

6.7   Synthetic Division

Streamlining Long Division   •   The Remainder Theorem

Streamlining Long Division

To divide a polynomial by a binomial of the type $x - a$, we can streamline
the usual procedure to develop a process called *synthetic division.*

Compare the following. In each stage, we attempt to write a bit less than
in the previous stage, while retaining enough essentials to solve the problem.
At the end, we will return to the usual polynomial notation.

Stage 1
When a polynomial is written in descending order, the coefficients provide
the essential information:

$$
\begin{array}{r}
4x^2 + 5x + 11 \\
x - 2 \overline{)\,4x^3 - 3x^2 + x + 7} \\
\underline{4x^3 - 8x^2} \\
5x^2 + x \\
\underline{5x^2 - 10x} \\
11x + 7 \\
\underline{11x - 22} \\
29
\end{array}
$$

$$
\begin{array}{r}
4 + 5 + 11 \\
1 - 2 \overline{)\,4 - 3 + 1 + 7} \\
\underline{4 - 8} \\
5 + 1 \\
\underline{5 - 10} \\
11 + 7 \\
\underline{11 - 22} \\
29
\end{array}
$$

Because the coefficient of x is 1 in the divisor, each time we multiply the divisor by a term in the answer, the leading coefficient of that product duplicates a coefficient in the answer. In the next stage, we don't bother to duplicate these numbers. We also show where -2 is used and drop the 1 from the divisor.

Stage 2

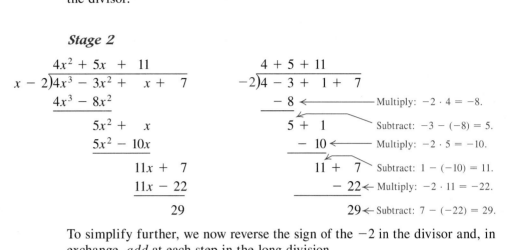

To simplify further, we now reverse the sign of the -2 in the divisor and, in exchange, *add* at each step in the long division.

Stage 3

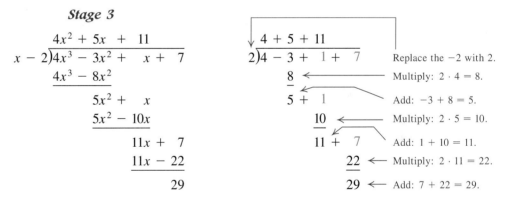

The blue numbers highlight some duplication that can be eliminated.

Stage 4

$$
\begin{array}{r}
4x^2 + 5x + 11 \\
x - 2\overline{)4x^3 - 3x^2 + x + 7} \\
\underline{4x^3 - 8x^2} \\
5x^2 + x \\
\underline{5x^2 - 10x} \\
11x + 7 \\
\underline{11x - 22} \\
29
\end{array}
$$

$$
\begin{array}{r}
4 \quad 5 \quad 11 \\
2\overline{)4 \quad -3 \quad 1 \quad 7} \\
\underline{8 \quad 10 \quad 22} \\
5 \quad 11 \quad 29
\end{array}
$$

Don't lose sight of how the products 8, 10, and 22 are found. Also, note that the 5 and 11 preceding the remainder 29 coincide with the 5 and 11 following the 4 on the top line. By writing a 4 to the left of 5 on the bottom line, we can eliminate the top line in stage 4 and read our answer from the bottom line. This final stage is commonly called **synthetic division**.

Stage 5

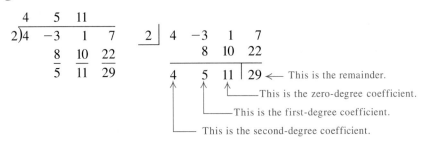

This is the remainder.
This is the zero-degree coefficient.
This is the first-degree coefficient.
This is the second-degree coefficient.

The quotient is $4x^2 + 5x + 11$. The remainder is 29.

> Remember that in order for this method to work, the divisor must be of the form $x - a$, that is, a variable minus a constant. The coefficient of the variable must be 1.

EXAMPLE 1   Use synthetic division to divide: $(x^3 + 6x^2 - x - 30) \div (x - 2)$.

SOLUTION

$$2\rfloor\ \ 1\ \ \ 6\ \ \ -1\ \ \ -30$$

Write the 2 of $x - 2$ and the coefficients of the dividend.

$$1$$

Bring down the first coefficient.

$$2\rfloor\ \ 1\ \ \ 6\ \ \ -1\ \ \ -30$$
$$2$$
$$1\ \ \ 8$$

Multiply 1 by 2 to get 2.

Add 6 and 2.

$$2\rfloor\ \ 1\ \ \ 6\ \ \ -1\ \ \ -30$$
$$2\ \ \ 16$$
$$1\ \ \ 8\ \ \ 15$$

Multiply 8 by 2.

Add -1 and 16.

$$2\rfloor\ \ 1\ \ \ 6\ \ \ -1\ \ \ -30$$
$$2\ \ \ 16\ \ \ 30$$
$$1\ \ \ 8\ \ \ 15\ \ \ 0$$

Multiply 15 by 2 and add.

The answer is $x^2 + 8x + 15$ with R0, or just $x^2 + 8x + 15$.

EXAMPLE 2   Use synthetic division to divide.

a) $(2x^3 + 7x^2 - 5) \div (x + 3)$ **b)** $(10x^2 - 13x + 3x^3 - 20) \div (4 + x)$

SOLUTION

a) $(2x^3 + 7x^2 - 5) \div (x + 3)$

The dividend has no x-term, so we must write a 0 for its coefficient of x. Note that $x + 3 = x - (-3)$, so we write -3 inside the $\rfloor$.

$$
\begin{array}{r|rrrr}
-3 & 2 & 7 & 0 & -5 \\
   &   & -6 & -3 & 9 \\
\hline
   & 2 & 1 & -3 & 4 \\
\end{array}
$$

The answer is $2x^2 + x - 3$, with R4, or $2x^2 + x - 3 + \dfrac{4}{x+3}$.

b) We first rewrite $(10x^2 - 13x + 3x^3 - 20) \div (4 + x)$ in descending order:

$$(3x^3 + 10x^2 - 13x - 20) \div (x + 4).$$

Next, we use synthetic division. Note that $x + 4 = x - (-4)$.

$$
\begin{array}{r|rrrr}
-4 & 3 & 10 & -13 & -20 \\
   &   & -12 & 8 & 20 \\
\hline
   & 3 & -2 & -5 & 0 \\
\end{array}
$$

The answer is $3x^2 - 2x - 5$.

TECHNOLOGY CONNECTION

6.7

In Example 1, the division by $x - 2$ gave a remainder of 0. The remainder theorem tells us that this means that when $x = 2$, the value of $x^3 + 6x^2 - x - 30$ is 0. Check this both graphically and algebraically (by substitution). Then perform a similar check for Example 2(b).

The Remainder Theorem

Example 1 shows that $(x^3 + 6x^2 - x - 30) \div (x - 2) = x^2 + 8x + 15$ or, equivalently,

$$x^3 + 6x^2 - x - 30 = (x - 2)(x^2 + 8x + 15).$$

Using this result and the principle of zero products, we know that if $f(x) = x^3 + 6x^2 - x - 30$, then $f(2) = 0$ (since $x - 2$ is a factor of $f(x)$). Similarly, Example 2(b) shows that if $g(x) = 10x^2 - 13x + 3x^3 - 20$, then $4 + x$ is a factor of $g(x)$ and $g(-4) = 0$. In both examples, the remainder from the division, 0, can serve as a function value. Remarkably, this pattern extends to nonzero remainders. To see this, note that the remainder in Example 2(a) is 4, and if $f(x) = 2x^3 + 7x^2 - 5$, then $f(-3)$ is also 4 (you should check this). The fact that the remainder and the function value coincide is predicted by the remainder theorem, which follows.

THE REMAINDER THEOREM

The remainder obtained by dividing $P(x)$ by $x - r$ is $P(r)$.

A proof of this result is outlined in Exercise 31.

EXAMPLE 3

Let $f(x) = 8x^5 - 6x^3 + x - 8$. Use synthetic division to find $f(2)$.

SOLUTION   The remainder theorem tells us that $f(2)$ is the remainder when $f(x)$ is divided by $x - 2$. We use synthetic division to find that remainder:

$$
\begin{array}{r}
2\,\rule[0.3ex]{0.4pt}{1.6ex}\ \ 8 \quad 0 \quad -6 \quad 0 \quad\ \ 1 \quad -8 \\
\ \ 16 \quad 32 \quad 52 \quad 104 \quad 210 \\
\hline
\ \ 8 \quad 16 \quad 26 \quad 52 \quad 105\ \rule[0.3ex]{0.4pt}{1.6ex}\ 202
\end{array}
$$

Although the bottom line can be used to find the quotient for the division $(8x^5 - 6x^3 + x - 8) \div (x - 2)$, what we are really interested in is the remainder. It tells us that $f(2) = 202$.

➡●

EXERCISE SET

6.7

Use synthetic division to divide.

1. $(x^3 - 2x^2 + 2x - 5) \div (x - 1)$

2. $(x^3 - 2x^2 + 2x - 5) \div (x + 1)$

3. $(a^2 + 11a - 19) \div (a + 4)$

4. $(a^2 + 11a - 19) \div (a - 4)$

5. $(x^3 - 7x^2 - 13x + 3) \div (x - 2)$

6. $(x^3 - 7x^2 - 13x + 3) \div (x + 2)$

7. $(3x^3 + 7x^2 - 4x + 3) \div (x + 3)$

8. $(3x^3 + 7x^2 - 4x + 3) \div (x - 3)$

9. $(y^3 - 3y + 10) \div (y - 2)$

10. $(x^3 - 2x^2 + 8) \div (x + 2)$

11. $(x^5 - 32) \div (x - 2)$

12. $(y^5 - 1) \div (y - 1)$

13. $(3x^3 + 1 - x + 7x^2) \div \left(x + \frac{1}{3}\right)$

14. $(8x^3 - 1 + 7x - 6x^2) \div \left(x - \frac{1}{2}\right)$

Use synthetic division to find the indicated function value.

15. $f(x) = 6x^4 + 15x^3 + 28x + 6;\ f(-3)$

16. $g(x) = 3x^4 - 25x^2 - 18;\ g(3)$

17. $P(x) = 2x^4 - x^3 - 5x^2 + x + 7;\ P(-1)$

18. $F(x) = 3x^4 + 8x^3 + 2x^2 - 7x - 4;\ F(-2)$

19. $f(x) = x^4 - x^3 - 19x^2 + 49x - 30;\ f(4)$

20. $p(x) = x^4 + 7x^3 + 11x^2 - 7x - 12;\ p(2)$

SKILL MAINTENANCE

Graph on a plane.

21. $2x - 3y < 6$

22. $5x + 3y \leqslant 15$

23. $y > 4$

24. $x \leqslant -2$

25. $y - 2 = \frac{3}{4}(x + 1)$

26. $y = -\frac{4}{3}x + 2$

SYNTHESIS

27. ◆ Explain how synthetic division could be useful when factoring a polynomial.

28. ◆ If $p(x) = 3x^5 - 7x^4 + x^3 - 4x^2 + 3x + 2$, what is the fastest way to calculate $p(4)$? Might your answer change if $p(x)$ were a different polynomial function? Why or why not?

29. ◆ Let $P(x)$ be a polynomial function with $p(x)$ a factor of $P(x)$. If $p(3) = 0$, does it follow that $P(3) = 0$? Why or why not? If $P(3) = 0$, does it follow that $p(3) = 0$? Why or why not?

30. ◆ What adjustments must be made if synthetic division is to be used to divide a polynomial by a binomial of the form $ax + b$, with $a > 1$?

31. To prove the remainder theorem, note that any polynomial $P(x)$ can be rewritten as $(x - r) \cdot Q(x) + R$, where $Q(x)$ is the quotient polynomial that arises when $P(x)$ is divided by $x - r$, and R is some constant (the remainder).

a) How do we know that R must be a constant?

b) Show that $P(r) = R$ (this says that $P(r)$ is the remainder when $P(x)$ is divided by $x - r$).

32. Let $f(x) = 4x^3 + 16x^2 - 3x - 45$. Find $f(-3)$ and then solve the equation $f(x) = 0$.

33. Let $f(x) = 6x^3 - 13x^2 - 79x + 140$. Find $f(4)$ and then solve the equation $f(x) = 0$.

34. ⛏ Use the TRACE feature on a grapher to check your answer to Exercise 32.

35. ⛏ Use the TRACE feature on a grapher to check your answer to Exercise 33.

Nested evaluation.   *One way to evaluate a polynomial function like* $P(x) = 3x^4 - 5x^3 + 4x^2 - 1$ *is to successively factor out x as shown:*

$$P(x) = x(x(x(3x - 5) + 4) + 0) - 1.$$

Computations are then performed using this "nested" form of P(x).

36. Use nested evaluation to find $f(-3)$ in Exercise 32. Note the similarities to the calculations performed with synthetic division.

37. Use nested evaluation to find $f(4)$ in Exercise 33. Note the similarities to the calculations performed with synthetic division.

6.8   Formulas and Applications

Solving for a Letter   •   Solving and Evaluating

Solving for a Letter

Formulas occur frequently as mathematical models. Many formulas contain rational expressions, and to solve such formulas for a specified letter, we proceed as when solving rational equations.

EXAMPLE 1

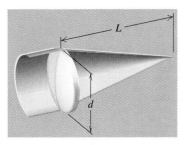

Optics.   The formula $f = L/d$ tells how to calculate a camera's "f-stop." In this formula, f is the f-stop, L is the focal length (approximately the distance from the lens to the film), and d is the diameter of the lens. Solve for d.

SOLUTION   We solve this equation as we did those in Section 6.4:

$$f = \frac{L}{d}$$

$$d \cdot f = d \cdot \frac{L}{d} \qquad \text{Multiplying by the LCD to clear fractions}$$

$$df = L$$

$$df \cdot \frac{1}{f} = L \cdot \frac{1}{f} \qquad \text{Multiplying by } \frac{1}{f} \text{ on both sides}$$

$$d = \frac{L}{f}. \qquad \text{Simplifying}$$

The formula $d = L/f$ can now be used to determine the diameter of a lens if we know the focal length and the f-stop.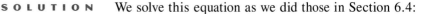

EXAMPLE 2

Astronomy.   The formula

$$L = \frac{dR}{D - d},$$

where D is the diameter of the sun, d is the diameter of the earth, R is the earth's distance from the sun, and L is some fixed distance, is used in calculating when lunar eclipses occur. Solve for D.

SOLUTION We first clear fractions by multiplying by the LCD, which is $D - d$:

$$(D - d)L = (D - d)\frac{dR}{D - d}$$

$$(D - d)L = dR.$$

We do *not* multiply the factors on the left since we wish to get D all alone. Instead we multiply both sides by $1/L$ and then add d:

$$D - d = \frac{dR}{L} \qquad \text{Multiplying on both sides by } \frac{1}{L}$$

$$D = \frac{dR}{L} + d. \qquad \text{Adding } d \text{ on both sides}$$

We now have D all alone on one side of the equation. Since D does not appear on the other side, we have solved the formula for D. ————●

EXAMPLE 3 *Acoustics (the Doppler Effect).* The formula

$$f = \frac{sg}{s + v}$$

is used to determine the frequency f of a sound that is moving at velocity v toward a listener who hears the sound as frequency g. Here s is the speed of sound in a particular medium. Solve for s.

SOLUTION We first clear fractions by multiplying by the LCD, $s + v$:

$$f \cdot (s + v) = \frac{sg}{s + v}(s + v)$$

$$fs + fv = sg. \qquad \text{Here, because } s \text{ } does \text{ appear on both sides,}$$
$$\text{we do distribute on the left side.}$$

Next, we must get all terms containing s on one side:

$$fv = sg - fs \qquad \text{Adding } -fs \text{ on both sides}$$

$$fv = s(g - f) \qquad \text{Factoring out } s$$

$$\frac{fv}{g - f} = s. \qquad \text{Multiplying by } \frac{1}{g - f} \text{ on both sides}$$

Since s is isolated on one side, we have solved for s. This last equation can be used to determine the speed of sound whenever f, v, and g are known.

Solving and Evaluating

EXAMPLE 4

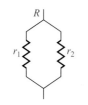

Resistance. The formula

$$\frac{1}{R} = \frac{1}{r_1} + \frac{1}{r_2}$$

relates the resistance R of two resistors r_1 and r_2 connected in parallel.

a) Solve for r_2.
b) Find r_2 when $R = 3.75$ ohms and $r_1 = 6$ ohms.

SOLUTION

a) To solve for r_2, we multiply by the LCD, Rr_1r_2:

$$\frac{1}{R} = \frac{1}{r_1} + \frac{1}{r_2}$$

$$Rr_1r_2 \cdot \frac{1}{R} = Rr_1r_2 \cdot \left[\frac{1}{r_1} + \frac{1}{r_2}\right] \qquad \text{Multiplying on both sides by the LCD}$$

$$Rr_1r_2 \cdot \frac{1}{R} = Rr_1r_2 \cdot \frac{1}{r_1} + Rr_1r_2 \cdot \frac{1}{r_2} \qquad \text{Using the distributive law}$$

$$\left.\begin{array}{l} \dfrac{R r_1 r_2}{R} = \dfrac{Rr_1r_2}{r_1} + \dfrac{Rr_1r_2}{r_2} \\[2mm] r_1r_2 = Rr_2 + Rr_1. \end{array}\right\} \quad \begin{array}{l}\text{Simplifying by removing} \\ \text{factors of 1:} \\ \frac{R}{R} = 1; \frac{r_1}{r_1} = 1; \frac{r_2}{r_2} = 1\end{array}$$

You might be tempted at this point to multiply by $1/r_1$ to get r_2 alone on the left, but note that r_2 also appears on the right. We must get all the terms involving r_2 on the *same side* of the equation:

$$r_1r_2 - Rr_2 = Rr_1 \qquad \text{Adding } -Rr_2 \text{ on both sides}$$

$$r_2(r_1 - R) = Rr_1 \qquad \text{Factoring out } r_2$$

$$r_2 = \frac{Rr_1}{r_1 - R}. \qquad \text{Multiplying by } \frac{1}{r_1 - R} \text{ on both sides}$$

b) We use the equation for r_2, replacing R with 3.75 and r_1 with 6:

$$r_2 = \frac{(3.75)6}{6 - 3.75} = 10.$$

The second resistor has a resistance of 10 ohms.

TO SOLVE A RATIONAL EQUATION FOR A SPECIFIC UNKNOWN:

1. If necessary, clear fractions.
2. Multiply, as needed, to remove parentheses.
3. Get all terms with the unknown alone on one side of the equation.
4. Factor out the unknown if it appears in more than one term.
5. Use the multiplication principle to get the unknown alone on one side.

EXERCISE SET

6.8

Solve the formula for the specified letter.

1. $\dfrac{W_1}{W_2} = \dfrac{d_1}{d_2}$; W_1

2. $\dfrac{W_1}{W_2} = \dfrac{d_1}{d_2}$; d_1

3. $s = \dfrac{(v_1 + v_2)t}{2}$; v_1

4. $s = \dfrac{(v_1 + v_2)t}{2}$; t

5. $\dfrac{1}{R} = \dfrac{1}{r_1} + \dfrac{1}{r_2}$; R

6. $\dfrac{1}{R} = \dfrac{1}{r_1} + \dfrac{1}{r_2}$; r_1

7. $R = \dfrac{gs}{g + s}$; g

8. $K = \dfrac{rt}{r - t}$; t

9. $I = \dfrac{2V}{R + 2r}$; R

10. $I = \dfrac{2V}{R + 2r}$; r

11. $\dfrac{1}{p} + \dfrac{1}{q} = \dfrac{1}{f}$; p

12. $\dfrac{1}{p} + \dfrac{1}{q} = \dfrac{1}{f}$; q

13. $I = \dfrac{nE}{R + nr}$; n

14. $I = \dfrac{nE}{R + nr}$; r

15. $S = \dfrac{H}{m(t_1 - t_2)}$; t_1

16. $S = \dfrac{H}{m(t_1 - t_2)}$; H

17. $\dfrac{E}{e} = \dfrac{R + r}{r}$; r

18. $\dfrac{E}{e} = \dfrac{R + r}{r}$; e

19. $S = \dfrac{a}{1 - r}$; r

20. $S = \dfrac{a - ar^n}{1 - r}$; a

Solve.

21. *Interest.* The formula

$$P = \dfrac{A}{1 + r}$$

is used to determine what principal P should be invested for one year at $(100 \cdot r)\%$ simple interest in order to have A dollars after a year. Solve for r.

22. *Average speed.* The formula

$$v = \dfrac{d_2 - d_1}{t_2 - t_1}$$

gives an object's average speed v when that object has traveled d_1 miles in t_1 hours and d_2 miles in t_2 hours. Solve for t_2.

23. *Average speed.* At what time will Enid's Taurus, averaging a speed of 60 mph, reach Philadelphia if Enid leaves New York at 2:00 A.M. and New York is 105 mi from Philadelphia? (See Exercise 22.)

24. *Interest.* At what yearly interest rate should Bernie invest $1600 if he wants it to grow to $1712 after one year? (See Exercise 21.)

25. *Earned run average.* The formula

$$A = 9 \cdot \dfrac{R}{I}$$

gives a pitcher's *earned run average,* where A is the earned run average, R is the number of earned runs, and I is the number of innings pitched. How many earned runs were given up if a pitcher's earned run average is 2.4 after 45 innings?

26. *Resistance.* Two resistors are connected in parallel. Their resistances are, respectively, 8 ohms and 15 ohms. What is the resistance of the combination? (See Example 4.)

27. *Resistance.* A resistor has a resistance of 50 ohms. What size resistor should be put with it, in parallel,

in order to obtain a resistance of 5 ohms?
(See Example 4.)

28. *Work rate.*   The formula

$$\frac{1}{t} = \frac{1}{a} + \frac{1}{b}$$

gives the total time t required for two workers to complete a job, if the workers' individual times are a and b. Solve for t.

29. *Area of a trapezoid.*   The area of a certain trapezoid is 25 cm². Its height is 5 cm and the length of one base is 4 cm. Find the length of the other base.

30. *Taxable interest.*   The formula

$$I_t = \frac{I_f}{1 - T}$$

gives the *taxable interest rate* I_t equivalent to the *tax-free interest rate* I_f for a person in the $(100 \cdot T)\%$ tax bracket. Solve for T.

31. *Escape velocity.*   The formula

$$\frac{V^2}{R^2} = \frac{2g}{R + h}$$

is used to find a satellite's *escape velocity* V, where R is a planet's radius, h is the satellite's height above the planet, and g is the planet's acceleration due to gravity. Solve for h.

32. *Astronomy.*   Solve the formula of Example 2 for d, the diameter of the earth.

33. *Semester average.*   The formula

$$A = \frac{2Tt + Qq}{2T + Q}$$

gives a student's average A after T tests and Q quizzes, where each test counts as 2 quizzes, t is the test average and q is the quiz average. Solve for Q.

34. *Average acceleration.*   The formula

$$a = \frac{v_2 - v_1}{t_2 - t_1}$$

gives a vehicle's *average acceleration* when its velocity changes from v_1 at time t_1 to v_2 at time t_2. Solve for t_1.

35. A line passing through two points has a slope of $-\frac{2}{5}$. If the coordinates of the points are $(x_1, 2)$ and $(2x_1, 8)$, find the coordinates of both points.

36. Nancy has a test average of 79 and a quiz average of 90. Her grade is calculated by using the formula found in Exercise 33. If Nancy's overall average is 84 and 5 quizzes were taken, how many tests were taken?

SKILL MAINTENANCE

37. Graph on a plane: $6x - y < 6$.

38. If $f(x) = x^3 - x$, find $f(2a)$.

39. Factor: $t^3 + 8b^3$.

40. Solve: $6x^2 = 11x + 35$.

SYNTHESIS

41. ◈ Solve both Exercise 9 on page 342 and Exercise 11 on page 366. In what ways are these exercises similar? How are they different?

42. ▦ *Escape velocity.*   A satellite's escape velocity is 6.5 mi/sec, the radius of the earth is 3960 mi, and the acceleration due to gravity is 32.2 ft/sec². How far is the satellite from the surface of the earth? (See Exercise 31.)

43. The *harmonic mean* of two numbers a and b is a number M such that the reciprocal of M is the average of the reciprocals of a and b. Find a formula for the harmonic mean.

44. Solve for x:

$$x^2\left(1 - \frac{2pq}{x}\right) = \frac{2p^2q^3 - pq^2x}{-q}.$$

45. *Average acceleration.*   The formula

$$a = \frac{\dfrac{d_4 - d_3}{t_4 - t_3} - \dfrac{d_2 - d_1}{t_2 - t_1}}{t_4 - t_2}$$

can be used to approximate average acceleration where the d's are distances and the t's are the corresponding times. Solve for t_1.

COLLABORATIVE C•O•R•N•E•R

Focus: Application of rational expressions

Time: 20–30 minutes

Group size: 2

Materials: Graph paper

Ginny is riding in the American Diabetes Association Tour de Cure 40-mi Bikeathon. For the first 20 mi, the route is primarily downhill so Ginny averages 18 mph. Over the final 20 mi, however, Ginny can average only 12 mph. Vince is also riding in the Bikeathon, but because his 40-mi route is flat, he averages a steady 15 mph over the entire route.

ACTIVITY

1. Assume that Ginny and Vince both begin riding at 9 A.M., and draw a graph representing the situation.
2. With one group member playing the role of Ginny

and the other playing the role of Vince, determine when each biker will cross the finish line.
3. *Question for Ginny*: If you knew in advance the speed at which Vince was biking, how could you have adjusted your speeds so that you would both finish together?
   Question for Vince: If you knew in advance the speeds at which Ginny was biking, how could you have adjusted your speed so that you would both finish together?
4. Assume now that both routes are only 20 mi long and that Ginny's speed drops from 18 mph to 12 mph at the 10-mi mark while Vince continues to ride at a steady 15 mph. Switch roles and again perform parts (2) and (3) above.
5. Suppose that Ginny rode the first half of a 10-mi route at 18 mph and the other half at 12 mph. What would be her average speed? What can you conclude, as a group, about Ginny's average speed?

SUMMARY AND REVIEW
6

KEY TERMS

IMPORTANT PROPERTIES AND FORMULAS

Addition: $\dfrac{A}{C} + \dfrac{B}{C} = \dfrac{A + B}{C}$

Subtraction: $\dfrac{A}{C} - \dfrac{B}{C} = \dfrac{A - B}{C}$

Multiplication: $\dfrac{A}{B} \cdot \dfrac{C}{D} = \dfrac{AC}{BD}$

Division: $\dfrac{A}{B} \div \dfrac{C}{D} = \dfrac{A}{B} \cdot \dfrac{D}{C}$

To find the least common multiple, LCM, use each factor the greatest number of times that it occurs in any one prime factorization.

Simplifying Complex Rational Expressions

I: By using multiplication by 1

1. Find the LCD of all expressions *within* the complex rational expression.
2. Multiply the complex rational expression by 1, using the LCD to form the expression for 1.
3. Distribute and simplify so that the numerator and the denominator of the complex rational expression are polynomials.
4. Factor and simplify, if possible.

II: By using division

1. Add or subtract, as necessary, to get one rational expression in the numerator.
2. Add or subtract, as necessary, to get one rational expression in the denominator.
3. Perform the indicated division (invert the divisor and multiply).
4. Simplify, if possible, by removing any factors equal to 1.

Modeling Work Problems

If

$a =$ the time needed for A to complete the work alone,

$b =$ the time needed for B to complete the work alone, and

$t =$ the time needed for A and B to complete the work together,

then

$$\dfrac{t}{a} + \dfrac{t}{b} = 1.$$

The Remainder Theorem

The remainder obtained by dividing $P(x)$ by $x - r$ is $P(r)$.

REVIEW EXERCISES

1. If

$$f(t) = \frac{t^2 - 3t + 2}{t^2 - 9},$$

find the following function values.

a) $f(0)$ **b)** $f(-1)$ **c)** $f(2)$

Find the LCD.

2. $\dfrac{7}{6x^3}, \quad \dfrac{y}{16x^2}$

3. $\dfrac{x + 8}{x^2 + x - 20}, \quad \dfrac{x}{x^2 + 3x - 10}$

Perform the indicated operations and simplify when possible.

4. $\dfrac{x^2}{x-3} - \dfrac{9}{x-3}$

5. $\dfrac{4x-2}{x^2-5x+4} - \dfrac{3x+2}{x^2-5x+4}$

6. $\dfrac{3a^2b^3}{5c^3d^2} \cdot \dfrac{15c^9d^4}{9a^7b}$

7. $\dfrac{5}{6m^2n^3p} + \dfrac{7}{9mn^4p^2}$

8. $\dfrac{y^2-64}{2y+10} \cdot \dfrac{y+5}{y+8}$

9. $\dfrac{x^3-8}{x^2-25} \cdot \dfrac{x^2+10x+25}{x^2+2x+4}$

10. $\dfrac{9a^2-1}{a^2-9} \div \dfrac{3a+1}{a+3}$

11. $\dfrac{x^3-64}{x^2-16} \div \dfrac{x^2+5x+6}{x^2-3x-18}$

12. $\dfrac{x}{x^2+5x+6} - \dfrac{2}{x^2+3x+2}$

13. $\dfrac{-4xy}{x^2-y^2} + \dfrac{x+y}{x-y}$

14. $\dfrac{2x^2}{x-y} + \dfrac{2y^2}{y-x}$

15. $\dfrac{3}{y+4} - \dfrac{y}{y-1} + \dfrac{y^2+3}{y^2+3y-4}$

Simplify.

16. $\dfrac{\dfrac{5}{x}-5}{\dfrac{7}{x}-7}$

17. $\dfrac{\dfrac{2}{a}+\dfrac{2}{b}}{\dfrac{4}{a^3}+\dfrac{4}{b^3}}$

18. $\dfrac{\dfrac{y^2+4y-77}{y^2-10y+25}}{\dfrac{y^2-5y-14}{y^2-25}}$

19. $\dfrac{\dfrac{5}{x^2-9}-\dfrac{3}{x+3}}{\dfrac{4}{x^2+6x+9}+\dfrac{2}{x-3}}$

Solve.

20. $\dfrac{6}{x} + \dfrac{4}{x} = 5$

21. $\dfrac{5}{3x+2} = \dfrac{3}{2x}$

22. $\dfrac{4x}{x+1} + \dfrac{4}{x} + 9 = \dfrac{4}{x^2+x}$

23. If

$$f(x) = \frac{2}{x-1} + \frac{2}{x+2},$$

find all a for which $f(a) = 1$.

Solve.

24. Kim can paint a garage in 12 hr. Kelly can paint the same garage in 9 hr. How long would it take them, working together, to paint the garage?

25. The Gold River's current is 6 mph. A boat travels 50 mi downstream in the same time that it takes to travel 30 mi upstream. What is the speed of the boat in still water?

26. A car and a motorcycle leave a rest area at the same time, with the car traveling 8 mph faster than the motorcycle. The car then travels 105 mi in the time it takes the motorcycle to travel 93 mi. Find the speed of each vehicle.

Divide.

27. $(20r^2s^3 + 15r^2s^2 - 10r^3s^3) \div (5r^2s)$

28. $(y^3 + 125) \div (y + 5)$

29. $(4x^3 + 3x^2 - 5x - 2) \div (x^2 + 1)$

30. Divide using synthetic division:
$(x^3 + 3x^2 + 2x - 6) \div (x - 3)$.

31. If $f(x) = 4x^3 - 6x^2 - 9$, use synthetic division to find $f(5)$.

32. Solve $R = \dfrac{gs}{g+s}$ for s.

33. Solve $S = \dfrac{H}{m(t_1 - t_2)}$ for m.

34. Solve $\dfrac{1}{ac} = \dfrac{2}{ab} - \dfrac{3}{bc}$ for c.

35. Solve $T = \dfrac{A}{v(t_2 - t_1)}$ for t_1.

S K I L L   M A I N T E N A N C E

Graph on a plane.

36. $y - 2x \geq 4$ **37.** $x > -3$

38. Factor: $125x^3 - 8y^3$.

39. Factor: $6x^2 + 29x - 42$.

40. Solve: $42 = 29x + 6x^2$.

41. If $f(x) = 7x^2 - 6x$, find $f(-2)$.

SYNTHESIS

42. ◆ Discuss at least three different uses of the LCD studied in this chapter.

43. ◆ Explain the difference between a rational expression and a rational equation.

Solve.

44. $\dfrac{5}{x - 13} - \dfrac{5}{x} = \dfrac{65}{x^2 - 13x}$

45. $\dfrac{\dfrac{x}{x^2 - 25} + \dfrac{2}{x - 5}}{\dfrac{3}{x - 5} - \dfrac{4}{x^2 - 10x + 25}} = 1$

46. One summer, Anna mowed 4 lawns for every 3 lawns mowed by her brother Franz. Together, they mowed 98 lawns. How many lawns did each mow?

CHAPTER TEST
6

Simplify.

1. $\dfrac{t - 1}{t + 3} \cdot \dfrac{3t + 9}{4t^2 - 4}$

2. $\dfrac{x^3 + 27}{x^2 - 16} \div \dfrac{x^2 + 8x + 15}{x^2 + x - 20}$

3. Find the LCD:
$$\dfrac{3x}{x^2 + 8x - 33}, \quad \dfrac{x + 1}{x^2 - 12x + 27}.$$

Perform the indicated operation and simplify when possible.

4. $\dfrac{25x}{x + 5} + \dfrac{x^3}{x + 5}$

5. $\dfrac{3a^2}{a - b} - \dfrac{3b^2 - 6ab}{b - a}$

6. $\dfrac{4ab}{a^2 - b^2} + \dfrac{a^2 + b^2}{a + b}$

7. $\dfrac{6}{x^3 - 64} - \dfrac{4}{x^2 - 16}$

8. $\dfrac{4}{y + 3} - \dfrac{y}{y - 2} + \dfrac{y^2 + 4}{y^2 + y - 6}$

Simplify.

9. $\dfrac{\dfrac{2}{a} + \dfrac{3}{b}}{\dfrac{5}{ab} + \dfrac{1}{a^2}}$

10. $\dfrac{\dfrac{x^2 - 5x - 36}{x^2 - 36}}{\dfrac{x^2 + x - 12}{x^2 - 12x + 36}}$

11. $\dfrac{\dfrac{4}{x + 3} - \dfrac{2}{x^2 - 3x + 2}}{\dfrac{3}{x - 2} + \dfrac{1}{x^2 + 2x - 3}}$

Solve.

12. $\dfrac{4}{2x - 5} = \dfrac{6}{5x + 3}$

13. $\dfrac{t + 11}{t^2 - t - 12} + \dfrac{1}{t - 4} = \dfrac{4}{t + 3}$

Let $f(x) = \dfrac{x + 3}{x - 1}.$

14. Find $f(2)$ and $f(-3)$.

15. Find all a for which $f(a) = 7$.

16. Kyla can cut and split a cord of wood in 3.5 hr. Ronson can cut and split a cord of wood in 4.5 hr. How long will it take them, working together, to cut and split a cord of wood?

Divide.

17. $(16ab^3c - 10ab^2c^2 + 12a^2b^2c) \div (4a^2b)$

18. $(y^2 - 20y + 64) \div (y - 6)$

19. $(6x^4 + 3x^2 + 5x + 4) \div (x^2 + 2)$

20. Divide using synthetic division:
$$(x^3 + 5x^2 + 4x - 7) \div (x - 4).$$

21. If $f(x) = 3x^4 - 5x^3 + 2x - 7$, use synthetic division to find $f(4)$.

22. Solve $A = \dfrac{h(b_1 + b_2)}{2}$ for b_1.

23. The product of the reciprocals of two consecutive integers is $\frac{1}{30}$. Find the integers.

24. Dodi bicycles 12 mph with no wind. Against the wind, Dodi bikes 8 mi in the same time that it takes to bike 14 mi with the wind. What is the speed of the wind?

SKILL MAINTENANCE

25. If $f(x) = x^2 - 3$, find $f(a + 1)$.

26. Factor: $16t^2 - 24t - 72$.

27. Solve: $16t^2 = 24t + 72$.

28. Graph on a plane: $2x + 5y > 10$.

SYNTHESIS

29. Let

$$f(x) = \frac{1}{x + 3} + \frac{5}{x - 2}.$$

Find all a for which $f(a) = f(a + 5)$.

30. Solve: $\dfrac{6}{x - 15} - \dfrac{6}{x} = \dfrac{90}{x^2 - 15x}$.

31. Find the x- and y-intercepts for the function given by

$$f(x) = \frac{\dfrac{5}{x + 4} - \dfrac{3}{x - 2}}{\dfrac{2}{x - 3} + \dfrac{1}{x + 4}}.$$

CUMULATIVE REVIEW 1–6

1. Evaluate

$$\frac{2x - y^2}{x + y}$$

for $x = 3$ and $y = -4$.

2. Convert to scientific notation: 5,760,000,000.

3. Determine the slope and the y-intercept for the line given by $7x - 4y = 12$.

4. Find an equation for the line that passes through the points $(-1, 7)$ and $(2, -3)$.

5. Solve the system

$$5x - 2y = -23,$$
$$3x + 4y = 7.$$

6. Solve the system

$$-3x + 4y + z = -5,$$
$$x - 3y - z = 6,$$
$$2x + 3y + 5z = -8.$$

7. Luigi's Pizzeria donated 45 pizzas for a charity event. Small pizzas sold for $7.00 each and large pizzas for $10.00 each. The total amount of funds raised from the sale of the pizzas was $402. How many of each size pizza were donated?

8. The sum of three numbers is 20. The first number is 3 less than twice the third number. The second number minus the third number is -7. What are the numbers?

9. If

$$f(x) = \frac{x - 2}{x - 5},$$

find **(a)** $f(3)$ and **(b)** the domain of f.

Solve.

10. $8x = 1 + 16x^2$

11. $625 = 49y^2$

12. $20 > 2 - 6x$

13. $\frac{1}{3}x - \frac{1}{5} \geq \frac{1}{5}x - \frac{1}{3}$

14. $-8 < x + 2 < 15$

15. $3x - 2 < -6 \; or \; x + 3 > 9$

16. $|x| > 6.4$

17. $|4x - 1| \leq 14$

18. $\frac{2}{n} - \frac{7}{n} = 3$

19. $\frac{6}{x - 5} = \frac{2}{2x}$

20. $\frac{3x}{x - 2} - \frac{6}{x + 2} = \frac{24}{x^2 - 4}$

21. $\frac{3x^2}{x + 2} + \frac{5x - 22}{x - 2} = \frac{-48}{x^2 - 4}$

22. Let $f(x) = |3x - 5|$. Find all values of x for which $f(x) = 2$.

23. Write the domain of f using interval notation if $f(x) = \sqrt{x - 7}$.

24. Solve $5m - 3n = 4m + 12$ for n.

25. Solve $P = \frac{3a}{a + b}$ for a.

Graph on a plane.

26. $4x \geq 5y + 20$ **27.** $y < -2$

Perform the indicated operations and simplify.

28. $(2x^2 - 3x + 1) + (6x - 3x^3 + 7x^2 - 4)$

29. $(5x^3y^2)(-3xy^2)$

30. $(3a + b - 2c) - (-4b + 3c - 2a)$

31. $(5x^2 - 2x + 1)(3x^2 + x - 2)$

32. $(2x^2 - y)^2$

33. $(2x^2 - y)(2x^2 + y)$

34. $(-5m^3n^2 - 3mn^3) + (-4m^2n + 4m^3n^2) - (2mn^3 - 3m^2n^2)$

35. $\frac{y^2 - 36}{2y + 8} \cdot \frac{y + 4}{y + 6}$

36. $\frac{x^4 - 1}{x^2 - x - 2} \div \frac{x^2 + 1}{x - 2}$

37. $\frac{5ab}{a^2 - b^2} + \frac{a + b}{a - b}$

38. $\frac{2}{m + 1} + \frac{3}{m - 5} - \frac{m^2 - 1}{m^2 - 4m - 5}$

39. $y - \frac{2}{3y}$

40. Simplify: $\dfrac{\dfrac{1}{x} - \dfrac{1}{y}}{x + y}$.

41. Divide: $(9x^3 + 5x^2 + 2) \div (x + 2)$.

Factor.

42. $4x^3 + 18x^2$

43. $x^2 + 8x - 84$

44. $16y^2 - 81$

45. $64x^3 + 8$

46. $t^2 - 16t + 64$

47. $x^6 - x^2$

48. $0.027b^3 - 0.008c^3$

49. $20x^2 + 7x - 3$

50. $3x^2 - 17x - 28$

51. $x^5 - x^3y + x^2y - y^2$

52. If $f(x) = x^2 - 4$ and $g(x) = x^2 - 7x + 10$, find the domain of f / g.

Solve.

53. Ed's tractor can plow a field in 3 hr. Nell's tractor can plow the same field in 1.5 hr. Working together, how long would it take them to plow the field?

54. The length of a rectangle is 3 ft longer than the width. The area is 54 ft^2. Find the perimeter of the rectangle.

55. The sum of the squares of three consecutive even integers is equal to 8 more than 3 times the square of the second number. Find the integers.

SYNTHESIS

56. Multiply: $(x - 4)^3$.

57. Find all roots for $f(x) = x^4 - 34x^2 + 225$.

Solve.

58. $4 \leq |3 - x| \leq 6$

59. $\dfrac{18}{x - 9} + \dfrac{10}{x + 5} = \dfrac{28x}{x^2 - 4x - 45}$

60. $16x^3 = x$

Exponents and Radicals

AN APPLICATION

Kit's cottage, which is 24 ft wide and 32 ft long, needs a new roof. By counting clapboards that are 4 in. apart, Kit determines that the peak of the roof is 6 ft higher than the sides. If one packet of shingles covers 100 square feet, how many packets will the job require?

THIS PROBLEM APPEARS AS EXERCISE 52 IN SECTION 7.7.

More information on roofing is available at **http://hepg.awl.com/be/inter_5**

*Y*ou need to know a lot of math to succeed in the construction business. We use geometry to calculate the number of squares of siding and roofing needed for a job. We also use math for calculating costs and job estimates.

CHERYL CARMER
Contractor
Indianapolis, IN

*I*n this chapter, we learn about square roots, cube roots, fourth roots, and so on. These roots are studied in connection with the manipulation of radical expressions and the solution of real-world applications. Fractional exponents are also studied and are used to ease much of our work with radicals. The chapter closes with an examination of the complex-number system.

7.1 Radical Expressions and Functions

Square Roots and Square Root Functions   •   Expressions of the Form $\sqrt{a^2}$ •   Cube Roots   •   Odd and Even nth Roots

---

In this section, we consider roots, such as square roots and cube roots. We look at the symbolism that is used and ways in which symbols can be manipulated to get equivalent expressions. All of this will be important in problem solving.

Square Roots and Square Root Functions

When a number is raised to the second power, the number is squared. Often we need to know what number was squared in order to produce some value a. If such a number can be found, we call that number a *square root* of a.

SQUARE ROOT

The number c is a *square root* of a if $c^2 = a$.

For example,

5 is a square root of 25 because $5^2 = 25$;

-5 is a square root of 25 because $(-5)^2 = 25$;

-4 does not have a real-number square root because there is no real number c such that $c^2 = -4$.

Later in this chapter, we will see that there is a number system, different from the real-number system, in which negative numbers do have square roots. Note that every positive number has two square roots, whereas 0 has only itself as a square root.

EXAMPLE 1   Find the two square roots of 64.

SOLUTION   The square roots are 8 and -8, because $8^2 = 64$ and $(-8)^2 = 64$.

PRINCIPAL SQUARE ROOT

The *principal square root* of a nonnegative number is its nonnegative square root. The symbol $\sqrt{a}$ represents the principal square root of a and is read "radical a," "the square root of a," or simply "root a." The negative square root of a is written $-\sqrt{a}$.

EXAMPLE 2   Simplify each of the following.

a) $\sqrt{25}$ **b)** $\sqrt{\dfrac{25}{64}}$ **c)** $-\sqrt{64}$ **d)** $\sqrt{0.0049}$

SOLUTION

a) $\sqrt{25} = 5$ $\sqrt{}$ indicates the principal square root.

b) $\sqrt{\dfrac{25}{64}} = \dfrac{5}{8}$ Since $\left(\dfrac{5}{8}\right)^2 = \dfrac{25}{64}$

c) $-\sqrt{64} = -8$ Since $\sqrt{64} = 8$, $-\sqrt{64} = -8$.

d) $\sqrt{0.0049} = 0.07$ $(0.07)(0.07) = 0.0049$

RADICAL NOTATION

The symbol $\sqrt{}$ is called a *radical sign* and the expression written under the radical sign is called the *radicand*. An expression written with a radical sign is called a *radical expression*.

The following are radical expressions:

$$\sqrt{5}, \qquad \sqrt{a}, \qquad -\sqrt{5x}, \qquad \sqrt{\dfrac{y^2 + 7}{\sqrt{x}}}.$$

All but the most basic calculators give values for square roots. These values are, for the most part, approximations. For example, on many calculators, if you enter 5 and then press $\boxed{\sqrt{}}$, a number like

2.23606798

appears, depending on how the calculator rounds. (On some calculators, the $\boxed{\sqrt{}}$ key is pressed first.) The exact value of $\sqrt{5}$ is not given by any repeating or terminating decimal. The same is true for the square root of any whole number that is not a perfect square. We discussed such *irrational numbers* in Chapter 1.

Since each nonnegative real number x has exactly one principal square root, $\sqrt{x}$, there is a square-root function, given by

$$f(x) = \sqrt{x}.$$

The domain of the square-root function is $[0, \infty)$. We can draw its graph by selecting convenient values for x and calculating the corresponding outputs. Once these ordered pairs have been graphed, a smooth curve can be drawn.

x	$\sqrt{x}$	$(x, f(x))$
0	0	(0, 0)
1	1	(1, 1)
4	2	(4, 2)
9	3	(9, 3)

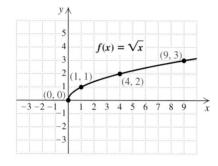

There is a second square-root function, given by

$$g(x) = -\sqrt{x}.$$

x	$-\sqrt{x}$	$(x, g(x))$
0	0	(0, 0)
1	−1	(1, −1)
4	−2	(4, −2)
9	−3	(9, −3)

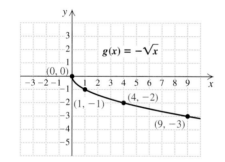

Years ago, the symbol $\sqrt{x}$ was used to represent both square roots of x. That usage has almost completely disappeared, because today we pay a lot more attention to functions. Recall that functions must have exactly one output for each member of the domain.

EXAMPLE   3

For each function, find the indicated function value.

a) $f(x) = \sqrt{3x - 2};\ f(1)$ **b)** $g(z) = -\sqrt{6z + 4};\ g(3)$

SOLUTION

a) $f(1) = \sqrt{3 \cdot 1 - 2}$ Substituting

$= \sqrt{1} = 1$ Simplifying and taking the square root

b) $g(3) = -\sqrt{6 \cdot 3 + 4}$ Substituting

$= -\sqrt{22}$ Simplifying

≈ -4.69041576 Using a calculator to approximate $\sqrt{22}$

Expressions of the Form $\sqrt{a^2}$

It is tempting to think that $\sqrt{x^2} = x$, but the next example shows that, in general, this is untrue.

EXAMPLE 4   Evaluate $\sqrt{x^2}$ for the following values: **(a)** 5; **(b)** 0; **(c)** -5.

SOLUTION

a) $\sqrt{5^2} = \sqrt{25} = 5$

b) $\sqrt{0^2} = \sqrt{0} = 0$

c) $\sqrt{(-5)^2} = \sqrt{25} = 5$ Note that $\sqrt{(-5)^2} \neq -5$.

This leads to the following rule for simplifying radical expressions.

SIMPLIFYING $\sqrt{a^2}$

For any real number a,
$$\sqrt{a^2} = |a|.$$
(The principal square root of a^2 is the absolute value of a.)

When a radicand consists of a perfect square, like $25x^2$ or $(m - 3)^2$, absolute-value signs are needed when simplifying. We use absolute-value signs unless we know that the quantities being squared are nonnegative.

EXAMPLE 5   Simplify each expression. Assume that the variable can represent any real number.

a) $\sqrt{(x + 1)^2}$ **b)** $\sqrt{x^2 - 8x + 16}$

c) $\sqrt{a^8}$ **d)** $\sqrt{t^6}$

SOLUTION

a) $\sqrt{(x + 1)^2} = |x + 1|$ Since $x + 1$ might be negative (for example, if $x = -3$), absolute-value notation is necessary.

b) $\sqrt{x^2 - 8x + 16} = \sqrt{(x - 4)^2} = |x - 4|$ Since $x - 4$ might be negative, absolute-value notation is necessary.

c) Note that $(a^4)^2 = a^8$ and that a^4 is never negative. Thus,
$$\sqrt{a^8} = a^4.$$ Absolute-value notation is unnecessary here.

d) Note that $(t^3)^2 = t^6$. Thus,
$$\sqrt{t^6} = |t^3|.$$ Since t^3 might be negative, absolute-value notation is necessary.

TECHNOLOGY CONNECTION

7.1A

To see the necessity of the absolute-value signs, let $y_1 = \sqrt{x^2}$, $y_2 = x$, and $y_3 = \text{ABS}(x)$. Then use a graph or table to show that $y_1 \neq y_2$ and $y_3 \neq y_2$, but $y_1 = y_3$.

EXAMPLE 6

Simplify each expression. Assume that no radicands were formed by raising negative quantities to even powers.

a) $\sqrt{y^2}$ **b)** $\sqrt{a^{10}}$ **c)** $\sqrt{9x^2 + 6x + 1}$

SOLUTION

a) $\sqrt{y^2} = y$ We are assuming that y is nonnegative, so no absolute-value notation is necessary. When y *is* negative, $\sqrt{y^2} \neq y$.

b) $\sqrt{a^{10}} = a^5$ Assuming that a^5 is nonnegative. Note that $(a^5)^2 = a^{10}$.

c) $\sqrt{9x^2 + 6x + 1} = \sqrt{(3x + 1)^2} = 3x + 1$ Assuming that $3x + 1$ is nonnegative

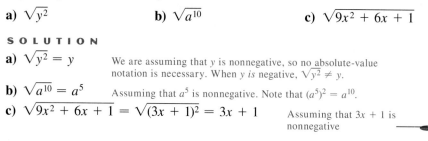

Cube Roots

We often need to know what number was cubed in order to produce a certain value. When such a number is found, we say that we have found a *cube root*.

CUBE ROOT

The number c is the *cube root* of a if $c^3 = a$. In symbols, we write $\sqrt[3]{a}$ to denote the cube root of a.

For example,

2 is the cube root of 8 because $2^3 = 2 \cdot 2 \cdot 2 = 8$;

-4 is the cube root of -64 because $(-4)^3 = (-4)(-4)(-4) = -64$.

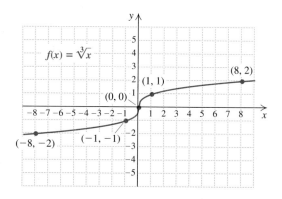

In the real-number system, every number has exactly one cube root. The cube root of a positive number is positive, and the cube root of a negative number is negative. In simplifying expressions involving cube roots, we do not use absolute-value signs.

EXAMPLE 7

For each function, find the indicated function value.

a) $f(y) = \sqrt[3]{y}$; $f(125)$ **b)** $g(x) = \sqrt[3]{x - 3}$; $g(-24)$

SOLUTION

a) $f(125) = \sqrt[3]{125} = 5$ Since $5 \cdot 5 \cdot 5 = 125$

b) $g(-24) = \sqrt[3]{-24 - 3}$
$\qquad\qquad = \sqrt[3]{-27}$
$\qquad\qquad = -3$ Since $(-3)(-3)(-3) = -27$

EXAMPLE 8   Simplify: $\sqrt[3]{-8y^3}$.

SOLUTION

$\quad \sqrt[3]{-8y^3} = -2y$ Since $(-2y)(-2y)(-2y) = -8y^3$

Odd and Even *n*th Roots

The fifth root of a number a is the number c for which $c^5 = a$. There are also 6th roots, 7th roots, and so on. We write $\sqrt[n]{a}$ for the *n*th root. The number n is called the *index* (plural, *indices*). When the index is 2, we do not write it.

If n is odd, we say that we are taking an *odd root*. Every number has just one real root when n is odd. Odd roots of positive numbers are positive and odd roots of negative numbers are negative. Absolute-value signs are not used when finding odd roots.

EXAMPLE 9   Find each of the following: **(a)** $\sqrt[5]{32}$; **(b)** $\sqrt[5]{-32}$; **(c)** $-\sqrt[5]{32}$; **(d)** $-\sqrt[5]{-32}$; **(e)** $\sqrt[7]{x^7}$; **(f)** $\sqrt[9]{(x-1)^9}$.

SOLUTION

a) $\sqrt[5]{32} = 2$ Since $2^5 = 32$

b) $\sqrt[5]{-32} = -2$ Since $(-2)^5 = -32$

c) $-\sqrt[5]{32} = -2$ Taking the opposite of $\sqrt[5]{32}$

d) $-\sqrt[5]{-32} = -(-2) = 2$ Taking the opposite of $\sqrt[5]{-32}$

e) $\sqrt[7]{x^7} = x$

f) $\sqrt[9]{(x-1)^9} = x - 1$

When the index n in $\sqrt[n]{a}$ is an even number, we say that we are taking an *even root*. Every positive real number has two real *n*th roots when n is even. One root is positive and one is negative. Negative numbers do not have real *n*th roots when n is even.

When n is even, the notation $\sqrt[n]{a}$ indicates the positive *n*th root. Thus, when we are finding even *n*th roots, absolute-value signs are often necessary.

EXAMPLE 10   Simplify each expression. Assume that variables can represent any real number.

a) $\sqrt[4]{16}$ **b)** $-\sqrt[4]{16}$ **c)** $\sqrt[4]{-16}$ **d)** $\sqrt[4]{81x^4}$ **e)** $\sqrt[6]{(y+7)^6}$

SOLUTION

a) $\sqrt[4]{16} = 2$ Since $2^4 = 16$

b) $-\sqrt[4]{16} = -2$ Taking the opposite of $\sqrt[4]{16}$

c) $\sqrt[4]{-16}$ cannot be simplified.   No real-number even root exists.

d) $\sqrt[4]{81x^4} = 3|x|$ Use absolute-value notation since x could represent a negative number.

e) $\sqrt[6]{(y+7)^6} = |y+7|$ Use absolute-value notation since $y+7$ could be negative.

SIMPLIFYING $\sqrt[n]{a^n}$

For any real number a:

a) $\sqrt[n]{a^n} = |a|$ when n is even. Unless a is known to be nonnegative, absolute-value notation is needed when n is even.

b) $\sqrt[n]{a^n} = a$ when n is odd. Absolute-value notation is not used when n is odd.

Recall from Section 4.2 that if $f(x) = \sqrt{x+2}$, then the domain of f is $\{x \mid x \geq -2\}$. This occurs because even roots of negative numbers are not real numbers.

EXAMPLE 11   Determine the domain of $g(x) = \sqrt[6]{7-3x}$.

SOLUTION   Since the index is even, the radicand, $7 - 3x$, must be nonnegative. We solve the inequality:

$$7 - 3x \geq 0$$ We cannot find the 6th root of a negative number.

$$-3x \geq -7$$

$$x \leq \tfrac{7}{3}.$$ Multiplying by $-\tfrac{1}{3}$ on both sides and reversing the inequality

Thus,

$$\text{Domain of } g = \left\{x \mid x \leq \tfrac{7}{3}\right\}$$
$$= \left(-\infty, \tfrac{7}{3}\right].$$

TECHNOLOGY CONNECTION

7.1B

Use a grapher to draw the graph of $f(x) = \sqrt{x+2}$. Then check graphically that the domain of f is $[-2, \infty)$.

EXERCISE SET

7.1

Find the square roots of each number.

1. 16

2. 225

3. 144

4. 9

5. 400

6. 81

7. 49

8. 900

Simplify.

9. $-\sqrt{\dfrac{49}{36}}$

10. $-\sqrt{\dfrac{361}{9}}$

11. $\sqrt{196}$

12. $\sqrt{441}$

13. $-\sqrt{\dfrac{16}{81}}$

14. $-\sqrt{\dfrac{81}{144}}$

15. $\sqrt{0.09}$

16. $\sqrt{0.36}$

17. $-\sqrt{0.0049}$

18. $\sqrt{0.0144}$

Identify the radicand and the index for each expression.

19. $5\sqrt{p^2 + 4}$

20. $-7\sqrt{y^2 - 8}$

21. $x^2 y^3 \sqrt[3]{\dfrac{x}{y + 4}}$

22. $a^2 b^3 \sqrt[3]{\dfrac{a}{a^2 - b}}$

For each function, find the specified function value, if it exists.

23. $f(y) = \sqrt{5y - 10}; \quad f(6), f(2), f(1), f(-1)$

24. $g(x) = \sqrt{x^2 - 25}; \quad g(-6), g(3), g(6), g(13)$

25. $t(x) = -\sqrt{2x + 1}; \quad t(4), t(0), t(-1), t\left(-\dfrac{1}{2}\right)$

26. $p(z) = \sqrt{2z^2 - 20}; \quad p(4), p(3), p(-5), p(0)$

27. $f(t) = \sqrt{t^2 + 1}; \quad f(0), f(-1), f(-10)$

28. $g(x) = -\sqrt{(x + 1)^2}; \quad g(-3), g(4), g(-5)$

29. $g(x) = \sqrt{x^3 + 9}; \quad g(-2), g(-3), g(3)$

30. $f(t) = \sqrt{t^3 - 10}; \quad f(2), f(3), f(4)$

Simplify. Assume that variables can represent any real number.

31. $\sqrt{25t^2}$

32. $\sqrt{16x^2}$

33. $\sqrt{(-6b)^2}$

34. $\sqrt{(-7c)^2}$

35. $\sqrt{(5 - b)^2}$

36. $\sqrt{(a + 1)^2}$

37. $\sqrt{y^2 + 16y + 64}$

38. $\sqrt{x^2 - 4x + 4}$

39. $\sqrt{9x^2 - 30x + 25}$

40. $\sqrt{4x^2 + 28x + 49}$

41. $-\sqrt[4]{256}$

42. $\sqrt[4]{625}$

43. $-\sqrt[5]{7^5}$

44. $\sqrt[5]{-1}$

45. $\sqrt[5]{-\dfrac{1}{32}}$

46. $\sqrt[5]{-\dfrac{32}{243}}$

47. $\sqrt[8]{y^8}$

48. $\sqrt[6]{x^6}$

49. $\sqrt[4]{(7b)^4}$

50. $\sqrt[4]{(5a)^4}$

51. $\sqrt[12]{(-10)^{12}}$

52. $\sqrt[10]{(-6)^{10}}$

53. $\sqrt[1976]{(2a + b)^{1976}}$

54. $\sqrt[414]{(a + b)^{414}}$

55. $\sqrt{x^{12}}$

56. $\sqrt{a^{22}}$

57. $\sqrt{a^{14}}$

58. $\sqrt{x^{16}}$

Simplify. Assume that no radicands were formed by raising negative quantities to even powers.

59. $\sqrt{25t^2}$

60. $\sqrt{16x^2}$

61. $\sqrt{(7c)^2}$

62. $\sqrt{(6b)^2}$

63. $\sqrt{(5 + b)^2}$

64. $\sqrt{(a + 1)^2}$

65. $\sqrt{9x^2 + 36x + 36}$

66. $\sqrt{4x^2 + 8x + 4}$

67. $\sqrt{25t^2 - 20t + 4}$

68. $\sqrt{9t^2 - 12t + 4}$

69. $-\sqrt[3]{64}$

70. $\sqrt[3]{27}$

71. $\sqrt[4]{81x^4}$

72. $\sqrt[4]{16x^4}$

73. $-\sqrt[5]{-100,000}$

74. $\sqrt[3]{-216}$

75. $-\sqrt[3]{-64x^3}$

76. $-\sqrt[3]{-125y^3}$

77. $\sqrt{a^{14}}$

78. $\sqrt{a^{22}}$

79. $\sqrt{(x + 3)^{10}}$

80. $\sqrt{(x - 2)^8}$

For each function, find the specified function value, if it exists.

81. $f(x) = \sqrt[3]{x + 1}; \quad f(7), f(26), f(-9), f(-65)$

82. $g(x) = -\sqrt[3]{2x - 1}; \quad g(0), g(-62), g(-13), g(63)$

83. $g(t) = \sqrt[4]{t - 3}; \quad g(19), g(-13), g(1), g(84)$

84. $f(t) = \sqrt[4]{t + 1}; \quad f(0), f(15), f(-82), f(80)$

Determine the domain of each function described.

85. $f(x) = \sqrt{x - 5}$

86. $g(x) = \sqrt{x + 8}$

87. $g(t) = \sqrt[4]{t + 3}$

88. $f(x) = \sqrt[4]{x - 7}$

89. $g(x) = \sqrt[4]{5 - x}$

90. $g(t) = \sqrt[3]{2t - 5}$

91. $f(t) = \sqrt[5]{2t + 9}$

92. $f(t) = \sqrt[6]{2t + 5}$

93. $h(z) = -\sqrt[6]{5z + 3}$

94. $d(x) = -\sqrt[4]{7x - 5}$

95. $f(t) = 4 + 2\sqrt[8]{3t - 7}$

96. $g(t) = 9 - 3\sqrt[8]{5t - 4}$

SKILL MAINTENANCE

Simplify.

97. $(a^3 b^2 c^5)^3$

98. $(5a^7 b^8)(2a^3 b)$

Multiply.

99. $(x - 3)(x + 3)$

100. $(a + bx)(a - bx)$

101. $(2x + 1)(x^2 - 3x + 1)$

102. $(5 - 2x + x^2)(x - 1)$

SYNTHESIS

103. ◈ Does the *n*th root of x^2 always exist? Why or why not?

104. ◈ If the domain of $f = \{x \mid x \le 4\}$, explain how to formulate an expression that could serve as $f(x)$.

105. ◈ If the domain of $g = \{x \mid x \geq 6\}$, explain how to formulate an expression that could serve as $g(x)$.

106. ◈ If the domain of $f = [1, \infty)$ and the range of $f = [2, \infty)$, find a possible expression for $f(x)$ and explain how such an expression is formulated.

107. *Spaces in a parking lot.* A parking lot has attendants to park the cars. The number N of stalls needed for waiting cars before attendants can get to them is given by the formula $N = 2.5\sqrt{A}$, where A is the number of arrivals in peak hours. Find the number of spaces needed for the given number of arrivals in peak hours: **(a)** 25; **(b)** 36; **(c)** 49; **(d)** 64.

Determine the domain of each function described. Then draw the graph of each function.

108. $f(x) = \sqrt{x + 5}$

109. $g(x) = \sqrt{x} + 5$

110. $g(x) = \sqrt{x} - 2$

111. $f(x) = \sqrt{x - 2}$

112. Find the domain of f if
$$f(x) = \frac{\sqrt{x + 3}}{\sqrt[4]{2 - x}}.$$

113. Find the domain of g if
$$g(x) = \frac{\sqrt[4]{5 - x}}{\sqrt[6]{x + 4}}.$$

114. Use a grapher to check your answers to Exercises 31, 39, and 49. On some graphers, a MATH key is needed to enter higher roots.

115. Use a grapher to check your answers to Exercises 112 and 113. (See Exercise 114.)

7.2   Rational Numbers as Exponents

Rational Exponents   •   Negative Rational Exponents   •   Laws of Exponents   •   Simplifying Radical Expressions

In Section 1.1, we considered the natural numbers as exponents. Our discussion of exponents was expanded to include all integers in Section 1.6. In this section, we expand the study still further—to include all rational numbers. This will give meaning to expressions like $a^{1/3}$, $7^{-1/2}$, and $(3x)^{4/5}$. Such notation will help us simplify certain radical expressions.

Rational Exponents

Consider $a^{1/2} \cdot a^{1/2}$. If we still want to add exponents when multiplying, it must follow that $a^{1/2} \cdot a^{1/2} = a^{1/2 + 1/2}$, or a^1. This suggests that $a^{1/2}$ is a square root of a. Similarly, $a^{1/3} \cdot a^{1/3} \cdot a^{1/3} = a^{1/3+1/3+1/3}$, or a^1, so $a^{1/3}$ should mean $\sqrt[3]{a}$.

$$a^{1/n} = \sqrt[n]{a}$$

$a^{1/n}$ means $\sqrt[n]{a}$. When a is nonnegative, n can be any index. When a is negative, n must be odd.

EXAMPLE 1   Rewrite without rational exponents: **(a)** $x^{1/2}$; **(b)** $(-8)^{1/3}$; **(c)** $(abc)^{1/5}$.

SOLUTION

a) $x^{1/2} = \sqrt{x}$

b) $(-8)^{1/3} = \sqrt[3]{-8} = -2$

c) $(abc)^{1/5} = \sqrt[5]{abc}$

EXAMPLE 2   Rewrite with rational exponents: **(a)** $\sqrt[5]{7xy}$; **(b)** $\sqrt[7]{x^3y/9}$.

SOLUTION   Parentheses are required to indicate the base.

a) $\sqrt[5]{7xy} = (7xy)^{1/5}$

b) $\sqrt[7]{\dfrac{x^3y}{9}} = \left(\dfrac{x^3y}{9}\right)^{1/7}$

How should we define $a^{2/3}$? If the property for multiplying exponents is to hold, we must have $a^{2/3} = (a^{1/3})^2$ and $a^{2/3} = (a^2)^{1/3}$. This would suggest that $a^{2/3} = (\sqrt[3]{a})^2$ and $a^{2/3} = \sqrt[3]{a^2}$. We make our definition accordingly.

POSITIVE RATIONAL EXPONENTS

For any natural numbers m and n ($n \neq 1$) and any real number a for which $\sqrt[n]{a}$ exists,

$$a^{m/n} \quad \text{means} \quad (\sqrt[n]{a})^m, \quad \text{or} \quad \sqrt[n]{a^m}.$$

EXAMPLE 3   Rewrite without rational exponents and simplify.

a) $27^{2/3}$ **b)** $25^{3/2}$

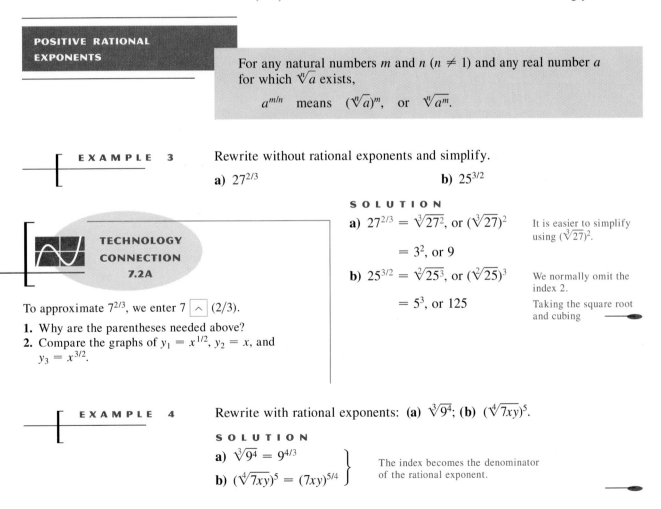

TECHNOLOGY CONNECTION
7.2A

To approximate $7^{2/3}$, we enter 7 $\boxed{\wedge}$ (2/3).

1. Why are the parentheses needed above?

2. Compare the graphs of $y_1 = x^{1/2}$, $y_2 = x$, and $y_3 = x^{3/2}$.

SOLUTION

a) $27^{2/3} = \sqrt[3]{27^2}$, or $(\sqrt[3]{27})^2$ It is easier to simplify using $(\sqrt[3]{27})^2$.

$= 3^2$, or 9

b) $25^{3/2} = \sqrt[2]{25^3}$, or $(\sqrt[2]{25})^3$ We normally omit the index 2.

$= 5^3$, or 125   Taking the square root and cubing

EXAMPLE 4   Rewrite with rational exponents: **(a)** $\sqrt[3]{9^4}$; **(b)** $(\sqrt[4]{7xy})^5$.

SOLUTION

a) $\sqrt[3]{9^4} = 9^{4/3}$

b) $(\sqrt[4]{7xy})^5 = (7xy)^{5/4}$

The index becomes the denominator of the rational exponent.

Negative Rational Exponents

Recall that $x^{-2} = 1/x^2$. Negative rational exponents behave similarly.

**NEGATIVE RATIONAL
EXPONENTS**

For any rational number m/n and any nonzero real number a for which $a^{m/n}$ exists,

$$a^{-m/n} \quad \text{means} \quad \frac{1}{a^{m/n}}.$$

> **CAUTION!**   A negative exponent does not indicate that the expression in which it appears is negative.

EXAMPLE 5

Rewrite with positive exponents and, if possible, simplify.

a) $9^{-1/2}$ **b)** $(5xy)^{-4/5}$ **c)** $64^{-2/3}$

d) $4x^{-2/3}y^{1/5}$ **e)** $\left(\dfrac{3r}{7s}\right)^{-5/2}$

SOLUTION

a) $9^{-1/2} = \dfrac{1}{9^{1/2}}$ $9^{-1/2}$ is the reciprocal of $9^{1/2}$.

Since $9^{1/2} = \sqrt{9} = 3$, the answer simplifies to $\dfrac{1}{3}$.

b) $(5xy)^{-4/5} = \dfrac{1}{(5xy)^{4/5}}$ $(5xy)^{-4/5}$ is the reciprocal of $(5xy)^{4/5}$.

c) $64^{-2/3} = \dfrac{1}{64^{2/3}}$ $64^{-2/3}$ is the reciprocal of $64^{2/3}$.

Since $64^{2/3} = (\sqrt[3]{64})^2 = 4^2 = 16$, the answer simplifies to $\dfrac{1}{16}$.

d) $4x^{-2/3}y^{1/5} = 4 \cdot \dfrac{1}{x^{2/3}} \cdot y^{1/5} = \dfrac{4y^{1/5}}{x^{2/3}}$

e) In Section 1.6, we found that $(a/b)^{-n} = (b/a)^n$. This property holds for *any* negative exponent:

$$\left(\frac{3r}{7s}\right)^{-5/2} = \left(\frac{7s}{3r}\right)^{5/2}.$$ Finding the reciprocal of the base and changing the sign of the exponent

Laws of Exponents

The same laws hold for rational exponents as for integer exponents.

LAWS OF EXPONENTS

For any real numbers a and b and any rational exponents m and n for which a^m, a^n, and b^m are defined:

1. $a^m \cdot a^n = a^{m+n}$ — In multiplying, add exponents if the bases are the same.

2. $\dfrac{a^m}{a^n} = a^{m-n}$ — In dividing, subtract exponents if the bases are the same. (Assume $a \neq 0$.)

3. $(a^m)^n = a^{m \cdot n}$ — To raise a power to a power, multiply the exponents.

4. $(ab)^m = a^m b^m$ — To raise a product to a power, raise each factor to the power and multiply.

EXAMPLE 6

Use the laws of exponents to simplify.

a) $3^{1/5} \cdot 3^{3/5}$ **b)** $a^{1/4}/a^{1/2}$ **c)** $(7.2^{2/3})^{3/4}$ **d)** $(a^{-1/3}b^{2/5})^{1/2}$

SOLUTION

a) $3^{1/5} \cdot 3^{3/5} = 3^{1/5+3/5} = 3^{4/5}$ Adding exponents

b) $\dfrac{a^{1/4}}{a^{1/2}} = a^{1/4-1/2} = a^{1/4-2/4}$ Subtracting exponents after finding a common denominator

$\qquad\qquad = a^{-1/4}$, or $\dfrac{1}{a^{1/4}}$

c) $(7.2^{2/3})^{3/4} = 7.2^{2/3 \cdot 3/4} = 7.2^{6/12}$ Multiplying exponents

$\qquad\qquad\qquad = 7.2^{1/2}$ Using arithmetic to simplify the exponent

d) $(a^{-1/3}b^{2/5})^{1/2} = a^{-1/3 \cdot 1/2} \cdot b^{2/5 \cdot 1/2}$ Raising a product to a power and multiplying exponents

$\qquad\qquad\qquad = a^{-1/6}b^{1/5}$, or $\dfrac{b^{1/5}}{a^{1/6}}$

Simplifying Radical Expressions

Many radical expressions can be simplified using rational exponents.

TO SIMPLIFY RADICAL EXPRESSIONS:

1. Convert radical expressions to exponential expressions.
2. Use arithmetic and the laws of exponents to simplify.
3. Convert back to radical notation when appropriate.

EXAMPLE 7

Use rational exponents to simplify.

a) $\sqrt[6]{(5x)^3}$ **b)** $\sqrt[5]{t^{20}}$ **c)** $(\sqrt[3]{ab^2c})^{12}$ **d)** $\sqrt{\sqrt[3]{x}}$

TECHNOLOGY CONNECTION 7.2B

One way to check Example 7(a) is to let $y_1 = (5x)^{3/6}$ and $y_2 = \sqrt{5x}$. Then see if y_1 and y_2 coincide. An alternative is to let $y_3 = y_2 - y_1$ and see if $y_3 = 0$. Check Example 7(a) using one of the checks just described.

1. Why are rational exponents especially useful when working on a grapher?

SOLUTION

a) $\sqrt[6]{(5x)^3} = (5x)^{3/6}$ Converting to exponential notation
$= (5x)^{1/2}$ Simplifying the exponent
$= \sqrt{5x}$ Returning to radical notation

b) $\sqrt[5]{t^{20}} = t^{20/5}$ Converting to exponential notation
$= t^4$ Simplifying the exponent

c) $(\sqrt[3]{ab^2c})^{12} = (ab^2c)^{12/3}$ Converting to exponential notation
$= (ab^2c)^4$ Simplifying the exponent
$= a^4b^8c^4$ Using the laws of exponents

d) $\sqrt{\sqrt[3]{x}} = \sqrt{x^{1/3}}$ Converting the radicand to exponential notation
$= (x^{1/3})^{1/2}$ Try to go directly to this step.
$= x^{1/6}$ Using the laws of exponents
$= \sqrt[6]{x}$ Returning to radical notation

EXERCISE SET

7.2

Note: Assume for all exercises that even roots are of nonnegative quantities and that all denominators are nonzero.

Rewrite without rational exponents and, if possible, simplify.

1. $x^{1/4}$
2. $y^{1/5}$
3. $16^{1/2}$
4. $8^{1/3}$
5. $81^{1/4}$
6. $64^{1/6}$
7. $9^{1/2}$
8. $25^{1/2}$
9. $(xyz)^{1/3}$
10. $(ab)^{1/4}$
11. $(a^2b^2)^{1/5}$
12. $(x^3y^3)^{1/4}$
13. $a^{2/3}$
14. $b^{3/2}$
15. $16^{3/4}$
16. $4^{7/2}$
17. $49^{3/2}$
18. $27^{4/3}$
19. $9^{5/2}$
20. $81^{3/2}$
21. $(81x)^{3/4}$
22. $(125a)^{2/3}$
23. $(25x^4)^{3/2}$
24. $(9y^6)^{3/2}$

Rewrite with rational exponents.

25. $\sqrt[3]{20}$
26. $\sqrt[3]{19}$
27. $\sqrt{17}$
28. $\sqrt{6}$
29. $\sqrt{x^3}$
30. $\sqrt{a^5}$
31. $\sqrt[5]{m^2}$
32. $\sqrt[5]{n^4}$
33. $\sqrt[4]{cd}$
34. $\sqrt[6]{xy}$
35. $\sqrt[5]{xy^2z}$
36. $\sqrt{x^3y^2z^2}$
37. $(\sqrt{3mn})^3$
38. $(\sqrt[3]{7xy})^4$
39. $(\sqrt{8x^2y})^5$

40. $(\sqrt[6]{2a^5b})^7$
41. $\dfrac{2x}{\sqrt[3]{z^2}}$
42. $\dfrac{3a}{\sqrt[5]{c^2}}$

Rewrite with positive rational exponents.

43. $x^{-1/3}$
44. $y^{-1/4}$
45. $(2rs)^{-3/4}$
46. $(5xy)^{-5/6}$
47. $\left(\dfrac{1}{10}\right)^{-2/3}$
48. $\left(\dfrac{1}{8}\right)^{-3/4}$
49. $\dfrac{1}{a^{-5/7}}$
50. $\dfrac{1}{a^{-3/5}}$
51. $2a^{3/4}b^{-1/2}c^{2/3}$
52. $5x^{-2/3}y^{4/5}z$
53. $2^{-1/3}x^4y^{-2/7}$
54. $3^{-5/2}a^3b^{-7/3}$
55. $\left(\dfrac{7x}{8yz}\right)^{-3/5}$
56. $\left(\dfrac{2ab}{3c}\right)^{-5/6}$
57. $\dfrac{7x}{\sqrt[3]{z}}$
58. $\dfrac{6a}{\sqrt[4]{b}}$
59. $\dfrac{5a}{3c^{-1/2}}$
60. $\dfrac{2z}{5x^{-1/3}}$

Use the laws of exponents to simplify. Do not use negative exponents in any answers.

61. $5^{3/4} \cdot 5^{1/8}$
62. $11^{2/3} \cdot 11^{1/2}$
63. $\dfrac{3^{5/8}}{3^{-1/8}}$

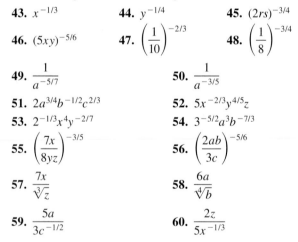

64. $\dfrac{8^{7/11}}{8^{-2/11}}$ **65.** $\dfrac{4.1^{-1/6}}{4.1^{-2/3}}$ **66.** $\dfrac{2.3^{-3/10}}{2.3^{-1/5}}$

67. $(10^{3/5})^{2/5}$ **68.** $(5^{5/4})^{3/7}$ **69.** $a^{2/3} \cdot a^{5/4}$

70. $x^{3/4} \cdot x^{2/3}$ **71.** $(x^{2/3})^{-3/7}$ **72.** $(a^{-3/2})^{2/9}$

73. $(m^{2/3}n^{-1/4})^{1/2}$ **74.** $(x^{-1/3}y^{2/5})^{1/4}$

Use rational exponents to simplify. Write the answer in radical notation when appropriate.

75. $\sqrt[6]{a^2}$ **76.** $\sqrt[6]{t^4}$ **77.** $\sqrt[3]{x^{15}}$

78. $\sqrt[4]{a^{12}}$ **79.** $\sqrt[6]{x^{18}}$ **80.** $\sqrt[5]{a^{10}}$

81. $(\sqrt[3]{ab})^{15}$ **82.** $(\sqrt[7]{xy})^{14}$ **83.** $\sqrt[8]{(3x)^2}$

84. $\sqrt[4]{(7a)^2}$ **85.** $(\sqrt[10]{3a})^5$ **86.** $(\sqrt[8]{2x})^6$

87. $\sqrt[4]{\sqrt{x}}$ **88.** $\sqrt[3]{\sqrt[6]{m}}$ **89.** $\sqrt{(ab)^6}$

90. $\sqrt[4]{(xy)^{12}}$ **91.** $(\sqrt[3]{x^2y^5})^{12}$ **92.** $(\sqrt[5]{a^2b^4})^{15}$

93. $\sqrt[3]{\sqrt[4]{xy}}$ **94.** $\sqrt[5]{\sqrt{2a}}$

SKILL MAINTENANCE

Solve.

95. $x^2 - 1 = 8$ **96.** $3x - 4 = 5x + 7$

97. $\dfrac{1}{x} + 2 = 5$ **98.** $x^2 = 49$

99. *Real estate taxes.*   For homes under \$100,000, the real-estate transfer tax in Vermont is 0.5% of the selling price. Find the selling price of a home that had a transfer tax of \$467.50.

100. What numbers are their own squares?

SYNTHESIS

101. ◆ If $f(x) = (x + 5)^{1/2}(x + 7)^{-1/2}$, find the domain of *f*. Explain how you found your answer.

102. ◆ Explain why $\sqrt[3]{x^6} = x^2$ for any value of *x*, whereas $\sqrt[9]{x^6} = x^3$ only when $x \geqslant 0$.

103. ◆ Let $f(x) = 5x^{-1/3}$. Under what condition will we have $f(x) > 0$? Why?

104. ◆ If $g(x) = x^{1/n}$, in what way does the domain of *g* depend on whether *n* is odd or even?

Use rational exponents to simplify.

105. $\sqrt[5]{x^2y\sqrt{xy}}$

106. $\sqrt{x\sqrt[3]{x^2}}$

107. $\sqrt[4]{\sqrt[3]{8x^3y^6}}$

108. $\sqrt[12]{p^2 + 2pq + q^2}$

109. 🖩 *Road pavement messages.*   In a psychological study, it was determined that the proper length *L* of the letters of a word printed on pavement is given by

$$L = \frac{0.000169d^{2.27}}{h},$$

where *d* is the distance of a car from the lettering and *h* is the height of the eye above the surface of the road. All units are in meters. This formula says that if a person is *h* meters above the surface of the road and is to be able to recognize a message *d* meters away, that message will be the most recognizable if the length of the letters is *L*. Find *L* to the nearest tenth of a meter, given *d* and *h*.

a) $h = 1$ m, $d = 60$ m
b) $h = 0.9906$ m, $d = 75$ m
c) $h = 2.4$ m, $d = 80$ m
d) $h = 1.1$ m, $d = 100$ m

110. *Dating fossils.*   The function $r(t) = 10^{-12}2^{-t/5700}$ expresses the ratio of carbon isotopes to carbon atoms in a fossil that is *t* years old. What ratio of carbon isotopes to carbon atoms would be present in a 1900-year-old bone?

111. *Physics.*   The equation $m = m_0(1 - v^2c^{-2})^{-1/2}$, developed by Albert Einstein, is used to determine the mass *m* of an object that is moving *v* meters per second and has mass m_0 before the motion begins. The constant *c* is the speed of light, approximately 3×10^8 m/sec. Suppose that a particle with mass 8 mg is accelerated to a speed of $\frac{9}{5} \times 10^8$ m/sec. Without using a calculator, find the new mass of the particle.

112. 〰 Use a grapher in the SIMULTANEOUS mode to graph

$$y_1 = x^{1/2}, \qquad y_2 = 3x^{2/5},$$
$$y_3 = x^{4/7}, \quad \text{and} \quad y_4 = \tfrac{1}{5}x^{3/4}.$$

Then, looking only at coordinates, match each graph with its equation.

COLLABORATIVE
C◆O◆R◆N◆E◆R

Focus: Functions and rational exponents

Time: 10–20 minutes

Group size: 3

Materials: Graph paper

In arithmetic, we have seen that $\frac{3}{5}$, $\frac{1}{10} \cdot 6$, and $6 \cdot \frac{1}{10}$ all represent the same number. Interestingly,

$$f(x) = x^{3/5},$$
$$g(x) = (x^{1/10})^6, \quad \text{and}$$
$$h(x) = (x^6)^{1/10}$$

represent three *different* functions.

ACTIVITY

1. Selecting a variety of values for x and using the definition of positive rational exponents, one group member should graph f, a second group member should graph g, and a third group member should graph h. Be sure to check whether negative x-values are in the domain of the function.

2. Compare the three graphs and check each other's work. How and why do the graphs differ?

3. Decide as a group which graph, if any, would best represent the graph of $k(x) = x^{6/10}$. Then be prepared to explain your reasoning to the entire class. (*Hint:* Study the definition of $a^{m/n}$ on page 385 carefully.)

7.3 Multiplying, Adding, and Subtracting Radical Expressions

Multiplying Radical Expressions   •   Simplifying by Factoring   •
Adding and Subtracting Radical Expressions

Multiplying Radical Expressions

Note that $\sqrt{4}\,\sqrt{25} = 2 \cdot 5 = 10$. Also $\sqrt{4 \cdot 25} = \sqrt{100} = 10$. Likewise,

$$\sqrt[3]{27}\,\sqrt[3]{8} = 3 \cdot 2 = 6 \quad \text{and} \quad \sqrt[3]{27 \cdot 8} = \sqrt[3]{216} = 6.$$

These examples suggest the following.

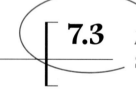

THE PRODUCT RULE
FOR RADICALS

For any real numbers $\sqrt[n]{a}$ and $\sqrt[n]{b}$,

$$\sqrt[n]{a} \cdot \sqrt[n]{b} = \sqrt[n]{a \cdot b}.$$

(To multiply, when the indices match, multiply the radicands.)

EXAMPLE   1

Multiply.

a) $\sqrt{3} \cdot \sqrt{5}$ **b)** $\sqrt{x+3} \sqrt{x-3}$ **c)** $\sqrt[3]{4} \cdot \sqrt[3]{5}$ **d)** $\sqrt[4]{\dfrac{y}{5}} \cdot \sqrt[4]{\dfrac{7}{x}}$

SOLUTION

a) $\sqrt{3} \cdot \sqrt{5} = \sqrt{3 \cdot 5} = \sqrt{15}$

b) $\sqrt{x+3} \sqrt{x-3} = \sqrt{(x+3)(x-3)}$
$$= \sqrt{x^2 - 9}$$

CAUTION!
$$\sqrt{x^2 - 9} \neq \sqrt{x^2} - \sqrt{9}.$$

c) $\sqrt[3]{4} \sqrt[3]{5} = \sqrt[3]{4 \cdot 5} = \sqrt[3]{20}$

d) $\sqrt[4]{\dfrac{y}{5}} \cdot \sqrt[4]{\dfrac{7}{x}} = \sqrt[4]{\dfrac{y}{5} \cdot \dfrac{7}{x}} = \sqrt[4]{\dfrac{7y}{5x}}$

**TECHNOLOGY
CONNECTION
7.3**

To check Example 1(b), let
$y_1 = \sqrt{x+3} \sqrt{x-3}$ and
$y_2 = \sqrt{x^2 - 9}$ and compare.

1. What should the graph of
$y = \sqrt{x^2 - 9} - \sqrt{x+3} \sqrt{x-3}$
look like?

It is important to remember that the product rule for radicals applies only when radicals have the same index. When indices differ, rational exponents can be useful. Study the steps in the following example carefully.

EXAMPLE   2

Multiply: $\sqrt{5x} \cdot \sqrt[4]{3y}$.

SOLUTION

$\sqrt{5x} \cdot \sqrt[4]{3y} = (5x)^{1/2}(3y)^{1/4}$ Converting to exponential notation

$\phantom{\sqrt{5x} \cdot \sqrt[4]{3y}} = (5x)^{2/4}(3y)^{1/4}$ Writing exponents with a common denominator; this is important.

$\phantom{\sqrt{5x} \cdot \sqrt[4]{3y}} = [(5x)^2(3y)]^{1/4}$ Using the laws of exponents

$\phantom{\sqrt{5x} \cdot \sqrt[4]{3y}} = \sqrt[4]{(25x^2)(3y)}$ Squaring $5x$ and returning to radical notation

$\phantom{\sqrt{5x} \cdot \sqrt[4]{3y}} = \sqrt[4]{75x^2 y}$ Multiplying under the radical

Simplifying by Factoring

An integer p is a *perfect square* if there exists an integer q for which $q^2 = p$. We say that p is a *perfect cube* if $q^3 = p$ for some integer q. In general, p is a *perfect nth power* if $q^n = p$ for some integer q. The product rule allows us to simplify $\sqrt[n]{ab}$ when a or b is a perfect nth power.

**USING THE PRODUCT
RULE TO SIMPLIFY**

$$\sqrt[n]{ab} = \sqrt[n]{a} \cdot \sqrt[n]{b}.$$

Thus to simplify $\sqrt{20}$, we note that 20 has 4, which is a perfect square, as a factor. Therefore,

$$\sqrt{20} = \sqrt{4 \cdot 5} \qquad \text{Factoring the radicand (4 is a perfect square)}$$
$$= \sqrt{4} \cdot \sqrt{5} \qquad \text{Factoring into two radicals}$$
$$= 2\sqrt{5}. \qquad \text{Taking the square root of 4}$$

TO SIMPLIFY A RADICAL EXPRESSION BY FACTORING:

1. Express the radicand as the product of two factors, one of which is a perfect nth power (where n is the index).
2. Take the nth root of each factor.
3. Simplification is complete when no radicand has a factor that is a perfect nth power (where n is the index).

EXAMPLE 3

Simplify by factoring: **(a)** $\sqrt{200}$; **(b)** $\sqrt[3]{32}$; **(c)** $\sqrt[4]{48}$; **(d)** $\sqrt{18x^2y}$.

SOLUTION

a) $\sqrt{200} = \sqrt{100 \cdot 2} = \sqrt{100} \cdot \sqrt{2} = 10\sqrt{2}$ This is the largest perfect-square factor of 200.

Note: Had we not seen that 100 is the largest perfect-square factor of 200, we could have used the prime factorization:

$$2 \cdot 2 \cdot 2 \cdot 5 \cdot 5.$$

Each pair of identical factors makes a square, so

$$\sqrt{2 \cdot 2 \cdot 2 \cdot 5 \cdot 5} = \sqrt{2^2 \cdot 5^2 \cdot 2}$$
$$= 2 \cdot 5 \cdot \sqrt{2} = 10\sqrt{2}$$

b) $\sqrt[3]{32} = \sqrt[3]{8 \cdot 4} = \sqrt[3]{8} \cdot \sqrt[3]{4} = 2\sqrt[3]{4}$ This is the largest perfect-cube (third power) factor of 32.

c) $\sqrt[4]{48} = \sqrt[4]{16 \cdot 3} = \sqrt[4]{16} \cdot \sqrt[4]{3} = 2\sqrt[4]{3}$ This is the largest fourth-power factor of 48.

d) $\sqrt{18x^2y} = \sqrt{9x^2 \cdot 2y}$ $9x^2$ is a perfect square.
$$= \sqrt{9x^2} \cdot \sqrt{2y} \qquad \text{Factoring into two radicals}$$
$$= |3x|\sqrt{2y}, \text{ or } 3|x|\sqrt{2y} \qquad \text{Taking the square root of } 9x^2$$

EXAMPLE 4

If $f(x) = \sqrt{3x^2 - 6x + 3}$, find a simplified form for $f(x)$.

SOLUTION

$$f(x) = \sqrt{3x^2 - 6x + 3}$$
$$= \sqrt{3(x^2 - 2x + 1)}$$
$$= \sqrt{3(x - 1)^2} \quad \Bigg\} \qquad \text{Factoring the radicand}$$
$$= \sqrt{(x - 1)^2} \cdot \sqrt{3} \qquad \text{Factoring into two radicals}$$
$$= |x - 1|\sqrt{3} \qquad \text{Taking the square root of } (x - 1)^2$$

The following computer-generated graphs of $y = \sqrt{3x^2 - 6x + 3}$ and $y = |x - 1|\sqrt{3}$ serve as a check for Example 4.

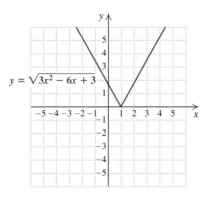

Since the graphs are identical, our simplification is correct. Note that the absolute-value symbol is important since the graph of $y = (x - 1)\sqrt{3}$ is different.

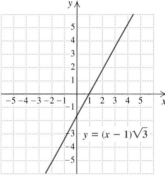

In many situations that do not involve functions, it is safe to assume that no radicands were formed by raising negative quantities to even powers. We now make this assumption and thus discontinue the use of absolute-value notation when taking even roots unless functions are involved.

EXAMPLE 5

Simplify: **(a)** $\sqrt{x^7 y^{11} z^9}$; **(b)** $\sqrt[3]{16a^7 b^{14}}$.

SOLUTION

a) There are many ways to factor $x^7 y^{11} z^9$. Because of the second (square) root, we want to use the largest exponents that are divisible by 2:

$$\sqrt{x^7 y^{11} z^9} = \sqrt{x^6 \cdot x \cdot y^{10} \cdot y \cdot z^8 \cdot z} \qquad \text{Identifying the largest even powers of } x, y, \text{ and } z$$

$$= \sqrt{x^6} \sqrt{y^{10}} \sqrt{z^8} \sqrt{xyz} \qquad \text{Factoring into several radicals}$$

$$= x^3 y^5 z^4 \sqrt{xyz}. \qquad \text{Note that } x^6 = (x^3)^2, y^{10} = (y^5)^2, \text{ and } z^8 = (z^4)^2. \text{ Assume } x, y, z \geqslant 0.$$

Check: $\quad (x^3 y^5 z^4 \sqrt{xyz})^2 = (x^3)^2 (y^5)^2 (z^4)^2 (\sqrt{xyz})^2$

$$= x^6 \cdot y^{10} \cdot z^8 \cdot xyz$$

$$= x^7 y^{11} z^9$$

Our check shows that $x^3 y^5 z^4 \sqrt{xyz}$ is the square root of $x^7 y^{11} z^9$.

b) There are many ways to factor $16a^7 b^{14}$. Because of the third (cube) root, we want to use the largest exponents that are divisible by 3:

$$\sqrt[3]{16a^7b^{14}} = \sqrt[3]{8 \cdot 2 \cdot a^6 \cdot a \cdot b^{12} \cdot b^2}$$ Identifying the largest perfect-cube factors

$$= \sqrt[3]{8} \cdot \sqrt[3]{a^6} \cdot \sqrt[3]{b^{12}} \cdot \sqrt[3]{2ab^2}$$ Factoring into several radicals

$$= 2a^2b^4\sqrt[3]{2ab^2}$$ Taking cube roots; note that $a^6 = (a^2)^3$ and $b^{12} = (b^4)^3$

Check: $(2a^2b^4\sqrt[3]{2ab^2})^3 = 2^3(a^2)^3(b^4)^3(\sqrt[3]{2ab^2})^3$

$$= 8 \cdot a^6 \cdot b^{12} \cdot 2ab^2 = 16a^7b^{14}$$

We see that $2a^2b^4\sqrt[3]{2ab^2}$ is the cube root of $16a^7b^{14}$.

Example 5 shows that to simplify an *n*th root, we search for factors in the radicand with exponents that are divisible by *n*.

Simplification can also be performed using rational exponents.

EXAMPLE 6   Simplify: $\sqrt[5]{32r^8t^{12}}$.

SOLUTION   We have

$$\sqrt[5]{32r^8t^{12}} = (2^5r^8t^{12})^{1/5}$$ Converting to exponential notation

$$= 2^1r^{8/5}t^{12/5}$$ Multiplying exponents

$$= 2 \cdot r^{1+3/5} \cdot t^{2+2/5}$$ Writing $\frac{8}{5}$ and $\frac{12}{5}$ as mixed numbers

$$= 2 \cdot r^1 \cdot r^{3/5} \cdot t^2 \cdot t^{2/5}$$ Factoring; study this carefully.

$$= 2rt^2(r^3t^2)^{1/5}$$ Using the laws of exponents

$$= 2rt^2\sqrt[5]{r^3t^2}.$$ Returning to radical notation

The check is left to the student.

Adding and Subtracting Radical Expressions

When two radical expressions have the same indices and radicands, they are said to be **like radicals.** Like radicals can be combined (added or subtracted) in much the same way that we combined like terms earlier in this text.

EXAMPLE 7   Simplify by combining like radical terms.

a) $6\sqrt{7} + 4\sqrt{7}$ **b)** $\sqrt[3]{2} - 7x\sqrt[3]{2} + 5\sqrt[3]{2}$

c) $6\sqrt[5]{4x} + 3\sqrt[5]{4x} - \sqrt[3]{4x}$

SOLUTION

a) $6\sqrt{7} + 4\sqrt{7} = (6 + 4)\sqrt{7}$ Using the distributive law (factoring out $\sqrt{7}$)

$$= 10\sqrt{7}$$ You can think: 6 square roots of 7 plus 4 square roots of 7 results in 10 square roots of 7.

b) $\sqrt[3]{2} - 7x\sqrt[3]{2} + 5\sqrt[3]{2} = (1 - 7x + 5)\sqrt[3]{2}$ Factoring out $\sqrt[3]{2}$

$$= (6 - 7x)\sqrt[3]{2}$$ These parentheses are important!

c) $6\sqrt[5]{4x} + 3\sqrt[5]{4x} - \sqrt[3]{4x} = (6 + 3)\sqrt[5]{4x} - \sqrt[3]{4x}$ Try to do this step mentally.

$$= 9\sqrt[5]{4x} - \sqrt[3]{4x}$$ Because the indices differ, we are done.

Our ability to simplify radical expressions can help us to find like radicals where, at first, it may appear that none exists.

EXAMPLE 8

Simplify by combining like radical terms, if possible.

a) $3\sqrt{8} - 5\sqrt{2}$

b) $9\sqrt{5} - 4\sqrt{3}$

c) $\sqrt[3]{2x^6y^4} + 7\sqrt[3]{2y}$

SOLUTION

a) $\left. \begin{aligned} 3\sqrt{8} - 5\sqrt{2} &= 3\sqrt{4 \cdot 2} - 5\sqrt{2} \\ &= 3\sqrt{4} \cdot \sqrt{2} - 5\sqrt{2} \\ &= 3 \cdot 2 \cdot \sqrt{2} - 5\sqrt{2} \end{aligned} \right\}$ Simplifying $\sqrt{8}$

$\qquad\qquad\quad = 6\sqrt{2} - 5\sqrt{2}$

$\qquad\qquad\quad = \sqrt{2}$ Combining like radicals

b) $9\sqrt{5} - 4\sqrt{3}$ cannot be simplified.

c) $\left. \begin{aligned} \sqrt[3]{2x^6y^4} + 7\sqrt[3]{2y} &= \sqrt[3]{x^6y^3 \cdot 2y} + 7\sqrt[3]{2y} \\ &= \sqrt[3]{x^6y^3} \cdot \sqrt[3]{2y} + 7\sqrt[3]{2y} \\ &= x^2y \cdot \sqrt[3]{2y} + 7\sqrt[3]{2y} \end{aligned} \right\}$ Simplifying $\sqrt[3]{2x^6y^4}$

$\qquad\qquad\qquad\quad = (x^2y + 7)\sqrt[3]{2y}$ Factoring to combine like radical terms

---

EXERCISE SET

7.3

Multiply.

1. $\sqrt{6}\,\sqrt{7}$

2. $\sqrt{5}\,\sqrt{7}$

3. $\sqrt[3]{2}\,\sqrt[3]{5}$

4. $\sqrt[3]{7}\,\sqrt[3]{2}$

5. $\sqrt[4]{8}\,\sqrt[4]{9}$

6. $\sqrt[4]{6}\,\sqrt[4]{3}$

7. $\sqrt{5a}\,\sqrt{3b}$

8. $\sqrt{2x}\,\sqrt{13y}$

9. $\sqrt[5]{9t^2}\,\sqrt[5]{2t}$

10. $\sqrt[5]{8y^3}\,\sqrt[5]{10y}$

11. $\sqrt{x-a}\,\sqrt{x+a}$

12. $\sqrt{y-b}\,\sqrt{y+b}$

13. $\sqrt[3]{0.5x}\,\sqrt[3]{0.2x}$

14. $\sqrt[3]{0.7y}\,\sqrt[3]{0.3y}$

15. $\sqrt[4]{x-1}\,\sqrt[4]{x^2+x+1}$

16. $\sqrt[5]{x-2}\,\sqrt[5]{(x-2)^2}$

17. $\sqrt{\dfrac{x}{5}}\,\sqrt{\dfrac{3}{y}}$

18. $\sqrt{\dfrac{7}{t}}\,\sqrt{\dfrac{s}{11}}$

19. $\sqrt[7]{\dfrac{x-3}{4}}\,\sqrt[7]{\dfrac{5}{x+2}}$

20. $\sqrt[6]{\dfrac{a}{b-2}}\,\sqrt[6]{\dfrac{3}{b+2}}$

Use rational exponents to write a single radical expression.

21. $\sqrt[3]{5}\,\sqrt{6}$

22. $\sqrt[3]{7} \cdot \sqrt[4]{5}$

23. $\sqrt{x}\,\sqrt[3]{7y}$

24. $\sqrt[3]{y}\,\sqrt[5]{3z}$

25. $\sqrt{x}\,\sqrt[3]{x-2}$

26. $\sqrt[4]{3x}\,\sqrt{y+4}$

27. $\sqrt[5]{yx^2}\,\sqrt{xy}$

28. $\sqrt{ab}\,\sqrt[5]{2a^2b^2}$

29. $\sqrt[4]{xy^2}\,\sqrt[3]{x^2y}$

30. $\sqrt[5]{a^2b^3}\,\sqrt[4]{a^2b}$

31. $\sqrt[4]{a^2bc^2}\,\sqrt[5]{a^2b^3c}$

32. $\sqrt[6]{x^2yz^3}\,\sqrt[5]{x^2yz^2}$

Simplify by factoring.

33. $\sqrt{27}$

34. $\sqrt{28}$

35. $\sqrt{12}$

36. $\sqrt{45}$

37. $\sqrt{8}$

38. $\sqrt{18}$

39. $\sqrt{44}$

40. $\sqrt{24}$

41. $\sqrt{36a^4b}$

42. $\sqrt{175y^8}$

43. $\sqrt[3]{8x^3y^2}$

44. $\sqrt[3]{27ab^6}$

45. $\sqrt[3]{-16x^6}$

46. $\sqrt[3]{-32a^6}$

Find a simplified form of f(x). Assume that x can be any real number.

47. $f(x) = \sqrt[3]{125x^5}$

48. $f(x) = \sqrt[3]{16x^6}$

49. $f(x) = \sqrt{49(x + 5)^2}$ **50.** $f(x) = \sqrt{81(x - 1)^2}$

51. $f(x) = \sqrt{5x^2 - 10x + 5}$

52. $f(x) = \sqrt{2x^2 + 8x + 8}$

Simplify. Assume that no radicands were formed by raising negative numbers to even powers.

53. $\sqrt{a^3 b^4}$ **54.** $\sqrt{x^6 y^9}$ **55.** $\sqrt[3]{x^5 y^6 z^{10}}$

56. $\sqrt[3]{a^6 b^7 c^{13}}$ **57.** $\sqrt[5]{-32 a^7 b^{11}}$ **58.** $\sqrt[4]{16 x^5 y^{11}}$

59. $\sqrt[5]{a^6 b^{12} c^7}$ **60.** $\sqrt[5]{x^{13} y^8 z^{17}}$

61. $\sqrt[4]{810 x^9}$ **62.** $\sqrt[3]{-80 a^{14}}$

Add or subtract. Simplify by combining like radical terms, if possible. Assume that all variables and radicands represent nonnegative numbers.

63. $3\sqrt{7} + 2\sqrt{7}$ **64.** $8\sqrt{5} + 9\sqrt{5}$

65. $9\sqrt[3]{5} - 6\sqrt[3]{5}$ **66.** $14\sqrt[5]{2} - 6\sqrt[5]{2}$

67. $4\sqrt[3]{y} + 9\sqrt[3]{y}$ **68.** $9\sqrt[4]{t} - 3\sqrt[4]{t}$

69. $8\sqrt{2} - 6\sqrt{2} + 5\sqrt{2}$ **70.** $2\sqrt{6} + 8\sqrt{6} - 3\sqrt{6}$

71. $9\sqrt[3]{7} - \sqrt{3} + 4\sqrt[3]{7} + 2\sqrt{3}$

72. $5\sqrt{7} - 8\sqrt[4]{11} + \sqrt{7} + 9\sqrt[4]{11}$

73. $8\sqrt{27} - 3\sqrt{3}$

74. $9\sqrt{50} - 4\sqrt{2}$

75. $3\sqrt{45} + 7\sqrt{20}$

76. $5\sqrt{12} + 16\sqrt{27}$

77. $3\sqrt[3]{16} + \sqrt[3]{54}$

78. $\sqrt[3]{27} - 5\sqrt[3]{8}$

79. $\sqrt{5a} + 2\sqrt{45a^3}$

80. $4\sqrt{3x^3} - \sqrt{12x}$

81. $\sqrt[3]{6x^4} + \sqrt[3]{48x}$

82. $\sqrt[3]{54x} - \sqrt[3]{2x^4}$

83. $\sqrt{4a - 4} + \sqrt{a - 1}$

84. $\sqrt{9y + 27} + \sqrt{y + 3}$

85. $\sqrt{x^3 - x^2} + \sqrt{9x - 9}$

86. $\sqrt{4x - 4} - \sqrt{x^3 - x^2}$

Find a simplified form for $f(x)$. Assume $x \geq 0$.

87. $f(x) = \sqrt{20x^2 + 4x^3} - 3x\sqrt{45 + 9x} + \sqrt{5x^2 + x^3}$

88. $f(x) = \sqrt{x^3 - x^2} + \sqrt{9x^3 - 9x^2} - \sqrt{4x^3 - 4x^2}$

89. $f(x) = \sqrt[4]{x^5 - x^4} + 3\sqrt[4]{x^9 - x^8}$

90. $f(x) = \sqrt[4]{16x^4 + 16x^5} - 2\sqrt[4]{x^8 + x^9}$

SKILL MAINTENANCE

91. During a one-hour television show, there were 12 commercials. Some of the commercials were 30 sec long and the others were 60 sec long. If the number of 30-sec commercials was 6 less than the total number of minutes of commercial time during the show, how many 60-sec commercials were used?

Add.

92. $(5x^3 - 4x^2 + x - 7) + (9x^2 - 4x - 9)$

93. $(7a^3 b^2 + 5a^2 b^2 - a^2 b) + (2a^3 b^2 - 7a^2 b + 3ab)$

94. Multiply: $(2x - 3)(2x + 3)$.

Factor.

95. $4x^2 - 49$ **96.** $2x^2 - 26x + 72$

SYNTHESIS

97. ◆ Why do we need to know how to multiply radical expressions before learning how to add them?

98. ◆ In what way(s) is combining like radical terms the same as combining like terms that are monomials?

99. ◆ Is the equation $\sqrt{(2x + 3)^8} = (2x + 3)^4$ always, sometimes, or never true? Why?

100. ▦ *Speed of a skidding car.* Police can estimate the speed at which a car was traveling by measuring its skid marks. The function

$$r(L) = 2\sqrt{5L}$$

can be used, where L is the length of a skid mark, in feet, and $r(L)$ is the speed, in miles per hour. Find the exact speed and an estimate (to the nearest tenth mile per hour) for the speed of a car that left skid marks **(a)** 20 ft long; **(b)** 70 ft long; **(c)** 90 ft long.

101. *Wind chill temperature.* When the temperature is T degrees Celsius and the wind speed is v meters per second, the *wind chill temperature, T_w,* is the temperature with no wind that it feels like. Here is a formula for finding wind chill temperature:

$$T_w = 33 - \frac{(10.45 + 10\sqrt{v} - v)(33 - T)}{22}.$$

Estimate the wind chill temperature (to the nearest tenth of a degree) for the given actual temperatures and wind speeds.

a) $T = 7°C$, $v = 8$ m/sec
b) $T = 0°C$, $v = 12$ m/sec
c) $T = -5°C$, $v = 14$ m/sec
d) $T = -23°C$, $v = 15$ m/sec

Simplify. Assume that all radicands are nonnegative.

102. $(\sqrt{r^3t})^7$

103. $(\sqrt[3]{25x^4})^4$

104. $\frac{1}{2}\sqrt{36a^5bc^4} - \frac{1}{2}\sqrt[3]{64a^4bc^6} + \frac{1}{6}\sqrt{144a^3bc^6}$

105. $7x\sqrt{(x + y)^3} - 5xy\sqrt{x + y} - 2y\sqrt{(x + y)^3}$

106. Graph the function given by
$$f(x) = \sqrt{(x - 2)^2}.$$
What is the domain of f?

107. Graph the function given by
$$g(x) = \sqrt{(x + 3)^2}.$$
What is the domain of g?

108. Use a grapher to check your answers to Exercises 15, 51, and 79.

109. Rony is puzzled. When he uses a grapher to graph $y = \sqrt{x} \cdot \sqrt{x}$, he gets the following screen. Explain why Rony did not get the complete line $y = x$.

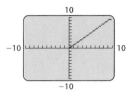

COLLABORATIVE
C◆O◆R◆N◆E◆R

Focus: Radical equations and problem solving

Time: 15–25 minutes

Group size: 2–3

Materials: Calculators or square-root tables

The faster a car is traveling, the more distance it needs to stop. Thus it is important for drivers to allow sufficient space between their vehicle and the vehicle in front of them. Police recommend that for each 10 mph of speed, a driver allow 1 car length. Thus a driver going 30 mph should have at least 3 car lengths between his or her vehicle and the one in front.

In Exercise 100, the function $r(L) = 2\sqrt{5L}$ was used to find the speed, in miles per hour, that a car was traveling when it left skid marks L feet long.

car's speed s, column 2 lists the minimum amount of space between cars traveling s miles per hour, as recommended by police. Column 3 is the speed that a vehicle *could* travel were it forced to stop in the distance listed in column 2, using the above function.

Column 1 s (in miles per hour)	Column 2 L(s) (in feet)	Column 3 r(L) (in miles per hour)
20		
30		
40		
50		
60		
70		

ACTIVITY

1. Each group member should estimate the length of a car in which he or she frequently travels. (Each should use a different length, if possible.)
2. Using a calculator as needed, each group member should complete the table below. Column 1 gives a

3. Determine whether there are any speeds at which the "1 car length per 10 mph" guideline might not suffice. On what reasoning do you base your answer? Compare tables to determine how car length affects the results. What recommendations would your group make to a new driver?

7.4 Multiplying, Dividing, and Simplifying Radical Expressions

---

Multiplying and Simplifying   •   Dividing and Simplifying

Multiplying and Simplifying

We have already used the product rule for radicals, $\sqrt[n]{a} \cdot \sqrt[n]{b} = \sqrt[n]{ab}$, to find certain products. The product rule was also used to simplify certain radical expressions. For some radical expressions, it is possible to do both: First find a product and then simplify.

EXAMPLE 1   Multiply and simplify.

a) $\sqrt{15} \, \sqrt{6}$ **b)** $3\sqrt[3]{25} \cdot 2\sqrt[3]{5}$ **c)** $\sqrt[4]{8x^3y^5} \, \sqrt[4]{4x^2y^3}$

SOLUTION

a) $\sqrt{15} \, \sqrt{6} = \sqrt{15 \cdot 6}$
$= \sqrt{90} = \sqrt{9 \cdot 10}$
$= 3\sqrt{10}$

b) $3\sqrt[3]{25} \cdot 2\sqrt[3]{5} = 6 \cdot \sqrt[3]{25 \cdot 5}$ Multiplying radicands
$= 6 \cdot \sqrt[3]{125}$ 125 is a perfect cube.
$= 6 \cdot 5,$ or 30

c) $\sqrt[4]{8x^3y^5} \, \sqrt[4]{4x^2y^3} = \sqrt[4]{32x^5y^8}$ Multiplying radicands
$= \sqrt[4]{16x^4y^8 \cdot 2x}$ Factoring the radicand
$= \sqrt[4]{16} \, \sqrt[4]{x^4} \, \sqrt[4]{y^8} \, \sqrt[4]{2x}$ Factoring into radicals
$= 2xy^2\sqrt[4]{2x}$ Finding the fourth roots; assume $x \geq 0$.

The checks are left to the student.

As we saw earlier, when indices differ, rational exponents are helpful.

EXAMPLE 2   Multiply and simplify: $\sqrt{x^3} \, \sqrt[3]{x}$.

SOLUTION

$\sqrt{x^3} \, \sqrt[3]{x} = x^{3/2} \cdot x^{1/3}$ Converting to exponential notation
$= x^{11/6}$ Adding exponents: $\frac{3}{2} + \frac{1}{3} = \frac{9}{6} + \frac{2}{6}$
$= x^{1+5/6}$ Writing 11/6 as a mixed number
$= x \cdot x^{5/6}$ Factoring
$= x\sqrt[6]{x^5}$ Converting back to radical notation

Dividing and Simplifying

Just as the root of a product can be expressed as the product of two roots, the root of a quotient can be expressed as the quotient of two roots. For example,

$$\sqrt[3]{\frac{27}{8}} = \frac{3}{2} \quad \text{and} \quad \frac{\sqrt[3]{27}}{\sqrt[3]{8}} = \frac{3}{2}.$$

This example suggests the following.

**THE QUOTIENT RULE
FOR RADICALS**

For any real numbers $\sqrt[n]{a}$ and $\sqrt[n]{b}$, $b \neq 0$,

$$\sqrt[n]{\frac{a}{b}} = \frac{\sqrt[n]{a}}{\sqrt[n]{b}}.$$

Remember that an nth root is simplified when its radicand has no factors that are perfect nth powers.

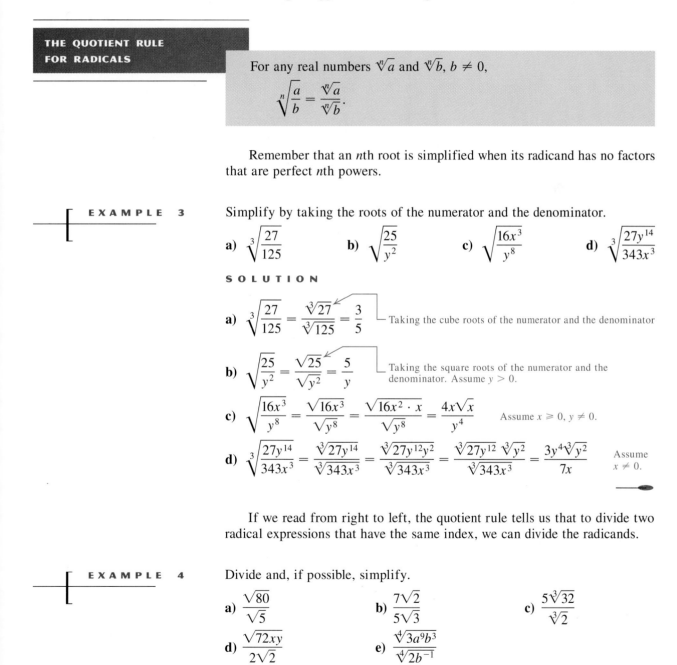

EXAMPLE 3   Simplify by taking the roots of the numerator and the denominator.

a) $\sqrt[3]{\dfrac{27}{125}}$ b) $\sqrt{\dfrac{25}{y^2}}$ c) $\sqrt{\dfrac{16x^3}{y^8}}$ d) $\sqrt[3]{\dfrac{27y^{14}}{343x^3}}$

SOLUTION

a) $\sqrt[3]{\dfrac{27}{125}} = \dfrac{\sqrt[3]{27}}{\sqrt[3]{125}} = \dfrac{3}{5}$ ⎯ Taking the cube roots of the numerator and the denominator

b) $\sqrt{\dfrac{25}{y^2}} = \dfrac{\sqrt{25}}{\sqrt{y^2}} = \dfrac{5}{y}$ ⎯ Taking the square roots of the numerator and the denominator. Assume $y > 0$.

c) $\sqrt{\dfrac{16x^3}{y^8}} = \dfrac{\sqrt{16x^3}}{\sqrt{y^8}} = \dfrac{\sqrt{16x^2 \cdot x}}{\sqrt{y^8}} = \dfrac{4x\sqrt{x}}{y^4}$ Assume $x \geq 0$, $y \neq 0$.

d) $\sqrt[3]{\dfrac{27y^{14}}{343x^3}} = \dfrac{\sqrt[3]{27y^{14}}}{\sqrt[3]{343x^3}} = \dfrac{\sqrt[3]{27y^{12}y^2}}{\sqrt[3]{343x^3}} = \dfrac{\sqrt[3]{27y^{12}}\,\sqrt[3]{y^2}}{\sqrt[3]{343x^3}} = \dfrac{3y^4\sqrt[3]{y^2}}{7x}$ Assume $x \neq 0$.

If we read from right to left, the quotient rule tells us that to divide two radical expressions that have the same index, we can divide the radicands.

EXAMPLE 4   Divide and, if possible, simplify.

a) $\dfrac{\sqrt{80}}{\sqrt{5}}$ b) $\dfrac{7\sqrt{2}}{5\sqrt{3}}$ c) $\dfrac{5\sqrt[3]{32}}{\sqrt[3]{2}}$

d) $\dfrac{\sqrt{72xy}}{2\sqrt{2}}$ e) $\dfrac{\sqrt[4]{3a^9b^3}}{\sqrt[4]{2b^{-1}}}$

SOLUTION

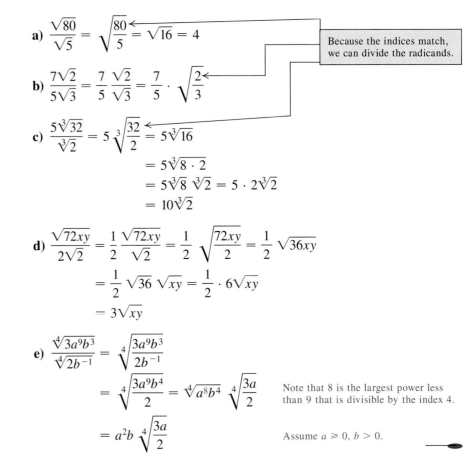

a) $\dfrac{\sqrt{80}}{\sqrt{5}} = \sqrt{\dfrac{80}{5}} = \sqrt{16} = 4$

> Because the indices match, we can divide the radicands.

b) $\dfrac{7\sqrt{2}}{5\sqrt{3}} = \dfrac{7}{5} \dfrac{\sqrt{2}}{\sqrt{3}} = \dfrac{7}{5} \cdot \sqrt{\dfrac{2}{3}}$

c) $\dfrac{5\sqrt[3]{32}}{\sqrt[3]{2}} = 5\sqrt[3]{\dfrac{32}{2}} = 5\sqrt[3]{16}$

$ = 5\sqrt[3]{8 \cdot 2}$

$ = 5\sqrt[3]{8}\,\sqrt[3]{2} = 5 \cdot 2\sqrt[3]{2}$

$ = 10\sqrt[3]{2}$

d) $\dfrac{\sqrt{72xy}}{2\sqrt{2}} = \dfrac{1}{2} \dfrac{\sqrt{72xy}}{\sqrt{2}} = \dfrac{1}{2}\sqrt{\dfrac{72xy}{2}} = \dfrac{1}{2}\sqrt{36xy}$

$ = \dfrac{1}{2}\sqrt{36}\,\sqrt{xy} = \dfrac{1}{2} \cdot 6\sqrt{xy}$

$ = 3\sqrt{xy}$

e) $\dfrac{\sqrt[4]{3a^9b^3}}{\sqrt[4]{2b^{-1}}} = \sqrt[4]{\dfrac{3a^9b^3}{2b^{-1}}}$

$ = \sqrt[4]{\dfrac{3a^9b^4}{2}} = \sqrt[4]{a^8b^4}\,\sqrt[4]{\dfrac{3a}{2}}$

> Note that 8 is the largest power less than 9 that is divisible by the index 4.

$ = a^2b\,\sqrt[4]{\dfrac{3a}{2}}$

> Assume $a \geqslant 0$, $b > 0$.

CAUTION! Before multiplying or dividing radicands, make sure that both radicals have the same index.

When indices differ, we use rational exponents.

EXAMPLE 5

Divide and, if possible, simplify.

a) $\dfrac{\sqrt[4]{(x+y)^3}}{\sqrt{x+y}}$ b) $\dfrac{\sqrt[3]{a^2b^4}}{\sqrt{ab}}$ c) $\dfrac{\sqrt[4]{x^3y^2}}{\sqrt[3]{x^2y}}$

SOLUTION

a) $\dfrac{\sqrt[4]{(x+y)^3}}{\sqrt{x+y}} = \dfrac{(x+y)^{3/4}}{(x+y)^{1/2}}$ Converting to exponential notation

$ = (x+y)^{3/4-1/2}$ Since the bases match, we can subtract exponents.

$\left.\begin{array}{l} = (x+y)^{1/4} \\ = \sqrt[4]{x+y} \end{array}\right\}$ Converting back to radical notation

b) $\dfrac{\sqrt[3]{a^2b^4}}{\sqrt{ab}} = \dfrac{(a^2b^4)^{1/3}}{(ab)^{1/2}}$ Converting to exponential notation

$= \dfrac{a^{2/3}b^{4/3}}{a^{1/2}b^{1/2}}$ Using the product and power rules

$= a^{2/3-1/2}b^{4/3-1/2}$ Subtracting exponents

$\left.\begin{array}{l} = a^{1/6}b^{5/6} \\ = (ab^5)^{1/6} \\ = \sqrt[6]{ab^5} \end{array}\right\}$ Converting back to radical notation

c) $\dfrac{\sqrt[4]{x^3y^2}}{\sqrt[3]{x^2y}} = \dfrac{(x^3y^2)^{1/4}}{(x^2y)^{1/3}}$ Converting to exponential notation

$= \dfrac{x^{3/4}y^{2/4}}{x^{2/3}y^{1/3}}$ Using the product and power rules

$= x^{3/4-2/3}y^{2/4-1/3}$ Subtracting exponents

$\left.\begin{array}{l} = x^{1/12}y^{2/12} \\ = (xy^2)^{1/12} \\ = \sqrt[12]{xy^2} \end{array}\right\}$ Converting back to radical notation

EXERCISE SET
7.4

Multiply and simplify. Write all answers in radical notation.

1. $\sqrt{10}\,\sqrt{5}$ **2.** $\sqrt{6}\,\sqrt{3}$

3. $\sqrt{6}\,\sqrt{14}$ **4.** $\sqrt{15}\,\sqrt{21}$

5. $\sqrt[3]{2}\,\sqrt[3]{4}$ **6.** $\sqrt[3]{9}\,\sqrt[3]{3}$

7. $\sqrt[3]{5a^2}\,\sqrt[3]{2a}$ **8.** $\sqrt[3]{7x}\,\sqrt[3]{3x^2}$

9. $\sqrt{3x^3}\,\sqrt{6x^5}$ **10.** $\sqrt{5a^7}\,\sqrt{15a^3}$

11. $\sqrt[3]{s^2t^4}\,\sqrt[3]{s^4t^6}$ **12.** $\sqrt[3]{x^2y^4}\,\sqrt[3]{x^2y^6}$

13. $\sqrt[3]{(x+5)^2}\,\sqrt[3]{(x+5)^4}$ **14.** $\sqrt[3]{(a-b)^5}\,\sqrt[3]{(a-b)^7}$

15. $\sqrt[4]{12a^3b^7}\,\sqrt[4]{4a^2b^5}$ **16.** $\sqrt[4]{9x^7y^2}\,\sqrt[4]{9x^2y^9}$

17. $\sqrt[5]{x^3(y+z)^4}\,\sqrt[5]{x^3(y+z)^6}$

18. $\sqrt[5]{a^3(b-c)^4}\,\sqrt[5]{a^7(b-c)^4}$

19. $\sqrt{a}\,\sqrt[4]{a^3}$ **20.** $\sqrt[3]{x^2}\,\sqrt[6]{x^5}$

21. $\sqrt[5]{b^2}\,\sqrt{b^3}$ **22.** $\sqrt[4]{a^3}\,\sqrt[3]{a^2}$

23. $\sqrt{xy^3}\,\sqrt[3]{x^2y}$ **24.** $\sqrt[5]{a^3b}\,\sqrt{ab}$

25. $\sqrt[4]{9ab^3}\,\sqrt{3a^4b}$ **26.** $\sqrt{2x^3y^3}\,\sqrt[4]{4xy^2}$

27. $\sqrt[3]{xy^2z}\,\sqrt{x^3yz^2}$ **28.** $\sqrt{a^4b^3c^4}\,\sqrt[3]{ab^2c}$

29. $\sqrt{27a^5(b+1)}\,\sqrt[3]{81a(b+1)^4}$

30. $\sqrt{8x(y+z)^5}\,\sqrt[3]{4x^2(y+z)^2}$

Simplify by taking the roots of the numerator and the denominator.

31. $\sqrt{\dfrac{25}{36}}$ **32.** $\sqrt{\dfrac{100}{81}}$ **33.** $\sqrt[3]{\dfrac{64}{27}}$

34. $\sqrt[3]{\dfrac{343}{1000}}$ **35.** $\sqrt{\dfrac{49}{y^2}}$ **36.** $\sqrt{\dfrac{121}{x^2}}$

37. $\sqrt{\dfrac{25y^3}{x^4}}$ **38.** $\sqrt{\dfrac{36a^5}{b^6}}$ **39.** $\sqrt[3]{\dfrac{27a^4}{8b^3}}$

40. $\sqrt[3]{\dfrac{64x^7}{216y^6}}$ **41.** $\sqrt[4]{\dfrac{16a^4}{b^4c^8}}$ **42.** $\sqrt[4]{\dfrac{81x^4}{y^8z^4}}$

43. $\sqrt[4]{\dfrac{a^5b^8}{c^{10}}}$ **44.** $\sqrt[4]{\dfrac{x^9y^{12}}{z^6}}$ **45.** $\sqrt[5]{\dfrac{32x^6}{y^{11}}}$

46. $\sqrt[5]{\dfrac{243a^9}{b^{13}}}$ **47.** $\sqrt[6]{\dfrac{x^6y^8}{z^{15}}}$ **48.** $\sqrt[6]{\dfrac{a^9b^{12}}{c^{13}}}$

Divide and, if possible, simplify.

49. $\dfrac{\sqrt{35x}}{\sqrt{7x}}$ **50.** $\dfrac{\sqrt{28y}}{\sqrt{4y}}$ **51.** $\dfrac{\sqrt[3]{270}}{\sqrt[3]{10}}$

52. $\dfrac{\sqrt[3]{40}}{\sqrt[3]{5}}$ **53.** $\dfrac{\sqrt{40xy^3}}{\sqrt{8x}}$ **54.** $\dfrac{\sqrt{56ab^3}}{\sqrt{7a}}$

55. $\dfrac{\sqrt[3]{96a^4b^2}}{\sqrt[3]{12a^2b}}$ **56.** $\dfrac{\sqrt[3]{189x^5y^7}}{\sqrt[3]{7x^2y^2}}$ **57.** $\dfrac{\sqrt{100ab}}{5\sqrt{2}}$

58. $\dfrac{\sqrt{75ab}}{3\sqrt{3}}$ **59.** $\dfrac{\sqrt[4]{48x^9y^{13}}}{\sqrt[4]{3xy^{-2}}}$ **60.** $\dfrac{\sqrt[5]{64a^{11}b^{28}}}{\sqrt[5]{2ab^{-2}}}$

61. $\dfrac{\sqrt[3]{x^3-y^3}}{\sqrt[3]{x-y}}$ **62.** $\dfrac{\sqrt[3]{r^3+s^3}}{\sqrt[3]{r+s}}$

Hint: Factor and then simplify.

63. $\dfrac{\sqrt[3]{a^2}}{\sqrt[4]{a}}$ **64.** $\dfrac{\sqrt[3]{x^2}}{\sqrt[5]{x}}$

65. $\dfrac{\sqrt[4]{x^2y^3}}{\sqrt[3]{xy}}$ **66.** $\dfrac{\sqrt[5]{a^4b^2}}{\sqrt[3]{ab^2}}$

67. $\dfrac{\sqrt{ab^3c}}{\sqrt[5]{a^2b^3c^{-1}}}$ **68.** $\dfrac{\sqrt[5]{x^3y^4z^9}}{\sqrt{xy^{-2}z}}$

69. $\dfrac{\sqrt[4]{(3x-1)^3}}{\sqrt[5]{(3x-1)^3}}$ **70.** $\dfrac{\sqrt[3]{(2+5x)^2}}{\sqrt[4]{2+5x}}$

71. $\dfrac{\sqrt[3]{(2x+1)^2}}{\sqrt[5]{(2x+1)^2}}$ **72.** $\dfrac{\sqrt[4]{(5-3x)^3}}{\sqrt[3]{(5-3x)^2}}$

SKILL MAINTENANCE

Solve.

73. $\dfrac{12x}{x-4}-\dfrac{3x^2}{x+4}=\dfrac{384}{x^2-16}$

74. $\dfrac{2}{3}+\dfrac{1}{t}=\dfrac{4}{5}$

75. The width of a rectangle is one-fourth the length. The area is twice the perimeter. Find the dimensions of the rectangle.

76. The sum of a number and its square is 20. Find the number.

77. Solve for a_1: $A=\dfrac{m}{a_2-a_1}$.

78. Solve for n: $3n+p=4(m-n)$.

SYNTHESIS

79. ◈ Explain how common denominators can be used when dividing one radical expression by another. (*Hint:* See Example 5.)

80. ◈ Is the quotient of two irrational numbers always an irrational number? Why or why not?

81. ◈ Ramon *incorrectly* writes
$$x^{2/5}\cdot x^{3/2}=\sqrt[5]{x^3}.$$
What mistake do you suspect he is making?

82. ◈ After examining Exercise 15, Dyan (correctly) concludes that a and b are both nonnegative. Explain how she could reach this conclusion.

83. ▤ *Pendulums.* The *period* of a pendulum is the time it takes to complete one cycle, swinging to and fro. For a pendulum that is L centimeters long, the period T is given by the formula
$$T=2\pi\sqrt{\dfrac{L}{980}},$$
where T is in seconds. Find, to the nearest hundredth of a second, the period of a pendulum of length **(a)** 65 cm; **(b)** 98 cm; **(c)** 120 cm. Use a calculator's $\boxed{\pi}$ key if possible.

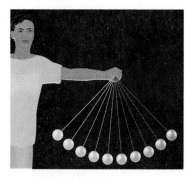

Perform the indicated operations.

84. $\dfrac{7\sqrt{a^2b}\ \sqrt{25xy}}{5\sqrt{a^{-4}b^{-1}}\ \sqrt{49x^{-1}y^{-3}}}$

85. $\dfrac{(\sqrt[3]{81mn^2})^2}{(\sqrt[3]{mn})^2}$

86. $\dfrac{\sqrt{44x^2y^9z}\ \sqrt{22y^9z^6}}{(\sqrt{11xy^8z^2})^2}$

87. $\dfrac{\sqrt{x^5 - 2x^4y} - \sqrt{xy^4 - 2y^5}}{\sqrt{xy^2 - 2y^3} + \sqrt{x^3 - 2x^2y}}$

88. Solve $\sqrt[3]{5x^{k+1}}\,\sqrt[3]{25x^k} = 5x^7$ for k.

89. Solve $\sqrt[5]{4a^{3k+2}}\,\sqrt[5]{8a^{6-k}} = 2a^4$ for k.

90. 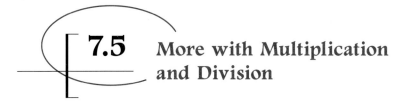 Use a grapher to check Example 2. Note that most graphers require that rational exponents be used when indices greater than 2 appear.

91. Use a grapher to check your answers to Exercises 9, 21, and 63.

7.5 More with Multiplication and Division

Multiplication with Two or More Radical Terms •
Rationalizing Denominators and Numerators

Multiplication with Two or More Radical Terms

Radical expressions often contain factors that have more than one term. The procedure for multiplying out such expressions is similar to finding products of polynomials.

EXAMPLE 1   Multiply.

a) $\sqrt{3}(x - \sqrt{5})$

b) $\sqrt[3]{y}(\sqrt[3]{y^2} + \sqrt[3]{2})$

c) $(4\sqrt{3} + \sqrt{2})(\sqrt{3} - 5\sqrt{2})$

d) $(\sqrt{a} + \sqrt{b})(\sqrt{a} - \sqrt{b})$

SOLUTION

a) $\sqrt{3}(x - \sqrt{5}) = \sqrt{3} \cdot x - \sqrt{3} \cdot \sqrt{5}$ Using the distributive law

$\qquad = x\sqrt{3} - \sqrt{15}$ Multiplying radicals

b) $\sqrt[3]{y}(\sqrt[3]{y^2} + \sqrt[3]{2}) = \sqrt[3]{y} \cdot \sqrt[3]{y^2} + \sqrt[3]{y} \cdot \sqrt[3]{2}$ Using the distributive law

$\qquad = \sqrt[3]{y^3} + \sqrt[3]{2y}$ Multiplying radicals

$\qquad = y + \sqrt[3]{2y}$ Simplifying $\sqrt[3]{y^3}$

c) $(4\sqrt{3} + \sqrt{2})(\sqrt{3} - 5\sqrt{2}) = \overset{F}{4(\sqrt{3})^2} - \overset{O}{20\sqrt{3} \cdot \sqrt{2}} + \overset{I}{\sqrt{2} \cdot \sqrt{3}} - \overset{L}{5(\sqrt{2})^2}$

$\qquad = 4 \cdot 3 - 20\sqrt{6} + \sqrt{6} - 5 \cdot 2$ Multiplying radicals

$\qquad = 12 - 20\sqrt{6} + \sqrt{6} - 10$

$\qquad = 2 - 19\sqrt{6}$ Combining like terms

d) $(\sqrt{a} + \sqrt{b})(\sqrt{a} - \sqrt{b}) = (\sqrt{a})^2 - (\sqrt{b})^2$ This is in the same form as a difference of two squares.

$\qquad = a - b$ We assume $a, b \geq 0$.

Note in part (d) above that the product of two radical expressions need not be a radical expression itself. Pairs of radical expressions like $\sqrt{a} + \sqrt{b}$ and $\sqrt{a} - \sqrt{b}$ or $5 - \sqrt{x}$ and $5 + \sqrt{x}$ are called **conjugates**. Conjugates will prove useful later in this section.

EXAMPLE 2

If $f(x) = x^2$, find $f(a + \sqrt{5})$.

SOLUTION

$$
\begin{aligned}
f(a + \sqrt{5}) &= (a + \sqrt{5})^2 \\
&= a^2 + 2a\sqrt{5} + (\sqrt{5})^2 \qquad \text{Squaring a binomial} \\
&= a^2 + 2a\sqrt{5} + 5
\end{aligned}
$$

As in our earlier work, when different indices appear, rational exponents are helpful.

EXAMPLE 3

If $f(x) = \sqrt[3]{x^2}$ and $g(x) = \sqrt{x} + \sqrt[4]{5x}$, find $(f \cdot g)(x)$.

SOLUTION   Recall from Section 2.6 that $(f \cdot g)(x) = f(x) \cdot g(x)$. Thus,

$$
\begin{aligned}
(f \cdot g)(x) &= \sqrt[3]{x^2}(\sqrt{x} + \sqrt[4]{5x}) & \text{We assume } x \geq 0. \\
&= x^{2/3}(x^{1/2} + (5x)^{1/4}) & \text{Converting to exponential notation} \\
&= x^{2/3}(x^{1/2} + 5^{1/4}x^{1/4}) & \text{Using the laws of exponents} \\
&= x^{2/3} \cdot x^{1/2} + x^{2/3} \cdot 5^{1/4}x^{1/4} & \text{Using the distributive law} \\
&= x^{2/3+1/2} + 5^{1/4}x^{2/3+1/4} & \text{Adding exponents} \\
&= x^{7/6} + 5^{3/12}x^{11/12} & \text{Finding a common denominator} \\
&= x^{1+1/6} + 5^{3/12}x^{11/12} & \text{Writing a mixed number} \\
&= x^1 x^{1/6} + (5^3 x^{11})^{1/12} & \text{Using the laws of exponents} \\
&= x\sqrt[6]{x} + \sqrt[12]{125x^{11}} & \text{Converting back to radical notation}
\end{aligned}
$$

Rationalizing Denominators and Numerators

When a radical expression appears in a denominator, it can be useful to find an equivalent expression in which the denominator no longer contains a radical.* The procedure for finding such an expression is called **rationalizing the denominator.** We carry this out by multiplying by 1 in either of two ways.

One way is to multiply by 1 *under* the radical to make the denominator of the radicand a perfect power.

EXAMPLE 4

Rationalize each denominator.

a) $\sqrt{\dfrac{7}{3}}$

b) $\sqrt[3]{\dfrac{5}{16}}$

---

*See Exercise 89 on p. 409.

SOLUTION

a) We multiply by 1 under the radical, using $\frac{3}{3}$. We do this so that the denominator of the radicand will be a perfect square:

$$\sqrt{\frac{7}{3}} = \sqrt{\frac{7}{3} \cdot \frac{3}{3}}$$

$$= \sqrt{\frac{21}{9}} \qquad \text{The denominator, 9, is now a perfect square.}$$

$$= \frac{\sqrt{21}}{\sqrt{9}} \qquad \text{Using the quotient rule for radicals}$$

$$= \frac{\sqrt{21}}{3}.$$

b) Note that $16 = 4^2$. Thus, to make the denominator a perfect cube, we multiply under the radical by $\frac{4}{4}$:

$$\sqrt[3]{\frac{5}{16}} = \sqrt[3]{\frac{5}{4 \cdot 4} \cdot \frac{4}{4}} \qquad \begin{array}{l}\text{Since the index is 3, we need 3 identical factors}\\ \text{in the denominator.}\end{array}$$

$$= \sqrt[3]{\frac{20}{4^3}} \qquad \text{The denominator is now a perfect cube.}$$

$$= \frac{\sqrt[3]{20}}{\sqrt[3]{4^3}}$$

$$= \frac{\sqrt[3]{20}}{4}.$$

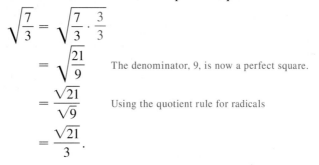

Another way to rationalize a denominator is to multiply by 1 *outside* the radical.

EXAMPLE   5

Rationalize each denominator.

a) $\sqrt{\dfrac{4}{5b}}$ **b)** $\dfrac{\sqrt[3]{a}}{\sqrt[3]{9x}}$ **c)** $\dfrac{3x}{\sqrt[5]{2x^2y^3}}$

SOLUTION

a) We rewrite the expression as a quotient of two radicals. Then we simplify and multiply by 1:

$$\sqrt{\frac{4}{5b}} = \frac{\sqrt{4}}{\sqrt{5b}} = \frac{2}{\sqrt{5b}} \qquad \text{We assume } b > 0.$$

$$= \frac{2}{\sqrt{5b}} \cdot \frac{\sqrt{5b}}{\sqrt{5b}} \qquad \text{Multiplying by 1}$$

$$= \frac{2\sqrt{5b}}{(\sqrt{5b})^2} \qquad \text{Try to do this step mentally.}$$

$$= \frac{2\sqrt{5b}}{5b}.$$

b) To rationalize the denominator $\sqrt[3]{9x}$, note that $9x$ is $3 \cdot 3 \cdot x$. In order for this radicand to be a cube, we need another factor of 3 and two more factors of x. Thus we multiply by 1, using $\sqrt[3]{3x^2}/\sqrt[3]{3x^2}$:

$$\frac{\sqrt[3]{a}}{\sqrt[3]{9x}} = \frac{\sqrt[3]{a}}{\sqrt[3]{9x}} \cdot \frac{\sqrt[3]{3x^2}}{\sqrt[3]{3x^2}} \qquad \text{Multiplying by 1}$$

$$= \frac{\sqrt[3]{3ax^2}}{\sqrt[3]{27x^3}} \longleftarrow \text{This radicand is now a perfect cube.}$$

$$= \frac{\sqrt[3]{3ax^2}}{3x}.$$

c) To change the radicand $2x^2y^3$ into a perfect fifth power, we need four more factors of 2, three more factors of x, and two more factors of y. Thus we multiply by 1, using $\sqrt[5]{2^4x^3y^2}/\sqrt[5]{2^4x^3y^2}$, or $\sqrt[5]{16x^3y^2}/\sqrt[5]{16x^3y^2}$:

$$\frac{3x}{\sqrt[5]{2x^2y^3}} = \frac{3x}{\sqrt[5]{2x^2y^3}} \cdot \frac{\sqrt[5]{16x^3y^2}}{\sqrt[5]{16x^3y^2}} \qquad \text{Multiplying by 1}$$

$$= \frac{3x\sqrt[5]{16x^3y^2}}{\sqrt[5]{32x^5y^5}} \longleftarrow \text{This radicand is now a perfect fifth power.}$$

$$= \frac{3x\sqrt[5]{16x^3y^2}}{2xy} = \frac{3\sqrt[5]{16x^3y^2}}{2y}. \qquad \text{Always simplify if possible.}$$

      Sometimes in calculus it is necessary to rationalize a numerator. To do so, we multiply by 1 to make the radicand in the *numerator* a perfect power.

EXAMPLE 6   Rationalize each numerator: **(a)** $\sqrt{\dfrac{7}{5}}$; **(b)** $\dfrac{\sqrt[3]{4a^2}}{\sqrt[3]{5b}}$.

SOLUTION

a) $\sqrt{\dfrac{7}{5}} = \sqrt{\dfrac{7}{5} \cdot \dfrac{7}{7}}$ Multiplying by 1 under the radical. We also could have multiplied by 1 outside the radical.

$$= \sqrt{\frac{49}{35}} \qquad \text{The numerator is now a perfect square.}$$

$$= \frac{\sqrt{49}}{\sqrt{35}} \qquad \text{Using the quotient rule for radicals}$$

$$= \frac{7}{\sqrt{35}}$$

b) $\dfrac{\sqrt[3]{4a^2}}{\sqrt[3]{5b}} = \dfrac{\sqrt[3]{4a^2}}{\sqrt[3]{5b}} \cdot \dfrac{\sqrt[3]{2a}}{\sqrt[3]{2a}}$ Multiplying by 1

$$= \frac{\sqrt[3]{8a^3}}{\sqrt[3]{10ba}} \longleftarrow \text{This radicand is now a perfect cube.}$$

$$= \frac{2a}{\sqrt[3]{10ab}}$$

Recall from Example 1(d) that when two radical expressions are conjugates, their product contains no radicals. When rationalizing a denominator or a numerator that has two terms, we proceed much as we did in Examples 4–6. The difference is that now we use conjugates to construct the symbol for 1.

EXAMPLE 7

Rationalize each denominator: **(a)** $\dfrac{4}{\sqrt{3}+x}$; **(b)** $\dfrac{4+\sqrt{2}}{\sqrt{5}-\sqrt{2}}$.

SOLUTION

a) $\dfrac{4}{\sqrt{3}+x} = \dfrac{4}{\sqrt{3}+x} \cdot \dfrac{\sqrt{3}-x}{\sqrt{3}-x}$ Multiplying by 1, using the conjugate of $\sqrt{3}+x$, which is $\sqrt{3}-x$

$= \dfrac{4(\sqrt{3}-x)}{(\sqrt{3}+x)(\sqrt{3}-x)}$ Multiplying numerators and denominators

$= \dfrac{4(\sqrt{3}-x)}{(\sqrt{3})^2-x^2}$ Using FOIL in the denominator

$= \dfrac{4\sqrt{3}-4x}{3-x^2}$ Simplifying

b) $\dfrac{4+\sqrt{2}}{\sqrt{5}-\sqrt{2}} = \dfrac{4+\sqrt{2}}{\sqrt{5}-\sqrt{2}} \cdot \dfrac{\sqrt{5}+\sqrt{2}}{\sqrt{5}+\sqrt{2}}$ Multiplying by 1, using the conjugate of $\sqrt{5}-\sqrt{2}$, which is $\sqrt{5}+\sqrt{2}$

$= \dfrac{(4+\sqrt{2})(\sqrt{5}+\sqrt{2})}{(\sqrt{5}-\sqrt{2})(\sqrt{5}+\sqrt{2})}$ Multiplying numerators and denominators

$= \dfrac{4\sqrt{5}+4\sqrt{2}+\sqrt{2}\sqrt{5}+(\sqrt{2})^2}{(\sqrt{5})^2-(\sqrt{2})^2}$ Using FOIL

$= \dfrac{4\sqrt{5}+4\sqrt{2}+\sqrt{10}+2}{5-2}$ Squaring in the denominator and the numerator

$= \dfrac{4\sqrt{5}+4\sqrt{2}+\sqrt{10}+2}{3}$

To rationalize a numerator with more than one term, we use the conjugate of the numerator.

EXAMPLE 8

Rationalize the numerator: $\dfrac{4+\sqrt{2}}{\sqrt{5}-\sqrt{2}}$.

SOLUTION

$\dfrac{4+\sqrt{2}}{\sqrt{5}-\sqrt{2}} = \dfrac{4+\sqrt{2}}{\sqrt{5}-\sqrt{2}} \cdot \dfrac{4-\sqrt{2}}{4-\sqrt{2}}$ Multiplying by 1, using the conjugate of $4+\sqrt{2}$, which is $4-\sqrt{2}$

$= \dfrac{16-(\sqrt{2})^2}{4\sqrt{5}-\sqrt{5}\sqrt{2}-4\sqrt{2}+(\sqrt{2})^2}$

$= \dfrac{14}{4\sqrt{5}-\sqrt{10}-4\sqrt{2}+2}$

EXERCISE SET

7.5

Multiply. Assume that all variables represent nonnegative real numbers. Note that Exercises 11–14 use conjugates.

1. $\sqrt{7}(3 - \sqrt{7})$

2. $\sqrt{3}(4 + \sqrt{3})$

3. $\sqrt{2}(\sqrt{3} - \sqrt{5})$

4. $\sqrt{5}(\sqrt{5} - \sqrt{2})$

5. $\sqrt{3}(2\sqrt{5} - 3\sqrt{4})$

6. $\sqrt{2}(3\sqrt{10} - 2\sqrt{2})$

7. $\sqrt[3]{2}(\sqrt[3]{4} - 2\sqrt[3]{32})$

8. $\sqrt[3]{3}(\sqrt[3]{9} - 4\sqrt[3]{21})$

9. $\sqrt[3]{a}(\sqrt[3]{a^2} + \sqrt[3]{24a^2})$

10. $\sqrt[3]{x}(\sqrt[3]{3x^2} - \sqrt[3]{81x^2})$

11. $(5 + \sqrt{6})(5 - \sqrt{6})$

12. $(2 - \sqrt{5})(2 + \sqrt{5})$

13. $(3 - 2\sqrt{7})(3 + 2\sqrt{7})$

14. $(4 + 3\sqrt{2})(4 - 3\sqrt{2})$

15. $(5 + \sqrt[3]{10})(3 - \sqrt[3]{10})$

16. $(\sqrt[3]{7} - 4)(\sqrt[3]{7} + 5)$

17. $(2\sqrt{7} - 4\sqrt{2})(3\sqrt{7} + 6\sqrt{2})$

18. $(4\sqrt{5} + 3\sqrt{3})(3\sqrt{5} - 4\sqrt{3})$

19. $(2\sqrt[3]{3} - \sqrt[3]{2})(\sqrt[3]{3} + 2\sqrt[3]{2})$

20. $(3\sqrt[4]{7} + \sqrt[4]{6})(2\sqrt[4]{9} - 3\sqrt[4]{6})$

21. $(\sqrt{3x} + \sqrt{y})^2$

22. $(\sqrt{t} - \sqrt{2r})^2$

23. $\sqrt[3]{x^2y}(\sqrt{xy} - \sqrt[5]{xy^3})$

24. $\sqrt[4]{a^2b}(\sqrt[3]{a^2b} - \sqrt[5]{a^2b^2})$

25. $(m + \sqrt[3]{n^2})(2m + \sqrt[4]{n})$

26. $(r - \sqrt[4]{s^3})(3r - \sqrt[5]{s})$

In Exercises 27–32, $f(x)$ and $g(x)$ are as given. Find $(f \cdot g)(x)$.

27. $f(x) = \sqrt[4]{x}, \ g(x) = \sqrt[4]{2x} - \sqrt[4]{x^{11}}$

28. $f(x) = \sqrt[4]{x^7} + \sqrt[4]{3x^2}, \ g(x) = \sqrt[4]{x}$

29. $f(x) = x + \sqrt{7}, \ g(x) = x - \sqrt{7}$

30. $f(x) = x - \sqrt{2}, \ g(x) = x + \sqrt{6}$

31. $f(x) = 2 - \sqrt{x}, \ g(x) = 1 - \sqrt{x}$

32. $f(x) = \sqrt{x} + \sqrt{3}, \ g(x) = \sqrt{x} + \sqrt{2}$

Let $f(x) = x^2$. Find each of the following.

33. $f(5 - \sqrt{2})$

34. $f(7 + \sqrt{3})$

35. $f(\sqrt{3} + \sqrt{5})$

36. $f(\sqrt{6} - \sqrt{3})$

37. $f(\sqrt{10} - \sqrt{5})$

38. $f(\sqrt{7} + \sqrt{8})$

Rationalize each denominator.

39. $\sqrt{\dfrac{5}{7}}$

40. $\sqrt{\dfrac{11}{6}}$

41. $\dfrac{6\sqrt{5}}{5\sqrt{3}}$

42. $\dfrac{4\sqrt{5}}{3\sqrt{2}}$

43. $\sqrt[3]{\dfrac{16}{9}}$

44. $\sqrt[3]{\dfrac{2}{9}}$

45. $\dfrac{\sqrt[3]{3a}}{\sqrt[3]{5c}}$

46. $\dfrac{\sqrt[3]{7x}}{\sqrt[3]{3y}}$

47. $\dfrac{\sqrt[3]{5y^4}}{\sqrt[3]{6x^4}}$

48. $\dfrac{\sqrt[3]{3a^4}}{\sqrt[3]{7b^2}}$

49. $\sqrt[3]{\dfrac{2}{x^2y}}$

50. $\sqrt[3]{\dfrac{5}{ab^2}}$

51. $\sqrt{\dfrac{7a}{18}}$

52. $\sqrt{\dfrac{3x}{10}}$

53. $\sqrt{\dfrac{9}{20x^2y}}$

54. $\sqrt{\dfrac{5}{32ab^2}}$

55. $\dfrac{5}{7 - \sqrt{2}}$

56. $\dfrac{3}{5 + \sqrt{6}}$

57. $\dfrac{\sqrt{x}}{\sqrt{x} + \sqrt{y}}$

58. $\dfrac{\sqrt{b}}{\sqrt{a} - \sqrt{b}}$

59. $\dfrac{\sqrt{3} + 4\sqrt{5}}{\sqrt{3} - 2\sqrt{6}}$

60. $\dfrac{10 - 2\sqrt{3}}{\sqrt{5} + 3\sqrt{3}}$

61. $\dfrac{5\sqrt{3} - 3\sqrt{2}}{3\sqrt{2} - 2\sqrt{3}}$

62. $\dfrac{7\sqrt{2} + 4\sqrt{3}}{4\sqrt{3} - 3\sqrt{2}}$

Rationalize each numerator.

63. $\dfrac{\sqrt{5}}{\sqrt{7x}}$

64. $\dfrac{\sqrt{10}}{\sqrt{3x}}$

65. $\sqrt{\dfrac{14}{21}}$

66. $\sqrt{\dfrac{12}{15}}$

67. $\dfrac{4\sqrt{13}}{3\sqrt{7}}$

68. $\dfrac{5\sqrt{21}}{2\sqrt{5}}$

69. $\dfrac{\sqrt[3]{7}}{\sqrt[3]{2}}$

70. $\dfrac{\sqrt[5]{5}}{\sqrt[3]{4}}$

71. $\sqrt{\dfrac{7x}{3y}}$

72. $\sqrt{\dfrac{6a}{5b}}$

73. $\sqrt[3]{\dfrac{2a^5}{5b}}$

74. $\sqrt[3]{\dfrac{2a^4}{7b}}$

75. $\sqrt{\dfrac{x^3y}{2}}$

76. $\sqrt{\dfrac{ab^5}{3}}$

77. $\dfrac{\sqrt{5} + 2}{6}$

78. $\dfrac{7 - \sqrt{3}}{4}$

79. $\dfrac{\sqrt{3} - 5}{\sqrt{2} + 5}$

80. $\dfrac{\sqrt{6} - 3}{\sqrt{3} + 7}$

81. $\dfrac{\sqrt{x} + \sqrt{y}}{\sqrt{x} - \sqrt{y}}$

82. $\dfrac{\sqrt{x} - \sqrt{y}}{\sqrt{x} + \sqrt{y}}$

83. $\dfrac{a\sqrt{b} - \sqrt{c}}{\sqrt{b} + \sqrt{c}}$

84. $\dfrac{\sqrt{a} + b\sqrt{c}}{\sqrt{a} - \sqrt{c}}$

SKILL MAINTENANCE

Solve.

85. $\dfrac{1}{2} - \dfrac{1}{3} = \dfrac{1}{t}$

86. $\dfrac{5}{x - 1} + \dfrac{9}{x^2 + x + 1} = \dfrac{15}{x^3 - 1}$

Divide and simplify.

87. $\dfrac{2x^2 - x - 6}{x^2 + 4x + 3} \div \dfrac{2x^2 + x - 3}{x^2 - 1}$

88. $\dfrac{1}{x^3 - y^3} \div \dfrac{1}{(x - y)(x^2 + xy + y^2)}$

SYNTHESIS

89. ◈ Explain why it is easier to approximate

$$\frac{\sqrt{2}}{2} \quad \text{than} \quad \frac{1}{\sqrt{2}}$$

if no calculator is available and $\sqrt{2} \approx 1.414213562$.

90. ◈ A student *incorrectly* claims that

$$\frac{5 + \sqrt{2}}{\sqrt{18}} = \frac{5 + \sqrt{1}}{\sqrt{9}} = \frac{5 + 1}{3}.$$

How could you convince the student that a mistake has been made? How would you explain the correct way of rationalizing the denominator?

Express each of the following as the product of two radical expressions.

91. $x - 5$ **92.** $y - 7$ **93.** $x - a$

Multiply.

94. $\sqrt{9 + 3\sqrt{5}} \ \sqrt{9 - 3\sqrt{5}}$

95. $(\sqrt{x + 2} - \sqrt{x - 2})^2$

For Exercises 96–99, assume that all radicands are positive and that no denominators are 0.

Rationalize each denominator.

96. $\dfrac{a - \sqrt{a + b}}{\sqrt{a + b} - b}$

97. $\dfrac{b + \sqrt{b}}{1 + b + \sqrt{b}}$

Rationalize each numerator.

98. $\dfrac{\sqrt{y + 18} - \sqrt{y}}{18}$

99. $\dfrac{\sqrt{x + 6} - 5}{\sqrt{x + 6} + 5}$

Simplify.

100. $\sqrt{a^2 - 3} - \dfrac{a^2}{\sqrt{a^2 - 3}}$

101. $5\sqrt{\dfrac{x}{y}} + 4\sqrt{\dfrac{y}{x}} - \dfrac{3}{\sqrt{xy}}$

102. $\dfrac{\dfrac{1}{\sqrt{w}} - \sqrt{w}}{\dfrac{\sqrt{w} + 1}{\sqrt{w}}}$

103. $\dfrac{1}{4 + \sqrt{3}} + \dfrac{1}{\sqrt{3}} + \dfrac{1}{\sqrt{3} - 4}$

104. ◈ Use what we know about factoring a difference of two cubes to present a method for rationalizing any denominator of the form $\sqrt[3]{a} - \sqrt[3]{b}$.

105. ◨ Use a grapher to check your answers to Exercises 9, 51, and 99.

7.6 Solving Radical Equations

The Principle of Powers • Equations with Two Radical Terms

The Principle of Powers

A **radical equation** has variables in one or more radicands. Examples are

$$\sqrt[3]{2x} + 1 = 5, \quad \sqrt{a} + \sqrt{a - 2} = 7, \quad \text{and} \quad 4 - \sqrt{3x + 1} = \sqrt{6 - x}.$$

To solve such equations, we need a new principle. Suppose an equation $a = b$ is true. If we square both sides, we get another true equation: $a^2 = b^2$. This can be generalized.

THE PRINCIPLE OF POWERS

If $a = b$, then $a^n = b^n$ for any exponent n.

Note that the principle of powers is an "if—then" statement. The statement obtained by interchanging the two parts of the sentence—"if $a^n = b^n$ for some exponent n, then $a = b$"—*is not always true.* For example, $3^2 = (-3)^2$ *is* true, but $3 = -3$ *is not* true. More generally, $3^n = (-3)^n$ is true for any even number n, whereas $3 = -3$ is false. For this reason, when raising both sides of an equation to an even power, we should check our answers in the original equation.

EXAMPLE 1

Solve: $\sqrt{x} - 3 = 4$.

SOLUTION

$$\sqrt{x} - 3 = 4$$
$$\sqrt{x} = 7 \qquad \text{Adding 3 on both sides to isolate the radical}$$
$$(\sqrt{x})^2 = 7^2 \qquad \text{Using the principle of powers}$$
$$x = 49$$

Check: $$\begin{array}{c|c} \sqrt{x} - 3 = 4 \\ \hline \sqrt{49} - 3 \;?\; 4 \\ 7 - 3 \\ 4 & 4 \quad \text{TRUE} \end{array}$$

The solution is 49.

EXAMPLE 2

Solve: $\sqrt{x} - 5 = -7$.

SOLUTION

$$\sqrt{x} - 5 = -7$$
$$\sqrt{x} = -2 \qquad \text{Adding 5 on both sides to isolate the radical}$$

The equation $\sqrt{x} = -2$ has no solution because the principal square root of a number is never negative. We continue as in Example 1 for comparison.

$$(\sqrt{x})^2 = (-2)^2 \qquad \text{Using the principle of powers (squaring)}$$
$$x = 4$$

Check:
$$\sqrt{x} - 5 = -7$$
$$\begin{array}{c|c}\sqrt{4} - 5 \ ? & -7 \\ 2 - 5 & \\ -3 & -7 \quad \text{FALSE}\end{array}$$

The number 4 does not check. Thus the equation $\sqrt{x} - 5 = -7$ has no real-number solution.

> **CAUTION!**   Raising both sides of an equation to an even power may not produce an equivalent equation. In this case, a check is essential.

Note in Example 2 that $x = 4$ has solution 4, but that $\sqrt{x} - 5 = -7$ has *no* solution. Thus the equations $x = 4$ and $\sqrt{x} - 5 = -7$ are *not* equivalent.

E X A M P L E   3   Solve: $x = \sqrt{x + 7} + 5$.

S O L U T I O N

$$x = \sqrt{x + 7} + 5$$
$$x - 5 = \sqrt{x + 7} \qquad \text{Adding } -5 \text{ on both sides}$$
$$\left.\begin{array}{c}(x - 5)^2 = (\sqrt{x + 7})^2 \\ x^2 - 10x + 25 = x + 7\end{array}\right\} \quad \begin{array}{l}\text{Using the principle of powers; squaring both sides}\end{array}$$
$$x^2 - 11x + 18 = 0 \qquad \begin{array}{l}\text{Adding } -x - 7 \text{ on both sides to write the quadratic equation in standard form}\end{array}$$
$$(x - 9)(x - 2) = 0 \qquad \text{Factoring}$$
$$x = 9 \quad or \quad x = 2 \qquad \begin{array}{l}\text{Using the principle of zero products}\end{array}$$

The possible solutions are 9 and 2. Let's check.

Check: For 9:
$$x = \sqrt{x + 7} + 5$$
$$\begin{array}{c|c}9 \ ? \ \sqrt{9 + 7} + 5 & \\ 9 & 9 \qquad \text{TRUE}\end{array}$$

For 2:
$$x = \sqrt{x + 7} + 5$$
$$\begin{array}{c|c}2 \ ? \ \sqrt{2 + 7} + 5 & \\ 2 & 8 \qquad \text{FALSE}\end{array}$$

Since 9 checks but 2 does not, the solution is 9.

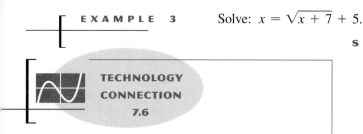

TECHNOLOGY CONNECTION
7.6

To solve Example 3 with a grapher, graph the curves $y_1 = x$ and $y_2 = (x + 7)^{1/2} + 5$ on the same set of axes.

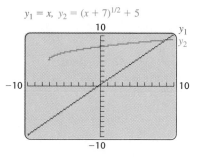

$$y_1 = x, \ y_2 = (x + 7)^{1/2} + 5$$

Using TRACE and ZOOM or the INTERSECT option, determine the point of intersection. The intersection should appear to occur when $x = 9$. Note that there is no intersection when $x = 2$, as predicted in the check of Example 3.

1. Use a grapher to solve Examples 1, 2, 4, 5, and 6. Compare your answers with those found using the algebraic methods shown.

It is important to isolate a radical term before using the principle of powers. Suppose in Example 3 that both sides of the equation were squared *before* isolating the radical. We then would have had the expression $(\sqrt{x + 7} + 5)^2$ or $x + 7 + 10\sqrt{x + 7} + 25$ on the right side, and the radical would have remained in the problem.

E X A M P L E   4

Solve: $(2x + 1)^{1/3} + 5 = 0$.

S O L U T I O N   We need not use radical notation to solve:

$$(2x + 1)^{1/3} + 5 = 0$$
$$(2x + 1)^{1/3} = -5 \qquad \text{Adding } -5 \text{ on both sides}$$
$$[(2x + 1)^{1/3}]^3 = (-5)^3 \qquad \text{Cubing both sides}$$
$$(2x + 1)^1 = (-5)^3 \qquad \text{Multiplying exponents; try to do this mentally.}$$
$$2x + 1 = -125$$
$$2x = -126 \qquad \text{Adding } -1 \text{ on both sides}$$
$$x = -63.$$

Because both sides were raised to an *odd* power, it is not essential that we check the answer. The student can show that -63 checks and is the solution.

Equations with Two Radical Terms

A strategy for solving equations with two or more radical terms is as follows.

TO SOLVE AN EQUATION WITH TWO OR MORE RADICAL TERMS:

1. Isolate one of the radical terms.
2. Use the principle of powers.
3. If a radical remains, perform steps (1) and (2) again.
4. Check possible solutions in the original equation.

E X A M P L E   5

Solve: $\sqrt{2x - 5} = 1 + \sqrt{x - 3}$.

S O L U T I O N

$$\sqrt{2x - 5} = 1 + \sqrt{x - 3}$$
$$(\sqrt{2x - 5})^2 = (1 + \sqrt{x - 3})^2 \qquad \text{One radical is already isolated; we square both sides.}$$

This is like squaring a binomial. We square 1, then find twice the product of 1 and $\sqrt{x - 3}$ and then the square of $\sqrt{x - 3}$.

$$2x - 5 = 1 + 2\sqrt{x - 3} + (\sqrt{x - 3})^2$$
$$2x - 5 = 1 + 2\sqrt{x - 3} + (x - 3)$$

Then

$$x - 3 = 2\sqrt{x - 3}$$ Isolating the remaining radical term

$$(x - 3)^2 = (2\sqrt{x - 3})^2$$ Squaring both sides

$$x^2 - 6x + 9 = 4(x - 3)$$ Remember to square both the 2 and the $\sqrt{x - 3}$ on the right side.

$$x^2 - 6x + 9 = 4x - 12$$

$$x^2 - 10x + 21 = 0$$

$$(x - 7)(x - 3) = 0$$ Factoring

$$x = 7 \quad or \quad x = 3.$$ Using the principle of zero products

We leave it to the student to show that 7 and 3 both check and are the solutions.

CAUTION!   A common error in solving equations like

$$\sqrt{2x - 5} = 1 + \sqrt{x - 3}$$

is to obtain $1 + (x - 3)$ as the square of the right side. This is wrong because $(A + B)^2 \neq A^2 + B^2$. For example,

$$(\sqrt{25} + \sqrt{9})^2 = (5 + 3)^2 = 8^2, \text{ or } 64,$$

whereas

$$(\sqrt{25})^2 + (\sqrt{9})^2 = 25 + 9, \text{ or } 34.$$

EXAMPLE 6

Let $f(x) = \sqrt{x + 5} - \sqrt{x - 3}$. Find all x-values for which $f(x) = 2$.

SOLUTION   We must have $f(x) = 2$, or

$$\sqrt{x + 5} - \sqrt{x - 3} = 2.$$ Substituting for $f(x)$

To solve, we isolate one radical term and square both sides:

$$\sqrt{x + 5} = 2 + \sqrt{x - 3}$$ Adding $\sqrt{x - 3}$; this isolates one of the radical terms.

$$(\sqrt{x + 5})^2 = (2 + \sqrt{x - 3})^2$$ Using the principle of powers (squaring both sides)

$$x + 5 = 4 + 4\sqrt{x - 3} + (x - 3)$$ Using $(A + B)^2 = A^2 + 2AB + B^2$

$$5 = 1 + 4\sqrt{x - 3}$$ Adding $-x$ and combining like terms

$$\left.\begin{array}{l} 4 = 4\sqrt{x - 3} \\ 1 = \sqrt{x - 3} \end{array}\right\}$$ Isolating the remaining radical term

$$1^2 = (\sqrt{x - 3})^2$$ Squaring both sides

$$1 = x - 3$$

$$4 = x.$$

Check: $f(4) = \sqrt{4 + 5} - \sqrt{4 - 3} = \sqrt{9} - \sqrt{1} = 3 - 1 = 2.$

We will have $f(x) = 2$ when $x = 4$.

EXERCISE SET

7.6

Solve.

1. $\sqrt{5x + 1} = 6$

2. $\sqrt{x + 3} = 6$

3. $\sqrt{3x + 1} = 7$

4. $\sqrt{2x - 1} = 7$

5. $\sqrt{y + 1} - 5 = 8$

6. $\sqrt{x - 2} - 7 = -4$

7. $\sqrt{x - 7} + 3 = 10$

8. $\sqrt{y + 4} + 6 = 7$

9. $\sqrt[3]{x + 5} = 2$

10. $\sqrt[3]{x - 2} = 3$

11. $\sqrt[4]{y - 3} = 2$

12. $\sqrt[4]{x + 3} = 3$

13. $3\sqrt{x} = x$

14. $8\sqrt{y} = y$

15. $2y^{1/2} - 7 = 9$

16. $3x^{1/2} + 12 = 9$

17. $\sqrt[3]{x} = -3$

18. $\sqrt[3]{y} = -4$

19. $t^{1/3} - 2 = 3$

20. $x^{1/4} - 2 = 1$

21. $(x + 2)^{1/2} = -4$

22. $(y - 3)^{1/2} = -2$

23. $\sqrt[4]{2x + 3} - 5 = -2$

24. $\sqrt[4]{3x + 1} - 4 = -1$

25. $(y - 7)^{1/4} = 3$

26. $(x + 5)^{1/3} = 4$

27. $\sqrt[3]{6x + 9} + 8 = 5$

28. $\sqrt[3]{3y + 6} + 2 = 3$

29. $\sqrt{2t - 7} = \sqrt{3t - 12}$

30. $\sqrt{3t + 4} = \sqrt{4t + 3}$

31. $2(1 - x)^{1/3} = 4^{1/3}$

32. $3(4 - t)^{1/4} = 6^{1/4}$

33. $x = \sqrt{x - 1} + 3$

34. $3 + \sqrt{5 - x} = x$

35. $3 + \sqrt{z - 6} = \sqrt{z + 9}$

36. $\sqrt{4x - 3} = 2 + \sqrt{2x - 5}$

37. $\sqrt{20 - x} + 8 = \sqrt{9 - x} + 11$

38. $4 + \sqrt{10 - x} = 6 + \sqrt{4 - x}$

39. $\sqrt{x + 2} + \sqrt{3x + 4} = 2$

40. $\sqrt{6x + 7} - \sqrt{3x + 3} = 1$

41. If $f(x) = \sqrt{x} + \sqrt{x - 9}$, find x such that $f(x) = 1$.

42. If $g(x) = \sqrt{x} + \sqrt{x - 5}$, find x such that $g(x) = 5$.

43. If $g(x) = \sqrt{2x + 7} - \sqrt{x + 15}$, find a such that $g(a) = -1$.

44. If $f(x) = \sqrt{x - 2} - \sqrt{4x + 1}$, find a such that $f(a) = -3$.

45. If $f(x) = \sqrt{2x - 3}$ and $g(x) = \sqrt{x + 7} - 2$, find x such that $f(x) = g(x)$.

46. If $f(x) = 2\sqrt{3x + 6}$ and $g(x) = 5 + \sqrt{4x + 9}$, find x such that $f(x) = g(x)$.

47. If $f(t) = 4 - \sqrt{t - 3}$ and $g(t) = (t + 5)^{1/2}$, find a such that $f(a) = g(a)$.

48. If $f(t) = 7 + \sqrt{2t - 5}$ and $g(t) = 3(t + 1)^{1/2}$, find a such that $f(a) = g(a)$.

SKILL MAINTENANCE

49. Solve:
$$\frac{3}{2x} + \frac{1}{x} = \frac{2x + 3.5}{3x}.$$

50. The base of a triangle is 2 in. longer than the height. The area is $31\frac{1}{2}$ in². Find the height and the base.

Graph.

51. $f(x) = \frac{2}{5}x - 7$

52. $y > -3x + 5$

SYNTHESIS

53. ◈ The principle of powers is an "if–then" statement that becomes false when the sentence parts are interchanged. Give an example of another such if–then statement.

54. ◈ Explain a method that could be used to solve
$$\sqrt{x} + \sqrt{2x - 1} - \sqrt{3x + 2} + \sqrt{2 + x} = 0.$$

55. ◈ Is checking essential when the principle of powers is used with an odd power n? Why or why not?

56. ◈ Outline a procedure that could be used to write radical equations that have no solution.

Escape velocity. *A formula for the escape velocity v of a satellite is*
$$v = \sqrt{2gr}\,\sqrt{\frac{h}{r + h}},$$

where g is the force of gravity, r is the planet or star's radius, and h is the height of the satellite above the planet or star's surface.

57. Solve for h.

58. Solve for r.

Sighting to the horizon. *The function* $D(h) = 1.2\sqrt{h}$ *can be used to approximate the distance D, in miles, that a person can see to the horizon from a height h, in feet.*

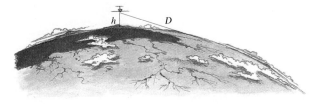

59. ▦ How far above sea level must a pilot fly in order to see a horizon that is 180 mi away?

60. ▦ How high above sea level must a sailor climb in order to see 10.2 mi out to sea?

Solve.

61. $\dfrac{x + \sqrt{x + 1}}{x - \sqrt{x + 1}} = \dfrac{5}{11}$

62. $\left(\dfrac{z}{4} - 5\right)^{2/3} = \dfrac{1}{25}$

63. $(z^2 + 17)^{3/4} = 27$

64. $\sqrt{\sqrt{y} + 49} = 7$

65. $x^2 - 5x - \sqrt{x^2 - 5x - 2} = 4$
(*Hint*: Let $u = x^2 - 5x - 2$.)

66. $\sqrt{8 - b} = b\sqrt{8 - b}$

Without graphing, determine the x-intercepts of the graphs given by each of the following.

67. $f(x) = \sqrt{x - 2} - \sqrt{x + 2} + 2$

68. $g(x) = 6x^{1/2} + 6x^{-1/2} - 37$

69. $f(x) = (x^2 + 30x)^{1/2} - x - (5x)^{1/2}$

70. 〰 Use a grapher to check your answers to Exercises 4, 10, and 26.

71. 〰 Use a grapher to check your answers to Exercises 21, 29, and 35.

72. ◈ 〰 Saul is trying to solve Exercise 62 using a grapher. Without resorting to trial and error, how can he determine a suitable viewing window for finding the solution?

◆ **COLLABORATIVE**
C ◆ O ◆ R ◆ N ◆ E ◆ R

Focus: Rational exponents, estimation, and solving equations

Time: 20–30 minutes

Group size: 2–3

Materials: Calculators are required.

Calistoga, California, has been a popular resort town for over 100 years, famous for its naturally occurring bubbling water. Thus residents were understandably unhappy with a proposal to raise the city water rates over a period of three years: 30% in the first year, 27% in the second, and 12% in the third.

ACTIVITY

1. Each group member should estimate the total amount of the percent increase in water rates at the end of the three-year period. Remember that the increase of 27% in the second year is applied to the

new, higher rates that would exist after the 30% rate hike. Similarly, the 12% increase would apply to the rates in effect after the first two increases.

2. Determine, as a group, what the *exact* percent increase would be for the three-year period. Then find a multiplier that could be used to predict a household's new bill after the proposed increases.

3. Each group member should estimate what three *equal* rate increases would produce the same total increase at the end of three years. Check how well each estimate works.

4. Using a calculator and the multiplier found in part (2) above, determine, as a group, a precise answer to part (3).

5. What would benefit consumers most: three equal rate hikes, as determined in part (4), or the three increases originally proposed? Share the reasoning behind your group's answer to this question with the entire class.

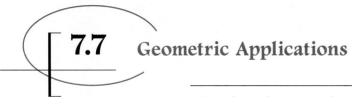

7.7   Geometric Applications

Using the Pythagorean Theorem   •   Two Special Triangles

Using the Pythagorean Theorem

There are many kinds of problems that involve powers and roots. Many also involve right triangles and the Pythagorean theorem, which we studied in Section 5.8.

EXAMPLE 1

Baseball.   A baseball diamond is actually a square 90 ft on a side. Suppose a catcher fields a ball along the third-base line 10 ft from home plate. How far would the catcher's throw to first base be? Give an exact answer and an approximation to three decimal places.

SOLUTION   We first make a drawing and let $d =$ the distance, in feet, to first base. Note that a right triangle is formed in which the length of the leg from home to first base is 90 ft. The length of the leg from home to where the catcher fields the ball is 10 ft. We substitute these values into the Pythagorean equation to find d:

$$d^2 = 90^2 + 10^2$$
$$d^2 = 8100 + 100$$
$$d^2 = 8200$$
$$d = \sqrt{8200}.*$$

Exact answer: $d = \sqrt{8200}$ ft
Approximation: $d \approx 90.554$ ft

Using a calculator

EXAMPLE 2

Guy Wires.   The base of a 40-ft-long guy wire is located 15 ft from the telephone pole that it is anchoring. How high up the pole does the guy wire reach? Give an exact answer and an approximation to three decimal places.

SOLUTION   We make a drawing and let h represent the height on the pole that the guy wire reaches. A right triangle is formed in which the length of one leg is 15 ft and the length of the hypotenuse is 40 ft. Using the

---

*Actually $d^2 = 8200$ also has $-\sqrt{8200}$ as a solution, but since the problems in this section all involve length, we concern ourselves only with positive answers.

Pythagorean equation, we have

$$h^2 + 15^2 = 40^2$$
$$h^2 + 225 = 1600$$
$$h^2 = 1375$$
$$h = \sqrt{1375}.$$

Exact answer:
$$h = \sqrt{1375} \text{ ft}$$

Approximation:
$$h \approx 37.081 \text{ ft} \quad \text{Using a calculator}$$

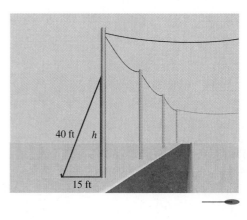

Two Special Triangles

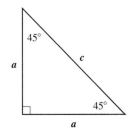

When both legs of a right triangle are the same size, we call the triangle an *isosceles right triangle*. If one leg of an isosceles right triangle has length a, we can find a formula for the length of the hypotenuse as follows:

$$c^2 = a^2 + b^2$$
$$c^2 = a^2 + a^2 \quad \text{Because the triangle is isosceles, both legs are the same size: } a = b.$$
$$c^2 = 2a^2 \quad \text{Combining like terms}$$
$$c = \sqrt{2a^2}$$
$$c = \sqrt{a^2 \cdot 2} = a\sqrt{2}. \quad \text{Try to remember this formula.}$$

E X A M P L E   3

One leg of an isosceles right triangle measures 7 cm. Find the length of the hypotenuse. Give an exact answer and an approximation to three decimal places.

S O L U T I O N   We substitute:

$$c = a\sqrt{2} \quad \text{This equation should be memorized.}$$
$$c = 7\sqrt{2}.$$

Exact answer: $c = 7\sqrt{2}$ cm
Approximation: $c \approx 9.899$ cm   Using a calculator

When the hypotenuse of an isosceles right triangle is known, the lengths of the legs can be found.

E X A M P L E   4

The hypotenuse of an isosceles right triangle is 5 ft long. Find the length of a leg. Give an exact answer and an approximation to three decimal places.

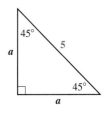

SOLUTION   We replace c with 5 and solve for a:

$$5 = a\sqrt{2} \qquad \text{Substituting 5 for } c \text{ in } c = a\sqrt{2}$$

$$\frac{5}{\sqrt{2}} = a \qquad \text{Dividing by } \sqrt{2} \text{ on both sides}$$

$$\frac{5\sqrt{2}}{2} = a. \qquad \text{Rationalize the denominator if necessary.}$$

Exact answer: $a = \dfrac{5}{\sqrt{2}}$ ft, or $\dfrac{5\sqrt{2}}{2}$ ft

Approximation: $a \approx 3.536$ ft   Using a calculator

A second special triangle is known as a 30°–60°–90° right triangle, so named because of the measures of its angles. Note that in an equilateral triangle, all sides have the same length and all angles are 60°. An altitude, drawn dashed in the figure, bisects, or splits, one angle and one side. Two 30°–60°–90° right triangles are thus formed.

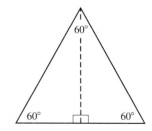

Because of the way in which the altitude is drawn, if a represents the length of the shorter leg in a 30°–60°–90° right triangle, then $2a$ represents the length of the hypotenuse. We have

$$a^2 + b^2 = (2a)^2 \qquad \text{Using the Pythagorean equation}$$

$$a^2 + b^2 = 4a^2$$

$$b^2 = 3a^2 \qquad \text{Adding } -a^2 \text{ on both sides}$$

$$b = \sqrt{3a^2}$$

$$= \sqrt{a^2 \cdot 3} = a\sqrt{3}.$$

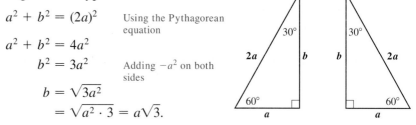

EXAMPLE 5

The shorter leg of a 30°–60°–90° right triangle measures 8 in. Find the lengths of the other sides. Give exact answers and, where appropriate, an approximation to three decimal places.

SOLUTION   The hypotenuse is twice as long as the shorter leg, so we have

$$c = 2a \qquad \text{This relationship should be memorized.}$$

$$= 2 \cdot 8 = 16 \text{ in.}$$

The length of the longer leg is the length of the shorter leg times $\sqrt{3}$. This gives us

$$b = a\sqrt{3} \qquad \text{This should also be memorized.}$$

$$= 8\sqrt{3} \text{ in.}$$

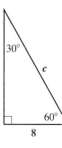

Exact answer: $c = 16$ in., $b = 8\sqrt{3}$ in.

Approximation: $b \approx 13.856$ in.

E X A M P L E   6

The length of the longer leg of a 30°–60°–90° right triangle is 14 cm. Find the length of the hypotenuse. Give an exact answer and an approximation to three decimal places.

S O L U T I O N   The length of the hypotenuse is twice the length of the shorter leg. We first find a, the length of the shorter leg, by using the length of the longer leg:

$$14 = a\sqrt{3} \qquad \text{Substituting 14 for } b \text{ in } b = a\sqrt{3}$$

$$\frac{14}{\sqrt{3}} = a. \qquad \text{Dividing by } \sqrt{3}$$

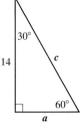

Since the hypotenuse is twice as long as the shorter leg, we have

$$c = 2a$$

$$= 2 \cdot \frac{14}{\sqrt{3}} \qquad \text{Substituting}$$

$$= \frac{28}{\sqrt{3}} \text{ cm.}$$

Exact answer: $c = \dfrac{28}{\sqrt{3}}$ cm, or $\dfrac{28\sqrt{3}}{3}$ cm if the denominator is rationalized.

Approximation: $c \approx 16.166$ cm

LENGTHS WITHIN ISOSCELES AND 30°–60°–90° RIGHT TRIANGLES

The length of the hypotenuse in an isosceles right triangle is the length of a leg times $\sqrt{2}$.

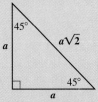

The length of the longer leg in a 30°–60°–90° right triangle is the length of the shorter leg times $\sqrt{3}$. The hypotenuse is twice as long as the shorter leg.

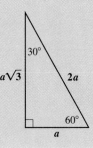

EXERCISE SET

7.7

In a right triangle, find the length of the side not given. Give an exact answer and, where appropriate, an approximation to three decimal places.

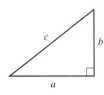

1. $a = 5, b = 3$
2. $a = 8, b = 10$
3. $a = 7, b = 7$
4. $a = 10, b = 10$
5. $b = 12, c = 13$
6. $a = 5, c = 12$
7. $c = 6, a = \sqrt{5}$
8. $c = 8, a = 4\sqrt{3}$
9. $b = 1, c = \sqrt{13}$
10. $a = 1, c = \sqrt{20}$
11. $a = 1, c = \sqrt{n}$
12. $c = 2, a = \sqrt{n}$

In Exercises 13–20, give an exact answer and, where appropriate, an approximation to three decimal places.

13. *Guy wire.* How long is a guy wire if it reaches from the top of a 15-ft pole to a point on the ground 10 ft from the pole?

14. *Softball.* A slow-pitch softball diamond is actually a square 65 ft on a side. How far is it from home to second base?

15. *Baseball.* Suppose the catcher in Example 1 makes a throw to second base. How far is that throw?

16. *Television sets.* What does it mean to refer to a 20-in. TV set or a 25-in. TV set? Such units refer to the diagonal of the screen. A 20-in. TV set has a width of 16 in. What is its height?

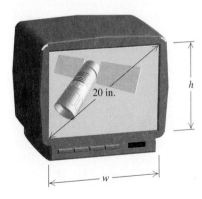

17. *Television sets.* A 25-in. TV set has a screen with a height of 15 in. What is its width? (See Exercise 16.)

18. *Speaker placement.* A stereo receiver is in a corner of a 12-ft by 14-ft room. Speaker wire will run under a rug, diagonally, to a speaker in the far corner. If 4 ft of slack is required on each end, how long a piece of wire should be purchased?

19. *Distance over water.* To determine the width of a pond, a surveyor locates two stakes at either end of the pond and uses instrumentation to place a third stake so that the distance across the pond is the length of a hypotenuse. If the third stake is 90 m from one stake and 70 m from the other, how wide is the pond?

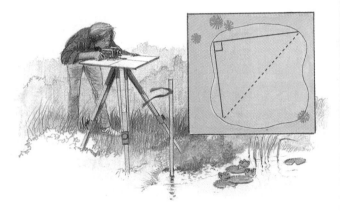

20. *Vegetable garden.* Benito and Dominique are planting a 30-ft by 40-ft vegetable garden and are laying it out using string. They would like to know the length of a diagonal to make sure that right angles are formed. Find the length of a diagonal.

For each triangle, find the missing length(s). Give an exact answer and, where appropriate, an approximation to three decimal places.

21.

22.

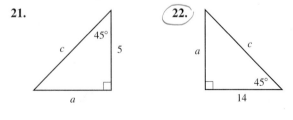

23.

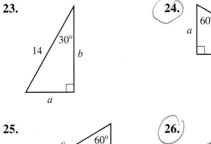

24.

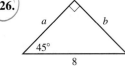

25.

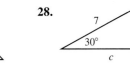

26.

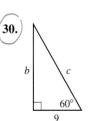

27.

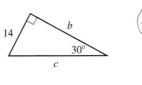

28.

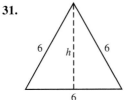

29.

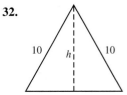

30.

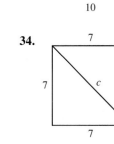

31.

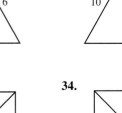

32.

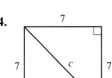

33.

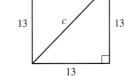

34.

35.

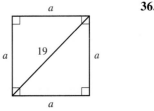

36.

In Exercises 37–44, give an exact answer and, where appropriate, an approximation to three decimal places.

37. *Bridge expansion.* During the summer heat, a 2-mi bridge expands 2 ft in length. If we assume that the bulge occurs straight up the middle, how high is the bulge? (The answer may surprise you. Most bridges have expansion spaces to avoid such buckling.)

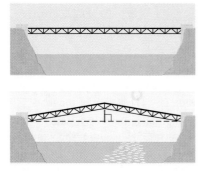

38. Triangle *ABC* has sides of lengths 25 ft, 25 ft, and 30 ft. Triangle *PQR* has sides of lengths 25 ft, 25 ft, and 40 ft. Which triangle has the greater area and by how much?

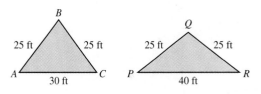

39. *Camping tent.* The entrance to a pup tent is the shape of an equilateral triangle. If the base of the tent is 4 ft wide, how tall is the tent?

40. Each side of a regular octagon has length *s*. Find a formula for the distance *d* between the parallel sides of the octagon.

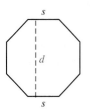

41. The diagonal of a square has length $8\sqrt{2}$ ft. Find the length of a side of the square.

42. The length and the width of a rectangle are given by consecutive integers. The area of the rectangle is 90 cm^2. Find the length of a diagonal of the rectangle.

43. Find all points on the *y*-axis of a Cartesian coordinate system that are 5 units from the point (3, 0).

44. Find all points on the *x*-axis of a Cartesian coordinate system that are 5 units from the point (0, 4).

SKILL MAINTENANCE

Solve.

45. $x^2 - 11x + 24 = 0$

46. $2x^2 + 11x - 21 = 0$

47. $|3x - 5| = 7$

48. $|2x - 3| = |x + 7|$

SYNTHESIS

49. ◈ Write a problem for a classmate to solve in which the solution is: "The height of the tepee is $5\sqrt{3}$ yd."

50. ◈ Write a problem for a classmate to solve in which the solution is: "The height of the window is $15\sqrt{3}$ ft."

51. A cube measures 5 cm on each side. How long is the diagonal that connects two opposite corners of the cube? Give an exact answer.

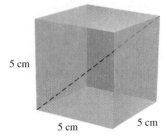

52. *Roofing.*   Kit's cottage, which is 24 ft wide and 32 ft long, needs a new roof. By counting clapboards that are 4 in. apart, Kit determines that the peak of the roof is 6 ft higher than the sides. If one packet of shingles covers 100 square feet, how many packets will the job require?

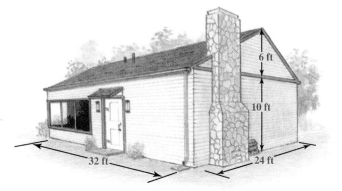

53. *Painting.*   (Refer to Exercise 52.) A gallon of paint covers about 275 square feet. If Kit's first floor is 10 ft high, how many gallons of paint should be bought to paint the house? What assumption(s) is made in your answer?

7.8   The Complex Numbers

Imaginary and Complex Numbers   •   Addition and Subtraction   •
Multiplication   •   Conjugates and Division   •   Powers of *i*

Imaginary and Complex Numbers

Negative numbers do not have square roots in the real-number system. However, a larger number system that contains the real-number system is designed so that negative numbers *do* have square roots. That system is called the **complex-number system,** and it makes use of a number that is a square root of −1. We call this new number *i*.

THE NUMBER i

We define the number i such that $i = \sqrt{-1}$ and $i^2 = -1$.

To express roots of negative numbers in terms of i, we can use the fact that in the complex numbers, $\sqrt{-p} = \sqrt{-1}\sqrt{p}$ when p is a positive real number.

EXAMPLE 1   Express in terms of i: **(a)** $\sqrt{-7}$; **(b)** $\sqrt{-16}$; **(c)** $-\sqrt{-13}$; **(d)** $-\sqrt{-50}$.

SOLUTION

a) $\sqrt{-7} = \sqrt{-1 \cdot 7} = \sqrt{-1} \cdot \sqrt{7} = i\sqrt{7}$, or $\sqrt{7}i$ $\boxed{i \text{ is } not \text{ under the radical.}}$

b) $\sqrt{-16} = \sqrt{-1 \cdot 16} = \sqrt{-1} \cdot \sqrt{16} = i \cdot 4 = 4i$

c) $-\sqrt{-13} = -\sqrt{-1 \cdot 13} = -\sqrt{-1} \cdot \sqrt{13} = -i\sqrt{13}$, or $-\sqrt{13}i$

d) $-\sqrt{-50} = -\sqrt{-1} \cdot \sqrt{25} \cdot \sqrt{2} = -i \cdot 5 \cdot \sqrt{2} = -5i\sqrt{2}$, or $-5\sqrt{2}i$

IMAGINARY NUMBERS

An *imaginary number* is a number that can be written $a + bi$, where a and b are real numbers and $b \neq 0$.

Don't let the name "imaginary" fool you. Imaginary numbers appear in such fields as engineering and the physical sciences. The following are examples of imaginary numbers:

$5 + 4i$,   Here $a = 5$, $b = 4$.

$\sqrt{5} - \pi i$,   Here $a = \sqrt{5}$, $b = -\pi$.

$17i$.   Here $a = 0$, $b = 17$.

When a and b are real numbers and b is allowed to be 0, the number $a + bi$ is said to be **complex**.

COMPLEX NUMBERS

A *complex number* is any number that can be written $a + bi$, where a and b are real numbers. (Note that a and b both can be 0.)

The following are examples of complex numbers:

$7 + 3i$ (here $a \neq 0$, $b \neq 0$); $4i$ (here $a = 0$, $b \neq 0$);

8 (here $a \neq 0$, $b = 0$); 0 (here $a = 0$, $b = 0$).

Complex numbers like $17i$ or $4i$, in which $a = 0$ and $b \neq 0$, are imaginary numbers with no real part. Such numbers are called *pure imaginary* numbers.

Note that when $b = 0$, $a + 0i = a$, so every real number is a complex number. The relationships among various real and complex numbers are shown below.

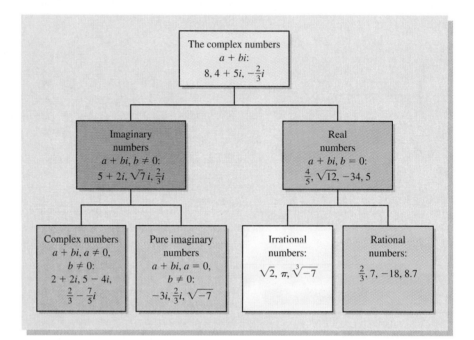

Observe that although $\sqrt{-7}$ and $\sqrt[3]{-7}$ are both complex numbers, $\sqrt{-7}$ is imaginary whereas $\sqrt[3]{-7}$ is real.

Addition and Subtraction

The complex numbers obey the commutative, associative, and distributive laws. Thus we can add and subtract them as we do binomials.

EXAMPLE 2

Add or subtract and simplify.

a) $(8 + 6i) + (3 + 2i)$ **b)** $(4 + 5i) - (6 - 3i)$

SOLUTION

a) $(8 + 6i) + (3 + 2i) = (8 + 3) + (6i + 2i)$ Combining the real parts and the imaginary parts

$\qquad = 11 + (6 + 2)i = 11 + 8i$

b) $(4 + 5i) - (6 - 3i) = (4 - 6) + [5i - (-3i)]$ Note that the 6 and the $-3i$ are both being subtracted.

$\qquad = -2 + 8i$

Multiplication

For complex numbers, the property $\sqrt{a}\,\sqrt{b} = \sqrt{ab}$ does *not* hold in general, but it does hold when $a = -1$ and b is nonnegative. To multiply square roots of negative real numbers, we first express them in terms of i. For example,

$$\sqrt{-2} \cdot \sqrt{-5} = \sqrt{-1} \cdot \sqrt{2} \cdot \sqrt{-1} \cdot \sqrt{5}$$
$$= i \cdot \sqrt{2} \cdot i \cdot \sqrt{5}$$
$$= i^2 \sqrt{10}$$
$$= -1\sqrt{10} = -\sqrt{10} \text{ is correct!}$$

CAUTION!   With complex numbers, simply multiplying radicands is *incorrect*: $\sqrt{-2} \cdot \sqrt{-5} \neq \sqrt{10}$.

With this in mind, we can now multiply complex numbers.

EXAMPLE 3

Multiply and simplify.

a) $\sqrt{-16} \cdot \sqrt{-25}$ **b)** $\sqrt{-5} \cdot \sqrt{-7}$ **c)** $-3i \cdot 8i$
d) $-4i(3 - 5i)$ **e)** $(1 + 2i)(1 + 3i)$

SOLUTION

a) $\sqrt{-16} \cdot \sqrt{-25} = \sqrt{-1} \cdot \sqrt{16} \cdot \sqrt{-1} \cdot \sqrt{25}$
$$= i \cdot 4 \cdot i \cdot 5$$
$$= i^2 \cdot 20$$
$$= -1 \cdot 20 \qquad i^2 = -1$$
$$= -20$$

b) $\sqrt{-5} \cdot \sqrt{-7} = \sqrt{-1} \cdot \sqrt{5} \cdot \sqrt{-1} \cdot \sqrt{7}$ Try to do this step mentally.
$$= i \cdot \sqrt{5} \cdot i \cdot \sqrt{7}$$
$$= i^2 \cdot \sqrt{35}$$
$$= -1 \cdot \sqrt{35} \qquad i^2 = -1$$
$$= -\sqrt{35}$$

c) $-3i \cdot 8i = -24 \cdot i^2$
$$= -24 \cdot (-1) \qquad i^2 = -1$$
$$= 24$$

d) $-4i(3 - 5i) = -4i \cdot 3 + (-4i)(-5i)$ Using the distributive law
$$= -12i + 20i^2$$
$$= -12i - 20 \qquad\qquad i^2 = -1$$
$$= -20 - 12i \qquad\qquad \text{Writing in the form } a + bi$$

e) $(1 + 2i)(1 + 3i) = 1 + 3i + 2i + 6i^2$ Multiplying each term of one number by every term of the other (FOIL)
$$= 1 + 3i + 2i - 6 \qquad i^2 = -1$$
$$= -5 + 5i \qquad\qquad \text{Combining like terms}$$

Conjugates and Division

Conjugates of complex numbers are defined as follows.

CONJUGATE OF A
COMPLEX NUMBER

The *conjugate* of a complex number $a + bi$ is $a - bi$, and the *conjugate* of $a - bi$ is $a + bi$.

EXAMPLE 4

Find the conjugate: **(a)** $-3 + 7i$; **(b)** $14 - 5i$; **(c)** $4i$.

SOLUTION

a) $-3 + 7i$ The conjugate is $-3 - 7i$.

b) $14 - 5i$ The conjugate is $14 + 5i$.

c) $4i$ The conjugate is $-4i$. Note that $4i = 0 + 4i$.

The product of a complex number and its conjugate is a real number.

EXAMPLE 5

Multiply: $(5 + 7i)(5 - 7i)$.

SOLUTION

$$
\begin{aligned}
(5 + 7i)(5 - 7i) &= 5^2 - (7i)^2 && \text{Using } (A + B)(A - B) = A^2 - B^2 \\
&= 25 - 49i^2 \\
&= 25 - 49(-1) && i^2 = -1 \\
&= 25 + 49 = 74
\end{aligned}
$$

Conjugates are used when dividing complex numbers. The procedure is much like that used to rationalize denominators in Section 7.5.

EXAMPLE 6

Divide and simplify to the form $a + bi$: **(a)** $\dfrac{-5 + 9i}{1 - 2i}$; **(b)** $\dfrac{7 + 3i}{5i}$.

SOLUTION

a) To divide and simplify $(-5 + 9i)/(1 - 2i)$, we multiply by 1, using the conjugate of the denominator to form 1:

$$
\begin{aligned}
\frac{-5 + 9i}{1 - 2i} &= \frac{-5 + 9i}{1 - 2i} \cdot \frac{1 + 2i}{1 + 2i} && \text{Multiplying by 1 using the conjugate of} \\
&&& \text{the denominator in the symbol for 1} \\[4pt]
&= \frac{(-5 + 9i)(1 + 2i)}{(1 - 2i)(1 + 2i)} \\[4pt]
&= \frac{-5 - 10i + 9i + 18i^2}{1^2 - 4i^2} && \text{Using FOIL} \\[4pt]
&= \frac{-5 - i - 18}{1 - 4(-1)} && i^2 = -1 \\[4pt]
&= \frac{-23 - i}{5} \\[4pt]
&= -\frac{23}{5} - \frac{1}{5}i && \text{Writing in the form } a + bi
\end{aligned}
$$

b) $\dfrac{7 + 3i}{5i} = \dfrac{7 + 3i}{5i} \cdot \dfrac{-5i}{-5i}$ Multiplying by 1 using the conjugate of $0 + 5i$ in the symbol for 1

$\qquad = \dfrac{-35i - 15i^2}{-25i^2}$ Multiplying

$\qquad = \dfrac{-35i - 15(-1)}{-25(-1)}$ $i^2 = -1$

$\qquad = \dfrac{15 - 35i}{25}$

$\qquad = \dfrac{15}{25} - \dfrac{35}{25}i$

$\qquad = \dfrac{3}{5} - \dfrac{7}{5}i$

Powers of *i*

Recall that -1 raised to an *even* power is 1, and -1 raised to an *odd* power is -1. Simplifying powers of *i* can then be done by using the fact that $i^2 = -1$ and expressing the given power of *i* in terms of i^2. Consider the following:

i, or $\sqrt{-1}$,

$i^2 = -1$,

$i^3 = i^2 \cdot i = (-1)i = -i$,

$i^4 = (i^2)^2 = (-1)^2 = 1$,

$i^5 = i^4 \cdot i = (i^2)^2 \cdot i = (-1)^2 \cdot i = i$,

$i^6 = (i^2)^3 = (-1)^3 = -1$.

Note that the powers of *i* cycle themselves through the values i, -1, $-i$, and 1.

EXAMPLE   7

Simplify: **(a)** i^{18}; **(b)** i^{24}; **(c)** i^{29}; **(d)** i^{75}.

SOLUTION

a) $i^{18} = (i^2)^9$

$\qquad = (-1)^9 = -1$

b) $i^{24} = (i^2)^{12}$

$\qquad = (-1)^{12} = 1$

c) $i^{29} = i^{28}i^1$

$\qquad = (i^2)^{14}i$

$\qquad = (-1)^{14}i$

$\qquad = 1 \cdot i = i$

d) $i^{75} = i^{74}i^1$

$\qquad = (i^2)^{37}i$

$\qquad = (-1)^{37}i$

$\qquad = -1 \cdot i = -i$

EXERCISE SET

7.8

Express in terms of i.

1. $\sqrt{-25}$ 2. $\sqrt{-36}$ 3. $\sqrt{-13}$

4. $\sqrt{-19}$ 5. $\sqrt{-18}$ 6. $\sqrt{-98}$

7. $\sqrt{-3}$ 8. $\sqrt{-4}$ 9. $\sqrt{-81}$

10. $\sqrt{-27}$ 11. $\sqrt{-300}$ 12. $-\sqrt{-75}$

13. $-\sqrt{-49}$ 14. $-\sqrt{-125}$

15. $4 - \sqrt{-60}$ 16. $6 - \sqrt{-84}$

17. $\sqrt{-4} + \sqrt{-12}$ 18. $-\sqrt{-76} + \sqrt{-125}$

Perform the indicated operation and simplify. Write each answer in the form a + bi.

19. $(4 + 7i) + (5 - 2i)$ 20. $(5 + 2i) + (7 - i)$

21. $(-2 + 8i) + (5 + 3i)$ 22. $(4 - 5i) + (3 + 9i)$

23. $(9 + 8i) - (5 + 3i)$ 24. $(9 + 7i) - (2 + 4i)$

25. $(8 - 3i) - (9 + 2i)$ 26. $(7 - 4i) - (5 - 3i)$

27. $(-2 + 6i) - (-7 + i)$ 28. $(-5 - i) - (7 + 4i)$

29. $6i \cdot 5i$ 30. $7i \cdot 6i$

31. $7i \cdot (-9i)$ 32. $(-4i)(-6i)$

33. $\sqrt{-49} \sqrt{-25}$ 34. $\sqrt{-36} \sqrt{-9}$

35. $\sqrt{-6} \sqrt{-7}$ 36. $\sqrt{-5} \sqrt{-2}$

37. $\sqrt{-15} \sqrt{-10}$ 38. $\sqrt{-6} \sqrt{-21}$

39. $2i(7 + 3i)$ 40. $5i(2 + 6i)$

41. $-4i(6 - 5i)$ 42. $-7i(3 - 4i)$

43. $(2 + 5i)(4 + 3i)$ 44. $(1 + i)(3 + 2i)$

45. $(5 - 6i)(2 + 5i)$ 46. $(6 - 5i)(3 + 4i)$

47. $(-4 + 5i)(3 - 4i)$ 48. $(7 - 2i)(2 - 6i)$

49. $(7 - 3i)(4 - 7i)$ 50. $(5 - 3i)(4 - 5i)$

51. $(-3 + 6i)(-3 + 4i)$ 52. $(-2 + 3i)(-2 + 5i)$

53. $(2 + 9i)(-3 - 5i)$ 54. $(-5 - 4i)(3 + 7i)$

55. $(5 - 2i)^2$ 56. $(3 - 2i)^2$

57. $(4 + 2i)^2$ 58. $(2 + 3i)^2$

59. $(-5 - 2i)^2$ 60. $(-2 + 3i)^2$

61. $\dfrac{7}{2 - i}$ 62. $\dfrac{4}{3 + i}$

63. $\dfrac{3i}{5 + 2i}$ 64. $\dfrac{4i}{5 - 3i}$

65. $\dfrac{8}{9i}$ 66. $\dfrac{5}{8i}$

67. $\dfrac{7 - 2i}{6i}$ 68. $\dfrac{3 + 8i}{9i}$

69. $\dfrac{4 + 5i}{3 - 7i}$ 70. $\dfrac{5 + 3i}{7 - 4i}$

71. $\dfrac{3 - 2i}{4 + 3i}$ 72. $\dfrac{5 - 2i}{3 + 6i}$

Simplify.

73. i^7 74. i^{11} 75. i^{24}

76. i^{35} 77. i^{42} 78. i^{64}

79. i^9 80. $(-i)^{71}$ 81. $(-i)^6$

82. $(-i)^4$ 83. $(5i)^3$ 84. $(-3i)^5$

85. $i^2 + i^4$ 86. $5i^5 + 4i^3$ 87. $i^5 + i^7$

88. $i^{84} - i^{100}$

SKILL MAINTENANCE

Solve.

89. $\dfrac{196}{x^2 - 7x + 49} - \dfrac{2x}{x + 7} = \dfrac{2058}{x^3 + 343}$

90. $\dfrac{5}{t} - \dfrac{3}{2} = \dfrac{4}{7}$

91. $28 = 3x^2 - 17x$

92. $|3x + 7| < 22$

SYNTHESIS

93. ◆ Is the product of two imaginary numbers always an imaginary number? Why or why not?

94. ◆ In what way(s) are conjugates of complex numbers similar to the conjugates used in Section 7.5?

95. ◆ Is the set of real numbers a subset of the complex numbers? Why or why not?

96. ◆ Is the union of the set of imaginary numbers and the set of real numbers the set of complex numbers? Why or why not?

97. A function g is given by

$$g(z) = \frac{z^4 - z^2}{z - 1}.$$

Find $g(2i)$; $g(i + 1)$; $g(2i - 1)$.

98. Evaluate

$$\frac{1}{w - w^2} \quad \text{for} \quad w = \frac{1 - i}{10}.$$

Simplify.

99. $\dfrac{i^5 + i^6 + i^7 + i^8}{(1 - i)^4}$

100. $(1 - i)^3(1 + i)^3$

101. $\dfrac{5 - \sqrt{5}i}{\sqrt{5}i}$

102. $\dfrac{6}{1 + \dfrac{3}{i}}$

103. $\left(\dfrac{1}{2} - \dfrac{1}{3}i\right)^2 - \left(\dfrac{1}{2} + \dfrac{1}{3}i\right)^2$

104. $\dfrac{i - i^{38}}{1 + i}$

⌈ SUMMARY AND REVIEW
7

KEY TERMS

Square root, p. 376	Rational exponent, p. 384	30°–60°–90° right triangle,
Principal square root, p. 377	Perfect square, p. 391	p. 418
Radical sign, p. 377	Perfect cube, p. 391	Imaginary number, p. 423
Radical expression, p. 377	Perfect nth power, p. 391	Complex number, p. 423
Radicand, p. 377	Like radicals, p. 394	Pure imaginary number,
Cube root, p. 380	Conjugates, p. 404	p. 424
nth root, p. 381	Rationalizing, p. 404	Conjugate of a complex
Index (pl., indices), p. 381	Radical equation, p. 409	number, p. 426
Odd root, p. 381	Isosceles right triangle,	
Even root, p. 381	p. 417	

IMPORTANT PROPERTIES AND FORMULAS

The number c is a square root of a if $c^2 = a$.

The number c is the cube root of a if $c^3 = a$.

For any real number a:

a) $\sqrt[n]{a^n} = |a|$ when n is even. Unless a is known to be nonnegative, absolute-value notation is needed when n is even.

b) $\sqrt[n]{a^n} = a$ when n is odd. Absolute-value notation is not used when n is odd.

$a^{1/n}$ means $\sqrt[n]{a}$. When a is nonnegative, n can be any index. When a is negative, n must be odd.

For any natural numbers m and n ($n \neq 1$), and any real number a for which $\sqrt[n]{a}$ exists,

$$a^{m/n} \quad \text{means} \quad (\sqrt[n]{a})^m \quad \text{or} \quad \sqrt[n]{a^m}.$$

For any rational number m/n and any nonzero real number a for which $a^{m/n}$ exists,

$$a^{-m/n} \quad \text{means} \quad \frac{1}{a^{m/n}}.$$

For any real numbers a and b and any rational exponents m and n for which a^m, a^n, and b^m are defined:

1. $a^m \cdot a^n = a^{m+n}$ — In multiplying, add exponents if the bases are the same.

2. $\dfrac{a^m}{a^n} = a^{m-n}$ — In dividing, subtract exponents if the bases are the same. (Assume $a \neq 0$.)

3. $(a^m)^n = a^{m \cdot n}$ — To raise a power to a power, multiply the exponents.

4. $(ab)^m = a^m b^m$ — To raise a product to a power, raise each factor to the power and multiply.

The Product Rule for Radicals

For any real numbers $\sqrt[n]{a}$ and $\sqrt[n]{b}$,

$$\sqrt[n]{a}\ \sqrt[n]{b} = \sqrt[n]{a \cdot b}.$$

The Quotient Rule for Radicals

For any real numbers $\sqrt[n]{a}$ and $\sqrt[n]{b}$, $b \neq 0$,

$$\sqrt[n]{\frac{a}{b}} = \frac{\sqrt[n]{a}}{\sqrt[n]{b}}.$$

Some Ways to Simplify Radical Expressions

1. *Simplifying by factoring.* Factor the radicand and look for factors raised to powers that are divisible by the index.

Example: $\sqrt[3]{a^6 b} = \sqrt[3]{a^6}\ \sqrt[3]{b} = a^2 \sqrt[3]{b}$

2. *Using rational exponents to simplify.* Convert to exponential notation and then use arithmetic and the laws of exponents to simplify the exponents. Then convert back to radical notation.

Example:
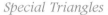
$$\begin{aligned} \sqrt[3]{p} \cdot \sqrt[4]{q^3} &= p^{1/3} \cdot q^{3/4} \\ &= p^{4/12} \cdot q^{9/12} \\ &= \sqrt[12]{p^4 q^9} \end{aligned}$$

3. *Combining like radical terms.*

Example:
$$\begin{aligned} \sqrt{8} + 3\sqrt{2} &= \sqrt{4} \cdot \sqrt{2} + 3\sqrt{2} \\ &= 2\sqrt{2} + 3\sqrt{2} = 5\sqrt{2} \end{aligned}$$

The Principle of Powers

If $a = b$, then $a^n = b^n$ for any exponent n.

A general strategy for solving equations with two or more radical terms is as follows.

1. Isolate one of the radical terms.
2. Use the principle of powers.
3. If a radical remains, repeat steps (1) and (2).
4. Check possible solutions in the original equation.

Special Triangles

The length of the hypotenuse in an isosceles right triangle is the length of a leg times $\sqrt{2}$.

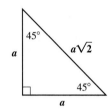

The length of the longer leg in a 30°–60°–90° right triangle is the length of the shorter leg times $\sqrt{3}$. The hypotenuse is twice as long as the shorter leg.

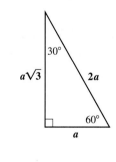

A complex number is any number that can be written $a + bi$, where a and b are real numbers and $i = \sqrt{-1}$.

REVIEW EXERCISES

Simplify.

1. $\sqrt{\dfrac{49}{36}}$

2. $\sqrt{0.25}$

Let $f(x) = \sqrt{2x - 7}$. Find the following.

3. $f(16)$

4. The domain of f

Simplify. Assume that each variable can represent any real number.

5. $\sqrt{49a^2}$

6. $\sqrt{(c + 8)^2}$

7. $\sqrt{x^2 - 6x + 9}$

8. $\sqrt{4x^2 + 4x + 1}$

9. $\sqrt[5]{-32}$

10. $\sqrt[3]{-\dfrac{64x^6}{27}}$

11. $\sqrt[4]{x^{12}y^8}$

12. $\sqrt[6]{64x^{12}}$

13. Rewrite with rational exponents: $(\sqrt[3]{5ab})^4$.

14. Rewrite without rational exponents: $(16a^8)^{3/4}$.

Use rational exponents to simplify. Assume $x, y \geq 0$.

15. $\sqrt{x^6 y^{10}}$

16. $(\sqrt[3]{a^2 b})^5$

Simplify. Do not use negative exponents in the answers.

17. $(x^{-2/3})^{3/5}$

18. $\dfrac{7^{-1/3}}{7^{-1/2}}$

19. If $f(x) = \sqrt{25(x - 3)^2}$, find a simplified form for $f(x)$.

Perform the indicated operation and, if possible, simplify. Write all answers using radical notation.

20. $\sqrt{5x}\,\sqrt{3y}$

21. $\sqrt[3]{a^5 b}\,\sqrt[3]{27b}$

22. $\sqrt[3]{ab}\,\sqrt[5]{a^3 b^2}$

23. $\dfrac{\sqrt[3]{60xy^3}}{\sqrt[3]{10x}}$

24. $\dfrac{\sqrt{75x}}{2\sqrt{3}}$

25. $\dfrac{\sqrt[3]{x^2}}{\sqrt[4]{x}}$

26. $5\sqrt[3]{x} + 2\sqrt[3]{x}$

27. $2\sqrt{75} - 7\sqrt{3}$

28. $\sqrt[3]{8x^4} + \sqrt[3]{xy^6}$

29. $\sqrt{50} + 2\sqrt{18} + \sqrt{32}$

30. $(\sqrt{5} - 3\sqrt{8})(\sqrt{5} + 2\sqrt{8})$

31. If $f(x) = x^2$, find $f(a - \sqrt{2})$.

32. Rationalize the denominator:

$$\dfrac{5\sqrt{12a}}{\sqrt{a} + \sqrt{b}}.$$

33. Rationalize the numerator of the expression in Exercise 32.

34. Solve: $1 + \sqrt{x} = \sqrt{3x - 3}$.

35. If $f(x) = \sqrt[4]{x + 2}$, find a such that $f(a) = 2$.

Solve. Give an exact answer and, where appropriate, an approximation to three decimal places.

36. The diagonal of a square has length $9\sqrt{2}$ cm. Find the length of a side of the square.

37. A bookcase is 5 ft tall and has a 7-ft diagonal brace, as shown. How wide is the bookcase?

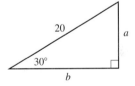

38. Find the missing lengths. Give exact answers and approximations to three decimal places.

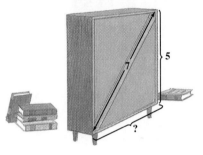

39. Express in terms of i and simplify: $-\sqrt{-8}$.

40. Add: $(-4 + 3i) + (2 - 12i)$.

41. Subtract: $(4 - 7i) - (3 - 8i)$.

Multiply.

42. $(2 + 5i)(2 - 5i)$

43. i^{13}

44. $(6 - 3i)(2 - i)$

45. Divide and simplify to the form $a + bi$:

$$\dfrac{7 - 2i}{3 + 4i}.$$

SKILL MAINTENANCE

46. Find three consecutive positive integers such that the product of the first and second integers is 26 less than the product of the second and third integers.

47. Solve:

$$\frac{7}{x+2} + \frac{5}{x^2 - 2x + 4} = \frac{84}{x^3 + 8}.$$

48. Solve:

$$2x^2 + 3x - 27 = 0.$$

49. Multiply and simplify:

$$\frac{x^2 + 3x}{x^2 - y^2} \cdot \frac{x^2 - xy + 2x - 2y}{x^2 - 9}.$$

SYNTHESIS

50. ◈ Explain why $\sqrt[n]{x^n} = |x|$ when n is even, but $\sqrt[n]{x^n} = x$ when n is odd.

51. ◈ What is the difference between real numbers and complex numbers?

52. Solve:

$$\sqrt{11x + \sqrt{6 + x}} = 6.$$

53. Simplify:

$$\frac{2}{1 - 3i} - \frac{3}{4 + 2i}.$$

CHAPTER TEST
7

Simplify. Assume that variables can represent any real number.

1. $\sqrt{75}$

2. $\sqrt[3]{-\dfrac{8}{x^6}}$

3. $\sqrt{100a^2}$

4. $\sqrt{x^2 - 8x + 16}$

5. $\sqrt[5]{x^{12}y^8}$

6. $\sqrt{\dfrac{25x^2}{36y^4}}$

7. $\sqrt[3]{2x} \ \sqrt[3]{5y^2}$

8. $\dfrac{\sqrt[5]{x^3y^4}}{\sqrt[5]{xy^2}}$

9. $\sqrt[4]{x^3y^2} \ \sqrt{xy}$

10. $\dfrac{\sqrt[5]{a^2}}{\sqrt[4]{a}}$

11. $7\sqrt{2} - 2\sqrt{2}$

12. $\sqrt{x^4y} + \sqrt{9y^3}$

13. $(7 + \sqrt{x})(2 - 3\sqrt{x})$

14. If $f(x) = \sqrt{8 - 4x}$, determine the domain of f.

15. If $f(x) = x^2$, find $f(5 + \sqrt{2})$.

16. Rationalize the denominator:

$$\frac{1 + \sqrt{2}}{3 - 5\sqrt{2}}.$$

17. Solve: $x = \sqrt{2x - 5} + 4.$

18. The shorter leg of a 30°–60°–90° right triangle measures 10 cm. Find the lengths of the other sides. Give exact answers and, where appropriate, approximations to three decimal places.

19. A referee jogs diagonally from one corner of a 50-ft by 90-ft basketball court to the far corner. How far does she jog? Give an exact answer and an approximation to three decimal places.

20. Express in terms of i and simplify: $\sqrt{-50}$.

21. Subtract: $(7 + 8i) - (-3 + 6i)$.

22. Multiply: $\sqrt{-16} \ \sqrt{-36}$.

23. Multiply. Write the answer in the form $a + bi$.

$$(4 - i)^2$$

24. Divide and simplify to the form $a + bi$:

$$\frac{-3 + i}{2 - 7i}.$$

25. Simplify: i^{37}.

SKILL MAINTENANCE

26. Solve: $6x^2 = 13x + 5.$

27. Divide and simplify:

$$\frac{x^3 - 27}{x^2 - 16} \div \frac{x^2 + 3x + 9}{x + 4}.$$

28. Solve:

$$\frac{11x}{x + 3} + \frac{33}{x} + 12 = \frac{99}{x^2 + 3x}.$$

29. Find two consecutive even integers whose product is 288.

SYNTHESIS

30. Solve:

$$\sqrt{2x - 2} + \sqrt{7x + 4} = \sqrt{13x + 10}.$$

31. Simplify:

$$\frac{1 - 4i}{4i(1 + 4i)^{-1}}.$$

Quadratic Functions and Equations

A river pilot uses math in a variety of ways from plotting a radar course when meeting other vessels to calculating the estimated time of arrival at a river location.

RANDY SVEUM
Licensed Pilot
Winona, MN

AN APPLICATION

The current in a typical Mississippi River shipping route flows at a rate of 4 mph. In order for a barge to travel 24 mi upriver and then return in a total of 5 hr, approximately how fast must the barge be able to travel in still water?

THIS PROBLEM APPEARS AS EXERCISE 9 IN SECTION 8.3.

More information on the Mississippi River is available at
http://hepg.awl.com/be/inter_5

*I*n *translating problem situations to mathematical language, we often obtain a function or equation containing a second-degree polynomial in one variable. Such functions or equations are said to be* quadratic. *In this chapter, we will study a variety of equations, inequalities, and applications for which we will need to solve quadratic equations or graph quadratic functions.*

8.1   Quadratic Equations

The Principle of Square Roots   •   Completing the Square   •
Problem Solving

In Section 5.8, we solved quadratic equations like $3x^2 = 2 - x$ by factoring. Let's review that procedure.

EXAMPLE 1   Solve: $3x^2 = 2 - x$.

SOLUTION   We have

$$3x^2 = 2 - x$$
$$3x^2 + x - 2 = 0 \qquad \text{Adding } -2 + x \text{ on both sides to obtain standard form}$$
$$(3x - 2)(x + 1) = 0 \qquad \text{Factoring}$$
$$3x - 2 = 0 \quad or \quad x + 1 = 0 \qquad \text{Using the principle of zero products}$$
$$3x = 2 \quad or \qquad x = -1$$
$$x = \tfrac{2}{3} \quad or \qquad x = -1.$$

Check: For -1:

$$\frac{3x^2 = 2 - x}{3(-1)^2 \ ? \ 2 - (-1)}$$
$$3 \cdot 1 \ \bigg| \ 2 + 1$$
$$3 \ \bigg| \ 3 \qquad \text{TRUE}$$

For $\tfrac{2}{3}$:

$$\frac{3x^2 = 2 - x}{3\left(\tfrac{2}{3}\right)^2 \ ? \ 2 - \tfrac{2}{3}}$$
$$3 \cdot \tfrac{4}{9} \ \bigg| \ \tfrac{6}{3} - \tfrac{2}{3}$$
$$\tfrac{4}{3} \ \bigg| \ \tfrac{4}{3} \qquad \text{TRUE}$$

The solutions are -1 and $\tfrac{2}{3}$.

EXAMPLE 2

Solve: $x^2 = 25$.

SOLUTION   We have

$$x^2 = 25$$
$$x^2 - 25 = 0 \qquad \text{Writing in standard form}$$
$$(x - 5)(x + 5) = 0 \qquad \text{Factoring}$$
$$x - 5 = 0 \quad or \quad x + 5 = 0 \qquad \text{Using the principle of zero products}$$
$$x = 5 \quad or \quad x = -5.$$

The solutions are 5 and -5. We leave the checks to the student.

The Principle of Square Roots

Consider the equation $x^2 = 25$ again. We know from Chapter 7 that the number 25 has two real-number square roots, namely, 5 and -5. Note that these are the solutions of the equation in Example 2. Thus square roots can provide a quick method for solving equations of the type $x^2 = k$.

THE PRINCIPLE OF SQUARE ROOTS

For any real number k, if $x^2 = k$, then $x = \sqrt{k}$ or $x = -\sqrt{k}$.

EXAMPLE 3

Solve: $3x^2 = 6$.

SOLUTION   We have

$$3x^2 = 6$$
$$x^2 = 2 \qquad \text{Multiplying by } \tfrac{1}{3}$$
$$x = \sqrt{2} \quad or \quad x = -\sqrt{2}. \qquad \text{Using the principle of square roots}$$

We often use the symbol $\pm\sqrt{2}$ to represent the two numbers $\sqrt{2}$ and $-\sqrt{2}$. We check as follows.

Check: For $\sqrt{2}$:

$$\begin{array}{c|c} 3x^2 = 6 \\ \hline 3(\sqrt{2})^2 \; ? \; 6 \\ 3 \cdot 2 \; \Big| \\ 6 \; \Big| \; 6 & \text{TRUE} \end{array}$$

For $-\sqrt{2}$:

$$\begin{array}{c|c} 3x^2 = 6 \\ \hline 3(-\sqrt{2})^2 \; ? \; 6 \\ 3 \cdot 2 \; \Big| \\ 6 \; \Big| \; 6 & \text{TRUE} \end{array}$$

The solutions are $\sqrt{2}$ and $-\sqrt{2}$, or $\pm \sqrt{2}$.

Sometimes we rationalize denominators to simplify answers.

EXAMPLE 4

Solve: $-5x^2 + 2 = 0$.

SOLUTION   We have

$$-5x^2 + 2 = 0$$

$$x^2 = \frac{2}{5} \qquad \text{Isolating } x^2$$

$$x = \sqrt{\frac{2}{5}} \quad or \quad x = -\sqrt{\frac{2}{5}}. \qquad \text{Using the principle of square roots}$$

The solutions are $\sqrt{\frac{2}{5}}$ and $-\sqrt{\frac{2}{5}}$. This can also be written as $\pm\sqrt{\frac{2}{5}}$, or, if we rationalize the denominator, $\pm\frac{\sqrt{10}}{5}$. We leave the checks to the student.

Sometimes we get solutions that are imaginary numbers.

EXAMPLE 5

Solve: $4x^2 + 9 = 0$.

SOLUTION   We have

$$4x^2 + 9 = 0$$

$$x^2 = -\frac{9}{4} \qquad \text{Isolating } x^2$$

$$x = \sqrt{-\frac{9}{4}} \qquad or \quad x = -\sqrt{-\frac{9}{4}} \qquad \text{Using the principle of square roots}$$

$$x = \sqrt{\frac{9}{4}}\sqrt{-1} \quad or \quad x = -\sqrt{\frac{9}{4}}\sqrt{-1}$$

$$x = \frac{3}{2}i \qquad\quad or \quad x = -\frac{3}{2}i.$$

Check: Since the solutions are opposites and the equation has an x^2-term and no x-term, we can check both solutions at once.

$$
\begin{array}{c|c}
4x^2 + 9 = 0 & \\
\hline
4\left(\pm\frac{3}{2}i\right)^2 + 9 & ?\ 0 \\
4 \cdot \frac{9}{4} \cdot i^2 + 9 & \\
-9 + 9 & \\
0 & 0 \quad \text{TRUE}
\end{array}
$$

The solutions are $\frac{3}{2}i$ and $-\frac{3}{2}i$, or $\pm\frac{3}{2}i$.

Equations like $(x - 2)^2 = 7$ can also be solved using the principle of square roots.

EXAMPLE 6

Let $f(x) = (x - 2)^2$. Find all x-values for which $f(x) = 7$.

SOLUTION We are asked to find all x-values for which

$$f(x) = 7,$$

or

$$(x - 2)^2 = 7. \qquad \text{Substituting } (x - 2)^2 \text{ for } f(x)$$

The principle of square roots gives us

$$x - 2 = \sqrt{7} \qquad or \quad x - 2 = -\sqrt{7} \qquad \text{Using the principle of square roots}$$
$$x = 2 + \sqrt{7} \quad or \qquad x = 2 - \sqrt{7}.$$

Check: $f(2 + \sqrt{7}) = (2 + \sqrt{7} - 2)^2 = (\sqrt{7})^2 = 7.$

Similarly,

$$f(2 - \sqrt{7}) = (2 - \sqrt{7} - 2)^2 = (-\sqrt{7})^2 = 7.$$

The solutions are $2 + \sqrt{7}$ and $2 - \sqrt{7}$, or $2 \pm \sqrt{7}$.

In Example 6, one side of the equation is the square of a binomial and the other side is a constant. Once an equation has been written in this form, we can proceed as we did in Example 6.

EXAMPLE 7

Solve: $x^2 + 6x + 9 = 2.$

SOLUTION We have

$$x^2 + 6x + 9 = 2 \qquad \text{The left side is the square of a binomial.}$$

$$(x + 3)^2 = 2$$
$$x + 3 = \sqrt{2} \qquad or \quad x + 3 = -\sqrt{2} \qquad \text{Using the principle of square roots}$$
$$x = -3 + \sqrt{2} \quad or \qquad x = -3 - \sqrt{2}.$$

The solutions are $-3 + \sqrt{2}$ and $-3 - \sqrt{2}$, or $-3 \pm \sqrt{2}$.

Completing the Square

By using a method called *completing the square*, we can use the principle of square roots to solve *any* quadratic equation.

EXAMPLE 8

Solve: $x^2 + 6x + 4 = 0.$

SOLUTION We have

$$x^2 + 6x + 4 = 0$$
$$x^2 + 6x = -4 \qquad \text{Adding } -4 \text{ on both sides}$$
$$x^2 + 6x + 9 = -4 + 9 \qquad \text{Adding 9 on both sides. We explain this shortly.}$$
$$(x + 3)^2 = 5 \qquad \text{Factoring the perfect-square trinomial}$$
$$x + 3 = \pm\sqrt{5} \qquad \text{Using the principle of square roots. Remember that } \pm\sqrt{5} \text{ represents two numbers.}$$
$$x = -3 \pm \sqrt{5}. \qquad \text{Adding } -3 \text{ on both sides}$$

Check: For $-3 + \sqrt{5}$:

$$x^2 + 6x + 4 = 0$$

$$\overline{(-3 + \sqrt{5})^2 + 6(-3 + \sqrt{5}) + 4 \ \overset{?}{\ } \ 0}$$

$$9 - 6\sqrt{5} + 5 - 18 + 6\sqrt{5} + 4$$

$$9 + 5 - 18 + 4 - 6\sqrt{5} + 6\sqrt{5}$$

$$0 \ \big| \ 0 \quad \text{TRUE}$$

For $-3 - \sqrt{5}$:

$$x^2 + 6x + 4 = 0$$

$$\overline{(-3 - \sqrt{5})^2 + 6(-3 - \sqrt{5}) + 4 \ \overset{?}{\ } \ 0}$$

$$9 + 6\sqrt{5} + 5 - 18 - 6\sqrt{5} + 4$$

$$9 + 5 - 18 + 4 + 6\sqrt{5} - 6\sqrt{5}$$

$$0 \ \big| \ 0 \quad \text{TRUE}$$

The solutions are $-3 + \sqrt{5}$ and $-3 - \sqrt{5}$, or $-3 \pm \sqrt{5}$. ⟶●

Let's examine how the above solutions were found. The decision to add 9 on both sides in Example 8 was not made arbitrarily. We chose 9 because it made the left side a perfect-square trinomial. The 9 was obtained by taking half of the coefficient of x and squaring it — that is,

$$\left(\tfrac{1}{2} \cdot 6\right)^2 = 3^2, \quad \text{or} \quad 9.$$

To help see why this procedure works, examine the following drawings.

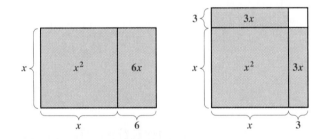

Note that both figures represent the same area, $x^2 + 6x$. However, only the figure on the right, in which the $6x$ is halved, can be converted into a square with the addition of a constant term. The constant term, 9, can be interpreted as the area of the "missing" piece of the diagram on the right. It *completes* the square.

EXAMPLE 9

Complete the square.

a) $x^2 + 14x$ **b)** $x^2 - 5x$ **c)** $x^2 + \tfrac{3}{4}x$

SOLUTION

a) We take half of the coefficient of x and square it.

$$x^2 + 14x$$

⟶ Half of 14 is 7, and $7^2 = 49$. We add 49.

Thus, $x^2 + 14x + 49$ is a perfect-square trinomial. It is equivalent to $(x + 7)^2$. We must add 49 in order for $x^2 + 14x$ to become a perfect-square trinomial.

b) We take half of the coefficient of x and square it:

$$x^2 - 5x$$

$$\longrightarrow \tfrac{1}{2} \cdot (-5) = -\tfrac{5}{2}, \quad \text{and} \quad \left(-\tfrac{5}{2}\right)^2 = \tfrac{25}{4}.$$

Thus, $x^2 - 5x + \frac{25}{4}$ is a perfect-square trinomial. It is equivalent to $\left(x - \frac{5}{2}\right)^2$. Note that for purposes of factoring, it is best to leave $\frac{25}{4}$ as a fraction.

c) We take half of the coefficient of x and square it:

$$x^2 + \tfrac{3}{4}x$$

$$\longrightarrow \tfrac{1}{2} \cdot \tfrac{3}{4} = \tfrac{3}{8}, \quad \text{and} \quad \left(\tfrac{3}{8}\right)^2 = \tfrac{9}{64}.$$

Thus, $x^2 + \frac{3}{4}x + \frac{9}{64}$ is a perfect-square trinomial. It is equivalent to $\left(x + \frac{3}{8}\right)^2$. ━━●

We can now use the method of completing the square to solve equations similar to Example 8.

EXAMPLE 10

Solve: **(a)** $x^2 - 8x - 7 = 0$; **(b)** $x^2 + 5x - 3 = 0$.

SOLUTION

a) $x^2 - 8x - 7 = 0$

$\qquad x^2 - 8x = 7$ Adding 7 on both sides. We can now complete the square on the left side.

$x^2 - 8x + 16 = 7 + 16$ Adding 16 on both sides to complete the square: $\frac{1}{2}(-8) = -4$, and $(-4)^2 = 16$

$\qquad (x - 4)^2 = 23$ Factoring

$\qquad x - 4 = \pm\sqrt{23}$ Using the principle of square roots

$\qquad x = 4 \pm \sqrt{23}$ Adding 4 on both sides

The solutions are $4 - \sqrt{23}$ and $4 + \sqrt{23}$, or $4 \pm \sqrt{23}$. The checks are left to the student.

b) $x^2 + 5x - 3 = 0$

$\qquad x^2 + 5x = 3$ Adding 3 on both sides

$x^2 + 5x + \dfrac{25}{4} = 3 + \dfrac{25}{4}$ Completing the square: $\frac{1}{2} \cdot 5 = \frac{5}{2}$, and $\left(\frac{5}{2}\right)^2 = \frac{25}{4}$

$\left(x + \dfrac{5}{2}\right)^2 = \dfrac{37}{4}$ Factoring and simplifying

$x + \dfrac{5}{2} = \pm\dfrac{\sqrt{37}}{2}$ Using the principle of square roots and the quotient rule for radicals

$x = \dfrac{-5 \pm \sqrt{37}}{2}$ Adding $-\frac{5}{2}$ on both sides

The checks are left to the student. The solutions are $(-5 - \sqrt{37})/2$ and $(-5 + \sqrt{37})/2$, or $(-5 \pm \sqrt{37})/2$. ━━●

Before we can complete the square, the coefficient of x^2 must be 1. When it is not 1, we divide both sides of the equation by the x^2-coefficient.

EXAMPLE 11   Solve: $3x^2 + 7x - 2 = 0$.

SOLUTION   We have

$$3x^2 + 7x - 2 = 0$$

$$3x^2 + 7x = 2 \qquad \text{Adding 2 on both sides}$$

$$x^2 + \frac{7}{3}x = \frac{2}{3} \qquad \text{Dividing on both sides by 3}$$

$$x^2 + \frac{7}{3}x + \frac{49}{36} = \frac{2}{3} + \frac{49}{36} \qquad \text{Completing the square: } \left(\frac{1}{2} \cdot \frac{7}{3}\right)^2 = \frac{49}{36}$$

$$\left(x + \frac{7}{6}\right)^2 = \frac{73}{36} \qquad \text{Factoring and simplifying}$$

$$x + \frac{7}{6} = \pm\frac{\sqrt{73}}{6} \qquad \begin{array}{l}\text{Using the principle of square roots}\\\text{and the quotient rule for radicals}\end{array}$$

$$x = \frac{-7 \pm \sqrt{73}}{6}. \qquad \text{Adding } -\frac{7}{6} \text{ on both sides}$$

The solutions are

$$\frac{-7 - \sqrt{73}}{6} \quad \text{and} \quad \frac{-7 + \sqrt{73}}{6}, \qquad \text{or} \qquad \frac{-7 \pm \sqrt{73}}{6}.$$

TECHNOLOGY CONNECTION
8.1

As we saw in Section 5.8, a grapher can be used to find approximate solutions of any quadratic equation that has real-number solutions.

To check Example 10(a), we graph $y = x^2 - 8x - 7$ and use the ROOT option of the CALC menu. When asked for a Lower and Upper Bound, we enter cursor positions to the left of and to the right of the root. A Guess between the bounds is entered and a value for the root appears.

1. Use a grapher to check the second solution of Example 10(a).
2. Use a grapher to confirm the solutions in Examples 8 and 10(b).
3. Can a grapher be used to find *exact* solutions in Example 11? Why or why not?

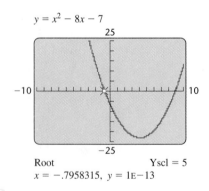

$y = x^2 - 8x - 7$

Root

$x = -.7958315, \ y = 1\text{E}{-}13$ Yscl = 5

4. Use a grapher to confirm that there are no real-number solutions in Example 5.

The procedure used in Example 11 can be used to solve *any* quadratic equation.

TO SOLVE A QUADRATIC EQUATION IN x BY COMPLETING THE SQUARE:

1. Isolate the terms with variables on one side of the equation, and arrange them in descending order.
2. Divide by the coefficient of x^2 on both sides if that coefficient is not 1.
3. Complete the square by taking half of the coefficient of x and adding its square on both sides.
4. Express one side as the square of a binomial and simplify the other side.
5. Use the principle of square roots.
6. Solve for x by adding appropriately on both sides.

Problem Solving

If you put money in a savings account, the bank will pay you interest. As interest is paid into your account, the bank will start paying you interest on both the original amount and the interest already earned. This is called **compounding interest**. If interest is paid yearly, we say that it is **compounded annually**.

THE COMPOUND-INTEREST FORMULA

If an amount of money P is invested at interest rate r, compounded annually, then in t years, it will grow to the amount A given by

$$A = P(1 + r)^t.$$

We can use quadratic equations to solve certain interest problems.

EXAMPLE 12   *Investment Growth.*   Rosa invested $4000 at interest rate r, compounded annually. In 2 yr, it grew to $4410. What was the interest rate?

SOLUTION

1. **Familiarize.** We are already familiar with the compound-interest formula. If we were not, we would need to consult an outside source.
2. **Translate.** The translation consists of substituting into the formula:

$$A = P(1 + r)^t$$
$$4410 = 4000(1 + r)^2.$$

3. Carry out. We solve for r:

$$4410 = 4000(1 + r)^2$$

$$\frac{4410}{4000} = (1 + r)^2 \qquad \text{Multiplying by } \tfrac{1}{4000} \text{ on both sides}$$

$$\frac{441}{400} = (1 + r)^2 \qquad \text{Simplifying}$$

$$\pm \sqrt{\frac{441}{400}} = 1 + r \qquad \text{Using the principle of square roots}$$

$$\pm \frac{21}{20} = 1 + r \qquad \text{Simplifying}$$

$$-\frac{20}{20} \pm \frac{21}{20} = r$$

$$\frac{1}{20} = r \quad or \quad -\frac{41}{20} = r.$$

4. Check. Since the interest rate cannot be negative, we need only check $\frac{1}{20}$, or 5%. If \$4000 were invested at 5% interest, compounded annually, then in 2 yr it would grow to $4000(1.05)^2$, or \$4410. The number 5% checks.

5. State. The interest rate was 5%. ————●

EXAMPLE   13

Freefalling Objects. The formula $s = 16t^2$ is used to approximate the distance s, in feet, that an object falls freely from rest in t seconds. The RCA Building in New York City is 850 ft tall. How long will it take an object to fall from the top?

SOLUTION

1. Familiarize. We make a sketch to help visualize the problem.

2. Translate. We substitute into the formula:

$$s = 16t^2$$
$$850 = 16t^2.$$

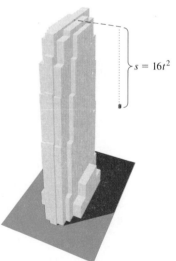

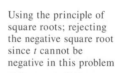

$s = 16t^2$

3. Carry out. We solve for t:

$$850 = 16t^2$$
$$\frac{850}{16} = t^2$$
$$53.125 = t^2$$
$$\sqrt{53.125} = t \qquad \text{Using the principle of square roots; rejecting the negative square root since } t \text{ cannot be negative in this problem}$$

$$7.3 \approx t. \qquad \text{Using a calculator to approximate the square root and rounding to the nearest tenth}$$

4. Check. Since $16(7.3)^2 = 852.64 \approx 850$, our answer checks.

5. State. It takes about 7.3 sec for an object to fall freely from the top of the RCA Building. ————●

EXERCISE SET
8.1

Solve.

1. $5x^2 = 15$

2. $7x^2 = 35$

3. $25x^2 + 4 = 0$

4. $9x^2 + 16 = 0$

5. $2x^2 - 3 = 0$

6. $3x^2 - 7 = 0$

7. $(x + 2)^2 = 49$

8. $(x - 1)^2 = 6$

9. $(a + 5)^2 = 8$

10. $(x - 13)^2 = 64$

11. $(x - 7)^2 = -4$

12. $(x + 1)^2 = -9$

13. $\left(x + \frac{3}{2}\right)^2 = \frac{7}{2}$

14. $\left(y + \frac{3}{4}\right)^2 = \frac{17}{16}$

15. $x^2 - 6x + 9 = 100$

16. $x^2 - 10x + 25 = 64$

17. Let $f(x) = (x - 7)^2$. Find x such that $f(x) = 16$.

18. Let $g(x) = (x - 2)^2$. Find x such that $g(x) = 25$.

19. Let $F(x) = (x - 3)^2$. Find x such that $F(x) = 13$.

20. Let $f(x) = (x + 3)^2$. Find x such that $f(x) = 17$.

21. Let $g(x) = x^2 + 14x + 49$. Find x such that $g(x) = 36$.

22. Let $F(x) = x^2 + 8x + 16$. Find x such that $F(x) = 9$.

Complete the square. Then write the trinomial square in factored form.

23. $x^2 + 10x$

24. $x^2 + 16x$

25. $x^2 - 6x$

26. $x^2 - 8x$

27. $x^2 - 24x$

28. $x^2 - 18x$

29. $x^2 + 9x$

30. $x^2 + 3x$

31. $x^2 - 3x$

32. $x^2 - 7x$

33. $x^2 + \frac{2}{3}x$

34. $x^2 + \frac{2}{5}x$

35. $x^2 - \frac{5}{6}x$

36. $x^2 - \frac{5}{3}x$

37. $x^2 + \frac{9}{5}x$

38. $x^2 + \frac{9}{4}x$

Solve by completing the square. Show your work.

39. $x^2 + 6x = 7$

40. $x^2 + 5x = -6$

41. $x^2 - 10x = 22$

42. $x^2 - 8x = -9$

43. $x^2 + 6x + 5 = 0$

44. $x^2 + 10x + 9 = 0$

45. $x^2 - 10x + 21 = 0$

46. $x^2 - 10x + 24 = 0$

47. $x^2 + 4x + 1 = 0$

48. $x^2 + 6x + 7 = 0$

49. $x^2 + 4 = 6x$

50. $x^2 + 23 = 10x$

51. $x^2 + 6x + 13 = 0$

52. $x^2 + 8x + 25 = 0$

53. $2x^2 - 5x - 3 = 0$

54. $3x^2 + 5x - 2 = 0$

55. $4x^2 + 8x + 3 = 0$

56. $9x^2 + 18x + 8 = 0$

57. $6x^2 - x = 15$

58. $6x^2 - x = 2$

59. $2x^2 + 4x + 1 = 0$

60. $2x^2 + 5x + 2 = 0$

61. $3x^2 - 5x - 3 = 0$

62. $4x^2 - 6x - 1 = 0$

Interest. Use $A = P(1 + r)^t$ to find the interest rate in Exercises 63–68. Refer to Example 12.

63. $2000 grows to $2420 in 2 yr

64. $2560 grows to $2890 in 2 yr

65. $1280 grows to $1805 in 2 yr

66. $1000 grows to $1440 in 2 yr

67. $6250 grows to $6760 in 2 yr

68. $6250 grows to $7290 in 2 yr

Free-falling objects. Use $s = 16t^2$ for Exercises 69–72. Refer to Example 13.

69. The CN Tower in Toronto, at 1815 ft, is the world's tallest self-supporting tower (no guy wires) (*Source: The Guinness Book of Records*). How long would it take an object to fall freely from the top?

70. Reaching 745 ft above the water, the towers of California's Golden Gate Bridge are the world's tallest bridge towers (*Source: The Guinness Book of Records*). How long would it take an object to fall freely from the top?

71. The Gateway Arch in St. Louis is 640 ft high. How long would it take an object to fall freely from the top?

72. The Sears Tower in Chicago is 1454 ft tall. How long would it take an object to fall freely from the top?

SKILL MAINTENANCE

Graph.

73. $f(x) = 5 - 2x$

74. $y = 7$

Simplify.

75. $\sqrt[3]{270}$

76. $\sqrt{80}$

Let $f(x) = \sqrt{3x - 5}$.

77. Find $f(10)$.

78. Find $f(18)$.

SYNTHESIS

79. ◈ Explain in your own words a sequence of steps that can be used to solve any quadratic equation in the quickest way.

80. ◈ Write an interest-rate problem for a classmate to solve. Devise the problem so that the solution is "The loan was made at 7% interest."

81. ◈ Write a problem involving a free-falling object for a classmate to solve (see Example 13). Devise the problem so that the solution is "The object takes about 4.5 sec to fall freely from the top of the structure."

82. ◈ What would be better: to receive 3% interest every 6 months, or to receive 6% interest every 12 months? Why?

Find b such that each trinomial is a square.

83. $x^2 + bx + 81$

84. $x^2 + bx + 49$

Solve.

85. $x(2x^2 - 9x - 56)(x^2 - 5) = 0$

86. $\left(x - \frac{1}{3}\right)\left(x^2 + \frac{1}{6}\right) + \left(x - \frac{1}{3}\right)\left(x^2 - \frac{2}{3}\right) = 0$

87. *Boating.* A barge and a fishing boat leave a dock at the same time, traveling at right angles to each other.

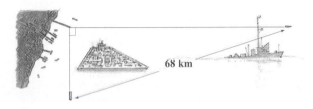

68 km

The barge travels 7 km/h slower than the fishing boat. After 4 hr, the boats are 68 km apart. Find the speed of each vessel.

88. Find three consecutive integers such that the square of the first plus the product of the other two is 67.

89. Exercises 17, 21, and 41 can be solved on a grapher without first rewriting in standard form. Simply let y_1 represent the left side of the equation and y_2 the right side. Then use a grapher to determine the x-coordinate of any point of intersection. Use a grapher to solve Exercises 17, 21, and 41 in this manner.

90. Use a grapher to check your answers to Exercises 5, 45, 59, and 61.

91. ◈ Example 12 can be solved with a grapher by graphing each side of

$$4410 = 4000(1 + r)^2.$$

How could you determine, from a reading of the problem, a suitable viewing window? What might that window be?

8.2 The Quadratic Formula

Solving Using the Quadratic Formula • Approximating Solutions

There are at least two reasons for learning to complete the square. One is to enhance your ability to graph certain equations that appear later in this and other chapters. The other is to prove a general formula for solving quadratic equations.

Solving Using the Quadratic Formula

Each time you solve by completing the square, the procedure is the same. In mathematics, when a procedure must be repeated many times, a formula is often developed to speed up our work.

Consider any quadratic equation in standard form:

$$ax^2 + bx + c = 0, \quad a > 0.$$

Let's solve by completing the square. As the steps are performed, compare them with Example 11 in Section 8.1.

$$ax^2 + bx = -c \qquad \text{Adding } -c \text{ on both sides}$$

$$x^2 + \frac{b}{a}x = -\frac{c}{a} \qquad \text{Dividing by } a \text{ on both sides}$$

Half of $\frac{b}{a}$ is $\frac{b}{2a}$ and $\left(\frac{b}{2a}\right)^2$ is $\frac{b^2}{4a^2}$. We add $\frac{b^2}{4a^2}$ on both sides:

$$x^2 + \frac{b}{a}x + \frac{b^2}{4a^2} = -\frac{c}{a} + \frac{b^2}{4a^2} \qquad \text{Adding } \frac{b^2}{4a^2} \text{ to complete the square}$$

$$\left(x + \frac{b}{2a}\right)^2 = -\frac{4ac}{4a^2} + \frac{b^2}{4a^2}$$

$$\left(x + \frac{b}{2a}\right)^2 = \frac{b^2 - 4ac}{4a^2}$$

Factoring on the left side; finding a common denominator on the right side

$$x + \frac{b}{2a} = \pm\frac{\sqrt{b^2 - 4ac}}{2a}$$

Using the principle of square roots and the quotient rule for radicals; since $a > 0$, $\sqrt{4a^2} = 2a$

$$x = \frac{-b \pm \sqrt{b^2 - 4ac}}{2a}. \qquad \text{Adding } -\frac{b}{2a} \text{ on both sides}$$

A similar derivation yields the same result when a is negative. It is important that you remember the quadratic formula and know how to use it.

THE QUADRATIC FORMULA

The solutions of $ax^2 + bx + c = 0$, $a \neq 0$, are given by

$$x = \frac{-b \pm \sqrt{b^2 - 4ac}}{2a}.$$

EXAMPLE 1 Solve $5x^2 + 8x = -3$ using the quadratic formula.

SOLUTION   We first find standard form and determine a, b, and c:

$$5x^2 + 8x + 3 = 0; \qquad \text{Adding 3 on both sides}$$
$$a = 5, \quad b = 8, \quad c = 3.$$

Next, we use the quadratic formula:

$$x = \frac{-b \pm \sqrt{b^2 - 4ac}}{2a}$$

$$x = \frac{-8 \pm \sqrt{8^2 - 4 \cdot 5 \cdot 3}}{2 \cdot 5} \qquad \text{Substituting}$$

$$x = \frac{-8 \pm \sqrt{64 - 60}}{10}$$

Be sure to write the fraction bar all the way across.

$$x = \frac{-8 \pm \sqrt{4}}{10} = \frac{-8 \pm 2}{10}.$$

Thus,

$$x = \frac{-8 + 2}{10} \quad or \quad x = \frac{-8 - 2}{10}$$

$$x = \frac{-6}{10} \quad or \quad x = \frac{-10}{10}$$

$$x = -\frac{3}{5} \quad or \quad x = -1.$$

The solutions are $-\frac{3}{5}$ and -1.

Because $5x^2 + 8x + 3$ can be factored as $(5x + 3)(x + 1)$, the quadratic formula may not have been the fastest way of solving Example 1. However, because the quadratic formula works for *any* quadratic equation, we need not spend much time struggling to solve a quadratic equation by factoring.

> **TO SOLVE A QUADRATIC EQUATION:**
>
> **1.** Check for the form $ax^2 = p$ or $(x + k)^2 = d$. If it is in either of these forms, use the principle of square roots as in Section 8.1.
> **2.** If it is not in the form of step (1), write it in standard form $ax^2 + bx + c = 0$.
> **3.** Try factoring and using the principle of zero products.
> **4.** If it is not possible to factor or factoring seems difficult, use the quadratic formula.
>
> The solutions of a quadratic equation can always be found using the quadratic formula. They cannot always be found by factoring.

Recall from Section 5.1 that a **quadratic function** is a second-degree polynomial function in one variable.

EXAMPLE 2   Given the quadratic function described by $f(x) = 5x^2 - 8x - 3$, find x such that $f(x) = 0$.

SOLUTION   We substitute and solve for x:

$$f(x) = 0$$
$$5x^2 - 8x - 3 = 0 \qquad \text{Substituting; this cannot be solved by factoring.}$$
$$a = 5, \quad b = -8, \quad c = -3.$$

We then substitute into the quadratic formula:

$$x = \frac{-(-8) \pm \sqrt{(-8)^2 - 4 \cdot 5 \cdot (-3)}}{2 \cdot 5}$$
$$= \frac{8 \pm \sqrt{64 + 60}}{10}.$$

Thus,

$$x = \frac{8 \pm \sqrt{124}}{10}$$ Note that 4 is a perfect-square factor of 124.

$$= \frac{8 \pm \sqrt{4 \cdot 31}}{10}$$

$$= \frac{8 \pm 2\sqrt{31}}{10}$$

$$= \frac{2(4 \pm \sqrt{31})}{2 \cdot 5} = \frac{4 \pm \sqrt{31}}{5}.$$ Removing a factor equal to 1: $\frac{2}{2} = 1$

CAUTION! To avoid a common error, *factor the numerator and the denominator* when removing a factor equal to 1.

The solutions are

$$\frac{4 + \sqrt{31}}{5} \quad \text{and} \quad \frac{4 - \sqrt{31}}{5}.$$

Some quadratic equations have solutions that are imaginary numbers.

TECHNOLOGY CONNECTION 8.2A

On many graphers, it is possible to check Example 2 by entering $5x^2 - 8x - 3$ as y_1 and then using the Y-VARS key to evaluate y_1 as shown. If 2nd ENTRY is pressed, y_1 can be evaluated quickly for a variety of inputs.

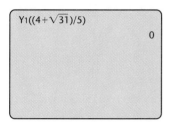

Y₁((4+√31)/5)

0

Use this approach to check the other solution of Example 2.

EXAMPLE 3

Solve: $x^2 + 2 = -x$.

SOLUTION We first find standard form:

$$x^2 + x + 2 = 0.$$ Adding x on both sides

Since we cannot solve by factoring, we use the quadratic formula with $a = 1$, $b = 1$, and $c = 2$:

$$x = \frac{-1 \pm \sqrt{1^2 - 4 \cdot 1 \cdot 2}}{2 \cdot 1}$$ Substituting

$$= \frac{-1 \pm \sqrt{1 - 8}}{2}$$

$$= \frac{-1 \pm \sqrt{-7}}{2}$$

$$= \frac{-1 \pm i\sqrt{7}}{2}.$$

The solutions are $\dfrac{-1 + i\sqrt{7}}{2}$ and $\dfrac{-1 - i\sqrt{7}}{2}$.

EXAMPLE 4

Solve: $2 + \dfrac{7}{x} = \dfrac{4}{x^2}$.

SOLUTION This is a rational equation similar to those found in Section 6.4. Note that $x \neq 0$. Since the LCD is x^2, we multiply by x^2 on both

**TECHNOLOGY
CONNECTION
8.2B**

We saw in Sections 5.8 and 8.1 how graphers can solve quadratic equations. To determine whether quadratic equations are solved more quickly on a grapher or by using the quadratic formula, solve Examples 2 and 4 both ways. Which method is faster? Which method is more precise? Why?

sides:

$$x^2\left(2 + \frac{7}{x}\right) = x^2 \cdot \frac{4}{x^2}$$

$$2x^2 + 7x = 4 \qquad \text{Simplifying}$$

$$2x^2 + 7x - 4 = 0. \qquad \text{Subtracting 4 on both sides}$$

We have

$$a = 2, \qquad b = 7, \quad \text{and} \quad c = -4.$$

Substituting then gives us

$$x = \frac{-7 \pm \sqrt{7^2 - 4 \cdot 2 \cdot (-4)}}{2 \cdot 2}$$

$$= \frac{-7 \pm \sqrt{49 + 32}}{4}$$

$$= \frac{-7 \pm \sqrt{81}}{4}$$

$$= \frac{-7 \pm 9}{4}$$

$$x = \frac{-7 + 9}{4} = \frac{1}{2} \quad or \quad x = \frac{-7 - 9}{4} = -4. \qquad \text{Both answers should check since } x \neq 0.$$

You can confirm that both $\frac{1}{2}$ and -4 check in the original equation. The solutions are $\frac{1}{2}$ and -4.

Checking the solutions of Examples 2 and 3 can be cumbersome. Fortunately, when the quadratic formula is used to solve a quadratic equation in standard form, the results will always be correct unless a careless error has been made. Thus checking for computational errors is usually sufficient.

Approximating Solutions

When the solution of an equation is irrational, a rational-number approximation is often useful. This is often the case in real-world applications similar to those found in Section 8.3.

EXAMPLE 5

Use a calculator to approximate the solutions of Example 2.

SOLUTION On most calculators, one of the following sequences of keystrokes can be used to approximate $(4 + \sqrt{31})/5$:

$$(\boxed{4} \boxed{+} \boxed{\sqrt{}} \boxed{31} \boxed{)} \boxed{\div} \boxed{5} \boxed{\text{ENTER}} \; ; \quad \text{or}$$

$$\boxed{31} \boxed{\sqrt{}} \boxed{+} \boxed{4} \boxed{=} \boxed{\div} \boxed{5} \boxed{=} .$$

Similar keystrokes can be used to approximate $(4 - \sqrt{31})/5$.

The solutions of Example 2 are approximately 1.913552873 and -0.3135528726.

EXERCISE SET
8.2

Solve.

1. $x^2 + 7x + 4 = 0$ **2.** $x^2 - 7x - 3 = 0$

3. $3p^2 = -8p - 5$ **4.** $3u^2 = 18u - 6$

5. $x^2 - x + 2 = 0$ **6.** $x^2 - x + 1 = 0$

7. $x^2 + 13 = 6x$ **8.** $x^2 + 13 = 4x$

9. $h^2 + 4 = 6h$ **10.** $r^2 + 3r = 8$

11. $3 + \dfrac{8}{x} = \dfrac{1}{x^2}$ **12.** $2 + \dfrac{5}{x^2} = \dfrac{9}{x}$

13. $3x + x(x - 2) = 0$ **14.** $4x + x(x - 3) = 0$

15. $14x^2 + 9x = 0$ **16.** $19x^2 + 8x = 0$

17. $25x^2 - 20x + 4 = 0$ **18.** $36x^2 + 84x + 49 = 0$

19. $7x(x + 2) + 6 = 3x(x + 1)$

20. $5x(x - 1) + 2 = 4x(x - 2)$

21. $14(x - 4) - (x + 2) = (x + 2)(x - 4)$

22. $11(x - 2) + (x - 5) = (x + 2)(x - 6)$

23. $5x^2 = 13x + 17$ **24.** $25x = 3x^2 + 28$

25. $x^2 + 9 = 4x$ **26.** $x^2 + 7 = 3x$

27. $x + \dfrac{1}{x} = \dfrac{13}{6}$ **28.** $\dfrac{3}{x} + \dfrac{x}{3} = \dfrac{5}{2}$

29. $x^3 - 8 = 0$ (*Hint*: Factor the difference of cubes. Then use the quadratic formula.)

30. $x^3 + 1 = 0$

31. Let $f(x) = 3x^2 - 5x - 1$. Find x such that $f(x) = 0$.

32. Let $g(x) = 4x^2 - 2x - 3$. Find x such that $g(x) = 0$.

33. Let
$$f(x) = \frac{7}{x} + \frac{7}{x + 4}.$$
Find x such that $f(x) = 1$.

34. Let
$$g(x) = \frac{2}{x} + \frac{2}{x + 3}.$$
Find x such that $g(x) = 1$.

35. Let
$$F(x) = \frac{x + 3}{x} \quad \text{and} \quad G(x) = \frac{x - 4}{3}.$$
Find x such that $F(x) = G(x)$.

36. Let
$$f(x) = x + 5 \quad \text{and} \quad g(x) = \frac{3}{x - 5}.$$
Find x such that $f(x) = g(x)$.

Solve. Use a calculator to approximate solutions as rational numbers.

37. $x^2 + 4x - 7 = 0$ **38.** $x^2 + 6x + 4 = 0$

39. $x^2 - 6x + 4 = 0$ **40.** $x^2 - 4x + 1 = 0$

41. $2x^2 - 3x - 7 = 0$ **42.** $3x^2 - 3x - 2 = 0$

SKILL MAINTENANCE

43. *Coffee beans.*   Twin Cities Roasters has Kenyan coffee worth $6.75 a pound and Peruvian coffee worth $11.25 a pound. How much of each kind should be mixed to obtain a 50-lb mixture that is worth $8.55 a pound?

Simplify.

44. $\sqrt{8a^3b} \cdot \sqrt{12ab^5}$

45. $\dfrac{\dfrac{3}{x - 1}}{\dfrac{1}{x + 1} + \dfrac{2}{x - 1}}$ **46.** $\dfrac{\dfrac{4}{a^2b}}{\dfrac{3}{a} - \dfrac{4}{b^2}}$

SYNTHESIS

47. ◆ Given the solutions of a quadratic equation, is it possible to reconstruct the original equation? Why or why not?

48. ◆ Are there any equations that can be solved by the quadratic formula but not by completing the square? Why or why not?

49. ◆ The list on p. 446 does not mention completing the square as a method of solving quadratic equations. Why not?

50. ◆ Explain how the quadratic formula can be used to factor a quadratic polynomial into two binomials.

Let $f(x) = \dfrac{x^2}{x - 2} + 1$ *and* $g(x) = \dfrac{4x - 2}{x - 2} + \dfrac{x + 4}{2}.$

51. Find the x-intercepts of the graph of g.

52. Find the x-intercepts of the graph of f.

53. Find x such that $f(x) = g(x)$.

Solve.

54. ▦ $x^2 - 0.75x - 0.5 = 0$

55. ▦ $z^2 + 0.84z - 0.4 = 0$

56. $\sqrt{2}x^2 + 5x + \sqrt{2} = 0$

57. $(1 + \sqrt{3})x^2 - (3 + 2\sqrt{3})x + 3 = 0$

58. $ix^2 - 2x + 1 = 0$

59. One solution of $kx^2 + 3x - k = 0$ is -2. Find the other.

60. 〰 Use a grapher to solve Exercises 3, 17, and 37.

61. 〰 Use a grapher to solve Exercises 9, 27, and 33. Use the method of graphing each side of the equation.

62. ◈ 〰 Can a grapher be used to solve *any* quadratic equation? Why or why not?

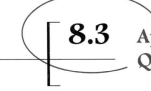

8.3 Applications Involving Quadratic Equations

Solving Problems • Solving Formulas

Solving Problems

As we found in Section 6.5, some problems translate to rational equations. The solution of such rational equations can involve quadratic equations.

EXAMPLE 1   *Motorcycle Travel.*   Makita's motorcycle traveled 300 mi at a certain speed. Had she gone 10 mph faster, the trip would have taken 1 hr less. Find the speed of the motorcycle.

SOLUTION

1. Familiarize. We make a drawing, labeling it with the known and unknown information. As in Section 6.5, we can organize the data in a table. We let r and t represent the rate, in miles per hour, and time, in hours, respectively, for Makita's trip.

300 miles

Time t Speed r

300 miles

Time $t - 1$ Speed $r + 10$

Distance	Speed	Time
300	r	t
300	$r + 10$	$t - 1$

Recall that the definition of speed, $r = d/t$, relates the three quantities.

2. **Translate.**   From the first line of the table, we obtain

$$r = \frac{300}{t}.$$

From the second line, we get

$$r + 10 = \frac{300}{t - 1}.$$

3. **Carry out.**   A system of equations has been formed. We substitute for r from the first equation into the second and solve the resulting equation:

$$\frac{300}{t} + 10 = \frac{300}{t - 1} \qquad \text{Substituting } 300/t \text{ for } r$$

$$t(t - 1) \cdot \left[\frac{300}{t} + 10 \right] = t(t - 1) \cdot \frac{300}{t - 1} \qquad \text{Multiplying by the LCD}$$

$$t(t - 1) \cdot \frac{300}{t} + t(t - 1) \cdot 10 = t(t-1) \cdot \frac{300}{t-1} \qquad \begin{array}{l} \text{Using the distributive} \\ \text{law and removing} \\ \text{factors that equal 1:} \\ \frac{t}{t} = 1; \frac{t-1}{t-1} = 1 \end{array}$$

$$300(t - 1) + 10(t^2 - t) = 300t$$

$$300t - 300 + 10t^2 - 10t = 300t$$

$$10t^2 - 10t - 300 = 0 \qquad \text{Standard form}$$

$$t^2 - t - 30 = 0 \qquad \text{Multiplying by } \frac{1}{10}$$

$$(t - 6)(t + 5) = 0 \qquad \text{Factoring}$$

$$t = 6 \quad or \quad t = -5. \qquad \begin{array}{l} \text{Principle of zero} \\ \text{products} \end{array}$$

4. **Check.**   Note that we have solved for t, not r as required. Since negative time has no meaning here, we disregard the -5 and use 6 hr to find r:

$$r = \frac{300}{6} = 50 \text{ mph.}$$

> **CAUTION!** Always make sure that you find the quantity asked for in the problem.

To see if 50 mph checks, we increase the speed 10 mph to 60 mph and see how long the trip would have taken at that speed:

$$t = \frac{d}{r} = \frac{300}{60} = 5 \text{ hr.}$$

This is 1 hr less than the trip actually took, so the answer checks.

5. **State.**   Makita's motorcycle traveled at a speed of 50 mph.

Solving Formulas

Recall that to solve a formula for a certain letter, we use the principles for solving equations to get that letter alone on one side.

EXAMPLE 2

Period of a Pendulum.   The time T required for a pendulum of length l to swing back and forth (complete one period) is given by the formula $T = 2\pi\sqrt{l/g}$, where g is the gravitational constant. Solve for l.

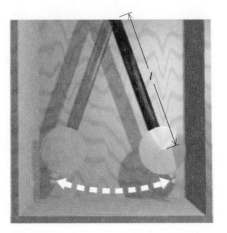

SOLUTION   We have

$$T = 2\pi\sqrt{\frac{l}{g}}$$ This is a radical equation (see Section 7.6).

$$T^2 = \left(2\pi\sqrt{\frac{l}{g}}\right)^2$$ Principle of powers (squaring)

$$T^2 = 2^2\pi^2\frac{l}{g}$$

$$gT^2 = 4\pi^2 l$$ Clearing fractions

$$\frac{gT^2}{4\pi^2} = l.$$ Multiplying by $\frac{1}{4\pi^2}$

We now have l alone on one side and l does not appear on the other side, so the formula is solved for l.

    In most formulas, variables represent nonnegative numbers, so we do not need to use absolute-value signs when taking square roots.

EXAMPLE 3

*Hang Time.** An athlete's *hang time* is the amount of time that the athlete can remain airborne when jumping. A formula relating an athlete's vertical leap V, in inches, to hang time T, in seconds, is $V = 48T^2$. Solve for T.

---

*This formula is taken from an article by Peter Brancazio, "The Mechanics of a Slam Dunk," *Popular Mechanics*, November 1991. Courtesy of Professor Peter Brancazio, Brooklyn College.

SOLUTION

$$48T^2 = V$$

$$T^2 = \frac{V}{48} \qquad \text{Multiplying by } \frac{1}{48} \text{ to get } T^2 \text{ alone}$$

$$T = \sqrt{\frac{V}{48}} \qquad \text{Using the principle of square roots; note that } T \geq 0.$$

$$T = \sqrt{\frac{V}{16 \cdot 3} \cdot \frac{3}{3}} \qquad \text{Factoring and multiplying by 1 to rationalize the denominator. Note that 16 is a perfect square.}$$

$$T = \sqrt{\frac{3V}{144}} = \frac{\sqrt{3V}}{12}$$

EXAMPLE 4

Falling Distance.   An object tossed downward with an initial speed (velocity) of v_0 will travel a distance of s meters, where $s = 4.9t^2 + v_0 t$ and t is measured in seconds. Solve for t.

SOLUTION   Since t is squared in one term and raised to the first power in the other term, the equation is quadratic in t.

$$4.9t^2 + v_0 t = s$$

$$4.9t^2 + v_0 t - s = 0 \qquad \text{Writing standard form}$$

$$a = 4.9, \quad b = v_0, \quad c = -s$$

$$t = \frac{-v_0 \pm \sqrt{v_0^2 - 4(4.9)(-s)}}{2(4.9)} \qquad \text{Using the quadratic formula}$$

Since the negative square root would yield a negative value for t, we use only the positive root:

$$t = \frac{-v_0 + \sqrt{v_0^2 + 19.6s}}{9.8}.$$

The following list of steps should help you when solving formulas for a given letter. Try to remember that when solving a formula, you do the same things you would do to solve any equation.

TO SOLVE A FORMULA FOR A LETTER—SAY, b:

1. Clear the fractions and use the principle of powers, as needed, until b does not appear in any radicand or denominator. (In some cases, you may clear the fractions first, and in some cases you may use the principle of powers first. Perform these steps until radicals containing b are gone and b is not in any denominator.)
2. Collect all terms with b^2 in them. Also collect all terms with b in them.
3. If b^2 does not appear, you can finish by using just the addition and multiplication principles as in Sections 1.5 and 6.8.
4. If b^2 appears but b does not, solve the equation for b^2. Then take square roots on both sides.
5. If there are terms containing both b and b^2, put the equation in standard form and use the quadratic formula.

EXERCISE SET
8.3

Solve.

1. *Car trips.* During the first part of a trip, Meira's Honda traveled 120 mi at a certain speed. Meira then drove another 100 mi at a speed that was 10 mph slower. If Meira's total trip time was 4 hr, what was her speed on each part of the trip?

2. *Canoeing.* During the first part of a canoe trip, Tim covered 60 km at a certain speed. He then traveled 24 km at a speed that was 4 km/h slower. If the total time for the trip was 8 hr, what was the speed on each part of the trip?

3. *Car trips.* Petra's Plymouth travels 200 mi at a certain speed. If the car had gone 10 mph faster, the trip would have taken 1 hr less. Find Petra's speed.

4. *Car trips.* Sandi's Subaru travels 280 mi at a certain speed. If the car had gone 5 mph faster, the trip would have taken 1 hr less. Find Sandi's speed.

5. *Air travel.* A Cessna flies 600 mi at a certain speed. A Beechcraft flies 1000 mi at a speed that is 50 mph faster, but takes 1 hr longer. Find the speed of each plane.

6. *Air travel.* A turbo-jet flies 50 mph faster than a super-prop plane. If a turbo-jet goes 2000 mi in 3 hr less time than it takes the super-prop to go 2800 mi, find the speed of each plane.

7. *Bicycling.* Naoki bikes the 40 mi to Hillsboro at a certain speed. The return trip is made at a speed that is 6 mph slower. Total time for the round trip is 14 hr. Find Naoki's speed on each part of the trip.

8. *Car speed.* On a sales trip, Gail drives the 600 mi to Richmond at a certain speed. The return trip is made at a speed that is 10 mph slower. Total time for the round trip was 22 hr. How fast did Gail travel on each part of the trip?

9. ▦ *Navigation.* The current in a typical Mississippi River shipping route flows at a rate of 4 mph. In order for a barge to travel 24 mi upriver and then return in a total of 5 hr, approximately how fast must the barge be able to travel in still water?

10. ▦ *Navigation.* The Hudson River flows at a rate of 3 mph. A patrol boat travels 60 mi upriver and returns in a total time of 9 hr. What is the speed of the boat in still water?

11. *Filling a pool.* Two hoses are connected to a swimming pool. Working together, they can fill the pool in 4 hr. The larger hose, working alone, can fill the pool in 6 hr less time than the smaller one. How long would the smaller one take, working alone, to fill the pool?

12. *Filling a tank.* Two pipes are connected to the same

tank. Working together, they can fill the tank in 2 hr. The larger pipe, working alone, can fill the tank in 3 hr less time than the smaller one. How long would the smaller one take, working alone, to fill the tank?

13. 🔲 *Rowing.*   Dan rows 10 km upstream and 10 km back in a total time of 3 hr. The speed of the river is 5 km/h. Find Dan's speed in still water.

14. 🔲 *Paddleboats.*   Ellen paddles 1 mi upstream and 1 mi back in a total time of 1 hr. The speed of the river is 2 mph. Find the speed of Ellen's paddleboat in still water.

Solve each formula for the indicated letter. Assume that all variables represent nonnegative numbers.

15. $A = 4\pi r^2$, for r
(Surface area of a sphere)

16. $A = 6s^2$, for s
(Surface area of a cube)

17. $A = 2\pi r^2 + 2\pi rh$, for r
(Surface area of a right cylindrical solid)

18. $F = \dfrac{Gm_1 m_2}{r^2}$, for r
(Law of gravity)

19. $N = \dfrac{kQ_1 Q_2}{s^2}$, for s
(Number of phone calls between two cities)

20. $A = \pi r^2$, for r
(Area of a circle)

21. $T = 2\pi \sqrt{\dfrac{l}{g}}$, for g
(A pendulum formula)

22. $a^2 + b^2 = c^2$, for b
(Pythagorean formula in two dimensions)

23. $a^2 + b^2 + c^2 = d^2$, for c
(Pythagorean formula in three dimensions)

24. $N = \dfrac{k^2 - 3k}{2}$, for k
(Number of diagonals of a polygon)

25. $s = v_0 t + \dfrac{gt^2}{2}$, for t
(A motion formula)

26. $A = \pi r^2 + \pi rs$, for r
(Surface area of a cone)

27. $N = \frac{1}{2}(n^2 - n)$, for n
(Number of games if n teams play each other once)

28. $A = A_0(1 - r)^2$, for r
(A business formula)

29. $V = 3.5\sqrt{h}$, for h
(Distance to horizon from a height)

30. $W = \sqrt{\dfrac{1}{LC}}$, for L
(An electricity formula)

31. $A = P_1(1 + r)^2 + P_2(1 + r)$, for r
(An investment formula)

32. $A = P_1\left(1 + \dfrac{r}{2}\right)^2 + P_2\left(1 + \dfrac{r}{2}\right)$, for r
(An investment formula)

Solve. Refer to Exercises 15–32 and Examples 2–4 for the appropriate formula.

33. *Falling distance.*

   a) An object is dropped 500 m from an airplane. How long does it take the object to reach the ground?
   b) An object is thrown downward 500 m from the plane at an initial velocity of 30 m/sec. How long does it take the object to reach the ground?
   c) How far will an object fall in 5 sec, when thrown downward at an initial velocity of 30 m/sec?

34. *Falling distance.*

   a) An object is dropped 75 m from an airplane. How long does it take the object to reach the ground?
   b) An object is thrown downward with an initial velocity of 30 m/sec from a plane 75 m above the ground. How long does it take the object to reach the ground?
   c) How far will an object fall in 2 sec, if thrown downward at an initial velocity of 30 m/sec?

35. *Bungee jumping.*   Jesse is tied to one end of a 40-m elasticized (bungee) cord. The other end of the cord is tied to the middle of a train trestle. If Jesse jumps off the bridge, for how long will he fall before the cord begins to stretch? (See Example 4 and let $v_0 = 0$.)

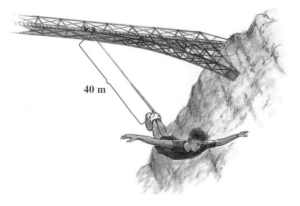

40 m

36. *Bungee jumping.* Sheila is tied to a bungee cord (see Exercise 35) and falls for 2.5 sec before her cord begins to stretch. How long is the bungee cord?

37. *Hang Time.* Anfernee Hardaway of the Orlando Magic has a vertical leap of about 36 in.* What is his hang time?

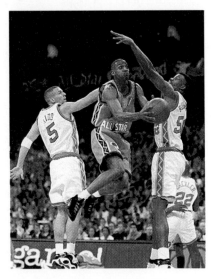

38. *League schedules.* In a volleyball league, each team plays each of the other teams once. If a total of 66 games is played, how many teams are in the league?

39. 📱 *Downward speed.* An object thrown downward from a 100-m cliff travels 51.6 m in 3 sec. What was the initial velocity of the object?

40. 📱 *Downward speed.* An object thrown downward from a 200-m cliff travels 91.2 m in 4 sec. What was the initial velocity of the object?

41. 📱 *Compound interest.* A firm invests $3000 in a savings account for 2 yr. At the beginning of the second year, an additional $1700 is invested. If a total of $5253.70 is in the account at the end of the second year, what is the annual interest rate? (*Hint*: See Exercise 31.)

42. 📱 *Compound interest.* A business invests $10,000 in a savings account for 2 yr. At the beginning of the second year, an additional $3500 is invested. If a total of $15,569.75 is in the account at the end of the second year, what is the annual interest rate? (*Hint*: See Exercise 31.)

SKILL MAINTENANCE

43. Solve: $\sqrt{3x + 1} = \sqrt{2x - 1} + 1$.

*Information provided by Orlando Magic media relations department.

44. Add: $\dfrac{1}{x - 1} + \dfrac{1}{x^2 - 3x + 2}$.

45. Multiply and simplify: $\sqrt[3]{18y^3}\,\sqrt[3]{4x^2}$.

46. Divide and simplify:

$$\frac{x - 3}{x^2 - 5x + 6} \div \frac{7x + 14}{x^2 - 4}.$$

SYNTHESIS

47. ◈ Write a problem for a classmate to solve. Devise the problem so that **(a)** the solution is found after solving a rational equation and **(b)** the solution is "The express train travels 90 mph."

48. ◈ Under what circumstances would a negative value for t, time, have meaning?

49. ◈ Marti is tied to a bungee cord that is twice as long as the cord tied to Pedro. Will Marti's fall take twice as long as Pedro's before their cords begin to stretch? Why or why not? (See Exercises 35 and 36.)

50. Find a when the reciprocal of $a - 1$ is $a + 1$.

51. *Purchasing.* A discount store bought a quantity of beach towels for $250 and sold all but 15 at a profit of $3.50 per towel. With the total amount received, the manager could buy 4 more than twice as many as were bought before. Find the cost per towel.

52. Solve for n:

$$mn^4 - r^2pm^3 - r^2n^2 + p = 0.$$

53. *The Golden Rectangle.* For over 2000 yr, the proportions of a "golden" rectangle have been considered visually appealing. A rectangle of width w and length l is considered "golden" if

$$\frac{w}{l} = \frac{l}{w + l}.$$

Solve for l.

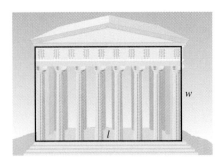

54. *Diagonal of a cube.* Find a formula that expresses the length of the three-dimensional diagonal of a cube as a function of the cube's surface area.

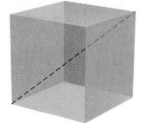

55. *Special relativity.* Einstein found that an object of mass m_0, traveling velocity v, has mass becomes

$$m = \frac{m_0}{\sqrt{1 - \dfrac{v^2}{c^2}}},$$

where c is the speed of light. Solve the formula for c.

56. *Surface area.* Find a formula that expresses the diameter of a right cylindrical solid as a function of its surface area and its height.

57. A sphere is inscribed in a cube as shown in the figure below. Express the surface area of the sphere as a function of the surface area S of the cube.

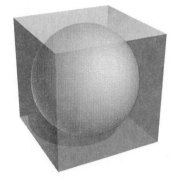

58. ◇ ⏁ Explain how Exercises 1–14 can be solved without factoring, completing the square, or using the quadratic formula.

8.4 Studying Solutions of Quadratic Equations

The Discriminant • Writing Equations from Solutions

The Discriminant

Sometimes in mathematics it is enough to know what *type* of number an equation will have for its solution(s), without actually solving the equation. To illustrate, suppose we want to know if the equation $4x^2 + 7x - 15$ has rational solutions (and thus can be solved by factoring). Using the quadratic formula, we would have

$$x = \frac{-b \pm \sqrt{b^2 - 4ac}}{2a} = \frac{-7 \pm \sqrt{7^2 - 4 \cdot 4 \cdot (-15)}}{2 \cdot 4}.$$

Note that the radicand, $7^2 - 4 \cdot 4 \cdot (-15)$, determines what type of number the solutions will be. Since $7^2 - 4 \cdot 4 \cdot (-15) = 49 - 16(-15) = 289$, and 289 is a perfect square ($\sqrt{289} = 17$), we know that the solutions of the equation will be two rational numbers. This means that $4x^2 + 7x - 15$ *can* be solved by factoring.

It is the expression $b^2 - 4ac$, known as the **discriminant**, that determines what type of number the solutions of a quadratic equation will be. When $b^2 - 4ac$ simplifies to 0, it doesn't matter if we use $+\sqrt{b^2 - 4ac}$ or $-\sqrt{b^2 - 4ac}$; we will get the same solution twice. Thus, when the discriminant is 0, there is one *repeated* solution and it will be rational.

Whenever the discriminant is positive, there will be two different real-number solutions. As we saw above, when $b^2 - 4ac$ is a perfect square, these solutions will be rational numbers. When $b^2 - 4ac$ is positive but not a perfect square, there will be two irrational solutions and they will be conjugates of each other (see p. 404).

Whenever the discriminant is negative, there will be two imaginary-number solutions and they will be complex conjugates of each other.

Discriminant $b^2 - 4ac$	Nature of Solutions
0	Only one solution; it is a rational number.

Positive
    Perfect square
    Not a perfect square
	Two different real-number solutions
	    Solutions are rational.
	    Solutions are irrational conjugates.

Negative	Two different imaginary-number solutions (complex conjugates)

EXAMPLE 1

TECHNOLOGY CONNECTION

8.4

Recall that the real-number solutions of $ax^2 + bx + c = 0$ are the x-intercepts of the graph of $y = ax^2 + bx + c$. Use a grapher to confirm that part (a) of Example 1 has one real solution, part (b) has no real solution, and part (c) has two real solutions.

For each equation, determine what type of number the solutions will be and how many solutions exist.

a) $9x^2 - 12x + 4 = 0$ **b)** $x^2 + 5x + 8 = 0$ **c)** $2x^2 + 7x - 3 = 0$

SOLUTION

a) For $9x^2 - 12x + 4 = 0$, we have

$$a = 9, \quad b = -12, \quad c = 4.$$

We substitute and compute the discriminant:

$$b^2 - 4ac = (-12)^2 - 4 \cdot 9 \cdot 4$$
$$= 144 - 144 = 0.$$

There is just one solution, and it is rational. This tells us that $9x^2 - 12x + 4 = 0$ can be solved by factoring.

b) For $x^2 + 5x + 8 = 0$, we have

$$a = 1, \quad b = 5, \quad c = 8.$$

We substitute and compute the discriminant:

$$b^2 - 4ac = 5^2 - 4 \cdot 1 \cdot 8$$
$$= 25 - 32 = -7.$$

Since the discriminant is negative, there are two imaginary-number solutions that are complex conjugates of each other.

c) For $2x^2 + 7x - 3 = 0$, we have

$$a = 2, \quad b = 7, \quad c = -3;$$
$$b^2 - 4ac = 7^2 - 4 \cdot 2(-3)$$
$$= 49 - (-24) = 73.$$

The discriminant is a positive number that is not a perfect square. Thus there are two irrational solutions that are conjugates of each other.

Writing Equations from Solutions

We know by the principle of zero products that $(x - 2)(x + 3) = 0$ has solutions 2 and −3. If we know the solutions of an equation, we can write an equation, using the principle in reverse.

EXAMPLE   2

Find an equation for the given solutions.

a) 3 and $-\frac{2}{5}$ **b)** $2i$ and $-2i$

c) $5\sqrt{7}$ and $-5\sqrt{7}$ **d)** $-4, 0$, and 1

SOLUTION

a)
$$x = 3 \quad or \quad x = -\tfrac{2}{5}$$
$$x - 3 = 0 \quad or \quad x + \tfrac{2}{5} = 0 \qquad \text{Getting 0's on one side}$$
$$(x - 3)\left(x + \tfrac{2}{5}\right) = 0 \qquad \text{Using the principle of zero products}$$
$$\qquad\qquad\qquad\qquad\qquad\qquad \text{(multiplying)}$$
$$x^2 + \tfrac{2}{5}x - 3x - 3 \cdot \tfrac{2}{5} = 0 \qquad \text{Multiplying}$$
$$x^2 - \tfrac{13}{5}x - \tfrac{6}{5} = 0 \qquad \text{Combining like terms}$$
$$5x^2 - 13x - 6 = 0 \qquad \text{Multiplying by 5 on both sides to clear}$$
$$\qquad\qquad\qquad\qquad\qquad \text{fractions}$$

b)
$$x = 2i \quad or \quad x = -2i$$
$$x - 2i = 0 \quad or \quad x + 2i = 0 \qquad \text{Getting 0's on one side}$$
$$(x - 2i)(x + 2i) = 0 \qquad \text{Using the principle of zero products}$$
$$\qquad\qquad\qquad\qquad\qquad \text{(multiplying)}$$
$$x^2 - (2i)^2 = 0 \qquad \text{Finding the product of a sum and difference}$$
$$x^2 - 4i^2 = 0$$
$$x^2 + 4 = 0 \qquad i^2 = -1$$

c)
$$x = 5\sqrt{7} \quad or \quad x = -5\sqrt{7}$$
$$x - 5\sqrt{7} = 0 \quad or \quad x + 5\sqrt{7} = 0 \qquad \text{Getting 0's on one side}$$
$$(x - 5\sqrt{7})(x + 5\sqrt{7}) = 0 \qquad \text{Using the principle of zero}$$
$$\qquad\qquad\qquad\qquad\qquad\qquad \text{products}$$
$$x^2 - (5\sqrt{7})^2 = 0 \qquad \text{Finding the product of a sum and}$$
$$\qquad\qquad\qquad\qquad\qquad \text{difference}$$
$$x^2 - 25 \cdot 7 = 0$$
$$x^2 - 175 = 0$$

d) $x = -4$ *or* $x = 0$ *or* $x = 1$

$x + 4 = 0$ *or* $x = 0$ *or* $x - 1 = 0$ Getting 0's on one side

$(x + 4)x(x - 1) = 0$ Using the principle of zero products

$x(x^2 + 3x - 4) = 0$ Multiplying

$x^3 + 3x^2 - 4x = 0$

To check any of these equations, we can simply substitute one or more of the given solutions. For example, in Example 2(d) above,

$$(-4)^3 + 3(-4)^2 - 4(-4) = -64 + 3 \cdot 16 + 16$$
$$= -64 + 48 + 16 = 0.$$

The other checks are left to the student.

Note that in Example 2(a) we multiplied on both sides by the LCD, 5. We normally perform this step and thus clear the equation of fractions. Had we preferred, we could have multiplied $x + \frac{2}{5} = 0$ by 5 on both sides, thus clearing fractions *before* using the principle of zero products.

EXERCISE SET 8.4

For each equation, determine what type of number the solutions are and how many solutions exist.

1. $x^2 - 4x + 3 = 0$ **2.** $x^2 + 6x + 5 = 0$

3. $x^2 + 5 = 0$ **4.** $x^2 + 3 = 0$

5. $x^2 - 2 = 0$ **6.** $x^2 - 5 = 0$

7. $4x^2 - 12x + 9 = 0$ **8.** $4x^2 + 8x - 5 = 0$

9. $x^2 - 2x + 4 = 0$ **10.** $x^2 + 4x + 6 = 0$

11. $a^2 + 11a + 28 = 0$ **12.** $t^2 - 8t + 16 = 0$

13. $6x^2 + 5x - 4 = 0$ **14.** $10x^2 - x - 2 = 0$

15. $9t^2 - 3t = 0$ **16.** $4m^2 + 7m = 0$

17. $x^2 + 5x = 7$ **18.** $x^2 + 4x = 7$

19. $2a^2 - 3a = -5$ **20.** $3a^2 + 5 = 7a$

21. $y^2 + \frac{9}{4} = 4y$ **22.** $x^2 = \frac{1}{2}x - \frac{3}{5}$

Write a quadratic equation having the given numbers as solutions.

23. $-7, 3$ **24.** $-6, 4$

25. 3, only solution (*Hint*: It must be a repeated solution.)   **26.** -5, only solution

27. $-2, -5$ **28.** $-1, -3$

29. $4, \frac{2}{3}$ **30.** $5, \frac{3}{4}$

31. $\frac{1}{2}, \frac{1}{3}$ **32.** $-\frac{1}{4}, -\frac{1}{2}$

33. $-0.6, 1.4$ **34.** $2.5, -0.4$

35. $-\sqrt{7}, \sqrt{7}$ **36.** $-\sqrt{3}, \sqrt{3}$

37. $3\sqrt{2}, -3\sqrt{2}$ **38.** $2\sqrt{5}, -2\sqrt{5}$

39. $3i, -3i$ **40.** $4i, -4i$

41. $5 - 2i, 5 + 2i$ **42.** $2 - 7i, 2 + 7i$

43. $2 - \sqrt{10}, 2 + \sqrt{10}$ **44.** $3 - \sqrt{14}, 3 + \sqrt{14}$

Write a third-degree equation having the given numbers as solutions.

45. $-3, 0, 4$ **46.** $-5, 0, 2$

47. $-1, 1, 2$ **48.** $-2, 2, 3$

SKILL MAINTENANCE

49. During a one-hour television show, there were 12 commercials. Some of the commercials were 30 sec long and the others were 60 sec long. The amount of time for 30-sec commercials was 6 min less than the total number of minutes of commercial time during the show. How many 30-sec commercials were used?

Graph.

50. $y = -\frac{3}{7}x + 4$

51. $f(x) = -x - 3$

52. $5x - 2y = 8$

SYNTHESIS

53. ◈ Describe a procedure that could be used to write an equation having the first seven natural numbers as solutions.

54. ◈ If we assume that a quadratic equation has integers for coefficients, will the product of the solutions always be a real number? Why or why not?

55. ◈ Can a fourth-degree equation have three irrational solutions? Why or why not?

56. Show that the product of the solutions of $ax^2 + bx + c = 0$ is c/a.

57. The graph of an equation of the form

$$y = ax^2 + bx + c$$

is a curve similar to the one shown below. Determine a, b, and c from the information given.

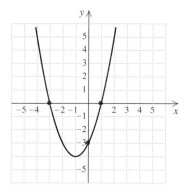

For each equation under the given condition, **(a)** *find k and* **(b)** *find the other solution.*

58. $kx^2 - 2x + k = 0$; one solution is -3

59. $x^2 - kx + 2 = 0$; one solution is $1 + i$

60. $x^2 - (6 + 3i)x + k = 0$; one solution is 3

61. Show that the sum of the solutions of $ax^2 + bx + c = 0$ is $-b/a$.

62. Find k for which

$$kx^2 - 4x + (2k - 1) = 0$$

and the product of the solutions is 3. (*Hint:* See Exercise 56.)

63. Find h and k, where $3x^2 - hx + 4k = 0$, the sum of the solutions is -12, and the product of the solutions is 20. (*Hint:* See Exercises 56 and 61.)

64. Suppose that $f(x) = ax^2 + bx + c$, with $f(-3) = 0$, $f\left(\frac{1}{2}\right) = 0$, and $f(0) = -12$. Find a, b, and c.

65. Find an equation for which $2 - \sqrt{3}$, $2 + \sqrt{3}$, $5 - 2i$, and $5 + 2i$ are solutions.

66. Find an equation for which $1 - \sqrt{5}$, $1 + \sqrt{5}$, $3 - 2i$, and $3 + 2i$ are solutions.

67. ◈ A discriminant that is a perfect square indicates that factoring can be used to solve the quadratic equation. Why?

68. ◈ Explain how the results in Exercises 56 and 61 could be used.

69. ◈ 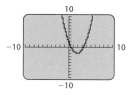 While solving a quadratic equation of the form $ax^2 + bx + c = 0$ with a grapher, Shawn-Marie gets the following screen.

How could the discriminant help her check the graph?

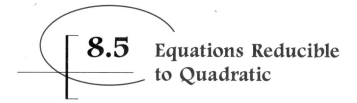

8.5   Equations Reducible to Quadratic

Recognizing Equations in Quadratic Form   •   Using Substitution to Solve

---

Certain equations that are not really quadratic can be thought of in such a way that they can be solved as quadratic. For example, if x^4 is regarded as $(x^2)^2$, the equation $x^4 - 9x^2 + 8 = 0$ is shown to be "quadratic in x^2":

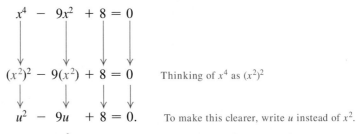

$$x^4 - 9x^2 + 8 = 0$$

$$(x^2)^2 - 9(x^2) + 8 = 0 \qquad \text{Thinking of } x^4 \text{ as } (x^2)^2$$

$$u^2 - 9u + 8 = 0. \qquad \text{To make this clearer, write } u \text{ instead of } x^2.$$

The equation $u^2 - 9u + 8 = 0$ can be solved by factoring or by the quadratic formula. Then, remembering that $u = x^2$, we can solve for x. Equations that can be solved like this are said to be *reducible to quadratic*, or *in quadratic form*.

EXAMPLE 1

Solve: $x^4 - 9x^2 + 8 = 0$.

SOLUTION   Let $u = x^2$. Then we solve by substituting u for x^2:

$$u^2 - 9u + 8 = 0$$

$$(u - 8)(u - 1) = 0 \qquad \text{Factoring}$$

$$u - 8 = 0 \quad or \quad u - 1 = 0 \qquad \text{Principle of zero products}$$

$$u = 8 \quad or \qquad u = 1.$$

> **CAUTION!**   A common error is to solve for u and then forget to solve for x. Remember that you must find values for the *original* variable!

We replace u with x^2 and solve these equations:

$$x^2 = 8 \qquad or \quad x^2 = 1$$

$$x = \pm\sqrt{8} \quad or \quad x = \pm 1$$

$$x = \pm 2\sqrt{2} \quad or \quad x = \pm 1.$$

To check, note that for $x = 2\sqrt{2}$, we have $x^2 = 8$ and $x^4 = 64$. Similarly, for $x = -2\sqrt{2}$, we have $x^2 = 8$ and $x^4 = 64$. When $x = 1$, we have $x^2 = 1$ and $x^4 = 1$, and when $x = -1$, we have $x^2 = 1$ and $x^4 = 1$. Thus instead of making four checks, we need make only two.

Check:

For $\pm 2\sqrt{2}$:

$$\begin{array}{c|c} x^4 - 9x^2 + 8 = 0 \\ \hline (\pm 2\sqrt{2})^4 - 9(\pm 2\sqrt{2})^2 + 8 \;?\; 0 \\ 64 - 9 \cdot 8 + 8 \\ 0 \;\big|\; 0 \text{ TRUE} \end{array}$$

For ± 1:

$$\begin{array}{c|c} x^4 - 9x^2 + 8 = 0 \\ \hline (\pm 1)^4 - 9(\pm 1)^2 + 8 \;?\; 0 \\ 1 - 9 + 8 \\ 0 \;\big|\; 0 \text{ TRUE} \end{array}$$

The solutions are $1, -1, 2\sqrt{2},$ and $-2\sqrt{2}$.

Example 1 can be solved directly by factoring:

$$x^4 - 9x^2 + 8 = 0$$
$$(x^2 - 1)(x^2 - 8) = 0$$
$$x^2 - 1 = 0 \quad or \quad x^2 - 8 = 0$$
$$x^2 = 1 \quad or \quad x^2 = 8$$
$$x = \pm 1 \quad or \quad x = \pm 2\sqrt{2}.$$

There is nothing wrong with this approach. However, in the examples that follow, you will note that it becomes increasingly difficult to solve the equation without first making a substitution.

Sometimes rational equations, radical equations, or equations containing fractional exponents are reducible to quadratic. It is especially important that answers to these equations be checked in the original equation.

EXAMPLE 2

Solve: $x - 3\sqrt{x} - 4 = 0$.

SOLUTION   This radical equation could be solved using the method discussed in Section 7.6. However, if we think of x as $(\sqrt{x})^2$, we can regard the equation as "quadratic in $\sqrt{x}$."

We let $u = \sqrt{x}$ and $u^2 = x$:

$$x - 3\sqrt{x} - 4 = 0$$
$$u^2 - 3u - 4 = 0 \qquad \text{Substituting}$$
$$(u - 4)(u + 1) = 0$$
$$u = 4 \quad or \quad u = -1.$$

Now we replace u with $\sqrt{x}$ and solve these equations:

$$\sqrt{x} = 4 \quad or \quad \sqrt{x} = -1.$$

Squaring gives us $x = 16$ or $x = 1$ and also makes checking essential.

Check: For 16:

$$\frac{x - 3\sqrt{x} - 4 = 0}{16 - 3\sqrt{16} - 4 \ ? \ 0}$$
$$16 - 3 \cdot 4 - 4 \ \Big|$$
$$0 \ \Big| \ 0 \quad \text{TRUE}$$

For 1:

$$\frac{x - 3\sqrt{x} - 4 = 0}{1 - 3\sqrt{1} - 4 \ ? \ 0}$$
$$1 - 3 \cdot 1 - 4 \ \Big|$$
$$-6 \ \Big| \ 0 \quad \text{FALSE}$$

The number 16 checks, but 1 does not. Had we noticed that $\sqrt{x} = -1$ has no solution (since principal roots are never negative), we could have solved only the equation $\sqrt{x} = 4$. The solution is 16.

EXAMPLE 3

Find the x-intercepts of the graph of $f(x) = (x^2 - 1)^2 - (x^2 - 1) - 2$.

SOLUTION   The x-intercepts occur where $f(x) = 0$ so we must have

$$(x^2 - 1)^2 - (x^2 - 1) - 2 = 0.$$

TECHNOLOGY CONNECTION 8.5

Check Example 3 with a grapher. Use the ROOT or INTERSECT option, if possible.

This equation is quadratic in $x^2 - 1$, so we let $u = x^2 - 1$:

$$u^2 - u - 2 = 0 \qquad \text{Substituting in } (x^2 - 1)^2 - (x^2 - 1) - 2 = 0$$
$$(u - 2)(u + 1) = 0$$
$$u = 2 \qquad or \qquad u = -1.$$

Now we replace u with $x^2 - 1$ and solve these equations:

$$\begin{array}{ccc} x^2 - 1 = 2 & or & x^2 - 1 = -1 \\ x^2 = 3 & or & x^2 = 0 \\ x = \pm\sqrt{3} & or & x = 0. \end{array}$$

The x-intercepts occur at $(-\sqrt{3}, 0)$, $(0, 0)$, and $(\sqrt{3}, 0)$.

Sometimes great care must be taken in deciding what substitution to make.

EXAMPLE 4

Solve: $m^{-2} - 6m^{-1} + 4 = 0$.

SOLUTION   We rewrite the equation using positive exponents:

$$\frac{1}{m^2} - \frac{6}{m} + 4 = 0.$$

If we let $u = 1/m$ and $u^2 = 1/m^2$, the equation is quadratic in $1/m$:

$$u^2 - 6u + 4 = 0 \qquad \text{Substituting}$$
$$u = \frac{-(-6) \pm \sqrt{(-6)^2 - 4 \cdot 1 \cdot 4}}{2 \cdot 1} \qquad \begin{array}{l}\text{Using the quadratic}\\\text{formula}\end{array}$$
$$\left.\begin{array}{l} u = \dfrac{6 \pm \sqrt{20}}{2} = \dfrac{2 \cdot 3 \pm 2\sqrt{5}}{2} \\[2mm] u = 3 \pm \sqrt{5}. \end{array}\right\} \qquad \text{Simplifying}$$

Now we replace u with $1/m$ and solve:

$$\frac{1}{m} = 3 \pm \sqrt{5}$$
$$1 = m(3 \pm \sqrt{5}) \qquad \text{Multiplying by } m \text{ on both sides}$$
$$\frac{1}{3 \pm \sqrt{5}} = m. \qquad \text{Dividing by } 3 \pm \sqrt{5} \text{ on both sides}$$

Check: For $1/(3 - \sqrt{5})$:

$$\begin{array}{c|c} \multicolumn{2}{c}{m^{-2} - 6m^{-1} + 4 = 0} \\ \hline \left(\dfrac{1}{3 - \sqrt{5}}\right)^{-2} - 6\left(\dfrac{1}{3 - \sqrt{5}}\right)^{-1} + 4 \;?\; 0 \\ (3 - \sqrt{5})^2 - 6(3 - \sqrt{5}) + 4 \\ 9 - 6\sqrt{5} + 5 - 18 + 6\sqrt{5} + 4 \\ 0 \;\Big|\; 0 \text{ TRUE} \end{array}$$

For $1/(3 + \sqrt{5})$:

$$\begin{array}{c} m^{-2} - 6m^{-1} + 4 = 0 \\ \hline \left(\dfrac{1}{3 + \sqrt{5}}\right)^{-2} - 6\left(\dfrac{1}{3 + \sqrt{5}}\right)^{-1} + 4 \ \text{?}\ 0 \\ (3 + \sqrt{5})^2 - 6(3 + \sqrt{5}) + 4 \ \Big| \\ 9 + 6\sqrt{5} + 5 - 18 - 6\sqrt{5} + 4 \ \Big| \\ 0 \ \Big|\ 0 \ \text{TRUE} \end{array}$$

Both numbers check. The solutions are $1/(3 - \sqrt{5})$ and $1/(3 + \sqrt{5})$, or approximately 1.309016994 and 0.1909830056. ⟶●

EXAMPLE 5   Solve: $t^{2/5} - t^{1/5} - 2 = 0$.

SOLUTION   Note that $t^{2/5}$ can be rewritten as $(t^{1/5})^2$. The equation can thus be written as $(t^{1/5})^2 - t^{1/5} - 2 = 0$. We let $u = t^{1/5}$ and solve the resulting equation:

$$\begin{aligned} u^2 - u - 2 &= 0 \qquad \text{Substituting} \\ (u - 2)(u + 1) &= 0 \\ u = 2 \quad &or \quad u = -1. \end{aligned}$$

Now we replace u with $t^{1/5}$ and solve:

$$\begin{aligned} t^{1/5} = 2 \quad &or \quad t^{1/5} = -1 \\ t = 32 \quad &or \quad t = -1. \qquad \text{Principle of powers; raising to the 5th power} \end{aligned}$$

Check:

For 32:

$$\begin{array}{c} t^{2/5} - t^{1/5} - 2 = 0 \\ \hline 32^{2/5} - 32^{1/5} - 2 \ \text{?}\ 0 \\ (32^{1/5})^2 - 32^{1/5} - 2 \ \Big| \\ 2^2 - 2 - 2 \ \Big| \\ 0 \ \Big|\ 0 \ \text{TRUE} \end{array}$$

For -1:

$$\begin{array}{c} t^{2/5} - t^{1/5} - 2 = 0 \\ \hline (-1)^{2/5} - (-1)^{1/5} - 2 \ \text{?}\ 0 \\ [(-1)^{1/5}]^2 - (-1)^{1/5} - 2 \ \Big| \\ (-1)^2 - (-1) - 2 \ \Big| \\ 0 \ \Big|\ 0 \ \text{TRUE} \end{array}$$

Both numbers check. The solutions are 32 and -1. ⟶●

The following tips may prove useful.

TO SOLVE AN EQUATION THAT IS REDUCIBLE TO QUADRATIC:

1. Determine if the equation is quadratic in form by identifying a term with a factor that is the square of a factor appearing in another term.
2. Write down any substitutions that you are making.
3. Whenever you make a substitution, be sure to solve for the variable that is used in the original equation.
4. Check possible answers in the original equation.

EXERCISE SET

8.5

Solve.

1. $x^4 - 5x^2 + 4 = 0$ **2.** $x^4 - 10x^2 + 9 = 0$

3. $x^4 - 12x^2 + 27 = 0$ **4.** $x^4 - 9x^2 + 20 = 0$

5. $4x^4 - 19x^2 + 12 = 0$ **6.** $9x^4 - 14x^2 + 5 = 0$

7. $x - 4\sqrt{x} - 1 = 0$ **8.** $x - 2\sqrt{x} - 6 = 0$

9. $(x^2 - 7)^2 - 3(x^2 - 7) + 2 = 0$

10. $(x^2 - 1)^2 - 5(x^2 - 1) + 6 = 0$

11. $(3 + \sqrt{x})^2 + 3(3 + \sqrt{x}) - 10 = 0$

12. $(1 + \sqrt{x})^2 + 5(1 + \sqrt{x}) + 6 = 0$

13. $x^{-2} - x^{-1} - 6 = 0$ **14.** $2x^{-2} - x^{-1} - 1 = 0$

15. $4x^{-2} + x^{-1} - 5 = 0$ **16.** $m^{-2} + 9m^{-1} - 10 = 0$

17. $t^{2/3} + t^{1/3} - 6 = 0$ (*Hint:* Let $u = t^{1/3}$.)

18. $w^{2/3} - 2w^{1/3} - 8 = 0$

19. $y^{1/3} - y^{1/6} - 6 = 0$ **20.** $t^{1/2} + 3t^{1/4} + 2 = 0$

21. $t^{1/3} + 2t^{1/6} = 3$ **22.** $m^{1/2} + 6 = 5m^{1/4}$

23. $(3 - \sqrt{x})^2 - 10(3 - \sqrt{x}) + 23 = 0$

24. $(5 + \sqrt{x})^2 - 12(5 + \sqrt{x}) + 33 = 0$

25. $16\left(\dfrac{x - 1}{x - 8}\right)^2 + 8\left(\dfrac{x - 1}{x - 8}\right) + 1 = 0$

26. $9\left(\dfrac{x + 2}{x + 3}\right)^2 - 6\left(\dfrac{x + 2}{x + 3}\right) + 1 = 0$

Find all x-intercepts of the given function f.

27. $f(x) = 5x + 13\sqrt{x} - 6$

28. $f(x) = 3x + 10\sqrt{x} - 8$

29. $f(x) = (x^2 - 3x)^2 - 10(x^2 - 3x) + 24$

30. $f(x) = (x^2 - 6x)^2 - 2(x^2 - 6x) - 35$

31. $f(x) = x^{2/5} + x^{1/5} - 6$

32. $f(x) = x^{1/2} - x^{1/4} - 6$

33. $f(x) = \left(\dfrac{x^2 - 2}{x}\right)^2 - 7\left(\dfrac{x^2 - 2}{x}\right) - 18$

34. $f(x) = \left(\dfrac{x^2 - 1}{x}\right)^2 - 4\left(\dfrac{x^2 - 1}{x}\right) - 12$

SKILL MAINTENANCE

35. Multiply and simplify: $\sqrt{3x^2}\sqrt{3x^3}$.

36. Solution A is 18% alcohol and solution B is 45% alcohol. How much of each should be mixed together to get 12 L of a solution that is 36% alcohol?

37. Subtract: $\dfrac{x + 1}{x - 1} - \dfrac{x + 1}{x^2 + x + 1}$.

38. If $g(x) = x^2 - x$, find $g(a + 1)$.

SYNTHESIS

39. ◆ Describe a procedure that could be used to solve any equation of the form $ax^4 + bx^2 + c = 0$.

40. ◆ Describe a procedure that could be used to write an equation that is quadratic in $3x^2 + 1$. Then explain how the procedure could be adjusted to write equations that are quadratic in $3x^2 + 1$ and have no real-number solution.

Solve.

41. $3x^4 + 5x^2 - 1 = 0$

42. $5x^4 - 7x^2 + 1 = 0$

43. $(x^2 - 5x - 1)^2 - 18(x^2 - 5x - 1) + 65 = 0$

44. $(x^2 - 4x - 2)^2 - 13(x^2 - 4x - 2) + 30 = 0$

45. $\dfrac{x}{x - 1} - 6\sqrt{\dfrac{x}{x - 1}} - 40 = 0$

46. $\left(\sqrt{\dfrac{x}{x - 3}}\right)^2 - 24 = 10\sqrt{\dfrac{x}{x - 3}}$

47. $a^5(a^2 - 25) + 13a^3(25 - a^2) + 36a(a^2 - 25) = 0$

48. $a^3 - 26a^{3/2} - 27 = 0$

49. $x^6 - 28x^3 + 27 = 0$

50. $x^6 + 7x^3 - 8 = 0$

51. 📈 Use a grapher to check your answers to Exercises 1, 3, 29, and 43.

52. 📈 Use a grapher to solve
$$x^4 - x^3 - 13x^2 + x + 12 = 0.$$

53. ◆ 📈 While trying to solve $0.05x^4 - 0.8 = 0$ with a grapher, Murray gets the following screen.

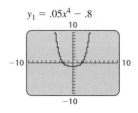

Can Murray solve this equation with a grapher? Why or why not?

8.6 Variation and Problem Solving

Direct Variation • Inverse Variation • Combined Variation

To extend our study of formulas and functions, we now examine three situations that frequently arise in problem solving: direct variation, inverse variation, and combined variation. These situations sometimes require the solution of a quadratic equation.

Direct Variation

A hair stylist earns $18 per hour. In 1 hr, $18 is earned. In 2 hr, $36 is earned. In 3 hr, $54 is earned, and so on. This gives rise to a set of ordered pairs of numbers:

(1, 18), (2, 36), (3, 54), (4, 72),   and so on.

The ratio of earnings to time is $\frac{18}{1}$ in every case.

Whenever a situation gives rise to pairs of numbers in which the ratio is constant, we say that there is **direct variation**. Here earnings *vary directly* as the time:

$E = 18t$ or, using function notation, $E(t) = 18t$.

DIRECT VARIATION

When a situation gives rise to a linear function of the form $f(x) = kx$, or $y = kx$, where k is a nonzero constant, we say that there is *direct variation*, that *y varies directly* as *x*, or that *y is proportional to x*. The number k is called the *variation constant*, or *constant of proportionality*.

The graph of $y = kx$, $k > 0$, always goes through the origin and rises from left to right. Note that as x increases, y increases.

The graph of a function describing direct variation is always a straight line.

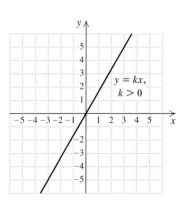

$y = kx,$
$k > 0$

EXAMPLE 1   Find the variation constant and an equation of variation if y varies directly as x, and $y = 32$ when $x = 2$.

SOLUTION   We know that $(2, 32)$ is a solution of $y = kx$. Therefore,

$$32 = k \cdot 2 \qquad \text{Substituting}$$

$$\frac{32}{2} = k, \quad \text{or} \quad k = 16. \qquad \text{Solving for } k$$

The variation constant is 16. The equation of variation is $y = 16x$. The notation $y(x) = 16x$ or $f(x) = 16x$ is also used.

EXAMPLE 2   *Water from Melting Snow.*   The number of centimeters W of water produced from melting snow varies directly as the number of centimeters S of snow. Meteorologists know that under certain conditions, 150 cm of snow will melt to 16.8 cm of water. How many centimeters of water will replace 200 cm of snow under these conditions?

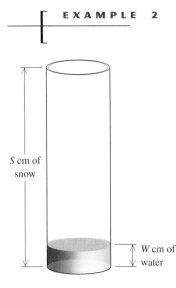

S cm of snow

W cm of water

SOLUTION

1. **Familiarize.**   Because of the phrase "W ... varies directly as ... S," we decide to express the amount of water as a function of the amount of snow. Thus, $W(S) = kS$, where k is the variation constant. Knowing that 150 cm of snow becomes 16.8 cm of water, we have $W(150) = 16.8$.

2. **Translate.**   We find the variation constant using the data and then find the equation of variation:

$$W(S) = kS$$
$$W(150) = k \cdot 150 \qquad \text{Replacing } S \text{ with } 150$$
$$16.8 = k \cdot 150 \qquad \text{Replacing } W(150) \text{ with } 16.8$$
$$\frac{16.8}{150} = k \qquad \text{Solving for } k$$
$$0.112 = k. \qquad \text{This is the variation constant.}$$

The equation of variation is $W(S) = 0.112S$. This is the translation.

3. **Carry out.**   To find how much water 200 cm of snow will become, we compute $W(200)$:

$$W(S) = 0.112S$$
$$W(200) = 0.112(200) \qquad \text{Substituting 200 for } S$$
$$= 22.4.$$

4. **Check.**   To check, we could reexamine all our calculations. Note that our answer seems reasonable since 200/22.4 and 150/16.8 are equal.

5. **State.**   200 cm of snow will melt into 22.4 cm of water.

Inverse Variation

To see what we mean by inverse variation, consider the following situation.

A bus is traveling a distance of 20 mi. At a speed of 20 mph, the trip will take 1 hr. At 40 mph, it will take $\frac{1}{2}$ hr. At 60 mph, it will take $\frac{1}{3}$ hr, and so on. This gives rise to a set of pairs of numbers, all having the same product:

$$(20, 1), \left(40, \tfrac{1}{2}\right), \left(60, \tfrac{1}{3}\right), \left(80, \tfrac{1}{4}\right), \quad \text{and so on.}$$

Whenever a situation gives rise to pairs of numbers whose product is constant, we say that there is **inverse variation**. The time t required for the bus to travel 20 mi at rate r is given by

$$t = \frac{20}{r} \quad \text{or, using function notation,} \quad t(r) = \frac{20}{r}.$$

INVERSE VARIATION

> When a situation gives rise to a function of the form $f(x) = k/x$, or $y = k/x$, where k is a nonzero constant, we say that there is *inverse variation*, that *y varies inversely as x*, or that *y is inversely proportional to x*. The number k is called the *variation constant*, or *constant of proportionality*.

Although we will not study such graphs until Chapter 10, it is helpful to look at the graph of $y = k/x$, for $k > 0$ and $x > 0$. The graph is like the one shown at right. Note that as x increases, y decreases.

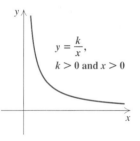

$y = \dfrac{k}{x}$,
$k > 0$ and $x > 0$

EXAMPLE 3   Find the variation constant and an equation of variation if y varies inversely as x, and $y = 32$ when $x = 0.2$.

SOLUTION   We know that $(0.2, 32)$ is a solution of

$$y = \frac{k}{x}.$$

Therefore,

$$32 = \frac{k}{0.2} \qquad \text{Substituting}$$

$$(0.2)32 = k$$

$$6.4 = k. \qquad \text{Solving for } k$$

The variation constant is 6.4. The equation of variation is

$$y = \frac{6.4}{x}.$$

There are many problems that translate to an equation of inverse variation.

EXAMPLE 4

Building a Shed. The time t required to do a job varies inversely as the number of people P who work on the job (assuming that all work at the same rate). It takes 4 hr for 12 people to build a woodshed. How long would it take 3 people to complete the same job?

SOLUTION

1. **Familiarize.** Because of the phrase "$t \ldots$ varies inversely as $\ldots P$," we express the amount of time required, in hours, as a function of the number of people working. Therefore we have $t(P) = k/P$. From the information given, we know that $t(12) = 4$. That is, 12 people take 4 hr to build the shed.

2. **Translate.** We find the variation constant using the data and then find the equation of variation:

$$t(P) = \frac{k}{P} \qquad \text{\small Using function notation}$$

$$t(12) = \frac{k}{12} \qquad \text{\small Replacing } P \text{ with } 12$$

$$4 = \frac{k}{12} \qquad \text{\small Replacing } t(12) \text{ with } 4$$

$$48 = k. \qquad \text{\small Solving for } k, \text{ the variation constant}$$

The equation of variation is $t(P) = 48/P$. This is the translation.

3. **Carry out.** To find how long it would take 3 people to complete the job, we compute $t(3)$:

$$t(P) = \frac{48}{P}$$

$$t(3) = \frac{48}{3} \qquad \text{\small Substituting 3 for } P$$

$$t = 16. \qquad \text{\small } t = 16 \text{ when } P = 3$$

4. Check. We could now recheck each step. Note that, as expected, as the number of people working goes *down*, the time required for the job goes *up*.

5. State.   It will take 3 people 16 hr to build a woodshed.   ——●

Combined Variation

Often one variable varies directly or inversely with more than one other variable. For example, in the formula for the volume of a right circular cylinder, $V = \pi r^2 h$, we say that V varies *jointly* as h and the square of r.

JOINT VARIATION

y varies *jointly* as x and z if, for some nonzero constant k, $y = kxz$.

EXAMPLE 5   Find an equation of variation if y varies jointly as x and z, and $y = 30$ when $x = 2$ and $z = 3$.

SOLUTION   We have

$$y = kxz,$$

so

$$30 = k \cdot 2 \cdot 3$$

$$k = 5. \qquad \text{The variation constant is 5.}$$

The equation of variation is $y = 5xz$.   ——●

EXAMPLE 6   Find an equation of variation if y varies jointly as x and z and inversely as the square of w, and $y = 105$ when $x = 3$, $z = 20$, and $w = 2$.

SOLUTION   The equation of variation is of the form

$$y = k \cdot \frac{xz}{w^2},$$

so, substituting, we have

$$105 = k \cdot \frac{3 \cdot 20}{2^2}$$

$$105 = k \cdot 15$$

$$k = 7.$$

Thus,

$$y = 7 \cdot \frac{xz}{w^2}.$$ ——●

EXAMPLE 7   *Volume of a Tree Trunk.* The volume of wood V in a tree trunk varies jointly as the height h and the square of the girth g (girth is distance around). If the volume is 35 ft^3 when the height is 20 ft and the girth is 5 ft, what is the girth when the volume is 85.75 ft^3 and the height is 25 ft?

SOLUTION

1. **Familiarize.** We make a table, including all given information and indicating the data we need to find.

	Volume of Wood	Height of Tree	Girth of Tree
Smaller Tree	35 ft^3	20 ft	5 ft
Larger Tree	85.75 ft^3	25 ft	g

Let h, g, and V represent the height, girth, and volume of a tree, respectively. We wish to determine g when V is 85.75 ft^3 and h is 25 ft. We know from the statement of the problem that in this situation the volume varies jointly as the height and the square of the girth.

2. **Translate.** First, we find k using the first set of data. Then we solve for g using the second set of data:

$$V = khg^2$$
$$35 = k \cdot 20 \cdot 5^2 \qquad \text{\footnotesize Using the given data}$$
$$0.07 = k. \qquad\qquad \text{\footnotesize This is the variation constant.}$$

The equation of variation is $V = 0.07hg^2$. This is the translation.

3. **Carry out.** We substitute and solve for g:

$$85.75 = 0.07 \cdot 25 \cdot g^2$$
$$\left. \begin{array}{r} 85.75 = 1.75g^2 \\ 49 = g^2 \\ \pm 7 = g. \end{array} \right\} \quad \text{\footnotesize Solving the quadratic equation}$$

We could have first solved the formula for g and then substituted; either approach is valid.

4. **Check.** We should now recheck all our calculations and perhaps make an estimate to see whether our answer is reasonable. This is left to the student. Since the girth of a tree must be positive, we accept only 7 as a solution. This seems to be a reasonable figure.

5. **State.** The girth of the 25-ft tree is 7 ft.

EXERCISE SET

8.6

Find the variation constant and an equation of variation if y varies directly as x and the following conditions exist.

1. $y = 28$ when $x = 7$

2. $y = 5$ when $x = 12$

3. $y = 3.4$ when $x = 2$

4. $y = 2$ when $x = 5$

5. $y = 30$ when $x = 8$

6. $y = 1$ when $x = \frac{1}{3}$

7. $y = 0.8$ when $x = 0.5$

8. $y = 0.6$ when $x = 0.4$

Solve.

9. *Ohm's law.* The electric current I, in amperes, in a circuit varies directly as the voltage V. When 15 volts are applied, the current is 5 amperes. What is the current when 18 volts are applied?

10. *Hooke's law.* Hooke's law states that the distance d that a spring is stretched by a hanging object varies directly as the mass m of the object. If the distance is 20 cm when the mass is 3 kg, what is the distance when the mass is 5 kg?

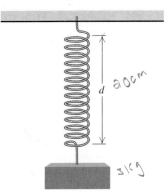

11. *Use of aluminum cans.* The number N of aluminum cans used each year varies directly as the number of people using the cans. If 250 people use 60,000 cans in one year, how many cans are used each year in Dallas, which has a population of 1,008,000?

12. *Weekly allowance.* According to Fidelity Investments *Investment Vision Magazine*, the average weekly allowance A of children varies directly as their grade level, G. In a recent year, the average allowance of a 9th-grade student was $9.66 per week. What was the average weekly allowance of a 4th-grade student?

13. *Mass of water in body.* The number of kilograms W of water in a human body varies directly as the mass of the body. A 96-kg person contains 64 kg of water. How many kilograms of water are in a 60-kg person?

14. *Weight on Mars.* The weight M of an object on Mars varies directly as its weight E on Earth. A person who weighs 95 lb on Earth weighs 38 lb on Mars. How much would a 100-lb person weigh on Mars?

15. *Lead pollution.* The average U.S. community of population 12,500 released about 385 tons of lead into the environment in a recent year.* How many tons were released nationally? Use 250,000,000 as the U.S. population.

16. *Relative aperture.* The relative aperture, or f-stop, of a 23.5-mm lens is directly proportional to the focal length F of the lens. If a lens with a 150-mm focal length has an f-stop of 6.3, find the f-stop of a 23.5-mm lens with a focal length of 80 mm.

Conservation Matters, Autumn 1995 issue. (Boston: Conservation Law Foundation), p. 30.

Find the variation constant and an equation of variation in which y varies inversely as x, and the following conditions exist.

17. $y = 6$ when $x = 10$

18. $y = 16$ when $x = 4$

19. $y = 4$ when $x = 3$

20. $y = 4$ when $x = 9$

21. $y = 12$ when $x = 3$

22. $y = 9$ when $x = 5$

23. $y = 27$ when $x = \frac{1}{3}$

24. $y = 81$ when $x = \frac{1}{9}$

Solve.

25. *Pumping rate.* The time t required to empty a tank varies inversely as the rate r of pumping. If a pump can empty a tank in 45 min at the rate of 600 kL/min, how long will it take the pump to empty the same tank at the rate of 1000 kL/min?

26. *Current and resistance.* The current I in an electrical conductor varies inversely as the resistance R of the conductor. If the current is $\frac{1}{2}$ ampere when the resistance is 240 ohms, what is the current when the resistance is 540 ohms?

27. *Volume and pressure.* The volume V of a gas varies inversely as the pressure P upon it. The volume of a gas is 200 cm^3 under a pressure of 32 kg/cm^2. What will be its volume under a pressure of 40 kg/cm^2?

28. *Work rate.* The time T required to do a job varies inversely as the number of people P working. It takes 5 hr for 7 bricklayers to build a park wall. How long will it take 10 bricklayers to complete the job?

29. *Wavelength and frequency.* The wavelength W of a radio wave varies inversely as its frequency F. A wave with a frequency of 1200 kilohertz has a length of 300 meters. What is the length of a wave with a frequency of 800 kilohertz?

30. *Rate of travel.* The time t required to drive a fixed distance varies inversely as the speed r. It takes 5 hr at a speed of 80 km/h to drive a fixed distance. How long will it take to drive the same distance at a speed of 70 km/h?

Find an equation of variation in which:

31. y varies directly as the square of x, and $y = 6$ when $x = 3$.

32. y varies directly as the square of x, and $y = 0.15$ when $x = 0.1$.

33. y varies inversely as the square of x, and $y = 6$ when $x = 3$.

34. y varies inversely as the square of x, and $y = 0.15$ when $x = 0.1$.

35. y varies jointly as x and z, and $y = 56$ when $x = 14$ and $z = 8$.

36. y varies directly as x and inversely as z, and $y = 4$ when $x = 12$ and $z = 15$.

37. y varies jointly as x and the square of z, and $y = 105$ when $x = 14$ and $z = 5$.

38. y varies jointly as x and z and inversely as w, and $y = \frac{3}{2}$ when $x = 2$, $z = 3$, and $w = 4$.

39. y varies jointly as w and the square of x and inversely as z, and $y = 49$ when $w = 3$, $x = 7$, and $z = 12$.

40. y varies directly as x and inversely as w and the square of z, and $y = 4.5$ when $x = 15$, $w = 5$, and $z = 2$.

41. y varies jointly as x and z and inversely as the product of w and p, and $y = \frac{3}{28}$ when $x = 3$, $z = 10$, $w = 7$, and $p = 8$.

42. y varies jointly as x and z and inversely as the square of w, and $y = \frac{12}{5}$ when $x = 16$, $z = 3$, and $w = 5$.

Solve.

43. *Stopping distance of a car.* The stopping distance d of a car after the brakes have been applied varies directly as the square of the speed r. If a car traveling 60 mph can stop in 200 ft, how fast can a car go and still stop in 72 ft?

44. *Volume of a gas.* The volume V of a given mass of a gas varies directly as the temperature T and inversely as the pressure P. If $V = 231$ cm^3 when $T = 42°$ and $P = 20$ kg/cm^2, what is the volume when $T = 30°$ and $P = 15$ kg/cm^2?

45. *Intensity of light.* The intensity I of light from a light bulb varies inversely as the square of the distance d from the bulb. Suppose I is 90 W/m^2 (watts per square meter) when the distance is 5 m. How much *further* would it be to a point where the intensity is 40 W/m^2?

46. *Intensity of a signal.* The intensity I of a television signal varies inversely as the square of the distance d from the transmitter. If the intensity is 25 W/m^2 at a distance of 2 km, how far from the transmitter are you when the intensity is 2.56 W/m^2?

47. *Weight of an astronaut.* The weight W of an object varies inversely as the square of the distance d from the center of the earth. At sea level (6400 km from the center of the earth), an astronaut weighs 100 lb. How far *above the earth* must the astronaut be in order to weigh 64 lb?

48. *Electrical resistance.* At a fixed temperature, the resistance R of a wire varies directly as the length l and inversely as the square of its diameter d. If the resistance is 0.1 ohm when the diameter is 1 mm and the length is 50 cm, what is the diameter when the resistance is 1 ohm and the length is 2000 cm?

49. ▦ *Atmospheric drag.* Wind resistance, or atmospheric drag, tends to slow down moving objects. Atmospheric drag varies jointly as an object's surface area A and velocity v. If a car traveling at a speed of 40 mph with a surface area of 37.8 ft^2 experiences a drag of 222 N (Newtons), how fast must a car with 51 ft^2 of surface area travel in order to experience a drag force of 430 N?

50. ▦ *Drag force.* The drag force F on a boat varies jointly as the wetted surface area A and the square of the velocity of the boat. If a boat going 6.5 mph experiences a drag force of 86 N when the wetted surface area is 41.2 ft^2, how fast must a boat with 28.5 ft^2 of wetted surface area go in order to experience a drag force of 94 N?

SKILL MAINTENANCE

51. Give an equation for a line with slope $-\frac{2}{3}$ and y-intercept $(0, -5)$.

52. Give an equation in point–slope form for the line passing through the point $(4, 7)$ with slope $-\frac{2}{7}$.

Simplify.

53. $\dfrac{\dfrac{1}{ab} - \dfrac{2}{bc}}{\dfrac{3}{ab} + \dfrac{4}{bc}}$

54. $\dfrac{\dfrac{3}{x^2 - 4}}{\dfrac{1}{3x + 6} - \dfrac{2}{4x - 8}}$

55. If $f(x) = x^3 - 2x^2$, find $f(3)$.

56. Multiply: $(3x - 2y)^2$.

SYNTHESIS

57. ◈ If y varies directly as x^2, explain why doubling x would not cause y to be doubled as well.

58. ◈ If y varies directly as x and x varies inversely as z, how does y vary with regard to z? Why?

59. ◈ Write a variation problem for a classmate to solve. Design the problem so that the answer is "When Simone studies for 8 hr a week, her quiz score is 92."

60. ◈ Suppose that the number of customer complaints is inversely proportional to the number of employees hired. Will a firm reduce the number of complaints more by expanding from 5 to 10 employees, or from 20 to 25? Explain. Consider using a graph to help justify your answer.

61. If y varies inversely as the cube of x and x is multiplied by 0.5, what is the effect on y?

Describe, in words, the variation given by the equation.

62. $Q = \dfrac{kp^2}{q^3}$

63. $W = \dfrac{km_1 M_1}{d^2}$

64. ▦ *Tension of a musical string.* The tension T on a string in a musical instrument varies jointly as the string's mass per unit length m, the square of its length l, and the square of its fundamental frequency f. A 2-m-long string of mass 5 gm/m with a fundamental frequency of 80 has a tension of 100 N. How long should the same string be if its tension is going to be changed to 72 N?

65. ▦ *The gravity model.* It has been determined that the average number of telephone calls in a day N, between two cities, is directly proportional to the populations P_1 and P_2 of the cities and inversely proportional to the square of the distance between the cities. This model is called the *gravity model* because the equation of variation resembles the equation that applies to Newton's law of gravitation.

a) In 1994, the population of Indianapolis was 752,279 and the population of Cincinnati was 358,170. The average number of daily phone calls between the two cities was 11,153. Find the value k and write the equation of variation given that the cities are 174 km apart.

b) In 1994, the average number of daily phone calls between Indianapolis and New York was 4270, and the population of New York was 7,333,153. Estimate the distance between Indianapolis and New York.

66. *Volume and cost.* A peanut butter jar in the shape of a right circular cylinder is 4 in. high and 3 in. in diameter and sells for $1.20. If we assume that cost is proportional to volume, how much should a jar 6 in. high and 6 in. in diameter cost?

67. *Golf distance finder.*   A device used in golf to estimate the distance d to a hole measures the size s that the 7-ft pin *appears* to be in a viewfinder. The viewfinder uses the principle, diagrammed here, that s gets bigger when d gets smaller. If $s = 0.56$ in. when $d = 50$ yd, find an equation of variation that expresses d as a function of s. What is d when $s = 0.40$ in.?

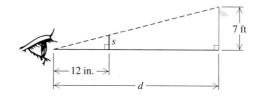

HOW IT WORKS:

Just sight the flagstick through the viewfinder…
fit flag between top dashed line and the solid line below…
…read the distance, 50 – 220 yards.

Nothing to focus.
•
Gives you exact distance that your ball
lies from the flagstick.
•
Choose proper club on every approach shot.
•
Figure new pin placement instantly.
•
Train your naked eye for formal and tournament play.
•
Eliminate the need to remember every stake,
tree, and bush on the course.

COLLABORATIVE
C•O•R•N•E•R

Focus: Direct variation and estimation

Time: 15 minutes

Group size: 2 or 3 and entire class

The National Park Service's estimates of crowd sizes for static (stationary) mass demonstrations vary directly as the area covered by the crowd. Park Service officials have found that at basic "shoulder-to-shoulder" demonstrations, 1 acre of land (about 45,000 ft^2) holds about 9000 people. Using aerial photographs, officials impose a grid to estimate the total area covered by the demonstrators. Once this has been accomplished, estimates of crowd size can be prepared.

ACTIVITY

1. In the grid imposed on the photograph on the following page, each square represents 10,000 ft^2. Estimate the size of the crowd photographed. Then compare your group's estimate with those of other groups. What might explain discrepancies between estimates? List ways in which your group's estimate could be made more accurate.

2. Park Service officials use an "acceptable margin of error" of no more than 20%. Using all estimates from part (1) above and allowing for error, find a range of values within which you feel certain that the actual crowd size lies.

3. The Million-Man March of 1995 was not a static demonstration because of a periodic turnover of people in attendance (many people stayed for only part of the day's festivities). How might you change your methodology to compensate for this complication?

8.7   Quadratic Functions and Their Graphs

Graphs of $f(x) = ax^2$ • **Graphs of** $f(x) = a(x - h)^2$ • **Graphs of** $f(x) = a(x - h)^2 + k$

We have already used some quadratic functions when we solved equations earlier in this chapter. In this section and the next, we develop techniques for graphing such functions.

Graphs of $f(x) = ax^2$

The most basic quadratic function is $f(x) = x^2$.

EXAMPLE 1   Graph: $f(x) = x^2$.

SOLUTION   We choose some values for x and compute $f(x)$ for each. Then we plot the ordered pairs and connect them with a smooth curve.

TECHNOLOGY CONNECTION 8.7A

To examine the effect of a when graphing $f(x) = ax^2$, draw the function $y_1 = x^2$ in a $[-5, 5, -10, 10]$ window. Without erasing this curve, graph the function $y_2 = 3x^2$. How do the graphs compare? Now include the graph of $y_3 = \frac{1}{3}x^2$. Find a rule that describes the effect of multiplying x^2 by a, when $a > 1$ and when $0 < a < 1$.

Clear the display and graph $y_1 = x^2$ again. Now include the graph of $y_2 = -x^2$. Note how it differs from $y_1 = x^2$. Next, graph $y_3 = \frac{2}{3}x^2$ and $y_4 = -\frac{2}{3}x^2$ and compare them. Find a rule that describes the effect of multiplying x^2 by a, when $a < -1$ and when $-1 < a < 0$.

x	$f(x) = x^2$	$(x, f(x))$
-3	9	$(-3, 9)$
-2	4	$(-2, 4)$
-1	1	$(-1, 1)$
0	0	$(0, 0)$
1	1	$(1, 1)$
2	4	$(2, 4)$
3	9	$(3, 9)$

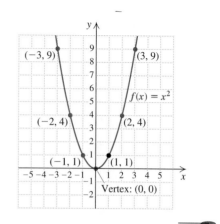

All quadratic functions have graphs similar to the one in Example 1. Such curves are called *parabolas*. They are cup-shaped curves that are symmetric with respect to a vertical line known as the parabola's *axis of symmetry*. In the graph of $f(x) = x^2$, the y-axis (or the line $x = 0$) is the axis of symmetry. Were the paper folded on this line, the two halves of the curve would match. The point $(0, 0)$ is known as the *vertex* of this parabola.

By plotting points, we can compare the graphs of $g(x) = \frac{1}{2}x^2$ and $h(x) = 2x^2$ with the graph of $f(x) = x^2$.

x	$h(x) = 2x^2$
-3	18
-2	8
-1	2
0	0
1	2
2	8
3	18

x	$g(x) = \frac{1}{2}x^2$
-3	$\frac{9}{2}$
-2	2
-1	$\frac{1}{2}$
0	0
1	$\frac{1}{2}$
2	2
3	$\frac{9}{2}$

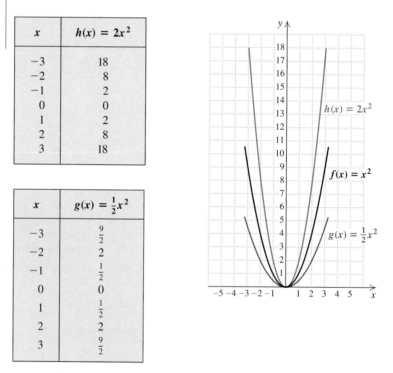

Note that the graph of $g(x) = \frac{1}{2}x^2$ is a flatter parabola than the graph of $f(x) = x^2$, and the graph of $h(x) = 2x^2$ is narrower. The vertex and the axis of symmetry, however, remain $(0, 0)$ and $x = 0$, respectively.

When we consider the graph of $k(x) = -\frac{1}{2}x^2$, we see that the parabola opens downward and is the same shape as the graph of $g(x) = \frac{1}{2}x^2$.

x	$k(x) = -\frac{1}{2}x^2$
-3	$-\frac{9}{2}$
-2	-2
-1	$-\frac{1}{2}$
0	0
1	$-\frac{1}{2}$
2	-2
3	$-\frac{9}{2}$

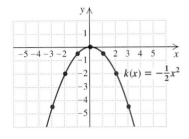

GRAPHING $f(x) = ax^2$

The graph of $f(x) = ax^2$ is a parabola with $x = 0$ as its axis of symmetry; its vertex is the origin.

For $a > 0$, the parabola opens upward; for $a < 0$, the parabola opens downward.

If $|a|$ is greater than 1, the parabola is narrower than $y = x^2$.

If $|a|$ is between 0 and 1, the parabola is wider than $y = x^2$.

Graphs of $f(x) = a(x - h)^2$

Why not now consider graphs of

$$f(x) = ax^2 + bx + c,$$

where b and c are not both 0? In effect, we will do that, but in a disguised form. It turns out to be convenient to first graph $f(x) = a(x - h)^2$.* This allows us to observe similarities to the graphs drawn above.

---

*The letters h and k are often used to name functions, in which case the notation $h(x)$ and $k(x)$ is used. When h and k appear in expressions like $f(x) = a(x - h)^2 + k$, assume that they represent constants.

TECHNOLOGY CONNECTION
8.7B

To investigate the effect of h on the graph of $f(x) = a(x - h)^2$, let $y_1 = 7x^2$ and $y_2 = 7(x - 1)^2$. Graph both y_1 and y_2 in the window $[-5, 5, -5, 5]$ and compare. If a TABLE feature is available, compare y-values, beginning at $x = 1$ and increasing x by one unit at a time.

Next, let $y_3 = 7(x - 2)^2$ and compare its graph and y-values with those of y_1 and y_2.

Finally, replace y_2 and y_3 with $y_2 = 7(x + 1)^2$ and $y_3 = 7(x + 2)^2$. Compare graphs and y-values and describe the effect of h on the graph of $f(x) = a(x - h)^2$.

EXAMPLE   2   Graph: $f(x) = (x - 3)^2$.

SOLUTION   We choose some values for x and compute $f(x)$. Then we plot the points and draw the curve. Comparing the pairs with those found in Example 1, we see that when an input here is 3 more than an input for Example 1, the outputs match.

x	$f(x) = (x - 3)^2$
-1	16
0	9
1	4
2	1
3	0
4	1
5	4
6	9

←——— Vertex

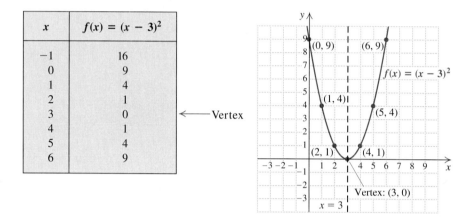

Note that $f(x) = 16$ when $x = -1$ and $f(x)$ gets larger as x gets more negative. Thus we use positive values to fill out the table. Note that the line $x = 3$ is now the axis of symmetry and the point $(3, 0)$ is the vertex. Had we known earlier that $x = 3$ is the axis of symmetry, we could have computed some values on one side, such as $(4, 1)$, $(5, 4)$, and $(6, 9)$, and then used symmetry to get their mirror images $(2, 1)$, $(1, 4)$, and $(0, 9)$ without further computation.

The graph of $f(x) = (x - 3)^2$ looks just like the graph of $f(x) = x^2$, except that it is moved, or *translated*, 3 units to the right.

GRAPHING $f(x) = a(x - h)^2$

The graph of $f(x) = a(x - h)^2$ has the same shape as the graph of $y = ax^2$.

If h is positive, the graph of $y = ax^2$ is shifted h units to the right.

If h is negative, the graph of $y = ax^2$ is shifted $|h|$ units to the left.

The vertex is $(h, 0)$ and the axis of symmetry is $x = h$.

EXAMPLE   3   Graph: $g(x) = -2(x + 3)^2$.

SOLUTION   We rewrite the equation as $g(x) = -2[x - (-3)]^2$. In this case, $a = -2$ and $h = -3$, so the graph looks like that of $y = 2x^2$ translated 3 units to the left and, since $-2 < 0$, flipped upside down. The vertex is $(-3, 0)$, and the axis of symmetry is $x = -3$. Plotting points as needed, we obtain the graph shown at the top of the next page.

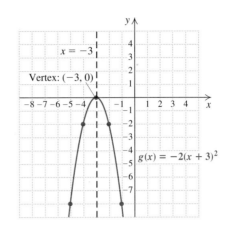

Graphs of $f(x) = a(x - h)^2 + k$

Given a graph of $f(x) = a(x - h)^2$, what happens if we add a constant k? Suppose that we add 2. This increases each function value $f(x)$ by 2, so the curve is moved up. If k is negative, the curve is moved down. The axis of symmetry for the parabola remains $x = h$, but the vertex will be at (h, k), or, equivalently, $(h, f(h))$.

Note that if a parabola opens upward ($a > 0$), the function value, or y-value, at the vertex is a least, or *minimum*, value. That is, it is less than the y-value at any other point on the graph. If the parabola opens downward ($a < 0$), the function value at the vertex is a greatest, or *maximum*, value.

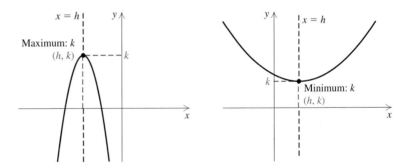

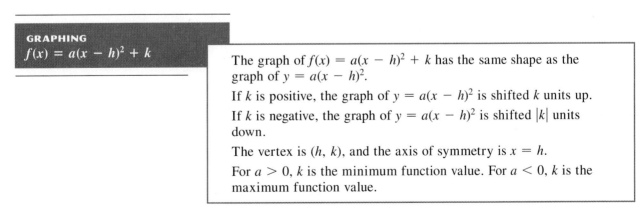

GRAPHING
$f(x) = a(x - h)^2 + k$

The graph of $f(x) = a(x - h)^2 + k$ has the same shape as the graph of $y = a(x - h)^2$.

If k is positive, the graph of $y = a(x - h)^2$ is shifted k units up.

If k is negative, the graph of $y = a(x - h)^2$ is shifted $|k|$ units down.

The vertex is (h, k), and the axis of symmetry is $x = h$.

For $a > 0$, k is the minimum function value. For $a < 0$, k is the maximum function value.

TECHNOLOGY CONNECTION 8.7C

To investigate the effect of k on the graph of $f(x) = a(x - h)^2 + k$, let $y_1 = 7(x - 1)^2$ and $y_2 = 7(x - 1)^2 + 2$. Graph both y_1 and y_2 in the window $[-5, 5, -5, 5]$ and compare. If possible, also compare y-values in a table.

Next, let $y_3 = 7(x - 1)^2 - 4$ and compare its graph and y-values with those of y_1 and y_2. Try other values of k, including decimals and fractions. Describe the effect of adding k to the right side of $f(x) = a(x - h)^2$.

E X A M P L E   4

Graph $g(x) = (x - 3)^2 - 5$, and find the minimum function value.

S O L U T I O N   The graph will look like that of $f(x) = (x - 3)^2$ (see Example 2) but shifted 5 units down. You can confirm this by plotting some points. For instance, $g(4) = (4 - 3)^2 - 5 = -4$, whereas in Example 2, $f(4) = (4 - 3)^2 = 1$.

The vertex is now $(3, -5)$, and the minimum function value is -5.

x	$g(x) = (x - 3)^2 - 5$
0	4
1	-1
2	-4
3	-5
4	-4
5	-1
6	4

← Vertex

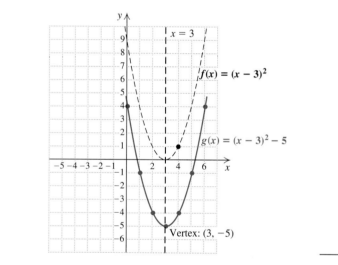

E X A M P L E   5

Graph $h(x) = \frac{1}{2}(x - 3)^2 + 5$, and find the minimum function value.

S O L U T I O N   The graph looks just like that of $f(x) = \frac{1}{2}x^2$ but moved 3 units to the right and 5 units up. The vertex is $(3, 5)$, and the axis of symmetry is $x = 3$. We draw $f(x) = \frac{1}{2}x^2$ and then shift the curve over and up. The minimum function value is 5. By plotting some points, we have a check.

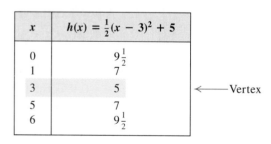

x	$h(x) = \frac{1}{2}(x - 3)^2 + 5$
0	$9\frac{1}{2}$
1	7
3	5
5	7
6	$9\frac{1}{2}$

←——Vertex

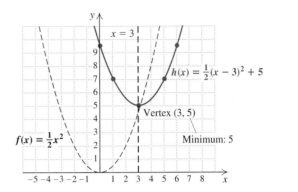

EXAMPLE 6   Graph $y = -2(x + 3)^2 + 5$. Find the vertex, the axis of symmetry, and the maximum or minimum value.

SOLUTION   We first express the equation in the equivalent form

$$y = -2[x - (-3)]^2 + 5.$$

The graph looks like that of $y = -2x^2$ translated 3 units to the left and 5 units up. The vertex is $(-3, 5)$, and the axis of symmetry is $x = -3$. Since $-2 < 0$, we know that 5, the second coordinate of the vertex, is the maximum y-value.

We compute a few points as needed. The graph is shown here.

x	$y = -2(x + 3)^2 + 5$
-4	3
-3	5
-2	3

← Vertex

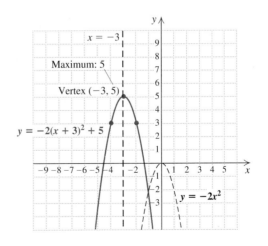

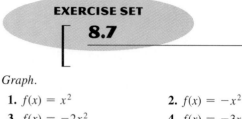

EXERCISE SET

8.7

Graph.

1. $f(x) = x^2$

2. $f(x) = -x^2$

3. $f(x) = -2x^2$

4. $f(x) = -3x^2$

5. $g(x) = \frac{1}{4}x^2$

6. $g(x) = \frac{1}{3}x^2$

7. $h(x) = -\frac{1}{3}x^2$

8. $h(x) = -\frac{1}{4}x^2$

9. $f(x) = \frac{3}{2}x^2$

10. $f(x) = \frac{5}{2}x^2$

For each of the following, graph the function, label the vertex, and draw the axis of symmetry.

11. $g(x) = (x + 1)^2$

12. $g(x) = (x + 4)^2$

13. $f(x) = (x - 2)^2$

14. $f(x) = (x - 1)^2$

15. $h(x) = (x - 3)^2$

16. $h(x) = (x - 4)^2$

17. $f(x) = -(x + 4)^2$

18. $f(x) = -(x - 2)^2$

19. $g(x) = -(x - 1)^2$

20. $g(x) = -(x + 5)^2$

21. $f(x) = 2(x + 1)^2$

22. $f(x) = 2(x + 4)^2$

23. $h(x) = -\frac{1}{2}(x - 3)^2$

24. $h(x) = -\frac{3}{2}(x - 2)^2$

25. $f(x) = \frac{1}{2}(x - 1)^2$

26. $f(x) = \frac{1}{3}(x + 2)^2$

27. $f(x) = -2(x + 5)^2$

28. $f(x) = 2(x + 7)^2$

29. $h(x) = -3\left(x - \frac{1}{2}\right)^2$

30. $h(x) = -2\left(x + \frac{1}{2}\right)^2$

For each of the following, graph the function and find the vertex, the axis of symmetry, and the maximum value or the minimum value.

31. $f(x) = (x - 5)^2 + 1$

32. $f(x) = (x + 3)^2 - 2$

33. $f(x) = (x + 1)^2 - 2$

34. $f(x) = (x - 1)^2 + 2$

35. $g(x) = (x + 4)^2 + 1$

36. $g(x) = -(x - 2)^2 - 4$

37. $h(x) = -2(x - 1)^2 - 3$

38. $h(x) = -2(x + 1)^2 + 4$

39. $f(x) = 2(x + 4)^2 + 1$

40. $f(x) = 2(x - 5)^2 - 3$

41. $g(x) = -\frac{3}{2}(x - 1)^2 + 2$

42. $g(x) = \frac{3}{2}(x + 2)^2 - 1$

Without graphing, find the vertex, the axis of symmetry, and the maximum value or the minimum value.

43. $f(x) = 8(x - 9)^2 + 5$

44. $f(x) = 10(x + 5)^2 - 8$

45. $h(x) = -\frac{2}{7}(x + 6)^2 + 11$

46. $h(x) = -\frac{3}{11}(x - 7)^2 - 9$

47. $f(x) = 5\left(x + \frac{1}{4}\right)^2 - 13$

48. $f(x) = 6\left(x - \frac{1}{4}\right)^2 + 19$

49. $f(x) = \sqrt{2}(x + 4.58)^2 + 65\pi$

50. $f(x) = 4\pi(x - 38.2)^2 - \sqrt{34}$

SKILL MAINTENANCE

Solve each system.

51. $3x + 4y = -19,$
$7x - 6y = -29$

52. $5x + 7y = 9,$
$3x - 4y = -11$

Complete the square.

53. $x^2 + 5x + \underline{\hspace{1cm}}$

54. $x^2 - 9x + \underline{\hspace{1cm}}$

SYNTHESIS

55. ◈ Explain, without plotting points, why the graph of $y = (x + 2)^2$ looks like the graph of $y = x^2$ translated 2 units to the left.

56. ◈ Explain, without plotting points, why the graph of $y = x^2 - 4$ looks like the graph of $y = x^2$ translated 4 units down.

57. ◈ Before graphing a quadratic function, Sophie always plots five points. First, she calculates and plots the coordinates of the vertex. Then she plots *four* more points after calculating *two* more ordered pairs. How is this possible?

58. ◈ If the graphs of $f(x) = a_1(x - h_1)^2 + k_1$ and $g(x) = a_2(x - h_2)^2 + k_2$ have the same shape, what, if anything, can you conclude about the a's, the h's, and the k's? Why?

For each of the following, write the equation of the parabola that has the shape of $f(x) = 2x^2$ or $g(x) = -2x^2$ and has a maximum or minimum value at the specified point.

59. Minimum: $(5, 0)$

60. Minimum: $(2, 0)$

61. Maximum: $(-4, 0)$

62. Maximum: $(0, 3)$

63. Maximum: $(3, 8)$

64. Minimum: $(-2, 3)$

Find an equation for a quadratic function F that satisfies the following conditions.

65. The graph of F is the same shape as the graph of f, where $f(x) = 3(x + 2)^2 + 7$, and $F(x)$ is a minimum at the same point that $g(x) = -2(x - 5)^2 + 1$ is a maximum.

66. The graph of F is the same shape as the graph of f, where $f(x) = -\frac{1}{3}(x - 2)^2 + 7$, and $F(x)$ is a maximum at the same point that $g(x) = 2(x + 4)^2 - 6$ is a minimum.

Functions other than parabolas can be translated. For a function $f(x)$, if we replace x with $x - h$, where h is a constant, the graph will be moved horizontally. If we add a constant k to a function $f(x)$, the graph will be moved vertically.

Use the graph of the function $y = f(x)$ below for Exercises 67–72.

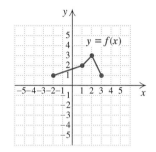

Draw a graph of each of the following.

67. $y = f(x - 1)$ **68.** $y = f(x + 2)$

69. $y = f(x) + 2$ **70.** $y = f(x) - 3$

71. $y = f(x + 3) - 2$ **72.** $y = f(x - 3) + 1$

73. 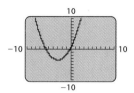 Use the TRACE and/or TABLE features of a grapher to confirm the maximum and minimum values given as answers to Exercises 43, 45, and 47. Be sure to adjust the window appropriately. On some graphers, a maximum or minimum option may be available by using a CALC key.

74. Use a grapher to check your graphs for Exercises 10, 20, and 40.

75. While trying to graph $y = -\frac{1}{2}x^2 + 3x + 1$, Omar gets the following screen.

How can Omar tell at a glance that a mistake has been made?

COLLABORATIVE
C◆O◆R◆N◆E◆R

Focus: Graphing quadratic functions

Time: 15–20 minutes

Group size: 6

ACTIVITY

1. On each of six scraps of paper, write one of the following equations:

$$y = \frac{1}{2}(x - 3)^2 + 1; \quad y = \frac{1}{2}(x - 1)^2 + 3;$$
$$y = \frac{1}{2}(x + 1)^2 - 3; \quad y = \frac{1}{2}(x + 3)^2 + 1;$$
$$y = \frac{1}{2}(x + 3)^2 - 1; \quad y = \frac{1}{2}(x + 1)^2 + 3.$$

2. Fold each scrap of paper and mix up the six scraps in a hat or bag. Then, one by one, each group member should select one of the equations.

3. Each group member should carefully graph the equation selected. Make the graph large enough so that when it is finished, it can be easily viewed by the rest of the group. Be sure to scale the axes, but **do not label the graph with the equation used.**

4. When all group members have drawn a graph, place the graphs in a pile. The group should then agree on the correct equation for each graph *with no help from the person who drew the graph.*

5. Compare your group's labeled graphs with those of other groups to reach consensus within the class on the correct label for each graph.

8.8 More About Graphing Quadratic Functions

Completing the Square   •   Finding Intercepts

Completing the Square

By *completing the square* (see Section 8.1), we can rewrite any polynomial $ax^2 + bx + c$ in the form $a(x - h)^2 + k$. Thus the procedures discussed in Section 8.7 enable us to graph any quadratic function.

EXAMPLE 1   Graph: $g(x) = x^2 - 6x + 4$.

SOLUTION   We have

$$g(x) = x^2 - 6x + 4$$
$$= (x^2 - 6x) + 4.$$

To complete the square inside the parentheses, we take half the x-coefficient, $\frac{1}{2} \cdot (-6) = -3$, and square it to get $(-3)^2 = 9$. Then we add $9 - 9$ inside the parentheses:

$$g(x) = (x^2 - 6x + 9 - 9) + 4 \qquad \text{The effect is of adding 0.}$$
$$= (x^2 - 6x + 9) + (-9 + 4) \qquad \text{Using the associative law of addition to regroup}$$
$$= (x - 3)^2 - 5. \qquad \text{Factoring and simplifying}$$

This equation was graphed in Example 4 of Section 8.7. The vertex is $(3, -5)$, and the axis of symmetry is $x = 3$.

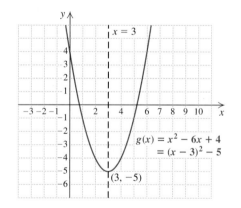

When the leading coefficient is not 1, we factor out that number from the first two terms. Then we complete the square.

EXAMPLE   2

Graph: $f(x) = 3x^2 + 12x + 13$.

SOLUTION   Since the coefficient of x^2 is not 1, we need to factor out that number—in this case, 3—from the first two terms. Remember that we want the form $f(x) = a(x - h)^2 + k$:

$$f(x) = 3x^2 + 12x + 13$$
$$= 3(x^2 + 4x) + 13.$$

Now we complete the square as before. We take half of the x-coefficient, $\frac{1}{2} \cdot 4 = 2$, and square it: $2^2 = 4$. Then we add $4 - 4$ inside the parentheses:

$$f(x) = 3(x^2 + 4x + 4 - 4) + 13. \qquad \text{Adding } 4 - 4, \text{ or } 0, \text{ inside the parentheses}$$

The distributive law allows us to separate the -4 from the perfect-square trinomial so long as it is multiplied by 3:

$$f(x) = 3(x^2 + 4x + 4) + 3(-4) + 13 \qquad \text{This leaves a perfect-square trinomial in the parentheses.}$$
$$= 3(x + 2)^2 + 1. \qquad \text{Factoring and simplifying}$$

The vertex is $(-2, 1)$, and the axis of symmetry is $x = -2$. The coefficient of x^2 is 3, so the graph is narrow and opens upward. We choose a few x-values on either side of the vertex, compute y-values, and then graph the parabola.

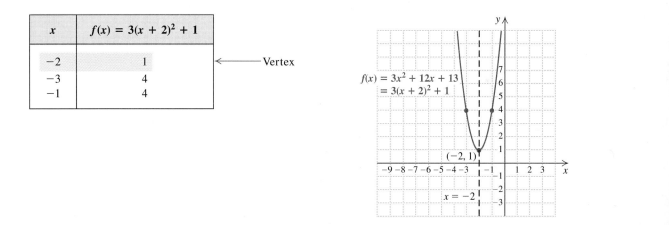

x	$f(x) = 3(x + 2)^2 + 1$
-2	1
-3	4
-1	4

← Vertex

$f(x) = 3x^2 + 12x + 13$
$= 3(x + 2)^2 + 1$

$(-2, 1)$

$x = -2$

EXAMPLE   3

Graph: $f(x) = -2x^2 + 10x - 7$.

SOLUTION   We first find the vertex by completing the square. To do so, we factor out -2 from the first two terms of the expression. This makes the coefficient of x^2 inside the parentheses 1:

$$f(x) = -2x^2 + 10x - 7$$
$$= -2(x^2 - 5x) - 7.$$

Now we complete the square as before. We take half of the x-coefficient and square it to get $\frac{25}{4}$. Then we add $\frac{25}{4} - \frac{25}{4}$ inside the parentheses:

$$f(x) = -2\left(x^2 - 5x + \frac{25}{4} - \frac{25}{4}\right) - 7$$

$$= -2\left(x^2 - 5x + \frac{25}{4}\right) + (-2)\left(-\frac{25}{4}\right) - 7 \quad \text{Multiplying by } -2, \text{ using the distributive law, and regrouping}$$

$$= -2\left(x - \frac{5}{2}\right)^2 + \frac{11}{2}. \quad \text{Factoring and simplifying}$$

The vertex is $\left(\frac{5}{2}, \frac{11}{2}\right)$, and the axis of symmetry is $x = \frac{5}{2}$. The coefficient of x^2, -2, is negative, so the graph opens downward. We plot a few points on either side of the vertex, including the y-intercept, $f(0)$, and graph the parabola.

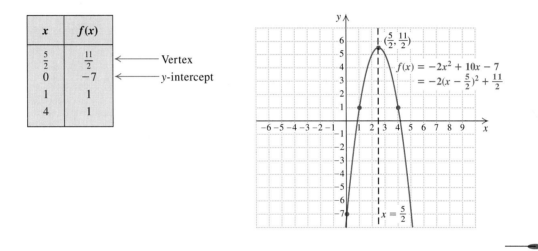

x	$f(x)$	
$\frac{5}{2}$	$\frac{11}{2}$	← Vertex
0	-7	← y-intercept
1	1	
4	1	

The method used in Examples 1–3 can be generalized to find a formula for locating the vertex. We complete the square as follows:

$$f(x) = ax^2 + bx + c$$

$$= a\left(x^2 + \frac{b}{a}x\right) + c. \quad \text{Factoring } a \text{ out of the first two terms. Check by multiplying.}$$

Half of the x-coefficient, $\frac{b}{a}$, is $\frac{b}{2a}$. We square it to get $\frac{b^2}{4a^2}$ and add $\frac{b^2}{4a^2} - \frac{b^2}{4a^2}$ inside the parentheses. Then we distribute the a and regroup terms:

$$f(x) = a\left(x^2 + \frac{b}{a}x + \frac{b^2}{4a^2} - \frac{b^2}{4a^2}\right) + c$$

$$= a\left(x^2 + \frac{b}{a}x + \frac{b^2}{4a^2}\right) + a\left(-\frac{b^2}{4a^2}\right) + c \quad \text{Using the distributive law}$$

$$= a\left(x + \frac{b}{2a}\right)^2 + \frac{-b^2}{4a} + \frac{4ac}{4a} \quad \text{Factoring and finding a common denominator}$$

$$= a\left[x - \left(-\frac{b}{2a}\right)\right]^2 + \frac{4ac - b^2}{4a}.$$

Thus we have the following.

THE VERTEX OF A PARABOLA

The vertex of the parabola given by $f(x) = ax^2 + bx + c$ is

$$\left(-\frac{b}{2a}, \frac{4ac - b^2}{4a}\right), \quad \text{or} \quad \left(-\frac{b}{2a}, f\left(-\frac{b}{2a}\right)\right).$$

The x-coordinate of the vertex is $-b/(2a)$. The axis of symmetry is $x = -b/(2a)$. The second coordinate of the vertex is most commonly found by computing $f\left(-\frac{b}{2a}\right)$.

Let us reexamine Example 3 to see how we could have found the vertex directly. From the formula above,

the x-coordinate of the vertex is $-\dfrac{b}{2a} = -\dfrac{10}{2(-2)} = \dfrac{5}{2}$.

Substituting $\frac{5}{2}$ into $f(x) = -2x^2 + 10x - 7$, we find the second coordinate of the vertex:

$$\begin{aligned}
f\left(\tfrac{5}{2}\right) &= -2\left(\tfrac{5}{2}\right)^2 + 10\left(\tfrac{5}{2}\right) - 7 \\
&= -2\left(\tfrac{25}{4}\right) + 25 - 7 \\
&= -\tfrac{25}{2} + 18 \\
&= -\tfrac{25}{2} + \tfrac{36}{2} = \tfrac{11}{2}.
\end{aligned}$$

The vertex is $\left(\tfrac{5}{2}, \tfrac{11}{2}\right)$. The axis of symmetry is $x = \tfrac{5}{2}$.

We have actually developed two methods for finding the vertex. One is by completing the square and the other is by using a formula. You should check with your instructor about which method to use.

Finding Intercepts

The points at which a graph crosses an axis are called intercepts. We saw in Chapter 2 and again in Example 3 that the y-intercept occurs at $f(0)$. For $f(x) = ax^2 + bx + c$, the y-intercept is simply $(0, c)$. To find x-intercepts, we look for points where $y = 0$ or $f(x) = 0$. To find the x-intercepts of a quadratic function $f(x) = ax^2 + bx + c$, we solve the equation

$$0 = ax^2 + bx + c.$$

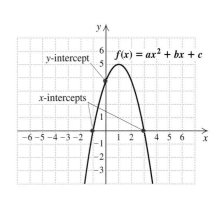

EXAMPLE 4

Find the x- and y-intercepts of the graph of $f(x) = x^2 - 2x - 2$.

SOLUTION The y-intercept is simply $(0, f(0))$, or $(0, -2)$. To find the x-intercepts, we solve the equation

$$0 = x^2 - 2x - 2.$$

We are unable to factor $x^2 - 2x - 2$, so we use the quadratic formula and get $x = 1 \pm \sqrt{3}$. Thus the x-intercepts are $(1 - \sqrt{3}, 0)$ and $(1 + \sqrt{3}, 0)$. If graphing, we would approximate, to get $(-0.7, 0)$ and $(2.7, 0)$.

The sign of the discriminant (see Section 8.4), $b^2 - 4ac$, indicates how many real-number solutions the equation $0 = ax^2 + bx + c$ has. This corresponds to the number of x-intercepts of the graph of $y = ax^2 + bx + c$. Compare the following.

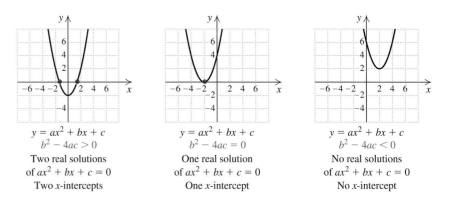

$y = ax^2 + bx + c$	$y = ax^2 + bx + c$	$y = ax^2 + bx + c$
$b^2 - 4ac > 0$	$b^2 - 4ac = 0$	$b^2 - 4ac < 0$
Two real solutions	One real solution	No real solutions
of $ax^2 + bx + c = 0$	of $ax^2 + bx + c = 0$	of $ax^2 + bx + c = 0$
Two x-intercepts	One x-intercept	No x-intercept

Similar sketches can be made for parabolas opening downward.

EXERCISE SET
8.8

*For each quadratic function, **(a)** find the vertex and the axis of symmetry and **(b)** graph the function.*

1. $f(x) = x^2 - 4x + 5$ **2.** $f(x) = x^2 + 2x - 5$

3. $g(x) = x^2 + 6x + 13$ **4.** $g(x) = x^2 + 4x + 5$

5. $f(x) = x^2 + 8x + 20$ **6.** $f(x) = x^2 - 10x + 21$

7. $h(x) = 2x^2 + 16x + 25$

8. $h(x) = 2x^2 - 16x + 23$

9. $f(x) = -x^2 + 2x + 5$

10. $f(x) = -x^2 - 2x + 7$

11. $g(x) = x^2 + 3x - 10$

12. $g(x) = x^2 + 5x + 4$

13. $f(x) = 3x^2 - 24x + 50$

14. $f(x) = 4x^2 + 8x - 3$

15. $h(x) = x^2 - 9x$

16. $h(x) = x^2 + x$

17. $f(x) = -2x^2 - 4x - 6$

18. $f(x) = -3x^2 + 6x + 2$

19. $g(x) = 2x^2 - 7x + 1$

20. $g(x) = 2x^2 + 5x - 1$

21. $f(x) = -3x^2 + 5x - 2$

22. $f(x) = -3x^2 - 7x + 2$

23. $h(x) = \frac{1}{2}x^2 + 4x + \frac{19}{3}$

24. $h(x) = \frac{1}{2}x^2 - 3x + 2$

Find the x- and y-intercepts. If no x-intercepts exist, state so.

25. $f(x) = x^2 - 6x + 3$ **26.** $f(x) = x^2 + 5x + 2$

27. $g(x) = -x^2 + 2x + 3$ **28.** $g(x) = x^2 - 6x + 9$

29. $f(x) = x^2 - 3x + 4$ **30.** $f(x) = x^2 - 7x - 2$

31. $h(x) = -x^2 + 4x - 4$ **32.** $h(x) = 2x^2 - 4x + 6$

33. $f(x) = 4x^2 - 12x + 3$ **34.** $f(x) = x^2 - x + 2$

SKILL MAINTENANCE

Solve.

35. $\sqrt{4x - 4} = \sqrt{x + 4} + 1$

36. $\sqrt{5x - 4} + \sqrt{13 - x} = 7$

37. Solve for n: $A = \dfrac{3Q + nT}{n}$.

38. Simplify:

$$\frac{\dfrac{3}{x} - \dfrac{2}{x - 1}}{\dfrac{5}{x^3 - x}}.$$

SYNTHESIS

39. ◈ Is it possible for the graph of a quadratic function to have only one x-intercept if the vertex is off the x-axis? Why or why not?

40. ◈ Does the graph of every quadratic function have a y-intercept? Why or why not?

41. ◈ Suppose that the graph of $f(x) = ax^2 + bx + c$ has $(x_1, 0)$ and $(x_2, 0)$ as x-intercepts. Explain why the graph of $g(x) = -ax^2 - bx - c$ will also have $(x_1, 0)$ and $(x_2, 0)$ as x-intercepts.

For each quadratic function, find **(a)** *the maximum or minimum value and* **(b)** *the x- and y-intercepts.*

42. ▦ $f(x) = 2.31x^2 - 3.135x - 5.89$

43. ▦ $f(x) = -18.8x^2 + 7.92x + 6.18$

44. Graph the function

$$f(x) = x^2 - x - 6.$$

Then use the graph to approximate solutions to each of the following equations.

a) $x^2 - x - 6 = 2$
b) $x^2 - x - 6 = -3$

45. Graph the function

$$f(x) = \frac{x^2}{8} + \frac{x}{4} - \frac{3}{8}.$$

Then use the graph to approximate solutions to each of the following equations.

a) $\dfrac{x^2}{8} + \dfrac{x}{4} - \dfrac{3}{8} = 0$

b) $\dfrac{x^2}{8} + \dfrac{x}{4} - \dfrac{3}{8} = 1$

c) $\dfrac{x^2}{8} + \dfrac{x}{4} - \dfrac{3}{8} = 2$

Find an equivalent equation of the type

$$f(x) = a(x - h)^2 + k.$$

46. $f(x) = mx^2 - nx + p$

47. $f(x) = 3x^2 + mx + m^2$

48. A quadratic function has $(-1, 0)$ as one of its intercepts and $(3, -5)$ as its vertex. Find an equation for the function.

49. A quadratic function has $(4, 0)$ as one of its intercepts and $(-1, 7)$ as its vertex. Find an equation for the function.

Graph.

50. $f(x) = |x^2 - 1|$ **51.** $f(x) = |x^2 - 3x - 4|$

52. $f(x) = |2(x - 3)^2 - 5|$

53. ▱ Use a grapher to check your answers to Exercises 9, 23, 33, 43, and 45.

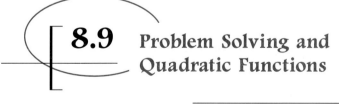

8.9 Problem Solving and Quadratic Functions

Maximum and Minimum Problems   •
Fitting Quadratic Functions to Data

Let's look now at some of the many situations in which quadratic functions are used for problem solving.

Maximum and Minimum Problems

We have seen that for any quadratic function f, the value of $f(x)$ at the vertex is either a maximum or a minimum. Thus problems in which a quantity must be maximized or minimized can often be solved by finding the coordinates of a vertex. This assumes that the problem can be modeled with a quadratic function.

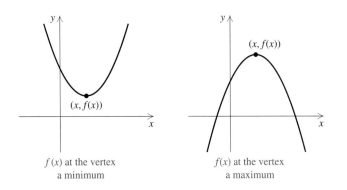

$f(x)$ at the vertex
a minimum

$f(x)$ at the vertex
a maximum

EXAMPLE 1

Fenced-in Land. What are the dimensions of the largest rectangular pen that a farmer can enclose with 64 m of fence?

SOLUTION

1. Familiarize. We make a drawing and label it. Recall these important formulas:

Perimeter: $2w + 2l$;
Area: $l \cdot w$.

To get a better feel for the problem, we can look at some possible dimensions for a rectangular pen that can be enclosed with 64 m of fence. All possibilities are chosen so that $2w + 2l = 64$.

l	w	A
22	10	220
20	12	240
18	14	252
.	.	.
.	.	.
.	.	.

What choice of l and w will maximize A?

2. **Translate.**   We have two equations: One guarantees that the perimeter is 64 m; the other expresses area in terms of length and width.

   $2w + 2l = 64$

   $A = l \cdot w$

3. **Carry out.**   We need to express A as a function of l or w but not both. To do so, we solve for l in the first equation to obtain $l = 32 - w$. Substituting for l in the second equation, we get a quadratic function, $A(w)$, or just A:

   $A = (32 - w)w$ Substituting for l

   $\quad = -w^2 + 32w.$ This is a parabola opening downward, so a maximum exists.

   Completing the square, we get

   $A = -(w^2 - 32w + 256 - 256)$

   $\quad = -(w - 16)^2 + 256.$

   The maximum function value, 256, occurs when $w = 16$ and $l = 32 - 16$, or 16.

4. **Check.**   Note that 256 is greater than any of the values for A found in the *Familiarize* step. To be more certain, we could check values other than those used in that step. For example, if $w = 15$, then $l = 32 - 15 = 17$, and $A = 15 \cdot 17 = 255$. The same area results if $w = 17$ and $l = 15$. Since 256 is greater than 255, it looks as though we have a maximum.

5. **State.**   The largest rectangular pen that can be enclosed is 16 m by 16 m.

EXAMPLE 2

What is the minimum product of two numbers that differ by 5? What are the numbers?

SOLUTION

1. **Familiarize.**   We try some pairs of numbers that differ by 5 and compute their products:

   $1 \cdot 6 = 6,$

   $0 \cdot 5 = 0,$

   $(-1) \cdot 4 = -4.$

   We suspect that one of the two numbers will be negative and the other positive. Let x represent the larger number and $x - 5$ the other number.

2. **Translate.**   We represent the product of the two numbers by

   $p = x(x - 5), \quad \text{or} \quad p(x) = x^2 - 5x.$

3. **Carry out.**   The function $p(x) = x^2 - 5x$ represents a parabola opening upward. Completing the square, we get

   $$p(x) = x^2 - 5x + \frac{25}{4} - \frac{25}{4} = \left(x - \frac{5}{2}\right)^2 - \frac{25}{4}.$$

   The minimum function value of $-\frac{25}{4}$ occurs when one number is $\frac{5}{2}$ and the other number is $\frac{5}{2} - 5$, or $-\frac{5}{2}$.

4. Check. To check, we see if other x-values, near $\frac{5}{2}$, yield smaller values of $x(x-5)$. Note that if $x = 2$, then $x - 5 = -3$ and $2(-3) = -6$. Also note that if $x = 3$, then $x - 5 = -2$ and $3(-2) = -6$. Thus, since $-\frac{25}{4} < -6$, it appears that $x(x-5)$ is minimized when x is $\frac{5}{2}$.

5. State. The minimum product of two numbers that differ by 5 is $-\frac{25}{4}$, or -6.25. This minimum occurs when $\frac{5}{2}$ and $-\frac{5}{2}$ are multiplied. ⎯⎯●

Fitting Quadratic Functions to Data

Whenever a certain quadratic function fits a situation, that function can be determined if three inputs and their outputs are known. Each of the given ordered pairs is called a *data point*.

E X A M P L E   3

Audio Equipment. The owner's manual for a top-rated cassette deck includes a table relating the time a tape has run to the counter reading.

Counter Reading	Time Tape Has Run (in minutes)
000	0
400	25
700	50

Find a quadratic function that fits the data. Then predict how long a tape has run when the counter reading is 200.

S O L U T I O N

1. Familiarize. The statement of the problem leads us to look for a function of the form

$$T(n) = an^2 + bn + c,$$

where $T(n)$ represents the time, in minutes, that the tape has run at counter reading n hundred. We need values for a, b, and c that fit the given data.

2. Translate. We substitute the given values of n and T:

$$0 = a \cdot 0^2 + b \cdot 0 + c,$$
$$25 = a \cdot 4^2 + b \cdot 4 + c,$$
$$50 = a \cdot 7^2 + b \cdot 7 + c.$$

After simplifying, we see that we need to solve the system

$$0 = c,$$
$$25 = 16a + 4b + c,$$
$$50 = 49a + 7b + c.$$

3. Carry out. Since $c = 0$, it suffices to solve the system

$$25 = 16a + 4b, \qquad (1)$$
$$50 = 49a + 7b. \qquad (2)$$

Multiplying the first equation by -7 and the second equation by 4, we obtain

$$-175 = -112a - 28b,$$
$$200 = 196a + 28b.$$

Thus,

$$25 = 84a \qquad \text{Adding equations}$$
$$\tfrac{25}{84} = a.$$

Next, we solve for b, using equation (1) above:

$$25 = 16 \cdot \tfrac{25}{84} + 4b \qquad \text{Substituting in equation (1)}$$
$$25 = \tfrac{400}{84} + 4b$$
$$\tfrac{1700}{84} = 4b \qquad\qquad \text{Note that } 25 = \tfrac{2100}{84}.$$
$$\tfrac{425}{84} = b.$$

Thus the function $T(n) = \tfrac{25}{84}n^2 + \tfrac{425}{84}n$ fits the given data. To find how long a tape has been running when the counter reading is 200, we find $T(2)$:

$$T(2) = \tfrac{25}{84} \cdot 2^2 + \tfrac{425}{84} \cdot 2$$
$$= \tfrac{50}{42} + \tfrac{425}{42}$$
$$= \tfrac{475}{42} \approx 11.3 \text{ min.} \qquad \text{Using a calculator}$$

4. **Check.**   Besides rechecking our calculations, we can check to see if the function we obtained will produce those values found in the table. As you should confirm.

$$T(0) = 0, \qquad T(4) = 25, \quad \text{and} \quad T(7) = 50.$$

A computer-generated graph can also be used as a check.

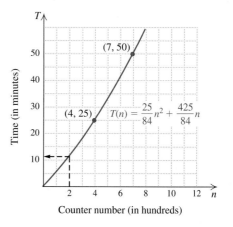

5. **State.**   When the counter reads 200, the tape has been running for about 11 min, 18 sec.

EXERCISE SET
8.9

Solve.

1. *Architecture.*   An architect is designing a rectangular family room with a perimeter of 56 ft. What dimensions will yield the maximum area? What is the maximum area?

2. *Stained-glass window design.*   An artist is designing a rectangular stained-glass window with a perimeter of 84 in. What dimensions will yield the maximum area?

3. What is the minimum product of two numbers that differ by 6? What are the numbers?

4. What is the maximum product of two numbers that add to 18? What numbers yield this product?

5. What is the maximum product of two numbers that add to 26? What numbers yield this product?

6. What is the minimum product of two numbers that differ by 8? What are the numbers?

7. What is the minimum product of two numbers that differ by 7? What are the numbers?

8. What is the maximum product of two numbers that add to −10? What numbers yield this product?

9. What is the maximum product of two numbers that add to −12? What numbers yield this product?

10. What is the minimum product of two numbers that differ by 9? What are the numbers?

11. *Patio design.*   A stone mason has enough stones to enclose a rectangular patio with 60 ft of perimeter, assuming that the attached house forms one side of the rectangle. What is the maximum area that the mason can enclose? What should the dimensions of the patio be in order to yield this area?

12. *Garden design.*   A farmer decides to enclose a rectangular garden, using the side of a barn as one side of the rectangle. What is the maximum area that the farmer can enclose with 40 ft of fence? What should the dimensions of the garden be in order to yield this area?

13. *Molding plastics.*   Economite Plastics plans to produce a one-compartment vertical file by bending the long side of an 8-in. by 14-in. sheet of plastic along two lines to form a U shape. How tall should the file be in order to maximize the volume that the file can hold?

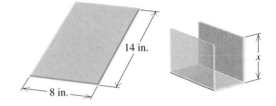

14. *Composting.*   A rectangular compost container is to be formed in a corner of a fenced yard, with 8 ft of chicken wire completing the other two sides of the rectangle. If the chicken wire is 3 ft high, what dimensions of the base will maximize the container's volume?

15. *Minimizing cost.*   Aki's Bicycle Designs has determined that when x hundred bicycles are built, the average cost per bicycle is given by

$$C(x) = 0.1x^2 - 0.7x + 2.425,$$

where $C(x)$ is in hundreds of dollars. How many bicycles should the shop build in order to minimize the average cost per bicycle?

16. *Corral design.*   A rancher needs to enclose two adjacent rectangular corrals, one for sheep and one for cattle. If a river forms one side of the corrals and 180 yd of fencing is available, what is the largest total area that can be enclosed?

Maximizing profit.   Recall that total profit P is the difference between total revenue R and total cost C. Given the following total-revenue and total-cost functions, find the total profit, the maximum value of the total profit, and the value of x at which it occurs.

17. $R(x) = 1000x - x^2$,
     $C(x) = 3000 + 20x$

18. $R(x) = 200x - x^2$,
     $C(x) = 5000 + 8x$

Find the quadratic function that fits the set of data points.

19. $(1, 4), (-1, -2), (2, 13)$

20. $(1, 4), (-1, 6), (-2, 16)$

21. $(2, 0), (4, 3), (12, -5)$

22. $(-3, -30), (3, 0), (6, 6)$

23. a) Find a quadratic function that fits the following data.

Travel Speed (in kilometers per hour)	Number of Nighttime Accidents (for every 200 million kilometers driven)
60	400
80	250
100	250

b) Use the function to estimate the number of nighttime accidents that occur at 50 km/h.

24. a) Find a quadratic function that fits the following data.

Travel Speed (in kilometers per hour)	Number of Daytime Accidents (for every 200 million kilometers driven)
60	100
80	130
100	200

b) Use the function to estimate the number of daytime accidents that occur at 50 km/h.

25. ▦ *Archery.*   The Olympic flame tower at the 1992 Summer Olympics was lit at a height of about 27 m by a flaming arrow that was launched about 63 m from the base of the tower. If the arrow landed about 63 m beyond the tower, find a quadratic function that expresses the height h of the arrow as a function of the distance d that it traveled horizontally.

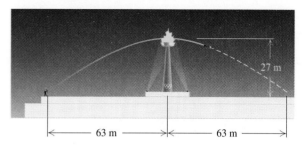

26. ▦ *Pizza prices.*   Pizza Unlimited has the following prices for pizzas.

Diameter	Price
8 in.	$ 6.00
12 in.	$ 8.50
16 in.	$11.50

Is price a quadratic function of diameter? It probably should be, because the price should be proportional to the area, and the area is a quadratic function of the diameter. (The area of a circular region is given by $A = \pi r^2$ or $(\pi/4) \cdot d^2$.)

a) Express price as a quadratic function of diameter using the data points (8, 6), (12, 8.50), and (16, 11.50).

b) Use the function to find the price of a 14-in. pizza.

SKILL MAINTENANCE

Simplify.

27. $\dfrac{x}{x^2 + 17x + 72} - \dfrac{8}{x^2 + 15x + 56}$

28. $\dfrac{x^2 - 9}{x^2 - 8x + 7} \div \dfrac{x^2 + 6x + 9}{x^2 - 1}$

SYNTHESIS

29. ◈ Explain how the leading coefficient of a quadratic function can be used to determine if a maximum or a minimum function value exists.

30. ◈ Explain what restrictions should be placed on the quadratic functions developed in Exercises 17 and 23 and why such restrictions are needed.

31. *Norman window.* A *Norman window* is a rectangle with a semicircle on top. Big Sky Windows is designing a Norman window that will require 24 ft of trim. What dimensions will allow the maximum amount of light to enter a house?

32. *Minimizing area.* A 36-in. piece of string is cut into two pieces. One piece is used to form a circle while the other is used to form a square. How should the string be cut so that the sum of the areas is a minimum?

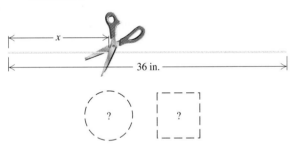

33. *Crop yield.* An orange grower finds that she gets an average yield of 40 bushels (bu) per tree when she plants 20 trees on an acre of ground. Each time she adds a tree to an acre, the yield per tree decreases by 1 bu, due to congestion. How many trees per acre should she plant for maximum yield?

34. *Cover charges.* When the owner of Sweet Sounds charges a $5 cover charge, an average of 100 people will attend a show. For each 25¢ increase in admission price, the average number attending decreases by 1. What should the owner charge in order to make the most money?

35. *Trajectory of a launched object.* The height above the ground of a launched object is a quadratic function of the time that it is in the air. Suppose that a flare is launched from a cliff 64 ft above sea level. If 3 sec after being launched the flare is again level with the cliff, and if 2 sec after that it lands in the sea, what is the maximum height that the flare will reach?

36. *Bridge design.* The cables supporting a straight-line suspension bridge are nearly parabolic in shape. Suppose that a suspension bridge is being designed with concrete supports 160 ft apart and with vertical cables 30 ft above road level at the midpoint of the bridge and 80 ft above road level at a point 50 ft from the midpoint of the bridge. How long are the longest vertical cables?

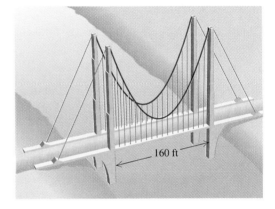

37. ◩ Many graphers have a QUADREG option as part of a STAT CALC menu. Such a feature can quickly fit a quadratic function to a LIST of ordered pairs. Use the QUAD REG feature to check your answers to Exercises 19, 23, and 25.

COLLABORATIVE
C ◆ O ◆ R ◆ N ◆ E ◆ R

Focus: Modeling quadratic functions

Time: 20–30 minutes

Group size: 3 or 4

Materials: Graphers are optional.

The Panasonic Portable Stereo System RX-DT680® has a counter for finding locations on an audio cassette. When a fully wound cassette with 45 min of music on a side begins to play, the counter is at 0. After 15 min of music has played, the counter reads 250 and after 35 min, it reads 487. When the 45-min side is finished playing, the counter reads 590.

ACTIVITY

1. The paragraph above describes four ordered pairs of the form (counter number, minutes played). Three pairs are enough to find a function of the form

$$T(n) = an^2 + bn + c,$$

where $T(n)$ represents the time, in minutes, that the tape has run at counter reading n hundred. Each group member should select a different set of three points from the four given and then fit a quadratic function to the data.

2. Of the 3 or 4 functions found in part (1) above, which fits the data "best"? One way to answer this is to see how well each function predicts other pairs. The same counter used above reads 432 after a 45-min tape has played for 30 min. Which function comes closest to predicting this?

3. If a grapher is available with a QUADREG option (see Exercise 37), what function does it fit to the four pairs originally listed?

4. If a class member has access to a Panasonic System RX-DT680, see how well the functions developed above predict the counter readings for a tape that has played for 5 or 10 min.

8.10   Polynomial and Rational Inequalities

Quadratic and Other Polynomial Inequalities   •   Rational Inequalities

Quadratic and Other Polynomial Inequalities

Inequalities like the following are called *polynomial inequalities*:

$$x^3 - 5x > x^2 + 7, \qquad 9x - 4x^2 < 0, \qquad 5x^2 - 3x + 2 \geq 0,$$
$$3x - 1 \leq 5.$$

Second-degree polynomial inequalities in one variable are called *quadratic inequalities*. To solve polynomial inequalities, we often focus attention on where the outputs of a polynomial function are positive and where they are negative.

E X A M P L E   1   Solve: $x^2 + 3x - 10 > 0$.

S O L U T I O N   Consider the "related" function $f(x) = x^2 + 3x - 10$ and its graph. Its graph opens upward since the leading coefficient is positive.

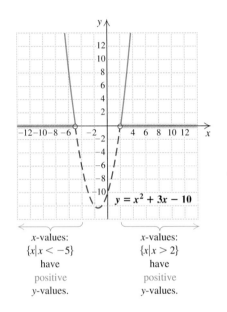

x-values: $\{x | x < -5\}$ have positive *y*-values.

x-values: $\{x | x > 2\}$ have positive *y*-values.

Values of *y* will be positive to the left and right of the *x*-intercepts, as shown. To find the intercepts, we set the polynomial equal to 0 and solve:

$$x^2 + 3x - 10 = 0$$
$$(x + 5)(x - 2) = 0$$
$$x + 5 = 0 \quad or \quad x - 2 = 0$$
$$x = -5 \quad or \quad x = 2.$$

Thus the solution set of the inequality is

$$\{x | \ x < -5 \ or \ x > 2\}, \quad or \quad (-\infty, -5) \cup (2, \infty).$$

Any inequality with 0 on one side can be solved by considering a graph of the related function and finding intercepts as in Example 1. Sometimes the quadratic formula is needed to find the intercepts.

E X A M P L E   2   Solve: $x^2 - 2x \leq 2$.

S O L U T I O N   We first find standard form with 0 on one side:

$$x^2 - 2x - 2 \leq 0. \qquad \text{This is equivalent to the original inequality.}$$

The graph of $f(x) = x^2 - 2x - 2$ is a parabola opening upward. Values of $f(x)$ are negative for x-values between the x-intercepts. We find the x-intercepts by solving $f(x) = 0$:

$$x = \frac{-b \pm \sqrt{b^2 - 4ac}}{2a}$$

$$= \frac{-(-2) \pm \sqrt{(-2)^2 - 4 \cdot 1(-2)}}{2 \cdot 1}$$

$$= \frac{2 \pm \sqrt{12}}{2} = \frac{2 \pm 2\sqrt{3}}{2} = 1 \pm \sqrt{3}.$$

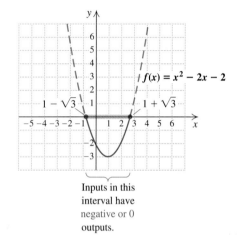

$f(x) = x^2 - 2x - 2$

$1 - \sqrt{3}$ $1 + \sqrt{3}$

Inputs in this interval have negative or 0 outputs.

At the x-intercepts, $1 - \sqrt{3}$ and $1 + \sqrt{3}$, the value of $f(x)$ is 0. Thus the solution set of the inequality is

$$[1 - \sqrt{3}, 1 + \sqrt{3}], \quad \text{or} \quad \{x \mid 1 - \sqrt{3} \leq x \leq 1 + \sqrt{3}\}.$$

Note that in Example 2 it was not essential to actually draw the graph. The important information came from the location of the x-intercepts and the sign of $f(x)$ on each side of those intercepts. In the next example, we solve a third-degree polynomial inequality, without graphing, by locating the x-intercepts, or **zeros**, of f and then using *test points* to determine the sign of $f(x)$ over each interval of the x-axis.

E X A M P L E   3

Solve: $5x^3 + 10x^2 - 15x > 0$.

S O L U T I O N   We first solve the related equation:

$$5x^3 + 10x^2 - 15x = 0$$
$$5x(x^2 + 2x - 3) = 0$$
$$5x(x + 3)(x - 1) = 0$$
$$5x = 0 \quad or \quad x + 3 = 0 \quad or \quad x - 1 = 0$$
$$x = 0 \quad or \qquad x = -3 \quad or \qquad x = 1.$$

We see that if $f(x) = 5x^3 + 10x^2 - 15x$, then the zeros of f are $-3, 0,$ and 1. These zeros divide the number line, or x-axis, into four intervals: A, B, C, and D.

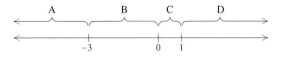

Next, using one convenient test value from each interval, we determine the sign of $f(x)$ for that interval. Using the factored form of $f(x)$ eases the computations:

$$f(x) = 5x(x + 3)(x - 1).$$

For interval A,

$$f(-4) = 5(-4)((-4) + 3)((-4) - 1)$$
$$= -20(-1)(-5)$$
$$= -100.$$

-4 is a convenient value in interval A.

$f(-4)$ is negative.

For interval B,

$$f(-1) = 5(-1)((-1) + 3)((-1) - 1)$$
$$= -5(2)(-2)$$
$$= 20.$$

-1 is a convenient value in interval B.

$f(-1)$ is positive.

For interval C,

$$f\left(\tfrac{1}{2}\right) = \underbrace{5 \cdot \tfrac{1}{2}}_{\text{Positive}} \cdot \underbrace{\left(\tfrac{1}{2} + 3\right)}_{\text{Positive}} \cdot \underbrace{\left(\tfrac{1}{2} - 1\right)}_{\text{Negative}}.$$

$\tfrac{1}{2}$ is a convenient value in interval C.

Only the sign is important.

The product is negative, so $f\left(\tfrac{1}{2}\right)$ is negative.

For interval D,

$$f(2) = \underbrace{5 \cdot 2}_{\text{Positive}} \cdot \underbrace{(2 + 3)}_{\text{Positive}} \cdot \underbrace{(2 - 1)}_{\text{Positive}}.$$

2 is a convenient value in interval D.

$f(2)$ is positive.

Recall that we are looking for all x for which $5x^3 + 10x^2 - 15x > 0$. The calculations above indicate that $f(x)$ is positive for any number in intervals B and D. The solution set of the original inequality is

$$(-3, 0) \cup (1, \infty), \quad \text{or} \quad \{x \mid -3 < x < 0 \text{ or } x > 1\}.$$

Note that the calculations in Example 3 were made simpler by using the factored form of the polynomial. The process was simplified further when, for intervals C and D, we concentrated on only the *sign* of $f(x)$. In the next example, we determine the sign of a polynomial function on each interval by tracking the sign of each factor. By looking at how many positive or negative factors are being multiplied, we will be able to determine the sign of the polynomial function.

EXAMPLE 4

Solve: $4x^3 - 4x \le 0$.

SOLUTION   We first solve the related equation:

$$4x^3 - 4x = 0$$
$$4x(x^2 - 1) = 0$$
$$4x(x + 1)(x - 1) = 0$$
$$4x = 0 \quad or \quad x + 1 = 0 \quad or \quad x - 1 = 0$$
$$x = 0 \quad or \quad x = -1 \quad or \quad x = 1.$$

The function $f(x) = 4x^3 - 4x$ has zeros at -1, 0, and 1. To determine the sign of $f(x)$ over each interval of the number line, we use the factorization $f(x) = 4x(x + 1)(x - 1)$. The product $4x(x + 1)(x - 1)$ is positive or negative, depending on the signs of the factors $4x$, $x + 1$, and $x - 1$. This is easily determined using a diagram.

Interval:	$(-\infty, -1)$	$(-1, 0)$	$(0, 1)$	$(1, \infty)$
Sign of $4x$:	−	−	+	+
Sign of $x + 1$:	−	+	+	+
Sign of $x - 1$:	−	−	−	+
Sign of product $4x(x + 1)(x - 1)$:	−	+	−	+

A product is negative when it has an odd number of negative factors. Since the $\le$ sign allows for equality, the endpoints -1, 0, and 1 are solutions. From the diagram, we see that the solution set is

$$(-\infty, -1] \cup [0, 1], \quad or \quad \{x \mid x \le -1 \ or \ 0 \le x \le 1\}.$$

TO SOLVE A POLYNOMIAL INEQUALITY USING FACTORS:

1. Get 0 on one side and solve the related polynomial equation by factoring.
2. Use the numbers found in step (1) to divide the number line into intervals.
3. Using a test value from each interval, determine the sign of each factor over that interval.
4. Determine the sign of the product of the factors over each interval. Remember that the product of an odd number of negative numbers is negative.
5. Select the interval(s) for which the inequality is satisfied and write set-builder notation or interval notation for the solution set. Include the endpoints when $\le$ or $\ge$ is used.

TECHNOLOGY CONNECTION
8.10

To solve $2.3x^2 \leq 9.11 - 2.94x$, we first rewrite the inequality in the form $2.3x^2 + 2.94x - 9.11 \leq 0$ and graph the function $f(x) = 2.3x^2 + 2.94x - 9.11$.

$f(x) = 2.3x^2 + 2.94x - 9.11$

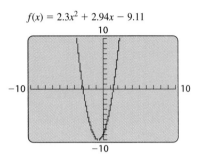

To find the values of x for which $f(x) \leq 0$, we focus on the region in which the graph lies *on or below* the x-axis.

From this graph, it appears that this region begins somewhere between -3 and -2, and continues to somewhere between 1 and 2. Using ZOOM and TRACE or the ROOT option of CALC, we can easily find the endpoints of this region. To two decimal places, the endpoints are -2.73 and 1.45. The solution set is approximately $\{x \mid -2.73 \leq x \leq 1.45\}$.

Had the inequality been $2.3x^2 > 9.11 - 2.94x$, we would look for portions of the graph that lie *above* the x-axis. An approximate solution set of such an inequality would be $\{x \mid x < -2.73 \ or \ x > 1.45\}$.

Use a grapher to solve each inequality to the nearest hundredth.

1. $4.32x^2 - 3.54x - 5.34 \leq 0$

2. $7.34x^2 - 16.55x - 3.89 \geq 0$

3. $10.85x^2 + 4.28x + 4.44 > 7.91x^2 + 7.43x + 13.03$

4. $5.79x^3 - 5.68x^2 + 10.68x > 2.11x^3 + 16.90x - 11.69$

Rational Inequalities

Inequalities involving rational expressions are called **rational inequalities.** Like polynomial inequalities, rational inequalities can be solved using test values.

EXAMPLE 5

Solve: $\dfrac{x - 3}{x + 4} \geq 2$.

SOLUTION   We write the related equation by changing the $\geq$ symbol to $=$:

$$\frac{x - 3}{x + 4} = 2.$$

Next, we solve this related equation:

$$(x + 4) \cdot \frac{x - 3}{x + 4} = (x + 4) \cdot 2 \qquad \text{Multiplying by the LCD, } x + 4, \text{ on both sides}$$

$$x - 3 = 2x + 8$$

$$-11 = x. \qquad \text{Solving for } x$$

In the case of rational inequalities, we also need to determine those values that make the denominator 0. We set the denominator equal to 0 and solve:

$$\left.\begin{array}{l} x + 4 = 0 \\ x = -4. \end{array}\right\} \quad \text{This tells us that } -4 \text{ is not in the domain of } f \text{ if } f(x) = \dfrac{x - 3}{x + 4}.$$

Now we use -11 and -4 to divide the number line into intervals:

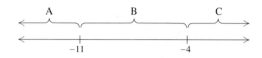

We test a number in each interval to see where the original inequality is satisfied:

$$\frac{x - 3}{x + 4} \geq 2.$$

A: Test -15, $\dfrac{-15 - 3}{-15 + 4} = \dfrac{-18}{-11}$

$$= \frac{18}{11} \not\geq 2 \qquad \begin{array}{l}-15 \text{ is } not \text{ a solution, so interval A} \\ \text{is not part of the solution set.}\end{array}$$

B: Test -8, $\dfrac{-8 - 3}{-8 + 4} = \dfrac{-11}{-4}$

$$= \frac{11}{4} \geq 2 \qquad \begin{array}{l}-8 \text{ } is \text{ a solution, so interval B is part} \\ \text{of the solution set.}\end{array}$$

C: Test 1, $\dfrac{1 - 3}{1 + 4} = \dfrac{-2}{5}$

$$= -\frac{2}{5} \not\geq 2 \qquad \begin{array}{l}1 \text{ is } not \text{ a solution, so interval C is not} \\ \text{part of the solution set.}\end{array}$$

The solution set includes the interval B. The endpoint -11 is included since the inequality symbol is $\geq$ and -11 is a solution of the related equation. The number -4 is *not* included since $(x - 3)/(x + 4)$ is undefined for $x = -4$. Thus the solution set of the original inequality is

$$[-11, -4), \quad \text{or} \quad \{x \mid -11 \leq x < -4\}.$$

There is an interesting visual interpretation of Example 5. If we graph the function $f(x) = (x - 3)/(x + 4)$, we see that the solutions of the inequality $(x - 3)/(x + 4) \geq 2$ can be found by inspection. We simply sketch the line $y = 2$ and locate all x-values for which $f(x) \geq 2$.

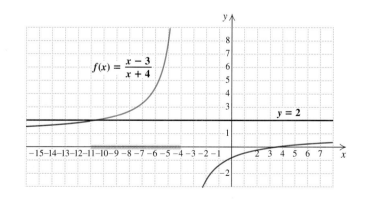

Because graphing rational functions can be very time-consuming, we normally just use test values.

**TO SOLVE A
RATIONAL INEQUALITY:**

1. Change the inequality symbol to an equals sign and solve the related equation.
2. Find any replacements for which the rational expression is undefined.
3. Use the numbers found in steps (1) and (2) to divide the number line into intervals.
4. Substitute a test value from each interval into the inequality. If the number is a solution, then the interval to which it belongs is part of the solution set.
5. Select the interval(s) for which the inequality is satisfied and write set-builder notation or interval notation for the solution set. If the inequality symbol is $\leq$ or $\geq$, then the solutions to step (1) are also included in the solution set. Those numbers found in step (2) should be excluded from the solution set.

**EXERCISE SET
8.10**

Solve.

1. $(x + 4)(x - 3) > 0$

2. $(x - 5)(x + 2) > 0$

3. $(x + 7)(x - 2) \leq 0$

4. $(x - 1)(x + 4) \leq 0$

5. $x^2 - x - 2 < 0$

6. $x^2 + x - 2 < 0$

7. $9 - x^2 \leq 0$

8. $4 - x^2 \geq 0$

9. $x^2 - 2x + 1 \geq 0$

10. $x^2 + 6x + 9 < 0$

11. $x^2 - 4x < 12$

12. $x^2 + 6x > -8$

13. $3x(x + 2)(x - 2) < 0$

14. $5x(x + 1)(x - 1) > 0$

15. $(x + 3)(x - 2)(x + 1) > 0$

16. $(x - 1)(x + 2)(x - 4) < 0$

17. $(x + 3)(x + 2)(x - 1) < 0$

18. $(x - 2)(x - 3)(x + 1) < 0$

19. $\dfrac{1}{x + 7} < 0$

20. $\dfrac{1}{x + 4} > 0$

21. $\dfrac{x + 1}{x - 3} \geq 0$

22. $\dfrac{x - 2}{x + 5} \leq 0$

23. $\dfrac{3x + 2}{x - 3} \leq 0$

24. $\dfrac{5 - 2x}{4x + 3} \leq 0$

25. $\dfrac{x + 1}{2x - 3} > 1$

26. $\dfrac{x - 1}{x - 2} < 1$

27. $\dfrac{(x - 2)(x + 1)}{x - 5} \leq 0$

28. $\dfrac{(x + 4)(x - 1)}{x + 3} \geq 0$

29. $\dfrac{x}{x + 3} \geq 0$

30. $\dfrac{x - 2}{x} \leq 0$

31. $\dfrac{x - 5}{x} < 1$

32. $\dfrac{x}{x - 1} > 2$

33. $\dfrac{x - 1}{(x - 3)(x + 4)} \leq 0$

34. $\dfrac{x + 2}{(x - 2)(x + 7)} \geq 0$

35. $4 < \dfrac{1}{x}$

36. $\dfrac{1}{x} \leq 5$

SKILL MAINTENANCE

37. Multiply and simplify: $\sqrt[5]{a^2 b}\, \sqrt[5]{ab^2}$.

38. The perimeter of an equilateral triangle is the same as that of a square. If the sides of the triangle are 2 units longer than the sides of the square, how long is each side of the square?

SYNTHESIS

39. ◈ Explain how any quadratic inequality can be solved by examining a parabola that opens upward.

40. ◈ Describe a method that could be used to create quadratic inequalities that have no solution.

Solve.

41. $x^2 + 2x > 4$

42. $x^4 + 2x^2 \geq 0$

43. $x^4 + 3x^2 \leq 0$

44. $\left|\dfrac{x + 2}{x - 1}\right| \leq 3$

45. *Total profit.* Derex, Inc., determines that its total-profit function is given by

$$P(x) = -3x^2 + 630x - 6000.$$

a) Find all values of x for which Derex makes a profit.

b) Find all values of x for which Derex loses money.

46. *Height of a thrown object.* The function

$$S(t) = -16t^2 + 32t + 1920$$

gives the height S, in feet, of an object thrown from a cliff that is 1920 ft high. Here t is the time, in seconds, that the object is in the air.

a) For what times does the height exceed 1920 ft?

b) For what times is the height less than 640 ft?

47. *Number of handshakes.* There are n people in a room. The number N of possible handshakes by the people is given by the function

$$N(n) = \frac{n(n - 1)}{2}.$$

For what number of people n is $66 \leq N \leq 300$?

48. *Number of diagonals.* A polygon with n sides has D diagonals, where D is given by the function

$$D(n) = \frac{n(n - 3)}{2}.$$

Find the number of sides n if

$$27 \leq D \leq 230.$$

📈 *Use a grapher to graph each function and find solutions of $f(x) = 0$. Then solve the inequalities $f(x) < 0$ and $f(x) > 0$.*

49. $f(x) = x^3 - 2x^2 - 5x + 6$

50. $f(x) = \dfrac{1}{3}x^3 - x + \dfrac{2}{3}$

51. $f(x) = x + \dfrac{1}{x}$

52. $f(x) = x - \sqrt{x}, x \geq 0$

53. $f(x) = x^4 - 4x^3 - x^2 + 16x - 12$

54. $f(x) = \dfrac{x^3 + x^2 - 2x}{x^2 + x - 6}$

55. 📈 Use a grapher to solve Exercises 11, 25, and 35 by drawing two curves, one for each side of the inequality.

SUMMARY AND REVIEW 8

KEY TERMS

IMPORTANT PROPERTIES AND FORMULAS

The Principle of Square Roots

For any real number k, if $x^2 = k$, then $x = \sqrt{k}$ or $x = -\sqrt{k}$.

To solve a quadratic equation in x by completing the square:

1. Isolate the terms with variables on one side of the equation, and arrange them in descending order.
2. Divide by the coefficient of x^2 on both sides if that coefficient is not 1.
3. Complete the square by taking half of the coefficient of x and adding its square on both sides.
4. Express one side as the square of a binomial and simplify the other side.
5. Use the principle of square roots.
6. Solve for x by adding appropriately on both sides.

The Quadratic Formula

The solutions of $ax^2 + bx + c = 0$, $a \neq 0$, are given by

$$x = \frac{-b \pm \sqrt{b^2 - 4ac}}{2a}.$$

To solve a quadratic equation:

1. Check for the form $ax^2 = p$ or $(x + k)^2 = d$. If it is in either of these forms, use the principle of square roots.
2. If it is not in the form of step (1), write it in standard form $ax^2 + bx + c = 0$.
3. Try factoring and using the principle of zero products.
4. If it is not possible to factor or factoring seems difficult, use the quadratic formula.

The solutions of a quadratic equation can always be found using the quadratic formula. They cannot always be found by factoring.

To solve a formula for a letter—say, b:

1. Clear the fractions and use the principle of powers, as needed, until b does not appear in any radicand or denominator. (In some cases you may clear the fractions first, and in some cases you may use the principle of powers first. Perform these steps until radicals containing b are gone and b is not in any denominator.)
2. Collect all terms with b^2 in them. Also collect all terms with b in them.
3. If b^2 does not appear, you can finish by using just the addition and multiplication principles as in Sections 1.5 and 6.8.
4. If b^2 appears but b does not, solve the equation for b^2. Then take square roots on both sides.
5. If there are terms containing both b and b^2, put the equation in standard form and use the quadratic formula.

Discriminant $b^2 - 4ac$	Nature of Solutions
0	Only one solution; it is a rational number
Positive Perfect square Not a perfect square	Two different real-number solutions Solutions are rational. Solutions are irrational conjugates.
Negative	Two different imaginary-number solutions (complex conjugates)

Variation

y varies directly as x if there is some nonzero constant k such that $y = kx$.

y varies inversely as x if there is some nonzero constant k such that $y = k/x$.

y varies jointly as x and z if there is some nonzero constant k such that $y = kxz$.

———————————

The graph of $g(x) = ax^2$ is a parabola with $x = 0$ as its axis of symmetry; its vertex is the origin.

For $a > 0$, the parabola opens upward; for $a < 0$, the parabola opens downward.

If $|a|$ is greater than 1, the parabola is narrower than $f(x) = x^2$.

If $|a|$ is between 0 and 1, the parabola is wider than $f(x) = x^2$.

———————————

The graph of $f(x) = a(x - h)^2$ has the same shape as the graph of $y = ax^2$.

If h is positive, the graph of $y = ax^2$ is shifted h units to the right.

If h is negative, the graph of $y = ax^2$ is shifted $|h|$ units to the left.

The vertex is $(h, 0)$, and the axis of symmetry is $x = h$.

———————————

The graph of $f(x) = a(x - h)^2 + k$ has the same shape as the graph of $y = a(x - h)^2$.

If k is positive, the graph of $y = a(x - h)^2$ is shifted k units up.

If k is negative, the graph of $y = a(x - h)^2$ is shifted $|k|$ units down.

The vertex is (h, k), and the axis of symmetry is $x = h$.

For $a > 0$, k is the minimum function value. For $a < 0$, k is the maximum function value.

———————————

The vertex of the parabola given by $f(x) = ax^2 + bx + c$ is

$$\left(-\frac{b}{2a}, \frac{4ac - b^2}{4a}\right), \quad \text{or} \quad \left(-\frac{b}{2a}, f\left(-\frac{b}{2a}\right)\right).$$

The x-coordinate of the vertex is $-b/(2a)$. The axis of symmetry is $x = -b/(2a)$.

———————————

To solve a polynomial inequality using factors:

1. Get 0 on one side and solve the related polynomial equation by factoring.
2. Use the numbers found in step (1) to divide the number line into intervals.
3. Using a test value from each interval, determine the sign of each factor over that interval.
4. Determine the sign of the product of the factors over each interval. Remember that the product of an odd number of negative numbers is negative.
5. Select the interval(s) for which the inequality is satisfied and write set-builder notation or interval notation for the solution set. Include the endpoints when ≤ or ≥ is used.

———————————

To solve a rational inequality:

1. Change the inequality symbol to an equals sign and solve the related equation.
2. Find any replacements for which the rational expression is undefined.
3. Use the numbers found in steps (1) and (2) to divide the number line into intervals.
4. Substitute a test value from each interval into the inequality. If the number is a solution, then the interval to which it belongs is part of the solution set.
5. Select the interval(s) for which the inequality is satisfied and write set-builder notation or interval notation for the solution set. If the inequality symbol is ≤ or ≥, then the solutions to step (1) are also included in the solution set. Those numbers found in step (2) should be excluded from the solution set.

REVIEW EXERCISES

Solve.

1. $2x^2 - 7 = 0$ **2.** $14x^2 + 5x = 0$

3. $x^2 - 12x + 36 = 9$ **4.** $x^2 - 5x + 9 = 0$

5. $x(3x + 4) = 4x(x - 1) + 15$

6. $x^2 - 5x - 2 = 0$. Use a calculator to approximate the solutions with rational numbers.

7. Let $f(x) = 4x^2 - 3x - 1$. Find x such that $f(x) = 0$.

Complete the square. Then write the perfect-square trinomial in factored form.

8. $x^2 - 12x$ **9.** $x^2 + \frac{3}{5}x$

Solve by completing the square. Show your work.

10. $x^2 + 2x - 8 = 0$ **11.** $x^2 - 6x + 1 = 0$

12. \$2500 grows to \$3025 in 2 yr. Use the formula $A = P(1 + r)^t$ to find the interest rate.

13. The Peachtree Center Plaza in Atlanta, Georgia, is 723 ft tall. Use the formula $s = 16t^2$ to approximate how long it would take an object to fall from the top.

Solve.

14. The Columbia River flows at a rate of 2 mph for the length of a popular boating route. In order for a motorized dinghy to travel 3 mi upriver and then return in a total of 4 hr, how fast must the boat be able to travel in still water?

15. Working together, Jean and Stacy can cut and split a cord of wood in 4 hr. Working alone, Jean takes 6 hr longer than Stacy. How long would it take Stacy to complete this job alone?

For each equation, determine what type of number the solutions will be.

16. $x^2 + 3x - 6 = 0$ **17.** $x^2 + 2x + 5 = 0$

18. Write a quadratic equation having the solutions 7 and −5.

19. Write a quadratic equation having −4 as its only solution.

20. Find all x-intercepts of the graph of
$$f(x) = x^4 - 13x^2 + 36.$$

Solve.

21. $15x^{-2} - 2x^{-1} - 1 = 0$

22. $(x^2 - 4)^2 - (x^2 - 4) - 6 = 0$

23. The power P expended by heat in an electric circuit of fixed resistance varies directly as the square of the current C in the circuit. A circuit expends 180 watts when a current of 6 amperes is flowing. What is the current in the circuit when it is expending 125 watts in heat?

24. A warning dye is used by people in lifeboats to aid searching airplanes. The radius r of the circle formed by the dye varies directly as the square root of the volume V. It is found that 4 L of dye will spread to a circle of radius 5 m. How much dye is needed to form a circle with a 20-m radius?

25. Find an equation of variation in which y varies jointly as x and z and inversely as w, and $y = 9$ when $x = 3$, $z = 2$, and $w = 5$.

26. a) Graph: $f(x) = -3(x + 2)^2 + 4$.
   b) Label the vertex.
   c) Draw the axis of symmetry.
   d) Find the maximum or the minimum value.

27. For the function $f(x) = 2x^2 - 12x + 23$:
   a) find the vertex and the axis of symmetry;
   b) graph the function.

28. Find the x- and y-intercepts of
$$f(x) = x^2 - 9x + 14.$$

29. Solve $N = 3\pi\sqrt{1/p}$ for p.

30. Solve $2A + T = 3T^2$ for T.

31. What is the minimum product of two numbers whose difference is 22? What numbers yield this product?

32. Find the quadratic function that fits the data points $(0, -2)$, $(1, 3)$, and $(3, 7)$.

Solve.

33. $x^3 - 3x > 2x^2$ **34.** $\dfrac{x - 5}{x + 3} \leq 0$

SKILL MAINTENANCE

35. Metal alloy A is 75% silver. Metal alloy B is 25% silver. How much of each should be mixed in order to produce 300 kg of an alloy that is 60% silver?

36. Solve: $\sqrt{5x - 1} + \sqrt{2x} = 5$.

37. Add: $\dfrac{x}{x^2 - 3x + 2} + \dfrac{2}{x^2 - 5x + 6}$.

38. Multiply and simplify: $\sqrt[3]{9t^6}\sqrt[3]{3s^4t^9}$.

SYNTHESIS

39. ◈ Explain how the x-intercepts of a quadratic function can be used to help find the maximum or minimum value of the function.

40. ◈ Suppose that the quadratic formula is used to solve a quadratic equation. If the discriminant is a perfect square, could factoring have been used to solve the equation? Why or why not?

41. ◈ What is the greatest number of solutions that an equation of the form $ax^4 + bx^2 + c = 0$ can have? Why?

42. ◈ Discuss two ways in which completing the square was used in this chapter.

43. A quadratic function has x-intercepts at -3 and 5. If the y-intercept is at -7, find an equation for the function.

44. Find h and k if, for $3x^2 - hx + 4k = 0$, the sum of the solutions is 20 and the product is 80.

45. The average of two positive integers is 171. One of the numbers is the square root of the other. Find the integers.

CHAPTER TEST

8

Solve.

1. $3x^2 - 16 = 0$

2. $4x(x - 2) - 3x(x + 1) = -18$

3. $x^2 + x + 1 = 0$

4. $x^{-2} - x^{-1} = \frac{3}{4}$

5. $x^2 + 3x = 5$. Use a calculator to approximate the solutions with rational numbers.

6. Let $f(x) = 12x^2 - 19x - 21$. Find x such that $f(x) = 0$.

Complete the square. Then write the perfect-square trinomial in factored form.

7. $x^2 + 14x$

8. $x^2 - \frac{2}{7}x$

Solve by completing the square. Show your work.

9. $x^2 + 3x - 18 = 0$

10. $x^2 + 10x + 15 = 0$

Solve.

11. The Connecticut River flows at a rate of 4 km/h for the length of a popular scenic route. In order for a cruiser to travel 60 km upriver and then return in a total of 8 hr, how fast must the boat be able to travel in still water?

12. Two pipes can fill a tank in $1\frac{1}{2}$ hr. One pipe requires 4 hr longer running alone to fill the tank than the other. How long would it take for the faster pipe, working alone, to fill the tank?

13. Determine the type of number that the solutions of $x^2 + 5x + 17 = 0$ will be.

14. Write a quadratic equation having solutions -2 and $\frac{1}{3}$.

15. Find all x-intercepts of the graph of
$$f(x) = (x^2 + 4x)^2 + 2(x^2 + 4x) - 3.$$

16. The surface area of a balloon varies directly as the square of its radius. The area is 3.4 in^2 when the radius is 5 in. What is the area when the radius is 7 in.?

17. a) Graph: $f(x) = 4(x - 3)^2 + 5$.
 b) Label the vertex.
 c) Draw the axis of symmetry.
 d) Find the maximum or the minimum function value.

18. For the function $f(x) = 2x^2 + 4x - 6$:
 a) find the vertex and the axis of symmetry;
 b) graph the function.

19. Find the x- and y-intercepts of
$$f(x) = x^2 - x - 6.$$

20. Solve $V = \frac{1}{3}\pi(R^2 + r^2)$ for r.

21. What is the minimum product of two numbers having a difference of 8?

22. Find the quadratic function that fits the data points $(0, 0)$, $(3, 0)$, and $(5, 2)$.

23. Solve: $x^2 + 5x \le 6$.

SKILL MAINTENANCE

24. Solve: $\sqrt{x + 3} = x - 3$.

25. Simplify: $\sqrt[4]{2a^2b^3}\,\sqrt[4]{a^4b}$. (Assume $a, b \ge 0$.)

26. Subtract and simplify:
$$\frac{x}{x^2 + 15x + 56} - \frac{7}{x^2 + 13x + 42}.$$

27. The perimeter of a hexagon with all six sides the same length is the same as the perimeter of a square. One side of the hexagon is 3 less than a side of the square. Find the perimeter of each polygon.

SYNTHESIS

28. One solution of $kx^2 + 3x - k = 0$ is -2. Find the other solution.

29. Find a fourth-degree polynomial equation, with integer coefficients, for which $2 - \sqrt{3}$ and $5 - i$ are solutions.

30. Find a polynomial equation, with integer coefficients, for which 5 is a repeated root and $\sqrt{2}$ and $\sqrt{3}$ are solutions.

Exponential and Logarithmic Functions

AN APPLICATION

Beginning in 1988, infestations of zebra mussels started spreading throughout North American waters. These mussels spread with such speed that water treatment facilities, power plants, and entire ecosystems have become threatened. The function $A(t) = 10 \cdot 34^t$ can be used to predict the number of square centimeters of lake bottom that will be covered with mussels t years after an infestation covering 10 cm^2 first appears. How many square centimeters of lake bottom will be covered 5 years later?

THIS PROBLEM APPEARS AS EXERCISE 37 IN SECTION 9.1.

More information on zebra mussels is available at
http://hepg.awl.com/be/inter_5

As an ecologist, I do a huge amount of statistics, most of which is based on linear algebra. The math helps me to understand more clearly what I observe in the field.

CHRISTOPHER MONZ
Research Scientist
Lander, WY

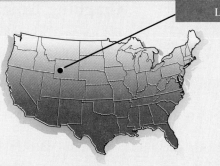

*T*he functions that we consider in this chapter are interesting not only from a purely intellectual point of view, but also for their rich applications to many fields. We will look at applications such as compound interest and population growth, to name just two.

The basis of the theory centers on functions having variable exponents (exponential functions). Results follow from those functions and their properties.

9.1   Exponential Functions

Graphing Exponential Functions   •   Equations with x and y Interchanged   •   Applications of Exponential Functions

Consider the graph below.* The rapidly rising curve approximates the graph of an *exponential function*. We now consider such functions and some of their applications.

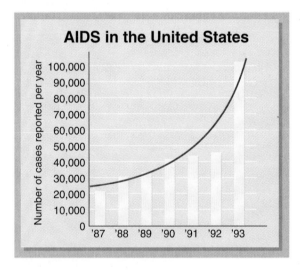

Graphing Exponential Functions

In Chapter 7, we studied exponential expressions with rational-number exponents, such as

$$5^{1/4}, \qquad 3^{-3/4}, \qquad 7^{2.34}, \qquad 5^{1.73}.$$

---

*Because of a change in the definition of AIDS, it is difficult to compare post-1993 figures with pre-1993 figures.

For example, $5^{1.73}$, or $5^{173/100}$, represents the 100th root of 5 raised to the 173rd power. What about expressions with irrational exponents, such as $5^{\sqrt{3}}$ or $7^{-\pi}$? To attach meaning to $5^{\sqrt{3}}$, consider a rational approximation, r, of $\sqrt{3}$. As r gets closer to $\sqrt{3}$, the value of 5^r gets closer to some real number p.

r closes in on $\sqrt{3}$.	5^r closes in on some real number p.
$1.7 < r < 1.8$	$15.426 \approx 5^{1.7} < p < 5^{1.8} \approx 18.119$
$1.73 < r < 1.74$	$16.189 \approx 5^{1.73} < p < 5^{1.74} \approx 16.452$
$1.732 < r < 1.733$	$16.241 \approx 5^{1.732} < p < 5^{1.733} \approx 16.267$

We define $5^{\sqrt{3}}$ to be the number p. To eight decimal places,

$$5^{\sqrt{3}} \approx 16.24245082.$$

Any positive irrational exponent can be defined in a similar way. Negative irrational exponents are then defined using reciprocals. Thus, so long as a is positive, a^x has meaning for *any* real number x. The general laws of exponents still hold, but we will not prove that here. We now define an *exponential function.*

EXPONENTIAL FUNCTION

> The function $f(x) = a^x$, where a is a positive constant, $a \neq 1$, is called the *exponential function,* base a.

We require the base a to be positive to avoid the imaginary numbers that would result from taking even roots of negative numbers. The restriction $a \neq 1$ is made to exclude the constant function $f(x) = 1^x$, or $f(x) = 1$.

The following are examples of exponential functions:

$$f(x) = 2^x, \qquad f(x) = \left(\tfrac{1}{3}\right)^x, \qquad f(x) = 5^{-3x}. \qquad \text{Note that } 5^{-3x} = (5^{-3})^x.$$

In contrast to polynomial functions like $f(x) = x^2$ or $g(x) = 5x^3 - 7$, exponential functions have a variable exponent. Because of this, graphs of exponential functions rise or fall dramatically.

EXAMPLE 1

Graph the exponential function $y = f(x) = 2^x$.

SOLUTION   We compute some function values, thinking of y as $f(x)$, and list the results in a table. It is a good idea to start by letting $x = 0$.

$$f(0) = 2^0 = 1; \qquad f(-1) = 2^{-1} = \frac{1}{2^1} = \frac{1}{2};$$
$$f(1) = 2^1 = 2;$$
$$f(2) = 2^2 = 4; \qquad f(-2) = 2^{-2} = \frac{1}{2^2} = \frac{1}{4};$$
$$f(3) = 2^3 = 8;$$
$$f(-3) = 2^{-3} = \frac{1}{2^3} = \frac{1}{8}.$$

x	y, or $f(x)$
0	1
1	2
2	4
3	8
-1	$\frac{1}{2}$
-2	$\frac{1}{4}$
-3	$\frac{1}{8}$

Next, we plot these points and connect them with a smooth curve.

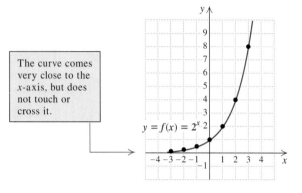

The curve comes very close to the x-axis, but does not touch or cross it.

Be sure to plot enough points to determine how steeply the curve rises.

Note that as x increases, the function values increase without bound. As x decreases, the function values decrease, getting very close to 0. The x-axis, or the line $y = 0$, is a horizontal *asymptote,* meaning that the curve gets closer and closer to this line the further we move to the left.

EXAMPLE 2

Graph the exponential function $y = f(x) = \left(\frac{1}{2}\right)^x$.

SOLUTION   We compute some function values, thinking of y as $f(x)$, and list the results in a table. Before we do this, note that

$$y = f(x) = \left(\tfrac{1}{2}\right)^x = (2^{-1})^x = 2^{-x}.$$

Then we have

$$f(0) = 2^{-0} = 1;$$

$$f(1) = 2^{-1} = \frac{1}{2^1} = \frac{1}{2};$$

$$f(2) = 2^{-2} = \frac{1}{2^2} = \frac{1}{4};$$

$$f(3) = 2^{-3} = \frac{1}{2^3} = \frac{1}{8};$$

$$f(-1) = 2^{-(-1)} = 2^1 = 2;$$

$$f(-2) = 2^{-(-2)} = 2^2 = 4;$$

$$f(-3) = 2^{-(-3)} = 2^3 = 8.$$

x	y, or $f(x)$
0	1
1	$\frac{1}{2}$
2	$\frac{1}{4}$
3	$\frac{1}{8}$
-1	2
-2	4
-3	8

Next, we plot these points and connect them with a smooth curve. Note that this curve is a mirror image, or *reflection*, of the above graph of $y = 2^x$ across the y-axis. The line $y = 0$ is again an asymptote.

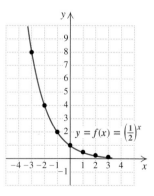

From Examples 1 and 2, we can make the following observations.

A. When $a > 1$, the graph of $f(x) = a^x$ increases from left to right. The greater the value of a, the steeper the curve. (See the figure on the left, below.)

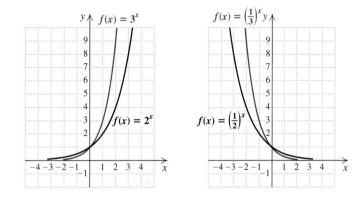

B. When $0 < a < 1$, the graph of $f(x) = a^x$ decreases from left to right. For smaller values of a, the curve becomes steeper. (See the figure on the right, above.)

C. All graphs of $f(x) = a^x$ go through the y-intercept $(0, 1)$.

E X A M P L E   3   Graph: $y = f(x) = 2^{x-2}$.

S O L U T I O N   We construct a table of values. Then we plot the points and connect them with a smooth curve. Here $x - 2$ is the *exponent*.

TECHNOLOGY CONNECTION

9.1A

Graphers are especially helpful when graphing exponential functions because bases may not always be whole numbers and function values can quickly become very large. For example, consider the function $y = 5000(1.075)^x$ shown at right. Because y-values will always be positive and will become quite large, an appropriate window for this function might be $[-10, 10, 0, 15000]$, where the *scale* of the y-axis is 1000.

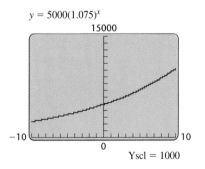

$y = 5000(1.075)^x$

Yscl = 1000

Use a grapher to draw the graph of each pair of functions. Select an appropriate viewing box and scale.

1. $y_1 = \left(\frac{5}{2}\right)^x$ and $y_2 = \left(\frac{2}{5}\right)^x$

2. $y_1 = 3.2^x$ and $y_2 = 3.2^{-x}$

3. $y_1 = 9.34^x$ and $y_2 = 9.34^{-x}$

4. $y_1 = \left(\frac{3}{7}\right)^x$ and $y_2 = \left(\frac{7}{3}\right)^x$

5. $y_1 = 5000(1.08)^x$ and $y_2 = 5000(1.08)^{x-3}$

6. $y_1 = 2000(1.09)^x$ and $y_2 = 2000(1.09)^{x+3}$

$$f(0) = 2^{0-2} = 2^{-2} = \frac{1}{2^2} = \frac{1}{4} \qquad f(-1) = 2^{-1-2} = 2^{-3} = \frac{1}{2^3} = \frac{1}{8}$$

$$f(1) = 2^{1-2} = 2^{-1} = \frac{1}{2^1} = \frac{1}{2} \qquad f(-2) = 2^{-2-2} = 2^{-4} = \frac{1}{2^4} = \frac{1}{16}$$

$$f(2) = 2^{2-2} = 2^0 = 1$$

$$f(3) = 2^{3-2} = 2^1 = 2$$

$$f(4) = 2^{4-2} = 2^2 = 4$$

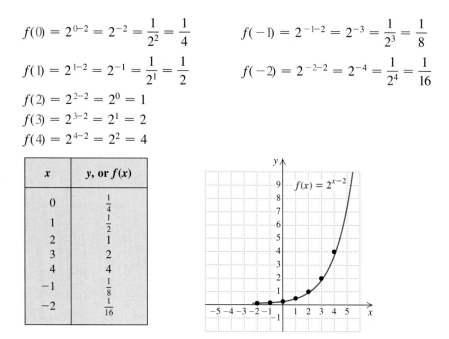

x	y, or $f(x)$
0	$\frac{1}{4}$
1	$\frac{1}{2}$
2	1
3	2
4	4
−1	$\frac{1}{8}$
−2	$\frac{1}{16}$

The graph looks just like the graph of $y = 2^x$, but it is translated 2 units to the right. The y-intercept of $y = 2^x$ is (0, 1). The y-intercept of $y = 2^{x-2}$ is $\left(0, \frac{1}{4}\right)$. The line $y = 0$ is again the asymptote.

Equations with x and y Interchanged

It will be helpful in later work to be able to graph an equation in which the x and the y in $y = a^x$ are interchanged.

EXAMPLE 4   Graph: $x = 2^y$.

SOLUTION   Note that x is alone on one side of the equation. To find ordered pairs that are solutions, we choose values for y and then compute values for x:

For $y = 0$, $x = 2^0 = 1$.

For $y = 1$, $x = 2^1 = 2$.

For $y = 2$, $x = 2^2 = 4$.

For $y = 3$, $x = 2^3 = 8$.

For $y = -1$, $x = 2^{-1} = \frac{1}{2^1} = \frac{1}{2}$.

For $y = -2$, $x = 2^{-2} = \frac{1}{2^2} = \frac{1}{4}$.

For $y = -3$, $x = 2^{-3} = \frac{1}{2^3} = \frac{1}{8}$.

x	y
1	0
2	1
4	2
8	3
$\frac{1}{2}$	−1
$\frac{1}{4}$	−2
$\frac{1}{8}$	−3

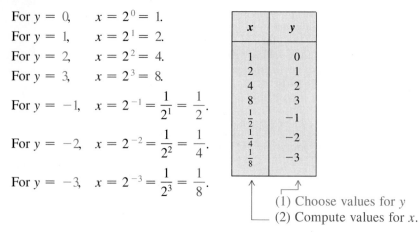

(1) Choose values for y

(2) Compute values for x.

We plot the points and connect them with a smooth curve.

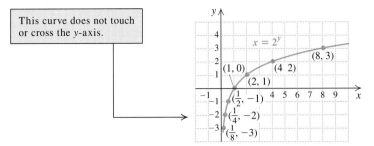

This curve does not touch or cross the y-axis.

Note too that this curve looks just like the graph of $y = 2^x$, except that it is reflected across the line $y = x$, as shown here.

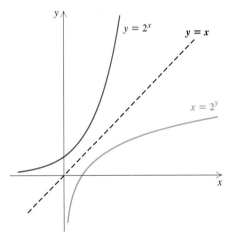

Applications of Exponential Functions

EXAMPLE 5

Interest Compounded Annually. The amount of money A that a principal P will be worth after t years at interest rate i, compounded annually, is given by the formula

$$A = P(1 + i)^t.$$ You might review Example 12 in Section 8.1.

Suppose that \$100,000 is invested at 8% interest, compounded annually.

a) Find a function for the amount in the account after t years.
b) Find the amount of money in the account at $t = 0$, $t = 4$, $t = 8$, and $t = 10$.
c) Graph the function.

S O L U T I O N

a) If $P = \$100,000$ and $i = 8\% = 0.08$, we can substitute these values and form the following function:

$$A(t) = \$100,000(1 + 0.08)^t$$
$$= \$100,000(1.08)^t.$$

TECHNOLOGY CONNECTION

9.1B

Graphers can quickly find many function values at the touch of a few keys. To see this, let $y_1 = 100,000(1.08)^x$. Then use either the TABLE feature (if one is available) or $y_1(\ )$ notation to check Example 5(b).

b) To find the function values, a calculator with a power key is helpful.

$$A(0) = \$100,000(1.08)^0$$
$$= \$100,000(1)$$
$$= \$100,000$$

$$A(4) = \$100,000(1.08)^4$$
$$= \$100,000(1.36048896)$$
$$\approx \$136,048.90$$

$$A(8) = \$100,000(1.08)^8$$
$$\approx \$100,000(1.85093021)$$
$$\approx \$185,093.02$$

$$A(10) = \$100,000(1.08)^{10}$$
$$\approx \$100,000(2.158924997)$$
$$\approx \$215,892.50$$

c) We use the function values computed in part (b), and others if we wish, to draw the graph as follows. Note that the axes are scaled differently because of the large numbers.

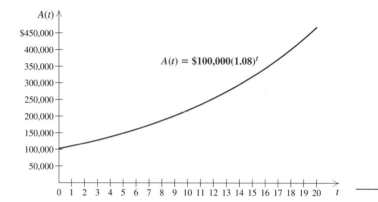

$$A(t) = \$100,000(1.08)^t$$

EXERCISE SET

9.1

Graph.

1. $y = f(x) = 2^x$

2. $y = f(x) = 3^x$

3. $y = 5^x$

4. $y = 6^x$

5. $y = 2^{x-1}$

6. $y = 2^{x+1}$

7. $y = 3^{x+2}$

8. $y = 3^{x-2}$

9. $y = 2^x - 1$

10. $y = 2^x + 3$

11. $y = 5^{x+3}$

12. $y = 6^{x-4}$

13. $y = \left(\frac{1}{2}\right)^x$

14. $y = \left(\frac{1}{3}\right)^x$

15. $y = \left(\frac{1}{5}\right)^x$

16. $y = \left(\frac{1}{4}\right)^x$

17. $y = 2^{2x-1}$

18. $y = 3^{4-x}$

19. $y = 2^{x-3} - 1$

20. $y = 2^{x+1} - 3$

21. $x = 3^y$

22. $x = 6^y$

23. $x = \left(\frac{1}{2}\right)^y$

24. $x = \left(\frac{1}{3}\right)^y$

25. $x = 5^y$

26. $x = 4^y$

27. $x = \left(\frac{3}{2}\right)^y$

28. $x = \left(\frac{4}{3}\right)^y$

Graph both equations using the same set of axes.

29. $y = 3^x, \ x = 3^y$

30. $y = 2^x, \ x = 2^y$

31. $y = \left(\frac{1}{2}\right)^x, \ x = \left(\frac{1}{2}\right)^y$

32. $y = \left(\frac{1}{4}\right)^x, \ x = \left(\frac{1}{4}\right)^y$

Solve.

33. 🔲 *Cases of AIDS.* The total number of Americans who have contracted AIDS, in thousands, can be approximated by the exponential function

$$N(t) = 100(1.4)^t,$$

where $t = 0$ corresponds to 1989.

a) According to the function, how many Americans had been infected as of 1993?

b) Estimate the total number of Americans who will have been infected as of 1998.

c) Graph the function.

34. █ *Growth of bacteria.* The bacteria *Escherichi coli* are commonly found in the human bladder. Suppose that 3000 of the bacteria are present at time $t = 0$. Then t minutes later, the number of bacteria present will be

$$N(t) = 3000(2)^{t/20}.$$

a) How many bacteria will be present after 10 min? 20 min? 30 min? 40 min? 60 min?

b) Graph the function.

35. █ *Recycling aluminum cans.* It is estimated that $\frac{2}{3}$ of all aluminum cans distributed will be recycled each year. A beverage company distributes 250,000 cans. The number still in use after time t, in years, is given by the exponential function

$$N(t) = 250{,}000\left(\tfrac{2}{3}\right)^t.$$

a) How many cans are still in use after 0 yr? 1 yr? 4 yr? 10 yr?

b) Graph the function.

36. █ *Salvage value.* A photocopier is purchased for $5200. Its value each year is about 80% of the value of the preceding year. Its value, in dollars, after t years is given by the exponential function

$$V(t) = 5200(0.8)^t.$$

a) Find the value of the machine after 0 yr, 1 yr, 2 yr, 5 yr, and 10 yr.

b) Graph the function.

37. █ *Spread of zebra mussels.* Beginning in 1988, infestations of zebra mussels started spreading throughout North American waters.* These mussels spread with such speed that water treatment facilities, power plants, and entire ecosystems can become threatened. The function

$$A(t) = 10 \cdot 34^t$$

can be used to estimate the number of square centimeters of lake bottom that will be covered with mussels t years after an infestation covering 10 cm^2 first occurs.

---

*Many thanks to Dr. Gerald Mackie of the Department of Zoology at the University of Guelph in Ontario for the background information for this exercise.

a) How many square centimeters of lake bottom will be covered with mussels 5 years after an infestation covering 10 cm^2 first appears? 7 years after the infestation first appears?

b) Graph the function.

38. █ *Cellular phones.* The number of cellular phones in use in the United States is increasing exponentially. The number N, in millions, in use is given by the exponential function

$$N(t) = 0.3(1.63)^t,$$

where t is the number of years after 1985.

a) Find the number of cellular phones in use in 1985, 1990, 1995, 2000, and 2005.

b) Graph the function.

SKILL MAINTENANCE

39. Multiply and simplify: $x^{-5} \cdot x^3$.

40. Simplify: $(x^{-3})^4$.

41. Divide and simplify: $\dfrac{x^{-3}}{x^4}$.

42. Simplify: 5^0.

SYNTHESIS

43. ◆ Suppose that $1000 is invested for 5 yr at 7% interest, compounded annually. In what year will the most interest be earned? Why?

44. ◆ Without using a calculator, explain why 2^π must be greater than 8 but less than 16.

45. ◆ Consider any exponential function of the form $f(x) = a^x$ with $a > 1$. Will it always follow that $f(3) - f(2) > f(2) - f(1)$, and, in general, $f(n + 2) - f(n + 1) > f(n + 1) - f(n)$? Why or why not? (*Hint:* Think graphically.)

46. ◆ Why was it necessary to discuss irrational exponents before graphing exponential functions?

Determine which of the two numbers is larger.

47. $\pi^{1.3}$ or $\pi^{2.4}$ 48. $\sqrt{8^3}$ or $8^{\sqrt{3}}$

Graph.

49. █ $f(x) = 3.8^x$ 50. █ $f(x) = 2.3^x$

51. $y = 2^x + 2^{-x}$ 52. $y = \left|\left(\tfrac{1}{2}\right)^x - 1\right|$

53. $y = |2^x - 2|$ 54. $y = 2^{-(x-1)^2}$

55. $y = |2^{x^2} - 1|$ 56. $y = 3^x + 3^{-x}$

Graph both equations using the same set of axes.

57. $y = 3^{-(x-1)}$, $x = 3^{-(y-1)}$

58. $y = 1^x$, $x = 1^y$

59. *Sales of color printers.*   As prices of color computer printers have been slashed, sales have soared. The graph below shows how sales have grown from 1993 to 1995.* Use a grapher with a STAT

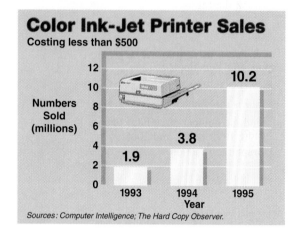

Color Ink-Jet Printer Sales
Costing less than $500

Numbers Sold (millions) — 1993: 1.9; 1994: 3.8; 1995: 10.2

Year

Sources: Computer Intelligence; The Hard Copy Observer.

The New York Times, 5/20/96. Story on p. D7 by Laurie Flynn (note sources on graph).

regression feature to find an exponential function that models the number of color printers costing under $500 that will be sold t years after 1993. Then use that function to predict the number of color printers costing under $500 that will be sold in 1998.

60. *Spread of AIDS.*   In 1985, a total of 8249 cases of AIDS was reported in the United States; in 1988, a total of 31,001 cases; in 1989, a total of 33,722 cases; and in 1990, a total of 41,595 cases. Use a REGRESSION feature (see Exercise 59) to find a model for $N(t)$, the number of AIDS cases in the United States t years after 1985. Then estimate the number of cases reported in 1987.

61. *Typing speed.*   Ali is studying typing. After he has studied for t hours, Ali's speed, in words per minute, is given by the exponential function

$$S(t) = 200[1 - (0.99)^t].$$

Use a graph and/or table of values to predict Ali's speed after studying for 10 hr, 40 hr, and 80 hr.

COLLABORATIVE
C◆O◆R◆N◆E◆R

Focus: Car loans and exponential functions

Time: 30 minutes

Group size: 2

Materials: Calculators with exponentiation keys

The formula

$$M = \frac{Pr}{1 - (1 + r)^{-n}}$$

is used to determine the payment size, M, when a loan of P dollars is to be repaid in n equally sized monthly payments. Here r represents the monthly interest rate. Loans repaid in this fashion are said to be *amortized* (spread out equally) over a period of n months.

ACTIVITY

1. Suppose one group member is selling the other a car for $2600, financed at 1% interest per month for 24 months. What should be the size of each monthly payment?

2. Suppose both group members are shopping for the same model new car. To save time, each group member visits a different dealer. One dealer offers the car for $13,000 at 10.5% interest (0.00875 monthly interest) for 60 months (no down payment). The other dealer offers the same car for $12,000, but at 12% interest (0.01 monthly interest) for 48 months (no down payment).

a) Determine the monthly payment size for each offer (remember to use the *monthly* interest rates). Then determine the total amount paid for the car under each offer. How much of each total is interest?

b) Work together to find the annual interest rate for which the total cost of 60 monthly payments for the $13,000 car would equal the total amount paid for the $12,000 car (as found in part a above).

9.2 Composite and Inverse Functions

Composite Functions • Inverses and One-to-One Functions •
Finding Formulas for Inverses • Graphing Functions and
Their Inverses • Inverse Functions and Composition

Composite Functions

In the real world, functions frequently occur in which some quantity depends on a variable that, in turn, depends on another variable. For instance, the number of employees hired by a firm may depend on the firm's profits, which may in turn depend on the number of items the firm produces. Functions like this are called **composite functions.**

For example, the function g that gives a correspondence between women's shoe sizes in the United States and those in Italy is given by $g(x) = 2x + 24$, where x is the U.S. size and $g(x)$ is the Italian size. Thus a U.S. size 4 corresponds to a shoe size of $g(4) = 2 \cdot 4 + 24$, or 32, in Italy.

There is also a function that gives a correspondence between women's shoe sizes in Italy and those in Britain. The function is given by $f(x) = \frac{1}{2}x - 14$, where x is the Italian size and $f(x)$ is the corresponding British size. Thus an Italian size 32 corresponds to a British size $f(32) = \frac{1}{2} \cdot 32 - 14$, or 2.

It seems reasonable to conclude that a shoe size of 4 in the United States corresponds to a size of 2 in Britain and that some function h describes this correspondence. Can we find a formula for h? If we look at the following tables, we might guess that such a formula is $h(x) = x - 2$, and that is indeed correct. But, for more complicated formulas, we would need to use algebra.

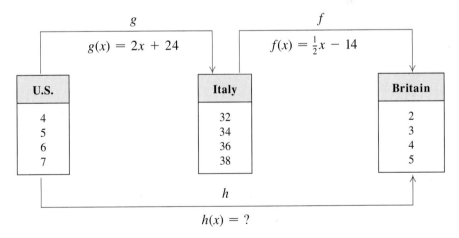

Size x shoes in the United States correspond to size $g(x)$ shoes in Italy, where

$$g(x) = 2x + 24.$$

Size n shoes in Italy correspond to size $f(n)$ shoes in Britain. Thus size $g(x)$ shoes in Italy correspond to size $f(g(x))$ shoes in Britain. Since the x in the expression $f(g(x))$ represents a U.S. shoe size, we can find the British shoe size that corresponds to a U.S. size x as follows:

$$f(g(x)) = f(2x + 24) = \tfrac{1}{2} \cdot (2x + 24) - 14 \qquad \text{Using } g(x) \text{ as an input}$$
$$= x + 12 - 14 = x - 2.$$

This gives a formula for h: $h(x) = x - 2$. Thus a shoe size of 4 in the United States corresponds to a shoe size of $h(4) = 4 - 2$, or 2, in Britain. The function h is called the *composition* of f and g and is denoted $f \circ g$ (read "f composed with g" or "f circle g").

COMPOSITION OF FUNCTIONS

The *composite function $f \circ g$*, the *composition* of f and g, is defined as

$$f \circ g(x) = f(g(x)).$$

We can visualize the composition of functions as follows.

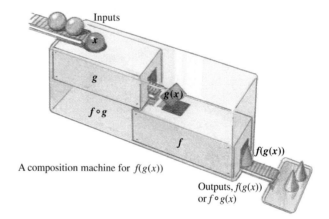

A composition machine for $f(g(x))$

EXAMPLE 1   Given $f(x) = 3x$ and $g(x) = 1 + x^2$:

a) Find $f \circ g(5)$ and $g \circ f(5)$.
b) Find $f \circ g(x)$ and $g \circ f(x)$.

SOLUTION   Consider each function separately:

$$f(x) = 3x \qquad \text{This function multiplies each input by 3.}$$

and

$$g(x) = 1 + x^2. \qquad \text{This function adds 1 to the square of each input.}$$

a) To find $f \circ g(5)$, we first find $g(5)$ by substituting in the formula for g: Square 5 and add 1, to get 26. We then use 26 as an input for f:

$$f \circ g(5) = f(g(5)) = f(1 + 5^2)$$
$$= f(26) = 3 \cdot 26 = 78.$$

To find $g \circ f(5)$, we first find $f(5)$ by substituting into the formula for f: Multiply 5 by 3, to get 15. We then use 15 as an input for g:

$$g \circ f(5) = g(f(5)) = g(3 \cdot 5) \qquad \text{Note that } f(5) = 3 \cdot 5 = 15.$$
$$= g(15) = 1 + 15^2 = 1 + 225 = 226.$$

b) We find $f \circ g(x)$ by substituting $g(x)$ for x in the equation for $f(x)$:

$$f \circ g(x) = f(g(x)) = f(1 + x^2) \qquad \text{Substituting } 1 + x^2 \text{ for } g(x)$$
$$= 3(1 + x^2) = 3 + 3x^2. \qquad \textit{These} \text{ parentheses indicate multiplication.}$$

To find $g \circ f(x)$, we substitute $f(x)$ for x in the equation for $g(x)$:

$$g \circ f(x) = g(f(x)) = g(3x) \qquad \text{Substituting } 3x \text{ for } f(x)$$
$$= 1 + (3x)^2 = 1 + 9x^2.$$

As a check, note that $g \circ f(5) = 1 + 9 \cdot 5^2 = 1 + 9 \cdot 25 = 226$, as expected from part (a) above. ⟶●

Example 1 shows that, in general, $f \circ g(5) \neq g \circ f(5)$ and $f \circ g(x) \neq g \circ f(x)$.

EXAMPLE 2   Given $f(x) = \sqrt{x}$ and $g(x) = x - 1$, find $f \circ g(x)$ and $g \circ f(x)$.

SOLUTION

$$f \circ g(x) = f(g(x)) = f(x - 1) = \sqrt{x - 1}$$
$$g \circ f(x) = g(f(x)) = g(\sqrt{x}) = \sqrt{x} - 1 \qquad ⟶●$$

In calculus, one needs to recognize how a function can be regarded as the composition of two "simpler" functions.

EXAMPLE 3   If $h(x) = (7x + 3)^2$, find $f(x)$ and $g(x)$ such that $h(x) = f \circ g(x)$.

SOLUTION   To find $h(x)$, we can think of two steps: forming $7x + 3$ and then squaring. This suggests that $g(x) = 7x + 3$ and $f(x) = x^2$. We check by forming the composition:

$$h(x) = f \circ g(x) = f(g(x))$$
$$= f(7x + 3) = (7x + 3)^2.$$

This is probably the most "obvious" answer to the question. There can be other less obvious answers. For example, if

$$f(x) = (x - 1)^2$$

and

$$g(x) = 7x + 4,$$

then

$$h(x) = f \circ g(x) = f(g(x)) = f(7x + 4)$$
$$= (7x + 4 - 1)^2 = (7x + 3)^2. \qquad ⟶●$$

TECHNOLOGY CONNECTION 9.2A

In Example 2, we saw that if $g(x) = x - 1$ and $f(x) = \sqrt{x}$, then $f(g(x)) = \sqrt{x - 1}$. One way to show this on a grapher is to let $y_1 = x - 1$ and $y_2 = \sqrt{y_1}$ (this is accomplished by using the Y-VARS key to enter y_1 as a variable). If we also let $y_3 = \sqrt{x - 1}$, graphs and tables can be used to show that $y_2 = y_3$.

Another approach is to let $y_1 = x - 1$ and $y_2 = \sqrt{x}$ and have $y_4 = y_2(y_1)$. Then show that if $y_3 = \sqrt{x - 1}$, we have $y_4 = y_3$.

1. Check Example 3 by using one of the above approaches. Use graphs and, if possible, tables.

Inverses and One~to~One Functions

Let us consider the following two functions. We think of them as relations, or correspondences.

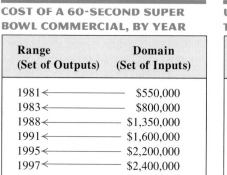

COST OF A 60-SECOND SUPER BOWL COMMERCIAL, BY YEAR

Domain (Set of Inputs)	Range (Set of Outputs)
1981	$550,000
1983	$800,000
1988	$1,350,000
1991	$1,600,000
1995	$2,200,000
1997	$2,400,000

U.S. SENATORS AND THEIR STATES

Domain (Set of Inputs)	Range (Set of Outputs)
Wellstone	Minnesota
Grams	
Mack	Florida
Graham	
Lautenberg	New Jersey
Torricelli	

Suppose we reverse the arrows. We obtain what is called the **inverse relation.** Are these inverse relations functions?

COST OF A 60-SECOND SUPER BOWL COMMERCIAL, BY YEAR

Range (Set of Outputs)	Domain (Set of Inputs)
1981	$550,000
1983	$800,000
1988	$1,350,000
1991	$1,600,000
1995	$2,200,000
1997	$2,400,000

U.S. SENATORS AND THEIR STATES

Range (Set of Outputs)	Domain (Set of Inputs)
Wellstone	Minnesota
Grams	
Mack	Florida
Graham	
Lautenberg	New Jersey
Torricelli	

We see that the inverse of the first correspondence is a function, but that the inverse of the second correspondence is not a function.

Recall that for each input, a function provides exactly one output. However, a function can have the same output for two or more different inputs. Thus it is possible for different inputs to correspond to the same output. Only when this possibility is *excluded* will the inverse be a function.

In the Super Bowl function, different inputs have different outputs. It is an example of a **one-to-one function.** In the U.S. Senator function, *Wellstone* and *Grams* are both paired with *Minnesota*. Thus the U.S. Senator function is not one-to-one.

ONE-TO-ONE FUNCTION

A function f is *one-to-one* if different inputs have different outputs. That is, if for any $a \neq b$, we have $f(a) \neq f(b)$, the function f is one-to-one. If a function is one-to-one, then its inverse correspondence is also a function.

How can we tell graphically whether a function is one-to-one?

EXAMPLE   4

Shown here is the graph of an exponential function. Determine whether the function is one-to-one and thus has an inverse that is a function.

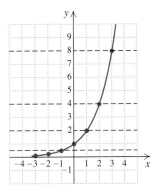

SOLUTION   A function is one-to-one if different inputs have different outputs. In other words, no two x-values will have the same y-value. For this function, we cannot find two x-values that have the same y-value. Note also that no horizontal line can be drawn so that it crosses the graph more than once. The function is one-to-one so its inverse is a function. ————

The graph of any function must pass the vertical-line test. In order for a function to have an inverse that is a function, it must pass the *horizontal-line test* as well.

THE HORIZONTAL-LINE TEST

A function is one-to-one, and thus has an inverse that is a function, if no horizontal line can cross its graph more than once.

EXAMPLE   5

Determine whether the function $f(x) = x^2$ is one-to-one and thus has an inverse that is a function.

SOLUTION   The graph of $f(x) = x^2$ is shown here. Many horizontal lines cross the graph more than once—in particular, the line $y = 4$. Note that where the line crosses, the first coordinates are -2 and 2. Although these are different inputs, they have the same output. That is, $-2 \neq 2$, but

$$f(-2) = (-2)^2 = 4 = 2^2 = f(2).$$

Thus the function is not one-to-one and no inverse function exists.

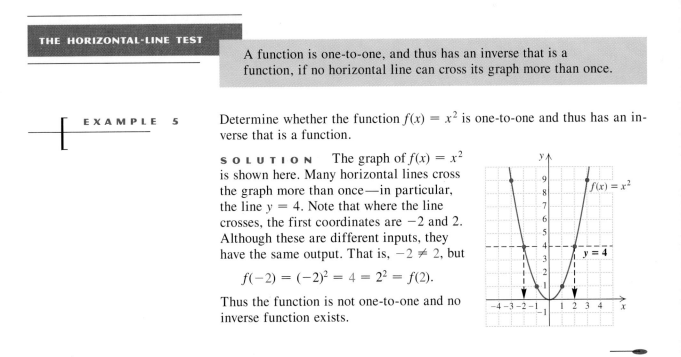

Finding Formulas for Inverses

When the inverse of f is also a function, it is denoted f^{-1} (read "f-inverse").

> **CAUTION!**   The -1 in f^{-1} is *not* an exponent!

Suppose a function is described by a formula. If it has an inverse that is a function, how do we find a formula for the inverse? For any equation in two variables, if we interchange the variables, we obtain an equation of the inverse correspondence. If it is a function, we proceed as follows to find a formula for f^{-1}.

TO FIND A FORMULA FOR f^{-1}:

First make sure that f is one-to-one. Then:

1. Replace $f(x)$ with y.
2. Interchange x and y. (This gives the inverse function.)
3. Solve for y.
4. Replace y with $f^{-1}(x)$. (This is inverse function notation.)

EXAMPLE 6   Determine if each function is one-to-one and if it is, find a formula for $f^{-1}(x)$:
(a) $f(x) = x + 2$; **(b)** $f(x) = 2x - 3$.

SOLUTION

a) The graph of $f(x) = x + 2$ is shown below. It passes the horizontal-line test, so it is one-to-one. Thus its inverse is a function.

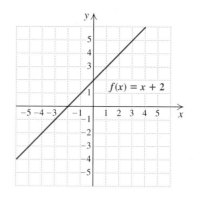

1. Replace $f(x)$ with y: $y = x + 2$.
2. Interchange x and y: $x = y + 2$.   This gives the inverse function.
3. Solve for y: $x - 2 = y$.
4. Replace y with $f^{-1}(x)$: $f^{-1}(x) = x - 2$.   We also "reversed" the equation.

In this case, the function f added 2 to all inputs. Thus, to "undo" f, the function f^{-1} must subtract 2 from its inputs.

b) The function $f(x) = 2x - 3$ is also linear. Any linear function that is not constant will pass the horizontal-line test. Thus, f is one-to-one.

1. Replace $f(x)$ with y: $y = 2x - 3.$

2. Interchange x and y: $x = 2y - 3.$

3. Solve for y: $x + 3 = 2y$

$$\frac{x + 3}{2} = y.$$

4. Replace y with $f^{-1}(x)$: $f^{-1}(x) = \dfrac{x + 3}{2}.$

Let us consider inverses of functions in terms of a function machine. Suppose that a one-to-one function f is programmed into a machine. If the machine has a reverse switch, when the switch is thrown, the machine performs the inverse function f^{-1}. Inputs then enter at the opposite end, and the entire process is reversed.

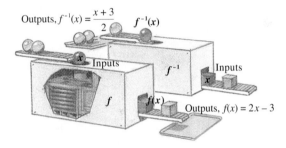

Consider $f(x) = 2x - 3$ and $f^{-1}(x) = \dfrac{x + 3}{2}$ from Example 6(b). For the input 5,

$$f(5) = 2 \cdot 5 - 3 = 10 - 3 = 7.$$

The output is 7. Now we use 7 for the input in the inverse:

$$f^{-1}(7) = \frac{7 + 3}{2} = \frac{10}{2} = 5.$$

The function f takes 5 to 7. The inverse function f^{-1} takes the number 7 back to 5.

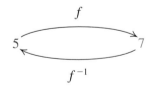

Graphing Functions and Their Inverses

How do the graphs of a function and its inverse compare?

EXAMPLE 7   Graph $f(x) = 2x - 3$ and $f^{-1}(x) = (x + 3)/2$ on the same set of axes. Then compare.

SOLUTION   The graph of each function follows. Note that the graph of f^{-1} can be drawn by reflecting the graph of f across the line $y = x$. That is, if we graph $f(x) = 2x - 3$ in wet ink and fold the paper along the line $y = x$, the graph of $f^{-1}(x) = (x + 3)/2$ will appear as the impression made by f.

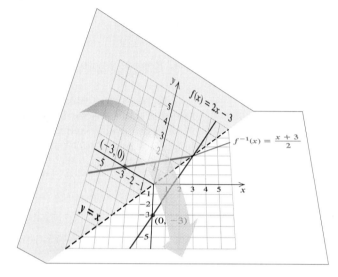

When x and y are interchanged to find a formula for the inverse, we are, in effect, flipping the graph of $f(x) = 2x - 3$ over the line $y = x$. For example, when the coordinates of the y-intercept of the graph of f, $(0, -3)$, are reversed, we get the x-intercept of the graph of f^{-1}, $(-3, 0)$.

VISUALIZING INVERSES

The graph of f^{-1} is a reflection of the graph of f across the line $y = x$.

EXAMPLE 8   Consider $g(x) = x^3 + 2$.

a) Determine whether the function is one-to-one.

b) If it is one-to-one, find a formula for its inverse.

c) Graph the inverse, if it exists.

SOLUTION

a) The graph of $g(x) = x^3 + 2$ is shown at right. It passes the horizontal-line test and thus has an inverse.

b) 1. Replace $g(x)$ with y: $y = x^3 + 2.$

   2. Interchange x and y: $x = y^3 + 2.$

   3. Solve for y: $x - 2 = y^3$

   $\sqrt[3]{x - 2} = y.$ Since a number has only one cube root, we can solve for y.

   4. Replace y with $g^{-1}(x)$: $g^{-1}(x) = \sqrt[3]{x - 2}.$

c) To find the graph, we reflect the graph of $g(x) = x^3 + 2$ across the line $y = x$, as we did in Example 7. It can also be found by substituting into $g^{-1}(x) = \sqrt[3]{x - 2}$ and plotting points. The graphs of g and g^{-1} are shown together at the bottom of the preceding page.   ———•

Inverse Functions and Composition

Suppose that we used some input x for the function f and found its output, $f(x)$. The function f^{-1} would then take that output back to x. Similarly, if we began with an input x for the function f^{-1} and found its output, $f^{-1}(x)$, the original function f would then take that output back to x. This is summarized as follows.

COMPOSITION AND INVERSES

If a function f is one-to-one, then f^{-1} is the unique function for which

$$f^{-1} \circ f(x) = x \quad \text{and} \quad f \circ f^{-1}(x) = x.$$

EXAMPLE   9   Let $f(x) = 2x + 1$. Show that

$$f^{-1}(x) = \frac{x - 1}{2}.$$

SOLUTION   We find $f^{-1} \circ f(x)$ and $f \circ f^{-1}(x)$ and check to see that each is x.

$$f^{-1} \circ f(x) = f^{-1}(f(x)) = f^{-1}(2x + 1)$$

$$= \frac{(2x + 1) - 1}{2}$$

$$= \frac{2x}{2} = x$$

$$f \circ f^{-1}(x) = f(f^{-1}(x)) = f\left(\frac{x - 1}{2}\right)$$

$$= 2 \cdot \frac{x - 1}{2} + 1$$

$$= x - 1 + 1 = x$$ ———•

TECHNOLOGY CONNECTION 9.2B   A grapher can provide a partial check of whether or not two functions are inverses of each other. First, we plot the functions believed to be inverses of each other and the line $y = x$, all on the same set of axes. Then, using the fact that a function and its inverse are reflections of each other across the line $y = x$, we can decide if the two functions may be inverses of each other.

Before we do this, however, there is a problem that must be addressed. Most graphers have rectangular, rather than square, displays. This means that a standard window, like $[-10, 10, -10, 10]$, distorts a graph because one unit in the x-direction is longer than one unit in the y-direction. To correct this, many graphers offer an option to "Square" axes, which guarantees that the unit length is the same in both the x- and y-directions.

To determine whether $y_1 = 2x + 6$ and $y_2 = \frac{1}{2}x - 3$ might be inverses of each other, we have drawn (to the right, above) both functions, along with the line $y = x$, on a "squared" set of axes. It *appears* that y_1 and y_2 are inverses of each other. For further verification, many graphers allow us to examine a table of values (see above right) in which $y_1 = 2x + 6$ and $y_2 = \frac{1}{2} \cdot y_1 - 3$. Note that y_2 "undoes" what y_1 "does."

A final, visual, check can be made on many graphers by graphing

$$y_1 = 2x + 6 \quad \text{and} \quad y_2 = \frac{1}{2}x - 3$$

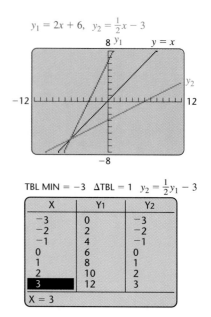

$y_1 = 2x + 6, \quad y_2 = \frac{1}{2}x - 3$

TBL MIN $= -3$ ΔTBL $= 1$ $y_2 = \frac{1}{2}y_1 - 3$

X	Y₁	Y₂
-3	0	-3
-2	2	-2
-1	4	-1
0	6	0
1	8	1
2	10	2
3	12	3

X = 3

and then pressing DRAW and selecting the DRAWINV option. Once this has been selected, we use Y-VARS to enter y_1. The resulting graph of the inverse of y_1 should coincide with y_2.

1. Use a grapher to check Examples 7, 8, and 9.
2. Will DRAWINV work for *any* choice of y_1? Why or why not?

EXERCISE SET

9.2

Find $f \circ g(x)$ and $g \circ f(x)$.

1. $f(x) = 3x^2 - 1$; $g(x) = 2x + 3$
2. $f(x) = 4x + 3$; $g(x) = 2x^2 - 5$
3. $f(x) = 4x^2 - 1$; $g(x) = 2/x$
4. $f(x) = 3/x$; $g(x) = 2x^2 + 3$
5. $f(x) = x^2 - 3$; $g(x) = x^2 + 1$
6. $f(x) = 1/x^2$; $g(x) = x + 2$

Find $f(x)$ and $g(x)$ such that $h(x) = f \circ g(x)$. Answers may vary.

7. $h(x) = (7 - 5x)^2$
8. $h(x) = 4(3x - 1)^2 + 9$
9. $h(x) = (3x^2 - 7)^5$
10. $h(x) = \sqrt{5x + 2}$

11. $h(x) = \dfrac{2}{x - 3}$
12. $h(x) = \dfrac{3}{x} + 4$
13. $h(x) = \dfrac{1}{\sqrt{7x + 2}}$
14. $h(x) = \sqrt{x - 7} - 3$
15. $h(x) = \dfrac{x^3 + 1}{x^3 - 1}$
16. $h(x) = (\sqrt{x} + 5)^4$

Determine whether each function is one-to-one.

17. $f(x) = x - 5$
18. $f(x) = 5 - 2x$
19. $f(x) = x^2 + 1$
20. $f(x) = 1 - x^2$
21. $g(x) = 3^x$
22. $g(x) = \left(\frac{1}{2}\right)^x$

23. $g(x) = |x|$ **24.** $h(x) = |x| - 1$

*For each function, **(a)** determine if it is one-to-one; **(b)** if it is one-to-one, find a formula for the inverse.*

25. $f(x) = x + 6$ **26.** $f(x) = x + 7$

27. $f(x) = 3 - x$ **28.** $f(x) = 9 - x$

29. $g(x) = x - 5$ **30.** $g(x) = x - 8$

31. $f(x) = 4x$ **32.** $f(x) = 7x$

33. $g(x) = 4x + 3$ **34.** $g(x) = 4x + 7$

35. $h(x) = 5$ **36.** $h(x) = -2$

37. $f(x) = \dfrac{1}{x}$ **38.** $f(x) = \dfrac{3}{x}$

39. $f(x) = \dfrac{2x + 1}{3}$ **40.** $f(x) = \dfrac{3x + 2}{5}$

41. $f(x) = x^3 - 5$ **42.** $f(x) = x^3 + 2$

43. $g(x) = (x - 2)^3$ **44.** $g(x) = (x + 7)^3$

45. $f(x) = \sqrt{x}$ **46.** $f(x) = \sqrt{x - 1}$

47. $f(x) = 2x^2 + 1, \ x \geq 0$

48. $f(x) = 3x^2 - 2, \ x \geq 0$

Graph each function and its inverse using the same set of axes.

49. $f(x) = \frac{1}{3}x - 2$ **50.** $g(x) = x + 4$

51. $f(x) = x^3$ **52.** $f(x) = x^3 - 1$

53. $y = 2^x$ **54.** $y = 3^x$

55. $y = \left(\frac{2}{3}\right)^x$ **56.** $y = \left(\frac{1}{2}\right)^x$

57. $f(x) = 3 - x^2, \ x \geq 0$

58. $f(x) = x^2 - 1, \ x \leq 0$

59. Let $f(x) = \frac{4}{5}x$. Show that
$$f^{-1}(x) = \tfrac{5}{4}x.$$

60. Let $f(x) = (x + 7)/3$. Show that
$$f^{-1}(x) = 3x - 7.$$

61. Let $f(x) = (1 - x)/x$. Show that
$$f^{-1}(x) = \frac{1}{x + 1}.$$

62. Let $f(x) = x^3 - 5$. Show that
$$f^{-1}(x) = \sqrt[3]{x + 5}.$$

63. *Dress sizes in the United States and France.* A size-6 dress in the United States is size 38 in France. A function that converts dress sizes in the United States to those in France is
$$f(x) = x + 32.$$

a) Find the dress sizes in France that correspond to sizes 8, 10, 14, and 18 in the United States.

b) Determine whether this function has an inverse that is a function. If so, find a formula for the inverse.

c) Use the inverse function to find dress sizes in the United States that correspond to sizes 40, 42, 46, and 50 in France.

64. *Dress sizes in the United States and Italy.* A size-6 dress in the United States is size 36 in Italy. A function that converts dress sizes in the United States to those in Italy is
$$f(x) = 2(x + 12).$$

a) Find the dress sizes in Italy that correspond to sizes 8, 10, 14, and 18 in the United States.

b) Determine whether this function has an inverse that is a function. If so, find a formula for the inverse. •

c) Use the inverse function to find dress sizes in the United States that correspond to sizes 40, 44, 52, and 60 in Italy.

SKILL MAINTENANCE

65. Find an equation of variation if y varies directly as x, and $y = 7.2$ when $x = 0.8$.

66. Find an equation of variation if y varies inversely as x, and $y = 3.5$ when $x = 6.1$.

Simplify.

67. $(a^3b^2)^5(a^2b^7)$ **68.** $(x^5y^3z^2)(x^2yz^2)^3$

SYNTHESIS

69. ◈ Mathematicians usually try to select "logical" words when forming definitions. Does the term "one-to-one" seem logical? Why or why not?

70. ◈ Does the constant function $f(x) = 4$ have an inverse that is a function? If so, find a formula. If not, explain why.

71. ◈ An organization determines that the cost per person of chartering a bus is given by the function
$$C(x) = \frac{100 + 5x}{x},$$
where $x =$ the number of people in the group and $C(x)$ is in dollars. Determine $C^{-1}(x)$ and explain how this inverse function could be used.

72. ◆ The function $V(t) = 750(1.2)^t$ is used to predict the value, $V(t)$, of a certain rare stamp t years from 1997. Do not calculate $V^{-1}(t)$, but explain how V^{-1} could be used.

73. *Dress sizes in France and Italy.* Use the information in Exercises 63 and 64 to find a function for the French dress size that corresponds to a size x dress in Italy.

74. ◆ Is function composition associative? That is, is it true that for any choices of f, g, and h,

$$f \circ (g \circ h) = (f \circ g) \circ h?$$

Why or why not?

 *In Exercises 75–78, use a grapher to help determine whether or not the given functions are inverses of each other.*

75. $f(x) = 0.75x^2 + 2$; $g(x) = \sqrt{\dfrac{4(x-2)}{3}}$

76. $f(x) = 1.4x^3 + 3.2$; $g(x) = \sqrt[3]{\dfrac{x - 3.2}{1.4}}$

77. $f(x) = \sqrt{2.5x + 9.25}$;
$g(x) = 0.4x^2 - 3.7$, $x \geq 0$

78. $f(x) = 0.8x^{1/2} + 5.23$;
$g(x) = 1.25(x^2 - 5.23)$, $x \geq 0$

79.  Use a grapher to help match each function in Column A with its inverse from Column B.

Column A

(1) $y = 5x^3 + 10$

(2) $y = (5x + 10)^3$

(3) $y = 5(x + 10)^3$

(4) $y = (5x)^3 + 10$

Column B

A. $y = \dfrac{\sqrt[3]{x} - 10}{5}$

B. $y = \sqrt[3]{\dfrac{x}{5}} - 10$

C. $y = \sqrt[3]{\dfrac{x - 10}{5}}$

D. $y = \dfrac{\sqrt[3]{x - 10}}{5}$

80. ◆  How could a grapher be used to determine if a function is one-to-one?

9.3 Logarithmic Functions

Graphs of Logarithmic Functions • Converting Exponential and Logarithmic Equations • Solving Certain Logarithmic Equations

---

TECHNOLOGY CONNECTION

9.3

To see that $f(x) = 10^x$ and $g(x) = \log_{10} x$ are inverses of each other, let $y_1 = 10^x$ and $y_2 = \log_{10} x = \log x$. Then, using a squared window, compare both graphs. Finally, let $y_3 = y_1(y_2)$ and $y_4 = y_2(y_1)$ to show, using a table or graphs, that $y_3 = y_4 = x$.

We are now ready to study inverses of exponential functions. These functions have many applications and are referred to as *logarithm*, or *logarithmic*, *functions*.

Graphs of Logarithmic Functions

Consider the exponential function $f(x) = 2^x$. Like all exponential functions, f is one-to-one. Can a formula for f^{-1} be found? To answer this, we use the method of Section 9.2:

1. Replace $f(x)$ with y:

$$y = 2^x.$$

2. Interchange x and y:

$$x = 2^y.$$

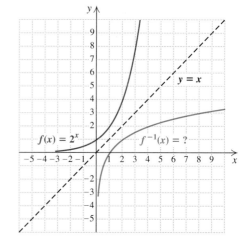

3. Solve for y: $y =$ the power to which we raise 2 to get x.

4. Replace y with $f^{-1}(x)$: $f^{-1}(x) =$ the power to which we raise 2 to get x.

We now define a new symbol to replace the words "the power to which we raise 2 to get x":

> $\log_2 x$, read "the logarithm, base 2, of x", or "log, base 2, of x," means "the power to which we raise 2 to get x."

Thus if $f(x) = 2^x$, then $f^{-1}(x) = \log_2 x$. Note that $f^{-1}(8) = \log_2 8 = 3$, because 3 is *the power to which we raise 2 to get* 8.

Although expressions like $\log_2 13$ can only be approximated, we must remember that $\log_2 13$ represents *the power to which we raise 2 to get* 13. That is, $2^{\log_2 13} = 13$.

For any exponential function $f(x) = a^x$, the inverse is called a **logarithmic function, base a.** The graph of the inverse can, of course, be drawn by reflecting the graph of $f(x) = a^x$ across the line $y = x$. It will be helpful to remember that the inverse of $f(x) = a^x$ is given by $f^{-1}(x) = \log_a x$. Normally, we use a number a that is greater than 1 for the logarithm base.

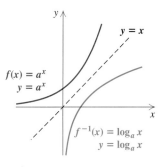

THE MEANING OF $\log_a x$

> For $x > 0$ and a a positive constant other than 1, $\log_a x$ is the number to which a is raised to get x. Thus,
>
> $$a^{\log_a x} = x \qquad \text{or equivalently,} \qquad \text{if } y = \log_a x, \text{ then } a^y = x.$$

It is important to remember that *the logarithm of a number is an exponent.* It might help to repeat to yourself several times: "The logarithm, base a, of a number x is the power to which a must be raised in order to get x."

EXAMPLE 1   Simplify: **(a)** $\log_{10} 1000$; **(b)** $7^{\log_7 13}$; **(c)** $\log_4 1$.

SOLUTION

a) Think of the meaning of $\log_{10} 1000$. It is the exponent to which we raise 10 to get 1000. That exponent is 3. Therefore, $\log_{10} 1000 = 3$.

b) It is important to remember what $\log_7 13$ is:

$\log_7 13$ is the power to which 7 is raised to get 13.

Thus, since $\log_7 13$ is the power to which 7 is raised to get 13,

$$7^{\log_7 13} = 13.$$

c) We ask ourselves: "To what power do we raise 4 in order to get 1?" That power is 0 (recall that $4^0 = 1$). Thus, $\log_4 1 = 0$.

The following is a comparison of exponential and logarithmic functions.

Exponential Function	Logarithmic Function
$y = a^x$	$x = a^y$
$f(x) = a^x$	$f(x) = \log_a x$
$a > 0, a \neq 1$	$a > 0, a \neq 1$
The input x can be any real number.	The output y can be any real number.
$y > 0$ (Outputs are positive.)	$x > 0$ (Inputs are positive.)

EXAMPLE 2

Graph: $y = f(x) = \log_5 x$.

SOLUTION   The equation $y = \log_5 x$ is equivalent to $5^y = x$. We can find ordered pairs that are solutions by choosing values for y and computing the x-values.

For $y = 0$, $x = 5^0 = 1$.
For $y = 1$, $x = 5^1 = 5$.
For $y = 2$, $x = 5^2 = 25$.
For $y = -1$, $x = 5^{-1} = \frac{1}{5}$.
For $y = -2$, $x = 5^{-2} = \frac{1}{25}$.

This table shows the following:

$\left.\begin{array}{l} \log_5 1 = 0; \\ \log_5 5 = 1; \\ \log_5 25 = 2; \\ \log_5 \frac{1}{5} = -1; \\ \log_5 \frac{1}{25} = -2. \end{array}\right\}$ These can all be checked using the equations above.

We plot the set of ordered pairs and connect the points with a smooth curve. The graph of $y = 5^x$ is shown only for reference.

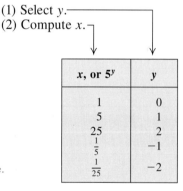

(1) Select y.
(2) Compute x.

x, or 5^y	y
1	0
5	1
25	2
$\frac{1}{5}$	-1
$\frac{1}{25}$	-2

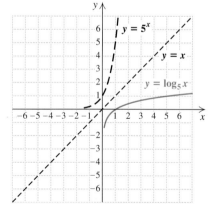

Converting Exponential and Logarithmic Equations

We use the definition of logarithm to convert from *exponential equations* to *logarithmic equations*:

$$y = \log_a x \quad \text{is equivalent to} \quad a^y = x.$$

> **CAUTION!** **Be sure to memorize this relationship!** It is probably the most important definition in the chapter. Many times this definition will serve as a justification for a property we are considering.

EXAMPLE 3   Convert each to a logarithmic equation: **(a)** $8 = 2^x$; **(b)** $y^{-1} = 4$; **(c)** $a^b = c$.

SOLUTION

a) $8 = 2^x$ is equivalent to $x = \log_2 8$ The exponent is the logarithm.

The base remains the same.

b) $y^{-1} = 4$ is equivalent to $-1 = \log_y 4$

c) $a^b = c$ is equivalent to $b = \log_a c$

We also use the definition of logarithm to convert from logarithmic equations to exponential equations.

EXAMPLE 4   Convert each to an exponential equation: **(a)** $y = \log_3 5$; **(b)** $-2 = \log_a 7$; **(c)** $a = \log_b d$.

SOLUTION

a) $y = \log_3 5$ is equivalent to $3^y = 5$ The logarithm is the exponent.

The base remains the same.

b) $-2 = \log_a 7$ is equivalent to $a^{-2} = 7$

c) $a = \log_b d$ is equivalent to $b^a = d$

Solving Certain Logarithmic Equations

Some logarithmic equations can be solved by converting to exponential equations.

EXAMPLE 5   Solve: **(a)** $\log_2 x = -3$; **(b)** $\log_x 16 = 2$.

SOLUTION

a) $\log_2 x = -3$

$\quad\quad 2^{-3} = x$ Converting to an exponential equation

$\quad\quad \frac{1}{8} = x$ Computing 2^{-3}

Check: $\log_2 \frac{1}{8}$ is the power to which 2 is raised to get $\frac{1}{8}$. Since that power is -3, we have a check. The solution is $\frac{1}{8}$.

b) $\log_x 16 = 2$

$$x^2 = 16 \qquad \text{Converting to an exponential equation}$$

$$x = 4 \quad or \quad x = -4 \qquad \text{Principle of square roots}$$

Check: $\log_4 16 = 2$ because $4^2 = 16$. Thus, 4 is a solution. Because all logarithm bases must be positive, -4 cannot be a solution. Logarithm bases must be positive because logarithms are defined using exponential functions that require positive bases. The solution is 4.   ━━●

One method for solving certain logarithmic and exponential equations relies on the following property, which results from the fact that exponential functions are one-to-one.

THE PRINCIPLE OF EXPONENTIAL EQUALITY

For any real number b, where $b \neq 0$ and $b \neq 1$,

$$b^x = b^y \quad \text{is equivalent to} \quad x = y.$$

EXAMPLE 6

Solve: **(a)** $\log_{10} 1000 = x$; **(b)** $\log_4 1 = t$.

SOLUTION

a) We convert $\log_{10} 1000 = x$ to exponential form and solve:

$$10^x = 1000 \qquad \text{Converting to an exponential equation}$$

$$10^x = 10^3 \qquad \text{Writing 1000 as a power of 10}$$

$$x = 3. \qquad \text{Equating exponents}$$

Check: This equation can also be solved using the definition of logarithm, exactly as we did in Example 1(a). Since in both cases we find that $\log_{10} 1000 = 3$, we have a check. The solution is 3.

b) We convert $\log_4 1 = t$ to exponential form and solve:

$$4^t = 1 \qquad \text{Converting to an exponential equation}$$

$$4^t = 4^0 \qquad \text{Writing 1 as a power of 4}$$

$$t = 0. \qquad \text{Equating exponents}$$

Check: As in part (a), this equation can be solved using the definition of logarithm. This is precisely what we did in Example 1(c). Since in both cases we find that $\log_4 1 = 0$, we have a check. The solution is 0.   ━━●

Example 6(b) illustrates an important property of logarithms.

$\log_a 1$

| The logarithm, base a, of 1 is always 0: $\log_a 1 = 0$. |

This follows from the fact that $a^0 = 1$ is equivalent to the logarithmic equation $\log_a 1 = 0$. Thus, $\log_{10} 1 = 0$, $\log_7 1 = 0$, and so on.

Another property results from the fact that $a^1 = a$. This is equivalent to the equation $\log_a a = 1$.

$\log_a a$

| The logarithm, base a, of a is always 1: $\log_a a = 1$. |

Thus, $\log_{10} 10 = 1$, $\log_8 8 = 1$, and so on.

EXERCISE SET

9.3

Graph.

1. $y = \log_2 x$ **2.** $y = \log_{10} x$

3. $y = \log_7 x$ **4.** $y = \log_3 x$

5. $f(x) = \log_4 x$ **6.** $f(x) = \log_6 x$

7. $f(x) = \log_{1/2} x$ **8.** $f(x) = \log_{2.5} x$

Graph both functions using the same set of axes.

9. $f(x) = 3^x$, $f^{-1}(x) = \log_3 x$

10. $f(x) = 4^x$, $f^{-1}(x) = \log_4 x$

Convert to logarithmic equations.

11. $10^4 = 10{,}000$ **12.** $10^2 = 100$

13. $5^{-3} = \frac{1}{125}$ **14.** $4^{-5} = \frac{1}{1024}$

15. $8^{1/3} = 2$ **16.** $16^{3/4} = 8$

17. $10^{0.3010} = 2$ **18.** $10^{0.4771} = 3$

19. $m^n = r$ **20.** $p^k = 3$

21. $Q^t = x$ **22.** $p^m = V$

23. $e^2 = 7.3891$ **24.** $e^3 = 20.0855$

25. $e^{-2} = 0.1353$ **26.** $e^{-4} = 0.0183$

Convert to exponential equations.

27. $t = \log_3 8$ **28.** $h = \log_7 10$

29. $\log_5 25 = 2$ **30.** $\log_6 6 = 1$

31. $\log_{10} 0.1 = -1$ **32.** $\log_{10} 0.01 = -2$

33. $\log_{10} 7 = 0.845$ **34.** $\log_{10} 3 = 0.4771$

35. $\log_c m = 17$ **36.** $\log_b n = 23$

37. $\log_t Q = k$ **38.** $\log_m P = a$

39. $\log_e 0.25 = -1.3863$

40. $\log_e 0.989 = -0.0111$

41. $\log_r T = -x$

42. $\log_c M = -w$

Solve.

43. $\log_3 x = 4$ **44.** $\log_4 x = 2$

45. $\log_x 125 = 3$ **46.** $\log_x 64 = 3$

47. $\log_2 16 = x$ **48.** $\log_5 25 = x$

49. $\log_3 27 = x$ **50.** $\log_4 16 = x$

51. $\log_x 8 = 1$ **52.** $\log_x 7 = 1$

53. $\log_6 x = 0$ **54.** $\log_9 x = 1$

55. $\log_2 x = -1$ **56.** $\log_3 x = -2$

57. $\log_8 x = \frac{2}{3}$ **58.** $\log_{32} x = \frac{2}{5}$

Find each of the following.

59. $\log_{10} 10{,}000$ **60.** $\log_{10} 100{,}000$

61. $\log_{10} 1$ **62.** $\log_{10} 10$

63. $\log_5 625$

64. $\log_6 1$

65. $\log_5 \frac{1}{25}$

66. $\log_4 64$

67. $\log_3 3$

68. $\log_2 \frac{1}{16}$

69. $\log_7 1$

70. $\log_2 2$

71. $6^{\log_6 15}$

72. $7^{\log_7 23}$

73. $\log_{27} 9$

74. $\log_8 2$

75. $\log_b b^7$

76. $\log_n n^8$

SKILL MAINTENANCE

Simplify.

77.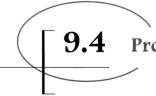

78. $\dfrac{\dfrac{4+x}{x^2+2x+1}}{\dfrac{3}{x+1} - \dfrac{2}{x+2}}$

Rename without using exponents.

79. 8^{-4}

80. $x^{1/5}$

81. $t^{-1/3}$

82. 5^1

SYNTHESIS

83. ◈ Express in words what number is represented by $\log_b c$.

84. ◈ Is it true that $2 = b^{\log_b 2}$? Why or why not?

85. ◈ Would a manufacturer be pleased or unhappy if sales of a product grew logarithmically? Why?

86. ◈ Explain why the number $\log_2 13$ must be between 3 and 4.

87. Graph both equations using the same set of axes:
$$y = \left(\tfrac{3}{2}\right)^x, \qquad y = \log_{3/2} x.$$

Graph.

88. $y = \log_2 (x - 1)$

89. $y = \log_3 |x + 1|$

Solve.

90. $|\log_3 x| = 2$

91. $\log_{125} x = \frac{2}{3}$

92. $\log_4 (3x - 2) = 2$

93. $\log_8 (2x + 1) = -1$

94. $\log_{10} (x^2 + 21x) = 2$

Simplify.

95. $\log_{1/4} \frac{1}{64}$

96. $\log_{1/5} 25$

97. $\log_{81} 3 \cdot \log_3 81$

98. $\log_{10} (\log_4 (\log_3 81))$

99. $\log_2 (\log_2 (\log_4 256))$

100. Show that $b^x = b^y$ is *not* equivalent to $x = y$ for $b = 0$ or $b = 1$.

9.4 Properties of Logarithmic Functions

Logarithms of Products • Logarithms of Powers •
Logarithms of Quotients • Using the Properties Together

Logarithmic functions are important in many applications and in more advanced mathematics. We now establish some basic properties that are useful in manipulating expressions involving logarithms. As their proofs reveal, the properties of logarithms are related to the properties of exponents.

Logarithms of Products

The first property we discuss is reminiscent of the property $a^m \cdot a^n = a^{m+n}$.

THE PRODUCT RULE FOR LOGARITHMS

For any positive numbers M, N, and a ($a \neq 1$),
$$\log_a MN = \log_a M + \log_a N.$$

(The logarithm of a product is the sum of the logarithms of the factors.)

EXAMPLE 1

Express as a sum of logarithms: $\log_2 (4 \cdot 16)$.

SOLUTION   We have

$$\log_2 (4 \cdot 16) = \log_2 4 + \log_2 16. \qquad \text{Using the product rule}$$

As a check, note that

$$\log_2 (4 \cdot 16) = \log_2 64 = 6$$

and that

$$\log_2 4 + \log_2 16 = 2 + 4 = 6.$$

EXAMPLE 2

Express as a single logarithm: $\log_{10} 0.01 + \log_{10} 1000$.

SOLUTION   We have

$$\log_{10} 0.01 + \log_{10} 1000 = \log_{10} (0.01 \times 1000) \qquad \text{Using the product rule}$$
$$= \log_{10} 10.$$

The check is left to the student.

A Proof of the Product Rule.   Let $\log_a M = x$ and $\log_a N = y$. Converting to exponential equations, we have $a^x = M$ and $a^y = N$.

Now we multiply the latter two equations, to obtain

$$MN = a^x \cdot a^y, \quad \text{or} \quad MN = a^{x+y}.$$

Converting back to a logarithmic equation, we get

$$\log_a MN = x + y.$$

Recalling what x and y represent, we get

$$\log_a MN = \log_a M + \log_a N.$$

Logarithms of Powers

The second basic property is related to the property $(a^m)^n = a^{mn}$.

THE POWER RULE FOR LOGARITHMS

For any positive numbers M and a ($a \neq 1$), and any real number p,

$$\log_a M^p = p \cdot \log_a M.$$

(The logarithm of a power of M is the exponent times the logarithm of M.)

EXAMPLE 3

Express as a product: **(a)** $\log_a 9^{-5}$; **(b)** $\log_7 \sqrt[3]{x}$.

SOLUTION

a) $\log_a 9^{-5} = -5 \log_a 9$ \qquad Using the power rule

b) $\log_7 \sqrt[3]{x} = \log_7 x^{1/3}$ \qquad Writing exponential notation

$\qquad\qquad = \frac{1}{3} \log_7 x$ \qquad Using the power rule

A Proof of the Power Rule.   Let $x = \log_a M$. We then convert to an exponential equation, to get $a^x = M$. Raising both sides to the pth power, we obtain

$$(a^x)^p = M^p, \quad \text{or} \quad a^{xp} = M^p.$$

Converting back to a logarithmic equation gives us

$$\log_a M^p = xp.$$

But $x = \log_a M$, so substituting, we have

$$\log_a M^p = (\log_a M)p = p \cdot \log_a M.$$

Logarithms of Quotients

The third property that we study is similar to the property $a^m/a^n = a^{m-n}$.

**THE QUOTIENT RULE
FOR LOGARITHMS**

For any positive numbers M, N, and a ($a \neq 1$),

$$\log_a \frac{M}{N} = \log_a M - \log_a N.$$

(The logarithm of a quotient is the logarithm of the dividend minus the logarithm of the divisor.)

EXAMPLE   4   Express as a difference of logarithms: $\log_t (6/U)$.

SOLUTION

$$\log_t \frac{6}{U} = \log_t 6 - \log_t U \qquad \text{Using the quotient rule}$$

EXAMPLE   5   Express as a single logarithm: $\log_b 17 - \log_b 27$.

SOLUTION

$$\log_b 17 - \log_b 27 = \log_b \frac{17}{27} \qquad \text{Using the quotient rule "in reverse"}$$

A Proof of the Quotient Rule.   Our proof uses both the product and power rules:

$$\log_a \frac{M}{N} = \log_a MN^{-1} \qquad \text{Rewriting } \frac{M}{N} \text{ with a negative exponent}$$

$$= \log_a M + \log_a N^{-1} \qquad \text{Using the product rule}$$

$$= \log_a M + (-1) \log_a N \qquad \text{Using the power rule}$$

$$= \log_a M - \log_a N.$$

Using the Properties Together

EXAMPLE 6

Express in terms of logarithms of x, y, and z.

a) $\log_b \dfrac{x^3}{yz}$

b) $\log_a \sqrt[4]{\dfrac{xy}{z^3}}$

SOLUTION

a) $\log_b \dfrac{x^3}{yz} = \log_b x^3 - \log_b yz$ Using the quotient rule

$\qquad = 3 \log_b x - \log_b yz$ Using the power rule

$\qquad = 3 \log_b x - (\log_b y + \log_b z)$ Using the product rule. Because of the subtraction, parentheses are essential!

$\qquad = 3 \log_b x - \log_b y - \log_b z$ Using the distributive law

b) $\log_a \sqrt[4]{\dfrac{xy}{z^3}} = \log_a \left(\dfrac{xy}{z^3}\right)^{1/4}$ Writing exponential notation

$\qquad = \dfrac{1}{4} \cdot \log_a \dfrac{xy}{z^3}$ Using the power rule

$\qquad = \dfrac{1}{4}(\log_a xy - \log_a z^3)$ Using the quotient rule. Parentheses are important.

$\qquad = \dfrac{1}{4}(\log_a x + \log_a y - 3 \log_a z)$ Using the product and power rules

---

CAUTION!   When subtraction or multiplication precedes use of the product or quotient rule, parentheses are needed, as in Example 6.

EXAMPLE 7

Express as a single logarithm.

a) $\dfrac{1}{2} \log_a x - 7 \log_a y + \log_a z$

b) $\log_a \dfrac{b}{\sqrt{x}} + \log_a \sqrt{bx}$

SOLUTION

a) $\dfrac{1}{2} \log_a x - 7 \log_a y + \log_a z$

$\qquad = \log_a x^{1/2} - \log_a y^7 + \log_a z$ Using the power rule

$\qquad = (\log_a \sqrt{x} - \log_a y^7) + \log_a z$ Using parentheses to emphasize the order of operations; $x^{1/2} = \sqrt{x}$

$\qquad = \log_a \dfrac{\sqrt{x}}{y^7} + \log_a z$ Using the quotient rule

$\qquad = \log_a \dfrac{z\sqrt{x}}{y^7}$ Using the product rule

b) $\log_a \dfrac{b}{\sqrt{x}} + \log_a \sqrt{bx} = \log_a \dfrac{b \ \sqrt{bx}}{\sqrt{x}}$ Using the product rule

$\qquad\qquad\qquad\qquad\quad = \log_a b\sqrt{b}$ Removing a factor equal to 1: $\dfrac{\sqrt{x}}{\sqrt{x}} = 1$

$\qquad\qquad\qquad\qquad\quad = \log_a b^{3/2}$, or $\dfrac{3}{2} \log_a b$ Since $b\sqrt{b} = b^1 \cdot b^{1/2}$

If we know the logarithms of two different numbers (to the same base), the properties allow us to calculate other logarithms.

EXAMPLE   8   Given $\log_a 2 = 0.301$ and $\log_a 3 = 0.477$, find each of the following.

a) $\log_a 6$ **b)** $\log_a \frac{2}{3}$ **c)** $\log_a 81$
d) $\log_a \frac{1}{3}$ **e)** $\log_a 2a$ **f)** $\log_a 5$

SOLUTION

a) $\log_a 6 = \log_a (2 \cdot 3) = \log_a 2 + \log_a 3$ Using the product rule
$\qquad\qquad\qquad\quad = 0.301 + 0.477 = 0.778$

b) $\log_a \frac{2}{3} = \log_a 2 - \log_a 3$ Using the quotient rule
$\qquad\quad = 0.301 - 0.477 = -0.176$

c) $\log_a 81 = \log_a 3^4 = 4 \log_a 3$ Using the power rule
$\qquad\qquad\quad = 4(0.477) = 1.908$

d) $\log_a \frac{1}{3} = \log_a 1 - \log_a 3$ Using the quotient rule
$\qquad\quad = 0 - 0.477 = -0.477$

e) $\log_a 2a = \log_a 2 + \log_a a$ Using the product rule
$\qquad\qquad = 0.301 + 1 = 1.301$

f) $\log_a 5$ *cannot be found using these properties.*
$(\log_a 5 \neq \log_a 2 + \log_a 3)$

A final property follows from the product rule: Since $\log_a a^k = k \log_a a$, and $\log_a a = 1$, we have $\log_a a^k = k$.

THE LOGARITHM OF THE BASE TO A POWER

For any base a,
$$\log_a a^k = k.$$
(The logarithm, base a, of a to a power is the power.)

This property also follows from the definition of logarithm: k is the power to which you raise a in order to get a^k.

EXAMPLE   9   Simplify: **(a)** $\log_3 3^7$; **(b)** $\log_{10} 10^{-5.2}$.

SOLUTION

a) $\log_3 3^7 = 7$ 7 is the power to which you raise 3 in order to get 3^7.

b) $\log_{10} 10^{-5.2} = -5.2$

We summarize the properties covered in this section as follows.

For any positive numbers M, N, and a ($a \neq 1$):

$$\log_a MN = \log_a M + \log_a N; \qquad \log_a M^p = p \cdot \log_a M;$$

$$\log_a \frac{M}{N} = \log_a M - \log_a N; \qquad \log_a a^k = k$$

CAUTION!   Keep in mind that, in general,

$$\log_a (M + N) \neq \log_a M + \log_a N, \qquad \log_a MN \neq (\log_a M)(\log_a N),$$

$$\log_a (M - N) \neq \log_a M - \log_a N, \qquad \log_a (M/N) \neq (\log_a M) \div (\log_a N)$$

EXERCISE SET
9.4

Express as a sum of logarithms.

1. $\log_3 (81 \cdot 27)$ **2.** $\log_2 (16 \cdot 32)$

3. $\log_4 (64 \cdot 16)$ **4.** $\log_5 (25 \cdot 125)$

5. $\log_c xyz$ **6.** $\log_t 3ab$

Express as a single logarithm.

7. $\log_a 5 + \log_a 14$ **8.** $\log_b 65 + \log_b 2$

9. $\log_c t + \log_c y$ **10.** $\log_t H + \log_t M$

Express as a product.

11. $\log_a t^7$ **12.** $\log_b t^5$ **13.** $\log_c y^6$

14. $\log_{10} y^7$ **15.** $\log_b C^{-3}$ **16.** $\log_c M^{-5}$

Express as a difference of logarithms.

17. $\log_2 \dfrac{64}{16}$ **18.** $\log_3 \dfrac{27}{9}$

19. $\log_b \dfrac{m}{n}$ **20.** $\log_a \dfrac{y}{x}$

Express as a single logarithm.

21. $\log_a 15 - \log_a 7$ **22.** $\log_b 42 - \log_b 7$

Express in terms of logarithms of w, x, y, and z.

23. $\log_a x^2 y^3 z$ **24.** $\log_a xy^4 z^3$ **25.** $\log_b \dfrac{xy^2}{z^3}$

26. $\log_b \dfrac{x^2 y^5}{w^4 z^7}$ **27.** $\log_a \dfrac{x^2}{y^3 z}$ **28.** $\log_a \dfrac{x^4}{yz^2}$

29. $\log_b \dfrac{xy^2}{wz^3}$ **30.** $\log_b \dfrac{w^2 x}{y^3 z}$

31. $\log_a \sqrt{\dfrac{x^6}{y^5 z^8}}$ **32.** $\log_c \sqrt[3]{\dfrac{x^4}{y^3 z^2}}$

33. $\log_a \sqrt[3]{\dfrac{x^6 y^3}{a^2 z^7}}$ **34.** $\log_a \sqrt[4]{\dfrac{x^8 y^{12}}{a^3 z^5}}$

Express as a single logarithm and, if possible, simplify.

35. $4 \log_a x + 3 \log_a y$ **36.** $2 \log_b m + \frac{1}{2} \log_b n$

37. $\log_a x^2 - 2 \log_a \sqrt{x}$ **38.** $\log_a \dfrac{a}{\sqrt{x}} - \log_a \sqrt{ax}$

39. $\frac{1}{2} \log_a x + 3 \log_a y - 2 \log_a x$

40. $\log_a 2x + 3(\log_a x - \log_a y)$

41. $\log_a (x^2 - 4) - \log_a (x - 2)$

42. $\log_a (2x + 10) - \log_a (x^2 - 25)$

Given $\log_b 3 = 1.099$ and $\log_b 5 = 1.609$. If possible, find each of the following.

43. $\log_b 15$ **44.** $\log_b \frac{5}{3}$ **45.** $\log_b \frac{3}{5}$

46. $\log_b \frac{1}{3}$ **47.** $\log_b \frac{1}{5}$ **48.** $\log_b \sqrt{b}$

49. $\log_b \sqrt{b^3}$ **50.** $\log_b 3b$ **51.** $\log_b 6$

52. $\log_b 45$ **53.** $\log_b 75$ **54.** $\log_b 20$

Simplify.

55. $\log_t t^9$ **56.** $\log_p p^4$

57. $\log_e e^m$ **58.** $\log_Q Q^{-2}$

SKILL MAINTENANCE

Compute and simplify. Express answers in the form $a + bi$, where $i^2 = -1$.

59. i^{29} **60.** $(2 + i)^2$ **61.** $5i(2 - i)$

62. i^{34} **63.** $(5 - 3i)^2$ **64.** $7i(i - 4)$

SYNTHESIS

65. ◆ Is it possible to express $\log_b \frac{x}{5}$ as a difference of two logarithms without using the quotient rule? Why or why not?

66. ◆ Is it true that $\log_a x + \log_b x = \log_{ab} x$? Why or why not?

67. ◆ A student *incorrectly* reasons that

$$\log_b \frac{1}{x} = \log_b \frac{x}{xx}$$

$$= \log_b x - \log_b x + \log_b x = \log_b x.$$

What mistake has the student made?

Express as a single logarithm and, if possible, simplify.

68. $\log_a (x^8 - y^8) - \log_a (x^2 + y^2)$

69. $\log_a (x + y) + \log_a (x^2 - xy + y^2)$

Express as a sum or difference of logarithms.

70. $\log_a \sqrt{1 - s^2}$ **71.** $\log_a \frac{c - d}{\sqrt{c^2 - d^2}}$

72. If $\log_a x = 2$, $\log_a y = 3$, and $\log_a z = 4$, what is

$$\log_a \frac{\sqrt[3]{x^2 z}}{\sqrt[3]{y^2 z^{-2}}}?$$

73. If $\log_a x = 2$, what is $\log_a (1/x)$?

74. If $\log_a x = 2$, what is $\log_{1/a} x$?

Classify each of the following as true or false. Assume a, x, P, and Q > 0.

75. $\log_a \left(\frac{P}{Q}\right)^x = x \log_a P - \log_a Q$

76. $\log_a (Q + Q^2) = \log_a Q + \log_a (Q + 1)$

77. ◪ Use graphs to show that
$$\log x^2 \neq \log x \cdot \log x. \quad \text{(Note: log means } \log_{10}.\text{)}$$

9.5 Common and Natural Logarithms

Common Logarithms on a Calculator   •   The Base e and Natural Logarithms on a Calculator   •   Changing Logarithmic Bases   •   Graphs of Exponential and Logarithmic Functions, Base e

---

Any positive number other than 1 can serve as the base of a logarithmic function. However, some numbers are easier to use than others, and there are logarithm bases that fit into certain applications more naturally than others. Base-10 logarithms, called **common logarithms,** are useful because they have the same base as our "commonly" used decimal system.

Before calculators became so widely available, common logarithms were helpful when performing tedious calculations. In fact, that is why logarithms were invented. Another logarithm base widely used today is an irrational number named *e*. We will consider *e* and base *e*, or *natural*, logarithms later in this section. First we examine common logarithms.

Common Logarithms on a Calculator

Before the invention of calculators, tables were developed to list common logarithms. Today we find common logarithms using calculators.

The abbreviation **log,** with no base written, is understood to mean logarithm base 10, or a common logarithm. Thus,

$$\log 17 \quad \text{means} \quad \log_{10} 17.$$

On scientific calculators, the key for common logarithms is usually marked $\boxed{\text{log}}$. To find the common logarithm of a number, we enter that number and press the $\boxed{\text{log}}$ key. On some calculators, we press $\boxed{\text{log}}$, the number, and then $\boxed{\text{ENTER}}$.

EXAMPLE 1

Use a calculator to find each number: **(a)** log 53,128; **(b)** log 0.000128.

SOLUTION

a) We enter 53,128 and then press the $\boxed{\text{log}}$ key. We find that

$$\log 53{,}128 \approx 4.7253. \qquad \text{Rounded to four decimal places}$$

b) We enter 0.000128 and then press the $\boxed{\text{log}}$ key. We find that

$$\log 0.000128 \approx -3.8928. \qquad \text{Rounded to four decimal places}$$

The inverse of a logarithmic function is an exponential function. Because of this, on many calculators the $\boxed{\text{log}}$ key doubles as the $\boxed{10^x}$ key after a $\boxed{\text{2nd}}$ or $\boxed{\text{SHIFT}}$ key is pressed.

EXAMPLE 2

Use a calculator to find $10^{3.417}$.

SOLUTION   We enter 3.417 and then press $\boxed{10^x}$. On some calculators, $\boxed{10^x}$ is pressed first, followed by 3.417 and $\boxed{\text{ENTER}}$. Since $10^{3.417}$ is irrational, our answer is approximate:

$$10^{3.417} \approx 2612.161354.$$

On calculators without a $\boxed{10^x}$ key, an exponentiation key, labeled $\boxed{x^y}$, $\boxed{a^x}$, or $\boxed{\wedge}$ may be available. Such a key can raise any nonnegative real number to any real-numbered power.

The Base *e* and Natural Logarithms on a Calculator

When interest is computed *n* times a year, the compound interest formula is

$$A = P\left(1 + \frac{r}{n}\right)^{nt},$$

where *A* is the amount that an initial investment *P* will be worth after *t* years at interest rate *r*. Suppose that $1 is invested at 100% interest for 1 year (no bank would pay this). The preceding formula becomes a function A defined in terms of the number of compounding periods n:

$$A(n) = \left(1 + \frac{1}{n}\right)^n.$$

Let us find some function values. We round to six decimal places, using a calculator.

n	$A(n) = \left(1 + \dfrac{1}{n}\right)^n$
1 (compounded annually)	$2.00
2 (compounded semiannually)	$2.25
3	$2.370370
4 (compounded quarterly)	$2.441406
5	$2.488320
100	$2.704814
365 (compounded daily)	$2.714567
8760 (compounded hourly)	$2.718127

The numbers in this table approach a very important number in mathematics, called e. Because e is irrational, its decimal representation does not terminate or repeat.

THE NUMBER e

$$e \approx 2.7182818284\ldots$$

Logarithms base e are called **natural logarithms,** or **Napierian logarithms,** in honor of John Napier (1550–1617), who first "discovered" logarithms.

The abbreviation "ln" is generally used with natural logarithms. Thus,

$$\ln 53 \quad \text{means} \quad \log_e 53.$$

To find the natural logarithm of a number, we enter that number and press $\boxed{\ln}$ on a calculator. On some calculators, we press $\boxed{\ln}$, the number, and then $\boxed{\text{ENTER}}$.

EXAMPLE 3   Use a calculator to find ln 4568.

SOLUTION   We enter 4568 and then press the $\boxed{\ln}$ key. We find that

$$\ln 4568 \approx 8.4268. \qquad \text{Rounded to four decimal places}$$

On many calculators, the $\boxed{\ln}$ key doubles as the $\boxed{e^x}$ key after a $\boxed{\text{2nd}}$ or $\boxed{\text{SHIFT}}$ key has been pressed.

EXAMPLE 4   Use a calculator to find $e^{-1.524}$.

SOLUTION   We enter -1.524 and then press $\boxed{e^x}$. On some calculators, $\boxed{e^x}$ is pressed first, followed by -1.524 and $\boxed{\text{ENTER}}$. Since $e^{-1.524}$ is irrational, our answer is approximate:

$$e^{-1.524} \approx 0.2178387868.$$

Changing Logarithmic Bases

Most calculators can find both common logarithms and natural logarithms. To find a logarithm with some other base, a conversion formula is needed.

THE CHANGE-OF-BASE FORMULA

For any logarithmic bases a and b, and any positive number M,

$$\log_b M = \frac{\log_a M}{\log_a b}.$$

Proof.   Let $x = \log_b M$. Then,

$$b^x = M \qquad \text{Rewriting } x = \log_b M \text{ in exponential form}$$

$$\log_a b^x = \log_a M \qquad \text{Taking the logarithm, base } a, \text{ on both sides}$$

$$x \log_a b = \log_a M \qquad \text{Using the power rule for logarithms}$$

$$x = \frac{\log_a M}{\log_a b}. \qquad \text{Solving for } x$$

But at the outset we stated that $x = \log_b M$. Thus, by substitution, we have

$$\log_b M = \frac{\log_a M}{\log_a b},$$

which is the change-of-base formula.

EXAMPLE 5   Find $\log_5 8$ using common logarithms.

SOLUTION   We use the change-of-base formula with $a = 10$, $b = 5$, and $M = 8$:

$$\log_5 8 = \frac{\log_{10} 8}{\log_{10} 5} \qquad \text{Substituting into } \log_b M = \frac{\log_a M}{\log_a b}$$

$$\approx \frac{0.9031}{0.6990} \qquad \text{Using the } \boxed{\text{LOG}} \text{ key twice}$$

$$\approx 1.2920. \qquad \text{When using a calculator, you need not round before dividing.}$$

To check, we use a calculator with an $\boxed{x^y}$ key to verify that $5^{1.2920} \approx 8$.

We can also use base e for a conversion.

EXAMPLE 6   Find $\log_4 31$ using natural logarithms.

SOLUTION   Substituting e for a, 4 for b, and 31 for M, we have

$$\log_4 31 = \frac{\log_e 31}{\log_e 4} \qquad \text{Substituting into } \log_b M = \frac{\log_a M}{\log_a b}$$

$$= \frac{\ln 31}{\ln 4} \approx \frac{3.4340}{1.3863} \qquad \text{Using the } \boxed{\text{LN}} \text{ key twice}$$

$$\approx 2.4771.$$

Graphs of Exponential and Logarithmic Functions, Base e

EXAMPLE 7   Graph $f(x) = e^x$ and $g(x) = e^{-x}$.

SOLUTION   We use a calculator with an $\boxed{e^x}$ key to find approximate values of e^x and e^{-x}. Using these values, we can graph the functions.

x	e^x	e^{-x}
0	1	1
1	2.7	0.4
2	7.4	0.1
−1	0.4	2.7
−2	0.1	7.4

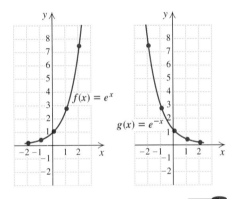

EXAMPLE 8   Graph: $f(x) = e^{-0.5x}$.

SOLUTION   We find some solutions with a calculator, plot them, and then draw the graph. For example,
$f(2) = e^{-0.5(2)} = e^{-1} \approx 0.4$.

x	$e^{-0.5x}$
0	1
1	0.6
2	0.4
3	0.2
−1	1.6
−2	2.7
−3	4.5

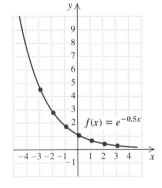

EXAMPLE 9

Graph: **(a)** $g(x) = \ln x$; **(b)** $f(x) = \ln (x + 3)$.

SOLUTION

a) We find some solutions with a calculator and then draw the graph. As expected, the graph is a reflection across the line $y = x$ of the graph of $y = e^x$.

TECHNOLOGY CONNECTION 9.5

Logarithmic functions with bases other than 10 or e can be easily drawn on a grapher, provided the change-of-base formula is used.

1. Graph $y = \log_5 x$.
2. Graph $y = \log_7 x$.
3. Graph $y = \log_5 (x + 2)$.
4. Graph $y = \log_7 x + 2$.

x	$\ln x$
1	0
4	1.4
7	1.9
0.5	-0.7

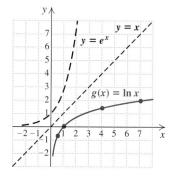

b) We find some solutions with a calculator, plot them, and draw the graph.

x	$\ln (x + 3)$
0	1.1
1	1.4
2	1.6
3	1.8
4	1.9
-1	0.7
-2	0
-2.5	-0.7

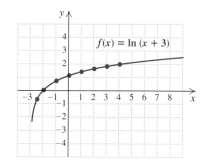

The graph of $y = \ln (x + 3)$ is the graph of $y = \ln x$ translated 3 units to the left.

EXERCISE SET

9.5

Use a calculator to find each of the following to the nearest ten-thousandth.

1. log 4   **2.** log 5   **3.** log 53.7

4. log 73.9   **5.** log 437   **6.** log 295

7. log 13,400   **8.** log 93,100   **9.** log 0.527

10. log 0.493   **11.** $10^{2.3}$ **12.** $10^{3.4}$

13. $10^{0.173}$ **14.** $10^{0.247}$ **15.** $10^{-2.9523}$

16. $10^{4.8982}$ **17.** ln 5   **18.** ln 2

19. ln 62   **20.** ln 30   **21.** ln 4365

22. ln 901.2   **23.** ln 0.0062   **24.** ln 0.00073

25. $e^{2.71}$ **26.** $e^{3.06}$ **27.** $e^{-3.49}$

28. $e^{-2.64}$ **29.** $e^{4.7}$ **30.** $e^{1.23}$

Find each of the following logarithms using the change-of-base formula.

31. $\log_6 100$ **32.** $\log_3 100$ **33.** $\log_2 100$

34. $\log_7 100$ **35.** $\log_7 65$ **36.** $\log_5 42$

37. $\log_{0.5} 5$ **38.** $\log_{0.1} 3$ **39.** $\log_2 0.2$

40. $\log_2 0.08$ **41.** $\log_\pi 58$ **42.** $\log_\pi 200$

Graph.

43. $f(x) = e^x$ **44.** $f(x) = e^{0.5x}$

45. $f(x) = e^{-0.5x}$ **46.** $f(x) = e^{-x}$

47. $f(x) = e^{x-1}$ **48.** $f(x) = e^{-x} - 3$

49. $f(x) = e^{x+5}$ **50.** $f(x) = e^{x-2}$

51. $f(x) = e^x + 3$ **52.** $f(x) = 2e^{0.5x}$

53. $f(x) = 2e^{-0.5x}$ **54.** $f(x) = \ln (x + 4)$

55. $f(x) = \ln (x + 1)$ **56.** $f(x) = \ln (x - 2)$

57. $f(x) = 2 \ln x$ **58.** $f(x) = 3 \ln x$

59. $f(x) = \ln x + 2$ **60.** $f(x) = \ln x - 3$

SKILL MAINTENANCE

Solve.

61. $4x^2 - 25 = 0$ **62.** $5x^2 - 7x = 0$

63. $17x - 15 = 0$ **64.** $9 - 13x = 0$

65. $x^{1/2} - 6x^{1/4} + 8 = 0$ **66.** $2y - 7\sqrt{y} + 3 = 0$

SYNTHESIS

67. ◈ Without referring to a calculator, explain why log 87 < ln 10 is a true statement.

68. ◈ Explain how the graph of $f(x) = e^x$ could be used to graph the function given by $g(x) = 1 + \ln x$.

69. ◈ Explain how the graph of $f(x) = \ln x$ could be used to graph the function given by $g(x) = e^{x-1}$.

70. ◈ Without drawing a graph or calculating any pairs, explain how the graphs of $f(x) = \ln |x|$ and $g(x) = |\ln x|$ differ.

71. Find a formula for converting common logarithms to natural logarithms.

72. Find a formula for converting natural logarithms to common logarithms.

Solve for x.

73. $\log (275x^2) = 38$ **74.** $\log (492x) = 5.728$

75. $\dfrac{3.01}{\ln x} = \dfrac{28}{4.31}$

76. $\log 692 + \log x = \log 3450$

For each function given below, **(a)** determine the domain, **(b)** set an appropriate window, and **(c)** draw the graph. Graphs may vary, depending on the scale used.

77. $f(x) = 7.4e^x \ln x$

78. $f(x) = 3.4 \ln x - 0.25e^x$

79. $f(x) = 5.3 \ln (x - 2.1)$

80. $f(x) = 2x^3 \ln x$

81. Use a grapher to check your answers to Exercises 45, 53, and 59.

82. Use a grapher to check your answers to Exercises 44, 52, and 60.

83. ◈ In an attempt to solve $\ln x = 1.5$, Emma gets the following graph.

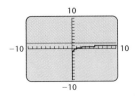

How can Emma tell at a glance that she has made a mistake?

9.6 Solving Exponential and Logarithmic Equations

Solving Exponential Equations • Solving Logarithmic Equations

Solving Exponential Equations

Equations with variables in exponents, such as $5^x = 12$ and $2^{7x} = 64$, are called **exponential equations.** In Section 9.3, we solved certain exponential equations by using the principle of exponential equality. We restate that principle below.

**THE PRINCIPLE OF
EXPONENTIAL EQUALITY**

For any real number b, where $b \neq -1$, 0, or 1,

$b^x = b^y$ is equivalent to $x = y$.

EXAMPLE 1

Solve: $4^{3x-5} = 16$.

SOLUTION   Note that $16 = 4^2$. Thus we can write each side as a power of the same number:

$4^{3x-5} = 4^2$.

Since the base is the same, 4, the exponents must be the same. Thus,

$$3x - 5 = 2 \qquad \text{Equating exponents}$$
$$3x = 7$$
$$x = \tfrac{7}{3}.$$

Check:

$$
\begin{array}{c|c}
4^{3x-5} = 16 & \\
\hline
4^{3 \cdot 7/3 - 5} \ ? \ 16 & \\
4^{7-5} & \\
4^2 & 16 \qquad \text{TRUE}
\end{array}
$$

The solution is $\tfrac{7}{3}$.

When it does not seem possible to write both sides of an equation as powers of the same base, we can use the following principle along with the properties developed in Section 9.4.

**THE PRINCIPLE OF
LOGARITHMIC EQUALITY**

For any logarithmic base a, and for $x, y > 0$,

$x = y$ is equivalent to $\log_a x = \log_a y$.

Because calculators can usually find only common or natural logarithms (without resorting to the change-of-base formula), we usually take the common or natural logarithm on both sides of the equation.

EXAMPLE 2

Solve: $5^x = 12$.

SOLUTION   We have

$$5^x = 12$$

$$\log 5^x = \log 12 \qquad \text{Using the principle of logarithmic equality to take the common logarithm on both sides. Natural logarithms also would work.}$$

$$x \log 5 = \log 12 \qquad \text{Using the power rule for logarithms}$$

$$x = \frac{\log 12}{\log 5} \quad \longleftarrow \quad \boxed{\textbf{CAUTION!} \ \ \text{This is not } \log 12 - \log 5!}$$

$$\approx \frac{1.0792}{0.6990} \approx 1.544. \qquad \text{Using a calculator and rounding to three decimal places}$$

Since $5^{1.544} \approx 12$, we have a check. The solution is $\log 12/\log 5$, or approximately 1.544.

EXAMPLE 3   Solve: $e^{0.06t} = 1500$.

SOLUTION   We take the natural logarithm on both sides:

$$\ln e^{0.06t} = \ln 1500 \qquad \text{Taking the natural logarithm on both sides}$$

$$0.06t = \ln 1500 \qquad \text{Finding the logarithm of the base to a power: } \log_a a^k = k$$

$$\left. \begin{array}{l} 0.06t \approx 7.3132 \\ t \approx 121.887. \end{array} \right\} \qquad \text{Using a calculator and rounding to three decimal places}$$

Solving Logarithmic Equations

Equations containing logarithmic expressions are called **logarithmic equations.** We saw in Section 9.3 that certain logarithmic equations can be solved by converting them into exponential equations.

EXAMPLE 4   Solve: $\log_4 (8x - 6) = 3$.

SOLUTION   We write an equivalent exponential equation:

$$4^3 = 8x - 6$$

$$64 = 8x - 6$$

$$70 = 8x \qquad \text{Adding 6 on both sides}$$

$$x = \frac{70}{8}, \text{ or } \frac{35}{4}.$$

The check is left to the student. The solution is $\frac{35}{4}$.

Often the properties for logarithms are needed.

EXAMPLE 5   Solve.

a) $\log x + \log (x - 3) = 1$
b) $\log_2 (x + 7) - \log_2 (x - 7) = 3$
c) $\log_7 (x + 1) + \log_7 (x - 1) = \log_7 8$

SOLUTION

a) As an aid in solving, we write in the base, 10.

$$\log_{10} x + \log_{10} (x - 3) = 1$$

$$\log_{10} [x(x - 3)] = 1 \qquad \text{Using the product rule for logarithms to obtain a single logarithm}$$

$$x(x - 3) = 10^1 \qquad \text{Writing an equivalent exponential equation}$$

$$x^2 - 3x = 10$$

$$x^2 - 3x - 10 = 0$$

$$(x + 2)(x - 5) = 0 \qquad \text{Factoring}$$

$$x + 2 = 0 \quad or \quad x - 5 = 0 \qquad \text{Using the principle of zero products}$$

$$x = -2 \quad or \quad x = 5$$

Check: For -2:

$$\frac{\log x + \log (x - 3) = 1}{\log (-2) + \log (-2 - 3) \ ? \ 1} \qquad \text{The number } -2 \text{ \textit{does not check} because negative numbers do not have logarithms.}$$

For 5:

$$\frac{\log x + \log (x - 3) = 1}{\begin{array}{c|c} \log 5 + \log (5 - 3) \ ? \ 1 & \\ \log 5 + \log 2 & \\ \log 10 & \\ 1 & 1 \quad \text{TRUE} \end{array}}$$

The solution is 5.

b) We have

$$\log_2 (x + 7) - \log_2 (x - 7) = 3$$

$$\log_2 \frac{x + 7}{x - 7} = 3 \qquad \text{Using the quotient rule for logarithms to obtain a single logarithm}$$

$$\frac{x + 7}{x - 7} = 2^3 \qquad \text{Writing an equivalent exponential equation}$$

$$\frac{x + 7}{x - 7} = 8$$

$$x + 7 = 8(x - 7) \qquad \text{Multiplying by the LCD, } x - 7$$

$$x + 7 = 8x - 56 \qquad \text{Using the distributive law}$$

$$63 = 7x$$

$$9 = x. \qquad \text{Dividing by 7}$$

Check:

$$\frac{\log_2 (x + 7) - \log_2 (x - 7) = 3}{\begin{array}{c|c} \log_2 (9 + 7) - \log_2 (9 - 7) \ ? \ 3 & \\ \log_2 16 - \log_2 2 & \\ 4 - 1 & \\ 3 & 3 \quad \text{TRUE} \end{array}}$$

The solution is 9.

c) We have

$$\log_7 (x + 1) + \log_7 (x - 1) = \log_7 8$$

$$\log_7 (x^2 - 1) = \log_7 8 \qquad \text{Using the product rule for logarithms: } (x + 1)(x - 1) = x^2 - 1$$

$$x^2 - 1 = 8 \qquad \text{Using the principle of logarithmic equality. Study this step carefully.}$$

$$x^2 - 9 = 0$$

$$(x - 3)(x + 3) = 0 \qquad \text{Solving the quadratic equation}$$

$$x = 3 \quad or \quad x = -3.$$

We leave it to the student to show that 3 checks but -3 does not. The solution is 3.

TECHNOLOGY CONNECTION 9.6

Exponential and logarithmic equations can be solved by graphing each side of the equation, using a window large enough to show all points of intersection. ZOOM, TRACE, and (on some graphers) INTERSECT can be used to determine the x-coordinate at each intersection.

For example, to solve $e^{0.5x} - 7 = 2x + 6$, we graph $y_1 = e^{0.5x} - 7$ and $y_2 = 2x + 6$ as shown at right. We then adjust the window and use TRACE to pinpoint any intersections. The x-coordinates at the intersections are approximately -6.48 and 6.52.

Use a grapher to find solutions, accurate to the nearest hundredth, for each of the following equations.

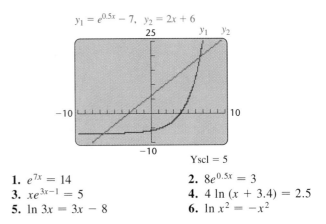

$y_1 = e^{0.5x} - 7, \ y_2 = 2x + 6$

Yscl = 5

1. $e^{7x} = 14$
2. $8e^{0.5x} = 3$
3. $xe^{3x-1} = 5$
4. $4 \ln (x + 3.4) = 2.5$
5. $\ln 3x = 3x - 8$
6. $\ln x^2 = -x^2$

EXERCISE SET 9.6

Solve. Where appropriate, include approximations to the nearest thousandth.

1. $3^x = 81$
2. $2^x = 8$
3. $4^x = 256$
4. $5^x = 125$
5. $2^{x+3} = 32$
6. $4^{3x} = 64$
7. $5^{3x} = 625$
8. $3^{5-x} = 27$
9. $4^{2x-1} = 64$
10. $5^{2x-3} = 25$
11. $3^{2x^2} \cdot 3^{5x} = 27$
12. $3^{4x} \cdot 3^{x^2} = \frac{1}{27}$

13. ▦ $2^x = 13$
14. ▦ $2^x = 19$
15. ▦ $4^x = 7$
16. ▦ $8^x = 10$
17. ▦ $e^t = 100$
18. ▦ $e^t = 1000$
19. ▦ $e^{-0.07t} = 0.08$
20. ▦ $e^{0.03t} = 5$
21. ▦ $2^x = 3^{x-1}$
22. ▦ $5^x = 3^{x+1}$
23. ▦ $4^{x+1} = 5^x$
24. ▦ $2^{x+3} = 7^x$
25. ▦ $20 - (1.7)^x = 0$
26. ▦ $125 - (4.5)^y = 0$
27. $\log_5 x = 4$
28. $\log_3 x = 3$

29. $\log_4 x = \frac{1}{2}$

30. $\log_2 x = -3$

31. $\log x = 3$

32. $\log x = 1$

33. $2 \log x = -6$

34. $4 \log x = -8$

35. $\ln x = 1$

36. $\ln x = 2$

37. $5 \ln x = -15$

38. $3 \ln x = -3$

39. $\log_2 (8 - 6x) = 5$

40. $\log_5 (2x - 7) = 3$

41. $\log (x + 9) + \log x = 1$

42. $\log (x - 9) + \log x = 1$

43. $\log x - \log (x + 7) = 1$

44. $\log x - \log (x + 3) = -1$

45. $\log_4 (x + 3) - \log_4 (x - 5) = 2$

46. $\log_2 (x + 3) + \log_2 (x - 3) = 4$

47. $\log_7 (x + 2) + \log_7 (x + 1) = \log_7 6$

48. $\log_6 (x + 3) + \log_6 (x + 2) = \log_6 20$

49. $\log_5 (x + 4) + \log_5 (x - 4) = 2$

50. $\log_{14} (x + 3) + \log_{14} (x - 2) = 1$

51. $\log_{12} (x - 4) - \log_{12} (x + 5) = \log_{12} 3$

52. $\log_6 (x + 7) - \log_6 (x - 2) = \log_6 5$

53. $\log_2 (x - 2) + \log_2 x = 3$

54. $\log_4 (x + 6) - \log_4 x = 2$

SKILL MAINTENANCE

Simplify.

55. $(125x^7 y^{-2} z^6)^{-2/3}$

56. i^{79}

57. $(3 + 5i)^2$

Solve.

58. $E = mc^2$, for c (Assume $E, m, c > 0$.)

59. $x^4 + 400 = 104x^2$

60. $x(x - 3) = 5$

SYNTHESIS

61. ◆ Can the principle of logarithmic equality be expanded to include all functions? That is, will the statement "$m = n$ is equivalent to $f(m) = f(n)$" be true for any function f? Why or why not?

62. ◆ Explain how Exercises 31–34 could be solved using the graph of $f(x) = \log x$.

63. ◆ In Example 3, we took the natural logarithm on both sides. What would have happened had we used common logarithms instead?

64. ◆ Christina finds that the solution of $\log_3 (x + 4) = 1$ is -1, but rejects -1 as an answer. What mistake is she making?

Solve.

65. $27^x = 81^{2x-3}$

66. $8^x = 16^{3x+9}$

67. $\log_x (\log_3 27) = 3$

68. $\log_6 (\log_2 x) = 0$

69. $x \log \frac{1}{8} = \log 8$

70. $\log_5 \sqrt{x^2 - 9} = 1$

71. $2^{x^2+4x} = \frac{1}{8}$

72. $\log (\log x) = 5$

73. $\log_5 |x| = 4$

74. $\log x^2 = (\log x)^2$

75. $\log \sqrt{2x} = \sqrt{\log 2x}$

76. $\log x^{\log x} = 25$

77. $3^{2x} - 8 \cdot 3^x + 15 = 0$

78. $(81^{x-2})(27^{x+1}) = 9^{2x-3}$

79. $3^{2x} - 3^{2x-1} = 18$

80. Given that $2^y = 16^{x-3}$ and $3^{y+2} = 27^x$, find the value of $x + y$.

81. If $x = (\log_{125} 5)^{\log_5 125}$, what is the value of $\log_3 x$?

82. 🖩 Find the value of x for which the natural logarithm is the same as the common logarithm.

83. 🖩 Use a grapher to check your answers to Exercises 3, 19, 25, 35, and 74.

9.7 Applications of Exponential and Logarithmic Functions

Applications of Logarithmic Functions •
Applications of Exponential Functions

We now consider applications of exponential and logarithmic functions.

Applications of Logarithmic Functions

EXAMPLE   1

Sound Levels.   To measure the "loudness" of any particular sound, the decibel scale is used. The loudness L, in decibels (dB), of a sound is given by

$$L = 10 \cdot \log \frac{I}{I_0},$$

where I is the intensity of the sound, in watts per square meter (W/m^2), and $I_0 = 10^{-12}$ W/m^2. (I_0 is approximately the intensity of the softest sound that can be heard.)

a) It is common for the intensity of sound at live performances of rock music to reach 10^{-1} W/m^2 (even higher close to the stage). How loud, in decibels, is this sound level?

b) Audiologists and physicians recommend that earplugs be worn when one is exposed to sounds in excess of 90 dB. What is the intensity of such sounds?

SOLUTION

a) To find the loudness, in decibels, we use the above formula:

$$
\begin{aligned}
L &= 10 \cdot \log \frac{I}{I_0} \\
&= 10 \cdot \log \frac{10^{-1}}{10^{-12}} \qquad \text{Substituting} \\
&= 10 \cdot \log 10^{11} \qquad \text{Subtracting exponents} \\
&= 10 \cdot 11 \qquad \qquad \log 10^a = a \\
&= 110.
\end{aligned}
$$

The volume of the music is 110 decibels.

b) We substitute and solve for I:

$$L = 10 \cdot \log \frac{I}{I_0}$$

$$90 = 10 \cdot \log \frac{I}{10^{-12}} \qquad \text{Substituting}$$

$$9 = \log \frac{I}{10^{-12}} \qquad \text{Dividing by 10 on both sides}$$

$$9 = \log I - \log 10^{-12} \qquad \text{Using the quotient rule for logarithms}$$

$$9 = \log I - (-12) \qquad \log 10^a = a$$

$$-3 = \log I \qquad \text{Adding } -12 \text{ on both sides}$$

$$10^{-3} = I. \qquad \text{Converting to an exponential equation}$$

Earplugs would be recommended for sounds with intensities exceeding 10^{-3} W/m^2.

EXAMPLE 2

Chemistry: pH of Liquids.   In chemistry the pH of a liquid is defined as follows:

$$pH = -\log [H^+],$$

where $[H^+]$ is the hydrogen ion concentration in moles per liter.

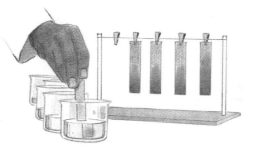

a) The hydrogen ion concentration of human blood is normally about 3.98×10^{-8} moles per liter. Find the pH.

b) The pH of seawater is about 8.3. Find the hydrogen ion concentration.

SOLUTION

a) To find the pH of blood, we use the above formula:

$$pH = -\log [H^+]$$

$$= -\log [3.98 \times 10^{-8}]$$

$$\approx -(-7.400117) \qquad \text{Using a calculator}$$

$$\approx 7.4.$$

The pH of human blood is normally about 7.4.

b) We substitute and solve for [H^+]:

$$8.3 = -\log [H^+] \qquad \text{Using pH} = -\log[H^+]$$
$$-8.3 = \log [H^+] \qquad \text{Dividing by } -1 \text{ on both sides}$$
$$10^{-8.3} = [H^+] \qquad \text{Converting to an exponential equation}$$
$$5.01 \times 10^{-9} \approx [H^+]. \qquad \text{Using a calculator; writing scientific notation}$$

The hydrogen ion concentration of seawater is about 5.01×10^{-9} moles per liter.

Applications of Exponential Functions

EXAMPLE 3

Interest Compounded Annually.   Suppose that $30,000 is invested at 8% interest, compounded annually. In t years, it will grow to the amount A given by the function

$$A(t) = 30{,}000(1.08)^t.$$

(See Example 5 in Section 9.1.)

a) How long will it take to accumulate $150,000 in the account?
b) Find the amount of time it takes for the $30,000 to double itself.

S O L U T I O N

a) We set $A(t) = 150{,}000$ and solve for t:

$$150{,}000 = 30{,}000(1.08)^t$$
$$\frac{150{,}000}{30{,}000} = 1.08^t \qquad \text{Dividing by 30,000 on both sides}$$
$$5 = 1.08^t$$
$$\log 5 = \log 1.08^t \qquad \text{Taking the common logarithm on both sides}$$
$$\log 5 = t \log 1.08 \qquad \text{Using the power rule for logarithms}$$
$$\frac{\log 5}{\log 1.08} = t \qquad \text{Dividing by log 1.08 on both sides}$$
$$20.9 \approx t. \qquad \text{Using a calculator}$$

Note: When doing a calculation like this on your calculator, it is best to wait until the end to round off. To do the last calculations on most scientific calculators, we would press

5 log $\div$ 1.08 log $=$.

It will take about 20.9 yr for the $30,000 to grow to $150,000.

b) To find the *doubling time T*, we replace $A(t)$ with 60,000, t with T, and solve for T:

$$60{,}000 = 30{,}000(1.08)^T$$
$$2 = (1.08)^T \qquad \text{Dividing by 30,000 on both sides}$$
$$\log 2 = \log (1.08)^T \qquad \text{Taking the common logarithm on both sides}$$
$$\log 2 = T \log 1.08 \qquad \text{Using the power rule for logarithms}$$
$$T = \frac{\log 2}{\log 1.08} \approx 9.0. \qquad \text{Using a calculator}$$

The doubling time is about 9 yr.

Like investments, populations often grow exponentially.

EXPONENTIAL GROWTH

An **exponential growth model** is a function of the form

$$P(t) = P_0 e^{kt}, \quad k > 0,$$

where P_0 is the population at time 0, $P(t)$ is the population at time t, and k is the **exponential growth rate** for the situation. The **doubling time** is the amount of time necessary for the population to double in size.

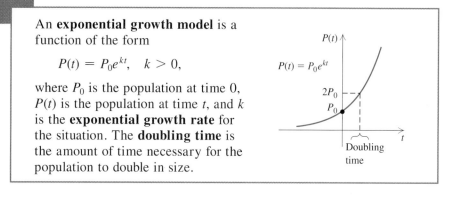

The exponential growth rate is the rate of growth of a population at any *instant* in time. Since the population is continually growing, the percent of total growth after one year will exceed the exponential growth rate.

EXAMPLE   4   *Growth of Zebra Mussel Populations.*   Zebra mussels, inadvertently imported from Europe, began fouling North American waters in 1988. These mussels are so prolific that lake and river bottoms, as well as water intake pipes, can become blanketed with them, altering an entire ecosystem. In 1997, a portion of the Mississippi River contained an average of 10 zebra mussels per square mile. The exponential growth rate was 340% per year.

a) Find the exponential growth model.
b) Predict the number of mussels per square mile in 2000.

SOLUTION

a) In 1997, at $t = 0$, the population was $10/\text{mi}^2$. We substitute 10 for P_0 and 340%, or 3.4, for k. This gives the exponential growth function

$$P(t) = 10e^{3.4t}.$$

b) In 2000, we have $t = 3$ (since 3 yr have passed since 1997). To find the population in 2000, we compute $P(3)$:

$$P(3) = 10e^{3.4(3)} \qquad \text{Using } P(t) = 10e^{3.4t} \text{ from part (a)}$$
$$= 10e^{10.2}$$
$$\approx 269{,}000. \qquad \text{Using a calculator}$$

The population of zebra mussels in the specified portion of the Mississippi River will reach approximately 269,000 per square mile in 2000. ⟶●

EXAMPLE 5

Spread of AIDS. The number of people $N(t)$ infected with a contagious disease at time t usually increases exponentially. Through 1985, 8249 cases of AIDS had been reported in the United States. By the end of 1994, the cumulative number had grown to 441,528. (*Source*: Centers for Disease Control)

a) Find the exponential growth rate and the exponential growth function.
b) Assuming exponential growth, predict the year in which the 1,000,000th case will occur.

SOLUTION

a) We use the equation $N(t) = N_0 e^{kt}$, where t is the number of years since 1985. In 1985, at $t = 0$, a total of 8249 cases had been reported. We substitute 8249 for N_0:

$$N(t) = 8249e^{kt}.$$

To find the exponential growth rate k, note that 9 yr later, at the end of 1994, the total number of cases had grown to 441,528:

$$\left.\begin{array}{l} N(9) = 8249e^{k \cdot 9} \\ 441{,}528 = 8249e^{9k} \end{array}\right\} \quad \text{Substituting}$$
$$53.525 \approx e^{9k} \qquad \text{Dividing by 8249 on both sides}$$
$$\ln 53.525 \approx \ln e^{9k} \qquad \text{Taking the natural logarithm on both sides}$$
$$3.9801 \approx 9k \qquad \ln e^a = a$$
$$0.442 \approx k. \qquad \text{Using a calculator}$$

The exponential growth function is $N(t) = 8249e^{0.442t}$, where t is measured in years since the end of 1985.

b) To predict when the 1,000,000th case will occur, we replace $N(t)$ with 1,000,000 and solve for t:

$$1{,}000{,}000 = 8249e^{0.442t}$$
$$121.227 \approx e^{0.442t} \qquad \text{Dividing by 8249 on both sides}$$
$$\ln 121.227 \approx \ln e^{0.442t} \qquad \text{Taking the natural logarithm on both sides}$$
$$4.798 \approx 0.442t \qquad \ln e^a = a$$
$$10.86 \approx t. \qquad \text{Using a calculator}$$

Rounding up to 11 yr, we see that, according to this model, by the end of 1985 + 11, or 1996, the 1,000,000th case occurred. (In reality, the spread of AIDS slowed in the mid-90s.) ⟶●

EXAMPLE   6

Interest Compounded Continuously.   Suppose that an amount of money P_0 is invested in a savings account at interest rate k, compounded continuously. That is, suppose that interest is computed every "instant" and added to the amount in the account. The balance $P(t)$, after t years, is given by the exponential growth model

$$P(t) = P_0 e^{kt}.$$

a) Suppose that \$30,000 is invested and grows to \$44,754.75 in 5 yr. Find the exponential growth function.
b) What is the doubling time?

SOLUTION

a) We have $P(0) = 30,000$. Thus the exponential growth function is

$$P(t) = 30,000 e^{kt}, \quad \text{where } k \text{ must still be determined.}$$

We know that at $t = 5$, $P(5) = 44,754.75$. We substitute and solve for k:

$$44,754.75 = 30,000 e^{k(5)} = 30,000 e^{5k}$$

$$\frac{44,754.75}{30,000} = e^{5k} \qquad \text{Dividing on both sides by 30,000}$$

$$1.491825 = e^{5k}$$

$$\ln 1.491825 = \ln e^{5k} \qquad \text{Taking the natural logarithm on both sides}$$

$$0.4 \approx 5k \qquad \text{Finding ln 1.491825 on a calculator and simplifying } \ln e^{5k}$$

$$\frac{0.4}{5} = 0.08 \approx k.$$

The interest rate is about 0.08, or 8%, compounded continuously. Note that since interest is being compounded continuously, the interest earned each year is more than 8%. The exponential growth function is

$$P(t) = 30,000 e^{0.08t}.$$

b) To find the doubling time T, we replace $P(T)$ with 60,000 and solve for T:

$$60,000 = 30,000 e^{0.08T}$$

$$2 = e^{0.08T} \qquad \text{Dividing by 30,000 on both sides}$$

$$\ln 2 = \ln e^{0.08T} \qquad \text{Taking the natural logarithm on both sides}$$

$$\ln 2 = 0.08T$$

$$\frac{\ln 2}{0.08} = T \qquad \text{Dividing}$$

$$\frac{0.693147}{0.08} \approx 8.7 \approx T.$$

Thus the original investment of \$30,000 will double in about 8.7 yr.

As the results of Examples 3(b) and 6(b) imply, for any specified interest rate, continuous compounding gives the highest yield and the shortest doubling time.

In some real-life situations, a quantity or population is *decreasing* or *decaying* exponentially.

EXPONENTIAL DECAY

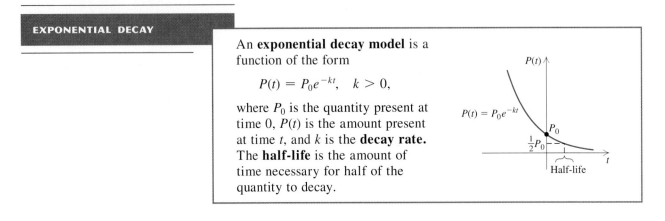

An **exponential decay model** is a function of the form

$$P(t) = P_0 e^{-kt}, \quad k > 0,$$

where P_0 is the quantity present at time 0, $P(t)$ is the amount present at time t, and k is the **decay rate.** The **half-life** is the amount of time necessary for half of the quantity to decay.

EXAMPLE 7

Carbon Dating. The radioactive element carbon-14 has a half-life of 5750 yr. The percentage of carbon-14 present in the remains of animal bones can be used to determine age. How old is an animal bone that has lost 40% of its carbon-14?

SOLUTION   We first find k. To do so, we use the concept of half-life. When $t = 5750$ (the half-life), $P(t)$ will be half of P_0. Then

$$0.5P_0 = P_0 e^{-k(5750)}$$

$$0.5 = e^{-5750k} \qquad \text{Dividing by } P_0 \text{ on both sides}$$

$$\ln 0.5 = \ln e^{-5750k} \qquad \text{Taking the natural logarithm on both sides}$$

$$\ln 0.5 = -5750k$$

$$\frac{\ln 0.5}{-5750} = k$$

$$0.00012 \approx k.$$

Now we have a function for the decay of carbon-14:

$$P(t) = P_0 e^{-0.00012t}. \qquad \text{This completes the first part of our solution.}$$

(*Note*: This equation can be used for any subsequent carbon-dating problem.) If an animal bone has lost 40% of its carbon-14 from an initial amount P_0, then 60% of P_0 is still present. To find the age t of the bone, we solve this equation for t:

$$0.6P_0 = P_0 e^{-0.00012t} \qquad \text{We want to find } t \text{ for which } P(t) = 0.6P_0.$$

$$0.6 = e^{-0.00012t} \qquad \text{Dividing by } P_0 \text{ on both sides}$$

$$\ln 0.6 = \ln e^{-0.00012t} \qquad \text{Taking the natural logarithm on both sides}$$

$$-0.5108 \approx -0.00012t$$

$$t \approx \frac{-0.5108}{-0.00012}$$

$$t \approx 4257.$$

The animal bone is about 4257 yr old.

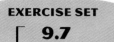

EXERCISE SET

9.7

Solve.

1. *Cellular phones.*   The number of cellular phones in use in the United States t years after 1985 can be approximated by

$$N(t) = 0.3(1.63)^t,$$

where $N(t)$ is the number of cellular phones in use, in millions.

  a) Determine the year in which 50 million cellular phones would be in use.
  b) What is the doubling time for the number of cellular phones in use?

2. *Compact discs.*   The number of compact discs N purchased each year, in millions, can be approximated by

$$N(t) = 150(1.28)^t,$$

where t is the number of years after 1988.

  a) After what amount of time will one billion compact discs be sold in a year?
  b) What is the doubling time for the number of compact discs sold in one year?

3. *Student loan repayment.*   A college loan of $29,000 is made at 8% interest, compounded annually. After t years, the amount due, A, is given by the function

$$A(t) = 29,000(1.08)^t.$$

  a) After what amount of time will the amount due reach $40,000?
  b) Find the doubling time.

4. *Spread of a rumor.*   The number of people who have heard a rumor increases exponentially. If all who hear a rumor repeat it to two people a day, and if 20 people start the rumor, the number of people N who have heard the rumor after t days is given by

$$N(t) = 20(3)^t.$$

  a) After what amount of time will 1000 people have heard the rumor?
  b) What is the doubling time for the number of people who have heard the rumor?

5. *Recycling aluminum cans.*   Approximately two thirds of all aluminum cans distributed will be recycled each year. A beverage company distributes 250,000 cans. The number still in use after t years is given by the function

$$N(t) = 250,000\left(\tfrac{2}{3}\right)^t.$$

  a) After how many years will 60,000 cans still be in use?
  b) After what amount of time will only 1000 cans still be in use?

6. *Salvage value.*   A color photocopier is purchased for $5200. Its value each year is about 80% of its value in the preceding year. Its value in dollars after t years is given by the exponential function

$$V(t) = 5200(0.8)^t.$$

  a) After what amount of time will the salvage value be $1200?
  b) After what amount of time will the salvage value be half the original value?

Use the pH formula given in Example 2 for Exercises 7–10.

7. *Chemistry.*   The hydrogen ion concentration of fresh-brewed coffee is about 1.3×10^{-5} moles per liter. Find the pH.

8. *Chemistry.*   The hydrogen ion concentration of milk is about 1.6×10^{-7} moles per liter. Find the pH.

9. *Medicine.*   When the pH of a patient's blood drops below 7.4, a condition called *acidosis* sets in. Acidosis can be deadly when the patient's pH reaches 7.0. What would the hydrogen ion concentration of the patient's blood be at that point?

10. *Medicine.*   When the pH of a patient's blood rises above 7.4, a condition called *alkalosis* sets in. Alkalosis can be deadly when the patient's pH reaches 7.8. What would the hydrogen ion concentration of the patient's blood be at that point?

Use the decibel formula given in Example 1 for Exercises 11–14.

11. *Acoustics.*   The intensity of sound in normal conversation is about 3.2×10^{-6} W/m². How loud in decibels is this sound level?

12. *Acoustics.* The intensity of sound of a riveter at work is about 3.2×10^{-3} W/m². How loud in decibels is this sound level?

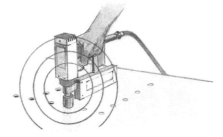

13. *Music.* The rock group Phish recently performed and sound measurements of 105 dB were recorded. What is the intensity of such sounds?

14. *Audiology.* Overexposure to excessive sound levels can diminish one's hearing to the point where the softest sound that is audible is 28 dB. What is the intensity of such a sound?

Use the compound-interest formula in Example 6 for Exercises 15 and 16.

15. *Interest compounded continuously.* Suppose that P_0 is invested in a savings account where interest is compounded continuously at 6% per year.

   a) Express $P(t)$ in terms of P_0 and 0.06.
   b) Suppose that $5000 is invested. What is the balance after 1 yr? after 2 yr?
   c) When will an investment of $5000 double itself?

16. *Interest compounded continuously.* Suppose that P_0 is invested in a savings account where interest is compounded continuously at 5% per year.

   a) Express $P(t)$ in terms of P_0 and 0.05.
   b) Suppose that $1000 is invested. What is the balance after 1 yr? after 2 yr?
   c) When will an investment of $1000 double itself?

17. *Population growth.* In 1995, the population of the United States was 264 million and the exponential growth rate was 1% per year.

   a) Find the exponential growth function.
   b) Predict the U.S. population in 2001.
   c) When will the U.S. population reach 300 million?

18. *World population growth.* In 1995, the world population was 5.7 billion and the exponential growth rate was 1.5% per year.

   a) Find the exponential growth function.
   b) Predict the world population in 2001.
   c) When will the world population be 8.0 billion?

19. *Growth of bacteria.* The bacteria *Escherichi coli* are commonly found in the human bladder. Suppose that 3000 of the bacteria are present at time $t = 0$. Then t minutes later, the number of bacteria present is

   $$N(t) = 3000(2)^{t/20}.$$

   a) After what amount of time will there be 60,000 bacteria?
   b) If 100,000,000 bacteria accumulate, a bladder infection can occur. What amount of time would have to pass in order for a possible infection to occur?
   c) What is the doubling time?

20. *Population growth.* The exponential growth rate of the population of Central America is 3.5% per year (one of the highest in the world). What is the doubling time?

21. *Advertising.* A model for advertising response is given by

   $$N(a) = 2000 + 500 \log a, \quad a \geq 1,$$

   where $N(a) =$ the number of units sold and $a =$ the amount spent on advertising, in thousands of dollars.

   a) How many units were sold after spending $1000 ($a = 1$) on advertising?
   b) How many units were sold after spending $8000?
   c) Graph the function.
   d) How much would have to be spent in order to sell 5000 units?

22. *Forgetting.* Students in an English class took a final exam. They took equivalent forms of the exam at monthly intervals thereafter. The average score $S(t)$, in percent, after t months was found to be given by

   $$S(t) = 68 - 20 \log (t + 1), \quad t \geq 0.$$

   a) What was the average score when they initially took the test, $t = 0$?
   b) What was the average score after 4 months? after 24 months?
   c) Graph the function.
   d) After what time t was the average score 50?

23. *Public health.* In 1995, an outbreak of Herpes infected 17 people in a large community. By 1996, the number of those infected had grown to 29.

   a) Find an exponential growth function that fits the data.
   b) Predict the number of people who will be infected in 2001.

24. *Heart transplants.* In 1967, Dr. Christiaan Barnard of South Africa stunned the world by performing the first heart transplant. There was 1 transplant in 1967. In 1987, there were 1418 such transplants.

a) Find an exponential growth function that fits the data.

b) Use the function to predict the number of heart transplants in 2002.

25. *Oil demand.* The exponential growth rate of the demand for oil in the United States is 10% per year. When will the demand be double that of 1995?

26. *Coal demand.* The exponential growth rate of the demand for coal in the world is 4% per year. When will the demand be double that of 1995?

27. *Decline of long-playing records.* The sales S of long-playing records has declined considerably in recent years because of the advent of the compact disc (see Exercise 2). There were 205 million LP records sold in 1983, and 1.2 million sold in 1993 (*Source*: Recording Industry Association of America). Assuming the sales are decreasing according to the exponential decay model:

a) Find the value k, and write an exponential function that describes the number of long-playing records sold after time t, in years.

b) Estimate the sales of LP records in the year 2001.

c) In what year (theoretically) will only 1 long-playing record be sold?

28. *Decline in beef consumption.* In 1985, the annual consumption of beef was about 80 lb per person. In 1996, it was about 67 lb per person. Assuming consumption is decreasing according to the exponential decay model:

a) Find the value k, and write an equation that describes beef consumption after time t, in years.

b) Estimate the consumption of beef in the year 2001.

c) In what year (theoretically) will the annual consumption of beef be 20 lb per person?

29. *Archaeology.* When archaeologists found the Dead Sea scrolls, they determined that the linen wrapping had lost 22.3% of its carbon-14. How old is the linen wrapping? (See Example 7.)

30. *Archaeology.* In 1996, researchers found an ivory tusk that had lost 18% of its carbon-14. How old was the tusk? (See Example 7.)

31. *Chemistry.* The exponential decay rate of iodine-131 is 9.6% per day. What is its half-life?

32. *Chemistry.* The decay rate of krypton-85 is 6.3% per year. What is its half-life?

33. *Home construction.* The chemical urea formaldehyde was found in some insulation used in houses built during the mid to late 60s. Unknown at the time was the fact that urea formaldehyde emitted toxic fumes as it decayed. The half-life of urea formaldehyde is 1 yr. What is its decay rate?

34. *Plumbing.* Lead pipes and solder are often found in older buildings. Unfortunately, as lead decays, toxic chemicals can get in the water resting in the pipes. The half-life of lead is 22 yr. What is its decay rate?

35. *Value of a sports card.* Because he objected to smoking, and because his first baseball card was issued in cigarette packs, the great shortstop Honus Wagner halted production of his card before many were produced. One of these cards was sold in 1991 for $451,000 and again in 1996 for $640,500. For the following questions, assume that the card's value increases exponentially.

WAGNER, PITTSBURG

a) Find an exponential function $V(t)$, if $V_0 = 451{,}000$.

b) Predict the card's value in 2002.

c) What is the doubling time for the value of the card?

d) In what year will the value of the card first exceed $1,000,000?

36. *Value of a Van Gogh painting.* The Van Gogh painting *Irises,* shown below, sold for $84,000 in 1947 and was sold again for $53,900,000 in 1987. Assume that the growth in the value V of the painting is exponential.

Van Gogh's *Irises*, a 28-in. by 32-in. oil on canvas.

a) Find the exponential growth rate k, and determine the exponential growth function, assuming $V_0 = 84,000$.
b) Estimate the value of the painting in 2001.
c) What is the doubling time for the value of the painting?
d) How long after 1947 will the value of the painting be $1 billion?

SKILL MAINTENANCE

Simplify.

37. $\dfrac{\dfrac{x-5}{x+3}}{\dfrac{x}{x-3}+\dfrac{2}{x+3}}$

38. $\dfrac{\dfrac{3}{a}+\dfrac{5}{b}}{\dfrac{2}{a^2}-\dfrac{4}{b^2}}$

39. $\dfrac{6a^3b^{-7}}{8a^{-5}b^{-10}}$

40. $\dfrac{a^{-1}+b^{-2}}{a^{-2}-b^{-1}}$

SYNTHESIS

41. ◆ By the end of 1996, the cumulative number of cases of AIDS in the United States had grown to approximately 607,308 (*Source*: Based on interviews with the Vermont AIDS hotline). Use the information in Example 5 to determine whether an exponential function or a linear function best fits the data. Then explain how you would predict the cumulative number of cases of AIDS in 2005.

42. ◆ *Atmospheric pressure.* Atmospheric pressure P at altitude a is given by

$$P = P_0 e^{-0.00005a},$$

where P_0 = the pressure at sea level ≈ 14.7 lb/in^2 (pounds per square inch). Explain how a barometer, or some other device for measuring atmospheric pressure, can be used to find the height of a skyscraper.

43. ◆ Examine the restriction on t in Exercise 22.

a) What upper limit might be placed on t?
b) In practice, would this upper limit ever be enforced? Why or why not?

44. ◆ Write a problem for a classmate to solve in which information is provided and the classmate is asked to find an exponential growth function. Make the problem as realistic as possible.

45. ▦ *Supply and demand.* The supply and demand for the sale of stereos by Sound Ideas are given by

$$S(x) = e^x \quad \text{and} \quad D(x) = 162{,}755e^{-x},$$

where $S(x)$ = the price at which the company is willing to supply x stereos and $D(x)$ = the demand price for a quantity of x stereos. Find the equilibrium point. (For reference, see Section 3.8.)

46. ▦ Use Exercises 1 and 17 to form a model for the percentage of U.S. residents owning a cellular phone t years after 1985.

47. ◆ Use the model developed in Exercise 46 to predict the percentage of U.S. residents who will own cellular phones in 2010. Does your prediction seem plausible? Why or why not?

COLLABORATIVE
C ♦ O ♦ R ♦ N ♦ E ♦ R

Focus: Exponential-growth models

Time: 30 minutes

Group size: 6

Prepaid calling cards have become a big business, not only as a convenient way to place a telephone call, but also as collectibles. Some telephone cards are worth far more than their original cost, particularly if they are unused.

Collectors often estimate the future value of their collections by examining the value's growth in the past. The value of some collectibles grows exponentially.

ACTIVITY

1. Suppose that in 1996, each group member were to invest $1200 in the telephone cards listed in the following table. Looking only at the approximate value

in 1996, each student should select $1200 worth of cards to buy. The card or cards chosen will become that student's portfolio.
2. Each group member should be assigned one of the cards. That person should then form an exponential growth model for the value of that card using the year of issue and the original cost of the card.
3. Using the models developed above, each group member should predict the value of his or her portfolio in 2005. Compare the values. Why, when buying a collector's item, is it important to consider its previous worth?
4. Look at the predicted value of each card in the year 2005. Does an exponential growth model seem appropriate? If possible, find the current value of some of the cards and compare those values with the predicted values.

Card	Approximate Value in 1996	Year of Issue	Original Cost
New York Telephone, 1992 Democratic National Convention	$500	1992	$1
McDonald's Extra-Value Meal Promotion	$600	1995	$3
Michigan Bell, University of Michigan	$1000	1987	$40
AT&T, America's Cup	$1200	1992	$50
Undecorated cards, Manning Prison, Columbia, SC	$100	1989	$20
Sprint, Coca-Cola collection	$200	1995	$17

Source: Phone Card Market Report, June 1996

SUMMARY AND REVIEW
9

KEY TERMS

Exponential function, p. 515
Asymptote, p. 516
Composite functions, p. 523
Inverse relation, p. 526
One-to-one function, p. 526
Horizontal-line test, p. 527
Logarithmic function, p. 535
Common logarithm, p. 546

Natural logarithm, p. 548
Exponential equation, p. 552
Logarithmic equation, p. 554
Exponential growth model,
   p. 561
Exponential growth rate,
   p. 561

Doubling time, p. 561
Exponential decay model,
   p. 564
Exponential decay rate,
   p. 564
Half-life, p. 564

IMPORTANT PROPERTIES AND FORMULAS

Exponential function: $f(x) = a^x$
Interest compounded annually: $A = P(1 + i)^t$
Composition of f and g: $f \circ g(x) = f(g(x))$

To Find a Formula for the Inverse of a Function

Make sure that the function f is one-to-one. Then:

1. Replace $f(x)$ with y.
2. Interchange x and y.
3. Solve for y.
4. Replace y with $f^{-1}(x)$.

For $x > 0$ and a a positive constant other than 1, $\log_a x$ is the number
to which a is raised to get x. Thus, $a^{\log_a x} = x$, or equivalently, if
$y = \log_a x$, then $a^y = x$.

The Principle of Exponential Equality

For any real number b, $b \neq -1, 0,$ or 1:

   $b^x = b^y$ is equivalent to $x = y$.

The Principle of Logarithmic Equality

For any logarithmic base a, and for x, $y > 0$:

$x = y$ is equivalent to $\log_a x = \log_a y$.

Properties of logarithms:

$$\log_a MN = \log_a M + \log_a N, \qquad \log_a \frac{M}{N} = \log_a M - \log_a N,$$

$$\log_a M^p = p \cdot \log_a M, \qquad \log_a 1 = 0,$$
$$\log_a a = 1, \qquad \log_a a^k = k,$$

$$\log M = \log_{10} M, \qquad \log_b M = \frac{\log_a M}{\log_a b}$$

$$\ln M = \log_e M,$$

$e \approx 2.7182818284\ldots$
Exponential growth: $\qquad\qquad\qquad P(t) = P_0 e^{kt}, k > 0$
Exponential decay: $\qquad\qquad\qquad\ \ P(t) = P_0 e^{-kt}, k > 0$
Interest compounded continuously: $\ P(t) = P_0 e^{kt}$, where P_0 is the principal invested for t years at interest rate k
Carbon dating: $\qquad\qquad\qquad\quad P(t) = P_0 e^{-0.00012t}$

REVIEW EXERCISES

Graph.

1. $f(x) = 3^{x-1}$

2. $x = 3^{y-1}$

3. $y = \log_3 x$

4. $f(x) = e^{x+1}$

5. Find $f \circ g(x)$ and $g \circ f(x)$ if $f(x) = x^2$ and $g(x) = 3x - 5$.

6. If $h(x) = \sqrt{4 - 7x}$, find $f(x)$ and $g(x)$ such that $h(x) = f \circ g(x)$. Answers may vary.

7. Determine whether $f(x) = 4 - x^2$ is one-to-one.

Find a formula for the inverse of each function.

8. $f(x) = x + 2$

9. $g(x) = \dfrac{2x - 3}{7}$

10. $f(x) = 8x^3$

Convert to exponential equations.

11. $\log_4 16 = x$

12. $\log_{1/2} 8 = -3$

Convert to logarithmic equations.

13. $10^4 = 10{,}000$

14. $25^{1/2} = 5$

Find each of the following.

15. $\log_3 9$

16. $\log_{10} \frac{1}{10}$

Express in terms of logarithms of x, y, and z.

17. $\log_a x^4 y^2 z^3$

18. $\log_a \dfrac{xy}{z^2}$

19. $\log \sqrt[4]{\dfrac{z^2}{x^3 y}}$

20. $\log_q \left(\dfrac{x^2 y^{1/3}}{z^4} \right)$

Express as a single logarithm.

21. $\log_a 8 + \log_a 15$

22. $\log_a 72 - \log_a 12$

23. $\frac{1}{2} \log a - \log b - 2 \log c$

24. $\frac{1}{3}[\log_a x - 2 \log_a y]$

Simplify.

25. $\log_m m$

26. $\log_m 1$

27. $\log_m m^{17}$

28. $\log_m m^{-7}$

Given $\log_a 2 = 1.8301$ *and* $\log_a 7 = 5.0999$, *find each of the following.*

29. $\log_a 14$ **30.** $\log_a \frac{2}{7}$ **31.** $\log_a 28$

32. $\log_a 3.5$ **33.** $\log_a \sqrt{7}$ **34.** $\log_a \frac{1}{4}$

Find each of the following using a calculator.

35. $\log 0.00627$ **36.** $\log 72{,}800{,}000$

37. $10^{1.789}$ **38.** $10^{-3.65}$

39. $\ln 23{,}912.2$ **40.** $\ln 0.06774$

41. $e^{-0.98}$ **42.** $e^{2.91}$

Find each of the following logarithms using the change-of-base formula.

43. $\log_5 2$ **44.** $\log_{12} 70$

Solve. Where appropriate, include approximations to the nearest ten-thousandth.

45. $\log_3 x = -2$ **46.** $\log_x 32 = 5$

47. $\log x = -4$ **48.** $3 \ln x = -6$

49. $4^{2x-5} = 16$ **50.** $2^{x^2} \cdot 2^{4x} = 32$

51. $4^x = 8.3$ **52.** $e^{-0.1t} = 0.03$

53. $\log_4 16 = x$

54. $\log_4 x + \log_4 (x - 6) = 2$

55. $\log x + \log (x - 15) = 2$

56. $\log_3 (x - 4) = 3 - \log_3 (x + 4)$

57. *Forgetting.* In a business class, students were tested at the end of the course on a final exam. They were tested again after 6 months. The forgetting formula was determined to be

$$S(t) = 62 - 18 \log (t + 1),$$

where t is the time, in months, after taking the first test.

a) Determine the average score when they first took the test (when $t = 0$).

b) What was the average score after 6 months?

c) After what time was the average score 34?

58. *Aeronautics.* There were 6 commercial space launches in 1994 and 12 in 1996.* Assume that the number is growing exponentially.

a) Find k and write the exponential growth function.

b) Predict the number of launches in 2003.

c) When will there be 192 annual scheduled launches?

59. The population of Riverton doubled in 16 yr. What was the exponential growth rate?

60. How long will it take $7600 to double itself if it is invested at 8.4%, compounded continuously?

61. How old is a skeleton that has lost 34% of its carbon-14? (Use $P(t) = P_0 e^{-0.00012t}$.)

62. What is the pH of a substance if its hydrogen ion concentration is 2.3×10^{-7} moles per liter? (Use pH $= -\log [\text{H}^+]$.)

63. The intensity of the sound of water at the foot of the Niagara Falls is about 10^{-3} W/m². * How loud in decibels is this sound level?

$$\left(\text{Use } L = 10 \cdot \log \frac{I}{I_0}.\right)$$

SKILL MAINTENANCE

64. Solve $aT^2 + bT = Q$ for T.

65. Solve: $x^4 + 80 = 21x^2$.

66. Divide: $\dfrac{4 - 5i}{1 + 3i}$.

67. Simplify:

$$\dfrac{\dfrac{1}{ab} - \dfrac{2}{bc}}{\dfrac{2}{ac} + \dfrac{3}{ab}}.$$

SYNTHESIS

68. Explain why negative numbers do not have logarithms.

69. Explain why taking the natural or common logarithm on each side of an equation produces an equivalent equation.

Solve.

70. $\ln (\ln x) = 3$

71. $2^{x^2+4x} = \frac{1}{8}$

72. $5^{x+y} = 25$,
$2^{2x-y} = 64$

Source: Telephone interview, U.S. Department of Transportation, Federal Aviation Administration, 1/6/97.

Sound and Hearing, Life Science Library. (New York: Time Incorporated, 1965), p. 173.

> # CHAPTER TEST
> ## 9

Graph.

1. $f(x) = 2^{x+3}$ **2.** $f(x) = \log_7 x$

3. Find $f \circ g(x)$ and $g \circ f(x)$ if $f(x) = x + x^2$ and $g(x) = 5x - 2$.

4. Determine whether $f(x) = 2 - |x|$ is one-to-one.

Find a formula for the inverse.

5. $f(x) = 4x - 3$ **6.** $f(x) = (x + 1)^3$

Convert to logarithmic equations.

7. $4^{-3} = x$ **8.** $256^{1/2} = 16$

Convert to exponential equations.

9. $\log_4 16 = 2$ **10.** $m = \log_7 49$

11. Express in terms of logarithms of a, b, and c:
$$\log \frac{a^3 b^{1/2}}{c^2}.$$

12. Express as a single logarithm:
$$\tfrac{1}{3} \log_a x + 2 \log_a z.$$

Simplify.

13. $\log_5 125$ **14.** $\log_t t^{23}$

15. $\log_p p$ **16.** $\log_c 1$

Given $\log_a 2 = 0.301$, $\log_a 6 = 0.778$, and $\log_a 7 = 0.845$, find each of the following.

17. $\log_a \frac{2}{7}$ **18.** $\log_a 12$ **19.** $\log_a 21$

Find each of the following using a calculator.

20. $\log 0.0123$ **21.** $10^{3.8}$

22. $\ln 0.01234$ **23.** $e^{4.68}$

24. Find $\log_{18} 31$ using the change-of-base formula.

Solve. Where appropriate, include approximations to the nearest ten-thousandth.

25. $\log_x 25 = 2$ **26.** $\log_4 x = \frac{1}{2}$

27. $\log x = 4$ **28.** $5^{4-3x} = 125$

29. 🔳 $7^x = 1.2$ **30.** 🔳 $\ln x = \frac{1}{4}$

31. $\log (x - 3) + \log (x + 1) = \log 5$

32. 🔳 The average walking speed R of people living in a city of population P, in thousands, is given by $R = 0.37 \ln P + 0.05$, where R is in feet per second.

a) The population of Akron, Ohio, is 225,000. Find the average walking speed.

b) A city has an average walking speed of 2.6 ft/sec. Find the population.

33. 🔳 The population of Canada was 30 million in 1996, and the exponential growth rate was 0.62% per year.

a) Write an exponential function describing the population of Canada.

b) What will the population be in 2003? in 2010?

c) When will the population be 50 million?

d) What is the doubling time?

34. 🔳 An investment with interest compounded continuously doubled itself in 15 yr. What is the interest rate?

35. 🔳 How old is an animal bone that has lost 43% of its carbon-14? (Use $P(t) = P_0 e^{-0.00012t}$.)

36. The sound of traffic at a busy intersection averages 75 dB. What is the intensity of such a sound?
$$\left(\text{Use } L = 10 \cdot \log \frac{I}{I_0}. \right)$$

37. The hydrogen ion concentration of water is 1.0×10^{-7} moles per liter. What is the pH? (Use pH $= -\log [\text{H}^+]$.)

S K I L L   M A I N T E N A N C E

38. Solve: $y - 9\sqrt{y} + 8 = 0$.

39. Solve $S = at^2 - bt$ for t.

40. Multiply: $(2 + 5i)(2 - 5i)$.

41. Simplify:
$$\frac{\dfrac{1}{x^2 - 4}}{\dfrac{1}{x + 2} + \dfrac{1}{x - 2}}.$$

S Y N T H E S I S

42. Solve: $\log_5 |2x - 7| = 4$.

43. If $\log_a x = 2$, $\log_a y = 3$, and $\log_a z = 4$, find
$$\log_a \frac{\sqrt[3]{x^2 z}}{\sqrt[3]{y^2 z^{-1}}}.$$

CUMULATIVE REVIEW
1-9

1. Evaluate $\dfrac{x^0 + y}{-z}$ for $x = 6$, $y = 9$, and $z = -5$.

Simplify.

2. $\left| -\dfrac{5}{2} + \left(-\dfrac{7}{2} \right) \right|$

3. $(-2x^2 y^{-3})^{-4}$

4. $(-5x^4 y^{-3} z^2)(-4x^2 y^2)$

5. $\dfrac{3x^4 y^6 z^{-2}}{-9x^4 y^2 z^3}$

6. $2x - 3 - 2[5 - 3(2 - x)]$

7. $3^3 + 2^2 - (32 \div 4 - 16 \div 8)$

Solve.

8. $8(2x - 3) = 6 - 4(2 - 3x)$

9. $4x - 3y = 15,$
   $3x + 5y = 4$

10. $x + y - 3z = -1,$
    $2x - y + z = 4,$
    $-x - y + z = 1$

11. $x(x - 3) = 10$

12. $\dfrac{7}{x^2 - 5x} - \dfrac{2}{x - 5} = \dfrac{4}{x}$

13. $\dfrac{8}{x + 1} + \dfrac{11}{x^2 - x + 1} = \dfrac{24}{x^3 + 1}$

14. $\sqrt{x - 1} = \sqrt{x + 4} - 1$

15. $\sqrt[3]{2x} = 1$

16. $3x^2 + 75 = 0$

17. $x - 8\sqrt{x} + 15 = 0$

18. $x^4 - 13x^2 + 36 = 0$

19. $\log_8 x = 1$

20. $\log_x 49 = 2$

21. $9^x = 27$

22. 🖩 $3^{5x} = 7$

23. $\log x - \log (x - 8) = 1$

24. $x^2 + 4x > 5$

25. If $f(x) = x^2 + 6x$, find a such that $f(a) = 11$.

26. If $f(x) = |2x - 3|$, find all x for which $f(x) \geq 9$.

Solve.

27. $D = \dfrac{ab}{b + a}$, for a

28. $\dfrac{1}{p} + \dfrac{1}{q} = \dfrac{1}{f}$, for q

29. $M = \dfrac{2}{3}(A + B)$, for B

Evaluate.

30. $\begin{vmatrix} 6 & -5 \\ 4 & -3 \end{vmatrix}$

31. $\begin{vmatrix} 7 & -6 & 0 \\ -2 & 1 & 2 \\ -1 & 1 & -1 \end{vmatrix}$

32. Find the domain of the function f given by

$$f(x) = \dfrac{-4}{3x^2 - 5x - 2}.$$

33. Lowell King, a farmer in the greater Indianapolis metropolitan area, is losing farmland to urban development. He farmed 1400 acres in 1994 and 950 in 1996 (*Source: Indianapolis Star,* 7/14/96). Let A represent the number of acres farmed and t the number of years since 1994.

    a) Find a linear function $A(t)$ that fits the data.
    b) Use the function of part (a) to predict the number of acres farmed in 2000.

Solve.

34. The perimeter of a rectangular garden is 112 m. The length is 16 m more than the width. Find the length and the width.

35. In triangle ABC, the measure of angle B is three times the measure of angle A. The measure of angle C is 105° greater than the measure of angle A. Find the angle measures.

36. Jenny can build a shed from a lumber kit in 10 hr. Phil can build the same shed in 12 hr. How long would it take Phil and Jenny, working together, to build the shed?

37. Swim Clean is 30% muriatic acid. Pure Swim is 80% muriatic acid. How many liters of each should be mixed together in order to get 100 L of a solution that is 50% muriatic acid?

38. A fishing boat with a trolling motor can move at a speed of 5 km/h in still water. The boat travels 42 km downstream in the same time that it takes to travel 12 km upstream. What is the speed of the stream?

39. What is the minimum product of two numbers whose difference is 14? What are the numbers that yield this product?

📷 *Students in a biology class just took a final exam. A formula for determining what the average exam grade will be t months later is*

$$S(t) = 78 - 15 \log (t + 1).$$

40. The average score when the students first took the test occurs when $t = 0$. Find the students' average score on the final exam.

41. What would the average score be on a retest after 4 months?

The population of the Ukraine was 52 million in 1996, and the exponential growth rate was 0.2% per year.

42. Write an exponential function describing the growth of the population of the Ukraine.

43. 📷 Predict what the population will be in 2001; in 2005.

44. y varies directly as the square of x and inversely as z, and $y = 2$ when $x = 5$ and $z = 100$. What is y when $x = 3$ and $z = 4$?

Perform the indicated operations and simplify.

45. $(5p^2q^3 - 4p^3q + 6pq - p^2 + 3)$
$\qquad\qquad + (2p^2q^3 + 2p^3q + p^2 - 5pq - 9)$

46. $(11x^2 - 6x - 3) - (3x^2 + 5x - 2)$

47. $(3x^2 - 2y)^2$

48. $(5a + 3b)(2a - 3b)$

49. $\dfrac{x^2 + 8x + 16}{2x + 6} \div \dfrac{x^2 + 3x - 4}{x^2 - 9}$

50. $\dfrac{1 + \dfrac{3}{x}}{x - 1 - \dfrac{12}{x}}$

51. $\dfrac{a^2 - a - 6}{a^3 - 27} \cdot \dfrac{a^2 + 3a + 9}{6}$

52. $\dfrac{3}{x + 6} - \dfrac{2}{x^2 - 36} + \dfrac{4}{x - 6}$

Factor.

53. $xy - 2xz + xw$

54. $1 - 125x^3$

55. $6x^2 + 8xy - 8y^2$

56. $x^4 - 4x^3 + 7x - 28$

57. $a^2 - 10a + 25 - 81b^2$

58. $2m^2 + 12mn + 18n^2$

59. $x^4 - 16y^4$

60. For the function described by
$$h(x) = -3x^2 + 4x + 8,$$
find $h(-2)$.

61. Divide: $(x^4 - 5x^3 + 2x^2 - 6) \div (x - 3)$.

62. Multiply $(5.2 \times 10^4)(3.5 \times 10^{-6})$. Write scientific notation for the answer.

63. Divide:
$$\frac{3.4 \times 10^5}{6.8 \times 10^{-9}}.$$
Write scientific notation for the answer.

For the radical expressions that follow, assume that all variables represent positive numbers.

64. Divide and simplify:
$$\frac{\sqrt[3]{40xy^8}}{\sqrt[3]{5xy}}.$$

65. Multiply and simplify: $\sqrt{7xy^3} \cdot \sqrt{28x^2y}$.

66. Rewrite without rational exponents: $(27a^6b)^{4/3}$.

67. Rationalize the denominator:
$$\frac{3 - \sqrt{y}}{2 - \sqrt{y}}.$$

68. Divide and simplify:
$$\frac{\sqrt{x + 5}}{\sqrt[5]{x + 5}}.$$

69. Multiply these complex numbers:
$$(1 + i\sqrt{3})(6 - 2i\sqrt{3}).$$

70. Divide these complex numbers:
$$\frac{3 - 2i}{4 - 3i}.$$

71. Find the inverse of f if $f(x) = 7 - 2x$.

72. Find a linear function with a graph that contains the points $(0, -3)$ and $(-1, 2)$.

73. Find an equation of the line containing the point $(-3, 5)$ and perpendicular to the line whose equation is $2x + y = 6$.

Graph.

74. $5x = 15 + 3y$

75. $y = 2x^2 - 4x - 1$

76. $y = \log_3 x$

77. $y = 3^x$

78. $-2x - 3y \leq 6$

79. Graph: $f(x) = 2(x + 3)^2 + 1$.

    a) Label the vertex.

    b) Draw the axis of symmetry.

    c) Find the maximum or minimum value.

80. Express in terms of logarithms of a, b, and c:
$$\log \left(\frac{a^2c^3}{b} \right).$$

81. Express as a single logarithm:

$$3 \log x - \tfrac{1}{2} \log y - 2 \log z \,.$$

82. Convert to an exponential equation: $\log_a 5 = x$.

83. Convert to a logarithmic equation: $x^3 = t$.

Find each of the following using a calculator.

84. log 0.05566   **85.** $10^{2.89}$

86. ln 12.78   **87.** $e^{-1.4}$

Solve.

88. $\dfrac{5}{3x - 3} + \dfrac{10}{3x + 6} = \dfrac{5x}{x^2 + x - 2}$

89. $\log \sqrt{3x} = \sqrt{\log 3x}$

90. A train travels 280 mi at a certain speed. If the speed had been increased by 5 mph, the trip could have been made in 1 hr less time. Find the actual speed.

Conic Sections

AN APPLICATION

Each side edge of a Burton® Twin snowboard is an arc of a circle. The Twin 53 has a "running length" of 1150 mm and a "sidecut depth" of 19.5 mm. What radius is used to manufacture the edge of this board?

THIS PROBLEM, ALONG WITH A DIAGRAM, APPEARS AS EXERCISE 95 IN SECTION 10.1.

More information on snowboarding is available at
http://hepg.awl.com/be/inter_5

The entire snowboard is designed by geometry from sidecut to nose length to the transition zones. The geometry is what controls the performance of the board.

G. SCOTT BARBIERI
Board Design Engineer
Burlington, VT

*T*he arcs described in the chapter opening are parts of a circle. A circle is one example of a conic section, *meaning that it can be regarded as a cross section of a cone. This chapter presents a variety of equations with graphs that are conic sections. We have already studied two conic sections,* lines *and* parabolas, *in some detail in Chapters 2 and 8. There are many applications involving conics, and we will consider some in this chapter.*

10.1 Conic Sections: Parabolas and Circles

Parabolas • The Distance and Midpoint Formulas • Circles

This section and the next two examine curves formed by cross sections of cones. These curves are graphs of second-degree equations in two variables. Some are shown below:

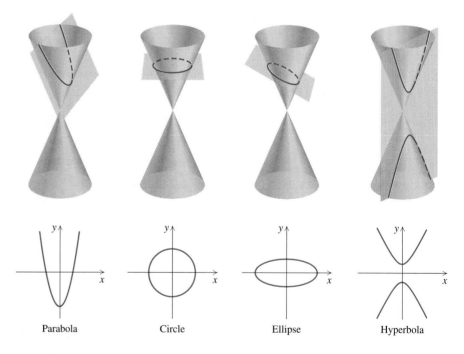

| Parabola | Circle | Ellipse | Hyperbola |

Parabolas

When a cone is cut as shown in the first figure above, the conic section formed is a **parabola**. Parabolas have many applications in electricity, mechanics, and optics. A cross section of a satellite dish is a parabola, and arches that support certain bridges are parabolas. (Free-hanging cables have a different shape, called a *catenary*.)

EQUATION OF A PARABOLA

Parabolas have equations as follows:

$y = ax^2 + bx + c$ (Axis of symmetry parallel to the y-axis);
$x = ay^2 + by + c$ (Axis of symmetry parallel to the x-axis).

Recall from Chapter 8 that the graph of $f(x) = ax^2 + bx + c$ (with $a \neq 0$) is a parabola.

EXAMPLE 1

Graph: $y = x^2 - 4x + 9$.

SOLUTION   To locate the vertex, we can use either of two approaches. One way is to complete the square:

$$y = (x^2 - 4x) + 9$$
$$= (x^2 - 4x + 4 - 4) + 9 \qquad \tfrac{1}{2}(-4) = -2; (-2)^2 = 4$$
$$= (x^2 - 4x + 4) + (-4 + 9) \qquad \text{Regrouping}$$
$$= (x - 2)^2 + 5. \qquad \text{Factoring and simplifying}$$

The vertex is $(2, 5)$.

A second way to find the vertex is to recall that *the x-coordinate of the vertex of the parabola given by* $y = ax^2 + bx + c$ *is* $-b/(2a)$:

$$x = -\frac{b}{2a} = -\frac{-4}{2(1)} = 2.$$

To find the y-coordinate of the vertex, we substitute 2 for x:

$$y = x^2 - 4x + 9 = 2^2 - 4(2) + 9 = 5.$$

Either way, the vertex is $(2, 5)$. Next, we calculate and plot some points on each side of the vertex. Since the x^2-coefficient, 1, is positive, the graph opens upward.

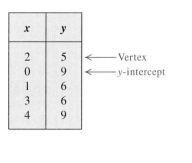

x	y	
2	5	← Vertex
0	9	← y-intercept
1	6	
3	6	
4	9	

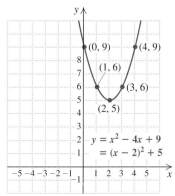

TO GRAPH AN EQUATION OF THE FORM
$y = ax^2 + bx + c$:

1. Find the vertex (h, k) either by completing the square to find an equivalent equation $y = a(x - h)^2 + k$, or by using $-b/(2a)$ to find the x-coordinate and substituting to find the y-coordinate.
2. Choose other values for x on each side of the vertex, and compute the corresponding y-values.
3. The graph opens upward for $a > 0$ and downward for $a < 0$.

Equations of the form $x = ay^2 + by + c$ represent horizontal parabolas. These parabolas open to the right for $a > 0$, open to the left for $a < 0$, and have axes of symmetry parallel to the x-axis.

EXAMPLE 2   Graph: $x = y^2 - 4y + 9$.

SOLUTION   This equation is like that in Example 1 except that x and y are interchanged. The vertex is $(5, 2)$ instead of $(2, 5)$. To find ordered pairs, we choose values for y on each side of the vertex. Then we compute values for x. Note that the x- and y-values of the table in Example 1 are now switched. You should confirm that, by completing the square, we get $x = (y - 2)^2 + 5$.

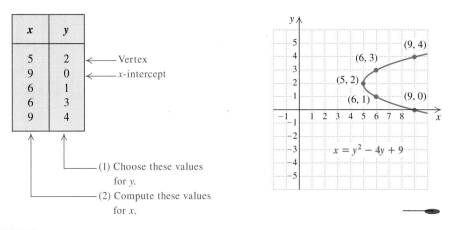

x	y	
5	2	← ——— Vertex
9	0	← ——— x-intercept
6	1	
6	3	
9	4	

(1) Choose these values for y.
(2) Compute these values for x.

TO GRAPH AN EQUATION OF THE FORM
$x = ay^2 + by + c$:

1. Find the vertex (h, k) either by completing the square to find an equivalent equation

    $$x = a(y - k)^2 + h,$$

    or by using $-b/(2a)$ to find the y-coordinate and substituting to find the x-coordinate.
2. Choose other values for y that are above and below the vertex, and compute the corresponding x-values.
3. The graph opens to the right if $a > 0$ and to the left if $a < 0$.

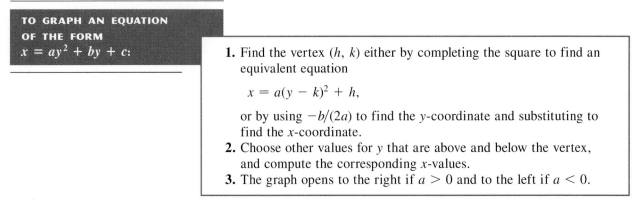

EXAMPLE   3

Graph: $x = -2y^2 + 10y - 7$.

SOLUTION   We use the method of completing the square:

$$x = -2y^2 + 10y - 7$$
$$= -2(y^2 - 5y) - 7$$
$$= -2\left(y^2 - 5y + \tfrac{25}{4}\right) - 7 - (-2)\tfrac{25}{4} \qquad \tfrac{1}{2}(-5) = \tfrac{-5}{2}; \left(\tfrac{-5}{2}\right)^2 = \tfrac{25}{4}; \text{ we add and}$$
$$\text{subtract } (-2)\tfrac{25}{4}.$$
$$= -2\left(y - \tfrac{5}{2}\right)^2 + \tfrac{11}{2}. \qquad\qquad \text{Factoring and simplifying}$$

The vertex is $\left(\tfrac{11}{2}, \tfrac{5}{2}\right)$.

For practice, we also find the vertex by first computing its y-coordinate, $-b/(2a)$, and then substituting to find the x-coordinate:

$$y = -\frac{b}{2a} = -\frac{10}{2(-2)} = \frac{5}{2}$$

$$x = -2y^2 + 10y - 7 = -2\left(\tfrac{5}{2}\right)^2 + 10\left(\tfrac{5}{2}\right) - 7$$
$$= \tfrac{11}{2}.$$

To find ordered pairs, we first choose values for y and then compute values for x. A table is shown below, together with the graph. The graph opens to the left because the y^2-coefficient, -2, is negative.

x	y	
$\frac{11}{2}$	$\frac{5}{2}$	← Vertex
-7	0	← x-intercept
5	2	
5	3	
1	1	
1	4	
-7	5	

(1) Choose these values for y.

(2) Compute these values for x.

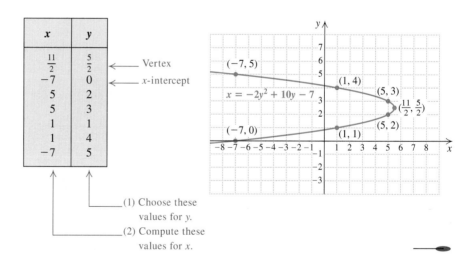

The Distance and Midpoint Formulas

Suppose that two points are on a horizontal line, and thus have the same second coordinate. We can find the distance between them by subtracting their first coordinates. This difference may be negative, depending on the order in which we subtract. So, to make sure we get a positive number, we take the absolute value of this difference. The distance between two points on a horizontal line (x_1, y_1) and (x_2, y_1) is thus $|x_2 - x_1|$. Similarly, the distance between two points on a vertical line (x_2, y_1) and (x_2, y_2) is $|y_2 - y_1|$.

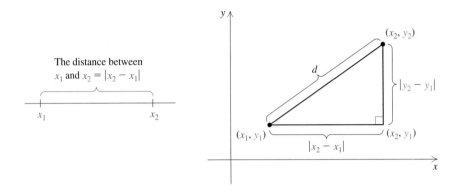

The distance between
x_1 and $x_2 = |x_2 - x_1|$

Now consider *any* two points (x_1, y_1) and (x_2, y_2). If $x_1 \neq x_2$ and $y_1 \neq y_2$, these points are vertices of a right triangle, as shown. The other vertex is then (x_2, y_1). The lengths of the legs are $|x_2 - x_1|$ and $|y_2 - y_1|$. We find d, the length of the hypotenuse, by using the Pythagorean theorem:

$$d^2 = |x_2 - x_1|^2 + |y_2 - y_1|^2.$$

Since the square of a number is the same as the square of its opposite, we don't really need these absolute-value signs. Thus,

$$d^2 = (x_2 - x_1)^2 + (y_2 - y_1)^2.$$

Taking the principal square root, we obtain the distance between two points.

THE DISTANCE FORMULA

The distance d between any two points (x_1, y_1) and (x_2, y_2) is given by

$$d = \sqrt{(x_2 - x_1)^2 + (y_2 - y_1)^2}.$$

E X A M P L E   4   Find the distance between $(5, -1)$ and $(-4, 6)$. Find an exact answer and an approximation to three decimal places.

S O L U T I O N   We substitute into the distance formula:

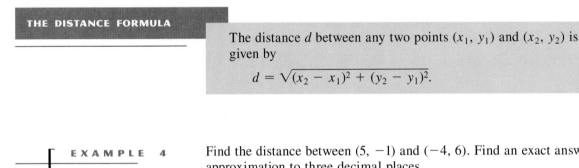

$$d = \sqrt{(-4 - 5)^2 + [6 - (-1)]^2} \qquad \text{Substituting}$$
$$= \sqrt{(-9)^2 + 7^2}$$
$$= \sqrt{130} \approx 11.402. \qquad \text{Using a calculator}$$

The distance formula is needed to develop the formula for a circle, which follows, and to verify certain properties of conic sections. It is also needed to verify a formula for the coordinates of the *midpoint* of a segment connecting two known endpoints. We state the midpoint formula and leave its proof to the exercises. Note that although the distance formula involves both subtraction and addition, the midpoint formula uses only addition.

If the endpoints of a segment are (x_1, y_1) and (x_2, y_2), then the coordinates of the midpoint are

$$\left(\frac{x_1 + x_2}{2}, \frac{y_1 + y_2}{2} \right).$$

(To locate the midpoint, determine the average of the x-coordinates and the average of the y-coordinates.)

EXAMPLE 5   Find the midpoint of the segment with endpoints $(-2, 3)$ and $(4, -6)$.

SOLUTION   Using the midpoint formula, we obtain

$$\left(\frac{-2 + 4}{2}, \frac{3 + (-6)}{2} \right), \quad \text{or} \quad \left(\frac{2}{2}, \frac{-3}{2} \right), \quad \text{or} \quad \left(1, -\frac{3}{2} \right).$$

Circles

One conic section, the **circle**, is a set of points in a plane that are a fixed distance r, called the **radius**, from a fixed point (h, k), called the **center**. If a point (x, y) is on the circle, then by the definition of a circle and the distance formula, it follows that

$$r = \sqrt{(x - h)^2 + (y - k)^2}.$$

Squaring both sides gives the equation of a circle in standard form.

The equation of a circle, centered at (h, k), with radius r, is given by

$$(x - h)^2 + (y - k)^2 = r^2.$$

Note that when $h = 0$ and $k = 0$, the circle is centered at the origin. Otherwise, the circle is translated $|h|$ units horizontally and $|k|$ units vertically.

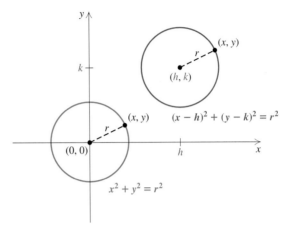

EXAMPLE 6

Find an equation of the circle having center $(4, 5)$ and radius 6.

SOLUTION   Using the standard form, we obtain

$$(x - 4)^2 + (y - 5)^2 = 6^2, \qquad \text{Using } (x - h)^2 + (y - k)^2 = r^2$$

or

$$(x - 4)^2 + (y - 5)^2 = 36.$$

EXAMPLE 7

Find the center and the radius and then graph each circle.

a) $(x - 2)^2 + (y + 3)^2 = 4^2$
b) $x^2 + y^2 + 8x - 2y + 15 = 0$

SOLUTION

a) We write standard form:

$$(x - 2)^2 + [y - (-3)]^2 = 4^2.$$

The center is $(2, -3)$ and the radius is 4. The graph is easily drawn using a compass.

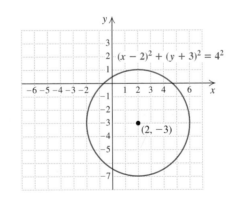

b) To write the equation $x^2 + y^2 + 8x - 2y + 15 = 0$ in standard form, we complete the square twice, once with $x^2 + 8x$ and once with $y^2 - 2y$:

$$x^2 + y^2 + 8x - 2y + 15 = 0$$

$$x^2 + 8x + y^2 - 2y + 15 = 0 \qquad \text{Grouping the } x\text{-terms and the } y\text{-terms}$$

$$x^2 + 8x + y^2 - 2y = -15 \qquad \text{Adding } -15 \text{ on both sides}$$

$$x^2 + 8x + 16 + y^2 - 2y + 1 = -15 + 16 + 1 \qquad \text{Adding } \left(\tfrac{8}{2}\right)^2, \text{ or 16, and } \left(-\tfrac{2}{2}\right)^2, \text{ or 1, on both sides}$$

$$(x + 4)^2 + (y - 1)^2 = 2 \qquad \text{Factoring}$$

$$[x - (-4)]^2 + (y - 1)^2 = (\sqrt{2})^2. \qquad \text{Writing standard form}$$

The center is $(-4, 1)$ and the radius is $\sqrt{2}$.

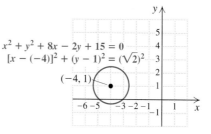

$x^2 + y^2 + 8x - 2y + 15 = 0$
$[x - (-4)]^2 + (y - 1)^2 = (\sqrt{2})^2$

$(-4, 1)$

TECHNOLOGY CONNECTION

10.1

Because most graphers can graph only functions, and because a circle is not a function, graphing a circle on a grapher requires two steps:

1. Solve the equation for y. The result will include a $\pm$ sign in front of a radical.

2. Graph two functions, one for the $+$ sign and the other for the $-$ sign, on the same set of axes.

For example, to graph $(x - 3)^2 + (y + 1)^2 = 16$, solve for $y + 1$ and then y:

$$(y + 1)^2 = 16 - (x - 3)^2$$

$$y + 1 = \pm\sqrt{16 - (x - 3)^2}$$

$$y = -1 \pm \sqrt{16 - (x - 3)^2},$$

or $$y_1 = -1 + \sqrt{16 - (x - 3)^2}$$

and $$y_2 = -1 - \sqrt{16 - (x - 3)^2}.$$

When both functions are graphed (in a "squared" window, to eliminate distortion), the result is similar to the graph shown here.

$y_1 = -1 + \sqrt{16 - (x - 3)^2}$,
$y_2 = -1 - \sqrt{16 - (x - 3)^2}$

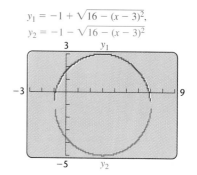

Some graphers will not connect the two pieces because of approximations made near each endpoint.

Use a grapher to graph each of the following equations.

1. $x^2 + y^2 - 16 = 0$

2. $4x^2 + 4y^2 = 100$

3. $x^2 + y^2 + 14x - 16y + 54 = 0$

4. $x^2 + y^2 - 10x - 11 = 0$

EXERCISE SET

10.1

Graph. Be sure to label each vertex.

1. $y = -x^2$

2. $x = y^2$

3. $x = y^2 - 4y + 1$

4. $y = x^2 + 2x + 3$

5. $y = -x^2 + 4x - 5$

6. $x = 4 - 3y - y^2$

7. $x = y^2 + 1$

8. $x = 2y^2$

9. $x = -\frac{1}{2}y^2$

10. $x = y^2 - 1$

11. $x = -y^2 - 3y$

12. $x = y^2 + y - 6$

13. $x = 8 - y - y^2$

14. $y = x^2 + 2x + 1$

15. $y = x^2 - 2x + 1$

16. $y = -\frac{1}{2}x^2$

17. $x = -y^2 + 2y - 1$

18. $x = -y^2 - 2y + 3$

19. $x = -2y^2 - 4y + 1$

20. $x = 2y^2 + 4y - 1$

Find the distance between each pair of points. Where appropriate, find an approximation to three decimal places.

21. $(2, 7)$ and $(6, 10)$

22. $(1, 10)$ and $(7, 2)$

23. $(0, -7)$ and $(3, -4)$

24. $(6, 2)$ and $(6, -8)$

25. $(-5, 5)$ and $(5, -5)$

26. $(5, 21)$ and $(-3, 1)$

27. $(8.6, -3.4)$ and $(-9.2, -3.4)$

28. $(5.9, 2)$ and $(3.7, -7.7)$

29. $\left(\frac{5}{7}, \frac{1}{14}\right)$ and $\left(\frac{1}{7}, \frac{11}{14}\right)$

30. $(0, \sqrt{7})$ and $(\sqrt{6}, 0)$

31. $(-\sqrt{6}, \sqrt{2})$ and $(0, 0)$

32. $(\sqrt{5}, -\sqrt{3})$ and $(0, 0)$

33. $(\sqrt{2}, -\sqrt{3})$ and $(-\sqrt{7}, \sqrt{5})$

34. $(\sqrt{8}, \sqrt{3})$ and $(-\sqrt{5}, -\sqrt{6})$

35. $(0, 0)$ and (s, t)

36. (p, q) and $(0, 0)$

Find the midpoint of each segment with the given endpoints.

37. $(2, 9)$ and $(6, -7)$

38. $(6, 7)$ and $(7, -9)$

39. $(8, 5)$ and $(-1, 2)$

40. $(-1, 2)$ and $(1, -3)$

41. $(-8, -5)$ and $(6, -1)$

42. $(8, -2)$ and $(-3, 4)$

43. $(-3.4, 8.1)$ and $(2.9, -8.7)$

44. $(4.1, 6.9)$ and $(5.2, -6.9)$

45. $\left(\frac{1}{6}, -\frac{3}{4}\right)$ and $\left(-\frac{1}{3}, \frac{5}{6}\right)$

46. $\left(-\frac{4}{5}, -\frac{2}{3}\right)$ and $\left(\frac{1}{8}, \frac{3}{4}\right)$

47. $(\sqrt{2}, -1)$ and $(\sqrt{3}, 4)$

48. $(9, 2\sqrt{3})$ and $(-4, 5\sqrt{3})$

Find an equation of the circle satisfying the given conditions.

49. Center $(0, 0)$, radius 6

50. Center $(0, 0)$, radius 5

51. Center $(7, 3)$, radius $\sqrt{5}$

52. Center $(5, 6)$, radius $\sqrt{2}$

53. Center $(-4, 3)$, radius $4\sqrt{3}$

54. Center $(-2, 7)$, radius $2\sqrt{5}$

55. Center $(-7, -2)$, radius $5\sqrt{2}$

56. Center $(-5, -8)$, radius $3\sqrt{2}$

57. Center $(0, 0)$, passing through $(-3, 4)$

58. Center $(3, -2)$, passing through $(11, -2)$

59. Center $(-4, 1)$, passing through $(-2, 5)$

60. Center $(-1, -3)$, passing through $(-4, 2)$

Find the center and the radius of each circle. Then graph the circle.

61. $x^2 + y^2 = 49$

62. $x^2 + y^2 = 36$

63. $(x + 1)^2 + (y + 3)^2 = 4$

64. $(x - 2)^2 + (y + 3)^2 = 1$

65. $(x - 4)^2 + (y + 3)^2 = 10$

66. $(x + 5)^2 + (y - 1)^2 = 15$

67. $x^2 + y^2 = 7$

68. $x^2 + y^2 = 8$

69. $(x - 5)^2 + y^2 = \frac{1}{4}$

70. $x^2 + (y - 1)^2 = \frac{1}{25}$

71. $x^2 + y^2 + 8x - 6y - 15 = 0$

72. $x^2 + y^2 + 6x - 4y - 15 = 0$

73. $x^2 + y^2 - 8x + 2y + 13 = 0$

74. $x^2 + y^2 + 6x + 4y + 12 = 0$

75. $x^2 + y^2 + 10y - 75 = 0$

76. $x^2 + y^2 - 8x - 84 = 0$

77. $x^2 + y^2 + 7x - 3y - 10 = 0$

78. $x^2 + y^2 - 21x - 33y + 17 = 0$

79. $36x^2 + 36y^2 = 1$

80. $4x^2 + 4y^2 = 1$

SKILL MAINTENANCE

81. A rectangle 10 in. long and 6 in. wide is bordered by a strip of uniform width. If the perimeter of the larger rectangle is twice that of the smaller rectangle, what is the width of the border?

82. One airplane flies 60 mph faster than another. To fly a certain distance, the faster plane takes 4 hr and the slower plane takes 4 hr and 24 min. What is the distance?

Solve each system.

83. $3x - 8y = 5,$
$2x + 6y = 5$

84. $4x - 5y = 9,$
$12x - 10y = 18$

SYNTHESIS

85. ◈ Is the center of a circle a part of the circle? Why or why not?

86. ◈ Outline a procedure that would use the distance formula to determine whether three points, (x_1, y_1), (x_2, y_2), and (x_3, y_3), are collinear (lie on the same line).

87. ◈ Describe a procedure that would use the distance formula to determine whether three points, (x_1, y_1), (x_2, y_2), and (x_3, y_3), are vertices of a right triangle.

88. ◈ Why does the discussion of the distance formula precede the discussion of circles?

Find an equation of a circle satisfying the given conditions.

89. Center $(3, -5)$ and tangent to (touching at one point) the y-axis

90. Center $(-7, -4)$ and tangent to the x-axis

91. The endpoints of a diameter are $(7, 3)$ and $(-1, -3)$.

92. Center $(-3, 5)$ with a circumference of 8π units

93. Find the point on the y-axis that is equidistant from $(2, 10)$ and $(6, 2)$.

94. Find the point on the x-axis that is equidistant from $(-1, 3)$ and $(-8, -4)$.

95. ▦ *Snowboarding.* Each side edge of a Burton® Twin snowboard is an arc of a circle. The Twin 53 has a "running length" of 1150 mm and a "sidecut depth" of 19.5 mm (see the figure at the top of the next column).

a) Using the coordinates shown, locate the center of the circle. (*Hint*: Equate distances.)

b) What radius is used for the edge of the board?

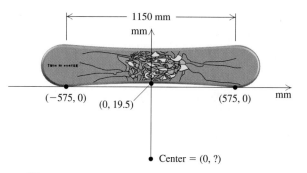

96. ▦ *Snowboarding.* The Burton® Twin 44 snowboard has a running length of 1070 mm and a sidecut depth of 17.5 mm (see Exercise 95). What radius is used for the edge of this snowboard?

97. ▦ *Skiing.* The Rossignol® Ski Corporation produces the Cut 10.4 ski. This ski, when lying flat and viewed from above, has edges that are arcs of a circle.*

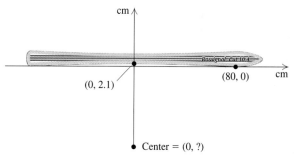

a) Using the coordinates shown, locate the center of the circle. (*Hint*: Equate distances.)

b) What radius is used for the arc passing through $(0, 2.1)$ and $(80, 0)$?

98. ▦ *Doorway construction.* Ace Carpentry needs to cut an arch for the top of an entranceway. The arch needs to be 8 ft wide and 2 ft high. To draw the arch, the carpenters will use a stretched string with chalk attached at an end as a compass.

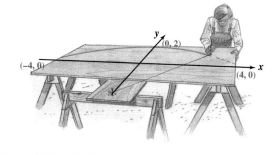

---

*Actually, each edge is made of two arcs of slightly different radii. The arc for the rear half of the ski edge has a slightly larger radius.

a) Using a coordinate system, locate the center of the circle.

b) What radius should the carpenters use to draw the arch?

99. ▦ *Archaeology.* During an archaeological dig, Martina finds the bowl fragment shown below. What was the original diameter of the bowl?

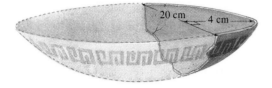

100. *Ferris wheel design.* A ferris wheel has a radius of 24.3 ft. Assuming that the center is 30.6 ft off the ground and that the origin is below the center, as in the following figure, find an equation of the circle.

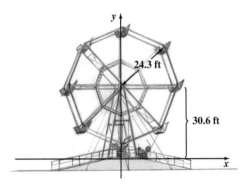

101. Use a graph of the equation $x = y^2 - y - 6$ to approximate the solutions of each of the following equations.

a) $y^2 - y - 6 = 2$ (*Hint:* Graph $x = 2$ on the same set of axes as the graph of $x = y^2 - y - 6$.)

b) $y^2 - y - 6 = -3$

102. *Power of a motor.* The horsepower of a certain kind of engine is given by the formula

$$H = \frac{D^2N}{2.5},$$

where N is the number of cylinders and D is the diameter, in inches, of each piston. Graph this equation, assuming that $N = 6$ (a six-cylinder engine). Let D run from 2.5 to 8.

103. Prove the midpoint formula by showing that

i) the distance from (x_1, y_1) to

$$\left(\frac{x_1 + x_2}{2}, \frac{y_1 + y_2}{2} \right)$$

equals the distance from (x_2, y_2) to

$$\left(\frac{x_1 + x_2}{2}, \frac{y_1 + y_2}{2} \right);$$

and

ii) the points

$$(x_1, y_1), \left(\frac{x_1 + x_2}{2}, \frac{y_1 + y_2}{2} \right),$$

and

$$(x_2, y_2)$$

lie on the same line (see Exercise 86).

104. ⌁ If the equation $x^2 + y^2 - 6x + 2y - 6 = 0$ is written as $y^2 + 2y + (x^2 - 6x - 6) = 0$, it can be regarded as quadratic in y.

a) Use the quadratic formula to solve for y.

b) Show that the graph of your answer to part (a) coincides with the graph in Technology Connection 10.1.

105. ◈ ⌁ How could a grapher best be used to help you sketch the graph of an equation of the form $x = ay^2 + by + c$?

106. ◈ ⌁ Why should a grapher's window be "squared" before graphing a circle?

10.2 Conic Sections: Ellipses

Ellipses Centered at (0, 0)   •   Ellipses Centered at (*h, k*)

When a cone is cut at an angle, as shown on the next page, the conic section formed is an *ellipse*. To draw an ellipse, stick two tacks in a piece of cardboard. Then tie a string to the tacks, place a pencil as shown, and draw.

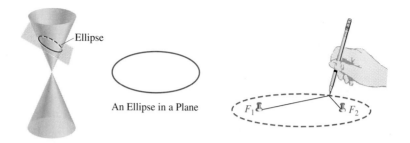

An Ellipse in a Plane

Ellipses Centered at (0, 0)

An **ellipse** is defined as the set of all points in a plane for which the *sum* of the distances from two fixed points F_1 and F_2 (called **foci**; singular, **focus**) is constant. In the figure above, the tacks are at the foci. The midpoint of the segment $F_1 F_2$ is the **center**. Ellipses have equations as follows. The proof is left to the exercises.

EQUATION OF AN ELLIPSE

The equation of an ellipse centered at the origin and symmetric with respect to both axes is

$$\frac{x^2}{a^2} + \frac{y^2}{b^2} = 1, \quad a, b > 0. \qquad \text{(Standard form)}$$

To graph an ellipse centered at the origin, it helps to first find the intercepts. If we replace x with 0, we can find the y-intercepts:

$$\frac{0^2}{a^2} + \frac{y^2}{b^2} = 1$$

$$\frac{y^2}{b^2} = 1$$

$$y^2 = b^2 \quad \text{or} \quad y = \pm b.$$

Thus the y-intercepts are $(0, b)$ and $(0, -b)$. Similarly, the x-intercepts are $(a, 0)$ and $(-a, 0)$. If $a > b$, the ellipse is horizontal and $(-a, 0)$ and $(a, 0)$ are the **vertices** (singular, **vertex**). If $b > a$, the ellipse is vertical and $(0, -b)$ and $(0, b)$ are the vertices.

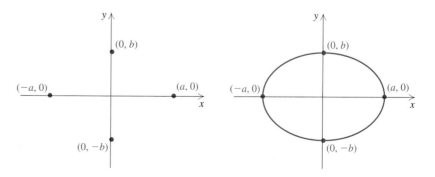

Plotting these points and filling in an oval-shaped curve, we get a graph of the ellipse. If a more precise graph is desired, we can plot more points.

USING a AND b TO GRAPH AN ELLIPSE

For the ellipse

$$\frac{x^2}{a^2} + \frac{y^2}{b^2} = 1,$$

the x-intercepts are $(-a, 0)$ and $(a, 0)$. The y-intercepts are $(0, -b)$ and $(0, b)$.

EXAMPLE   1   Graph the ellipse

$$\frac{x^2}{4} + \frac{y^2}{9} = 1.$$

SOLUTION   Note that

$$\frac{x^2}{4} + \frac{y^2}{9} = \frac{x^2}{2^2} + \frac{y^2}{3^2}. \qquad \text{Identifying } a \text{ and } b$$

Thus the x-intercepts are $(-2, 0)$ and $(2, 0)$, and the y-intercepts are $(0, 3)$ and $(0, -3)$. We plot these points and connect them with an oval-shaped curve. To plot some other points, we let $x = 1$ and solve for y:

$$\frac{1^2}{4} + \frac{y^2}{9} = 1$$

$$36\left(\frac{1}{4} + \frac{y^2}{9}\right) = 36 \cdot 1$$

$$36 \cdot \frac{1}{4} + 36 \cdot \frac{y^2}{9} = 36$$

$$9 + 4y^2 = 36$$

$$4y^2 = 27$$

$$y^2 = \frac{27}{4}$$

$$y = \pm\sqrt{\frac{27}{4}}$$

$$y \approx \pm 2.6.$$

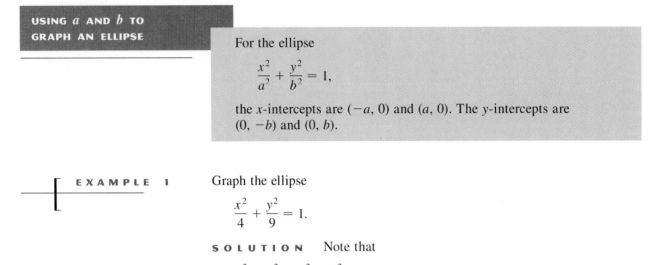

Thus, $(1, 2.6)$ and $(1, -2.6)$ can also be used to draw the graph. Similarly, the points $(-1, 2.6)$ and $(-1, -2.6)$ can also be computed and plotted.

EXAMPLE 2   Graph: $4x^2 + 25y^2 = 100$.

SOLUTION   To write the equation in standard form, we multiply on both sides by $\frac{1}{100}$:

$$\frac{1}{100}(4x^2 + 25y^2) = \frac{1}{100}(100)$$

Multiplying by $\frac{1}{100}$ to get 1 on the right side.

$$\left.\begin{array}{l}\dfrac{1}{100}(4x^2) + \dfrac{1}{100}(25y^2) = 1 \\[2mm] \dfrac{x^2}{25} + \dfrac{y^2}{4} = 1\end{array}\right\}$$

Simplifying

$$\frac{x^2}{5^2} + \frac{y^2}{2^2} = 1.$$ $a = 5, b = 2$

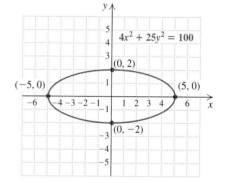

The x-intercepts are $(-5, 0)$ and $(5, 0)$, and the y-intercepts are $(0, -2)$ and $(0, 2)$. We plot the intercepts and connect them with an oval-shaped curve. Other points can also be computed and plotted.

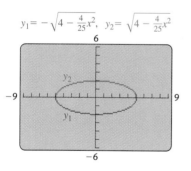

TECHNOLOGY CONNECTION

10.2

Graphing an ellipse on a grapher is much like graphing a circle: We graph it in two pieces after solving for y. To illustrate, check Example 2:

$$4x^2 + 25y^2 = 100$$
$$25y^2 = 100 - 4x^2$$
$$y^2 = 4 - \frac{4}{25}x^2$$
$$y = \pm\sqrt{4 - \frac{4}{25}x^2}.$$

Using a squared window, we have our check:

$$y_1 = -\sqrt{4 - \frac{4}{25}x^2}, \quad y_2 = \sqrt{4 - \frac{4}{25}x^2}$$

Ellipses Centered at (*h*, *k*)

Horizontal and vertical translations, similar to those used in Chapter 8, can be used to graph ellipses that are not centered at the origin.

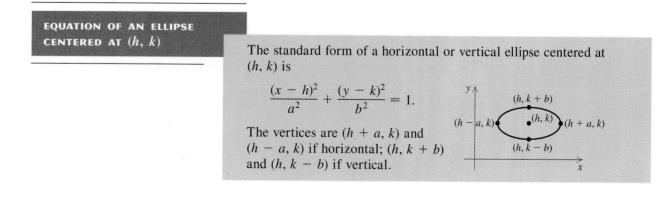

EQUATION OF AN ELLIPSE CENTERED AT (h, k)

The standard form of a horizontal or vertical ellipse centered at (h, k) is

$$\frac{(x - h)^2}{a^2} + \frac{(y - k)^2}{b^2} = 1.$$

The vertices are $(h + a, k)$ and $(h - a, k)$ if horizontal; $(h, k + b)$ and $(h, k - b)$ if vertical.

EXAMPLE 3   Graph the ellipse

$$\frac{(x-1)^2}{4} + \frac{(y+5)^2}{9} = 1.$$

SOLUTION   Note that

$$\frac{(x-1)^2}{4} + \frac{(y+5)^2}{9} = \frac{(x-1)^2}{2^2} + \frac{(y+5)^2}{3^2}.$$

Thus, $a = 2$ and $b = 3$. To determine the center of the ellipse, (h, k), note that

$$\frac{(x-1)^2}{2^2} + \frac{(y+5)^2}{3^2} = \frac{(x-1)^2}{2^2} + \frac{(y-(-5))^2}{3^2}.$$

Thus the center is $(1, -5)$. We locate $(1, -5)$ and then plot $(1 + 2, -5)$, $(1 - 2, -5)$, $(1, -5 + 3)$, and $(1, -5 - 3)$. These are the points $(3, -5)$, $(-1, -5)$, $(1, -2)$, and $(1, -8)$.

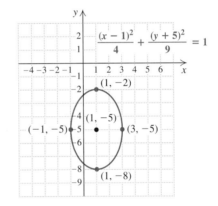

Note that this ellipse is the same as the ellipse in Example 1 but translated 1 unit to the right and 5 units down.

Ellipses have many applications. Communications satellites move in elliptical orbits with the earth as a focus while the earth itself follows an elliptical path around the sun. A recent medical innovation, the lithotripter, uses shock waves originating at one focus to crush a kidney stone located at the other focus.

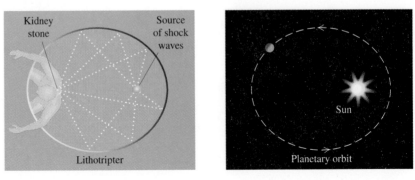

Lithotripter

Planetary orbit

In some buildings, an ellipsoidal ceiling creates a "whispering gallery" in which a person at one focus can whisper and still be heard clearly at the other focus. This happens because sound waves coming from one focus are all reflected to the other focus. Similarly, light waves bouncing off an ellipsoidal mirror are used in a dentist's reflector light. The light source is located at one focus while the patient's mouth is at the other.

EXERCISE SET
10.2

Graph each ellipse.

1. $\dfrac{x^2}{4} + \dfrac{y^2}{1} = 1$

2. $\dfrac{x^2}{1} + \dfrac{y^2}{4} = 1$

3. $\dfrac{x^2}{9} + \dfrac{y^2}{25} = 1$

4. $\dfrac{x^2}{25} + \dfrac{y^2}{16} = 1$

5. $4x^2 + 9y^2 = 36$

6. $9x^2 + 4y^2 = 36$

7. $16x^2 + 9y^2 = 144$

8. $9x^2 + 16y^2 = 144$

9. $2x^2 + 3y^2 = 6$

10. $5x^2 + 7y^2 = 35$

11. $12x^2 + 5y^2 = 120$

12. $8x^2 + 5y^2 = 80$

13. $3x^2 + 7y^2 - 63 = 0$

14. $3x^2 + 8y^2 - 72 = 0$

15. $8x^2 = 96 - 3y^2$

16. $6y^2 = 24 - 8x^2$

17. $16x^2 + 25y^2 = 1$

18. $9x^2 + 4y^2 = 1$

19. $\dfrac{(x-2)^2}{9} + \dfrac{(y-1)^2}{25} = 1$

20. $\dfrac{(x-3)^2}{25} + \dfrac{(y-4)^2}{9} = 1$

21. $\dfrac{(x+4)^2}{16} + \dfrac{(y-3)^2}{49} = 1$

22. $\dfrac{(x+5)^2}{4} + \dfrac{(y-2)^2}{36} = 1$

23. $12(x-1)^2 + 3(y+4)^2 = 48$

24. $4(x-6)^2 + 9(y+2)^2 = 36$

25. $(x+4)^2 + 4(y+1)^2 - 10 = 90$

26. $9(x+6)^2 + (y+2)^2 - 20 = 61$

SKILL MAINTENANCE

Solve.

27. $3x^2 - 2x + 7 = 0$

28. $9x^2 + 4 = 0$

29. $\dfrac{1}{x-5} = x + 5$

30. $9 - \sqrt{2x+1} = 7$

SYNTHESIS

31. ◆ An eccentric person builds a pool table in the shape of an ellipse with a hole at one focus and a tiny dot at the other. Guests are amazed at how many bank shots the owner of the pool table makes. Explain why this occurs.

32. ◆ Can a circle be considered a special type of ellipse? Why or why not?

Find an equation of an ellipse that contains the following points.

33. $(-9, 0)$, $(9, 0)$, $(0, -11)$, and $(0, 11)$

34. $(-7, 0)$, $(7, 0)$, $(0, -5)$, and $(0, 5)$

35. $(-2, -1)$, $(6, -1)$, $(2, -4)$, and $(2, 2)$

36. $(-6, 3)$, $(4, 3)$, $(-1, 7)$, and $(-1, -1)$

37. *Planetary motion.* The maximum distance of the planet Mars from the sun is 2.48×10^8 mi. The minimum distance is 3.46×10^7 mi. The sun is at one focus of the elliptical orbit. Find the distance from the sun to the other focus.

38. Let $(-c, 0)$ and $(c, 0)$ be the foci of an ellipse. Any point $P(x, y)$ is on the ellipse if the sum of the distances from the foci to P is some constant. Use $2a$ to represent this constant.

a) Show that an equation for the ellipse is given by
$$\frac{x^2}{a^2} + \frac{y^2}{a^2 - c^2} = 1.$$

b) Substitute b^2 for $a^2 - c^2$ to get standard form.

39. *President's office.* The Oval Office of the President of the United States is an ellipse 31 ft wide and 38 ft long. Show in a sketch precisely where the President and an adviser could sit to best hear each other using the room's acoustics. (*Hint:* See Exercise 38(b) and the discussion following Example 3.)

40. *Dentistry.* The light source in a dental lamp shines against a reflector that is shaped like a portion of an ellipse in which the light source is one focus of the ellipse. Reflected light enters a patient's mouth at the other focus of the ellipse. If the ellipse from which the reflector was formed is 2 ft wide and 6 ft long, how far should the patient's mouth be from the light source? (*Hint:* See Exercise 38(b).)

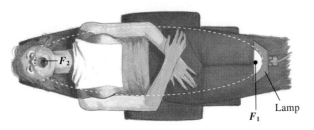

For each of the following equations, complete the square as needed and find an equivalent equation in standard form. Then graph the ellipse.

41. $x^2 - 4x + 4y^2 + 8y - 8 = 0$

42. $4x^2 + 24x + y^2 - 2y - 63 = 0$

43. Use a grapher to check your answers to Exercises 3, 17, 21, and 25.

COLLABORATIVE
C ◆ O ◆ R ◆ N ◆ E ◆ R

Focus: Ellipses

Time: 20–30 minutes

Group size: 2

Materials: Scientific calculators

In March 1996, the comet Hyakutake came within 21 million mi of the sun, and closer to Earth than any comet in over 500 yr (*Source:* Associated Press newspaper story, 3/20/96). Hyakutake is traveling in an elliptical orbit with the sun at one focus. The comet's average speed is about 100,000 mph (it actually goes much faster near its foci and slower as it gets further from the foci) and one orbit takes about 15,000 yr. (Astronomers estimate the time at 10,000–20,000 yr.)

ACTIVITY

1. The elliptical orbit of Hyakutake is so elongated that the distance traveled in one orbit can be estimated by $4a$ (see the following figure). Use the information above to estimate the distance, in millions of miles, traveled in one orbit. Then determine a.

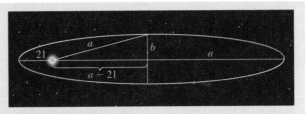

Units are millions of miles.

2. Using the figure above, express b^2 as a function of a. Then solve for b using the value found for a in part (1).

3. Approximately how far will Hyakutake be from the sun at the most distant part of its orbit?

4. Repeat parts (1)–(3), with one group member using the lower estimate of orbit time (10,000 yr) and the other using the upper estimate of orbit time (20,000 yr). By how much do the three answers to part (3) vary?

10.3 Conic Sections: Hyperbolas

Hyperbolas • Hyperbolas (Nonstandard Form) •
Classifying Graphs of Equations

Hyperbolas

A **hyperbola** looks like a pair of parabolas, but the actual shapes are different. A hyperbola has two **vertices** and the line through the vertices is known as an **axis**. The point halfway between the vertices is called the **center**.

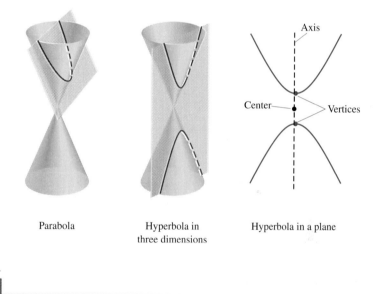

Parabola

Hyperbola in
three dimensions

Hyperbola in a plane

EQUATION OF A HYPERBOLA

Hyperbolas with their centers at the origin* have equations as follows:

$$\frac{x^2}{a^2} - \frac{y^2}{b^2} = 1 \qquad \text{(Axis horizontal)};$$

$$\frac{y^2}{b^2} - \frac{x^2}{a^2} = 1 \qquad \text{(Axis vertical)}.$$

Note that both equations have a 1 on the right-hand side and a subtraction symbol between the terms.

---

*Hyperbolas with horizontal or vertical axes and centers *not* at the origin are discussed in Exercises 47–52.

To graph a hyperbola, it helps to begin by graphing two lines called **asymptotes**. Although the asymptotes themselves are not part of the graph, they serve as guidelines for an accurate sketch.

ASYMPTOTES OF A HYPERBOLA

For hyperbolas with equations as given in the preceding box, the asymptotes are the lines

$$y = \frac{b}{a}x \quad \text{and} \quad y = -\frac{b}{a}x.$$

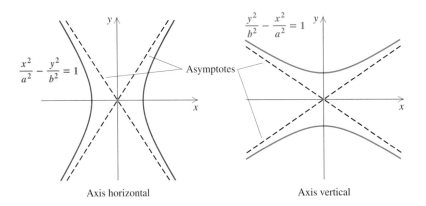

Axis horizontal   Axis vertical

As a hyperbola gets further away from the origin, it gets closer and closer to its asymptotes. The larger $|x|$ gets, the closer the graph gets to an asymptote. The asymptotes act to "constrain" the graph of a hyperbola. Parabolas are *not* constrained by any asymptotes.

The next thing to do after sketching asymptotes is to plot vertices. Then it is easy to sketch the curve.

EXAMPLE 1

Graph: $\dfrac{x^2}{4} - \dfrac{y^2}{9} = 1$.

SOLUTION   Note that

$$\frac{x^2}{4} - \frac{y^2}{9} = \frac{x^2}{2^2} - \frac{y^2}{3^2}, \qquad \text{Identifying } a \text{ and } b$$

so $a = 2$ and $b = 3$. The asymptotes are thus

$$y = \frac{3}{2}x \quad \text{and} \quad y = -\frac{3}{2}x.$$

We sketch them, as shown in the graph on the left on the next page.

For horizontal or vertical hyperbolas centered at the origin, the vertices also serve as intercepts. Since this hyperbola is horizontal, we replace y with 0 and solve for x. We see that $x^2/2^2 = 1$ when $x = \pm 2$. The intercepts are $(2, 0)$ and $(-2, 0)$. You can check that no y-intercepts exist.

Finally, we plot the intercepts and sketch the graph. Through each intercept, we draw a smooth curve that approaches the asymptotes closely, as shown.

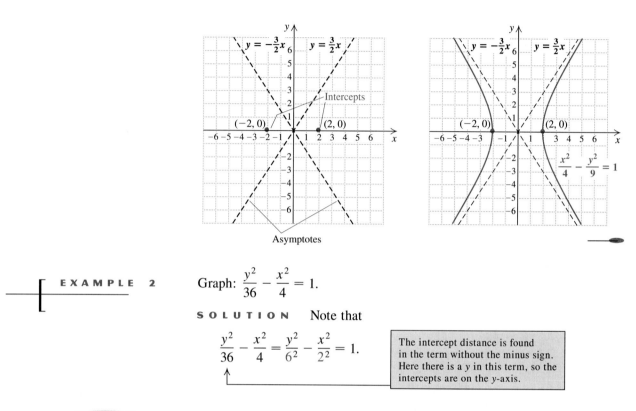

Asymptotes

EXAMPLE 2

Graph: $\dfrac{y^2}{36} - \dfrac{x^2}{4} = 1$.

SOLUTION   Note that

$$\frac{y^2}{36} - \frac{x^2}{4} = \frac{y^2}{6^2} - \frac{x^2}{2^2} = 1.$$

> The intercept distance is found in the term without the minus sign. Here there is a y in this term, so the intercepts are on the y-axis.

TECHNOLOGY CONNECTION
10.3

The procedure used to graph a hyperbola in standard form on a grapher is similar to that used to draw a circle or ellipse. Consider the graph of the hyperbola given by the equation

$$\frac{x^2}{25} - \frac{y^2}{49} = 1.$$

Solving for y gives us

$$y_1 = \frac{\sqrt{49x^2 - 1225}}{5} = \frac{7}{5}\sqrt{x^2 - 25}$$

and $y_2 = \dfrac{-\sqrt{49x^2 - 1225}}{5} = -\dfrac{7}{5}\sqrt{x^2 - 25}$,

or $\ y_2 = -y_1$.

When the two pieces are drawn on the same squared axes, the result is as shown. Again, note the problem that the grapher has at the points where the graph is nearly vertical.

$$y_1 = \frac{7}{5}\sqrt{x^2 - 25}, \ \ y_2 = -\frac{7}{5}\sqrt{x^2 - 25}$$

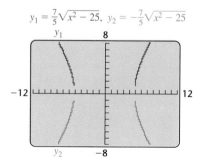

Use a grapher to draw the graph of each hyperbola. Use squared axes so that the shapes are not distorted.

1. $\dfrac{x^2}{16} - \dfrac{y^2}{60} = 1$ **2.** $16x^2 - 3y^2 = 64$

3. $\dfrac{y^2}{20} - \dfrac{x^2}{64} = 1$ **4.** $45y^2 - 9x^2 = 441$

The asymptotes are thus $y = \frac{6}{2}x$ and $y = -\frac{6}{2}x$, or $y = 3x$ and $y = -3x$.

The numbers 6 and 2 can be used to sketch a rectangle that helps with graphing. Using ±2 as x-coordinates and ±6 as y-coordinates, we form all possible ordered pairs: $(2, 6)$, $(2, -6)$, $(-2, 6)$, and $(-2, -6)$. We plot these pairs and lightly sketch a rectangle through them. The asymptotes pass through the corners (see the figure on the left below). Since the hyperbola is vertical, we plot its y-intercepts, $(0, 6)$ and $(0, -6)$. Finally, we draw curves through the intercepts toward the asymptotes, as shown below.

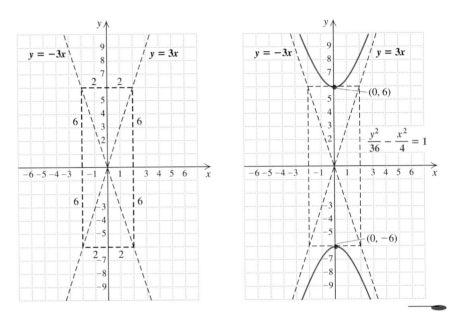

Hyperbolas (Nonstandard Form)

The equations for hyperbolas just examined are the standard ones, but there are other hyperbolas. We consider some of them.

EQUATION OF A HYPERBOLA IN NONSTANDARD FORM

Hyperbolas having the x- and y-axes as asymptotes have equations as follows:

$$xy = c, \quad \text{where } c \text{ is a nonzero constant.}$$

EXAMPLE 3   Graph: $xy = -8$.

SOLUTION   We first solve for y:

$$y = -\frac{8}{x}. \qquad \text{Dividing by } x \text{ on both sides. Note that } x \neq 0.$$

Next, we find some solutions, keeping the results in a table. Note that x cannot be 0 and that for large values of $|x|$, y will be close to 0. Thus the x- and y-axes serve as asymptotes. We plot the points and draw two curves.

x	y
2	-4
-2	4
4	-2
-4	2
1	-8
-1	8
8	-1
-8	1

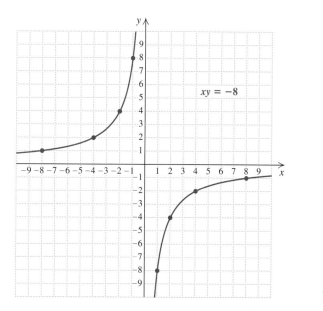

$xy = -8$

Hyperbolas have many applications. A jet breaking the sound barrier creates a sonic boom with a wave front the shape of a cone. The intersection of the cone with the ground is one branch of a hyperbola. Some comets travel in hyperbolic orbits, and a cross section of certain lenses may be hyperbolic in shape.

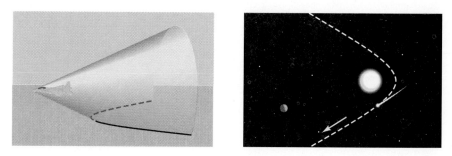

Classifying Graphs of Equations

We summarize the equations and the graphs of the conic sections studied.

PARABOLA

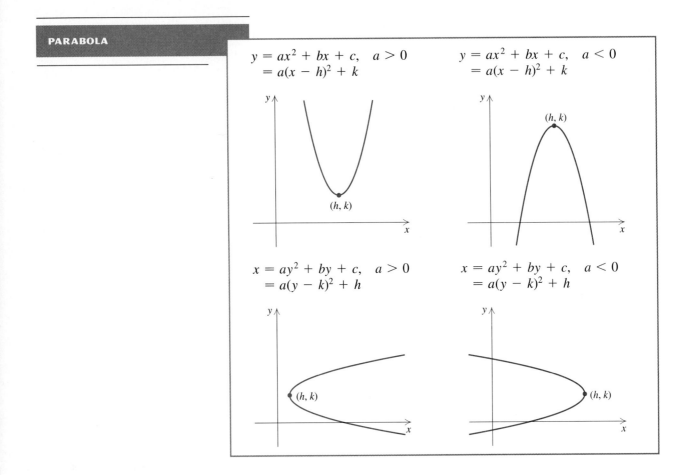

$y = ax^2 + bx + c, \quad a > 0$
$\quad = a(x - h)^2 + k$

$y = ax^2 + bx + c, \quad a < 0$
$\quad = a(x - h)^2 + k$

$x = ay^2 + by + c, \quad a > 0$
$\quad = a(y - k)^2 + h$

$x = ay^2 + by + c, \quad a < 0$
$\quad = a(y - k)^2 + h$

CIRCLE

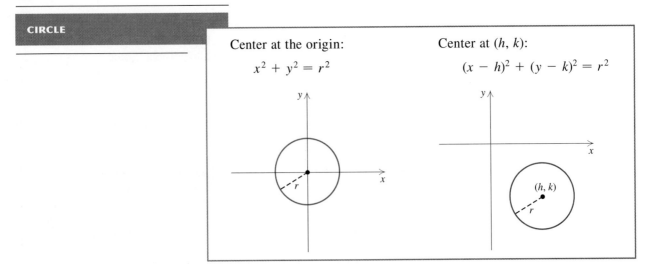

Center at the origin:

$$x^2 + y^2 = r^2$$

Center at (h, k):

$$(x - h)^2 + (y - k)^2 = r^2$$

ELLIPSE

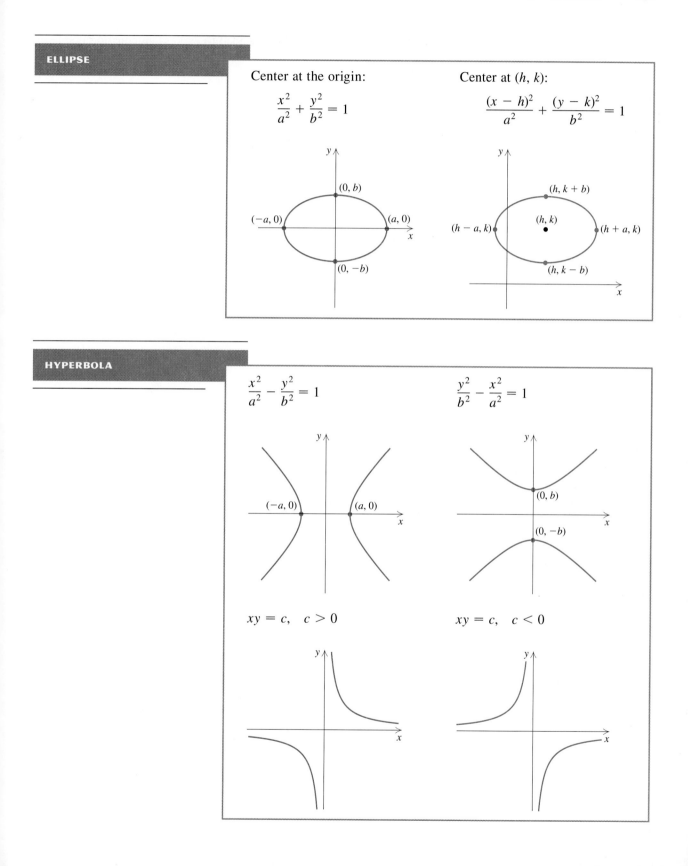

Center at the origin:

$$\frac{x^2}{a^2} + \frac{y^2}{b^2} = 1$$

Center at (h, k):

$$\frac{(x - h)^2}{a^2} + \frac{(y - k)^2}{b^2} = 1$$

HYPERBOLA

$$\frac{x^2}{a^2} - \frac{y^2}{b^2} = 1$$

$$\frac{y^2}{b^2} - \frac{x^2}{a^2} = 1$$

$$xy = c, \quad c > 0$$

$$xy = c, \quad c < 0$$

Algebraic manipulations may be needed to express an equation in one of the preceding forms.

EXAMPLE 4

Classify the graph of each equation as a circle, an ellipse, a parabola, or a hyperbola.

a) $5x^2 = 20 - 5y^2$
b) $x + 3 + 8y = y^2$
c) $x^2 = y^2 + 4$
d) $x^2 = 16 - 4y^2$

SOLUTION

a) We get the terms with variables on one side by adding $5y^2$:

$$5x^2 + 5y^2 = 20.$$

Since x and y are *both* squared, we do not have a parabola. The fact that the squared terms are *added* tells us that we do not have a hyperbola. Do we have a circle? To find out, we need to get $x^2 + y^2$ by itself. We can do that by factoring the 5 out of both terms on the left and then dividing by 5:

$$5(x^2 + y^2) = 20 \qquad \text{Factoring out 5}$$
$$x^2 + y^2 = 4 \qquad \text{Dividing on both sides by 5}$$
$$x^2 + y^2 = 2^2.$$

We can see that the graph is a circle with center at the origin and radius 2.

b) The equation $x + 3 + 8y = y^2$ has only one variable squared, so we solve for the other variable:

$$x = y^2 - 8y - 3.$$

The graph is a horizontal parabola that opens to the right.

c) In $x^2 = y^2 + 4$, both variables are squared, so the graph is not a parabola. We subtract y^2 on both sides and divide by 4 to obtain

$$\frac{x^2}{2^2} - \frac{y^2}{2^2} = 1.$$

The minus sign here indicates that the graph of this equation is a hyperbola.

d) In $x^2 = 16 - 4y^2$, both variables are squared, so the graph cannot be a parabola. We obtain the following equivalent equation:

$$x^2 + 4y^2 = 16.$$

If the coefficients of the terms were the same, we would have the graph of a circle, as in part (a), but they are not. Dividing by 16 on both sides yields

$$\frac{x^2}{16} + \frac{y^2}{4} = 1.$$

The graph of this equation is an ellipse.

EXERCISE SET

10.3

Graph each hyperbola. Label all vertices and sketch all asymptotes.

1. $\dfrac{y^2}{9} - \dfrac{x^2}{9} = 1$

2. $\dfrac{x^2}{16} - \dfrac{y^2}{16} = 1$

3. $\dfrac{x^2}{4} - \dfrac{y^2}{25} = 1$

4. $\dfrac{y^2}{16} - \dfrac{x^2}{9} = 1$

5. $\dfrac{y^2}{36} - \dfrac{x^2}{9} = 1$

6. $\dfrac{x^2}{25} - \dfrac{y^2}{36} = 1$

7. $y^2 - x^2 = 25$

8. $x^2 - y^2 = 4$

9. $25x^2 - 16y^2 = 400$

10. $4y^2 - 9x^2 = 36$

Graph.

11. $xy = -4$

12. $xy = 6$

13. $xy = 3$

14. $xy = -9$

15. $xy = -2$

16. $xy = -1$

17. $xy = 1$

18. $xy = 2$

Classify each of the following as the equation of a circle, an ellipse, a parabola, or a hyperbola.

19. $x^2 + y^2 - 10x + 8y - 40 = 0$

20. $y + 5 = 2x^2$

21. $9x^2 + 4y^2 - 36 = 0$

22. $1 + 3y = 2y^2 - x$

23. $4x^2 - 9y^2 - 100 = 0$

24. $y^2 + x^2 = 7$

25. $x^2 + y^2 = 2x + 4y + 4$

26. $2y + 13 + x^2 = 8x - y^2$

27. $4x^2 = 64 - y^2$

28. $y = \dfrac{1}{x}$

29. $x - \dfrac{3}{y} = 0$

30. $x - 4 = y^2 - 3y$

31. $y + 6x = x^2 + 6$

32. $x^2 = 16 + y^2$

33. $9y^2 = 36 + 4x^2$

34. $3x^2 + 5y^2 + x^2 = y^2 + 49$

35. $3x^2 + y^2 - x = 2x^2 - 9x + 10y + 40$

36. $4y^2 + 20x^2 + 1 = 8y - 5x^2$

37. $16x^2 + 5y^2 - 12x^2 + 8y^2 - 3x + 4y = 568$

38. $56x^2 - 17y^2 = 234 - 13x^2 - 38y^2$

SKILL MAINTENANCE

39. Simplify: $\sqrt[3]{125t^{15}}$.

40. Solve: $2x^2 + 10 = 0$.

41. Rationalize the denominator:

$$\dfrac{4\sqrt{2} - 5\sqrt{3}}{6\sqrt{3} - 8\sqrt{2}}.$$

42. An airplane travels 500 mi at a certain speed. A larger plane travels 1620 mi at a speed that is 320 mph faster, but takes 1 hr longer. Find the speed of each plane.

SYNTHESIS

43. ◈ Is it possible for a hyperbola to represent the graph of a function? Why or why not?

44. ◈ If, in

$$\dfrac{x^2}{a^2} - \dfrac{y^2}{b^2} = 1,$$

$a = b$, what are the asymptotes of the graph? Why?

Find an equation of a hyperbola satisfying the given conditions.

45. Having intercepts $(0, 6)$ and $(0, -6)$ and asymptotes $y = 3x$ and $y = -3x$

46. Having intercepts $(8, 0)$ and $(-8, 0)$ and asymptotes $y = 4x$ and $y = -4x$

The standard equations for horizontal or vertical hyperbolas centered at (h, k) are as follows:

$$\dfrac{(x - h)^2}{a^2} - \dfrac{(y - k)^2}{b^2} = 1$$

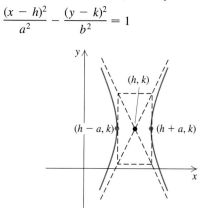

$$\frac{(y-k)^2}{b^2} - \frac{(x-h)^2}{a^2} = 1$$

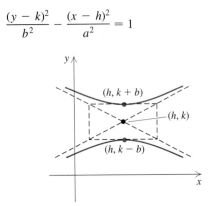

The vertices are as labeled and the asymptotes are

$$y - k = \frac{b}{a}(x - h) \quad and \quad y - k = -\frac{b}{a}(x - h).$$

For each of the following equations of hyperbolas, complete the square, if necessary, and write in standard form. Find the center, the vertices, and the asymptotes. Then graph the hyperbola.

47. $\dfrac{(x-5)^2}{36} - \dfrac{(y-2)^2}{25} = 1$

48. $\dfrac{(x-2)^2}{9} - \dfrac{(y-1)^2}{4} = 1$

49. $8(y+3)^2 - 2(x-4)^2 = 32$

50. $25(x-4)^2 - 4(y+5)^2 = 100$

51. $4x^2 - y^2 + 24x + 4y + 28 = 0$

52. $4y^2 - 25x^2 - 8y - 100x - 196 = 0$

53. Use a grapher to check your answers to Exercises 5, 17, 23, and 47.

10.4 Nonlinear Systems of Equations

Systems Involving One Nonlinear Equation • Systems of Two Nonlinear Equations • Problem Solving

The equations appearing in systems of two equations have thus far all been linear. We now consider systems of two equations in which at least one equation is nonlinear.

Systems Involving One Nonlinear Equation

Suppose that a system consists of an equation of a circle and an equation of a line. In what ways can the circle and the line intersect? The figures below represent three ways in which this situation can occur. We see that such a system will have 0, 1, or 2 real solutions.

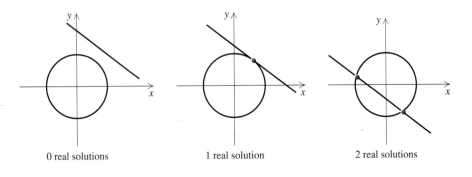

0 real solutions   1 real solution   2 real solutions

Recall that we used both *elimination* and *substitution* to solve systems of linear equations. When solving systems in which one equation is of first degree and one is of second degree, it is preferable to use the *substitution* method.

EXAMPLE 1

Solve the system

$$x^2 + y^2 = 25, \quad (1) \qquad \text{(The graph is a circle.)}$$
$$3x - 4y = 0. \quad (2) \qquad \text{(The graph is a line.)}$$

SOLUTION First, we solve the linear equation, (2), for x:

$$x = \tfrac{4}{3}y. \qquad (3)$$

Then we substitute $\tfrac{4}{3}y$ for x in equation (1) and solve for y:

$$\left(\tfrac{4}{3}y\right)^2 + y^2 = 25$$
$$\tfrac{16}{9}y^2 + y^2 = 25$$
$$\tfrac{25}{9}y^2 = 25$$
$$y^2 = 9 \qquad \text{Multiplying by } \tfrac{9}{25} \text{ on both sides}$$
$$y = \pm 3. \qquad \text{Using the principle of square roots}$$

Now we substitute these numbers for y in equation (3) and solve for x:

for $y = 3$, $x = \tfrac{4}{3}(3) = 4$;   for $y = -3$, $x = \tfrac{4}{3}(-3) = -4$.

Check: For (4, 3):

$$\begin{array}{c|c}
x^2 + y^2 = 25 & \\
\hline
4^2 + 3^2 \;?\; 25 & \\
16 + 9 & \\
25 & 25 \;\; \text{TRUE}
\end{array}
\qquad
\begin{array}{c|c}
3x - 4y = 0 & \\
\hline
3(4) - 4(3) \;?\; 0 & \\
12 - 12 & \\
0 & 0 \;\; \text{TRUE}
\end{array}$$

It is left to the student to confirm that $(-4, -3)$ also checks in both equations.

The pairs (4, 3) and $(-4, -3)$ check, so they are solutions. We can see the solutions in the graph. Intersections occur at (4, 3) and $(-4, -3)$.

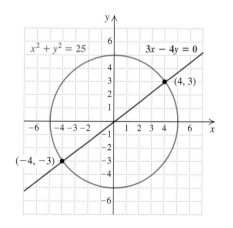

EXAMPLE 2   Solve the system

$$y + 3 = 2x, \qquad (1)$$
$$x^2 + 2xy = -1. \qquad (2)$$

SOLUTION   First, we solve the linear equation (1) for y:

$$y = 2x - 3. \qquad (3)$$

Then we substitute $2x - 3$ for y in equation (2) and solve for x:

$$x^2 + 2x(2x - 3) = -1$$
$$x^2 + 4x^2 - 6x = -1$$
$$5x^2 - 6x + 1 = 0$$
$$(5x - 1)(x - 1) = 0 \qquad \text{Factoring}$$
$$5x - 1 = 0 \quad or \quad x - 1 = 0 \qquad \text{Using the principle of zero products}$$
$$x = \tfrac{1}{5} \quad or \qquad x = 1.$$

Now we substitute these numbers for x in equation (3) and solve for y:

$$\text{for } x = \tfrac{1}{5}, \quad y = 2\left(\tfrac{1}{5}\right) - 3 = -\tfrac{13}{5};$$
$$\text{for } x = 1, \quad y = 2(1) - 3 = -1.$$

You can confirm that $\left(\tfrac{1}{5}, -\tfrac{13}{5}\right)$ and $(1, -1)$ check, so they are solutions.

TECHNOLOGY CONNECTION
10.4A

Because the algebra is often difficult, finding the solution(s) to systems of nonlinear equations provides an excellent opportunity to use a grapher. As with systems of linear equations, we use ZOOM and TRACE or INTERSECT to find the coordinates of any points of intersection. Using a grapher restricts solutions to real numbers since few graphers can find imaginary solutions.

For example, to solve Example 2,

$$y + 3 = 2x,$$
$$x^2 + 2xy = -1,$$

we first solve each equation for y:

$$\left.\begin{array}{l} y_1 = 2x - 3, \\ y_2 = \dfrac{-1 - x^2}{2x}. \end{array}\right\} \quad \text{Note that } x, y \neq 0.$$

Both equations are then graphed.

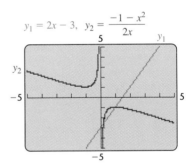

$$y_1 = 2x - 3, \quad y_2 = \frac{-1 - x^2}{2x}$$

Using INTERSECT or ZOOM and TRACE, we find the coordinates to be $(0.2, -2.6)$ and $(1, -1)$.

Use a grapher to solve each system. Round all values to two decimal places.

1. $4xy - 7 = 0,$
    $x - 3y - 2 = 0$

2. $x^2 + y^2 = 14,$
    $16x + 7y^2 = 0$

EXAMPLE   3   Solve the system

$$x + y = 5, \qquad (1) \qquad \text{(The graph is a line.)}$$
$$y = 3 - x^2. \qquad (2) \qquad \text{(The graph is a parabola.)}$$

SOLUTION   We substitute $3 - x^2$ for y in the first equation:

$$x + 3 - x^2 = 5$$
$$-x^2 + x - 2 = 0 \qquad \text{Adding } -5 \text{ on both sides and rearranging}$$
$$x^2 - x + 2 = 0. \qquad \text{Multiplying by } -1 \text{ on both sides}$$

To solve this equation, we need the quadratic formula:

$$x = \frac{-b \pm \sqrt{b^2 - 4ac}}{2a}$$

$$= \frac{-(-1) \pm \sqrt{(-1)^2 - 4 \cdot 1 \cdot 2}}{2(1)} \qquad \text{Substituting}$$

$$= \frac{1 \pm \sqrt{1 - 8}}{2} = \frac{1 \pm \sqrt{-7}}{2} = \frac{1}{2} \pm \frac{\sqrt{7}}{2} i.$$

Solving equation (1) for y gives us $y = 5 - x$. Substituting values for x gives

$$y = 5 - \left(\frac{1}{2} + \frac{\sqrt{7}}{2} i \right) = \frac{9}{2} - \frac{\sqrt{7}}{2} i \quad \text{and}$$

$$y = 5 - \left(\frac{1}{2} - \frac{\sqrt{7}}{2} i \right) = \frac{9}{2} + \frac{\sqrt{7}}{2} i.$$

The solutions are

$$\left(\frac{1}{2} + \frac{\sqrt{7}}{2} i, \frac{9}{2} - \frac{\sqrt{7}}{2} i \right) \quad \text{and} \quad \left(\frac{1}{2} - \frac{\sqrt{7}}{2} i, \frac{9}{2} + \frac{\sqrt{7}}{2} i \right).$$

There are no real-number solutions. Note in the figure at right that the graphs do not intersect. Getting only nonreal solutions tells us that the graphs do not intersect.

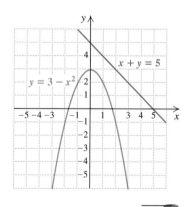

Systems of Two Nonlinear Equations

We now consider systems of two second-degree equations. Graphs of such systems can involve any two conic sections. The following figure shows some ways in which a circle and a hyperbola can intersect.

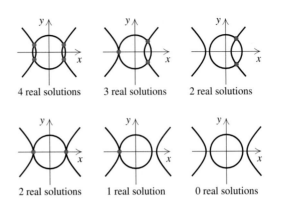

4 real solutions   3 real solutions   2 real solutions

2 real solutions   1 real solution   0 real solutions

To solve systems of two second-degree equations, we can use either substitution or elimination. The elimination method is generally better when both equations are of the form $Ax^2 + By^2 = C$. Then we can eliminate an x^2- or y^2-term in a manner similar to the procedure used in Chapter 3.

EXAMPLE 4   Solve the system

$$2x^2 + 5y^2 = 22, \qquad (1)$$
$$3x^2 - y^2 = -1. \qquad (2)$$

SOLUTION   Here we multiply equation (2) by 5 and then add:

$$
\begin{array}{rl}
2x^2 + 5y^2 = & 22 \\
\underline{15x^2 - 5y^2 = -5} & \quad \text{Multiplying by 5 on both sides of equation (2)} \\
17x^2 \qquad\;\; = & 17 \quad \text{Adding} \\
x^2 = & 1 \\
x = & \pm 1.
\end{array}
$$

If $x = 1$, $x^2 = 1$, and if $x = -1$, $x^2 = 1$, so substituting 1 or -1 for x in equation (2), we have

$$
\begin{aligned}
3 \cdot (\pm 1)^2 - y^2 &= -1 \\
3 - y^2 &= -1 \\
-y^2 &= -4 \\
y^2 = 4 \quad &\text{or} \quad y = \pm 2.
\end{aligned}
$$

Thus, if $x = 1$, $y = 2$ or $y = -2$; and if $x = -1$, $y = 2$ or $y = -2$. The four possible solutions are $(1, 2)$, $(1, -2)$, $(-1, 2)$, and $(-1, -2)$.

Check: Since $(2)^2 = 4$, $(-2)^2 = 4$, $(1)^2 = 1$, and $(-1)^2 = 1$, we can check all four pairs at once.

$$
\begin{array}{c|c}
\underline{2x^2 + 5y^2 = 22} & \underline{3x^2 - y^2 = -1} \\
2(\pm 1)^2 + 5(\pm 2)^2 \;?\; 22 & 3(\pm 1)^2 - (\pm 2)^2 \;?\; -1 \\
2 + 20 & 3 - 4 \\
22 \;\bigm|\; 22 \;\; \text{TRUE} & -1 \;\bigm|\; -1 \;\; \text{TRUE}
\end{array}
$$

The solutions are $(1, 2)$, $(1, -2)$, $(-1, 2)$, and $(-1, -2)$.

When a product of variables is in one equation and the other equation is of the form $Ax^2 + By^2 = C$, we often solve for a variable in the equation with the product and then use substitution.

EXAMPLE 5 Solve the system

$$x^2 + 4y^2 = 20, \qquad (1)$$
$$xy = 4. \qquad (2)$$

SOLUTION First, we solve equation (2) for y:

$$y = \frac{4}{x}.$$ Dividing by x on both sides.
Note that $x \neq 0$.

Then we substitute $4/x$ for y in equation (1) and solve for x:

$$x^2 + 4\left(\frac{4}{x}\right)^2 = 20$$

$$x^2 + \frac{64}{x^2} = 20$$

$$x^4 + 64 = 20x^2 \qquad \text{Multiplying by } x^2$$

$$x^4 - 20x^2 + 64 = 0 \qquad \text{Obtaining standard form. This equation is reducible to quadratic.}$$

$$(x^2 - 4)(x^2 - 16) = 0 \qquad \text{Factoring. If you prefer, let } u = x^2 \text{ and substitute.}$$

$$(x - 2)(x + 2)(x - 4)(x + 4) = 0 \qquad \text{Factoring}$$

$$x = 2 \quad or \quad x = -2 \quad or \quad x = 4 \quad or \quad x = -4. \qquad \text{Using the principle of zero products}$$

Since $y = 4/x$, for $x = 2$, we have $y = 4/2$, or 2. Thus, $(2, 2)$ is a solution. Similarly, $(-2, -2)$, $(4, 1)$, and $(-4, -1)$ are solutions. You can show that all four pairs check.

TECHNOLOGY CONNECTION
10.4B

A grapher can be used to check Example 4. To do so, each equation in the system must first be solved for y. When this has been done, we have $y_1 = \sqrt{(22 - 2x^2)/5}$ and $y_2 = -\sqrt{(22 - 2x^2)/5}$ for equation (1) and $y_3 = \sqrt{3x^2 + 1}$ and $y_4 = -\sqrt{3x^2 + 1}$ for equation (2). The graph verifies the solutions found algebraically.

1. Use a grapher to provide a visual check for Example 5.

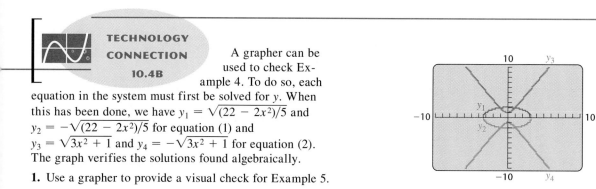

Problem Solving

We now consider applications in which the translation is to a system of equations in which at least one is not linear.

EXAMPLE 6

Architecture. For a college gymnasium, an architect wants to lay out a rectangular piece of land that has a perimeter of 204 m and an area of 2565 m². Find the dimensions of the piece of land.

SOLUTION

1. **Familiarize.** We draw and label a sketch, letting l = the length and w = the width, both in meters.

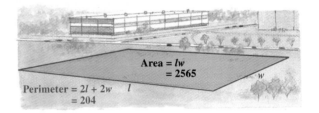

Area = lw
= 2565

Perimeter = $2l + 2w$
= 204

2. **Translate.** We then have the following translation:

Perimeter: $2w + 2l = 204$;
Area: $lw = 2565.$

3. **Carry out.** We solve the system

$$2w + 2l = 204,$$
$$lw = 2565.$$

Solving the second equation for l gives us $l = 2565/w$. Then we substitute $2565/w$ for l in the first equation and solve for w:

$$2w + 2\left(\frac{2565}{w}\right) = 204$$
$$2w^2 + 2(2565) = 204w \qquad \text{Multiplying by } w$$
$$2w^2 - 204w + 2(2565) = 0 \qquad \text{Standard form}$$
$$w^2 - 102w + 2565 = 0 \qquad \text{Multiplying by } \tfrac{1}{2}$$

> Factoring could be used instead of the quadratic formula, but the numbers are quite large.

$$w = \frac{-(-102) \pm \sqrt{(-102)^2 - 4 \cdot 1 \cdot 2565}}{2 \cdot 1}$$
$$w = \frac{102 \pm \sqrt{144}}{2} = \frac{102 \pm 12}{2}$$
$$w = 57 \quad or \quad w = 45.$$

If $w = 57$, then $l = 2565/w = 2565/57 = 45$. If $w = 45$, then $l = 2565/w = 2565/45 = 57$. Since length is usually considered to be longer than width, we have the solution $l = 57$ and $w = 45$, or (57, 45).

4. **Check.**   If $l = 57$ and $w = 45$, the perimeter is $2 \cdot 57 + 2 \cdot 45$, or 204. The area is $57 \cdot 45$, or 2565. The numbers check.

5. **State.**   The length is 57 m and the width is 45 m.

EXAMPLE 7

Design of a Van.   The cargo area of a delivery van must be 60 ft^2, and the length of a diagonal must accommodate a 13-ft board. Find the dimensions of the cargo area.

SOLUTION

1. **Familiarize.**   We draw a picture and label it. Note that there is a right triangle in the figure. We let l = the length and w = the width, both in feet.

2. **Translate.**   We translate to a system of equations:

   $l^2 + w^2 = 13^2$,   Using the Pythagorean theorem

   $lw = 60$.   Using the formula for the area of a rectangle

3. **Carry out.**   We solve the system

   $l^2 + w^2 = 169$,   You should complete the solution of this system.

   $lw = 60$

   to get $(12, 5)$, $(5, 12)$, $(-12, -5)$, and $(-5, -12)$.

4. **Check.**   Measurements cannot be negative and length is usually greater than width, so we check only $(12, 5)$. In the right triangle, $12^2 + 5^2 = 144 + 25 = 169$, which is 13^2. The area is $12 \cdot 5$, or 60, so we have a solution.

5. **State.**   The length is 12 ft and the width is 5 ft.

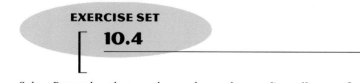

EXERCISE SET

10.4

Solve. Remember that graphs can be used to confirm all real solutions.

1. $x^2 + y^2 = 100,$
$\quad y - x = 2$

2. $x^2 + y^2 = 25,$
$\quad y - x = 1$

3. $9x^2 + 4y^2 = 36,$
$\quad 3x + 2y = 6$

4. $4x^2 + 9y^2 = 36,$
$\quad 3y + 2x = 6$

5. $y = x^2,$
$\quad 3x = y + 2$

6. $y^2 = x + 3,$
$\quad 2y = x + 4$

7. $2y^2 + xy + x^2 = 7$,
$x - 2y = 5$

8. $x^2 - xy + 3y^2 = 27$,
$x - y = 2$

9. $y^2 - x^2 = 16$,
$2x - y = 1$

10. $x^2 + 4y^2 = 25$,
$x + 2y = 7$

11. $m^2 + 3n^2 = 10$,
$m - n = 2$

12. $x^2 - xy + 3y^2 = 5$,
$x - y = 2$

13. $2y^2 + xy = 5$,
$4y + x = 7$

14. $3x + y = 7$,
$4x^2 + 5y = 24$

15. $p + q = -6$,
$pq = -7$

16. $a + b = 7$,
$ab = 4$

17. $4x^2 + 9y^2 = 36$,
$x + 3y = 3$

18. $2a + b = 1$,
$b = 4 - a^2$

19. $xy = 4$,
$x + y = 5$

20. $a^2 + b^2 = 89$,
$a - b = 3$

21. $y = x^2$,
$x = y^2$

22. $x^2 + y^2 = 25$,
$y^2 = x + 5$

23. $x^2 + y^2 = 9$,
$x^2 - y^2 = 9$

24. $y^2 - 4x^2 = 4$,
$4x^2 + y^2 = 4$

25. $x^2 + y^2 = 25$,
$xy = 12$

26. $x^2 - y^2 = 16$,
$x + y^2 = 4$

27. $x^2 + y^2 = 4$,
$16x^2 + 9y^2 = 144$

28. $x^2 + y^2 = 9$,
$25x^2 + 16y^2 = 400$

29. $x^2 + y^2 = 16$,
$y^2 - 2x^2 = 10$

30. $x^2 + y^2 = 14$,
$x^2 - y^2 = 4$

31. $x^2 + y^2 = 5$,
$xy = 2$

32. $x^2 + y^2 = 20$,
$xy = 8$

33. $x^2 + y^2 = 13$,
$xy = 6$

34. $x^2 + 4y^2 = 20$,
$xy = 4$

35. $3xy + x^2 = 34$,
$2xy - 3x^2 = 8$

36. $2xy + 3y^2 = 7$,
$3xy - 2y^2 = 4$

37. $xy - y^2 = 2$,
$2xy - 3y^2 = 0$

38. $4a^2 - 25b^2 = 0$,
$2a^2 - 10b^2 = 3b + 4$

39. $x^2 - y = 5$,
$x^2 + y^2 = 25$

40. $ab - b^2 = -4$,
$ab - 2b^2 = -6$

Solve.

41. *Computer parts.* Dataport Electronics needs a rectangular memory board that has a perimeter of 28 cm and a diagonal of length 10 cm. What should the dimensions of the board be?

42. *Tile design.* The New World tile company wants to make a new rectangular tile that has a perimeter of

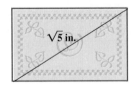

$\sqrt{5}$ in.

6 in. and a diagonal of length $\sqrt{5}$ in. What should the dimensions of the tile be?

43. *Geometry.* A rectangle has an area of 20 in² and a perimeter of 18 in. Find its dimensions.

44. *Geometry.* A rectangle has an area of 2 yd² and a perimeter of 6 yd. Find its dimensions.

45. *Fencing.* It will take 210 yd of fencing to enclose a rectangular field. The area of the field is 2250 yd². What are the dimensions of the field?

46. *Dimensions of a rug.* The diagonal of a Persian rug is 25 ft. The area of the rug is 300 ft². Find the length and the width of the rug.

47. The product of the lengths of the legs of a right triangle is 156. The hypotenuse has length $\sqrt{313}$. Find the lengths of the legs.

48. The product of two numbers is 60. The sum of their squares is 136. Find the numbers.

49. *Garden design.* A garden contains two square peanut beds. Find the length of each bed if the sum of their areas is 832 ft² and the difference of their areas is 320 ft².

50. *Investments.* A certain amount of money saved for 1 yr at a certain interest rate yielded $225 in interest. If $750 more had been invested and the rate had been 1% less, the interest would have been the same. Find the principal and the rate.

51. The area of a rectangle is $\sqrt{3}$ m², and the length of a diagonal is 2 m. Find the dimensions.

52. The area of a rectangle is $\sqrt{2}$ m², and the length of a diagonal is $\sqrt{3}$ m. Find the dimensions.

S K I L L   M A I N T E N A N C E

Simplify.

53. $\sqrt{48}$

54. $\sqrt[4]{32a^{24}d^9}$

55. A boat travels 4 mi upstream and 4 mi back downstream. The total time for the trip is 3 hr. The speed of the stream is 2 mph. Find the speed of the boat in still water.

56. Rationalize the denominator:

$$\frac{\sqrt{x} - \sqrt{h}}{\sqrt{x} + \sqrt{h}}.$$

S Y N T H E S I S

57. ◆ Write a problem that translates to a system of two equations. Design the problem so that at least one equation is nonlinear and so that no real solution exists.

58. ◆ Write a problem for a classmate to solve. Devise the problem so that a system of two nonlinear equations with exactly one real solution is solved.

59. A piece of wire 100 cm long is to be cut into two pieces and those pieces are each to be bent to make a square. The area of one square is to be 144 cm² greater than that of the other. How should the wire be cut?

60. Find the equation of a circle that passes through $(-2, 3)$ and $(-4, 1)$ and whose center is on the line $5x + 8y = -2$.

61. Find the equation of an ellipse centered at the origin that passes through the points $(2, -3)$ and $(1, \sqrt{13})$.

62. *Box design.* Four squares with sides 5 in. long are cut from the corners of a rectangular metal sheet that has an area of 340 in². The edges are bent up to form an open box with a volume of 350 in³. Find the dimensions of the box.

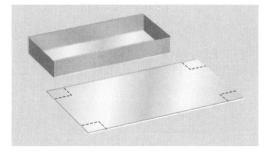

63. ▤ *Railing sales.* Fireside Castings finds that the total revenue R from the sale of x units of railing is given by

$$R = 100x + x^2.$$

Fireside also finds that the total cost C of producing x units of the same product is given by

$$C = 80x + 1500.$$

A break-even point is a value of x for which total revenue is the same as total cost; that is, $R = C$. How many units must be sold to break even?

Solve.

64. $p^2 + q^2 = 13,$
$\dfrac{1}{pq} = -\dfrac{1}{6}$

65. $a + b = \dfrac{5}{6},$
$\dfrac{a}{b} + \dfrac{b}{a} = \dfrac{13}{6}$

66. ▱ Use a grapher to check your answers to Exercises 5, 17, and 39.

┌ **SUMMARY AND REVIEW**
└   **10**

┌─────────────────────────────

KEY TERMS

IMPORTANT PROPERTIES AND FORMULAS

The Distance Formula

The distance d between any two points (x_1, y_1) and (x_2, y_2) is given by

$$d = \sqrt{(x_2 - x_1)^2 + (y_2 - y_1)^2}.$$

The Midpoint Formula

If the endpoints of a segment are (x_1, y_1) and (x_2, y_2), then the coordinates of the midpoint are

$$\left(\frac{x_1 + x_2}{2}, \frac{y_1 + y_2}{2}\right).$$

(See the summary of graphs near the end of Section 10.3.)

Parabola

Vertical with vertex at (h, k):

$$\begin{aligned} y &= ax^2 + bx + c \\ &= a(x - h)^2 + k \end{aligned}$$

Horizontal with vertex at (h, k):

$$\begin{aligned} x &= ay^2 + by + c \\ &= a(y - k)^2 + h \end{aligned}$$

Circle

Center at the origin:

$$x^2 + y^2 = r^2$$

Center at (h, k):

$$(x - h)^2 + (y - k)^2 = r^2$$

Ellipse

Center at the origin:

$$\frac{x^2}{a^2} + \frac{y^2}{b^2} = 1$$

Center at (h, k):

$$\frac{(x - h)^2}{a^2} + \frac{(y - k)^2}{b^2} = 1$$

Hyperbola

Center at the origin

Axis horizontal: $\dfrac{x^2}{a^2} - \dfrac{y^2}{b^2} = 1$ Axis vertical: $\dfrac{y^2}{b^2} - \dfrac{x^2}{a^2} = 1$

With x- and y-axes as asymptotes: $xy = c$

REVIEW EXERCISES

Find the distance between each pair of points. Where appropriate, find an approximation to three decimal places.

1. (2, 6) and (6, 6)

2. (−1, 1) and (−5, 4)

3. (1.4, 3.6) and (4.7, −5.3)

4. (2, 3a) and (−1, a)

Find the midpoint of the segment with the given endpoints.

5. (1, 6) and (7, 6)

6. (−1, 1) and (−5, 4)

7. $(1, \sqrt{3})$ and $\left(\frac{1}{2}, -\sqrt{2}\right)$

8. (2, 3a) and (−1, a)

Find the center and the radius of each circle.

9. $(x + 2)^2 + (y - 3)^2 = 2$

10. $(x - 5)^2 + y^2 = 49$

11. $x^2 + y^2 - 6x - 2y + 1 = 0$

12. $x^2 + y^2 + 8x - 6y - 10 = 0$

13. Find an equation of the circle with center (−4, 3) and radius $4\sqrt{3}$.

14. Find an equation of the circle with center (7, −2) and radius $2\sqrt{5}$.

Classify each equation as a circle, an ellipse, a parabola, or a hyperbola. Then graph.

15. $4x^2 + 4y^2 = 100$

16. $9x^2 + 2y^2 = 18$

17. $y = -x^2 + 2x - 3$

18. $\dfrac{y^2}{9} - \dfrac{x^2}{4} = 1$

19. $xy = 9$

20. $x = y^2 + 2y - 2$

21. $\dfrac{(x + 1)^2}{3} + (y - 3)^2 = 1$

22. $x^2 + y^2 + 6x - 8y - 39 = 0$

Solve.

23. $x^2 - y^2 = 33,$
$x + y = 11$

24. $x^2 - 2x + 2y^2 = 8,$
$2x + y = 6$

25. $x^2 - y = 3,$
$2x - y = 3$

26. $x^2 + y^2 = 25,$
$x^2 - y^2 = 7$

27. $x^2 - y^2 = 3,$
$y = x^2 - 3$

28. $x^2 + y^2 = 18,$
$2x + y = 3$

29. $x^2 + y^2 = 100,$
$2x^2 - 3y^2 = -120$

30. $x^2 + 2y^2 = 12,$
$xy = 4$

31. A rectangle has a perimeter of 38 m and an area of 84 m². What are the dimensions of the rectangle?

32. Find two positive integers whose sum is 12 and the sum of whose reciprocals is $\frac{3}{8}$.

33. The perimeter of a square is 12 cm more than the perimeter of another square. Its area exceeds the area of the other by 39 cm². Find the perimeter of each square.

34. The sum of the areas of two circles is 130π ft². The difference of the circumferences is 16π ft. Find the radius of each circle.

SKILL MAINTENANCE

35. Simplify: $\sqrt[3]{81a^8b^{10}}$.

36. Solve: $x^2 + 2x + 5 = 0$.

37. Rationalize the numerator:

$$\frac{4 - \sqrt{a}}{2 + \sqrt{a}}.$$

38. The speed of a moving sidewalk at an airport is 5 ft/sec. Bea can walk 55 ft forward on the moving sidewalk in the same time it takes to walk 5 ft in the opposite direction. At what rate would she walk on a nonmoving sidewalk?

SYNTHESIS

39. ◈ How does the graph of a hyperbola differ from the graph of a parabola?

40. ◈ Explain why function notation is not used in this chapter, and list the graphs discussed for which function notation could be used.

41. Solve:

$$4x^2 - x - 3y^2 = 9,$$
$$-x^2 + x + y^2 = 2$$

42. Find the points whose distance from (8, 0) and from (−8, 0) is 10.

43. Find an equation of the circle that passes through (−2, −4), (5, −5), and (6, 2).

44. Find an equation of the ellipse with the following intercepts: (−7, 0), (7, 0), (0, −3), and (0, 3).

45. Find the point on the *x*-axis that is equidistant from (−3, 4) and (5, 6).

CHAPTER TEST
10

Find the distance between each pair of points. Where appropriate, find an approximation to three decimal places.

1. $(4, -1)$ and $(-5, 8)$ **2.** $(3, -a)$ and $(-3, a)$

Find the midpoint of the segment with the given endpoints.

3. $(4, -1)$ and $(-5, 8)$ **4.** $(3, -a)$ and $(-3, a)$

Find the center and the radius of the circle.

5. $(x + 2)^2 + (y - 3)^2 = 64$

6. $x^2 + y^2 + 4x - 6y + 4 = 0$

Classify the equation as a circle, an ellipse, a parabola, or a hyperbola. Then graph.

7. $y = x^2 - 4x - 1$

8. $x^2 + y^2 + 2x + 6y + 6 = 0$

9. $\dfrac{x^2}{9} - \dfrac{y^2}{4} = 1$ **10.** $16x^2 + 4y^2 = 64$

11. $xy = -5$ **12.** $x = -y^2 + 4y$

Solve.

13. $\dfrac{x^2}{16} + \dfrac{y^2}{9} = 1,$ **14.** $x^2 + y^2 = 16,$
$3x + 4y = 12$ $\dfrac{x^2}{16} - \dfrac{y^2}{9} = 1$

15. In a rational expression, the sum of the values of the numerator and the denominator is 23. The product of their values is 120. Find the values of the numerator and the denominator.

16. A rectangle with diagonal of length $5\sqrt{5}$ has an area of 22. Find the dimensions of the rectangle.

17. Two squares are such that the sum of their areas is 8 m^2 and the difference of their areas is 2 m^2. Find the length of a side of each square.

18. A rectangle has a diagonal of length 20 ft and a perimeter of 56 ft. Find the dimensions of the rectangle.

SKILL MAINTENANCE

19. Solve: $x^2 + 2x = 5$.

20. Simplify: $\sqrt[3]{48a^5b^{18}}$.

21. Rationalize the denominator:
$$\frac{4 - \sqrt{a}}{2 + \sqrt{a}}.$$

22. A boat travels 6 mi upstream in the same time it takes to travel 30 mi downstream. The speed of the stream is 4 mph. Find the speed of the boat in still water.

SYNTHESIS

23. Find an equation of the ellipse passing through $(6, 0)$ and $(6, 6)$ with vertices at $(1, 3)$ and $(11, 3)$.

24. Find the point on the y-axis that is equidistant from $(-3, -5)$ and $(4, -7)$.

25. The sum of two numbers is 36, and the product is 4. Find the sum of the reciprocals of the numbers.

Sequences, Series, and the Binomial Theorem

AN APPLICATION

The Sanders Amphitheater has 20 seats in the first row, 22 seats in the second row, 24 in the third, and so on, for 19 rows. How many seats are in the amphitheater?

THIS PROBLEM APPEARS AS EXERCISE 43 IN SECTION 11.2.

More information on theater construction is available at **http://hepg.awl.com/be/inter_5**

We use math every day to calculate slope, grading, area, volume, cost estimates, maximum capacity, and exit widths, and to draw buildings to scale. One of the most necessary skills in design is translating words into the numbers needed to yield a final answer.

ELLEN MILLS
Architect
Chicago, IL

The first three sections of this chapter are devoted to sequences *and* series. *A sequence is simply an ordered list. For example, when a baseball coach writes a batting order, a sequence is being formed. When the members of a sequence are numbers, we can discuss their sum. Such a sum is called a* series.

Section 11.4 presents the binomial theorem, *which is used to expand expressions of the form* $(a + b)^n$. *Such an expansion is itself a series.*

11.1 Sequences and Series

Sequences • Finding the General Term • Sums and Series • Sigma Notation

Sequences

Suppose that $1000 is invested at 8%, compounded annually. The amounts to which the account will grow after 1 year, 2 years, 3 years, and so on, are as follows:

①   ②   ③   ④
↓   ↓   ↓   ↓
$1080.00, $1166.40, $1259.71, $1360.49, ...

We can regard this as a function that pairs 1 with $1080.00, 2 with $1166.40, 3 with $1259.71, and so on. A **sequence** is thus a function, where the domain is a set of consecutive positive integers beginning with 1.

If we continue computing the amounts in the account forever, we obtain an **infinite sequence,** with function values

$1080.00, $1166.40, $1259.71, $1360.49, $1469.33, $1586.87, ...

The three dots at the end indicate that the sequence goes on without stopping. If we stop after a certain number of years, we obtain a **finite sequence:**

$1080.00, $1166.40, $1259.71, $1360.49.

SEQUENCES

An *infinite sequence* is a function having for its domain the set of positive integers: $\{1, 2, 3, 4, 5, \ldots\}$.

A *finite sequence* is a function having for its domain a set of positive integers $\{1, 2, 3, 4, 5, \ldots, n\}$, for some positive integer n.

As another example, consider the sequence given by

$$a(n) = 2^n, \quad \text{or} \quad a_n = 2^n.$$

The notation a_n means the same as $a(n)$ but is used more commonly with sequences. Some function values (also called *terms* of the sequence) follow:

$$a_1 = 2^1 = 2,$$
$$a_2 = 2^2 = 4,$$
$$a_3 = 2^3 = 8,$$
$$a_6 = 2^6 = 64.$$

The first term of the sequence is a_1, the fifth term is a_5, and the nth term, or **general term,** is a_n. This sequence can also be denoted in the following ways:

$$2, 4, 8, \ldots; \text{ or}$$
$$2, 4, 8, \ldots, 2^n, \ldots.$$ The 2^n emphasizes that the nth term of this sequence is found by raising 2 to the nth power.

EXAMPLE   1

Find the first four terms and the 57th term of the sequence for which the general term is given by $a_n = (-1)^n/(n + 1)$.

SOLUTION   We have

$$a_1 = \frac{(-1)^1}{1 + 1} = -\frac{1}{2},$$
$$a_2 = \frac{(-1)^2}{2 + 1} = \frac{1}{3},$$
$$a_3 = \frac{(-1)^3}{3 + 1} = -\frac{1}{4},$$
$$a_4 = \frac{(-1)^4}{4 + 1} = \frac{1}{5},$$
$$a_{57} = \frac{(-1)^{57}}{57 + 1} = -\frac{1}{58}.$$

Note that the expression $(-1)^n$ causes the signs of the terms to alternate between positive and negative, depending on whether n is even or odd.

TECHNOLOGY CONNECTION

11.1

Sequences are entered and graphed much like functions. The difference is that the SEQUENCE MODE must be selected. You can then enter U_n or V_n using n (the 2nd function of 9 on some keypads) as the variable. Use this approach to check Example 1 with a table of values for the sequence.

Finding the General Term

When only the first few terms of a sequence are known, it is impossible to be certain what the general term is, but a prediction can be made by looking for a pattern.

EXAMPLE   2

For each sequence, predict the general term.

a) 1, 4, 9, 16, 25, ...   **b)** $-1, 2, -4, 8, -16, \ldots$

c) 2, 4, 8, ...

SOLUTION

a) 1, 4, 9, 16, 25, . . .
These are squares of consecutive positive integers, so the general term may be n^2.

b) −1, 2, −4, 8, −16, . . .
These are powers of 2 with alternating signs, so the general term may be $(-1)^n[2^{n-1}]$. To check, note that 8 is the fourth term, and $(-1)^4[2^{4-1}] = 1 \cdot 2^3 = 8$.

c) 2, 4, 8, . . .
We regard the pattern as powers of 2, so 16 is the next term and 2^n is the general term. The sequence is then written with more terms as

$$2, 4, 8, 16, 32, 64, 128, \ldots .$$

In part (c) above, suppose that the second term is found by adding 2, the third term by adding 4, the next term by adding 6, and so on. In this case, 14 would be the next term and the sequence would be

$$2, 4, 8, 14, 22, 32, 44, 58, \ldots .$$

This illustrates that the fewer terms we are given, the greater the uncertainty about the nth term.

Sums and Series

SERIES

Given the infinite sequence

$$a_1, a_2, a_3, a_4, \ldots, a_n, \ldots,$$

the sum of the terms

$$a_1 + a_2 + a_3 + \cdots + a_n + \cdots$$

is called an *infinite series*. A *partial sum* is the sum of the first n terms:

$$a_1 + a_2 + a_3 + \cdots + a_n.$$

A partial sum is also called a *finite series* and is denoted S_n.

EXAMPLE 3   For the sequence −2, 4, −6, 8, −10, 12, −14, find: **(a)** S_1; **(b)** S_3; **(c)** S_6.

SOLUTION

a) $S_1 = -2$

b) $S_3 = -2 + 4 + (-6) = -4$

c) $S_6 = -2 + 4 + (-6) + 8 + (-10) + 12 = 6$

Sigma Notation

When the general term of a sequence is known, the Greek letter Σ (sigma) can be used to write a series. For example, the sum of the first four terms of the sequence $3, 5, 7, 9, \ldots, 2k + 1, \ldots$ can be named as follows, using *sigma notation,* or *summation notation*:

$$\sum_{k=1}^{4} (2k + 1).$$

This is read "the sum as k goes from 1 to 4 of $(2k + 1)$." The letter k is called the *index of summation*. The index of summation need not start at 1.

EXAMPLE 4

Find and evaluate each sum.

a) $\displaystyle\sum_{k=1}^{5} k^2$ **b)** $\displaystyle\sum_{k=4}^{6} (-1)^k (2k)$ **c)** $\displaystyle\sum_{k=0}^{3} (2^k + 5)$

SOLUTION

a) $\displaystyle\sum_{k=1}^{5} k^2 = 1^2 + 2^2 + 3^2 + 4^2 + 5^2 = 1 + 4 + 9 + 16 + 25 = 55$

Evaluate k^2 for all integers from 1 through 5. Then add.

b) $\displaystyle\sum_{k=4}^{6} (-1)^k (2k) = (-1)^4 (2 \cdot 4) + (-1)^5 (2 \cdot 5) + (-1)^6 (2 \cdot 6)$

$$= 8 - 10 + 12 = 10$$

c) $\displaystyle\sum_{k=0}^{3} (2^k + 5) = (2^0 + 5) + (2^1 + 5) + (2^2 + 5) + (2^3 + 5)$

$$= 6 + 7 + 9 + 13 = 35$$

EXAMPLE 5

Write sigma notation for each sum.

a) $1 + 4 + 9 + 16 + 25$ **b)** $-1 + 3 - 5 + 7$
c) $3 + 9 + 27 + 81 + \cdots$

SOLUTION

a) $1 + 4 + 9 + 16 + 25$
This is a sum of squares, $1^2 + 2^2 + 3^2 + 4^2 + 5^2$, so the general term is k^2. Sigma notation is

$$\sum_{k=1}^{5} k^2.$$ The sum starts with 1^2 and ends with 5^2.

Answers may vary here. For example, another—perhaps less obvious—way of writing $1 + 4 + 9 + 16 + 25$ is

$$\sum_{k=2}^{6} (k - 1)^2.$$

b) $-1 + 3 - 5 + 7$

Except for the alternating signs, this is the sum of the first four positive odd numbers. Note that $2k - 1$ is a formula for the kth positive odd number, and $(-1)^k = 1$ when k is even and $(-1)^k = -1$ when k is odd. The general term is thus $(-1)^k(2k - 1)$, beginning with $k = 1$. Sigma notation is

$$\sum_{k=1}^{4} (-1)^k(2k - 1).$$

To check, we can evaluate $(-1)^k(2k - 1)$ using 1, 2, 3, and 4. Then we can write the sum of the four terms. We leave this to the student.

c) $3 + 9 + 27 + 81 + \cdots$

This is a sum of powers of 3, and it is also an infinite series. We use the symbol ∞ for infinity and write the series using sigma notation:

$$\sum_{k=1}^{\infty} 3^k.$$

EXERCISE SET 11.1

In each of the following, the nth term of a sequence is given. In each case, find the first 4 terms; the 10th term, a_{10}; and the 15th term, a_{15}.

1. $a_n = 5n - 3$

2. $a_n = 2n + 5$

3. $a_n = \dfrac{n}{n + 2}$

4. $a_n = n^2 + 1$

5. $a_n = n^2 - 2n$

6. $a_n = \dfrac{n^2 - 1}{n^2 + 1}$

7. $a_n = n + \dfrac{1}{n}$

8. $a_n = \left(-\dfrac{1}{2}\right)^{n-1}$

9. $a_n = (-1)^n n^2$

10. $a_n = (-1)^n(n + 3)$

11. $a_n = (-1)^{n+1}(3n - 5)$

12. $a_n = (-1)^n(n^3 - 1)$

Find the indicated term of each sequence.

13. $a_n = 3n - 5$; a_7

14. $a_n = 5n + 2$; a_8

15. $a_n = (3n + 4)(2n - 5)$; a_9

16. $a_n = (3n + 2)^2$; a_6

17. $a_n = (-1)^{n-1}(3.4n - 17.3)$; a_{12}

18. $a_n = (-2)^{n-2}(45.68 - 1.2n)$; a_{23}

19. $a_n = 3n^2(9n - 100)$; a_{11}

20. $a_n = 4n^2(2n - 39)$; a_{22}

21. $a_n = \left(1 + \dfrac{1}{n}\right)^2$; a_{20}

22. $a_n = \left(1 - \dfrac{1}{n}\right)^3$; a_{15}

23. $a_n = \log 10^n$; a_{43}

24. $a_n = \ln e^n$; a_{67}

Predict the general term, or nth term, a_n, of each sequence. Answers may vary.

25. $1, 3, 5, 7, 9, \ldots$

26. $3, 9, 27, 81, 243, \ldots$

27. $-2, 6, -18, 54, \ldots$

28. $-2, 3, 8, 13, 18, \ldots$

29. $\dfrac{1}{2}, \dfrac{2}{3}, \dfrac{3}{4}, \dfrac{4}{5}, \dfrac{5}{6}, \ldots$

30. $1, \sqrt{3}, \sqrt{5}, \sqrt{7}, 3, \ldots$

31. $\sqrt{3}, 3, 3\sqrt{3}, 9, 9\sqrt{3}, \ldots$

32. $1 \cdot 2, 2 \cdot 3, 3 \cdot 4, 4 \cdot 5, \ldots$

33. $-1, -4, -7, -10, -13, \ldots$

34. $\log 1, \log 10, \log 100, \log 1000, \ldots$

Find the indicated partial sum for each sequence.

35. $1, -2, 3, -4, 5, -6, \ldots$; S_7

36. $1, -3, 5, -7, 9, -11, \ldots$; S_8

37. $2, 4, 6, 8, \ldots ; S_5$

38. $1, \frac{1}{4}, \frac{1}{9}, \frac{1}{16}, \frac{1}{25}, \ldots ; S_5$

Rename and evaluate each sum.

39. $\displaystyle\sum_{k=1}^{5} \frac{1}{2k}$

40. $\displaystyle\sum_{k=1}^{6} \frac{1}{2k - 1}$

41. $\displaystyle\sum_{k=0}^{4} 3^k$

42. $\displaystyle\sum_{k=4}^{7} \sqrt{2k + 1}$

43. $\displaystyle\sum_{k=1}^{8} \frac{k}{k + 1}$

44. $\displaystyle\sum_{k=1}^{4} \frac{k - 2}{k + 3}$

45. $\displaystyle\sum_{k=1}^{5} (-1)^k$

46. $\displaystyle\sum_{k=1}^{5} (-1)^{k+1}$

47. $\displaystyle\sum_{k=1}^{8} (-1)^{k+1} 2^k$

48. $\displaystyle\sum_{k=1}^{7} (-1)^k 4^{k+1}$

49. $\displaystyle\sum_{k=0}^{5} (k^2 - 2k + 3)$

50. $\displaystyle\sum_{k=0}^{5} (k^2 - 3k + 4)$

51. $\displaystyle\sum_{k=3}^{5} \frac{(-1)^k}{k(k + 1)}$

52. $\displaystyle\sum_{k=3}^{7} \frac{k}{2^k}$

Rewrite each sum using sigma notation. Answers may vary.

53. $\dfrac{2}{3} + \dfrac{3}{4} + \dfrac{4}{5} + \dfrac{5}{6} + \dfrac{6}{7}$

54. $3 + 6 + 9 + 12 + 15$

55. $1 + 4 + 9 + 16 + 25 + 36$

56. $\dfrac{1}{1^2} + \dfrac{1}{2^2} + \dfrac{1}{3^2} + \dfrac{1}{4^2} + \dfrac{1}{5^2}$

57. $4 - 9 + 16 - 25 + \cdots + (-1)^n n^2$

58. $9 - 16 + 25 + \cdots + (-1)^{n+1} n^2$

59. $5 + 10 + 15 + 20 + 25 + \cdots$

60. $7 + 14 + 21 + 28 + 35 + \cdots$

61. $\dfrac{1}{1 \cdot 2} + \dfrac{1}{2 \cdot 3} + \dfrac{1}{3 \cdot 4} + \dfrac{1}{4 \cdot 5} + \cdots$

62. $\dfrac{1}{1 \cdot 2^2} + \dfrac{1}{2 \cdot 3^2} + \dfrac{1}{3 \cdot 4^2} + \dfrac{1}{4 \cdot 5^2} + \cdots$

SKILL MAINTENANCE

Simplify.

63. $6^{\log_6 29}$

64. $9^{\log_9 43}$

65. $\log_3 3$

66. $\log_3 1$

67. $\log_3 3^7$

68. $\log_c c$

SYNTHESIS

69. ◈ The sequence $1, 4, 9, 16, \ldots$ can be written as $f(x) = x^2$ with the domain of $f = \{x \mid x$ is an integer

and $x > 0\}$. Explain how the graph of f would compare with the graph of $y = x^2$.

70. ◈ Explain why the equation

$$\sum_{k=1}^{n} (a_k + b_k) = \sum_{k=1}^{n} a_k + \sum_{k=1}^{n} b_k$$

is true for any positive integer n. What laws are used to justify this result?

71. ◈ Consider the sums

$$\sum_{k=1}^{5} 3k^2 \quad \text{and} \quad 3\sum_{k=1}^{5} k^2.$$

a) Which is easier to evaluate and why?

b) Is it true that

$$\sum_{k=1}^{n} ca_k = c \sum_{k=1}^{n} a_k?$$

Why or why not?

Some sequences are given by a recursive definition. The value of the first term, a_1, is given, and then we are told how to find any subsequent term from the term preceding it. Find the first six terms of each of the following recursively defined sequences.

72. $a_1 = 1, \ a_{n+1} = 5a_n - 2$

73. $a_1 = 0, \ a_{n+1} = a_n^2 + 3$

74. *Cell biology.* A single cell of bacterium divides into two every 15 min. Suppose that the same rate of division is maintained for 4 hr. Give a sequence that lists the number of cells after successive 15-min periods.

75. *Value of a copier.* The value of a color photocopier is $5200. Its scrap value each year is 75% of its value the year before. Give a sequence that lists the scrap value of the machine at the start of each year for a 10-yr period.

76. *Hourly wages.* Katrina gets $8.20 an hour for filling orders for River's Bend Publishing. Each year she gets a $0.40 hourly raise. Give a sequence that lists Katrina's hourly salary over a 10-yr period.

Find the first five terms of each sequence; then find S_5.

77. $a_n = \dfrac{1}{2^n} \log 1000^n$

78. $a_n = i^n, \ i = \sqrt{-1}$

79. Find all values for x that solve the following:

$$\sum_{k=1}^{x} i^k = -1.$$

80. ▦ The nth term of a sequence is given by

$$a_n = n^5 - 14n^4 + 6n^3 + 416n^2 - 655n - 1050.$$

Use a grapher with a TABLE feature to determine what term in the sequence is 6144.

81. [graph icon] To define a sequence recursively on a grapher (see Exercises 72 and 73), the SEQ MODE is used. The general term U_n or V_n can often be expressed in terms of U_{n-1} or V_{n-1} by pressing [2nd] [7] or [2nd] [8]. The starting values of U_n, V_n, and n are set as one of the WINDOW variables.

Use recursion to determine how many handshakes will occur if a group of 50 people shake hands with one another. To develop the recursion formula, begin with a group of 2 and determine how many additional handshakes occur with the arrival of each new group member. (See the Collaborative Corner following Exercise Set 5.1 on p. 258.)

11.2 Arithmetic Sequences and Series

Arithmetic Sequences   •   Sum of the First n Terms of an
Arithmetic Sequence   •   Problem Solving

In this section, we concentrate on sequences and series that are said to be arithmetic (pronounced ar-ith-MET-ik).

Arithmetic Sequences

In an **arithmetic sequence,** any term (other than the first) can be found by adding the same number to its preceding term. For example, the sequence 2, 5, 8, 11, 14, 17, . . . is arithmetic because adding 3 to any term produces the next term. In other words, the difference between any term and the preceding one is 3. Arithmetic sequences are also called *arithmetic progressions.*

ARITHMETIC SEQUENCE

A sequence is *arithmetic* if there exists a number d, called the *common difference,* such that $a_{n+1} = a_n + d$ for any integer $n \geq 1$.

EXAMPLE 1   For each arithmetic sequence, identify the first term, a_1, and the common difference, d.

a) 4, 9, 14, 19, 24, . . .   **b)** 27, 20, 13, 6, −1, −8, . . .

SOLUTION   To find a_1, we simply use the first term listed. To find d, we choose any term beyond the first and subtract the preceding term from it.

Sequence	*First Term, a_1*	*Common Difference, d*
a) 4, 9, 14, 19, 24, . . .	4	5 ⟵ 9 − 4 = 5
b) 27, 20, 13, 6, −1, −8, . . .	27	−7 ⟵ 20 − 27 = −7

To find the common difference, we subtracted a_1 from a_2. Had we subtracted

a_2 from a_3 or a_3 from a_4 we would have found the same values for d. As a check, simply add d to any term and see if the next term results.

Check: **a)** $4 + 5 = 9$, $9 + 5 = 14$, $14 + 5 = 19$, $19 + 5 = 24$

b) $27 + (-7) = 20$, $20 + (-7) = 13$, $13 + (-7) = 6$,
$6 + (-7) = -1$, $-1 + (-7) = -8$

To find a formula for the general, or nth, term of any arithmetic sequence, we denote the common difference by d and write out the first few terms:

$a_1,$

$a_2 = a_1 + d,$

$a_3 = a_2 + d = (a_1 + d) + d = a_1 + 2d,$ Substituting for a_2

$a_4 = a_3 + d = (a_1 + 2d) + d = a_1 + 3d.$ Substituting for a_3

Note that the coefficient of d in each case is 1 less than the subscript.

Generalizing, we obtain the following formula.

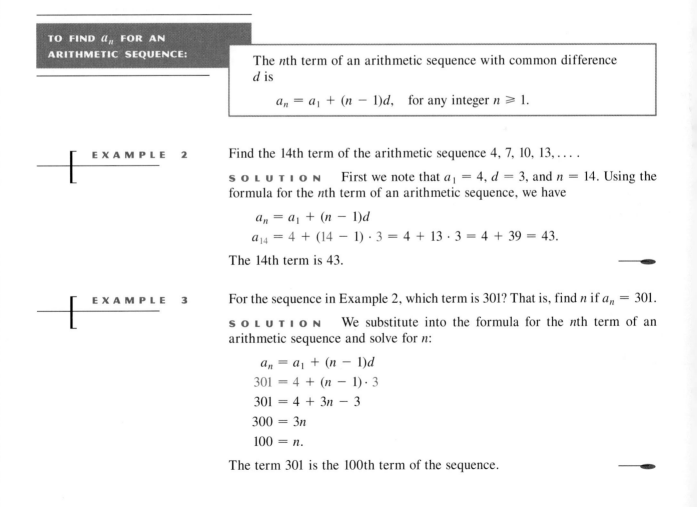

TO FIND a_n FOR AN ARITHMETIC SEQUENCE:

The nth term of an arithmetic sequence with common difference d is

$$a_n = a_1 + (n - 1)d, \quad \text{for any integer } n \geqslant 1.$$

EXAMPLE 2   Find the 14th term of the arithmetic sequence $4, 7, 10, 13, \ldots$.

SOLUTION   First we note that $a_1 = 4$, $d = 3$, and $n = 14$. Using the formula for the nth term of an arithmetic sequence, we have

$$a_n = a_1 + (n - 1)d$$
$$a_{14} = 4 + (14 - 1) \cdot 3 = 4 + 13 \cdot 3 = 4 + 39 = 43.$$

The 14th term is 43.

EXAMPLE 3   For the sequence in Example 2, which term is 301? That is, find n if $a_n = 301$.

SOLUTION   We substitute into the formula for the nth term of an arithmetic sequence and solve for n:

$$a_n = a_1 + (n - 1)d$$
$$301 = 4 + (n - 1) \cdot 3$$
$$301 = 4 + 3n - 3$$
$$300 = 3n$$
$$100 = n.$$

The term 301 is the 100th term of the sequence.

Given two terms and their places in an arithmetic sequence, we can construct the sequence.

E X A M P L E   4   The 3rd term of an arithmetic sequence is 8, and the 16th term is 47. Find a_1 and d and construct the sequence.

S O L U T I O N   We know that $a_3 = 8$ and $a_{16} = 47$. Thus we would have to add d thirteen times to get from 8 to 47. That is,

$$8 + 13d = 47. \qquad \text{a_3 and a_{16} are 13 terms apart.}$$

Solving $8 + 13d = 47$, we obtain

$$13d = 39$$
$$d = 3.$$

We subtract d twice from a_3 to get to a_1. Thus,

$$a_1 = 8 - 2 \cdot 3 = 2. \qquad \text{a_1 and a_3 are 2 terms apart.}$$

The sequence is $2, 5, 8, 11, \ldots$. Note that we could have subtracted d 15 times from a_{16} in order to find a_1.   ———●

In general, d should be subtracted $(n - 1)$ times from a_n in order to find a_1.

Sum of the First n Terms of an Arithmetic Sequence

When the terms of an arithmetic sequence are added, an **arithmetic series** is formed. To find a formula for computing S_n when the series is arithmetic, we denote the first n terms as follows:

This is the next-to-last term. If you add d to this term, the result is a_n.

$$a_1, (a_1 + d), (a_1 + 2d), \ldots, (a_n - 2d), (a_n - d), a_n$$

This term is two terms back from the last. If you add d to this term, you get the next-to-last term, $a_n - d$.

Thus, S_n is given by

$$S_n = a_1 + (a_1 + d) + (a_1 + 2d) + \cdots + (a_n - 2d) + (a_n - d) + a_n.$$

Reversing the order of addition, we have

$$S_n = a_n + (a_n - d) + (a_n - 2d) + \cdots + (a_1 + 2d) + (a_1 + d) + a_1.$$

Adding corresponding terms on each side of the above equations, we get

$$2S_n = [a_1 + a_n] + [(a_1 + d) + (a_n - d)] + [(a_1 + 2d) + (a_n - 2d)]$$
$$+ \cdots + [(a_n - 2d) + (a_1 + 2d)] + [(a_n - d) + (a_1 + d)]$$
$$+ [a_n + a_1].$$

This simplifies to

$$2S_n = [a_1 + a_n] + [a_1 + a_n] + [a_1 + a_n]$$
$$+ \cdots + [a_n + a_1] + [a_n + a_1] + [a_n + a_1].$$

Since $(a_1 + a_n)$ is being added n times, it follows that

$$2S_n = n(a_1 + a_n).$$

This leads to the following formula.

TO FIND S_n FOR AN ARITHMETIC SEQUENCE:

The sum of the first n terms of an arithmetic sequence is given by

$$S_n = \frac{n}{2}(a_1 + a_n).$$

EXAMPLE 5   Find the sum of the first 100 positive even numbers.

SOLUTION   The sum is

$$2 + 4 + 6 + \cdots + 198 + 200.$$

This is the sum of the first 100 terms of the arithmetic sequence for which

$$a_1 = 2, \quad n = 100, \quad \text{and} \quad a_n = 200.$$

Substituting in the formula

$$S_n = \frac{n}{2}(a_1 + a_n),$$

we get

$$S_{100} = \frac{100}{2}(2 + 200)$$
$$= 50(202) = 10{,}100.$$

The above formula is useful when we know the first and last terms, a_1 and a_n. To find S_n when a_n is unknown, but a_1, n, and d are known, we must first calculate a_n.

EXAMPLE 6   Find the sum of the first 15 terms of the arithmetic sequence 4, 7, 10, 13, ...

SOLUTION   Note that

$$a_1 = 4, \quad n = 15, \quad \text{and} \quad d = 3.$$

Before using the formula for S_n, we find a_{15}:

$$a_{15} = 4 + (15 - 1)3 \qquad \text{Substituting into the formula for } a_n$$
$$= 4 + 14 \cdot 3 = 46.$$

Thus, knowing that $a_{15} = 46$, we have

$$S_{15} = \frac{15}{2}(4 + 46) \qquad \text{Using the formula for } S_n$$

$$= \frac{15}{2}(50) = 375.$$

Problem Solving

For some problem-solving situations, the translation may involve sequences or series. We look at some examples.

EXAMPLE 7

Hourly Wages. Chris takes a job, starting with an hourly wage of $14.25, and is promised a raise of 15¢ per hour every 2 months for 5 years. At the end of 5 years, what will be Chris's hourly wage?

SOLUTION

1. **Familiarize.** It helps to write down the hourly wage for several two-month time periods.

   Beginning: 14.25,
   After two months: 14.40,
   After four months: 14.55,
   and so on.

   What appears is a sequence of numbers: 14.25, 14.40, 14.55, ... Since the same amount is added each time, the sequence is arithmetic.
       We ask ourselves what we know about arithmetic sequences. The pertinent formulas are

   $$a_n = a_1 + (n - 1)d$$

   and

   $$S_n = \frac{n}{2}(a_1 + a_n).$$

   In this case, we are not looking for a sum, so it is probably the first formula that will give us our answer. We want to determine the last term in a sequence. To do so, we need to know a_1, n, and d. From our list above, we see that

   $$a_1 = 14.25 \quad \text{and} \quad d = 0.15.$$

   What is n? That is, how many terms are in the sequence? Each year there are 6 raises, since Chris gets a raise every 2 months. There are 5 years, so the total number of raises will be $5 \cdot 6$, or 30. There will be 31 terms: the original wage and 30 increased rates.

2. **Translate.** We want to find a_n for the arithmetic sequence in which $a_1 = 14.25$, $n = 31$, and $d = 0.15$.

3. **Carry out.** Substituting in the formula for a_n gives us

   $$a_{31} = 14.25 + (31 - 1) \cdot 0.15$$

   $$= 18.75.$$

4. Check. We can check by redoing the calculations or we can calculate in a slightly different way for another check. For example, at the end of a year, there will be 6 raises, for a total raise of $0.90. At the end of 5 years, the total raise will be 5 × $0.90, or $4.50. If we add that to the original wage of $14.25, we obtain $18.75. The answer checks.

5. State. At the end of 5 years, Chris's hourly wage will be $18.75.

Example 7 is one in which the calculations or the translation could be done in a number of ways. There is often a variety of ways in which a problem can be solved. You should use the one that is best or easiest for you. In this chapter, however, we will try to emphasize sequences and series and their related formulas.

E X A M P L E   8

Telephone Pole Storage.   A stack of telephone poles has 30 poles in the bottom row. There are 29 poles in the second row, 28 in the next row, and so on. How many poles are in the stack if there are 5 poles in the top row?

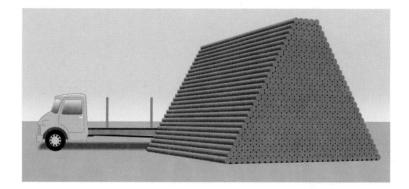

S O L U T I O N

1. Familiarize. A picture will help in this case. The following figure shows the ends of the poles and the way in which they stack. There are 30 poles on the bottom, and we see that there will be one fewer in each succeeding row. How many rows will there be?

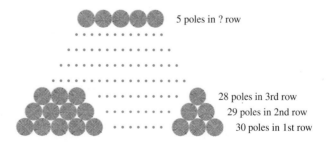

5 poles in ? row

28 poles in 3rd row
29 poles in 2nd row
30 poles in 1st row

Note that there are 30 − 1 = 29 logs in the 2nd row, 30 − 2 = 28 logs in the 3rd row, 30 − 3 = 27 logs in the 4th row, and so on. The pattern leads to 30 − 25 = 5 logs in the 26th row.

The situation is represented by the equation

$$30 + 29 + 28 + \cdots + 5. \qquad \text{There are 26 terms in this series.}$$

Thus we have an arithmetic series. We recall the formula

$$S_n = \frac{n}{2}(a_1 + a_n).$$

2. **Translate.** We want to find the sum of the first 26 terms of an arithmetic sequence in which $a_1 = 30$ and $a_{26} = 5$.

3. **Carry out.** Substituting into the above formula gives us

$$S_{26} = \frac{26}{2}(30 + 5)$$
$$= 13 \cdot 35 = 455.$$

4. **Check.** In this case, we can check the calculations by doing them again. A longer, harder way would be to do the entire addition:

$$30 + 29 + 28 + \cdots + 5.$$

5. **State.** There are 455 poles in the stack.

EXERCISE SET
11.2

Find the first term and the common difference.

1. 3, 8, 13, 18, . . .

2. 1.06, 1.12, 1.18, 1.24, . . .

3. 6, 2, −2, −6, . . .

4. −9, −6, −3, 0, . . .

5. $\frac{3}{2}, \frac{9}{4}, 3, \frac{15}{4}, \ldots$

6. $\frac{3}{5}, \frac{1}{10}, -\frac{2}{5}, \ldots$

7. \$2.12, \$2.24, \$2.36, \$2.48, . . .

8. \$214, \$211, \$208, \$205, . . .

9. Find the 12th term of the arithmetic sequence 3, 7, 11, ...

10. Find the 11th term of the arithmetic sequence 0.07, 0.12, 0.17, ...

11. Find the 17th term of the arithmetic sequence 7, 4, 1, ...

12. Find the 14th term of the arithmetic sequence $3, \frac{7}{3}, \frac{5}{3}, \ldots$.

13. Find the 13th term of the arithmetic sequence \$1200, \$964.32, \$728.64, ...

14. Find the 10th term of the arithmetic sequence \$2345.78, \$2967.54, \$3589.30, ...

15. In the sequence of Exercise 9, what term is 107?

16. In the sequence of Exercise 10, what term is 1.67?

17. In the sequence of Exercise 11, what term is −296?

18. In the sequence of Exercise 12, what term is −27?

19. Find a_{17} when $a_1 = 2$ and $d = 5$.

20. Find a_{20} when $a_1 = 14$ and $d = -3$.

21. Find a_1 when $d = 4$ and $a_8 = 33$.

22. Find a_1 when $d = 8$ and $a_{11} = 26$.

23. Find n when $a_1 = 5$, $d = -3$, and $a_n = -76$.

24. Find n when $a_1 = 25$, $d = -14$, and $a_n = -507$.

25. For an arithmetic sequence in which $a_{17} = -40$ and $a_{28} = -73$, find a_1 and d. Write the first five terms of the sequence.

26. In an arithmetic sequence, $a_{17} = \frac{25}{3}$ and $a_{32} = \frac{95}{6}$. Find a_1 and d. Write the first five terms of the sequence.

27. Find the sum of the first 20 terms of the arithmetic series $1 + 5 + 9 + 13 + \cdots$.

28. Find the sum of the first 14 terms of the arithmetic series $11 + 7 + 3 + \cdots$.

29. Find the sum of the first 300 natural numbers.

30. Find the sum of the first 400 natural numbers.

31. Find the sum of the even numbers from 2 to 100, inclusive.

32. Find the sum of the odd numbers from 1 to 99, inclusive.

33. Find the sum of all multiples of 6 from 6 to 102, inclusive.

34. Find the sum of all multiples of 4 that are between 14 and 523.

35. An arithmetic series has $a_1 = 2$ and $d = 5$. Find S_{20}.

36. An arithmetic series has $a_1 = 7$ and $d = -3$. Find S_{32}.

Solve.

37. *Band formations.* The Duxbury marching band has 14 marchers in the front row, 16 in the second row, 18 in the third row, and so on, for 15 rows. How many marchers are in the last row? How many marchers are there altogether?

38. *Gardening.* A gardener is making a triangular planting, with 39 plants in the front row, 35 in the second row, 31 in the third row, and so on. If the pattern is consistent, how many plants will be in the last row? How many plants will there be altogether?

39. *Telephone pole piles.* How many poles will be in a pile of telephone poles if there are 50 in the first layer, 49 in the second, and so on, until there are 6 in the last layer?

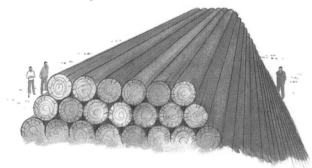

40. *Accumulated savings.* If 10¢ is saved on October 1, another 20¢ on October 2, another 30¢ on October 3, and so on, how much is saved during October? (October has 31 days.)

41. *Accumulated savings.* Renata saves money in an arithmetic sequence: $600 the first year, another $700 the second, and so on, for 20 yr. How much does she save in all (disregarding interest)?

42. *Spending.* Jacob spent $30 on August 1, $50 on August 2, $70 on August 3, and so on. How much did Jacob spend in August? (August has 31 days.)

43. *Auditorium design.* Theaters are often built with more seats per row as the rows move toward the back. The Sanders Amphitheater has 20 seats in the first row, 22 in the second, 24 in the third, and so on, for 19 rows. How many seats are in the amphitheater?

44. *Accumulated savings.* Shirley sets up an investment such that it will return $5000 the first year, $6125 the second year, $7250 the third year, and so on, for 25 yr. How much in all is received from the investment?

SKILL MAINTENANCE

Convert to an exponential equation.

45. $\log_a P = k$ **46.** $\ln t = a$

Find an equation of the circle satisfying the given conditions.

47. Center $(0, 0)$, radius 9

48. Center $(-2, 5)$, radius $3\sqrt{2}$

SYNTHESIS

49. ◈ It is said that as a young child, the mathematician Karl F. Gauss (1777–1855) was able to compute the sum $1 + 2 + 3 + \cdots + 100$ very quickly in his head. Explain how Gauss might have done this and present a formula for the sum of the first n natural numbers. (*Hint*: $1 + 99 = 100$.)

50. ◈ The sum of the first n terms of an arithmetic sequence is given by

$$S_n = \frac{n}{2}[2a_1 + (n-1)d].$$

Use the formulas for a_n and S_n to explain how this equation was developed.

51. Find a formula for the sum of the first n consecutive odd numbers starting with 1:

$$1 + 3 + 5 + \cdots + (2n - 1).$$

52. Find three numbers in an arithmetic sequence for which the sum of the first and third is 10 and the product of the first and second is 15.

53. ▦ In an arithmetic sequence, $a_1 = \$8760$ and $d = -\$798.23$. Find the first 10 terms of the sequence.

54. ▦ Find the sum of the first 10 terms of the sequence given in Exercise 53.

55. Prove that if p, m, and q are consecutive terms in an arithmetic sequence, then

$$m = \frac{p + q}{2}.$$

56. *Straight-line depreciation.* A company buys a color copier for $5200 on January 1 of a given year. The machine is expected to last for 8 yr, at the end of which time its *trade-in,* or *salvage, value* will be $1100. If the company figures the decline in value to be the same each year, then the trade-in values, after t years, $0 \le t \le 8$, form an arithmetic sequence given by

$$a_t = C - t\left(\frac{C - S}{N}\right),$$

where $C =$ the original cost of the item, $N =$ the years of expected life, and $S =$ the salvage value.

a) Find the formula for a_t for the straight-line depreciation of the copier.
b) Find the salvage value after 0 yr, 1 yr, 2 yr, 3 yr, 4 yr, 7 yr, and 8 yr.
c) Find a formula that expresses a_t recursively.

11.3 Geometric Sequences and Series

Geometric Sequences • Sum of the First n Terms of a Geometric Sequence • Infinite Geometric Series • Problem Solving

In an arithmetic sequence, a certain number is added to each term to get the next term. When each term in a sequence is *multiplied* by a certain number to get the next term, the sequence is **geometric**. In this section, we examine geometric sequences (or progressions) and *geometric series.*

Geometric Sequences

Consider the sequence

2, 6, 18, 54, 162, ...

If we multiply each term by 3, we obtain the next term. The multiplier is called the *common ratio* because it is found by dividing any term by the preceding term.

GEOMETRIC SEQUENCE

A sequence is *geometric* if there exists a number r, called the *common ratio,* for which

$$\frac{a_{n+1}}{a_n} = r, \quad \text{or} \quad a_{n+1} = a_n \cdot r \quad \text{for any integer } n \ge 1.$$

EXAMPLE 1   For each geometric sequence, find the common ratio.

a) 3, 6, 12, 24, 48, . . .   **b)** 3, −6, 12, −24, 48, −96, . . .
c) $5200, $3900, $2925, $2193.75, . . .

SOLUTION

Sequence	*Common Ratio*	
a) 3, 6, 12, 24, 48, . . .	2	$\frac{6}{3} = 2, \frac{12}{6} = 2$, and so on
b) 3, −6, 12, −24, 48, −96, . . .	−2	$\frac{-6}{3} = -2, \frac{12}{-6} = -2$, and so on
c) $5200, $3900, $2925, $2193.75, . . .	0.75	$\frac{\$3900}{\$5200} = 0.75, \frac{\$2925}{\$3900} = 0.75$

To develop a formula for the general, or nth, term of a geometric sequence, let a_1 be the first term and let r be the common ratio. We write out the first few terms as follows:

$a_1,$

$a_2 = a_1 r,$

$a_3 = a_2 r = (a_1 r)r = a_1 r^2,$ Substituting $a_1 r$ for a_2

$a_4 = a_3 r = (a_1 r^2)r = a_1 r^3.$ Substituting $a_1 r^2$ for a_3

Note that the exponent is 1 less than the subscript.

Generalizing, we obtain the following.

TO FIND a_n FOR A GEOMETRIC SEQUENCE:

The nth term of a geometric sequence is given by
$$a_n = a_1 r^{n-1}, \quad \text{for any integer } n \geq 1.$$

EXAMPLE 2   Find the 7th term of the geometric sequence 4, 20, 100, ...

SOLUTION   First we note that

$a_1 = 4$ and $n = 7.$

To find the common ratio, we can divide any term (other than the first) by the term preceding it. Since the second term is 20 and the first is 4,

$$r = \frac{20}{4}, \quad \text{or } 5.$$

The formula

$$a_n = a_1 r^{n-1}$$

gives us

$$a_7 = 4 \cdot 5^{7-1} = 4 \cdot 5^6 = 4 \cdot 15,625 = 62,500.$$

EXAMPLE 3   Find the 10th term of the geometric sequence

$$64, \ -32, \ 16, \ -8, \dots .$$

SOLUTION   First we note that

$$a_1 = 64, \qquad n = 10, \quad \text{and} \quad r = \frac{-32}{64} = -\frac{1}{2}.$$

Then, using the formula for the nth term of a geometric series, we have

$$a_{10} = 64 \cdot \left(-\frac{1}{2}\right)^{10-1} = 64 \cdot \left(-\frac{1}{2}\right)^{9} = 2^6 \cdot \left(-\frac{1}{2^9}\right) = -\frac{1}{2^3} = -\frac{1}{8}.$$

Sum of the First n Terms of a Geometric Sequence

We next develop a formula for S_n when a sequence is geometric:

$$a_1, \ a_1 r, \ a_1 r^2, \ a_1 r^3, \dots, \ a_1 r^{n-1}, \ \dots .$$

The **geometric series** S_n is given by

$$S_n = a_1 + a_1 r + a_1 r^2 + \dots + a_1 r^{n-2} + a_1 r^{n-1}. \tag{1}$$

Multiplying by r on both sides gives us

$$r S_n = a_1 r + a_1 r^2 + a_1 r^3 + \dots + a_1 r^{n-1} + a_1 r^n. \tag{2}$$

When we subtract corresponding sides of equation (2) from equation (1), the color terms drop out, leaving

$$S_n - r S_n = a_1 - a_1 r^n,$$

or

$$S_n(1 - r) = a_1(1 - r^n). \qquad \text{Factoring}$$

Dividing on both sides by $1 - r$ gives us the following formula.

TO FIND S_n FOR A GEOMETRIC SEQUENCE:

The sum of the first n terms of a geometric sequence is given by

$$S_n = \frac{a_1(1 - r^n)}{1 - r}, \quad \text{for any } r \neq 1.$$

EXAMPLE 4   Find the sum of the first seven terms of the geometric sequence 3, 15, 75, 375, ...

SOLUTION   First we note that

$$a_1 = 3, \qquad n = 7, \quad \text{and} \quad r = \frac{15}{3} = 5.$$

Then, substituting in the formula $S_n = \dfrac{a_1(1 - r^n)}{1 - r}$, we have

$$S_7 = \frac{3(1 - 5^7)}{1 - 5} = \frac{3(1 - 78{,}125)}{-4}$$

$$= \frac{3(-78{,}124)}{-4}$$

$$= 58{,}593.$$

Infinite Geometric Series

Suppose we consider the sum of the terms of an infinite geometric sequence, such as $2, 4, 8, 16, 32, \ldots$. We get what is called an **infinite geometric series:**

$$2 + 4 + 8 + 16 + 32 + \cdots.$$

Here, as n grows larger and larger, the sum of the first n terms, S_n, becomes larger and larger without bound. There are also infinite series that get closer and closer to some specific number. Here is an example:

$$\frac{1}{2} + \frac{1}{4} + \frac{1}{8} + \frac{1}{16} + \cdots + \frac{1}{2^n} + \cdots.$$

Let's consider S_n for the first five values of n:

$$S_1 = \tfrac{1}{2} \qquad\qquad\quad = \tfrac{1}{2} = 0.5,$$
$$S_2 = \tfrac{1}{2} + \tfrac{1}{4} \qquad\quad = \tfrac{3}{4} = 0.75,$$
$$S_3 = \tfrac{1}{2} + \tfrac{1}{4} + \tfrac{1}{8} \qquad = \tfrac{7}{8} = 0.875,$$
$$S_4 = \tfrac{1}{2} + \tfrac{1}{4} + \tfrac{1}{8} + \tfrac{1}{16} = \tfrac{15}{16} = 0.9375.$$

> The denominator of the sum is 2^n, where n is the subscript of S. The numerator is $2^n - 1$.

Thus, for this particular series, we have

$$S_n = \frac{2^n - 1}{2^n} = \frac{2^n}{2^n} - \frac{1}{2^n} = 1 - \frac{1}{2^n}.$$

Note that the value of S_n is less than 1 for any value of n, but as n gets larger and larger, the values of S_n get closer and closer to 1. We say that 1 is the *limit* of S_n and that 1 is the sum of this infinite geometric sequence. An infinite geometric series is denoted S_∞. It can be shown (but we will not do it here) that the sum of the terms of an infinite geometric sequence exists if and only if $|r| < 1$ (that is, the absolute value of the common ratio is less than 1).

To find a formula for the sum of an infinite geometric sequence, we first consider the sum of the first n terms:

$$S_n = \frac{a_1(1 - r^n)}{1 - r} = \frac{a_1 - a_1 r^n}{1 - r}. \qquad \text{Using the distributive law}$$

For $|r| < 1$, it follows that values of r^n get closer and closer to 0 as n gets larger. (Check this by selecting a number between -1 and 1 and finding larger and larger powers on a calculator.) As r^n gets closer and closer to 0, so does $a_1 r^n$. Thus, S_n gets closer and closer to $a_1/(1 - r)$.

THE LIMIT OF AN INFINITE GEOMETRIC SERIES

When $|r| < 1$, the limit of an infinite geometric series is given by

$$S_\infty = \frac{a_1}{1 - r}. \qquad \text{(For } |r| \geq 1, \text{ no limit exists.)}$$

EXAMPLE 5 Determine whether each series has a limit. If one exists, find it.

a) $1 + 3 + 9 + 27 + \cdots$ **b)** $-2 + 1 - \frac{1}{2} + \frac{1}{4} - \frac{1}{8} + \cdots$

SOLUTION

a) Here $r = 3$, so $|r| = |3| = 3$. Since $|r| \not< 1$, the series does *not* have a limit.

b) Here $r = -\frac{1}{2}$, so $|r| = |-\frac{1}{2}| = \frac{1}{2}$. Since $|r| < 1$, the series *does* have a limit. We find the limit by substituting into the formula for S_∞:

$$S_\infty = \frac{-2}{1 - \left(-\frac{1}{2}\right)} = \frac{-2}{\frac{3}{2}} = -\frac{4}{3}.$$

EXAMPLE 6 Find fractional notation for $0.63636363\ldots$.

SOLUTION We can express this as

$$0.63 + 0.0063 + 0.000063 + \cdots.$$

This is an infinite geometric series, where $a_1 = 0.63$ and $r = 0.01$. Since $|r| < 1$, this series has a limit:

$$S_\infty = \frac{a_1}{1 - r} = \frac{0.63}{1 - 0.01} = \frac{0.63}{0.99} = \frac{63}{99}.$$

Thus fractional notation for $0.63636363\ldots$ is $\frac{63}{99}$, or $\frac{7}{11}$.

Problem Solving

For some problem-solving situations, the translation may involve geometric sequences or series.

EXAMPLE 7 *Daily Wages.* Suppose someone offered you a job for the month of September (30 days) under the following conditions. You will be paid \$0.01 for the first day, \$0.02 for the second, \$0.04 for the third, and so on, doubling your previous day's salary each day. How much would you earn? (Would you take the job? Make a guess before reading further.)

SOLUTION

1. **Familiarize.** You earn $0.01 the first day, $0.01(2) the second day, $0.01(2)(2) the third day, and so on. Since each day's wages are a constant multiple of the previous day's wage, a geometric sequence is formed.

2. **Translate.** The amount earned is the geometric series

$$\$0.01 + \$0.01(2) + \$0.01(2^2) + \$0.01(2^3) + \cdots + \$0.01(2^{29}),$$

where

$$a_1 = \$0.01, \qquad n = 30, \quad \text{and} \quad r = 2.$$

3. **Carry out.** Using the formula

$$S_n = \frac{a_1(1 - r^n)}{1 - r},$$

we have

$$S_{30} = \frac{\$0.01(1 - 2^{30})}{1 - 2}$$

$$= \frac{\$0.01(-1{,}073{,}741{,}823)}{-1} \qquad \text{Using a calculator}$$

$$= \$10{,}737{,}418.23.$$

4. **Check.** The calculations can be repeated as a check.

5. **State.** The pay exceeds $10.7 million for the month. Most people would probably take the job!

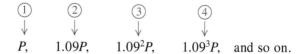

EXAMPLE   8

Loan Repayment.   A student loan is in the amount of $6000. Interest is to be 9% compounded annually, and the entire amount is to be paid after 10 yr. How much is to be paid back?

SOLUTION

1. **Familiarize.** Suppose we let P represent any principal amount. At the end of one year, the amount owed will be $P + 0.09P$, or $1.09P$. That amount will be the principal for the second year. The amount owed at the end of the second year will be $1.09 \times$ New principal $= 1.09(1.09P)$, or 1.09^2P. Thus the amount owed at the beginning of successive years is as follows:

①	②	③	④
↓	↓	↓	↓
P,	$1.09P$,	1.09^2P,	1.09^3P, and so on.

We have a geometric sequence. The amount owed at the beginning of the 11th year will be the amount owed at the end of the 10th year.

2. **Translate.** We have a geometric sequence with $a_1 = 6000$, $r = 1.09$, and $n = 11$. The appropriate formula is

$$a_n = a_1 r^{n-1}.$$

3. Carry out. We substitute and calculate:

$$a_{11} = \$6000(1.09)^{11-1} = \$6000(1.09)^{10}$$
$$\approx \$6000(2.3673637) \quad \text{Using a calculator to approximate } 1.09^{10}$$
$$\approx \$14,204.18. \quad \text{Rounded to the nearest hundredth}$$

4. Check. A check, by repeating the calculations, is left to the student.

5. State. A total of \$14,204.18 is to be paid back at the end of 10 yr.

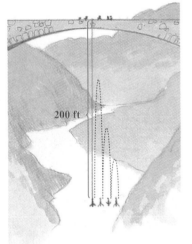

EXAMPLE 9

Bungee Jumping. A bungee jumper rebounds 60% of the height jumped. A bungee jump is made using a cord that stretches to 200 ft.

a) After jumping and then rebounding 9 times, how far has a bungee jumper traveled upward (the total rebound distance)?

b) Approximately how far will a jumper have traveled upward (bounced) before coming to rest?

SOLUTION

1. Familiarize. Let's do some calculations and look for a pattern.

First fall:	200 ft
First rebound:	0.6×200, or 120 ft
Second fall:	120 ft, or 0.6×200
Second rebound:	0.6×120, or $0.6(0.6 \times 200)$, which is 72 ft
Third fall:	72 ft, or $0.6(0.6 \times 200)$
Third rebound:	0.6×72, or $0.6(0.6(0.6 \times 200))$, which is 43.2 ft

The rebound distances form a geometric sequence:

$$\underset{\substack{\downarrow \\ 120,}}{①} \quad \underset{\substack{\downarrow \\ 0.6 \times 120,}}{②} \quad \underset{\substack{\downarrow \\ 0.6^2 \times 120,}}{③} \quad \underset{\substack{\downarrow \\ 0.6^3 \times 120, \dots}}{④}$$

2. Translate.

a) The total rebound distance after 9 bounces is the sum of a geometric sequence. The first term is 120 and the common ratio is 0.6. There will be 9 terms, so we can use the formula

$$S_n = \frac{a_1(1 - r^n)}{1 - r}.$$

b) Theoretically, the jumper will never stop bouncing. Realistically, the bouncing will eventually stop. To approximate the actual distance bounced, we consider an infinite number of bounces and use the formula

$$S_\infty = \frac{a_1}{1 - r}.$$

3. Carry out.

a) We substitute into the formula and calculate:

$$S_9 = \frac{120[1 - (0.6)^9]}{1 - 0.6} \approx 297. \quad \text{Using a calculator}$$

b) We substitute and calculate:

$$S_\infty = \frac{120}{1 - 0.6} = 300.$$

4. Check. We can do the calculations again.

5. State.

a) In 9 bounces, the bungee jumper will have traveled upward a total distance of about 297 ft.

b) The jumper will travel upward about 300 ft before coming to rest.

EXERCISE SET
11.3

Find the common ratio for each geometric sequence.

1. 5, 10, 20, 40, ...

2. 2, 6, 18, 54, ...

3. 5, −5, 5, −5, ...

4. −5, −0.5, −0.05, −0.005, ...

5. $\frac{1}{2}, -\frac{1}{4}, \frac{1}{8}, -\frac{1}{16}, \ldots$

6. $\frac{2}{3}, -\frac{4}{3}, \frac{8}{3}, -\frac{16}{3}, \ldots$

7. 75, 15, 3, $\frac{3}{5}$, ...

8. 12, −4, $\frac{4}{3}$, $-\frac{4}{9}$, ...

9. $\frac{1}{m}, \frac{3}{m^2}, \frac{9}{m^3}, \frac{27}{m^4}, \ldots$

10. $4, \frac{4m}{5}, \frac{4m^2}{25}, \frac{4m^3}{125}, \ldots$

Find the indicated term for each geometric sequence.

11. 5, 10, 20, ... ; the 7th term

12. 2, 8, 32, ... ; the 9th term

13. 3, $3\sqrt{2}$, 6, ... ; the 9th term

14. 4, $4\sqrt{3}$, 12, ... ; the 8th term

15. ▦ $-\frac{8}{243}, \frac{8}{81}, -\frac{8}{27}, \ldots$; the 10th term

16. ▦ $\frac{7}{625}, \frac{-7}{125}, \frac{7}{25}, \ldots$; the 13th term

17. ▦ $1000, $1080, $1166.40, ... ; the 12th term

18. ▦ $1000, $1070, $1144.90, ... ; the 11th term

Find the nth, or general, term for each geometric sequence.

19. 1, 3, 9, ...

20. 25, 5, 1, ...

21. 1, −1, 1, −1, ...

22. 2, 4, 8, ...

23. $\frac{1}{x}, \frac{1}{x^2}, \frac{1}{x^3}, \ldots$

24. $5, \frac{5m}{2}, \frac{5m^2}{4}, \ldots$

For Exercises 25–32, use the formula for S_n to find the indicated sum.

25. S_7 for the geometric series $7 + 14 + 28 + \cdots$

26. S_6 for the geometric series $16 - 8 + 4 - \cdots$

27. S_7 for the geometric series $\frac{1}{18} - \frac{1}{6} + \frac{1}{2} - \cdots$

28. S_5 for the geometric series $6 + 0.6 + 0.06 + \cdots$

29. S_8 for the series $1 + x + x^2 + x^3 + \cdots$

30. S_{10} for the series $1 + x^2 + x^4 + x^6 + \cdots$

31. ▦ S_{16} for the geometric sequence
$200, $200(1.06), $200(1.06)^2, \ldots$

32. ▦ S_{23} for the geometric sequence
$1000, $1000(1.08), $1000(1.08)^2, \ldots$

Determine whether each infinite geometric series has a limit. If a limit exists, find it.

33. $9 + 3 + 1 + \cdots$

34. $8 + 4 + 2 + \cdots$

35. $7 + 3 + \frac{9}{7} + \cdots$

36. $12 + 9 + \frac{27}{4} + \cdots$

37. $3 + 15 + 75 + \cdots$

38. $2 + 3 + \frac{9}{2} + \cdots$

39. $4 - 6 + 9 - \frac{27}{2} + \cdots$

40. $-6 + 3 - \frac{3}{2} + \frac{3}{4} - \cdots$

41. $0.43 + 0.0043 + 0.000043 + \cdots$

42. $0.37 + 0.0037 + 0.000037 + \cdots$

43. $\$500(1.02)^{-1} + \$500(1.02)^{-2} + \$500(1.02)^{-3} + \cdots$

44. $\$1000(1.08)^{-1} + \$1000(1.08)^{-2} + \$1000(1.08)^{-3} + \cdots$

Find fractional notation for each infinite sum. (These are geometric series.)

45. $0.7777\ldots$

46. $0.2222\ldots$

47. $8.3838\ldots$

48. $7.4747\ldots$

49. $0.15151515\ldots$

50. $0.12121212\ldots$

Solve. Use a calculator as needed for evaluating formulas.

51. *Rebound distance.* A ping-pong ball is dropped from a height of 20 ft and always rebounds one fourth of the distance fallen. How high does it rebound the 6th time?

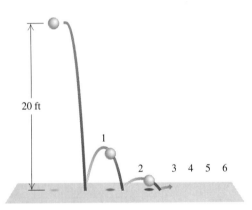

52. *Rebound distance.* Approximate the total of the rebound heights of the ball in Exercise 51.

53. *Population growth.* Yorktown has a current population of 100,000, and the population is increasing by 3% each year. What will the population be in 15 yr?

54. *Doubling time.* How long will it take for the population of Yorktown to double? (See Exercise 53.)

55. *Amount owed.* Gilberto borrows $15,000. The loan is to be repaid in 13 yr at 8.5% interest, compounded annually. How much will be repaid at the end of 13 yr?

56. *Shrinking population.* A population of 5000 fruit flies is dying off at a rate of 4% per minute. How many flies will be alive after 15 min?

57. *Shrinking population.* For the population of fruit flies in Exercise 56, how long will it take for only

1800 fruit flies to remain alive? (See Exercise 56 and use logarithms.) Round to the nearest minute.

58. *Investing.* Leslie is saving money in a retirement account. At the beginning of each year, she invests $1000 at 7%, compounded annually. How much will be in the retirement fund at the end of 40 yr?

59. *Rebound distance.* A superball dropped from the top of the Washington Monument (556 ft high) rebounds three fourths of the distance fallen. How far (up and down) will the ball have traveled when it hits the ground for the 6th time?

60. *Rebound distance.* Approximate the total distance that the ball of Exercise 59 will have traveled when it comes to rest.

61. *Stacking paper.* Construction paper is about 0.02 in. thick. Beginning with just one piece, a stack is doubled again and again 10 times. Find the height of the final stack.

62. *Monthly earnings.* Suppose you accepted a job for the month of February (28 days) under the following conditions. You will be paid $0.01 the first day, $0.02 the second, $0.04 the third, and so on, doubling your previous day's salary each day. How much would you earn?

SKILL MAINTENANCE

Solve the system.

63. $5x - 2y = -3,$
$2x + 5y = -24$

64. $x - 2y + 3z = 4,$
$2x - y + z = -1,$
$4x + y + z = 1$

SYNTHESIS

65. ◆ Write a problem for a classmate to solve. Devise the problem so that a geometric series is involved and the solution is "The total amount in the bank is
$\$900(1.08)^{40},$
or about $19,550."

66. ◆ The infinite series
$$S_\infty = 2 + \frac{1}{2} + \frac{1}{2 \cdot 3} + \frac{1}{2 \cdot 3 \cdot 4} + \frac{1}{2 \cdot 3 \cdot 4 \cdot 5} + \frac{1}{2 \cdot 3 \cdot 4 \cdot 5 \cdot 6} + \cdots$$
is not geometric, but it does have a sum. Using S_1, S_2, S_3, S_4, S_5, and S_6, make a conjecture about the value of S_∞ and explain your reasoning.

67. ◆ When r is negative, a series is said to be *alternating.* Why do you suppose this terminology is used?

68. Find the sum of the first n terms of
$$x^2 - x^3 + x^4 - x^5 + \cdots.$$

69. Find the sum of the first n terms of
$$1 + x + x^2 + x^3 + \cdots.$$

70. The sides of a square are each 16 cm long. A second square is inscribed by joining the midpoints of the sides, successively. In the second square we repeat

the process, inscribing a third square. If this process is continued indefinitely, what is the sum of all of the areas of all the squares? (*Hint*: Use an infinite geometric series.)

71. ◆ To compare the *graphs* of an arithmetic and a geometric sequence, we plot n on the horizontal axis and a_n on the vertical axis. Graph Example 1(a) of Section 11.2 and Example 1(a) of Section 11.3 on the same set of axes. How do the graphs of geometric sequences differ from the graphs of arithmetic sequences?

72. ◆ 〰 Using Example 5 and Exercises 33–44, explain how the graph of a geometric sequence can be used to determine whether a geometric series has a limit.

COLLABORATIVE
C ◆ O ◆ R ◆ N ◆ E ◆ R

Focus: Geometric series

Time: 30 minutes

Group size: 2

Materials: Graphing calculators are optional.

ACTIVITY*

1. One group member ("the seller") has a car for sale and is asking \$3500. The second ("the buyer") offers \$1500. The seller splits the difference (\$2000 ÷ 2 = \$1000) and lowers the price to \$2500. The buyer then splits the difference again (\$1000 ÷ 2 = \$500) and counters with \$2000. Continue in this manner and stop when you are able to agree on the car's selling price to the nearest penny.

2. What should the buyer's initial offer be in order to achieve a purchase price of \$2000? (Check several guesses to find the appropriate initial offer.)

---

*This activity is based on the article, "Bargaining Theory, or Zeno's Used Cars," by James C. Kirby, *The College Mathematics Journal*, **27**(4), September 1996.

3. The seller's price in the bargaining above can be modeled recursively (see Exercises 72, 73, and 81 in Section 11.1) by the sequence

$$a_1 = 3500, \qquad a_n = a_{n-1} - \frac{d}{2^{2n-3}},$$

where d is the difference between the initial price and the first offer. Use this recursively defined sequence to solve parts (1) and (2) above either manually or with the SEQUENTIAL FUNCTION mode and the TABLE feature of a grapher.

4. The first four terms in the sequence in part (3) can be written as

$$a_0, \quad a_0 - \frac{d}{2}, \quad a_0 - \frac{d}{2} - \frac{d}{8},$$

$$a_0 - \frac{d}{2} - \frac{d}{8} - \frac{d}{32}.$$

Use the formula for the limit of an infinite geometric series to find a simple algebraic formula for the eventual sale price, P, when the bargaining process from above is followed. Verify the formula by using it to solve parts (1) and (2) above.

11.4   The Binomial Theorem

Binomial Expansion Using Pascal's Triangle   •
Binomial Expansion Using Factorial Notation

Binomial Expansion Using Pascal's Triangle

Consider the following expanded powers of $(a + b)^n$:

$$(a + b)^0 = 1$$
$$(a + b)^1 = a + b$$
$$(a + b)^2 = a^2 + 2a^1b^1 + b^2$$
$$(a + b)^3 = a^3 + 3a^2b^1 + 3a^1b^2 + b^3$$
$$(a + b)^4 = a^4 + 4a^3b^1 + 6a^2b^2 + 4a^1b^3 + b^4$$
$$(a + b)^5 = a^5 + 5a^4b^1 + 10a^3b^2 + 10a^2b^3 + 5a^1b^4 + b^5.$$

Each expansion is a polynomial. There are some patterns to be noted:

1. There is one more term than the power of the binomial, n. That is, there are $n + 1$ terms in the expansion of $(a + b)^n$.
2. In each term, the sum of the exponents is the power to which the binomial is raised.
3. The exponents of a start with n, the power of the binomial, and decrease to 0. The last term has no factor of a. The first term has no factor of b, so powers of b start with 0 and increase to n.
4. The coefficients start at 1, increase through certain values, and then decrease through these same values back to 1. Let's study the coefficients further.

Suppose we wish to expand $(a + b)^8$. The patterns we noticed above indicate 9 terms in the expansion:

$$a^8 + c_1a^7b + c_2a^6b^2 + c_3a^5b^3 + c_4a^4b^4 + c_5a^3b^5 + c_6a^2b^6 + c_7ab^7 + b^8.$$

How can we determine the values for the c's? One method seems to be the easiest, but is not always. It involves writing down the coefficients in a triangular array as follows. We form what is known as **Pascal's triangle:**

$$
\begin{array}{cccccccccccc}
(a + b)^0: & & & & & & 1 & & & & & \\
(a + b)^1: & & & & & 1 & & 1 & & & & \\
(a + b)^2: & & & & 1 & & 2 & & 1 & & & \\
(a + b)^3: & & & 1 & & 3 & & 3 & & 1 & & \\
(a + b)^4: & & 1 & & 4 & & 6 & & 4 & & 1 & \\
(a + b)^5: & 1 & & 5 & & 10 & & 10 & & 5 & & 1
\end{array}
$$

There are many patterns in the triangle. Find as many as you can.

Perhaps you discovered a way to write the next row of numbers, given the numbers in the row above it. There are always 1's on the outside. Each remaining number is the sum of the two numbers above:

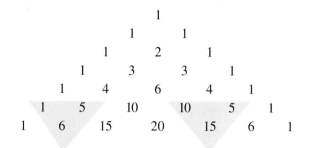

We see that in the bottom (seventh) row

the 1st and last numbers are 1;
the 2nd number is 1 + 5, or 6;
the 3rd number is 5 + 10, or 15;
the 4th number is 10 + 10, or 20;
the 5th number is 10 + 5, or 15; and
the 6th number is 5 + 1, or 6.

Thus the expansion of $(a + b)^6$ is

$$(a + b)^6 = 1a^6 + 6a^5b + 15a^4b^2 + 20a^3b^3 + 15a^2b^4 + 6ab^5 + 1b^6.$$

To expand $(a + b)^8$, we complete two more rows of Pascal's triangle:

$$
\begin{array}{ccccccccccccccc}
 & & & & & & & 1 \\
 & & & & & & 1 & & 1 \\
 & & & & & 1 & & 2 & & 1 \\
 & & & & 1 & & 3 & & 3 & & 1 \\
 & & & 1 & & 4 & & 6 & & 4 & & 1 \\
 & & 1 & & 5 & & 10 & & 10 & & 5 & & 1 \\
 & 1 & & 6 & & 15 & & 20 & & 15 & & 6 & & 1 \\
1 & & 7 & & 21 & & 35 & & 35 & & 21 & & 7 & & 1 \\
\end{array}
$$

1 8 28 56 70 56 28 8 1

Thus the expansion of $(a + b)^8$ is

$$(a + b)^8 = 1a^8 + 8a^7b + 28a^6b^2 + 56a^5b^3 + 70a^4b^4 + 56a^3b^5$$
$$+ 28a^2b^6 + 8ab^7 + 1b^8.$$

We can generalize our results as follows:

THE BINOMIAL THEOREM (FORM 1)

For any binomial $a + b$ and any natural number n,

$$(a+b)^n = c_0 a^n b^0 + c_1 a^{n-1} b^1 + c_2 a^{n-2} b^2 + \cdots + c_{n-1} a^1 b^{n-1} + c_n a^0 b^n,$$

where the numbers $c_0, c_1, c_2, \ldots, c_n$ are from the $(n + 1)$st row of Pascal's triangle.

EXAMPLE 1

Expand: $(u - v)^5$.

SOLUTION   Using the binomial theorem, we have $a = u$, $b = -v$, and $n = 5$. We use the 6th row of Pascal's triangle: 1 5 10 10 5 1. Thus,

$$(u - v)^5 = [u + (-v)]^5 \qquad \text{Rewriting } u - v \text{ as a sum}$$
$$= 1(u)^5 + 5(u)^4(-v)^1 + 10(u)^3(-v)^2 + 10(u)^2(-v)^3$$
$$+ 5(u)^1(-v)^4 + 1(-v)^5$$
$$= u^5 - 5u^4v + 10u^3v^2 - 10u^2v^3 + 5uv^4 - v^5.$$

Note that the signs of the terms alternate between $+$ and $-$. When $-v$ is raised to an odd power, the sign is $-$.   ━━●

EXAMPLE 2

Expand: $\left(2t + \dfrac{3}{t}\right)^6$.

SOLUTION   Note that $a = 2t$, $b = 3/t$, and $n = 6$. We use the 7th row of Pascal's triangle: 1 6 15 20 15 6 1. Thus,

$$\left(2t + \frac{3}{t}\right)^6 = 1(2t)^6 + 6(2t)^5\left(\frac{3}{t}\right)^1 + 15(2t)^4\left(\frac{3}{t}\right)^2 + 20(2t)^3\left(\frac{3}{t}\right)^3$$
$$+ 15(2t)^2\left(\frac{3}{t}\right)^4 + 6(2t)^1\left(\frac{3}{t}\right)^5 + 1\left(\frac{3}{t}\right)^6$$
$$= 64t^6 + 6(32t^5)\left(\frac{3}{t}\right) + 15(16t^4)\left(\frac{9}{t^2}\right) + 20(8t^3)\left(\frac{27}{t^3}\right)$$
$$+ 15(4t^2)\left(\frac{81}{t^4}\right) + 6(2t)\left(\frac{243}{t^5}\right) + \frac{729}{t^6}$$
$$= 64t^6 + 576t^4 + 2160t^2 + 4320 + 4860t^{-2} + 2916t^{-4}$$
$$+ 729t^{-6}.$$ ━━●

Binomial Expansion Using Factorial Notation

The disadvantage in using Pascal's triangle is that we must compute all the preceding rows in the table to obtain the row needed for the expansion. The following method avoids this difficulty. It will also enable us to find a specific term—say, the 8th term—without computing all the other terms in the expansion. This method is useful in such courses as finite mathematics, calculus, and statistics.

To develop the method, we need some new notation. Products of successive natural numbers, such as $6 \cdot 5 \cdot 4 \cdot 3 \cdot 2 \cdot 1$ and $8 \cdot 7 \cdot 6 \cdot 5 \cdot 4 \cdot 3 \cdot 2 \cdot 1$, have a special notation. For the product $6 \cdot 5 \cdot 4 \cdot 3 \cdot 2 \cdot 1$, we write 6!, read "6 factorial."

FACTORIAL NOTATION

For any natural number n,

$$n! = n(n - 1)(n - 2) \cdots (3)(2)(1).$$

Here are some examples:

$$6! = 6 \cdot 5 \cdot 4 \cdot 3 \cdot 2 \cdot 1 = 720,$$
$$5! = 5 \cdot 4 \cdot 3 \cdot 2 \cdot 1 = 120,$$
$$4! = 4 \cdot 3 \cdot 2 \cdot 1 = 24,$$
$$3! = 3 \cdot 2 \cdot 1 = 6,$$
$$2! = 2 \cdot 1 = 2,$$
$$1! = 1 = 1.$$

We also define 0! to be 1 for reasons explained shortly.

To simplify expressions like

$$\frac{8!}{5! \, 3!},$$

note that

$$8! = 8 \cdot 7 \cdot 6 \cdot 5 \cdot 4 \cdot 3 \cdot 2 \cdot 1 = 8 \cdot 7! = 8 \cdot 7 \cdot 6! = 8 \cdot 7 \cdot 6 \cdot 5!,$$

and so on.

EXAMPLE 3

Simplify: $\dfrac{8!}{5! \, 3!}$.

SOLUTION

$$\frac{8!}{5! \, 3!} = \frac{8 \cdot 7 \cdot 6 \cdot 5!}{5! \cdot 3 \cdot 2 \cdot 1} = 8 \cdot 7 \qquad \text{Removing a factor equal to 1: } \frac{6 \cdot 5!}{5! \cdot 3 \cdot 2} = 1$$
$$= 56.$$

The following notation is used in our second formulation of the binomial theorem.

$\binom{n}{r}$ NOTATION

For n, r nonnegative integers with $n \geq r$,

$$\binom{n}{r}, \quad \text{read "n choose r," means} \quad \frac{n!}{(n - r)! \, r!}.$$

EXAMPLE   4

Simplify: **(a)** $\begin{pmatrix} 7 \\ 2 \end{pmatrix}$; **(b)** $\begin{pmatrix} 9 \\ 6 \end{pmatrix}$; **(c)** $\begin{pmatrix} 6 \\ 6 \end{pmatrix}$.

SOLUTION

a) $\begin{pmatrix} 7 \\ 2 \end{pmatrix} = \dfrac{7!}{(7-2)!\,2!}$

$= \dfrac{7!}{5!\,2!} = \dfrac{7 \cdot 6 \cdot 5!}{5! \cdot 2 \cdot 1} = \dfrac{7 \cdot 6}{2}$

$= 7 \cdot 3$

$= 21$

b) $\begin{pmatrix} 9 \\ 6 \end{pmatrix} = \dfrac{9!}{3!\,6!}$

$= \dfrac{9 \cdot 8 \cdot 7 \cdot 6!}{3 \cdot 2 \cdot 1 \cdot 6!} = \dfrac{9 \cdot 8 \cdot 7}{3 \cdot 2}$

$= 3 \cdot 4 \cdot 7$

$= 84$

c) $\begin{pmatrix} 6 \\ 6 \end{pmatrix} = \dfrac{6!}{0!\,6!} = \dfrac{6!}{1 \cdot 6!}$ Since $0! = 1$

$= \dfrac{6!}{6!}$

$= 1$

Now we can restate the binomial theorem using our new notation.

THE BINOMIAL THEOREM (FORM 2)

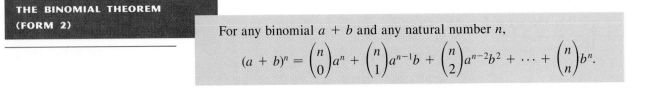

For any binomial $a + b$ and any natural number n,

$$(a + b)^n = \begin{pmatrix} n \\ 0 \end{pmatrix} a^n + \begin{pmatrix} n \\ 1 \end{pmatrix} a^{n-1}b + \begin{pmatrix} n \\ 2 \end{pmatrix} a^{n-2}b^2 + \cdots + \begin{pmatrix} n \\ n \end{pmatrix} b^n.$$

EXAMPLE   5

Expand: $(3x + y)^4$.

SOLUTION   We use the binomial theorem (Form 2) with $a = 3x$, $b = y$, and $n = 4$:

$$(3x + y)^4 = \begin{pmatrix} 4 \\ 0 \end{pmatrix}(3x)^4 + \begin{pmatrix} 4 \\ 1 \end{pmatrix}(3x)^3 y + \begin{pmatrix} 4 \\ 2 \end{pmatrix}(3x)^2 y^2 + \begin{pmatrix} 4 \\ 3 \end{pmatrix}(3x)y^3 + \begin{pmatrix} 4 \\ 4 \end{pmatrix} y^4$$

$$= \frac{4!}{4!\,0!} 3^4 x^4 + \frac{4!}{3!\,1!} 3^3 x^3 y + \frac{4!}{2!\,2!} 3^2 x^2 y^2 + \frac{4!}{1!\,3!} 3xy^3 + \frac{4!}{0!\,4!} y^4$$

$$= 81x^4 + 108x^3 y + 54x^2 y^2 + 12xy^3 + y^4. \qquad \text{Simplifying}$$

EXAMPLE 6   Expand: $(x^2 - 2y)^5$.

SOLUTION   In this case, $a = x^2$, $b = -2y$, and $n = 5$:

$$(x^2 - 2y)^5 = \binom{5}{0}(x^2)^5 + \binom{5}{1}(x^2)^4(-2y) + \binom{5}{2}(x^2)^3(-2y)^2$$

$$+ \binom{5}{3}(x^2)^2(-2y)^3 + \binom{5}{4}(x^2)(-2y)^4 + \binom{5}{5}(-2y)^5$$

$$= \frac{5!}{5!\,0!}x^{10} + \frac{5!}{4!\,1!}x^8(-2y) + \frac{5!}{3!\,2!}x^6(-2y)^2 + \frac{5!}{2!\,3!}x^4(-2y)^3$$

$$+ \frac{5!}{1!\,4!}x^2(-2y)^4 + \frac{5!}{0!\,5!}(-2y)^5$$

$$= x^{10} - 10x^8y + 40x^6y^2 - 80x^4y^3 + 80x^2y^4 - 32y^5.$$

Note that in the binomial theorem (Form 2), $\binom{n}{0}a^n b^0$ gives us the first term, $\binom{n}{1}a^{n-1}b^1$ gives us the second term, $\binom{n}{2}a^{n-2}b^2$ gives us the third term, and so on. This can be generalized to give a method for finding a specific term without writing the entire expansion.

FINDING A SPECIFIC TERM

The $(r + 1)$st term of $(a + b)^n$ is

$$\binom{n}{r}a^{n-r}b^r.$$

EXAMPLE 7   Find the 5th term in the expansion of $(2x - 3y)^7$.

SOLUTION   First, we note that $5 = 4 + 1$. Thus, $r = 4$, $a = 2x$, $b = -3y$, and $n = 7$. Then the 5th term of the expansion is

$$\binom{7}{4}(2x)^{7-4}(-3y)^4, \quad \text{or} \quad \frac{7!}{3!\,4!}(2x)^3(-3y)^4, \quad \text{or} \quad 22{,}680x^3y^4.$$

It is because of the binomial theorem that $\binom{n}{r}$ is called a *binomial coefficient*. We can now explain why 0! is defined to be 1. In the binomial expansion, we want $\binom{n}{0}$ to equal 1 and we also want the definition

$$\binom{n}{r} = \frac{n!}{(n-r)!\,r!}$$

to hold for all whole numbers n and r. Thus we must have

$$\binom{n}{0} = \frac{n!}{(n-0)!\,0!} = \frac{n!}{n!\,0!} = 1.$$

This is satisfied only if 0! is defined to be 1.

EXERCISE SET

11.4

Simplify.

1. 8!

2. 9!

3. 10!

4. 11!

5. $\dfrac{7!}{4!}$

6. $\dfrac{8!}{6!}$

7. $\dfrac{10!}{7!}$

8. $\dfrac{9!}{5!}$

9. $\dbinom{8}{2}$

10. $\dbinom{7}{4}$

11. $\dbinom{10}{6}$

12. $\dbinom{9}{5}$

13. $\dbinom{20}{18}$

14. $\dbinom{30}{3}$

15. $\dbinom{35}{2}$

16. $\dbinom{40}{38}$

Expand. Use both of the methods shown in this section.

17. $(m + n)^5$

18. $(a - b)^4$

19. $(x - y)^6$

20. $(p + q)^7$

21. $(x^2 - 3y)^5$

22. $(3c - d)^7$

23. $(3c - d)^6$

24. $(t^{-2} + 2)^6$

25. $(x - y)^3$

26. $(x - y)^5$

27. $\left(x + \dfrac{2}{y}\right)^9$

28. $\left(3s + \dfrac{1}{t}\right)^9$

29. $(a^2 - b^3)^5$

30. $(x^3 - 2y)^5$

31. $(\sqrt{3} - t)^4$

32. $(\sqrt{5} + t)^6$

33. $(x^{-2} + x^2)^4$

34. $\left(\dfrac{1}{\sqrt{x}} - \sqrt{x}\right)^6$

Find the indicated term for each binomial expression.

35. 3rd, $(a + b)^6$

36. 6th, $(x + y)^7$

37. 12th, $(a - 3)^{14}$

38. 11th, $(x - 2)^{12}$

39. 5th, $(2x^3 + \sqrt{y})^8$

40. 4th, $\left(\dfrac{1}{b^2} + c\right)^7$

41. Middle, $(2u - 3v^2)^{10}$

42. Middle two, $(\sqrt{x} + \sqrt{3})^5$

SKILL MAINTENANCE

Solve.

43. $\log_2 x + \log_2(x - 2) = 3$

44. $\log_3(x + 2) - \log_3(x - 2) = 2$

45. $e^t = 280$

46. $\log_5 x^2 = 2$

SYNTHESIS

47. ◈ Devise two problems requiring the use of the binomial theorem. Design the problems so that one is solved more easily using Form 1 and the other is solved more easily using Form 2. Then explain what makes one form easier to use than the other in each case.

48. Show that there are exactly $\dbinom{5}{3}$ ways of forming a subset of size 3 from a set of 5 elements.

49. ▦ *Baseball.* At one point in a recent season, Mike Piazza of the Los Angeles Dodgers had a batting average of 0.313. Suppose he came to bat 5 times in a game. The probability of his getting exactly 3 hits is the 3rd term of the binomial expansion of $(0.313 + 0.687)^5$. Find that term and use a calculator to estimate the probability.

50. ▦ *Widows or divorcees.* The probability that a woman will be either widowed or divorced is 85%. Suppose 8 women are interviewed. The probability that exactly 5 of them will be either widowed or divorced in their lifetime is the 6th term of the binomial expansion of $(0.15 + 0.85)^8$. Find that term and use a calculator to estimate the probability.

51. ▦ *Baseball.* In reference to Exercise 49, the probability that Piazza will get *at most* 3 hits is found by adding the last 4 terms of the binomial expansion of $(0.313 + 0.687)^5$. Find these terms and use a calculator to estimate the probability.

52. ▦ *Widows or divorcees.* In reference to Exercise 50, the probability that *at least* 6 of the women will be widowed or divorced is found by adding the last three terms of the binomial expansion of $(0.15 + 0.85)^8$. Find these terms and use a calculator to estimate the probability.

53. Prove that

$$\binom{n}{r} = \binom{n}{n-r}$$

for any whole numbers n and r. Assume $r \le n$.

54. Find the term of

$$\left(\frac{3x^2}{2} - \frac{1}{3x}\right)^{12}$$

that does not contain x.

55. Find the middle term of $(x^2 - 6y^{3/2})^6$.

56. Find the ratio of the 4th term of

$$\left(p^2 - \frac{1}{2}p\sqrt[3]{q}\right)^5$$

to the 3rd term.

57. Find the term containing $\dfrac{1}{x^{1/6}}$ of

$$\left(\sqrt[3]{x} - \frac{1}{\sqrt{x}}\right)^7.$$

58. What is the degree of $(x^2 + 3)^4$?

SUMMARY AND REVIEW 11

KEY TERMS

IMPORTANT PROPERTIES AND FORMULAS

Arithmetic sequence: $\qquad a_{n+1} = a_n + d$

nth term of an arithmetic sequence: $\qquad a_n = a_1 + (n-1)d$

Sum of the first n terms of an arithmetic sequence: $\qquad S_n = \dfrac{n}{2}(a_1 + a_n)$

Geometric sequence: $\qquad a_{n+1} = a_n \cdot r$

nth term of a geometric sequence: $\quad a_n = a_1 r^{n-1}$

Sum of the first n terms of a geometric sequence: $\quad S_n = \dfrac{a_1(1 - r^n)}{1 - r}$

Limit of an infinite geometric series: $\quad S_\infty = \dfrac{a_1}{1 - r}, \quad |r| < 1$

Factorial notation: $\quad n! = n(n - 1)(n - 2) \cdots 3 \cdot 2 \cdot 1$

Binomial coefficient: $\quad \dbinom{n}{r} = \dfrac{n!}{(n - r)!\, r!}$

Binomial theorem: $\quad (a + b)^n = \dbinom{n}{0}a^n + \dbinom{n}{1}a^{n-1}b + \dbinom{n}{2}a^{n-2}b^2 + \cdots + \dbinom{n}{n}b^n$

$(r + 1)$st term of $(a + b)^n$: $\quad \dbinom{n}{r}a^{n-r}b^r$

REVIEW EXERCISES

Find the first four terms; the 8th term, a_8; and the 12th term, a_{12}.

1. $a_n = 4n - 3$

2. $a_n = \dfrac{n - 1}{n^2 + 1}$

Predict the general term. Answers may vary.

3. $-2, -4, -6, -8, -10, \ldots$

4. $1, 4, 9, 16, 25, \ldots$

Rename and evaluate each sum.

5. $\displaystyle\sum_{k=1}^{5} (-2)^k$

6. $\displaystyle\sum_{k=2}^{7} (1 - 2k)$

Rewrite using sigma notation.

7. $4 + 8 + 12 + 16 + 20$

8. $\dfrac{-1}{2} + \dfrac{1}{4} + \dfrac{-1}{8} + \dfrac{1}{16} + \dfrac{-1}{32}$

9. Find the 14th term of the arithmetic sequence $-6, 1, 8, \ldots$.

10. Find d when $a_1 = 11$ and $a_{10} = 35$. Assume an arithmetic sequence.

11. Find a_1 and d when $a_{12} = 25$ and $a_{24} = 40$. Assume an arithmetic sequence.

12. Find the sum of the first 17 terms of the arithmetic series $-8 + (-11) + (-14) + \cdots$.

13. Find the sum of all the multiples of 6 from 12 to 318, inclusive.

14. Find the 20th term of the geometric sequence $2, 2\sqrt{2}, 4, \ldots$.

15. Find the common ratio of the geometric sequence $2, \dfrac{4}{3}, \dfrac{8}{9}, \ldots$.

16. Find the nth term of the geometric sequence $-2, 2, -2, \ldots$.

17. Find the nth term of the geometric sequence $3, \dfrac{3}{4}x, \dfrac{3}{16}x^2, \ldots$.

18. Find S_6 for the geometric series
$$3 + 12 + 48 + \cdots.$$

19. Find S_{12} for the geometric series
$$3x - 6x + 12x - \cdots.$$

Determine whether each infinite geometric series has a limit. If a limit exists, find it.

20. $6 + 3 + 1.5 + 0.75 + \cdots$

21. $7 - 4 + \dfrac{16}{7} - \cdots$

22. $2 + (-2) + 2 + (-2) + \cdots$

23. $0.04 + 0.08 + 0.16 + 0.32 + \cdots$

24. $\$2000 + \$1900 + \$1805 + \$1714.75 + \cdots$

25. Find fractional notation for $0.555555\ldots$.

26. Find fractional notation for $1.39393939\ldots$.

Solve.

27. You take a job, starting with an hourly wage of $\$11.40$. You are promised a raise of 20¢ per hour every 3 mos for 8 yr. At the end of 8 yr, what will be your hourly wage?

28. A stack of logs has 42 poles in the bottom row. There are 41 poles in the second row, 40 poles in the third row, and so on, ending with 1 pole in the top row. How many poles are in the stack?

29. A student loan is in the amount of $10,000. Interest is 7%, compounded annually, and the amount is to be paid off in 12 yr. How much is to be paid back?

30. Find the total rebound distance of a ball, given that it is dropped from a height of 12 m and each rebound is one third of the preceding one.

Simplify.

31. 8!

32. $\binom{8}{3}$

33. Find the 3rd term of $(a + b)^{20}$.

34. Expand: $(x - 2y)^4$.

SKILL MAINTENANCE

Solve.

35. $3x - y = 7,$
$2x + 3y = 5$

36. $\log (x + 5) - \log x = 1$

37. Simplify: $7^{\log_7 13}$.

38. Find an equation of the circle with center $(3, -1)$ and radius $\sqrt{3}$.

SYNTHESIS

39. ◈ What happens to the terms of a geometric sequence with $|r| < 1$ as n gets larger? Why?

40. ◈ Compare the two forms of the binomial theorem given in the text. Under what circumstances would one be more useful than the other?

41. Find the sum of the first n terms of the geometric series $1 - x + x^2 - x^3 + \cdots$.

42. Expand: $(x^{-3} + x^3)^5$.

CUMULATIVE REVIEW
1-11

1. Find the first five terms and the 16th term of a sequence with general term $a_n = 6n - 5$.

2. Predict the general term of the sequence
$$\frac{4}{3}, \frac{4}{9}, \frac{4}{27}, \ldots .$$

3. Rename and evaluate:
$$\sum_{k=1}^{5} (3 - 2^k).$$

4. Rewrite using sigma notation:
$$1 + 8 + 27 + 64 + 125.$$

5. Find the 12th term, a_{12}, of the arithmetic sequence $9, 4, -1, \ldots .$

Assume arithmetic sequences for Questions 6 and 7.

6. Find the common difference d when $a_1 = 9$ and $a_7 = 11\frac{1}{4}$.

7. Find a_1 and d when $a_5 = 16$ and $a_{10} = -3$.

8. Find the sum of all the multiples of 12 from 24 to 240, inclusive.

9. Find the 6th term of the geometric sequence $72, 18, 4\frac{1}{2}, \ldots .$

10. Find the common ratio of the geometric sequence $22\frac{1}{2}, 15, 10, \ldots .$

11. Find the nth term of the geometric sequence $3, -9, 27, \ldots .$

12. Find the sum of the first nine terms of the geometric series
$$(1 + x) + (2 + 2x) + (4 + 4x) + \cdots .$$

Determine whether each infinite geometric series has a limit. If a limit exists, find it.

13. $0.5 + 0.25 + 0.125 + \cdots$

14. $0.5 + 1 + 2 + 4 + \cdots$

15. $\$1000 + \$80 + \$6.40 + \cdots$

16. Find fractional notation for $0.85858585 \ldots .$

17. An auditorium has 31 seats in the first row, 33 seats in the second row, 35 seats in the third row, and so on, for 18 rows. How many seats are in the 17th row?

18. A stack of poles has 52 poles in the bottom row. There are 51 poles in the second row, 50 poles in the third row, and so on, ending with 1 pole in the top row. How many poles are in the stack?

19. Each week the price of a $15,000 boat will be reduced 5% of the previous week's price. If we assume that it is not sold, what will be the price after 10 weeks?

20. Find the total rebound distance of a ball that is dropped from a height of 18 m, with each rebound two thirds of the preceding one.

21. Simplify: $\begin{pmatrix} 13 \\ 11 \end{pmatrix}$.

22. Expand: $(x^2 - 3y)^5$.

23. Find the 4th term in the expansion of $(a + x)^{12}$.

SKILL MAINTENANCE

24. Solve: $4^{2x-3} = 64$.

25. Find an equation of the circle with center $(1, -2)$ and radius $3\sqrt{3}$.

26. Solve:
$$y = 3x + 5,$$
$$2x + 5y = 8.$$

27. Convert to an exponential equation: $\log x = 1.5$.

SYNTHESIS

28. Find a formula for the sum of the first n even natural numbers:
$$2 + 4 + 6 + \cdots + 2n.$$

29. Find the sum of the first n terms of
$$1 + \frac{1}{x} + \frac{1}{x^2} + \frac{1}{x^3} + \cdots.$$

CUMULATIVE REVIEW
1–11

Simplify.

1. $(-9x^2y^3)(5x^4y^{-7})$

2. $|-3.5 + 9.8|$

3. $2y - [3 - 4(5 - 2y) - 3y]$

4. $(10 \cdot 8 - 9 \cdot 7)^2 - 54 \div 9 - 3$

5. Evaluate
$$\frac{ab - ac}{bc}$$
for $a = -2$, $b = 3$, and $c = -4$.

Perform the indicated operations and simplify.

6. $(5a^2 - 3ab - 7b^2) - (2a^2 + 5ab + 8b^2)$

7. $(-3x^2 + 4x^3 - 5x - 1) + (9x^3 - 4x^2 + 7 - x)$

8. $(2a - 1)(3a + 5)$

9. $(3a^2 - 5y)^2$

10. $\dfrac{1}{x - 2} - \dfrac{4}{x^2 - 4} + \dfrac{3}{x + 2}$

11. $\dfrac{x^2 - 6x + 8}{3x + 9} \cdot \dfrac{x + 3}{x^2 - 4}$

12. $\dfrac{3x + 3y}{5x - 5y} \div \dfrac{3x^2 + 3y^2}{5x^3 - 5y^3}$

13. $\dfrac{x - \dfrac{a^2}{x}}{1 + \dfrac{a}{x}}$

Factor.

14. $4x^2 - 12x + 9$

15. $27a^3 - 8$

16. $a^3 + 3a^2 - ab - 3b$

17. $15y^4 + 33y^2 - 36$

18. For the function described by
$$f(x) = 3x^2 - 4x,$$
find $f(-2)$.

19. Divide:
$$(7x^4 - 5x^3 + x^2 - 4) \div (x - 2).$$

Solve.

20. $9(x - 1) - 3(x - 2) = 1$

21. $\dfrac{6}{x} + \dfrac{6}{x + 2} = \dfrac{5}{2}$

22. $2x + 1 > 5 \ or \ x - 7 \leqslant 3$

23. $5x + 3y = 2,$
$3x + 5y = -2$

24. $x + y - z = 0,$
$3x + y + z = 6,$
$x - y + 2z = 5$

25. $3\sqrt{x - 1} = 5 - x$

26. $x^4 - 29x^2 + 100 = 0$

27. $x^2 + y^2 = 8,$
$x^2 - y^2 = 2$

28. ▦ $5^x = 8$

29. $\log (x^2 - 25) - \log (x + 5) = 3$

30. $\log_4 x = -2$

31. $7^{2x+3} = 49$

32. $|2x - 1| \leqslant 5$

33. $7x^2 + 14 = 0$

34. $x^2 + 4x = 3$

35. $y^2 + 3y > 10$

36. Let $f(x) = x^2 - 2x$. Find a such that $f(a) = 48$.

Solve.

37. The perimeter of a rectangle is 34 ft. The length of a diagonal is 13 ft. Find the dimensions of the rectangle.

38. A music club offers two types of membership. Limited members pay a fee of $10 a year and can buy CDs for $10 each. Preferred members pay $20 a year and can buy CDs for $7.50 each. For what numbers of annual CD purchases would it be less expensive to be a preferred member?

39. Find three consecutive integers whose sum is 198.

40. A pentagon with all five sides the same size has a perimeter equal to that of an octagon in which all eight sides are the same size. One side of the pentagon is 2 less than 3 times one side of the octagon. What is the perimeter of each figure?

41. A chemist has two solutions of ammonia and water. Solution A is 6% ammonia and solution B is 2% ammonia. How many liters of each solution are needed in order to obtain 80 L of a solution that is 3.2% ammonia?

42. An airplane can fly 190 mi with the wind in the same time it takes to fly 160 mi against the wind. The speed of the wind is 30 mph. How fast can the plane fly in still air?

43. Bianca can do a certain job in 21 min. Dahlia can do the same job in 14 min. How long would it take to do the job if the two worked together?

44. The centripetal force F of an object moving in a circle varies directly as the square of the velocity v and inversely as the radius r of the circle. If $F = 8$ when $v = 1$ and $r = 10$, what is F when $v = 2$ and $r = 16$?

45. A farmer wants to fence in a rectangular area next to a river. (Note that no fence will be needed along the river.) What is the area of the largest region that can be fenced in with 100 ft of fencing?

Graph.

46. $3x - y = 6$

47. $\dfrac{x^2}{25} + \dfrac{y^2}{4} = 1$

48. $y = \log_2 x$

49. $2x - 3y < -6$

50. Graph: $f(x) = -2(x - 3)^2 + 1$.

   a) Label the vertex.
   b) Draw the axis of symmetry.
   c) Find the maximum or minimum value.

51. Solve $V = P - Prt$ for r.

52. Solve $I = \dfrac{R}{R + r}$ for R.

53. Find a linear equation whose graph contains the point $(-1, 4)$ and is perpendicular to the line whose equation is $3x - y = 6$.

Find the domain of each function.

54. $f(x) = \sqrt{5 - 3x}$

55. $g(x) = \dfrac{x - 4}{x^2 - 2x + 1}$

56. Multiply $(8.9 \times 10^{-17})(7.6 \times 10^4)$. Write scientific notation for the answer.

57. Multiply and simplify: $\sqrt{8x}\,\sqrt{8x^3y}$.

58. Simplify: $(25x^{4/3}y^{1/2})^{3/2}$.

59. Divide and simplify:
$$\dfrac{\sqrt[3]{15x}}{\sqrt[3]{3y^2}}.$$

60. Rationalize the denominator:
$$\dfrac{1 - \sqrt{x}}{1 + \sqrt{x}}.$$

61. Write a single radical expression:
$$\dfrac{\sqrt[3]{(x + 1)^5}}{\sqrt{(x + 1)^3}}.$$

62. Multiply these complex numbers:
$$(3 + 2i)(4 - 7i).$$

63. Write a quadratic equation whose solutions are $5\sqrt{2}$ and $-5\sqrt{2}$.

64. Find the center and the radius of the circle
$$x^2 + y^2 - 4x + 6y - 23 = 0.$$

65. Express as a single logarithm:
$$\tfrac{2}{3}\log_a x - \tfrac{1}{2}\log_a y + 5\log_a z.$$

66. Convert to an exponential equation: $\log_a c = 5$.

Find each of the following using a calculator or table.

67. $\log 5677.2$

68. $10^{-3.587}$

69. $\ln 5677.2$

70. $e^{-3.587}$

Calling cards. *The sale of prepaid long-distance telephone cards brought in $75 million in 1993 and $800 million in 1995 (Source: Associated Press article in the* Indianapolis Star, *7/14/96). Assume that the business is growing exponentially.*

71. Write an exponential function describing the sales of the cards t years after 1993.

72. Predict the amount of sales in the year 2000.

73. Find the distance between the points $(-1, -5)$ and $(2, -1)$.

74. Find the 21st term of the arithmetic sequence $19, 12, 5, \ldots$.

75. Find the sum of the first 25 terms of the arithmetic series $-1 + 2 + 5 + \cdots$.

76. Find the general term of the geometric sequence $16, 4, 1, \ldots$.

77. Find the 7th term of $(a - 2b)^{10}$.

78. Find the sum of the first nine terms of the geometric series $x + 1.5x + 2.25x + \cdots$.

79. On Mark's 9th birthday, his grandmother opened a savings account for him with $100. The account draws 6% interest, compounded annually. If Mark neither adds to nor withdraws any money from the bank, how much will be in the account on his 18th birthday?

SYNTHESIS

Solve.

80. $\dfrac{9}{x} - \dfrac{9}{x+12} = \dfrac{108}{x^2+12x}$

81. $\log_2 (\log_3 x) = 2$

82. y varies directly as the cube of x and x is multiplied by 0.5. What is the effect on y?

83. Divide these complex numbers:
$$\dfrac{2\sqrt{6} + 4\sqrt{5}i}{2\sqrt{6} - 4\sqrt{5}i}.$$

84. Diaphantos, a famous mathematician, spent $\frac{1}{6}$ of his life as a child, $\frac{1}{12}$ as an adolescent, and $\frac{1}{7}$ as a bachelor. Five years after he was married, he had a son who died 4 years before his father at half his father's final age. How long did Diaphantos live?

The Graphing Calculator

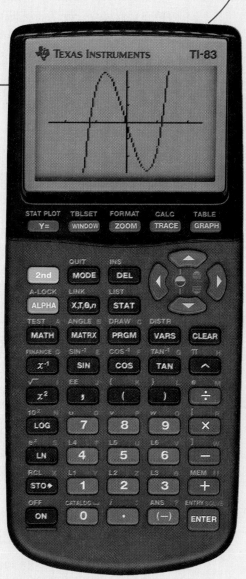

*G*raphing calculators and computer graphing software, often referred to collectively as graphers, *make possible the quick graphing of equations. This appendix discusses the graphing calculator. Sections I.1 and I.2 introduce the calculator and the other section numbers in this appendix correspond to section numbers in the text. Topics discussed in the text are not repeated in the appendix.*

Different calculators may require different keystrokes to use the same feature. In this appendix, exact keystrokes referenced and screens shown apply to the Texas Instruments TI-83® calculator. If your calculator does not have a key by the name shown, consult the user's manual that came with your calculator to learn how to perform the procedure.

I.1 Introduction to the Graphing Calculator

A graphing calculator does not look exactly like the scientific calculator to which you may be accustomed. The screen is much larger; a typical screen may display up to 8 lines of text and may be 16 columns wide. Screens are composed of *pixels*, or dots. After the calculator has been turned on, a blinking rectangle or line appears on the screen. This rectangle or line is called the *cursor*. It indicates your current position on the screen. In the graphing mode, the cursor may appear as a cross. The four arrow keys on the grapher's keyboard are used to move the cursor.

The *contrast* controls how dark the characters appear on the screen. Once the grapher is on, you can adjust the contrast by pressing ⟨2nd⟩ and either the up or down arrow key.

Entering and Editing Expressions

The large screen allows room for graphs and lets you view an instruction in its entirety before it is executed. Pressing the operation keys ($+$, $-$, $\times$, $\div$) does not automatically perform the operation; you must also press an Execute or ENTER key.

EXAMPLE 1

Calculate: $4 + 3(7 - 10)$.

SOLUTION Enter the entire expression as it is written. Then you can check your typing. If it is correct, pressing ENTER will perform all the operations in the correct order. The answer appears on the right side of the screen.

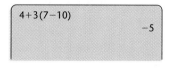

```
4+3(7−10)
                    −5
```

Thus, $4 + 3(7 - 10) = -5$.

If, before pressing ENTER , you see that your typing is not correct, use the arrow keys to move through the expression and correct any errors. Pressing the DEL key will delete the character under the cursor. To replace the character under the cursor with another character, simply press the desired key. If you left out a character, press 2nd INS . The cursor will change to a line, and characters can then be inserted to the left of the cursor.

After pressing ENTER , you can still edit, or change, the previous expression. To do so, press 2nd ENTRY to copy the previous expression on the screen. You can then change it as described above.

A graphing calculator has many more features than those written on the keys. Written above each key are secondary features, which are performed by first pressing 2nd and then the key below the desired feature. Above many keys, there is also an *alpha character.* These are accessed by first pressing ALPHA and then the key below the character. The ALPHA key may be locked on by pressing 2nd A-LOCK . Press ALPHA again to unlock the key.

Other features are available from *menus.* When certain keys are pressed, a menu, or list of further options, appears on the screen. You can select an item from the menu by pressing the number of the item or by using the cursor keys to highlight the item and then pressing ENTER . Sometimes a menu has a *secondary menu* from which to choose. For example, pressing MATH gives you the menu shown at left. The secondary menu is listed across the top of the screen.

```
MATH NUM CPX PRB
1:▶Frac
2:▶Dec
3: 3
4: ³√(
5: ˣ√
6: fMin(
7↓fMax(
```

The screen in which you perform calculations is called the *home screen.* To return to the home screen from a menu or other type of screen, press 2nd QUIT .

Functions and Grouping Symbols

A grapher will perform the functions of a scientific calculator, although the exact keystrokes may be different. For example, on most scientific calculators, the square root of a number is found by entering the number first and then pressing the square root key. On most graphers, the square root is found by first pressing the square root key, then entering the number, and then pressing the ENTER key. You will find that most operations can be entered in the order in which you would write them on paper, except that you may need to add parentheses. For example, a fraction bar is entered as a division symbol, /, and does not serve as a grouping symbol, so numerators and denominators may need to be enclosed in parentheses. It is important to remember that the subtraction symbol and the negative symbol are two different keys. To enter a negative number, we press $(-)$ and to subtract, we press $-$.

EXAMPLE 2

Calculate:

$$\sqrt{3 \cdot 5 - 3(6 - 4)} \quad \text{and} \quad \frac{4 + 1.5}{3.2 - 1}.$$

SOLUTION   The keystrokes and the results of the calculations can be seen in the screen at left.

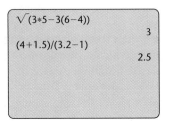

$\sqrt{(3*5-3(6-4))}$
                    3
$(4+1.5)/(3.2-1)$
                    2.5

Other functions, like absolute value (ABS), work in a similar way.

Any number can be squared by typing the number, pressing the x^2 key, and then pressing ENTER . You may be able to cube an expression in a similar way (see the MATH menu pictured on page 659). In general, for exponents other than 2, use the $\wedge$ key, followed by the exponent. If the exponent is itself an expression, enclose it within parentheses. An exponent that is a single negative number does not need to be enclosed in parentheses. But a rational exponent, in fractional notation, must be in parentheses. (For more on rational exponents, see Chapter 7.)

EXAMPLE 3

Calculate: 1.4^2, $(3 + 5)^{-2}$, and 7^{8-1}.

SOLUTION   The results of the calculations are shown in the screen below.

1.4^2
                    1.96
$(3+5)^{-2}$
                    .015625
$7^{(8-1)}$
                    823543

EXERCISE SET

I.1

Short exercise sets appear throughout this appendix.
These provide an opportunity to practice the skills just
discussed. Answers can be found at the end of the
appendix.

Evaluate each expression using a graphing calculator.

1. $\sqrt{-7 + 2(10 - 2)}$

2. $|-9|$

3. $|9 - (5 + (-3(6.4 - 19)))|$

4. $\dfrac{-4 - 3|13 - 5.3| + \sqrt{2.25}}{5.6 \div 7 - 55 \div 10}$

5. $(3.4 - 5.6(7.3 - 8.79))^2$

6. $2.6^{2(3-1)}$

7. $\left(\dfrac{3}{4}\right)^{-2}$

I.2   Variables and Functions

Graphers have a large amount of memory available in which to store values of variables, equations of functions, and programs. (A *program* is a series of instructions that can be performed repeatedly. Consult your user's manual for information about storing and running programs.)

To store a numerical value for later use, type the number, press $\boxed{\text{STO▸}}$ and then type the name of the variable using an alpha character. To recall that value, press $\boxed{\text{2nd}}$ $\boxed{\text{RCL}}$ and the variable name or simply type the variable name to recall its value. This can be done in the middle of a calculation.

To store the equation of a function, press $\boxed{\text{Y=}}$. You can enter an equation the same way in which you would write it on paper. Always let the variable be x. You can enter an x using the X, T, Θ, n key.

EXAMPLE 1

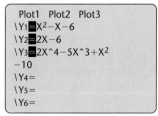

```
Plot1  Plot2  Plot3
\Y1■X²−X−6
\Y2■2X−6
\Y3■2X^4−5X^3+X²
−10
\Y4=
\Y5=
\Y6=
```

Enter the equations

$$y_1 = x^2 - x - 6, \qquad y_2 = 2x - 6, \quad \text{and} \quad y_3 = 2x^4 - 5x^3 + x^2 - 10.$$

SOLUTION   Press $\boxed{\text{Y=}}$ $\boxed{\text{X, T, } \theta, n}$ $\boxed{x^2}$ $\boxed{-}$ $\boxed{\text{X, T, } \theta, n}$ $\boxed{-}$ $\boxed{6}$ $\boxed{\text{ENTER}}$ to enter the first equation. You will now be on the line for Y2 and can press $\boxed{2}$ $\boxed{\text{X, T, } \theta, n}$ $\boxed{-}$ $\boxed{6}$ $\boxed{\text{ENTER}}$. Enter y_3 in a similar way. Your screen should look like the one at left.

Remember: After you have worked on a screen other than the home screen, press $\boxed{\text{2nd}}$ $\boxed{\text{QUIT}}$ to return to the home screen. ⟶●

Once entered, a function is available for graphing or solving. Graphing functions and solving equations are discussed throughout the text and in the remainder of this appendix.

To find the function value for a particular value of x, you can use function notation or store the value to the variable X. Shown on the following page are the keystrokes for each method of finding $f(2)$, where $f(x) = x^2 - x - 6$, which was y_1 in Example 1. (On the TI-82®, 2nd Y-VARS should be used in place of VARS .)

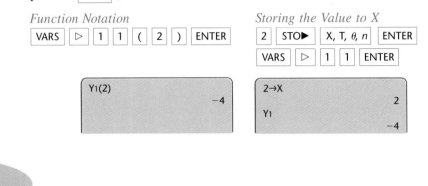

Function Notation

VARS ▷ 1 1 (2) ENTER

Storing the Value to X

2 STO▶ X, T, θ, n ENTER

VARS ▷ 1 1 ENTER

Y₁(2)
                                    −4

2→X
                                     2
Y₁
                                    −4

EXERCISE SET
1.2

1. Find the circumference and area of a circle with radius 10.3452 cm using the following steps.

   a) Store the value 10.3452 to the variable R.
   b) Calculate the circumference using the formula $C = 2\pi R$. Use the π key and the stored value for R. (You can type simply $2\pi R$ and press ENTER .) Round to four decimal places.
   c) Calculate the area using the formula $A = \pi R^2$. Round to four decimal places.

2. The formula $I = PRT$ gives the amount of simple interest earned on a principal P at an interest rate R (in decimal form) for a length of time T (in years).

You have \$1452.39 to invest for 7 months. Store the values for P and T and use them to calculate the interest earned when the interest rate is each of the following.

a) 10%   b) 6%
c) 4.375%   d) 8.65%

Evaluate each polynomial for the given values.

3. $y = 3x^4 - 2x^3 + 9x - 17$, for $x = 4, -1, 36$, and 98.
4. $y = 1.5x^6 - 2.8x^4 + 0.1x^2$, for $x = 1.23, -1.23, 107$, and -107.

2.1   Graphs

When you draw a graph by hand, you first draw a set of axes and label them with an appropriate scale. A grapher will do this as well. The part of the graph shown by the grapher is in a *window*. The window is described by four numbers: the minimum x-value shown (Xmin), the maximum x-value (Xmax), the minimum y-value (Ymin), and the maximum y-value (Ymax). These dimensions are often abbreviated [L, R, B, T], or Left and Right endpoints of the x-axis and Bottom and Top endpoints of the y-axis. The number of markings, or tick marks, on each axis is determined by setting the x-scale (Xscl) and the y-scale (Yscl). The *scale* is the distance between the tick marks. The graphs in this appendix are labeled to indicate the size of the viewing window used. If the Xscl or Yscl is not 1, it is also indicated.

Choosing appropriate dimensions for a viewing window is important. The window dimensions should be big enough to include important features of the graph. However, if the dimensions are too large, it may be difficult to see the curvature of the graph.

E X A M P L E   1

Graph $y = x^3 + 3x^2 - x + 1$ using the windows

$[0, 10, 0, 10]$, $[-100, 100, -100, 100]$,   and $[-10, 10, -10, 10]$.

S O L U T I O N   First enter the equation using the Y= key and then set the window dimensions using the WINDOW key. You can choose the Xscl and Yscl. A good choice of scale might be 1 for the first and last graphs and 10 for the second graph. Graph the equation by pressing GRAPH after setting each window.

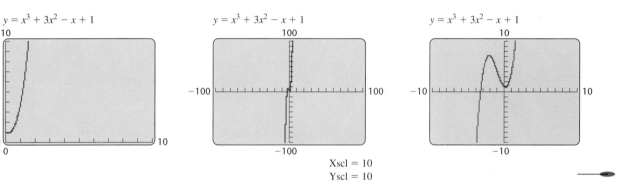

Xscl = 10
Yscl = 10

Which viewing window is the best? The answer depends on what you need to see from the graph. In Example 1, if you wanted to see where the graph crossed the x-axis, the first window would not show that point, and you could not be sure from the second window that it crossed the x-axis only once. The third window would be the best choice for determining where the x-axis is crossed.

On most graphers, two or more equations can be graphed one after the other (*sequentially*) or at the same time (*simultaneously*). If equations are graphed simultaneously, all y-values are calculated and plotted for one x-value, then all y-values are calculated for the next x-value, and so on. Often a grapher can be set to the SEQUENTIAL or SIMULTANEOUS mode using the MODE key. To *select* the equation(s) that is graphed, locate the cursor on the equals sign of the equation you wish to graph, and press ENTER .

EXERCISE SET

2.1

Graph each equation or pair of equations using the given viewing windows. Use an Xscl and a Yscl of 1 unless otherwise indicated.

1. $y = 2x - 5$

  a) $[-10, 10, -10, 10]$
  b) $[0, 10, 0, 10]$

2. $y = 2/x$

  a) $[-100, 100, -100, 100]$
        Xscl = 10   Yscl = 10
  b) $[-10, 10, -10, 10]$

3. $y_1 = 3x - 2, y_2 = x^2 + x + 1$
       $[-10, 10, -10, 10]$

2.4 Another Look at Linear Graphs

Most graphers can graph only functions; they cannot graph a vertical line using its equation. The DRAW menu allows you to draw a vertical line on a graph.

EXAMPLE 1

Graph the equations $y = 5$ and $x = -7$.

SOLUTION   The graph of $y = 5$ is a horizontal line. One way to graph it on a grapher is to enter $y_1 = 5$ and press $\boxed{\text{GRAPH}}$. The graph of $x = -7$ cannot be graphed in this manner. Both equations can be graphed using options in the DRAW menu. For example, to graph $y = 5$, select option 3: Horizontal. When the word "Horizontal" appears on the screen, type 5 and press $\boxed{\text{ENTER}}$. To graph $x = -7$ from the home screen, press $\boxed{\text{2nd}}$ $\boxed{\text{DRAW}}$ $\boxed{4}$ $\boxed{(-)}$ $\boxed{7}$.

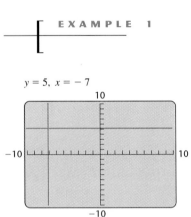

$y = 5,\ x = -7$

The graph of an equation can be used to visually check what type of equation it represents. If the graph is obviously not a straight line, the equation is not linear. If the graph appears to be a straight line, the equation *is* probably linear. However, a line may appear to be straight in the window used but have a different shape when viewed using a different window. Algebraic checks for linearity are more accurate than graphing.

EXAMPLE 2

Determine whether $y = 3x^2 + 5x - 1$ is linear.

SOLUTION   For the sake of illustration, the graph of the equation is shown using two different viewing windows. Notice that in the figure on the left, the graph appears to be a straight line.

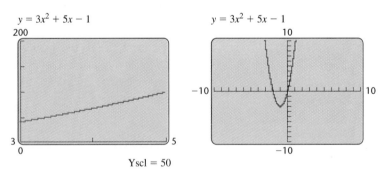

$y = 3x^2 + 5x - 1$ $y = 3x^2 + 5x - 1$

Yscl = 50

From the figure on the right, we can see that the equation is not linear. Algebraically, it is not linear because it cannot be written in the standard form of a linear equation.

1. Graph $x = 3.8$ and $y = -2.7$ on the same set of axes.
2. Determine whether each equation is linear.
   a) $y = 2.98x + 1.307$
   b) $y = |3x - 6.5|$
   c) $y = 0.002x^3$

4.4 Inequalities in Two Variables

You can use a grapher to graph a linear inequality on the TI-83 by selecting the appropriate graph style in the Y= editing screen. The graph style icon immediately precedes the Y-variable. If you position the cursor on the icon and press ENTER, the icon will rotate through seven styles. Two of the styles shade above or below the graph of the corresponding equation. There are also four different shading patterns: vertical lines, horizontal lines, positively sloped diagonal lines, and negatively sloped diagonal lines.

EXAMPLE 1

Graph: $y < 4x - 1$.

SOLUTION Enter $4x - 1$ as y_1. Then move the cursor to the icon directly to the left of y_1. The region below the line $y = 4x - 1$ is the graph of the inequality. Press ENTER until the icon showing shading below a line appears, as shown on the left below.

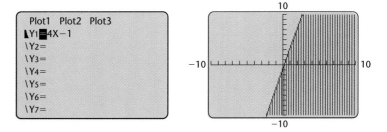

Now press GRAPH. The graph of the inequality is shown on the right above.

Because the TI-83 uses a variety of shading patterns, it is well suited to graphing systems of inequalities.

Graph the system

$$2.65x - 1.3y < 7.3,$$
$$4.8x + 2.9y < 3.78.$$

SOLUTION   Solving for y, we can write the systems as

$$y_1 > (2.65/1.3)x - (7.3/1.3)$$

and

$$y_2 < (-4.8/2.9)x + (3.78/2.9).$$

We enter the related equations, and shade above y_1 and below y_2. The grapher will use vertical lines for the first shading and horizontal lines for the second shading. The graph of the system is the intersection of the individual solution sets, or the region that is shaded by both vertical and horizontal lines.

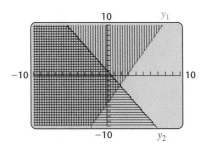

EXERCISE SET
4.4

Graph each inequality.

1. $y > 2.7x + 4$

2. $y \leq 4 - x$

3. $2.8x - 4.6 \geq 0.5y$

4. Graph the system

$$7.4x + 3.8y > 1.9,$$
$$4.3x - 1.5y < -2.8.$$

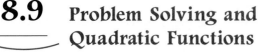

8.9   Problem Solving and Quadratic Functions

The point at which a maximum or minimum value of a quadratic function occurs can often be seen directly from its graph. To help identify this value, many graphers have a feature that calculates the maximum or minimum value of a function over a specified interval.

EXAMPLE 1

Estimate the minimum value of the function $g(x) = 1.65x^2 - 3.7x - 2.95$.

S O L U T I O N   You can see from the graph that $g(x)$ has a minimum value somewhere near $x = 1$.

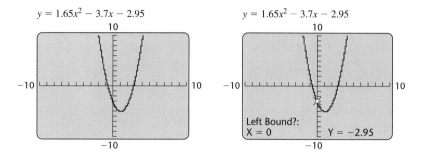

The CALC menu, which is the 2nd function of the TRACE key, lists some information that can be calculated from the graph. When we select option 3: minimum, the graph is displayed, along with the question "Left Bound?" as shown in the figure on the right above. Position the cursor on the graph somewhere to the left of the minimum and press $\boxed{\text{ENTER}}$.

Now you will be asked for the right bound. Position the cursor to the *right* of the minimum and press $\boxed{\text{ENTER}}$.

Finally, the grapher asks for a guess. Move the cursor closer to the minimum and press $\boxed{\text{ENTER}}$.

At the bottom of the graph, you can now read the coordinates of the minimum point on the graph. To the nearest hundredth, the minimum value of the function is -5.02.

Many graphers have the capability of fitting functions to data. These curve-fitting options are usually listed under the statistics, or STAT, menu. You can enter the data points, graph them, fit a function to them, and graph the function.

EXAMPLE 2

Find a quadratic equation that fits the data $(1, 2.3)$, $(2.7, 3.6)$, $(4.8, -1.5)$. Graph the points and the function.

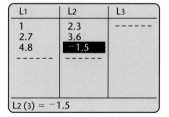

S O L U T I O N   To enter the data points, press the STAT key and look at the EDIT submenu. Choose the option ClrList to erase any previous data and then press $\boxed{\text{L}_1}$ $\boxed{,}$ $\boxed{\text{L}_2}$ $\boxed{\text{ENTER}}$ (L1 and L2 are 2nd functions of the number keys 1 and 2).

Now select the Edit option in the same menu. Under L1, enter the first coordinates of the data points by typing each number and pressing $\boxed{\text{ENTER}}$. Then move to L2 using the right arrow key, and enter the second coordinates of the data points.

To find the quadratic function $f(x) = ax^2 + bx + c$ that fits the data, choose the CALC submenu under the STAT menu, select the QuadReg option, and then press $\boxed{\text{ENTER}}$. The coefficients a, b, and c are calculated for you. Rounding these to the nearest thousandth, we find that the function is given by

$$f(x) = -0.840x^2 + 3.874x - 0.734.$$

The grapher can copy this function directly to Y1. To do so, press $\boxed{\text{Y=}}$ and clear y_1. Press $\boxed{\text{VARS}}$ $\boxed{5}$ (for statistics), and then select the EQ submenu.

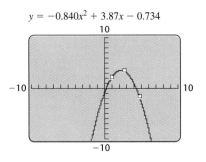

Select the RegEQ option to copy the quadratic function to y_1, as shown on the right above.

To plot the data points, press $\boxed{\text{STAT PLOT}}$ (the 2nd function of the Y= key) and turn Plot 1 on. Now press $\boxed{\text{GRAPH}}$. The points will be plotted first, then the quadratic function, as shown below.

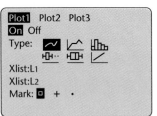

$$y = -0.840x^2 + 3.87x - 0.734$$

Find the maximum or minimum value of each function and the value of x at which it occurs. Round to the nearest hundredth.

1. $f(x) = -2.15x^2 + 3.8x - 6.9$

2. $f(x) = 8.3x^2 + 6.7x + 3.9$

3. $g(x) = 0.016 + 2.3x + 0.003x^2$

Find a quadratic function that fits the data. Graph the points and the function.

4. (2.6, 3.8), (6.9, 8.7), (8.4, 9.3)

5. (−254, 341), (16, 896), (189, 12)

A grapher can fit a quadratic function to more than three data points. The graph of this function may not go through the points, but it is considered a best fit. Find and graph the quadratic function that best fits the data.

6. (0.3, 0.1), (3.5, 5.5), (3.8, 9.7), (4.5, 8.7), (6.8, 3.2)

7. (−2.5, 9.8), (0.2, 9.7), (1.6, 4.3), (3.8, 0.1), (5.7, 3.6), (8.7, 4.9)

Appendix Answers

EXERCISE SET I.1
1. 3   **2.** 9   **3.** 33.8   **4.** 5.446808511   **5.** 137.921536
6. 45.6976   **7.** 1.777777778

EXERCISE SET I.2
1. $C = 65.0008$ cm; $A = 336.2232$ cm^2
2. (a) \$84.72; (b) \$50.83; (c) \$37.07; (d) \$73.29
3. 659; -21; 4,945,843; 274,828,929
4. -1.06329696; -1.06329696; $2.250728506 \times 10^{12}$;
$2.250728506 \times 10^{12}$

EXERCISE SET 2.1
1. (a)

$y = 2x - 5$

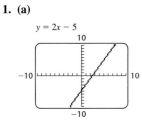

(b)

$y = 2x - 5$

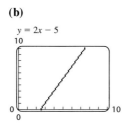

2. (a)

$y = \dfrac{2}{x}$

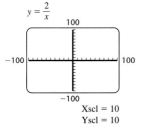

Xscl = 10
Yscl = 10

(b)

$y = \dfrac{2}{x}$

3.

$y_1 = 3x - 2; \ y_2 = x^2 + x + 1$

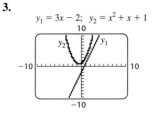

EXERCISE SET 2.4
1. $x = 3.8; \ y = -2.7$

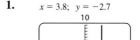

2. (a) Yes; (b) no; (c) no

EXERCISE SET 4.4
1. $y > 2.7x + 4$

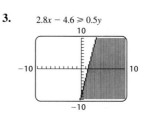

2. $y \leq 4 - x$

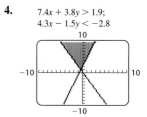

3. $2.8x - 4.6 \geq 0.5y$

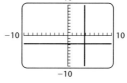

4. $7.4x + 3.8y > 1.9;$
$4.3x - 1.5y < -2.8$

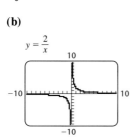

EXERCISE SET 8.9
1. Maximum of -5.22 at $x = 0.88$
2. Minimum of 2.55 at $x = -0.40$
3. Minimum of -440.82 at $x = -383.33$
4. $f(x) = -0.128x^2 + 2.351x - 1.450$

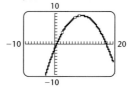

5. $f(x) = -0.016x^2 - 1.794x + 928.845$

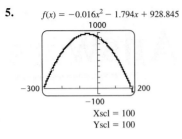

Xscl = 100
Yscl = 100

6. $f(x) = -0.590x^2 + 4.730x - 1.427$

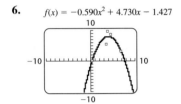

7. $f(x) = 0.143x^2 - 1.478x + 6.568$

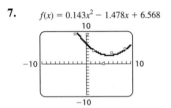

Answers

CHAPTER 1

EXERCISE SET 1.1, PP. 9–11

1. Let n represent the number; $7 + n$, or $n + 7$
3. Let t represent the number; $12t$ **5.** Let x represent the number; $0.65x$, or $\frac{65}{100}x$ **7.** Let y represent the number; $2y - 9$ **9.** Let s represent the number; $0.1s + 8$
11. Let m and n represent the numbers; $m - n - 1$
13. $90 \div 4$, or $\frac{90}{4}$ **15.** 10   **17.** 16   **19.** 23   **21.** 387
23. 5   **25.** 4   **27.** 57   **29.** 17.5 sq ft   **31.** 6.4 sq m
33. {a, e, i, o, u} or {a, e, i, o, u, y}   **35.** {1, 3, 5, 7, ...}
37. {7, 14, 21, 28, ...}   **39.** $\{x \mid x$ is an odd number between 10 and 30$\}$ **41.** $\{x \mid x$ is a whole number less than 5$\}$ **43.** $\{n \mid n$ is a multiple of 5 between 7 and 79$\}$
45. False   **47.** True   **49.** False   **51.** True   **53.** True
55. True   **57.** ◆   **59.** ◆   **61.** Let a and b represent the numbers; $(a + b)(a - b)$ **63.** Let r and s represent the numbers; $\frac{1}{2}(r - s)$, or $\frac{r - s}{2}$ **65.** {0}
67. {5, 10, 15, 20, ...}   **69.** {..., −4, −2, 0, 2, 4, ...}
71.

EXERCISE SET 1.2, PP. 19–20

1. 8   **3.** 9   **5.** 6.2   **7.** 0   **9.** $1\frac{7}{8}$ **11.** 4.21
13. −8 is less than or equal to −2; true
15. −7 is greater than 1; false   **17.** 3 is greater than or equal to −5; true   **19.** −9 is less than −4; true
21. −4 is greater than or equal to −4; true
23. −5 is less than −5; false   **25.** 17   **27.** −11
29. −3.2   **31.** $-\frac{11}{35}$ **33.** −8.5   **35.** $\frac{5}{9}$ **37.** −4.5
39. 0   **41.** −6.4   **43.** −7.29   **45.** $4\frac{1}{3}$ **47.** 0
49. −7   **51.** 2.7   **53.** −1.79   **55.** 0   **57.** 2
59. −5   **61.** 4   **63.** −17   **65.** −3.1   **67.** $-\frac{11}{10}$
69. 2.9   **71.** 7.9   **73.** −28   **75.** 24   **77.** −21
79. $-\frac{3}{7}$ **81.** 0   **83.** 5.44   **85.** 5   **87.** −5
89. −73   **91.** 0   **93.** $\frac{1}{5}$ **95.** $-\frac{1}{9}$ **97.** $\frac{3}{2}$ **99.** $-\frac{11}{3}$

101. $\frac{5}{6}$ **103.** $-\frac{6}{5}$ **105.** $\frac{1}{36}$ **107.** $-\frac{12}{7}$ **109.** 27
111. $-\frac{6}{11}$ **113.** $-\frac{65}{7}$ **115.** 117   **117.** 28   **119.** −5
121. $\frac{62}{3}$ **123.** $8y + 3x$; $x3 + 8y$; $3x + y8$
125. $y(7x)$; $(x7)y$ **127.** $3(xy)$ **129.** $(x + 2y) + 5$
131. $3a + 3$ **133.** $4x - 4y$ **135.** $-10a - 15b$
137. $2ab - 2ac + 2ad$ **139.** $5(x + y)$ **141.** $3(p - 3)$
143. $7(x - 3y)$ **145.** $2(x - y + z)$
147. $0.7x - 5$, or $\frac{70}{100}x - 5$ **149.** ◆   **151.** ◆
153. $(3 - 8)^2 + 9 = 34$ **155.** $5 \cdot 2^3 \div (3 - 4)^4 = 40$
157. −6.2

EXERCISE SET 1.3, PP. 26–27

1. Equivalent   **3.** Equivalent   **5.** Not equivalent
7. Not equivalent   **9.** 14.6   **11.** 8   **13.** 18   **15.** 5
17. 24   **19.** $9a$ **21.** $-2rt$ **23.** $9x^2$ **25.** $11a$
27. $-8t$ **29.** $10x$ **31.** $8x - 2x^2$ **33.** $2a + 9a^2$
35. $22x + 18$ **37.** $-5t^2 + 2t + 4t^3$ **39.** $-a - 5$
41. $m + 1$ **43.** $5d - 12$ **45.** $-7x + 14$
47. $-9x + 21$ **49.** $44a - 22$ **51.** $-100a - 90$
53. $-12y - 145$ **55.** 8   **57.** 21   **59.** 2   **61.** 2
63. 2   **65.** 7   **67.** 5   **69.** 2   **71.** 2   **73.** $\frac{49}{9}$ **75.** $\frac{4}{5}$
77. 5   **79.** $-\frac{1}{2}$ **81.** ∅; contradiction   **83.** $\mathbb{R}$; identity
85. {0}; conditional   **87.** $\mathbb{R}$; identity
89. {1, 2, 3, 4, 5, 6, 7, 8, 9}; $\{x \mid x$ is a positive integer less than 10}   **91.** ◆   **93.** ◆   **95.** −4.176190476
97. 8   **99.** $\frac{224}{29}$

EXERCISE SET 1.4, PP. 34–36

1. Let x and $x + 7$ represent the numbers; $x + (x + 7) = 65$
3. Let $t = $ the time, in seconds, it takes Alida to walk the length of the sidewalk; $9t = 300$ **5.** Let $t = $ the time, in hours, required for the plane to travel 725 km into the wind; $325t = 725$ **7.** Let $x, x + 1,$ and $x + 2$ be the angle measures; $x + (x + 1) + (x + 2) = 180$ **9.** Let t be the number of minutes climbing; $3500t = 29{,}000 - 8000$.
11. Let x be the measure of the second angle; $3x + x + (2x - 12) = 180$ **13.** Let n be the first odd number; $n + 2(n + 2) + 3(n + 4) = 70$ **15.** Let s be the length of a side of the smaller square; $4s + 4 \cdot 2s = 100$

17. Let x be the first number; $x + (3x - 6) +$
$\left[\frac{2}{3}(3x - 6) + 2\right] = 172$ **19.** Let x be the score on the next
test; $\dfrac{93 + 89 + 72 + 80 + 96 + x}{6} = 88$ **21.** $\frac{5}{2}$
23. 156.7   **25.** 1368   **27.** 63   **29.** Width: 3.5 m;
length: 7 m   **31.** $\frac{2}{3}$ hr   **33.** 14 and 16   **35.** A 6-m piece
and a 4-m piece   **37.** \$1.16   **39.** 20   **41.** $\mathbb{R}$ **43.** ◆
45. ◆   **47.** 10   **49.** About \$110,000

EXERCISE SET 1.5, PP. 41–42

1. $t = \dfrac{d}{r}$ **3.** $a = \dfrac{F}{m}$ **5.** $E = \dfrac{W}{I}$ **7.** $h = \dfrac{V}{lw}$

9. $k = Ld^2$ **11.** $n = \dfrac{G - w}{150}$ **13.** $l = p - 2w - 2h$

15. $y = \dfrac{C - Ax}{B}$ **17.** $F = \frac{9}{5}C + 32$ **19.** $r^3 = \dfrac{3V}{4\pi}$

21. $b_2 = \dfrac{2A}{h} - b_1$ **23.** $m = \dfrac{Fr}{v^2}$ **25.** $n = \dfrac{q_1 + q_2 + q_3}{A}$

27. $t = \dfrac{d_2 - d_1}{v}$ **29.** $d_1 = d_2 - vt$ **31.** $m = \dfrac{r}{1 + np}$

33. $a = \dfrac{y}{b - c^2}$ **35.** \$1571.43   **37.** 6 cm

39. About 1504.6 g   **41.** About 8.5 cm   **43.** 9 ft
45. 7 yr   **47.** About 7.7 hr   **49.** 72%
51. -58.44 **53.** ◆   **55.** ◆   **57.** About 7.4 cm
59. $a = \dfrac{2s - 2v_i t}{t^2}$ **61.** $T_2 = \dfrac{T_1 P_2 V_2}{P_1 V_1}$ **63.** $b = \dfrac{ac}{1 + c}$

TECHNOLOGY CONNECTION 1.6, P. 46

1. Answers may vary; $\boxed{2}$ $\boxed{x^y}$ $\boxed{5}$ $\boxed{-}$ $\boxed{=}$, $\boxed{2}$ $\boxed{\wedge}$ $\boxed{(-)}$
$\boxed{5}$ $\boxed{\text{ENTER}}$, $\boxed{2}$ $\boxed{x^y}$ $\boxed{-x}$ $\boxed{5}$ $\boxed{\text{ENTER}}$ **2.** Compute $\dfrac{1}{2^5}$.

EXERCISE SET 1.6, PP. 50–51

1. 7^7 **3.** 5^9 **5.** a^3 **7.** $18x^7$ **9.** $21m^{13}$ **11.** $x^{10}y^{10}$
13. a^6 **15.** $2x^3$ **17.** m^5n^4 **19.** $5x^6y^4$ **21.** $-7a^3b^{10}$
23. $-3x^4y^6z^6$ **25.** 81   **27.** -81 **29.** $\dfrac{1}{25}$ **31.** $-\dfrac{1}{25}$

33. $-\dfrac{1}{32}$ **35.** $-\dfrac{1}{64}$ **37.** $\dfrac{1}{a^3}$ **39.** $\dfrac{1}{(5x)^3}$ **41.** $\dfrac{x^2}{y^3}$

43. $\dfrac{x^2}{y^2}$ **45.** $\dfrac{1}{x^2y^5}$ **47.** $\dfrac{x^3}{y^5}$ **49.** $\dfrac{y^7z^4}{x^2}$ **51.** 3^{-4}

53. $(-16)^{-2}$ **55.** $\dfrac{1}{x^{-5}}$ **57.** $\dfrac{6}{x^{-2}}$ **59.** $(5y)^{-3}$ **61.** $\dfrac{y^{-4}}{3}$

63. 8^{-6}, or $\dfrac{1}{8^6}$ **65.** b^{-3}, or $\dfrac{1}{b^3}$ **67.** a^3 **69.** $-28m^5n^5$

71. $-14x^{-11}$, or $\dfrac{-14}{x^{11}}$ **73.** $10a^{-6}b^{-2}$, or $\dfrac{10}{a^6b^2}$

75. 10^{-9}, or $\dfrac{1}{10^9}$ **77.** 2^{-2}, or $\dfrac{1}{2^2}$, or $\dfrac{1}{4}$ **79.** y^9

81. $-3ab^2$ **83.** $-\dfrac{7a^{-4}b^2}{4}$, or $-\dfrac{7b^2}{4a^4}$

85. $-\dfrac{x^3y^{-2}z^{10}}{6}$, or $-\dfrac{x^3z^{10}}{6y^2}$ **87.** x^{12} **89.** 9^{-12}, or $\dfrac{1}{9^{12}}$

91. 7^{40} **93.** $36x^2y^2$ **95.** $a^{12}b^4$ **97.** $5x^6y^6$

99. $\dfrac{x^{-6}y^8}{49}$, or $\dfrac{y^8}{49x^6}$ **101.** $\dfrac{x^{16}y^{-20}z^{-8}}{4096}$, or $\dfrac{x^{16}}{4096y^{20}z^8}$

103. $\dfrac{9x^8y^9}{2}$ **105.** $\dfrac{625x^{-20}y^{24}}{256}$, or $\dfrac{625y^{24}}{256x^{20}}$

107. $9a^{-4}b^{-10}$, or $\dfrac{9}{a^4b^{10}}$ **109.** 1   **111.** 19

113. 59, 61, and 63   **115.** ◆   **117.** ◆
119. $3a^{-x-4}$ **121.** y^4 **123.** $3a^{2+2a}$ **125.** $2x^{a+2}y^{b-2}$
127. $2^{-2a-2b+ab}$ **129.** $\dfrac{2}{27}$

EXERCISE SET 1.7, PP. 57–58

1. 4.7×10^{10} **3.** 8.63×10^{17} **5.** 1.6×10^{-8}
7. 7×10^{-11} **9.** 4.07×10^{11} **11.** 6.03×10^{-7}
13. 4.927×10^{11} **15.** 0.0004   **17.** 673,000,000
19. 0.0000000008923   **21.** 90,300,000,000
23. 0.00000004037   **25.** 8,007,000,000,000
27. 9.7×10^{-5} **29.** 1.3×10^{-11} **31.** 8.3×10^{10}
33. 2.0×10^3 **35.** 1.51×10^{-12} **37.** 1.5×10^3
39. 3.0×10^{-5} **41.** 4.0×10^{-16} **43.** 3.00×10^{-22}
45. 2.00×10^{26} **47.** 1×10^{11} **49.** 8.3×10^{23}
51. 6.79×10^8 km   **53.** 2.2×10^{-3} lb
55. 8.00 light years   **57.** 3.08×10^{26} Å
59. 1×10^{22} cu Å   **61.** 7.90×10^7
63. 4.49×10^4 km/h   **65.** $-\dfrac{1}{12}$ **67.** $x - 6$
69. ◆   **71.** ◆   **73.** $8 \cdot 10^{-90}$ is larger by
7.1×10^{-90} **75.** 8   **77.** 8×10^{18} grains

REVIEW EXERCISES: CHAPTER 1, PP. 61–63

1. [1.1] Let x and y represent the numbers; $\dfrac{x}{y} - 5$

2. [1.1] 22   **3.** [1.1] $\{2, 4, 6, 8, 10, 12\}$; $\{x \mid x$ is an even
number between 1 and 13$\}$ **4.** [1.1] 3150 sq cm
5. [1.2] 7.3   **6.** [1.2] 4.09   **7.** [1.2] 0   **8.** [1.2] -13.1
9. [1.2] $-\dfrac{23}{35}$ **10.** [1.2] $\dfrac{7}{15}$ **11.** [1.2] -11.5
12. [1.2] $-\dfrac{1}{6}$ **13.** [1.2] -5.4 **14.** [1.2] 6.3
15. [1.2] $-\dfrac{5}{12}$ **16.** [1.2] -4.8 **17.** [1.2] 5
18. [1.2] -9.1 **19.** [1.2] $-\dfrac{21}{4}$ **20.** [1.2] 4.01
21. [1.2] $a + 5$ **22.** [1.2] $y7$ **23.** [1.2] $x5 + y$, or
$y + 5x$ **24.** [1.2] $4 + (a + b)$ **25.** [1.2] $x(y7)$
26. [1.2] $7m(n + 2)$ **27.** [1.3] $6x^3 - 8x^2 + 2$
28. [1.3] $47x - 60$ **29.** [1.3] 6.6   **30.** [1.3] $\dfrac{27}{2}$

31. [1.3] $-\frac{4}{11}$ **32.** [1.3] $\mathbb{R}$; identity
33. [1.3] $\varnothing$; contradiction   **34.** [1.4] $2x - 13 = 21$
35. [1.4] 49   **36.** [1.4] 90°, 30°, 60°   **37.** [1.5] $m = PS$
38. [1.5] $x = \dfrac{c}{m - r}$ **39.** [1.5] 4 cm   **40.** [1.6] $-10a^5b^8$
41. [1.6] $4xy^6$ **42.** [1.6] 1, 28.09, -28.09
43. [1.6] 3^3, or 27   **44.** [1.6] 5^3a^6, or $125a^6$
45. [1.6] $-\dfrac{a^9}{8b^6}$ **46.** [1.6] $\dfrac{z^8}{x^4y^6}$ **47.** [1.6] $\dfrac{b^{16}}{16a^{20}}$
48. [1.2] $-\dfrac{16}{7}$ **49.** [1.2] 0   **50.** [1.7] 1.03×10^{-7}
51. [1.7] 3.086×10^{13} **52.** [1.7] 3.7×10^7
53. [1.7] 2.0×10^{-6} **54.** [1.7] 1.4×10^4 mm³, or
1.4×10^{-5} m³   **55.** [1.3] ◈ To write an equation that has
no solution, begin with a simple equation that is false for
any value of x, such as $x = x + 1$. Then add or multiply by
the same quantities on both sides of the equation to
construct a more complicated equation with no solution.
56. [1.3] ◈ Use the distributive law to rewrite a sum of
like terms as a single term by first writing the sum as a
product. For example, $2a + 5a = (2 + 5)a = 7a$.
57. [1.7] 0.0000003%   **58.** [1.1], [1.6] $-\dfrac{23}{24}$
59. [1.4], [1.5] The 17-in. pizza is a better deal. It costs
about 5¢ per square inch; the 13-in. pizza costs about 6¢ per
square inch.   **60.** [1.5] 729 cm³
61. [1.5] $z = y - \dfrac{x}{m}$ **62.** [1.6] $3^{-2a+2b-8ab}$
63. [1.4] $88.\overline{3}$ **64.** [1.3] -39 **65.** [1.3] $-40x$
66. [1.2] $a2 + cb + cd + ad = ad + a2 + cb + cd = a(d + 2) + c(b + d)$ **67.** [1.1] $\sqrt{5}/4$; answers may vary

TEST: CHAPTER I, P. 63

1. [1.1] Let m and n represent the numbers; $mn + 3$
2. [1.1] -47 **3.** [1.1] 3.75 sq cm   **4.** [1.2] -41
5. [1.2] -3.7 **6.** [1.2] -2.11 **7.** [1.2] -14.2
8. [1.2] -43.2 **9.** [1.2] -33.92 **10.** [1.2] $-\dfrac{19}{12}$
11. [1.2] $\dfrac{5}{49}$ **12.** [1.2] 6   **13.** [1.2] $-\dfrac{4}{3}$ **14.** [1.2] $-\dfrac{3}{2}$
15. [1.2] $y + 7x$; $x7 + y$; answers may vary
16. [1.3] $8y$ **17.** [1.3] $a^2b - 4ab^2 + 2$
18. [1.3] $3x + 8$ **19.** [1.3] -2 **20.** [1.3] $\mathbb{R}$; identity
21. [1.5] $P_2 = \dfrac{P_1 V_1 T_2}{T_1 V_2}$ **22.** [1.4] 94   **23.** [1.4] 17, 19, 21
24. [1.3] $-8x - 1$ **25.** [1.3] $24b - 9$ **26.** [1.6] $-\dfrac{72}{x^{10}y^6}$
27. [1.6] $-\dfrac{1}{3^2}$, or $-\dfrac{1}{9}$ **28.** [1.6] $\dfrac{y^8}{36x^4}$ **29.** [1.6] $\dfrac{x^6}{4y^8}$
30. [1.6] 1   **31.** [1.7] 2.01×10^{-7} **32.** [1.7] 2.0×10^{10}
33. [1.7] 3.8×10^2 **34.** [1.7] 4.2×10^8 mi
35. [1.6] $16^c x^{6ac} y^{2bc+2c}$ **36.** [1.6] $-9a^3$ **37.** [1.6] $\dfrac{4}{7y^2}$

CHAPTER 2

TECHNOLOGY CONNECTION 2.1, P. 71

1. $y = 5x - 3$
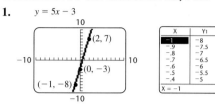

2. $y = x^2 - 4x + 3$
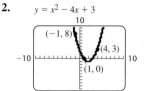
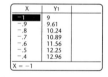

3. $y = (x + 4)^2$
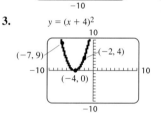

4. $y = \sqrt{x + 2}$
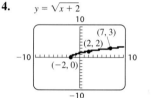
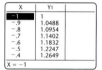

5. $y = |x + 2|$
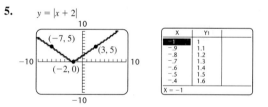

EXERCISE SET 2.1, PP. 72–74

1.
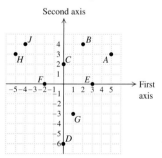

3.

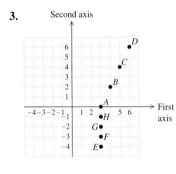

5.   Triangle; 21 units^2

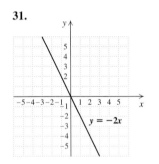

7. III   **9.** II   **11.** I   **13.** IV   **15.** Yes   **17.** No
19. Yes   **21.** Yes   **23.** Yes   **25.** No   **27.** Yes
29. No

31.

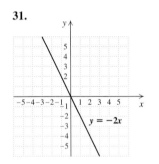

33.

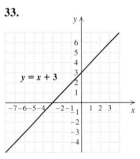

35.

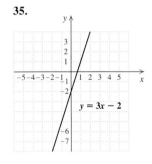

37.

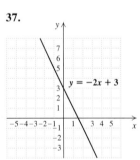

39.

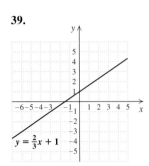

41.

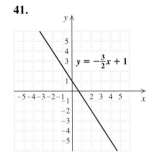

43.

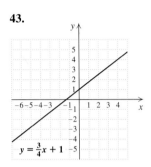

45.

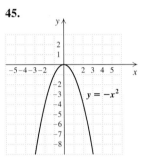

47.

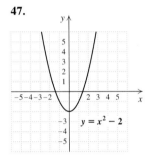

49.

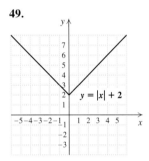

51.

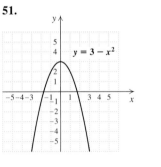

53.

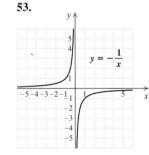

55. 25 ft   **57.** −1.4   **59.** ◈   **61.** ◈   **63.** ◈

65.

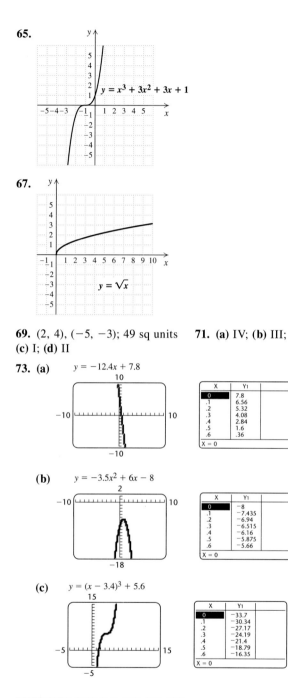

$y = x^3 + 3x^2 + 3x + 1$

67.

$y = \sqrt{x}$

69. $(2, 4)$, $(-5, -3)$; 49 sq units   **71. (a)** IV; **(b)** III;
(c) I; **(d)** II

73. (a) $y = -12.4x + 7.8$

X	Y₁
0	7.8
.1	6.56
.2	5.32
.3	4.08
.4	2.84
.5	1.6
.6	.36
X = 0	

(b) $y = -3.5x^2 + 6x - 8$

X	Y₁
0	-8
.1	-7.435
.2	-6.94
.3	-6.515
.4	-6.16
.5	-5.875
.6	-5.66
X = 0	

(c) $y = (x - 3.4)^3 + 5.6$

X	Y₁
0	-33.7
.1	-30.34
.2	-27.17
.3	-24.19
.4	-21.4
.5	-18.79
.6	-16.35
X = 0	

EXERCISE SET 2.2, PP. 82–86

1. No   **3.** Yes   **5.** Yes   **7.** Function
9. A relation but not a function   **11.** Function
13. (a) 3; **(b)** $\{-4, -3, -2, -1, 0, 1, 2\}$; **(c)** $-2, 0$;
(d) $\{1, 2, 3, 4\}$ **15. (a)** About $\frac{5}{2}$; **(b)** $\{x|-3 \leq x \leq 5\}$;
(c) About $\frac{7}{3}$; **(d)** $\{y| 1 \leq y \leq 4\}$ **17. (a)** About $\frac{9}{4}$;
(b) $\{x| -4 \leq x \leq 3\}$; **(c)** About 0; **(d)** $\{y| -5 \leq y \leq 4\}$
19. (a) About $\frac{3}{2}$; **(b)** $\{x| -5 \leq x \leq 2\}$;

(c) About $\frac{7}{6}$; **(d)** $\{y| -3 \leq y \leq 5\}$ **21. (a)** 2;
(b) $\{x| -4 \leq x \leq 4\}$; **(c)** $\{x| 0 < x \leq 2\}$; **(d)** $\{1, 2, 3, 4\}$
23. Yes   **25.** Yes   **27.** No   **29.** No
31. (a) 1; **(b)** -3; **(c)** -6; **(d)** 9; **(e)** $a + 3$
33. (a) 0; **(b)** 1; **(c)** 57; **(d)** $5t^2 + 4t$; **(e)** $20a^2 + 8a$
35. (a) $\frac{3}{5}$; **(b)** $\frac{1}{3}$; **(c)** $\frac{4}{7}$; **(d)** 0; **(e)** $\dfrac{x - 1}{2x - 1}$
37. (a) $\{x| x$ is a real number and $x \neq 3\}$;
(b) $\{x| x$ is a real number and $x \neq 5\}$; **(c)** $\mathbb{R}$; **(d)** $\mathbb{R}$;
(e) $\left\{x| x$ is a real number and $x \neq \frac{5}{2}\right\}$; **(f)** $\mathbb{R}$
39. $4\sqrt{3}$ cm$^2 \approx 6.93$ cm^2 **41.** 36π in$^2 \approx 113.10$ in^2
43. $14°$F   **45.** 159.48 cm   **47.** 75   **49.** 1,000,000
51. 3.5 drinks

53. About 2.5 hr

55. About 64,000

57. About $313,000

59. 18, 20, 22   **61.** $l = \dfrac{S - 2wh}{2h + 2w}$

63. $3x^2 - 13x - 14$ **65.** ◈   **67.** ◈   **69.** 26; 99

71. About 22 mm   **73.** ◈

75.

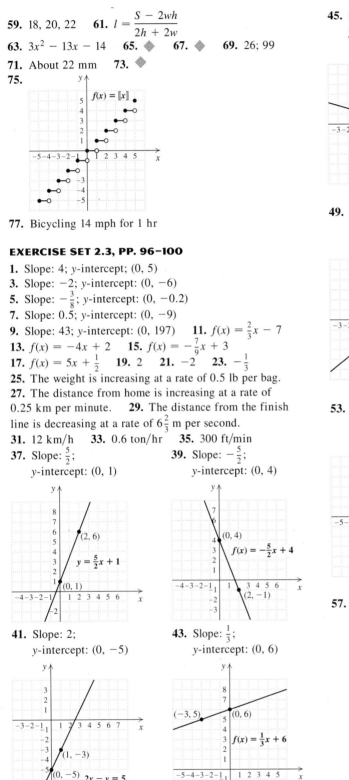

77. Bicycling 14 mph for 1 hr

EXERCISE SET 2.3, PP. 96–100

1. Slope: 4; y-intercept; (0, 5)
3. Slope: -2; y-intercept: (0, -6)
5. Slope: $-\frac{3}{8}$; y-intercept: (0, -0.2)
7. Slope: 0.5; y-intercept: (0, -9)
9. Slope: 43; y-intercept: (0, 197)   **11.** $f(x) = \frac{2}{3}x - 7$
13. $f(x) = -4x + 2$ **15.** $f(x) = -\frac{7}{9}x + 3$
17. $f(x) = 5x + \frac{1}{2}$ **19.** 2   **21.** -2 **23.** $-\frac{1}{3}$
25. The weight is increasing at a rate of 0.5 lb per bag.
27. The distance from home is increasing at a rate of
0.25 km per minute.   **29.** The distance from the finish
line is decreasing at a rate of $6\frac{2}{3}$ m per second.
31. 12 km/h   **33.** 0.6 ton/hr   **35.** 300 ft/min
37. Slope: $\frac{5}{2}$;
y-intercept: (0, 1)
39. Slope: $-\frac{5}{2}$;
y-intercept: (0, 4)

41. Slope: 2;
y-intercept: (0, -5)
43. Slope: $\frac{1}{3}$;
y-intercept: (0, 6)

45. Slope: $-\frac{2}{7}$;
y-intercept: (0, 1)

47. Slope: -0.25;
y-intercept: (0, 2)

49. Slope: $\frac{4}{5}$;
y-intercept: (0, -2)

51. Slope: $\frac{5}{4}$;
y-intercept: (0, -2)

53. Slope: $-\frac{3}{4}$;
y-intercept: (0, 3)

55. (a)

(b) 4 chirps per minute

57. $145

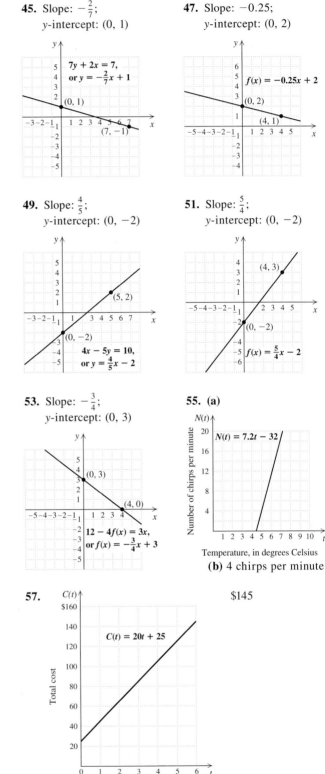

59.

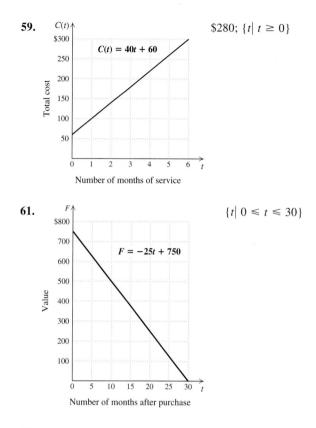

$280; \{t \mid t \geq 0\}$

61.

$\{t \mid 0 \leq t \leq 30\}$

63. 0.75 signifies that the cost per mile is $0.75; 2 signifies that the minimum cost of a taxi ride is $2.
65. 2.6 signifies that sales increase $2.6 billion per year after 1975; 17.8 signifies that sales in 1975 were $17.8 billion.   **67.** $\frac{3}{20}$ signifies that the life expectancy of American women increases $\frac{3}{20}$ yr per year after 1950; 72 signifies that the life expectancy of American women was 72 yr in 1950.   **69. (a)** II; **(b)** IV; **(c)** I; **(d)** III
71. $45x + 54$ **73.** $25x^6y^8$ **75.** ◈   **77.** ◈

79. Slope: $-\dfrac{r}{p}$; y-intercept: $\left(0, \dfrac{s}{p}\right)$ **81.** Since (x_1, y_1) and (x_2, y_2) are two points on the graph of $y = mx + b$, then $y_1 = mx_1 + b$ and $y_2 = mx_2 + b$. Using the definition of slope, we have

$$\begin{aligned}
\text{Slope} &= \frac{y_2 - y_1}{x_2 - x_1} \\
&= \frac{(mx_2 + b) - (mx_1 + b)}{x_2 - x_1} \\
&= \frac{m(x_2 - x_1)}{x_2 - x_1} \\
&= m.
\end{aligned}$$

83. False   **85.** False

87.

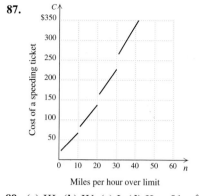

89. (a) III; **(b)** IV; **(c)** I; **(d)** II   **91.** ◈

EXERCISE SET 2.4, PP. 107–109
1. Undefined   **3.** 3   **5.** 0   **7.** $\frac{5}{7}$ **9.** Undefined
11. $\frac{1}{2}$ **13.** Undefined   **15.** -2 **17.** 0   **19.** 0   **21.** 3
23.

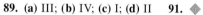

25.

27.

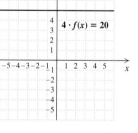

29.

31.

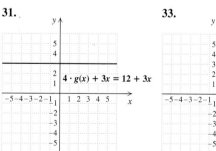

33.

35.

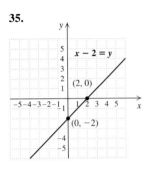

37.

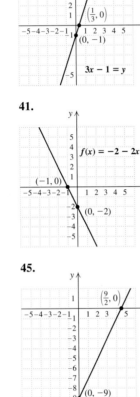

39.

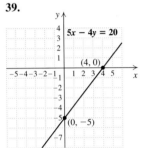

41.

43.

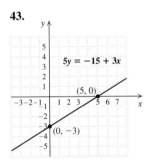

45.

47.

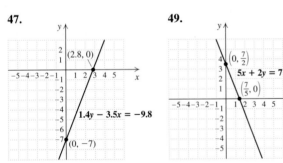

49.

51. 7 **53.** 3 **55.** 9 **57.** 1 **59.** 6 **61.** $\frac{4}{3}$
63. Linear; $\frac{5}{3}$ **65.** Linear; 0 **67.** Not linear
69. Not linear **71.** Not linear **73.** Not linear
75. $F = \dfrac{f(c - v_s)}{c - v_0}$ **77.** $3(3x - 5y)$ **79.** ◆
81. ◆ **83.** $4x - 5y = 20$ **85.** Linear **87.** Linear
89. The slope of equation B is $\frac{1}{2}$ the slope of equation A.
91. $a = 7, b = -3$ **93.** $0.\overline{6}$ **95.** 2.6

TECHNOLOGY CONNECTION 2.5, P. 113

1. $y_1 = \frac{3}{4}x + 2;\ y_2 = -\frac{4}{3}x - 1$ **2.** $y_1 = -\frac{2}{5}x - 4;\ y_2 = \frac{5}{2}x + 3$

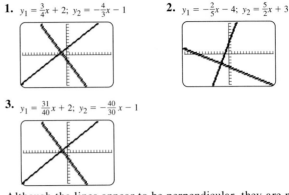

3. $y_1 = \frac{31}{40}x + 2;\ y_2 = -\frac{40}{30}x - 1$

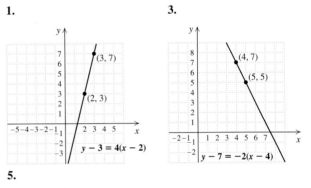

Although the lines appear to be perpendicular, they are not because the product of their slopes is not -1:

$$\frac{31}{40}\left(-\frac{40}{30}\right) = -\frac{1240}{1200} \neq -1.$$

EXERCISE SET 2.5, PP. 114–117

1.

3.

5.

7.

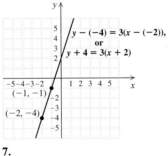

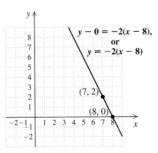

9.

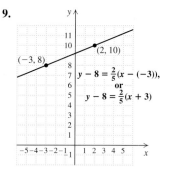

$y - 8 = \frac{2}{5}(x - (-3)),$
or
$y - 8 = \frac{2}{5}(x + 3)$

11. $\frac{2}{7}$; $(1, 3)$ **13.** -5; $(7, -2)$ **15.** $-\frac{5}{3}$; $(-2, 1)$
17. $f(x) = 5x - 13$ **19.** $f(x) = -\frac{2}{3}x - \frac{13}{3}$
21. $f(x) = -0.6x - 5.8$ **23.** $f(x) = \frac{1}{2}x + \frac{7}{2}$
25. $f(x) = 1.5x - 6.75$ **27.** $f(x) = \frac{1}{2}x - 2$
29. $f(x) = \frac{3}{2}x$ **31.** (a) $R(t) = -0.075t + 46.8$;
(b) 41.625 sec; 41.4 sec; (c) 2021
33. (a) $A(t) = 7.65t + 132.7$; (b) \$255.1 million
35. (a) $N(t) = \frac{121}{30}t + 32.9$; (b) about 77.3 million tons
37. (a) $A(t) = -0.375t + 76.4$; (b) 71.9 million acres
39. (a) $E(t) = 0.18t + 78.8$; (b) 81.32 yr **41.** Yes
43. No **45.** Yes **47.** $y = -\frac{1}{2}x + \frac{17}{2}$ **49.** $y = \frac{5}{7}x - \frac{17}{7}$
51. $y = \frac{1}{3}x + 4$ **53.** $y = -\frac{3}{2}x - \frac{13}{2}$ **55.** Yes **57.** No
59. $y = \frac{1}{2}x + 4$ **61.** $y = \frac{4}{3}x - 6$ **63.** $y = \frac{5}{2}x + 9$
65. $y = -\frac{5}{3}x - \frac{41}{3}$ **67.** \$35 **69.** $82\frac{2}{3}$ **71.** ◆
73. ◆ **75.** \$1940 **77.** \$11,000 **79.** 21.1°C
81. (a) $g(x) = x - 8$; (b) -10; (c) 83 **83.** 7 **85.** ◆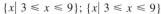

EXERCISE SET 2.6, PP. 123–125
1. 1 **3.** -41 **5.** 12 **7.** $\frac{13}{18}$ **9.** 5 **11.** 2
13. $x^2 - x + 1$ **15.** 21 **17.** 5 **19.** 42
21. $-\frac{3}{4}$ **23.** $\frac{1}{6}$ **25.** 1550; in January 1993, there were
1550 new cases of AIDS diagnosed among people born in
1960 or later. **27.** $T(6) \approx 4800$, $G(6) \approx 3550$,
$F(6) \approx 1250$; $(G + F)(6) = G(6) + F(6) \approx 3550 +$
$1250 \approx 4800 \approx T(6)$. **29.** 44; a total of about 44 million
passengers used Newark and LaGuardia airports in 1992.
31. 8; about 8 million more passengers used Kennedy
airport than LaGuardia airport in 1992. **33.** 72; a total of
about 72 million passengers used Newark, LaGuardia, and
Kennedy airports in 1993. **35.** $\mathbb{R}$
37. $\{x \mid x$ is a real number and $x \neq 2\}$
39. $\{x \mid x$ is a real number and $x \neq 0\}$
41. $\{x \mid x$ is a real number and $x \neq 1\}$
43. $\{x \mid x$ is a real number and $x \neq 2$ and $x \neq 4\}$
45. $\{x \mid x$ is a real number and $x \neq 3\}$
47. $\{x \mid x$ is a real number and $x \neq 4\}$
49. $\{x \mid x$ is a real number and $x \neq 4$ and $x \neq 5\}$
51. $\{x \mid x$ is a real number and $x \neq -2.5$ and $x \neq -1\}$
53. 4; 3 **55.** $\{x \mid 0 \leq x \leq 9\}$; $\{x \mid 3 \leq x \leq 10\}$;
$\{x \mid 3 \leq x \leq 9\}$; $\{x \mid 3 \leq x \leq 9\}$

57.

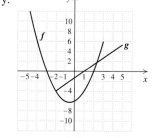

$F + G$

59. $\frac{7}{5}$ **61.** 86
63. ◆

65. $\{x \mid x$ is a real number and $-1 < x < 5$ and $x \neq \frac{3}{2}\}$
67. 5; 15; $\frac{2}{3}$ **69.** $\{x \mid x$ is a real number and $x \neq -\frac{5}{2}$ and
$x \neq -3$ and $x \neq -1$ and $x \neq 1\}$
71. Answers may vary.

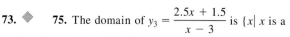

73. ◆ **75.** The domain of $y_3 = \dfrac{2.5x + 1.5}{x - 3}$ is $\{x \mid x$ is a

real number and $x \neq 3\}$. The CONNECTED mode graph
contains the line $x = 3$, whereas the DOT mode graph
contains no points having 3 as the first coordinate. Thus the
DOT mode graph represents y_3 more accurately.

REVIEW EXERCISES: CHAPTER 2, PP. 127–129
1. [2.1] Yes **2.** [2.1] No **3.** [2.1] Yes **4.** [2.1] No
5. [2.1], [2.3] **6.** [2.1]

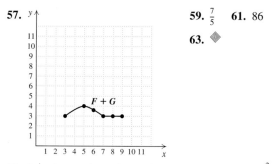

$y = -3x + 2$ $(0, 2)$ $(1, -1)$ $8x + 32 = 0$ $y = -x^2 + 1$

7. [2.4] **8.** [2.4]

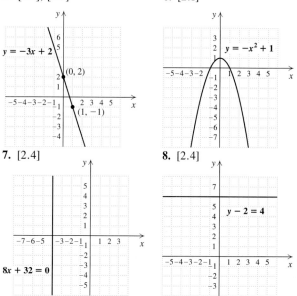

$y - 2 = 4$

9. [2.2] **(a)** 3; **(b)** $\{x \mid -2 \le x \le 4\}$; **(c)** -1; **(d)** $\{y \mid 1 \le y \le 5\}$ **10.** [2.2] $18,185
11. [2.3] Slope: -4; y-intercept: $(0, -9)$
12. [2.3] Slope: $\frac{1}{3}$; y-intercept: $\left(0, -\frac{7}{6}\right)$ **13.** [2.3] $\frac{4}{7}$
14. [2.4] Undefined   **15.** [2.3] $2500 of personal income per year since high school   **16.** [2.3] $f(x) = \frac{2}{7}x - 6$

17. [2.4]

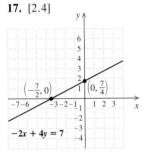

18. [2.4] Yes   **19.** [2.4] Yes   **20.** [2.4] No
21. [2.4] No   **22.** [2.5] $y - 4 = -2(x + 3)$
23. [2.5] $f(x) = \frac{4}{3}x + \frac{7}{3}$ **24.** [2.5] Perpendicular
25. [2.5] Parallel   **26.** [2.5] **(a)** $W(t) = 0.0875t + 0.75$;
(b) about $5.13   **27.** [2.5] $y = \frac{3}{5}x - \frac{31}{5}$
28. [2.5] $y = -\frac{5}{3}x - \frac{5}{3}$ **29.** [2.2] -6 **30.** [2.2] 26
31. [2.6] 102   **32.** [2.6] -17 **33.** [2.6] $-\frac{9}{2}$
34. [2.2] $3a + 3b - 6$ **35.** [2.6] $\mathbb{R}$ **36.** [2.6] $\{x \mid x$ is a real number and $x \ne 2\}$ **37.** [1.2] $\frac{2}{15}$ **38.** [1.6] $25a^6b^2$
39. [1.3] -26 **40.** [1.7] 5.28×10^6 **41.** [2.2] ◆ For any function, each member of the domain corresponds to *exactly one* member of the range. Thus, for any function, each member of the domain corresponds to *at least one* member of the range. Therefore, a function is a relation. In a relation, every member of the domain corresponds to *at least one*, but not necessarily *exactly one*, member of the range. Therefore, a relation may or may not be a function.   **42.** [2.4] ◆ The slope of a line is the rise between two points on the line divided by the run between those points. For a vertical line, there is no run between any two points, and division by 0 is undefined; therefore, the slope is undefined. For a horizontal line, there is no rise between any two points, so the slope is 0/run, or 0.

43. [1.6], [2.4] -9 **44.** [2.5] $-\frac{9}{2}$
45. [2.5] $f(x) = 3.09x + 3.75$ **46.** [2.3] **(a)** III; **(b)** IV; **(c)** I; **(d)** II

TEST: CHAPTER 2, PP. 129–130
1. [2.1] Yes   **2.** [2.1] No

3. [2.1], [2.3]   **4.** [2.1]

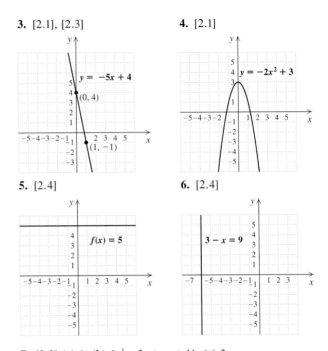

5. [2.4]   **6.** [2.4]

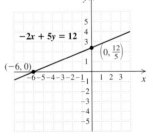

7. [2.2] **(a)** 1; **(b)** $\{x \mid -3 \le x \le 4\}$; **(c)** 3;
(d) $\{y \mid -1 \le y \le 2\}$ **8.** [2.2] $33.4 billion
9. [2.3] Slope: $-\frac{3}{5}$; y-intercept: $(0, 12)$
10. [2.3] Slope: $-\frac{2}{5}$; y-intercept: $\left(0, -\frac{7}{5}\right)$ **11.** [2.3] $\frac{5}{8}$
12. [2.4] 0   **13.** [2.3] $f(x) = -5x - 1$
14. [2.4]

15. [2.4] **(a)**, **(c)**   **16.** [2.5] $y + 4 = 4(x + 2)$
17. [2.5] $f(x) = -x + 2$ **18.** [2.5] Parallel
19. [2.5] Perpendicular   **20.** [2.5] $y = \frac{2}{5}x + \frac{16}{5}$
21. [2.5] $y = -\frac{5}{2}x - \frac{11}{2}$ **22.** [2.2], [2.6] **(a)** 5; **(b)** -130;
(c) $\left\{x \mid x$ is a real number and $x \ne -\frac{4}{3}\right\}$
23. [2.5] **(a)** $C(m) = 0.3m + 25$; **(b)** $175   **24.** [1.2] $-\frac{14}{15}$
25. [1.6] a^6b^{12} **26.** [1.3] 3   **27.** [1.7] 6.7×10^{-6}
28. [2.2], [2.3] **(a)** 30 mi; **(b)** 15 mph
29. [2.5] $s = -\frac{3}{2}r + \frac{27}{2}$ **30.** [2.6] $h(x) = 7x - 2$

CHAPTER 3

TECHNOLOGY CONNECTION 3.1, P. 137
1. $(1.53, 2.58)$ **2.** $(-0.26, 57.06)$ **3.** $(2.23, 1.14)$
4. $(0.87, -0.32)$

EXERCISE SET 3.1, PP. 138–141

1. Let x = the larger number and y = the smaller number; $x - y = 11$, $3x + 2y = 123$ **3.** Let x = the number of $8.50 brushes sold and y = the number of $9.75 brushes sold; $x + y = 45$, $8.50x + 9.75y = 398.75$ **5.** Let x and y represent the angles; $x + y = 180$, $x = 2y - 3$ **7.** Let x = the number of field goals and y = the number of free throws; $x + y = 18$, $2x + y = 30$ **9.** Let h = the number of vials of Humulin sold and n = the number of vials of Novolin; $h + n = 65$, $15.75h + 12.95n = 959.35$ **11.** Let l and w represent the length and width of the court, respectively; $2l + 2w = 228$, $w = l - 42$ **13.** Let l = the number of units of lumber produced and p = the number of units of plywood produced; $p = 2l$, $25l + 40p = 10,920$ **15.** Let w = the number of wins and t = the number of ties; $2w + t = 60$, $w = t + 9$ **17.** Let x = the number of ounces of lemon juice and y = the number of ounces of linseed oil; $y = 2x$, $x + y = 32$ **19.** Yes   **21.** No   **23.** Yes   **25.** Yes   **27.** $(4, 1)$ **29.** $(2, -1)$ **31.** $(4, 3)$ **33.** $(-3, -2)$ **35.** $\left(\frac{5}{2}, -2\right)$ **37.** $(-3, 2)$ **39.** $(7, 2)$ **41.** $(4, 0)$ **43.** No solution   **45.** $\{(x, y) \mid y = 3 - x\}$ **47.** All except 43   **49.** 45   **51.** -3 **53.** $\frac{9}{20}$ **55.** $3(x - 7)$ **57.** ◆   **59.** ◆   **61.** 1989   **63.** Answers may vary. **(a)** $x + y = 6$, $x - y = 4$; **(b)** $x + y = 1$, $2x + 2y = 3$; **(c)** $x + y = 1$, $2x + 2y = 2$ **65.** $A = -\frac{17}{4}$, $B = -\frac{12}{5}$ **67.** Let x and y represent the number of years that Lou and Juanita have taught at the university, respectively; $x + y = 46$, $x - 2 = 2.5(y - 2)$ **69.** Let s and v represent the number of ounces of baking soda and vinegar needed, respectively; $s = 4v$, $s + v = 16$ **71.** $(0, 0)$, $(1, 1)$ **73.** $(0.07, -7.95)$ **75.** $(0.00, 1.25)$

EXERCISE SET 3.2, PP. 147–148

1. $(-4, 3)$ **3.** $(-3, -15)$ **5.** $(2, -7)$ **7.** $(4, -1)$ **9.** $(3, -2)$ **11.** $\left(\frac{25}{23}, -\frac{11}{23}\right)$ **13.** No solution   **15.** $(1, 2)$ **17.** $(3, 0)$ **19.** $\left(\frac{1}{2}, -5\right)$ **21.** $\left(\frac{10}{21}, \frac{11}{14}\right)$ **23.** $(1, 3)$ **25.** $(2, 3)$ **27.** $(-2, 3)$ **29.** $(12, 15)$ **31.** $\{(x, y) \mid 3x + 5y = 7\}$ **33.** $(-2, -9)$ **35.** $(30, 6)$ **37.** $\{(x, y) \mid x = 2 + 3y\}$ **39.** No solution   **41.** $(140, 60)$ **43.** 11%   **45.** $\mathbb{R}$ **47.** ◆   **49.** ◆   **51.** $m = -\frac{1}{2}$, $b = \frac{5}{2}$ **53.** $a = 5$, $b = 2$ **55.** $\left(-\frac{32}{17}, \frac{38}{17}\right)$ **57.** $\left(-\frac{1}{5}, \frac{1}{10}\right)$ **59.** ◆ 〰

EXERCISE SET 3.3, PP. 158–161

1. 29, 18   **3.** 32 at $8.50, 13 at $9.75   **5.** 119°, 61°   **7.** 12 field goals, 6 free throws   **9.** Humulin: 42; Novolin: 23   **11.** Length: 78 ft; width: 36 ft   **13.** Lumber: 104; plywood: 208   **15.** 23 wins, 14 ties   **17.** Lemon juice: $10\frac{2}{3}$ oz; linseed oil: $21\frac{1}{3}$ oz   **19.** 18 scientific calculators, 27 graphing calculators

21. Kenyan: 8 lb; Sumatran: 12 lb   **23.** Deep Thought: 12 lb; Oat Dream: 8 lb   **25.** 4 L of 25%, 6 L of 50%   **27.** $7500 at 6%, $4500 at 9%   **29.** Whole milk: $169\frac{3}{13}$ lb; cream: $30\frac{10}{13}$ lb   **31.** Length: 265 ft; width: 165 ft   **33.** 7 $5 bills, 15 $1 bills   **35.** 375 km   **37.** 14 km/h   **39.** About 1489 mi   **41.** $x + 29$ **43.** $y = -\frac{3}{4}x - \frac{7}{2}$ **45.** $f(x) = \frac{3}{2}x - 5$ **47.** ◆   **49.** ◆   **51.** Burl: 40; son: 20   **53.** Width: $\frac{102}{5}$ in.; length: $\frac{288}{5}$ in.   **55.** $4\frac{4}{7}$ L   **57.** 180   **59.** First train: 36 km/h; second train: 54 km/h   **61.** Brown: 0.8 gal; neutral: 0.2 gal   **63.** $P(x) = \frac{0.1 + x}{1.5}$ (This expresses the percent as a decimal quantity.)

EXERCISE SET 3.4, PP. 167–169

1. No   **3.** $(4, 0, 2)$ **5.** $(2, -2, 2)$ **7.** $(3, -2, 1)$ **9.** $(7, -3, -4)$ **11.** $(2, 1, 3)$ **13.** $(2, -5, 6)$ **15.** The equations are dependent.   **17.** $(3, -5, 8)$ **19.** $\left(\frac{3}{5}, \frac{2}{3}, -3\right)$ **21.** $\left(4, \frac{1}{2}, -\frac{1}{2}\right)$ **23.** $(17, 9, 79)$ **25.** $\left(\frac{1}{4}, -\frac{1}{2}, -\frac{1}{4}\right)$ **27.** $\left(\frac{98}{5}, \frac{304}{5}, \frac{498}{5}\right)$ **29.** No solution   **31.** The equations are dependent.   **33.** $2a + 9$ **35.** $b = a - \frac{2K}{t}$, or $\frac{at - 2K}{t}$ **37.** ◆   **39.** ◆   **41.** $(1, -1, 2)$ **43.** $(-3, -1, 0, 4)$ **45.** $\left(-\frac{1}{2}, -1, -\frac{1}{3}\right)$ **47.** 14   **49.** $z = 8 - 2x - 4y$

EXERCISE SET 3.5, PP. 172–174

1. 16, 19, 22   **3.** 8, 21, -3 **5.** 32°, 96°, 52°   **7.** Automatic transmission: $865; power door locks: $520; air conditioning: $375   **9.** A: 1500; B: 1900; C: 2300   **11.** 8 10-oz cups, 20 14-oz cups, 6 20-oz cups   **13.** First fund: $45,000; second fund: $10,000; third fund: $25,000   **15.** Roast beef: 2; baked potato: 1; broccoli: 2   **17.** Asian-American: 385; African-American: 200; Caucasian: 154   **19.** 32 2-point field goals, 5 3-point field goals, 13 foul shots   **21.** 2   **23.** $\frac{a^3}{b}$ **25.** $\{x \mid x$ is a real number and $x \neq -7\}$ **27.** ◆   **29.** 464   **31.** 5 adults, 1 student, 94 children   **33.** 180°

EXERCISE SET 3.6, PP. 178–179

1. $(2, -1)$ **3.** $(-4, 3)$ **5.** $\left(\frac{3}{2}, \frac{5}{2}\right)$ **7.** $\left(\frac{3}{2}, -4, 3\right)$ **9.** $(2, -2, 1)$ **11.** $\left(4, \frac{1}{2}, -\frac{1}{2}\right)$ **13.** $(1, -3, -2, -1)$ **15.** 4 dimes, 30 nickels   **17.** 5 lb of $4.05-per-pound granola, 10 lb of $2.70-per-pound granola   **19.** $400 at 7%, $500 at 8%, $1600 at 9%   **21.** $\frac{69}{10}$, or 6.9   **23.** -32 **25.** ◆   **27.** 1324

EXERCISE SET 3.7, PP. 183–184

1. 18   **3.** 36   **5.** 27   **7.** -3 **9.** -5 **11.** $(-3, 2)$
13. $\left(\frac{9}{19}, \frac{51}{38}\right)$ **15.** $\left(-1, -\frac{6}{7}, \frac{11}{7}\right)$ **17.** $(2, -1, 4)$
19. $(1, 2, 3)$ **21.** $\frac{333}{245}$ **23.** One piece, 20.8 ft; the other
piece, 12 ft   **25.** 3   **27.** An equation of the line through
(x_1, y_1) and (x_2, y_2) is

$$y - y_1 = \frac{y_2 - y_1}{x_2 - x_1}(x - x_1),$$

which is equivalent to

$$yx_2 - yx_1 - y_1x_2 + y_1x_1 = y_2x - y_2x_1 - y_1x + y_1x_1$$

or

$$y_2x_1 + y_1x - y_2x + yx_2 - yx_1 - y_1x_2 = 0. \qquad (1)$$

$$\begin{vmatrix} x & y & 1 \\ x_1 & y_1 & 1 \\ x_2 & y_2 & 1 \end{vmatrix} = 0$$

is equivalent to

$$x(y_1 - y_2) - x_1(y - y_2) + x_2(y - y_1) = 0$$

or

$$xy_1 - xy_2 - x_1y + x_1y_2 + x_2y - x_2y_1 = 0. \qquad (2)$$

Equations (1) and (2) are equivalent.

EXERCISE SET 3.8, PP. 188–190

1. (a) $P(x) = 45x - 270,000$; (b) 6000 units
3. (a) $P(x) = 50x - 120,000$; (b) 2400 units
5. (a) $P(x) = 80x - 10,000$; (b) 125 units
7. (a) $P(x) = 18x - 16,000$; (b) 889 units
9. (a) $P(x) = 75x - 195,000$; (b) 2600 units
11. ($70, 300)   **13.** ($22, 474)   **15.** ($50, 6250)
17. ($10, 1070)   **19.** (a) $C(x) = 22,500 + 40x$;
(b) $R(x) = 85x$; (c) $P(x) = 45x - 22,500$;
(d) $112,500 profit, $4500 loss; (e) (500 lamps, $42,500)
21. (a) $C(x) = 16,404 + 6x$; (b) $R(x) = 18x$;
(c) $P(x) = 12x - 16,404$; (d) $19,596 profit, $4404 loss;
(e) (1367 dozen caps, $24,606)
23.

25. $\frac{5}{2}$ **27.** ◈
29. 308 pairs
31. (a) $8.74;
(b) 24,509 units

REVIEW EXERCISES: CHAPTER 3, PP. 192–193

1. [3.1] $(-2, 1)$ **2.** [3.1] $(3, 2)$ **3.** [3.2] $\left(-\frac{4}{5}, \frac{2}{5}\right)$
4. [3.2] No solution   **5.** [3.2] $\left(-\frac{11}{15}, -\frac{43}{30}\right)$

6. [3.2] $\left(\frac{37}{19}, \frac{53}{19}\right)$ **7.** [3.2] $\left(\frac{76}{17}, -\frac{2}{119}\right)$ **8.** [3.2] $(2, 2)$
9. [3.2] $\{(x, y) \mid 3x + 4y = 6\}$ **10.** [3.3] CD: $14;
cassette: $9   **11.** [3.3] 4 hr   **12.** [3.3] 5 L of each
13. [3.4] $(4, -8, 10)$ **14.** [3.4] The equations
are dependent.   **15.** [3.4] $(2, 0, 4)$
16. [3.2] No solution   **17.** [3.4] $\left(\frac{8}{9}, -\frac{2}{3}, \frac{10}{9}\right)$
18. [3.5] A: 90°, B: 67.5°, C: 22.5°   **19.** [3.5] 641
20. [3.5] 7 $20 bills, 3 $5 bills, 4 $1 bills
21. [3.6] $\left(55, -\frac{89}{2}\right)$ **22.** [3.6] $(-1, 1, 3)$ **23.** [3.7] 2
24. [3.7] 9   **25.** [3.7] $(6, -2)$ **26.** [3.7] $(-3, 0, 4)$
27. [3.8] ($3, 81)   **28.** [3.8] (a) $C(x) = 175x + 35,000$;
(b) $R(x) = 300x$; (c) $P(x) = 125x - 35,000$; (d) $115,000
profit, $10,000 loss; (e) 280 beds   **29.** [1.3] $-\frac{4}{3}$
30. [1.5] $t = \dfrac{Q}{a - 4}$ **31.** [2.5] $y = -\frac{7}{3}x + 12$
32. [1.4] 69, 76, 83   **33.** [3.5] ◈ To solve a problem
involving four variables, go through the *Familiarize* and
Translate steps as usual. The resulting system of equations
can be solved using the elimination method just as for three
variables but likely with more steps.   **34.** [3.4] ◈ A
system of equations can be both dependent and inconsistent
if it is equivalent to a system with fewer equations that has
no solution. An example is a system of three equations in
three unknowns in which two of the equations represent the
same plane, and the third represents a parallel plane.
35. [3.1] $(0, 2), (1, 3)$ **36.** [3.5] $a = -\frac{2}{3}$, $b = -\frac{4}{3}$, $c = 3$;
$f(x) = -\frac{2}{3}x^2 - \frac{4}{3}x + 3$

TEST: CHAPTER 3, PP. 193–194

1. [3.2] $\left(3, -\frac{11}{3}\right)$ **2.** [3.2] $\left(\frac{15}{7}, -\frac{18}{7}\right)$
3. [3.2] $\left(-\frac{3}{2}, -\frac{3}{2}\right)$ **4.** [3.2] No solution
5. [3.3] Length: 30; width: 18   **6.** [3.5] Mortgage:
$74,000; car loan: $600; credit card bill: $700
7. [3.4] The equations are dependent.
8. [3.4] $\left(2, -\frac{1}{2}, -1\right)$ **9.** [3.4] No solution
10. [3.4] $(0, 1, 0)$ **11.** [3.6] $\left(\frac{34}{107}, -\frac{104}{107}\right)$
12. [3.6] $(3, 1, -2)$ **13.** [3.7] 34   **14.** [3.7] 133
15. [3.7] $\left(\frac{13}{18}, \frac{7}{27}\right)$ **16.** [3.5] 3.5 hr   **17.** [3.8] ($3, 55)
18. [3.8] (a) $C(x) = 40,000 + 30x$; (b) $R(x) = 80x$;
(c) $P(x) = 50x - 40,000$; (d) $35,000 profit;
(e) (800 rackets, $64,000)   **19.** [1.4] $35
20. [1.5] $a = \dfrac{P + 3b}{4}$ **21.** [1.3] $\frac{9}{5}$
22. [2.5] $y + 1 = -\frac{5}{8}(x - 5)$, or $y = -\frac{5}{8}x + \frac{17}{8}$
23. [2.3], [3.3] $m = 7$, $b = 10$ **24.** [3.5] Adult: 1346;
senior citizen: 335; child: 1651

CUMULATIVE REVIEW: 1–3

1. [1.3] 14.87   **2.** [1.3] -22 **3.** [1.3] -42.9
4. [1.3] 20   **5.** [1.3] $-\frac{21}{4}$ **6.** [1.3] 6   **7.** [1.3] -5
8. [1.3] $\frac{10}{9}$ **9.** [1.3] $-\frac{32}{5}$ **10.** [1.3] $\frac{18}{17}$ **11.** [1.6] x^{11}

12. [1.6] $-\dfrac{40x}{y^5}$ **13.** [1.6] $-288x^4y^{18}$ **14.** [1.6] y^{10}

15. [1.6] $-\dfrac{2a^{11}}{5b^{33}}$ **16.** [1.6] $\dfrac{81x^{36}}{256y^8}$ **17.** [1.7] 1.12×10^6

18. [1.7] 4.00×10^6 **19.** [1.5] $b = \dfrac{2A}{h} - t$, or $\dfrac{2A - ht}{h}$

20. [2.1] Yes

21. [2.3]   **22.** [2.1]

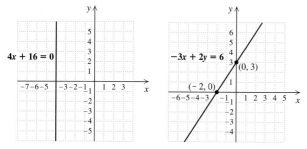

23. [2.4]   **24.** [2.3]

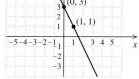

25. [2.3] Slope: $\dfrac{9}{4}$; y-intercept: $(0, -3)$ **26.** [2.3] $\dfrac{4}{3}$
27. [2.5] $y = -3x - 5$ **28.** [2.5] $y = -\dfrac{1}{10}x + \dfrac{12}{5}$
29. [2.5] Parallel   **30.** [2.5] $y = -2x + 5$
31. [2.2] $\{-5, -3, -1, 1, 3\}$; $\{-3, -2, 1, 4, 5\}$; -2; 3
32. [2.2] $\left\{x \mid x \text{ is a real number and } x \neq \frac{1}{2}\right\}$ **33.** [2.2] -31
34. [2.2] 3 **35.** [2.6] 7 **36.** [2.6] $8a^2 + 4a - 4$
37. [3.2] $(1, 1)$ **38.** [3.2] $(-2, 3)$ **39.** [3.2] $\left(-3, \frac{2}{5}\right)$
40. [3.4] $(-3, 2, -4)$ **41.** [3.4] $(0, -1, 2)$ **42.** [3.7] 14
43. [3.7] 0 **44.** [3.3] $8, 18$ **45.** [3.3] Soakem: $48\frac{8}{9}$ oz;
Rinsem: $71\frac{1}{9}$ oz   **46.** [1.4] $3, 5, 7$ **47.** [1.4] 90
48. [3.3] $l = 10$ cm, $w = 6$ cm   **49.** [3.3] 19 nickels,
15 dimes   **50.** [3.5] \$120   **51.** [3.5] 23 wins, 33 losses,
8 ties   **52.** [3.5] Cookie: 90; banana: 80; yogurt: 165
53. [1.6] $-12x^{2a}y^{b+y+3}$ **54.** [2.5] \$151,000
55. [2.5], [3.3] $m = -\dfrac{5}{9}$, $b = -\dfrac{2}{9}$

CHAPTER 4

EXERCISE SET 4.1, PP. 206–209

1. No, no, no, yes   **3.** No, yes, yes, no

5.

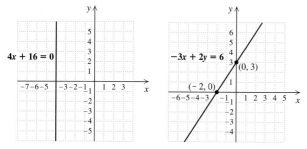

$\{y \mid y < 5\}$, $(-\infty, 5)$

7.

$\{x \mid x \geq -4\}$, $[-4, \infty)$

9.

$\{t \mid t > -2\}$, $(-2, \infty)$

11.

$\{x \mid x \leq -5\}$, $(-\infty, -5]$

13. $\{x \mid x > -5\}$, or $(-5, \infty)$
15. $\{a \mid a \leq -20\}$, or $(-\infty, -20]$
17. $\{x \mid x \leq 19\}$, or $(-\infty, 19]$
19. $\{y \mid y > -9\}$, or $(-9, \infty)$
21. $\{y \mid y \leq 14\}$, or $(-\infty, 14]$
23. $\{t \mid t < -9\}$, or $(-\infty, -9)$
25. $\{x \mid x < 50\}$, or $(-\infty, 50)$
27. $\{y \mid y \geq -0.4\}$, or $[-0.4, \infty)$
29. $\left\{y \mid y \geq \frac{9}{10}\right\}$, or $\left[\frac{9}{10}, \infty\right)$
31. $\{y \mid y > 3\}$, or $(3, \infty)$
33. $\{x \mid x \leq 4\}$, or $(-\infty, 4]$
35. $\left\{x \mid x < -\frac{2}{5}\right\}$, or $\left(-\infty, -\frac{2}{5}\right)$
37. $\{x \mid x \geq 11.25\}$, or $[11.25, \infty)$
39. $\left\{y \mid y \leq -\frac{53}{6}\right\}$, or $\left(-\infty, -\frac{53}{6}\right]$ **41.** $\left\{x \mid x > -\frac{2}{17}\right\}$, or
$\left(-\frac{2}{17}, \infty\right)$ **43.** $\left\{m \mid m > \frac{7}{3}\right\}$, or $\left(\frac{7}{3}, \infty\right)$ **45.** $\{x \mid x \geq 2\}$,
or $[2, \infty)$ **47.** $\{y \mid y < 5\}$, or $(-\infty, 5)$ **49.** $\left\{x \mid x \leq \frac{4}{7}\right\}$, or
$\left(-\infty, \frac{4}{7}\right]$ **51.** Mileages greater than 65
53. \$11,500 or more   **55.** More than 4.25 hr
57. More than 25   **59.** More than \$1850   **61.** \$5000
63. (a) $\left\{x \mid x < 8181\frac{9}{11}\right\}$; (b) $\left\{x \mid x > 8181\frac{9}{11}\right\}$
65.

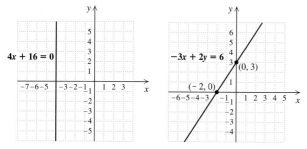

$f(x) = 2x - 1$

67. -2 **69.** 16
71. ◈   **73.** ◈
75. $\left\{y \mid y \geq -\dfrac{10}{b+4}\right\}$
77. $\left\{x \mid x < \dfrac{4c + 3d}{6c - 2d}\right\}$
79. $\left\{x \mid x < \dfrac{-3a + 2d}{c - (4a + 5d)}\right\}$

81. False; $-3 < -2$, but $9 > 4$.   **83.** ◈   **85.** ∅
87. (a) $\{x \mid x < 4\}$, or $(-\infty, 4)$; (b) $\{x \mid x \geq 2\}$, or $[2, \infty)$;
(c) $\{x \mid x \geq 3.2\}$, or $[3.2, \infty)$

TECHNOLOGY CONNECTION 4.2, P. 215

1. Domain of $y_1 = \{x \mid x \le 3\}$, or $(-\infty, 3]$; domain of $y_2 = \{x \mid x \ge -1\}$, or $[-1, \infty)$ **2.** Domain of $y_1 + y_2 =$ domain of $y_1 - y_2 =$ domain of $y_2 - y_1 =$ domain of $y_1 \cdot y_2 = \{x \mid -1 \le x \le 3\}$, or $[-1, 3]$

EXERCISE SET 4.2, PP. 216–218

1. $\{9, 11\}$ **3.** $\{1, 5, 10, 15, 20\}$ **5.** $\{b\}$ **7.** $\{r, s, t, u, v\}$
9. $\{5, 7\}$ **11.** $\{3, 5, 7\}$
13.

;

$(2, 7)$

15.

;

$[-6, -2]$

17.

;

$(-\infty, -2) \cup (1, \infty)$

19.

;

$(-\infty, -1] \cup (4, \infty)$

21.

;

$(-5, 3]$

23.

;

$(-2, 4)$

25.

;

$(-\infty, 5) \cup (7, \infty)$

27.

;

$(-\infty, -4] \cup [5, \infty)$

29.

;

$[-6, 5)$

31.

;

$[3, 7)$

33.

;

$(-\infty, 5)$

35.

;

$(-\infty, \infty)$

37.

;

$(7, \infty)$

39. $\{t \mid -5 < t < 5\}$, or $(-5, 5)$

41. $\{x \mid -1 < x \le 4\}$, or $(-1, 4]$

43. $\{a \mid -2 \le a < 2\}$, or $[-2, 2)$

45. $\{x \mid x \le -9 \text{ or } x \ge -2\}$, or $(-\infty, -9] \cup [-2, \infty)$

47. $\{x \mid 1 \le x \le 3\}$, or $[1, 3]$

49. $\left\{x \mid -\frac{7}{2} < x \le \frac{11}{2}\right\}$, or $\left(-\frac{7}{2}, \frac{11}{2}\right]$

51. $\{x \mid x \le 1 \text{ or } x \ge 3\}$, or $(-\infty, 1] \cup [3, \infty)$

53. $\{x \mid x < 3 \text{ or } x > 4\}$, or $(-\infty, 3) \cup (4, \infty)$

55. $\left\{a \mid a < \frac{7}{2}\right\}$, or $\left(-\infty, \frac{7}{2}\right)$

57. $\{a \mid a < -5\}$, or $(-\infty, -5)$

59. $\mathbb{R}$, or $(-\infty, \infty)$

61. $\{t \mid t \le 6\}$, or $(-\infty, 6]$

63. $(-\infty, 5) \cup (5, \infty)$ **65.** $[-4, \infty)$
67. $\left(-\infty, \frac{5}{2}\right) \cup \left(\frac{5}{2}, \infty\right)$ **69.** $(-\infty, 4]$ **71.** $\left(-\frac{16}{13}, -\frac{41}{13}\right)$
73. $-\frac{27}{7}$
75.

$f(x) = 5$

77. ◈ **79.** ◈
81. $\{x \mid -1 < x < 6\}$, or $(-1, 6)$ **83.** More than 1 and less than 5
85. $0 \text{ ft} \le d \le 198 \text{ ft}$
87. From 2005 through 2010

89. $\left\{a \mid -\frac{3}{2} \le a \le 1\right\}$, or $\left[-\frac{3}{2}, 1\right]$;

91. $\{x \mid -4 < x \le 1\}$, or $(-4, 1]$;

93. True **95.** False **97.** $\left[-\frac{5}{2}, 1\right) \cup (1, \infty)$ **99.**
101.

TECHNOLOGY CONNECTION 4.3A, P. 220

1. The graphs of $y_1 = \text{abs}(4 - 7x)$ and $y_2 = -8$ do not intersect.

EXERCISE SET 4.3, PP. 225–227

1. $\{-7, 7\}$ **3.** $\varnothing$ **5.** $\{-8.6, 8.6\}$ **7.** $\{0\}$
9. $\left\{-1, \frac{1}{5}\right\}$ **11.** $\varnothing$ **13.** $\{-9, 9\}$ **15.** $\{-2, 2\}$
17. $\left\{-\frac{38}{5}, \frac{46}{5}\right\}$ **19.** $\{6, 8\}$ **21.** $\left\{\frac{3}{2}, \frac{7}{2}\right\}$ **23.** $\left\{-\frac{4}{3}, 4\right\}$
25. $\{-8.3, 8.3\}$ **27.** $\left\{-\frac{8}{3}, 4\right\}$ **29.** 11 **31.** 33
33. 34 **35.** $\{1, 11\}$ **37.** $\left\{\frac{3}{2}\right\}$ **39.** $\left\{-4, -\frac{10}{9}\right\}$ **41.** $\mathbb{R}$
43. $\{1\}$ **45.** $\left\{\frac{8}{3}, 32\right\}$
47. $\{a \mid -6 \le a \le 6\}$, or $[-6, 6]$

49. $\{x \mid x < -7 \text{ or } x > 7\}$, or $(-\infty, -7) \cup (7, \infty)$

51. $\{t \mid t < 0 \text{ or } t > 0\}$, or $(-\infty, 0) \cup (0, \infty)$

53. $\{x \mid -2 < x < 8\}$, or $(-2, 8)$

55. $\{x \mid -7 \le x \le 3\}$, or $[-7, 3]$;

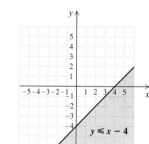

57. $\{x \mid x < -2 \text{ or } x > 8\}$, or $(-\infty, -2) \cup (8, \infty)$

59. $\{y \mid y \text{ is a real number}\}$, or $\mathbb{R}$, or $(-\infty, \infty)$;

61. $\left\{a \mid a \le -\frac{1}{3} \text{ or } a \ge 3\right\}$, or $\left(-\infty, -\frac{1}{3}\right] \cup [3, \infty)$;

63. $\{y \mid -9 < y < 15\}$, or $(-9, 15)$;

65. $\{x \mid x \le -8 \text{ or } x \ge 0\}$, or $(-\infty, -8] \cup [0, \infty)$

67. $\left\{y \mid y < -\frac{4}{3} \text{ or } y > 4\right\}$, or $\left(-\infty, -\frac{4}{3}\right) \cup (4, \infty)$;

69. $\varnothing$

71. $\left\{x \mid x \le -\frac{2}{15} \text{ or } x \ge \frac{14}{15}\right\}$, or $\left(-\infty, -\frac{2}{15}\right] \cup \left[\frac{14}{15}, \infty\right)$;

73. $\{m \mid -12 \le m \le 2\}$, or $[-12, 2]$;

75. $\{a \mid -6 < a < 0\}$, or $(-6, 0)$

77. $\left\{x \mid -\frac{1}{2} \le x \le \frac{7}{2}\right\}$, or $\left[-\frac{1}{2}, \frac{7}{2}\right]$;

79. $\left\{x \mid x \le -\frac{7}{3} \text{ or } x \ge 5\right\}$, or $\left(-\infty, -\frac{7}{3}\right] \cup [5, \infty)$

81. $\{x \mid -4 < x < 5\}$, or $(-4, 5)$

83. 118 child's, 132 adult's   **85.** 68   **87.** ◈   **89.** ◈
91. $\left\{x \mid x \ge \frac{5}{2}\right\}$, or $\left[\frac{5}{2}, \infty\right)$ **93.** $\left\{-\frac{1}{4}, 1\right\}$
95. $\{x \mid -4 \le x \le -1 \text{ or } 3 \le x \le 6\}$, or $[-4, -1] \cup [3, 6]$
97. $|y| \le 5$ **99.** $|x| > 4$ **101.** $|x + 2| < 3$
103. $|x - 5| > 1$, or $|5 - x| > 1$
105. $\{x \mid 1 \le x \le 5\}$, or $[1, 5]$ **107.** ◈   **109.** ◺◹

TECHNOLOGY CONNECTION 4.4, P. 231

1. $y > x + 3.5$
2. $7y \le 2x + 5$
3. $8x - 2y < 11$
4. $11x + 13y + 4 \ge 0$

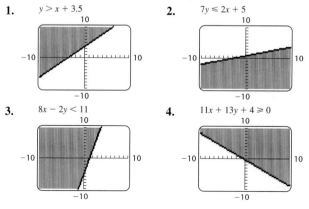

EXERCISE SET 4.4, PP. 234–235
1. Yes   **3.** No

5.

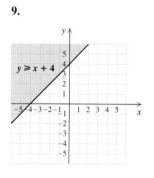
$y < \frac{1}{2}x$

7.

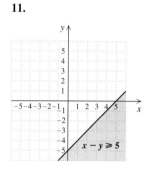
$y \le x - 4$

9.

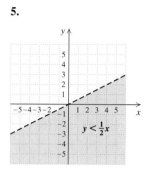
$y \ge x + 4$

11.

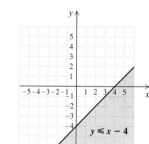
$x - y \ge 5$

13.

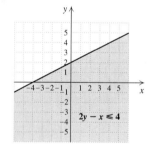
$2x + 3y < 6$

15.

$2y - x \le 4$

17.

$2x - 2y \ge 8 + 2y$

19.

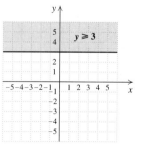
$y \ge 3$

21.

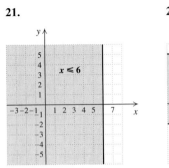

23.

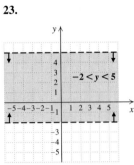

37.

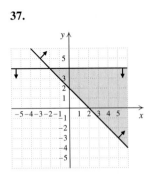

39.

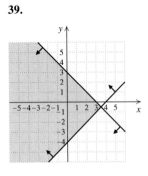

25.

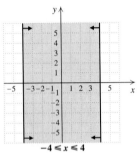

27.

41.

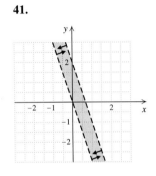

43.

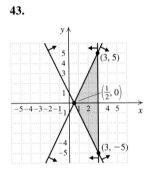

29.

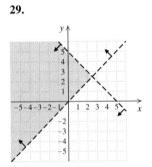

31.

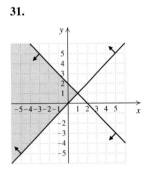

45.

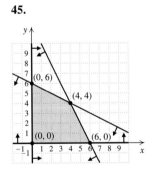

33.

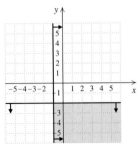

35.

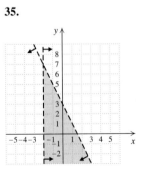

47.

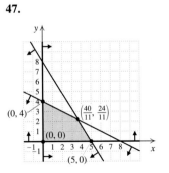

49.

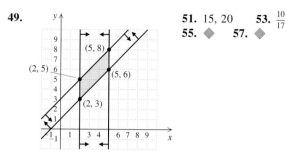

51. 15, 20   **53.** $\frac{10}{17}$
55. ◈   **57.** ◈

(c)

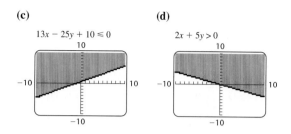

$13x - 25y + 10 \le 0$

(d)

$2x + 5y > 0$

59.

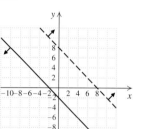

61.

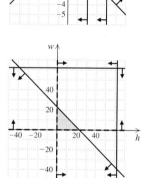

63. $0 < w \le 62$,
$0 < h \le 62$,
$62 + 2w + 2h \le 108$,
or $w + h \le 23$

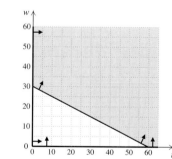

65. $2w + t \ge 60$,
$w \ge 0$,
$t \ge 0$

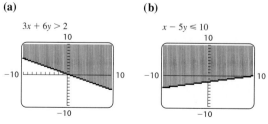

67. (a)

$3x + 6y > 2$

(b)

$x - 5y \le 10$

EXERCISE SET 4.5, PP. 241–242

1. Maximum 168, when $x = 0$, $y = 6$; minimum 0, when
$x = 0$, $y = 0$ **3.** Maximum 152, when $x = 7$, $y = 0$;
minimum 32, when $x = 0$, $y = 4$ **5.** Maximum 5, when
$x = 3$, $y = 7$; minimum -15, when $x = 3$, $y = -3$
7. 40 orders of chili, 50 burritos   **9.** 60 motorcycles,
100 bicycles   **11.** 8 short-answer, 10 word problems; 102
13. Corn: 80 acres; oats: 160 acres
15. City Bank: $7000; State Bank: $15,000; $1395
17. 2 knit, 4 worsted   **19.** ◈   **21.** 30 P-2's, 15 P-3's

REVIEW EXERCISES: CHAPTER 4, PP. 244–245

1. [4.1] $\{x|\ x \le -4\}$,
or $(-\infty, -4]$;

2. [4.1] $\{a|\ a \le -21\}$,
or $(-\infty, -21]$;

3. [4.1] $\{y|\ y \ge -7\}$,
or $[-7, \infty)$;

4. [4.1] $\left\{y|\ y > -\frac{15}{4}\right\}$,
or $\left(-\frac{15}{4}, \infty\right)$;

5. [4.1] $\{y|\ y > -30\}$,
or $(-30, \infty)$;

6. [4.1] $\{x|\ x > -3\}$,
or $(-3, \infty)$;

7. [4.1] $\{x|\ x < -3\}$,
or $(-\infty, -3)$;

8. [4.1] $\left\{y|\ y > -\frac{220}{23}\right\}$,
or $\left(-\frac{220}{23}, \infty\right)$;

9. [4.1] $\left\{x|\ x \le -\frac{5}{2}\right\}$,
or $\left(-\infty, -\frac{5}{2}\right]$;

10. [4.1] $\{x|\ x \le 4\}$, or $(-\infty, 4]$ **11.** [4.1] More than
125 hours   **12.** [4.1] $1500   **13.** [4.2] $\{1, 5, 9\}$
14. [4.2] $\{1, 2, 3, 5, 6, 9\}$
15. [4.2]   ; $(-5, 3]$
16. [4.2]   ; $(-\infty, \infty)$

17. [4.2] $\{x|\ -7 < x \le 2\}$, or $(-7, 2]$

18. [4.2] $\left\{x|\ -\frac{5}{4} < x < \frac{5}{2}\right\}$, or $\left(-\frac{5}{4}, \frac{5}{2}\right)$

19. [4.2] $\{x|\ x < -3\ or\ x > 1\}$, or $(-\infty, -3) \cup (1, \infty)$

20. [4.2] $\{x|\ x < -11\ or\ x \ge -6\}$, or $(-\infty, -11) \cup [-6, \infty)$

21. [4.2] $\{x|\ x \le -6\ or\ x \ge 8\}$, or $(-\infty, -6] \cup [8, \infty)$

22. [4.2] $\left\{x|\ x < -\frac{2}{5}\ or\ x > \frac{8}{5}\right\}$, or $\left(-\infty, -\frac{2}{5}\right) \cup \left(\frac{8}{5}, \infty\right)$

23. [4.2] $[3, \infty)$ **24.** [4.3] $\{-6, 6\}$
25. [4.3] $\{x|\ x \le -3.5\ or\ x \ge 3.5\}$, or $(-\infty, -3.5] \cup [3.5, \infty)$ **26.** [4.3] $\{-5, 9\}$
27. [4.3] $\left\{x|\ -\frac{17}{2} < x < \frac{7}{2}\right\}$, or $\left(-\frac{17}{2}, \frac{7}{2}\right)$
28. [4.3] $\left\{x|\ x \le -\frac{11}{3}\ or\ x \ge \frac{19}{3}\right\}$, or $\left(-\infty, -\frac{11}{3}\right] \cup \left[\frac{19}{3}, \infty\right)$
29. [4.3] $\left\{-14, \frac{4}{3}\right\}$ **30.** [4.3] $\varnothing$
31. [4.3] $\{x|\ -12 \le x \le 4\}$, or $[-12, 4]$
32. [4.3] $\{x|\ x < 0\ or\ x > 10\}$, or $(-\infty, 0) \cup (10, \infty)$
33. [4.3] $\varnothing$
34. [4.4]

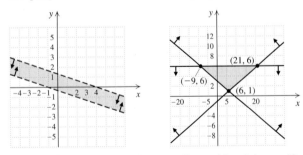

$x - 2y \ge 6$

35. [4.4]

36. [4.4]

(21, 6)

(−9, 6)   (6, 1)

37. [4.5] Maximum 40, when $x = 7$, $y = 15$; minimum 10, when $x = 1$, $y = 3$ **38.** [4.5] 120 from the East plant, 40 from the West plant   **39.** [3.2] $\left(0, \frac{5}{7}\right)$ **40.** [1.3] -22

41. [2.3]

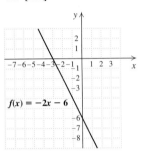

$f(x) = -2x - 6$

42. [3.3] 38,592 ft^2

43. [4.3] ◈ The equation $|x| = p$ has two solutions when p is positive because x can be either p or $-p$. The same equation has no solution when p is negative because no number has a negative absolute value.   **44.** [4.4] ◈ The solution set of a system of inequalities is all ordered pairs that make *all* the individual inequalities true. This consists of ordered pairs that are common to all the individual solution sets, or the intersection of the graphs.
45. [4.3] $\left\{x|\ -\frac{8}{3} \le x \le -2\right\}$, or $\left[-\frac{8}{3}, -2\right]$
46. [4.1] False; $-4 < 3$ is true, but $(-4)^2 < 9$ is false.
47. [4.3] $|d - 1.1| \le 0.03$

TEST: CHAPTER 4, PP. 245–246
1. [4.1] $\{x|\ x < 14\}$, or $(-\infty, 14)$

2. [4.1] $\{y|\ y > -50\}$, or $(-50, \infty)$

3. [4.1] $\{y|\ y \le -2\}$, or $(-\infty, -2]$

4. [4.1] $\left\{a|\ a \le \frac{11}{5}\right\}$, or $\left(-\infty, \frac{11}{5}\right]$

5. [4.1] $\left\{x|\ x > \frac{5}{2}\right\}$, or $\left(\frac{5}{2}, \infty\right)$

6. [4.1] $\left\{x|\ x \le \frac{7}{4}\right\}$, or $\left(-\infty, \frac{7}{4}\right]$

7. [4.1] $\{x|\ x > 1\}$, or $(1, \infty)$
8. [4.1] More than 166 mi   **9.** [4.1] Less than or equal to 2.5 hr   **10.** [4.2] $\{3, 5\}$ **11.** [4.2] $\{1, 3, 5, 7, 9, 11, 13\}$
12. [4.2] $(-\infty, 7]$

13. [4.2] $\{x|-1 < x < 6\}$, or $(-1, 6)$

14. [4.2] $\left\{x\left|-\frac{2}{5} < x \leq \frac{9}{5}\right.\right\}$, or $\left(-\frac{2}{5}, \frac{9}{5}\right]$

15. [4.2] $\{x| x < 3 \text{ or } x > 6\}$, or $(-\infty, 3) \cup (6, \infty)$

16. [4.2] $\left\{x\left| x < -4 \text{ or } x > -\frac{5}{2}\right.\right\}$, or $(-\infty, -4) \cup \left(-\frac{5}{2}, \infty\right)$

17. [4.2] $\left\{x\left| 4 \leq x < \frac{15}{2}\right.\right\}$, or $\left[4, \frac{15}{2}\right)$

18. [4.3] $\{-9, 9\}$

19. [4.3] $\{x| x < -3 \text{ or } x > 3\}$, or $(-\infty, -3) \cup (3, \infty)$

20. [4.3] $\left\{x\left|-\frac{7}{8} < x < \frac{11}{8}\right.\right\}$, or $\left(-\frac{7}{8}, \frac{11}{8}\right)$

21. [4.3] $\left\{x\left| x \leq -\frac{13}{5} \text{ or } x \geq \frac{7}{5}\right.\right\}$, or $\left(-\infty, -\frac{13}{5}\right) \cup \left[\frac{7}{5}, \infty\right)$

22. [4.3] $\varnothing$

23. [4.2] $\left\{x\left| x < \frac{1}{2} \text{ or } x > \frac{7}{2}\right.\right\}$, or $\left(-\infty, \frac{1}{2}\right) \cup \left(\frac{7}{2}, \infty\right)$

24. [4.3] $\{1\}$

25. [4.4]   **26.** [4.4]

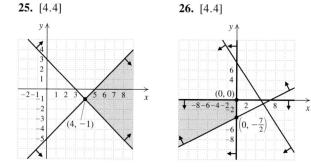

27. [4.5] Maximum 57, when $x = 6$, $y = 9$; minimum 5, when $x = 1$, $y = 0$ **28.** [4.5] $40,000 in municipal bonds, $20,000 in mutual funds; $6800   **29.** [1.3] $-\frac{33}{2}$
30. [3.2] $\left(-\frac{127}{8}, \frac{89}{8}\right)$

31. [2.4]

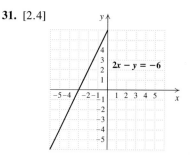

32. [3.3] 30-sec: 10; 60-sec: 5   **33.** [4.3] $[-1, 0] \cup [4, 6]$
34. [4.2] $\left(\frac{1}{5}, \frac{4}{5}\right)$

CHAPTER 5

TECHNOLOGY CONNECTION 5.1A, PP. 250–251

1.

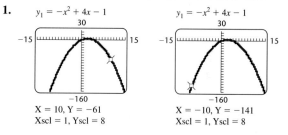

2. This approach is faster than using the TRACE feature, and it is possibly more accurate.

3.   **4.**

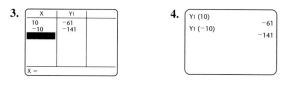

TECHNOLOGY CONNECTION 5.1B, P. 254

1. The graphs appear to coincide in the given window. Moving the cursor from curve to curve does not change the coordinates.
2. A quick visual inspection can be performed more quickly and easily with this procedure.

EXERCISE SET 5.1, PP. 255–258

1. 5, 3, 2, 1, 0; 5   **3.** 3, 7, 6, 0; 7   **5.** 5, 6, 2, 1, 0; 6
7. $-4y^3 - 6y^2 + 7y + 15$; $-4y^3$; -4
9. $3x^7 + 5x^2 - x + 12$; $3x^7$; 3
11. $-a^7 + 8a^5 + 5a^3 - 19a^2 + a$; $-a^7$; -1
13. $-9 + 7x - 5x^2 + 3x^4$
15. $3xy^3 + x^2y^2 - 9x^3y + 2x^4$
17. $-7ab + 4ax - 7ax^2 + 4x^6$ **19.** 45; 5
21. -45; $-8\frac{19}{27}$ **23.** -23 **25.** -12 **27.** 9   **29.** 1320
31. 144 ft   **33.** $18,750   **35.** $8375   **37.** About 399
39. About 340 mg   **41.** $M(5) \approx 65$ **43.** 56.5 in^2

45. $-2a + 6 + 2a^3$ **47.** $-6a^2b - 2b^2$
49. $9x^2 + 2xy + 15y^2$ **51.** $9a + 4b - c$
53. $-4a^2 - b^2 + 3c^2$ **55.** $-2x^2 + x - xy - 1$
57. $5x^2y - 4xy^2 + 5xy$ **59.** $9r^2 + 9r - 9$
61. $-\frac{5}{24}xy - \frac{27}{20}x^3y^2 + 1.4y^3$
63. $-(5x^3 - 7x^2 + 3x - 6),\ -5x^3 + 7x^2 - 3x + 6$
65. $-(-12y^5 + 4ay^4 - 7by^2),\ 12y^5 - 4ay^4 + 7by^2$
67. $13x - 6$ **69.** $-4x^2 - 3x + 13$ **71.** $2a - 4b + 3c$
73. $-2x^2 + 6x$ **75.** $-4a^2 + 8ab - 5b^2$
77. $8a^2b + 16ab + 3ab^2$ **79.** $x^4 - x^2 - 1$ **81.** $9700

83.

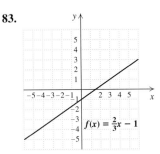

$f(x) = \frac{2}{3}x - 1$

85. $3y - 6$ **87.** ◈ **89.** ◈
91. $68x^5 - 81x^4 - 22x^3 + 52x^2 + 2x + 250$
93. $45x^5 - 8x^4 + 208x^3 - 176x^2 + 116x - 25$
95. $494.55\ \text{cm}^3$
97. $200\pi rh + 20,000\pi r^2\ \text{cm}^2$, or $0.02\pi rh + 2\pi r^2\ \text{m}^2$
99. $x^{6a} - 5x^{5a} - 4x^{4a} + x^{3a} - 2x^{2a} + 8$ **101.**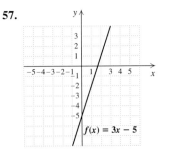
103. $p(0) \ne 5$

TECHNOLOGY CONNECTION 5.2B, P. 264

1. A screen like the one below will be produced for each example.

EXERCISE SET 5.2, PP. 265–266

1. $21a^3$ **3.** $-20x^3y$ **5.** $-10x^5y^6$ **7.** $16x - 8x^2$
9. $15c^3d^2 - 25c^2d^3$ **11.** $6x^2 + 7x - 20$
13. $m^2 - mn - 6n^2$ **15.** $2y^2 + 9xy - 56x^2$
17. $a^4 - 5a^2b^2 + 6b^4$ **19.** $x^3 - 64$ **21.** $x^3 + y^3$
23. $a^4 + 5a^3 - 2a^2 - 9a + 5$
25. $4a^3b^2 + 4a^3b - 10a^2b^2 - 2a^2b + 3ab^3 + 7ab^2 - 6b^3$
27. $x^2 - \frac{3}{4}x + \frac{1}{8}$ **29.** $3x^2 - 1.5xy - 15y^2$
31. $12x^4 - 21x^3 - 14x^2 + 35x - 10$ **33.** $a^2 + 13a + 40$
35. $y^2 - y - 12$ **37.** $x^2 + 6x + 9$
39. $x^2 - 4xy + 4y^2$ **41.** $2x^2 + 13x + 18$
43. $100a^2 - 2.4ab + 0.0144b^2$ **45.** $4x^2 - 4xy - 3y^2$
47. $4a^2 + \frac{4}{3}a + \frac{1}{9}$ **49.** $4x^6 - 12x^3y^2 + 9y^4$
51. $a^4b^4 + 2a^2b^2 + 1$ **53.** $16x^2 - 8x + 1$ **55.** $c^2 - 4$

57. $4a^2 - 1$ **59.** $9m^2 - 4n^2$ **61.** $x^6 - y^2z^2$
63. $-m^2n^2 + m^4$ **65.** $x^4 - 1$ **67.** $a^4 - 2a^2b^2 + b^4$
69. $a^2 + 2ab + b^2 - 1$ **71.** $4x^2 + 12xy + 9y^2 - 16$
73. $A = P + 2Pi + Pi^2$ **75.** **(a)** $t^2 + 3t - 4$;
(b) $5h + 2ah + h^2$ **77.** 8 **79.** A: 75; B: 84; C: 63
81. ◈ **83.** ◈ **85.** $a^{2n}b^{6n2}$ **87.** z^{5n5}
89. $a^{xw+xz}b^{yw+yz}$
91. $-a^4 - 2a^3b + 25a^2 + 2ab^3 - 25b^2 + b^4$
93. $y^{12} - 6y^{10} + 15y^8 - 20y^6 + 15y^4 - 6y^2 + 1$
95. $\frac{4}{9}x^2 - \frac{1}{9}y^2 - \frac{2}{3}y - 1$ **97.** $16x^4 + 4x^2y^2 + y^4$
99. $x^{4a} - y^{4b}$ **101.** $x^{a^2-b^2}$

103.

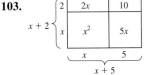

105. (b) and (c) are identities.

TECHNOLOGY CONNECTION 5.3, P. 269

1. A table should show that $y_3 = 0$ for any value of x.

EXERCISE SET 5.3, PP. 270–272

1. $2t(4t + 1)$ **3.** $y(y - 5)$ **5.** $y^2(y + 9)$
7. $5x^2(1 - 3x^2)$ **9.** $4xy(x - 3y)$ **11.** $3(y^2 - y - 3)$
13. $2a(3b - 2d + 6c)$ **15.** $3x^2y^4z^2(3xy^2 - 4x^2z^2 + 5yz)$
17. $-5(x - 3)$ **19.** $-6(y + 12)$ **21.** $-2(x^2 - 2x + 6)$
23. $-3(y^2 - 8x)$ **25.** $-3y(y^2 - 4y + 5)$
27. $-(x^2 - 5x + 9)$ **29.** $-a(a^3 - 2a^2 + 13)$
31. **(a)** $h(t) = -16t(t - 6)$; **(b)** $h(2) = 128$ ft
33. $N(x) = \frac{1}{6}x(x^2 + 3x + 2)$ **35.** $\pi r(2h + r)$
37. $R(x) = 0.4x(700 - x)$ **39.** $(b - 2)(a + c)$
41. $(x + 7)(2x - 3)$ **43.** $(x - y)(a^2 - 5)$
45. $(c + d)(a + b)$ **47.** $(b - 1)(b^2 + 2)$
49. $(a - 3)(a^2 - 2)$ **51.** $12x(2x^2 - 3x + 6)$
53. $x^3(x - 1)(x^2 + 1)$ **55.** $(y^2 + 3)(2y^2 + 5)$

57.

$f(x) = 3x - 5$

59. 9.375 lb of Countryside; 15.625 lb of Mystic
61. ◈ **63.** ◈ **65.** $x^5y^4 + x^4y^6 = x^3y(x^2y^3 + xy^5)$
67. $(x^2 - x + 5)(r + s)$ **69.** $x(5 - y)(x + 2 + 3z)$
71. $2x^a(x^{2a} + 4 + 2x^a)$ **73.** $x^a(4x^b + 7x^{-b})$ **75.**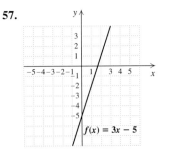

TECHNOLOGY CONNECTION 5.4, P. 275

1. They should coincide.   **2.** The x-axis

3. Let $y_1 = x^3 - x^2 - 30x$, $y_2 = x(x + 5)(x - 6)$, and $y_3 = y_2 - y_1$. The graphs of y_1 and y_2 should coincide; the graph of y_3 should be the x-axis.

4. Let $y_1 = 2x^2 + x - 15$, $y_2 = (2x + 5)(x - 3)$, and $y_3 = y_2 - y_1$. The graphs of y_1 and y_2 do not coincide; the graph of y_3 is not the x-axis.

EXERCISE SET 5.4, PP. 280–281

1. $(x + 2)(x + 6)$ **3.** $(t - 5)(t - 3)$
5. $(x - 9)(x + 3)$ **7.** $2(n - 5)(n - 5)$, or $2(n - 5)^2$
9. $a(a + 9)(a - 8)$ **11.** $(x + 9)(x + 5)$
13. $(y + 9)(y - 7)$ **15.** $(t - 9)(t - 5)$
17. $(x + 5)(x - 2)$ **19.** $3(x + 2)(x + 3)$
21. $(8 - x)(7 + x)$ **23.** $y(8 - y)(4 + y)$
25. $(x^2 + 16)(x^2 - 5)$ **27.** Prime
29. $(p - 8q)(p + 3q)$
31. $(y + 4z)(y + 4z)$, or $(y + 4z)^2$
33. $(p^2 + 79)(p^2 + 1)$ **35.** $(x^4 - 5)(x^4 - 2)$
37. $(3x + 5)(2x - 5)$ **39.** $y(5y + 4)(2y - 3)$
41. $2(4a - 1)(3a - 1)$ **43.** $(5y + 2)(7y + 4)$
45. $2(5t - 3)(t + 1)$ **47.** $4(2x + 1)(x - 4)$
49. $(a^3 - 3)(a^3 + 2)$ **51.** $x^2(7x + 1)(2x - 3)$
53. $4(3a - 4)(a + 1)$ **55.** $(3x + 1)(3x + 4)$
57. $(4 + 3z)(2 - 3z)$ **59.** $-2(2t - 3)(2t + 5)$
61. $xy(6y + 5)(3y - 2)$ **63.** $(24x + 1)(x - 2)$
65. $3x(7x + 3)(3x + 4)$ **67.** $(6x^2 + 5)(4x^2 - 3)$
69. $(4a - 3b)(3a - 2b)$ **71.** $(2x - 3y)(x + 2y)$
73. $(2x - 7y)(3x - 4y)$
75. $(3x - 5y)(3x - 5y)$, or $(3x - 5y)^2$
77. $(9xy - 4)(xy + 1)$
79. 224 ft; 288 ft; 320 ft; 288 ft; 128 ft   **81.** -24
83. ◈   **85.** ◈   **87.** $(2a^2b^3 + 5)(a^2b^3 - 4)$
89. $\left(x + \frac{4}{5}\right)\left(x - \frac{1}{5}\right)$ **91.** $(y - 0.1)(y + 0.5)$
93. $(x^a + 8)(x^a - 3)$ **95.** $(bx + a)(dx + c)$
97. $a^2(p^a + 2)(p^a - 1)$
99. $[3(x - 7) - 1][2(x - 7) + 5]$, or $(3x - 22)(2x - 9)$
101. $31, -31, 14, -14, 4, -4$
103. Since $ax^2 + bx + c = (mx + r)(nx + s)$, from FOIL we know that $a = mn$, $c = rs$, and $b = ms + rn$. If $P = ms$ and $Q = rn$, then $b = P + Q$. Since $ac = mnrs = msrn$, we have $ac = PQ$.   **105.** ◺◹

EXERCISE SET 5.5, PP. 285–286

1. $(x + 4)^2$ **3.** $(a - 8)^2$ **5.** $2(a + 2)^2$ **7.** $(y - 6)^2$
9. $a(a + 12)^2$ **11.** $2(4x + 3)^2$ **13.** $(5y - 8)^2$
15. $(a^2 - 5)^2$ **17.** $(0.5x + 0.3)^2$ **19.** $(p - q)^2$
21. $(5a - 3b)^2$ **23.** $(t^4 + s^4)^2$ **25.** $(x + 4)(x - 4)$
27. $(p + 7)(p - 7)$ **29.** $(ab + 9)(ab - 9)$
31. $6(x + y)(x - y)$ **33.** $7x(y^2 + z^2)(y + z)(y - z)$
35. $a(2a + 7)(2a - 7)$

37. $3(x^4 + y^4)(x^2 + y^2)(x + y)(x - y)$
39. $a^2(3a + 5b^2)(3a - 5b^2)$ **41.** $\left(\frac{1}{5} + x\right)\left(\frac{1}{5} - x\right)$
43. $(a + b + 3)(a + b - 3)$ **45.** $(x - 3 + y)(x - 3 - y)$
47. $(m - n + 5)(m - n - 5)$
49. $(6 + x + y)(6 - x - y)$
51. $(r - 1 - 2s)(r - 1 + 2s)$
53. $(4 + a - b)(4 - a + b)$
55. $(m - 7)(m + 2)(m - 2)$ **57.** $(a - 2)(a + b)(a - b)$
59. $(2, -1, 3)$ **61.** $\left\{x \mid -\frac{4}{7} \le x \le 2\right\}$, or $\left[-\frac{4}{7}, 2\right]$
63. ◈   **65.** ◈   **67.** $-3\left(\frac{1}{2}p - \frac{2}{5}\right)^2$ **69.** $\left(\frac{1}{6}x^4 + \frac{2}{3}\right)^2$
71. $(a + b + c - 3)(a + b - c + 3)$
73. $(x^a + y)(x^a - y)$ **75.** $4(y^{2a} + 5)^2$ **77.** $8(a - 7)^2$
79. $5(c^{50} + 4d^{50})(c^{25} + 2d^{25})(c^{25} - 2d^{25})$
81. $c(c^w + 1)^2$ **83.** $h(2a + h)(2a^2 + 2ah + h^2)$
85. ◺◹

EXERCISE SET 5.6, PP. 289–290

1. $(t - 2)(t^2 + 2t + 4)$ **3.** $(x + 3)(x^2 - 3x + 9)$
5. $(m - 4)(m^2 + 4m + 16)$ **7.** $(2a + 1)(4a^2 - 2a + 1)$
9. $(2 - 3b)(4 + 6b + 9b^2)$ **11.** $(2x + 3)(4x^2 - 6x + 9)$
13. $(y - z)(y^2 + yz + z^2)$ **15.** $\left(x + \frac{1}{3}\right)\left(x^2 - \frac{1}{3}x + \frac{1}{9}\right)$
17. $2(y - 4)(y^2 + 4y + 16)$
19. $8(a + 5)(a^2 - 5a + 25)$ **21.** $r(s + 4)(s^2 - 4s + 16)$
23. $2(y - 3z)(y^2 + 3yz + 9z^2)$
25. $(y + 0.5)(y^2 - 0.5y + 0.25)$
27. $(5c^2 - 2d^2)(25c^4 + 10c^2d^2 + 4d^4)$
29. $3z^2(z - 1)(z^2 + z + 1)$ **31.** $(t^2 + 1)(t^4 - t^2 + 1)$
33. $(p + q)(p^2 - pq + q^2)(p - q)(p^2 + pq + q^2)$
35. $(a^3 + b^4c^5)(a^6 - a^3b^4c^5 + b^8c^{10})$ **37.** 228 ft^2
39. Slope: $\frac{4}{3}$; y-intercept: $\left(0, -\frac{8}{3}\right)$
41. $\left\{x \mid -\frac{33}{5} \le x \le 9\right\}$, or $\left[-\frac{33}{5}, 9\right]$ **43.** ◈   **45.** ◈
47. $2(x^a + 2y^b)(x^{2a} - 2x^ay^b + 4y^{2b})$
49. $\frac{1}{2}\left(\frac{1}{2}x^a + y^{2a}z^{3b}\right)\left(\frac{1}{4}x^{2a} - \frac{1}{2}x^ay^{2a}z^{3b} + y^{4a}z^{6b}\right)$ or $\frac{1}{16}(x^a + 2y^{2a}z^{3b})(x^{2a} - 2x^ay^{2a}z^{3b} + 4y^{4a}z^{6b})$
51. $-y(3x^2 + 3xy + y^2)$ **53.** $-(3x^{4a} - 3x^{2a} + 1)$
55. $h(3a^2 + 3ah + h^2)$
57.

$h(x) = (x - 2)^3$
$f(x) = x^3$
$g(x) = x^3 - 8$

59. ◺◹

EXERCISE SET 5.7, PP. 293–294

1. $5(m^2 + 2)(m^2 - 2)$ **3.** $(a + 9)(a - 9)$
5. $(4x + 1)(2x - 5)$ **7.** $(a + 5)^2$ **9.** $3(x + 12)(x - 7)$
11. $(3x + 5y)(3x - 5y)$

13. $(m + 1)(m^2 - m + 1)(m - 1)(m^2 + m + 1)$
15. $(x + y + 3)(x - y + 3)$
17. $(7x + 3y)(49x^2 - 21xy + 9y^2)$
19. $(m^3 + 10)(m^3 - 2)$ **21.** $(a + d)(c - b)$
23. $(2c - d)^2$ **25.** $(8 + 3t^2)(3 + t)$
27. $2(x + 3)(x + 2)(x - 2)$
29. $2(3a - 2b)(9a^2 + 6ab + 4b^2)$ **31.** $(6y - 5)(6y + 7)$
33. $(a^4 + b^4)(a^2 + b^2)(a + b)(a - b)$
35. $ab(a + 4b)(a - 4b)$ **37.** $2(a - 3)(a + 3)$
39. $7a(a^3 - 2a^2 + 3a - 1)$ **41.** $(9ab + 2)(3ab + 4)$
43. $p(1 + 4p)(1 - 4p + 16p^2)$
45. $(a - b - 3)(a + b + 3)$

47.

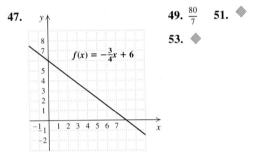

$f(x) = -\frac{3}{4}x + 6$

49. $\frac{80}{7}$ **51.** ◈
53. ◈

55. $(12x + 5y^2)(5x + 6y^2)$ **57.** $(11 + y^2)(2 + y)(2 - y)$
59. $(a + b)(a - b)(a + 7b)(a - 7b)$
61. $(3x^{2s} + 4y^t)(9x^{4s} - 12x^{2s}y^t + 16y^{2t})$
63. $(2x + y - r + 3s)(2x + y + r - 3s)$
65. $6(2t^a + 1)(2t^a - 1)$ **67.** $\left(\frac{x^9}{10} - 1\right)\left(\frac{x^{18}}{100} + \frac{x^9}{10} + 1\right)$
69. $3(x - 3)(x + 2)$ **71.** $3(a + 7)^2$ **73.** 0

TECHNOLOGY CONNECTION 5.8, P. 298

1. $\{-5.00, 2.00\}$ **2.** $\{-4.00, 6.00\}$ **3.** No solution
4. $\{-8.98, -4.56\}$ **5.** $\{-1.21, 3.45, 8.98\}$
6. $\{-1.50, 3.97, 4.43\}$ **7.** $\{-3.68, 0.97\}$
8. $\{0.51, 4.08\}$

EXERCISE SET 5.8, PP. 302–305

1. $\{-9, 5\}$ **3.** $\{1\}$ **5.** $\{6\}$ **7.** $\{-5, -4\}$ **9.** $\{0, 8\}$
11. $\{-5, 0, 8\}$ **13.** $\{-4, 4\}$ **15.** $\{-9, 9\}$ **17.** $\left\{\frac{2}{3}, 2\right\}$
19. $\left\{-2, -\frac{3}{4}, 0\right\}$ **21.** $\{-4, 5\}$ **23.** $\left\{-\frac{7}{4}, \frac{4}{3}\right\}$
25. $\left\{-\frac{1}{8}, \frac{1}{8}\right\}$ **27.** $\{-5, -1, 1, 5\}$ **29.** $\{-8, -4\}$
31. $\left\{-4, \frac{3}{2}\right\}$ **33.** $\{-9, -3\}$
35. $\{x \mid x \in \mathbb{R} \text{ and } x \neq -1 \text{ and } x \neq 5\}$
37. $\{x \mid x \in \mathbb{R} \text{ and } x \neq -3 \text{ and } x \neq 3\}$
39. $\left\{x \mid x \in \mathbb{R} \text{ and } x \neq 0 \text{ and } x \neq \frac{1}{2}\right\}$
41. $\{x \mid x \in \mathbb{R} \text{ and } x \neq 0 \text{ and } x \neq 2 \text{ and } x \neq 5\}$
43. -13 or 12 **45.** Length: 12 cm; width: 7 cm
47. 6 cm   **49.** 3 cm   **51.** 5 ft
53. Height: 6 ft; base: 4 ft   **55.** 16, 18, 20
57. 150 ft by 200 ft   **59.** 41 ft
61. Length: 100 m; width: 75 m   **63.** 11 sec   **65.** 6

67. Faster car: 54 mph; slower car: 39 mph
69. $\left\{x \mid x > \frac{10}{3}\right\}$, or $\left(\frac{10}{3}, \infty\right)$ **71.** ◈   **73.** ◈
75. $\left\{-\frac{11}{8}, -\frac{1}{4}, \frac{2}{3}\right\}$ **77.** $\{1\}$
79. $\{-3, 1\}$; $\{x \mid -4 \leq x \leq 2\}$, or $[-4, 2]$
81. Answers may vary. $g(x) = 3x^3 - 9x^2 - 39x + 45$
83. 14 cm by 28 cm   **85.** About 5.7 sec   **87.** ⌇
89. $\{-4.00, 1.00\}$ **91.** $\{-3.33, 5.15\}$ **93.** ◈ ⌇

REVIEW EXERCISES: CHAPTER 5, PP. 307–308

1. [5.1] 7, 11, 3, 0; 11
2. [5.1] $-5x^3 + 2x^2 + 4x - 7$; $-5x^3$; -5
3. [5.1] $-3x^2 + 2x^3 + 3x^6y - 7x^8y^3$
4. [5.1] 0; -6 **5.** [5.1] 4   **6.** [5.1] $-x^2y - 2xy^2$
7. [5.1] $ab + 4 + 12ab^2$ **8.** [5.1] $-x^3 + 2x^2 + 5x + 2$
9. [5.1] $-3x^4 + 3x^3 - x + 16$
10. [5.1] $-8xy^2 + 4xy + 13x^2y$ **11.** [5.1] $9x - 7$
12. [5.1] $-2a + 6b + 7c$ **13.** [5.1] $6x^2 - 7xy + 3y^2$
14. [5.2] $-18x^3y^4$ **15.** [5.2] $x^8 - x^6 + 5x^2 - 3$
16. [5.2] $8a^2b^2 + 2abc - 3c^2$ **17.** [5.2] $4x^2 - 25y^2$
18. [5.2] $4x^2 - 20xy + 25y^2$ **19.** [5.2] $2x^2 + 5x - 3$
20. [5.2] $x^4 + 8x^2y^3 + 16y^6$ **21.** [5.2] $x^3 - 125$
22. [5.2] $x^2 - \frac{1}{2}x + \frac{1}{18}$ **23.** [5.3] $x(6x + 5)$
24. [5.3] $3y^2(3y^2 - 1)$ **25.** [5.3] $3x(5x^3 - 6x^2 + 7x - 3)$
26. [5.4] $(a - 9)(a - 3)$ **27.** [5.4] $(3m + 2)(m + 4)$
28. [5.5] $(5x + 2)^2$ **29.** [5.5] $4(y + 2)(y - 2)$
30. [5.4] $x(x - 2)(x + 7)$ **31.** [5.3] $(a + 2b)(x - y)$
32. [5.3] $(y + 2)(3y^2 - 5)$
33. [5.5] $(a^2 + 9)(a + 3)(a - 3)$
34. [5.3] $4(x^4 + x^2 + 5)$
35. [5.6] $(3x - 2)(9x^2 + 6x + 4)$
36. [5.6] $(0.4b - 0.5c)(0.16b^2 + 0.2bc + 0.25c^2)$
37. [5.3] $y(y^4 + 1)$ **38.** [5.3] $2z^6(z^2 - 8)$
39. [5.6] $2y(3x^2 - 1)(9x^4 + 3x^2 + 1)$
40. [5.5] $4(3x - 5)^2$ **41.** [5.4] $(3t + p)(2t + 5p)$
42. [5.5] $(x + 3)(x - 3)(x + 2)$
43. [5.5] $(a - b + 2t)(a - b - 2t)$ **44.** [5.8] $\{10\}$
45. [5.8] $\left\{\frac{2}{3}, \frac{3}{2}\right\}$ **46.** [5.8] $\left\{0, \frac{7}{4}\right\}$ **47.** [5.8] $\{-4, 4\}$
48. [5.8] $\{-4, 11\}$
49. [5.8] $\left\{x \mid x \in \mathbb{R} \text{ and } x \neq -7 \text{ and } x \neq \frac{2}{3}\right\}$ **50.** [5.8] 5
51. [5.8] 3, 5, 7; $-7, -5, -3$ **52.** [5.8] 5 in. by 8 in.
53. [3.4] $(0, -2, 7)$ **54.** [4.1] $\left\{x \mid x > -\frac{7}{3}\right\}$, or $\left(-\frac{7}{3}, \infty\right)$
55. [4.3] $\left\{x \mid -\frac{4}{3} \leq x \leq 8\right\}$, or $\left[-\frac{4}{3}, 8\right]$
56. [2.3]

$f(x) = \frac{2}{3}x + 2$

57. [3.5] 31 multiple-choice, 26 true–false, 13 fill-in
58. [5.8] ◈ The roots of a polynomial function are the x-coordinates of the points at which the graph of the function crosses the x-axis.
59. [5.8] ◈ The principle of zero products states that if a product is equal to 0, at least one of the factors must be 0. If a product is nonzero, we cannot conclude that any one of the factors is a particular value.
60. [5.6] $2(2x - y)(4x^2 + 2xy + y^2)(2x + y) \cdot$ $(4x^2 - 2xy + y^2)$ **61.** [5.6] $-2(3x^2 + 1)$
62. [5.2], [5.6] $a^3 - b^3 + 3b^2 - 3b + 1$ **63.** [5.2] z^{5n^5}
64. [5.8] $\left\{0, \frac{1}{8}, -\frac{1}{8}\right\}$

TEST: CHAPTER 5, PP. 308–309

1. [5.1] 9   **2.** [5.1] $5x^5y^4 - 2x^4y - 4x^2y + 3xy^3$
3. [5.1] $-4a^3$ **4.** [5.1] 4; 2   **5.** [5.2] $2ah + h^2 - 5h$
6. [5.1] $3xy + 3xy^2$ **7.** [5.1] $-3x^3 + 3x^2 - 6y - 7y^2$
8. [5.1] $7m^3 + 2m^2n + 3mn^2 - 7n^3$ **9.** [5.1] $6a - 8b$
10. [5.1] $2y^2 + 5y + y^3$ **11.** [5.2] $64x^3y^3$
12. [5.2] $12a^2 - 4ab - 5b^2$ **13.** [5.2] $x^3 - 2x^2y + y^3$
14. [5.2] $-3m^4 - 13m^3 + 5m^2 + 26m - 10$
15. [5.2] $16y^2 - 72y + 81$ **16.** [5.2] $x^2 - 4y^2$
17. [5.3] $5x^2(3 - x^2)$ **18.** [5.5] $(y + 5)(y + 2)(y - 2)$
19. [5.4] $(p - 14)(p + 2)$ **20.** [5.4] $(6m + 1)(2m + 3)$
21. [5.5] $(3y + 5)(3y - 5)$
22. [5.6] $3(r - 1)(r^2 + r + 1)$ **23.** [5.5] $(3x - 5)^2$
24. [5.5] $(x^4 + y^4)(x^2 + y^2)(x + y)(x - y)$
25. [5.5] $(y + 4 + 10t)(y + 4 - 10t)$
26. [5.5] $5(2a - b)(2a + b)$ **27.** [5.4] $2(4x - 1)(3x - 5)$
28. [5.6] $2ab(2a^2 + 3b^2)(4a^4 - 6a^2b^2 + 9b^4)$
29. [5.3] $4xy(y^3 + 9x + 2x^2y - 4)$ **30.** [5.8] $\{6, -3\}$
31. [5.8] $\{-5, 5\}$ **32.** [5.8] $\left\{-\frac{3}{2}, -7\right\}$
33. [5.8] $\left\{-\frac{1}{3}, 0\right\}$ **34.** [5.8] $\{0, 5\}$
35. [5.8] $\{x \mid x \in \mathbb{R} \text{ and } x \neq -1\}$
36. [5.8] Length: 8 cm; width: 5 cm
37. [4.3] $\left\{x \mid -6 < x < \frac{2}{3}\right\}$, or $\left(-6, \frac{2}{3}\right)$
38. [4.1] $\{x \mid x > 1\}$, or $(1, \infty)$ **39.** [3.4] $(5, -1, -2)$
40. [2.3] Slope: $-\frac{1}{3}$; y-intercept: $\left(0, \frac{8}{3}\right)$
41. **(a)** [5.2] $x^5 + x + 1$;
(b) [5.2], [5.7] $(x^2 + x + 1)(x^3 - x^2 + 1)$
42. [5.4] $(3x^n + 4)(2x^n - 5)$

CHAPTER 6

TECHNOLOGY CONNECTION 6.1, P. 315

1. Let $y_1 = (7x^2 + 21x)/(14x)$, $y_2 = (x + 3)/2$, and $y_3 = y_1 - y_2$ (or $y_2 - y_1$). A table or the TRACE feature can be used to show that, except when $x = 0$, y_3 is always 0. As an alternative, let $y_1 = (7x^2 + 21x)/(14x) - (x + 3)/2$ and show that, except when $x = 0$, y_1 is always 0.
2. Let $y_1 = (x + 3)/x$, $y_2 = 3$, and $y_3 = y_1 - y_2$ (or $y_2 - y_1$). Use a table or the TRACE feature to show that y_3 is not always 0. As an alternative, let $y_1 = (x + 3)/x - 3$ and

show that y_1 is not always 0.

EXERCISE SET 6.1, PP. 318–320

1. $\frac{2}{3}$; $\frac{23}{6}$; $\frac{163}{10}$ **3.** 0; -184; does not exist   **5.** $-\frac{9}{4}$; does not exist; $-\frac{11}{9}$ **7.** $\frac{5x(x - 3)}{5x(x + 2)}$ **9.** $\frac{(t - 2)(-1)}{(t + 3)(-1)}$ **11.** $\frac{3}{x}$
13. $\frac{2}{3t^4}$ **15.** $a - 5$ **17.** $\frac{3}{5a - 6}$ **19.** $\frac{x - 4}{x + 5}$ **21.** $\frac{5}{x}$
23. $-\frac{1}{2}$ **25.** $\frac{8}{t + 2}$ **27.** $-\frac{1}{1 + 2t}$ **29.** $-\frac{6}{5}$
31. $\frac{a - 5}{a + 5}$ **33.** $\frac{x + 8}{x - 4}$ **35.** $\frac{4 + t}{4 - t}$ **37.** $\frac{7b^2}{6a^4}$ **39.** $\frac{3x^2}{25}$
41. $\frac{y + 4}{2}$ **43.** $\frac{(x + 4)(x - 4)}{x(x + 3)}$ **45.** $-\frac{a + 1}{2 + a}$
47. $\frac{-2}{t^4(t + 3)(t + 2)}$ **49.** $\frac{(x + 5)(2x + 3)}{7x}$
51. $c(c - 2)$ **53.** $\frac{a^2 + ab + b^2}{3(a + 2b)}$ **55.** $\frac{1}{2x + 3y}$
57. $6x^4y^7$ **59.** $\frac{6}{x^5}$ **61.** $\frac{(x + 2)(x + 4)}{x^7}$
63. $-\frac{5x + 2}{x - 3}$ **65.** $-\frac{1}{y^3}$ **67.** $\frac{(x + 4)(x + 2)}{3(x - 5)}$
69. $\frac{y(y^2 + 3)}{(y + 3)(y - 2)}$ **71.** $\frac{x^2 + 4x + 16}{(x + 4)^2}$ **73.** $\frac{(2a + b)^2}{2(a + b)}$
75. $\left(\frac{7}{2}, \frac{5}{2}\right)$ **77.** $\frac{16}{3}$ **79.** ◈   **81.** ◈

83.

y or $f(x)$

$f(x) = \dfrac{x^2 - 9}{x - 3}$

85. $\frac{(d - 1)(d - 5)}{5d(d + 5)}$ **87.** $\frac{-5}{x + 2}$ **89.** $\frac{a^2 + 2}{a^2 - 3}$
91. $\frac{(u^2 - uv + v^2)^2}{u - v}$ **93. (a)** $\frac{16(x + 1)}{(x - 1)^2(x^2 + x + 1)}$;
(b) $\frac{x^2 + x + 1}{(x + 1)^3}$; **(c)** $\frac{(x + 1)^3}{x^2 + x + 1}$ **95.** ⊠   **97.** ◈ ⊠

EXERCISE SET 6.2, PP. 327–329

1. $\frac{4}{a}$ **3.** $-\frac{1}{a^2b}$ **5.** 2   **7.** $\frac{3y + 5}{y - 2}$ **9.** $\frac{1}{x - 4}$
11. $\frac{-1}{a + 5}$ **13.** $a + b$ **15.** $\frac{11}{x}$ **17.** $\frac{2}{x + 4}$

19. $\dfrac{1}{t^2 + 4}$ **21.** $\dfrac{1}{m^2 + mn + n^2}$ **23.** $\dfrac{2a^2 - a + 14}{(a - 4)(a + 3)}$

25. $\dfrac{3x - 1}{x + 1}$ **27.** $\dfrac{x + y}{x - y}$ **29.** $\dfrac{3x^2 + 7x + 14}{(2x - 5)(x - 1)(x + 2)}$

31. $\dfrac{8x + 1}{(x + 1)(x - 1)}$ **33.** $\dfrac{-x + 34}{20(x + 2)}$

35. $\dfrac{-a^2 + 7ab - b^2}{(a - b)(a + b)}$ **37.** $\dfrac{x - 5}{(x + 5)(x + 3)}$

39. $\dfrac{y}{(y - 2)(y - 3)}$ **41.** $\dfrac{7x + 1}{x - y}$

43. $\dfrac{3y^2 - 3y - 29}{(y - 3)(y + 8)(y - 4)}$ **45.** $\dfrac{-2a^2 - 7a - 10}{(a + 4)(a - 4)(a + 6)}$

47. $\dfrac{10a - 16}{(a^2 - 5a + 4)(a^2 - 4)}$ **49.** $\dfrac{4t^2 - 2t - 14}{(t + 2)(t - 2)}$

51. $\dfrac{-y}{(y + 3)(y - 1)}$ **53.** $\dfrac{-14y^2 - 3y + 3}{(2y + 1)(2y - 1)}$

55. $\dfrac{-6x + 42}{(x - 3)(x - 2)(x + 1)(x + 3)}$ **57.** $\dfrac{3y^6 z^7}{7x^5}$

59. 2 rolls of dimes, 5 rolls of nickels, 5 rolls of quarters

61. ◈   **63.** ◈   **65.** 12

67. $x^4(x^2 + 1)(x + 1)(x - 1)(x^2 + x + 1)(x^2 - x + 1)$

69. $8a^4, 8a^4 b, 8a^4 b^2, 8a^4 b^3, 8a^4 b^4, 8a^4 b^5, 8a^4 b^6, 8a^4 b^7$

71. $\dfrac{x^4 + 6x^3 + 2x^2}{(x + 2)(x - 2)(x + 5)}$ **73.** $\dfrac{x^5}{(x^2 - 4)(x^2 + 3x - 10)}$

75. $\dfrac{9x^2 + 28x + 15}{(x - 3)(x + 3)^2}$ **77.** $\dfrac{1}{2x(x - 5)}$ **79.** $-4t^4$

81.

TECHNOLOGY CONNECTION 6.3, P. 333

1. $-1, 0, \dfrac{2}{3}, 2$

EXERCISE SET 6.3, PP. 335–337

1. $\dfrac{5a + 1}{1 - 2a}$ **3.** $\dfrac{x^2 - 1}{x^2 + 1}$ **5.** $\dfrac{3y + 4x}{4y - 3x}$ **7.** $\dfrac{x + y}{x}$

9. $\dfrac{x^2(3 - y)}{y^2(2x - 1)}$ **11.** $\dfrac{1}{a - b}$ **13.** $-\dfrac{1}{x(x + h)}$

15. $\dfrac{(a - 2)(a - 7)}{(a + 1)(a - 6)}$ **17.** $\dfrac{3y - 1}{7y - 5}$ **19.** $\dfrac{a^2 - 3a - 6}{a^2 - 2a - 3}$

21. $\dfrac{x + 2}{x + 3}$ **23.** $\dfrac{2a - 3}{a + 1}$ **25.** $\dfrac{7 - 3x}{3 - 2x}$ **27.** $\dfrac{1}{y + 3}$

29. $-y$ **31.** $\dfrac{2(5a^2 + 4a + 12)}{5(a^2 + 6a + 18)}$ **33.** $\dfrac{(2x + 1)(x + 2)}{2x(x - 1)}$

35. $\dfrac{(2a - 3)(a + 5)}{2(a - 3)(a + 2)}$ **37.** $\dfrac{-y - 1}{y + 8}$ **39.** $\dfrac{2x^2 - 11x - 27}{2x^2 + 21x + 13}$

41. 22

43.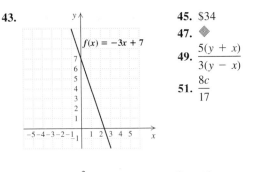

45. \$34

47. ◈

49. $\dfrac{5(y + x)}{3(y - x)}$

51. $\dfrac{8c}{17}$

53. $\dfrac{x^2}{x^4 + x^3 + x^2 + x + 1}$ **55.** $\dfrac{2a - 1}{-a + 5}$ **57.** $\dfrac{-5}{x(x + h)}$

59. $\dfrac{1}{(1 + x + h)(1 + x)}$ **61.** $\{x \mid x$ is a real number and $x \neq 1$ and $x \neq -1$ and $x \neq 4$ and $x \neq -4$ and $x \neq 5$ and $x \neq -5\}$ **63.** ⬚   **65.** ◈ ⬚

EXERCISE SET 6.4, PP. 342–343

1. $\dfrac{51}{5}$ **3.** 144   **5.** $\dfrac{40}{9}$ **7.** 5   **9.** No solution

11. $-4, -1$ **13.** No solution   **15.** 11   **17.** 2

19. 2, 3   **21.** -23 **23.** -1 **25.** $-\dfrac{5}{2}, 3$ **27.** 14

29. $\dfrac{3}{4}$ **31.** 4   **33.** 3   **35.** No solution   **37.** $-\dfrac{7}{3}$

39. $\dfrac{1}{7}$ **41.** $(9x^2 + y^2)(3x + y)(3x - y)$

43. $\{x \mid x < -1 \text{ or } x > 5\}$, or $(-\infty, -1) \cup (5, \infty)$

45. 16 and 18, -18 and -16 **47.** ◈   **49.** ◈

51. -8 **53.** $\{a \mid a$ is a real number and $a \neq -1$ and $a \neq 1\}$ **55.** 0.0854697   **57.** Yes   **59.** ⬚   **61.** ⬚

EXERCISE SET 6.5, PP. 350–352

1. 2   **3.** $-3, -2$ **5.** 6 and 7, -7 and -6 **7.** $1\dfrac{5}{7}$ hr

9. $8\dfrac{4}{7}$ hr   **11.** 2.475 hr   **13.** 6 hr   **15.** Sara: 6 hr; Kate: 3 hr   **17.** Zsuzanna: $\dfrac{4}{3}$ hr; Stan: 4 hr

19. 300 min, or 5 hr   **21.** New machine: $22\dfrac{1}{2}$ hr; old machine: 45 hr   **23.** 12 mph   **25.** 5.2 ft/sec

27. Freight: 66 mph; passenger: 80 mph

29. Local: 35 mph; express: 42 mph   **31.** 9 km/h

33. Jaime: 23 km/h; Mara: 15 km/h   **35.** 5 m per minute

37. 20 mph   **39.** $\{x \mid x$ is a real number and $x \neq 5$ and $x \neq -1\}$ **41.** $2y - 8xy^2 + 6xy$ **43.** ◈   **45.** ◈

47. 40 min   **49.** 2250   **51.** $\dfrac{119}{8}$ mi   **53.** 30 mi

55. $8\dfrac{2}{11}$ min after 10:30   **57.** $51\dfrac{3}{7}$ mph

EXERCISE SET 6.6, PP. 357–358

1. $4x^4 + 3x^3 - 6$ **3.** $4a^2 + a - \dfrac{3}{7} - \dfrac{2}{a}$

5. $13y - \dfrac{9}{2} - \dfrac{4}{y}$ **7.** $-6x^5 + 9x^2 + 1$

9. $1 - ab^2 - a^3 b^4$ **11.** $-2pq + 3p - 4q$ **13.** $x + 7$

15. $a - 12 + \dfrac{32}{a + 4}$ **17.** $x - 6 + \dfrac{-7}{x - 5}$ **19.** $y - 5$

21. $y^2 - 2y - 1 + \dfrac{-8}{y - 2}$ **23.** $2x^2 - x + 1 + \dfrac{-5}{x + 2}$

25. $a^2 + 4a + 15 + \dfrac{72}{a - 4}$ **27.** $2y^2 + 2y - 1 + \dfrac{8}{5y - 2}$

29. $2x^2 - x - 9 + \dfrac{3x + 12}{x^2 + 2}$ **31.** $16x^2 + 8x + 4, x \neq \frac{1}{2}$

33. $2x - 5, x \neq -\frac{2}{3}$ **35.** $x^2 + 6, x \neq -3, x \neq 3$

37. $2x^3 - 3x^2 + 5, x \neq -1, x \neq 1$ **39.** $\{0, 5\}$

41. 12, 13, 14   **43.** $\{-2, 5\}$ **45.** ◈

47. ◈   **49.** $x^2 + 2y$

51. $a^6 - a^5b + a^4b^2 - a^3b^3 + a^2b^4 - ab^5 + b^6$

53. $-\frac{3}{2}$ **55.** ◈   **57.** 📈

EXERCISE SET 6.7, PP. 362–363

1. $x^2 - x + 1$, R -4, or $x^2 - x + 1 + \dfrac{-4}{x - 1}$

3. $a + 7$, R -47, or $a + 7 + \dfrac{-47}{a + 4}$

5. $x^2 - 5x - 23$, R -43, or $x^2 - 5x - 23 + \dfrac{-43}{x - 2}$

7. $3x^2 - 2x + 2$, R -3, or $3x^2 - 2x + 2 + \dfrac{-3}{x + 3}$

9. $y^2 + 2y + 1$, R 12, or $y^2 + 2y + 1 + \dfrac{12}{y - 2}$

11. $x^4 + 2x^3 + 4x^2 + 8x + 16$

13. $3x^2 + 6x - 3$, R 2, or $3x^2 + 6x - 3 + \dfrac{2}{x + \frac{1}{3}}$ **15.** 3

17. 4   **19.** 54

21.

$2x - 3y < 6$

23.

$y > 4$

25.

$y - 2 = \frac{3}{4}(x + 1)$

27. ◈   **29.** ◈   **31. (a)** The degree of R must be less than 1, the degree of $x - r$; **(b)** Let $x = r$. Then

$$\begin{aligned} P(r) &= (r - r) \cdot Q(r) + R \\ &= 0 \cdot Q(r) + R \\ &= R. \end{aligned}$$

33. $0; -\frac{7}{2}, \frac{5}{3}, 4$ **35.** 📈   **37.** 0

EXERCISE SET 6.8, PP. 366–367

1. $W_1 = \dfrac{d_1 W_2}{d_2}$ **3.** $v_1 = \dfrac{2s}{t} - v_2$, or $\dfrac{2s - tv_2}{t}$

5. $R = \dfrac{r_1 r_2}{r_2 + r_1}$ **7.** $g = \dfrac{Rs}{s - R}$ **9.** $R = \dfrac{2V}{I} - 2r$, or

$\dfrac{2V - 2Ir}{I}$ **11.** $p = \dfrac{qf}{q - f}$ **13.** $n = \dfrac{IR}{E - Ir}$

15. $t_1 = \dfrac{H}{Sm} + t_2$, or $\dfrac{H + Smt_2}{Sm}$ **17.** $r = \dfrac{Re}{E - e}$

19. $r = 1 - \dfrac{a}{S}$, or $\dfrac{S - a}{S}$ **21.** $r = \dfrac{A}{P} - 1$, or

$\dfrac{A - P}{P}$ **23.** 3:45 A.M.   **25.** 12   **27.** $5\frac{5}{9}$ ohms

29. 6 cm   **31.** $h = \dfrac{2gR^2}{V^2} - R$, or $\dfrac{2gR^2 - RV^2}{V^2}$

33. $Q = \dfrac{2Tt - 2AT}{A - q}$ **35.** $(-15, 2), (-30, 8)$

37.

$6x - y < 6$

39. $(t + 2b)(t^2 - 2tb + 4b^2)$ **41.** ◈

43. $M = \dfrac{2ab}{b + a}$

45. $t_1 = t_2 + \dfrac{(d_2 - d_1)(t_4 - t_3)}{a(t_4 - t_2)(t_4 - t_3) + d_3 - d_4}$

REVIEW EXERCISES: CHAPTER 6, PP. 369–371

1. [6.1] **(a)** $-\dfrac{2}{9}$; **(b)** $-\dfrac{3}{4}$; **(c)** 0   **2.** [6.2] $48x^3$

3. [6.2] $(x + 5)(x - 2)(x - 4)$ **4.** [6.2] $x + 3$

5. [6.2] $\dfrac{1}{x - 1}$ **6.** [6.1] $\dfrac{b^2 c^6 d^2}{a^5}$ **7.** [6.2] $\dfrac{15np + 14m}{18m^2 n^4 p^2}$

8. [6.1] $\dfrac{y - 8}{2}$ **9.** [6.1] $\dfrac{(x - 2)(x + 5)}{x - 5}$

10. [6.1] $\dfrac{3a-1}{a-3}$ **11.** [6.1] $\dfrac{(x^2+4x+16)(x-6)}{(x+4)(x+2)}$

12. [6.2] $\dfrac{x-3}{(x+1)(x+3)}$ **13.** [6.2] $\dfrac{x-y}{x+y}$

14. [6.2] $2(x+y)$ **15.** [6.2] $\dfrac{-y}{(y+4)(y-1)}$

16. [6.3] $\dfrac{5}{7}$ **17.** [6.3] $\dfrac{a^2b^2}{2(b^2-ba+a^2)}$

18. [6.3] $\dfrac{(y+11)(y+5)}{(y-5)(y+2)}$ **19.** [6.3] $\dfrac{(14-3x)(x+3)}{2x^2+16x+6}$

20. [6.4] 2   **21.** [6.4] 6   **22.** [6.4] No solution

23. [6.4] $-1, 4$ **24.** [6.5] $5\frac{1}{7}$ hr   **25.** [6.5] 24 mph

26. [6.5] Motorcycle: 62 mph; car: 70 mph

27. [6.6] $4s^2+3s-2rs^2$ **28.** [6.6] $y^2-5y+25$

29. [6.6] $4x+3+\dfrac{-9x-5}{x^2+1}$

30. [6.7] $x^2+6x+20+\dfrac{54}{x-3}$ **31.** [6.7] 341

32. [6.8] $s=\dfrac{Rg}{g-R}$ **33.** [6.8] $m=\dfrac{H}{S(t_1-t_2)}$

34. [6.8] $c=\dfrac{b+3a}{2}$ **35.** [6.8] $t_1=\dfrac{-A}{vT}+t_2$, or

$\dfrac{-A+vTt_2}{vT}$

36. [4.4]   **37.** [4.4]

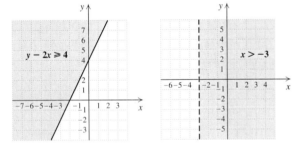

38. [5.6] $(5x-2y)(25x^2+10xy+4y^2)$

39. [5.4] $(x+6)(6x-7)$ **40.** [5.8] $-6, \frac{7}{6}$ **41.** [2.2] 40

42. [6.2], [6.3], [6.4] ◈ The least common denominator was used to add and subtract rational expressions, to simplify complex rational expressions, and to solve rational equations.   **43.** [6.1], [6.4] ◈ A rational *expression* is a quotient of two polynomials. Expressions can be simplified, multiplied, or added, but they cannot be solved for a variable. A rational *equation* is an equation containing rational expressions. In a rational equation, we often can solve for a variable.   **44.** [6.4] All real numbers except 0 and 13   **45.** [6.3], [6.4] 45   **46.** [6.5] Anna: 56; Franz: 42

TEST: CHAPTER 6, PP. 371–372

1. [6.1] $\dfrac{3}{4(t+1)}$ **2.** [6.1] $\dfrac{x^2-3x+9}{x+4}$

3. [6.2] $(x-3)(x+11)(x-9)$ **4.** [6.2] $\dfrac{25x+x^3}{x+5}$

5. [6.2] $3(a-b)$ **6.** [6.2] $\dfrac{a^3-a^2b+4ab+ab^2-b^3}{(a-b)(a+b)}$

7. [6.2] $\dfrac{-2(2x^2+5x+20)}{(x-4)(x+4)(x^2+4x+16)}$

8. [6.2] $\dfrac{y-4}{(y+3)(y-2)}$ **9.** [6.3] $\dfrac{a(2b+3a)}{5a+b}$

10. [6.3] $\dfrac{(x-9)(x-6)}{(x+6)(x-3)}$ **11.** [6.3] $\dfrac{4x^2-14x+2}{3x^2+7x-11}$

12. [6.4] $-\dfrac{21}{4}$ **13.** [6.4] 15   **14.** [6.1] 5; 0

15. [6.4] $\dfrac{5}{3}$ **16.** [6.5] $1\frac{31}{32}$ hr

17. [6.6] $\dfrac{4b^2c}{a}-\dfrac{5bc^2}{2a}+3bc$ **18.** [6.6] $y-14+\dfrac{-20}{y-6}$

19. [6.6] $6x^2-9+\dfrac{5x+22}{x^2+2}$

20. [6.7] $x^2+9x+40+\dfrac{153}{x-4}$ **21.** [6.7] 449

22. [6.8] $b_1=\dfrac{2A}{h}-b_2$, or $\dfrac{2A-b_2h}{h}$ **23.** [6.5] 5 and 6;

-6 and -5 **24.** [6.5] $3\frac{3}{11}$ mph   **25.** [2.2] a^2+2a-2

26. [5.4] $8(2t+3)(t-3)$ **27.** [5.8] $-\frac{3}{2}, 3$

28. [4.4]

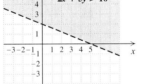

29. [6.4] $-\dfrac{19}{3}$ **30.** [6.4] $\{x\mid x$ is a real number and $x\neq 0$ and $x\neq 15\}$ **31.** [6.1], [6.3], [6.4] x-intercept: (11, 0); y-intercept: $\left(0, -\dfrac{33}{5}\right)$

CUMULATIVE REVIEW: 1–6

1. [1.1], [1.2] 10   **2.** [1.7] 5.76×10^9 **3.** [2.3] Slope is $\dfrac{7}{4}$; y-intercept is $(0, -3)$ **4.** [2.5] $y=-\dfrac{10}{3}x+\dfrac{11}{3}$

5. [3.2] $(-3, 4)$ **6.** [3.4] $(-2, -3, 1)$ **7.** [3.3] 16 small, 29 large   **8.** [3.5] $12, \frac{1}{2}, 7\frac{1}{2}$ **9. (a)** [6.1] $-\frac{1}{2}$;

(b) [2.2] $\{x\mid x$ is a real number and $x\neq 5\}$ **10.** [5.8] $\frac{1}{4}$

11. [5.8] $-\dfrac{25}{7}, \dfrac{25}{7}$ **12.** [4.1] $\{x\mid x>-3\}$, or $(-3, \infty)$

13. [4.1] $\{x \mid x \geq -1\}$, or $[-1, \infty)$
14. [4.2] $\{x \mid -10 < x < 13\}$, or $(-10, 13)$
15. [4.2] $\{x \mid x < -\frac{4}{3}$ or $x > 6\}$, or $\left(-\infty, -\frac{4}{3}\right) \cup (6, \infty)$
16. [4.3] $\{x \mid x < -6.4$ or $x > 6.4\}$, or
$(-\infty, -6.4) \cup (6.4, \infty)$ **17.** [4.3] $\{x \mid -\frac{13}{4} \leq x \leq \frac{15}{4}\}$, or
$\left[-\frac{13}{4}, \frac{15}{4}\right]$ **18.** [6.4] $-\frac{5}{3}$ **19.** [6.4] -1 **20.** [6.4] No
solution   **21.** [6.4] $\frac{1}{3}$ **22.** [4.3] $1, \frac{7}{3}$ **23.** [4.2] $[7, \infty)$
24. [1.5] $n = \dfrac{m - 12}{3}$ **25.** [6.8] $a = \dfrac{Pb}{3 - P}$
26. [4.4]   **27.** [4.4]

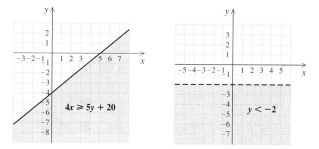

28. [5.1] $-3x^3 + 9x^2 + 3x - 3$ **29.** [5.2] $-15x^4y^4$
30. [5.1] $5a + 5b - 5c$
31. [5.2] $15x^4 - x^3 - 9x^2 + 5x - 2$
32. [5.2] $4x^4 - 4x^2y + y^2$
33. [5.2] $4x^4 - y^2$ **34.** [5.1] $-m^3n^2 - m^2n^2 - 5mn^3$
35. [6.1] $\dfrac{y - 6}{2}$ **36.** [6.1] $x - 1$
37. [6.2] $\dfrac{a^2 + 7ab + b^2}{(a - b)(a + b)}$ **38.** [6.2] $\dfrac{-m^2 + 5m - 6}{(m + 1)(m - 5)}$
39. [6.2] $\dfrac{3y^2 - 2}{3y}$ **40.** [6.3] $\dfrac{y - x}{xy(x + y)}$
41. [6.6] $9x^2 - 13x + 26 + \dfrac{-50}{x + 2}$
42. [5.3] $2x^2(2x + 9)$ **43.** [5.4] $(x - 6)(x + 14)$
44. [5.5] $(4y - 9)(4y + 9)$
45. [5.6] $8(2x + 1)(4x^2 - 2x + 1)$ **46.** [5.5] $(t - 8)^2$
47. [5.5] $x^2(x - 1)(x + 1)(x^2 + 1)$
48. [5.6] $(0.3b - 0.2c)(0.09b^2 + 0.06bc + 0.04c^2)$
49. [5.4] $(4x - 1)(5x + 3)$ **50.** [5.4] $(3x + 4)(x - 7)$
51. [5.3] $(x^3 + y)(x^2 - y)$ **52.** [2.6], [5.8] $\{x \mid x$ is a real
number and $x \neq 2$ and $x \neq 5\}$ **53.** [6.5] 1 hr
54. [5.8] 30 ft   **55.** [5.8] All such sets of even integers
satisfy this condition.   **56.** [5.2] $x^3 - 12x^2 + 48x - 64$
57. [5.8] $-3, 3, -5, 5$ **58.** [4.2], [4.3] $\{x \mid -3 \leq x \leq -1$
or $7 \leq x \leq 9\}$, or $[-3, -1] \cup [7, 9]$ **59.** [6.4] All real
numbers except 9 and -5 **60.** [5.8] $0, \frac{1}{4}, -\frac{1}{4}$

CHAPTER 7

EXERCISE SET 7.1, PP. 382–384

1. $4, -4$ **3.** $12, -12$ **5.** $20, -20$ **7.** $7, -7$ **9.** $-\frac{7}{6}$
11. 14 **13.** $-\frac{4}{9}$ **15.** 0.3 **17.** -0.07 **19.** $p^2 + 4$; 2
21. $\dfrac{x}{y + 4}$; 3   **23.** $\sqrt{20}$; 0; does not exist; does not exist
25. -3; -1; does not exist; 0   **27.** 1; $\sqrt{2}$; $\sqrt{101}$
29. 1; does not exist; 6   **31.** $5|t|$ **33.** $6|b|$ **35.** $|5 - b|$
37. $|y + 8|$ **39.** $|3x - 5|$ **41.** -4 **43.** -7 **45.** $-\frac{1}{2}$
47. $|y|$ **49.** $7|b|$ **51.** 10   **53.** $|2a + b|$ **55.** x^6
57. $|a^7|$ **59.** $5t$ **61.** $7c$ **63.** $5 + b$ **65.** $3(x + 2)$,
or $3x + 6$ **67.** $5t - 2$ **69.** -4 **71.** $3x$ **73.** 10
75. $4x$ **77.** a^7 **79.** $(x + 3)^5$ **81.** 2; 3; -2; -4
83. 2; does not exist; does not exist; 3
85. $\{x \mid x \geq 5\}$, or $[5, \infty)$ **87.** $\{t \mid t \geq -3\}$, or $[-3, \infty)$
89. $\{x \mid x \leq 5\}$, or $(-\infty, 5]$ **91.** $\mathbb{R}$
93. $\left\{z \mid z \geq -\frac{3}{5}\right\}$, or $\left[-\frac{3}{5}, \infty\right)$ **95.** $\left\{t \mid t \geq \frac{7}{3}\right\}$, or $\left[\frac{7}{3}, \infty\right)$
97. $a^9b^6c^{15}$ **99.** $x^2 - 9$ **101.** $2x^3 - 5x^2 - x + 1$
103. ◈   **105.** ◈   **107.** (a) 13; (b) 15; (c) 18; (d) 20
109. $\{x \mid x \geq 0\}$, or $[0, \infty)$;   **111.** $\{x \mid x \geq 2\}$, or $[2, \infty)$;

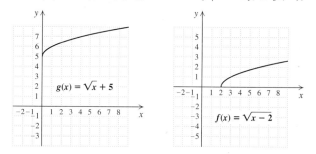

113. $\{x \mid -4 < x \leq 5\}$, or $(-4, 5]$ **115.** ◰

TECHNOLOGY CONNECTION 7.2A, P. 385

1. Without parentheses, the expression entered would be $\dfrac{7^2}{3}$.

2. For $x = 0$ or $x = 1$, $y_1 = y_2 = y_3$; on $(0, 1)$,
$y_1 > y_2 > y_3$; on $(1, \infty)$, $y_1 < y_2 < y_3$.

TECHNOLOGY CONNECTION 7.2B, P. 388

1. Most graphers do not have keys for radicals of index 3
or higher. On those graphers that offer $\sqrt[x]{}$ in a MATH menu,
rational exponents still require fewer keystrokes.

EXERCISE SET 7.2, PP. 388–389

1. $\sqrt[4]{x}$ **3.** 4   **5.** 3   **7.** 3   **9.** $\sqrt[3]{xyz}$ **11.** $\sqrt[5]{a^2b^2}$
13. $\sqrt[5]{a^2}$ **15.** 8   **17.** 343   **19.** 243
21. $\sqrt[4]{81^3x^3}$, or $27\sqrt[4]{x^3}$ **23.** $125x^6$ **25.** $20^{1/3}$ **27.** $17^{1/2}$
29. $x^{3/2}$ **31.** $m^{2/5}$ **33.** $(cd)^{1/4}$ **35.** $(xy^2z)^{1/5}$

37. $(3mn)^{3/2}$ **39.** $(8x^2y)^{5/7}$ **41.** $\dfrac{2x}{z^{2/3}}$ **43.** $\dfrac{1}{x^{1/3}}$

45. $\dfrac{1}{(2rs)^{3/4}}$ **47.** $10^{2/3}$ **49.** $a^{5/7}$ **51.** $\dfrac{2a^{3/4}c^{2/3}}{b^{1/2}}$

53. $\dfrac{x^4}{2^{1/3}y^{2/7}}$ **55.** $\left(\dfrac{8yz}{7x}\right)^{3/5}$ **57.** $\dfrac{7x}{z^{1/3}}$ **59.** $\dfrac{5ac^{1/2}}{3}$

61. $5^{7/8}$ **63.** $3^{3/4}$ **65.** $4.1^{1/2}$ **67.** $10^{6/25}$ **69.** $a^{23/12}$

71. $\dfrac{1}{x^{2/7}}$ **73.** $\dfrac{m^{1/3}}{n^{1/8}}$ **75.** $\sqrt[3]{a}$ **77.** x^5 **79.** x^3

81. a^5b^5 **83.** $\sqrt[4]{3x}$ **85.** $\sqrt{3a}$ **87.** $\sqrt[8]{x}$ **89.** a^3b^3

91. x^8y^{20} **93.** $\sqrt[12]{xy}$ **95.** $-3, 3$ **97.** $\dfrac{1}{3}$ **99.** \$93,500

101. ◈   **103.** ◈   **105.** $\sqrt[10]{x^5y^3}$ **107.** $\sqrt[4]{2xy^2}$

109. (a) 1.8 m; (b) 3.1 m; (c) 1.5 m; (d) 5.3 m

111. 10 mg

TECHNOLOGY CONNECTION 7.3, P. 391

1. The graph of $y = 0$ (the x-axis)

EXERCISE SET 7.3, PP. 395–397

1. $\sqrt{42}$ **3.** $\sqrt[3]{10}$ **5.** $\sqrt[4]{72}$ **7.** $\sqrt{15ab}$ **9.** $\sqrt[5]{18t^3}$

11. $\sqrt{x^2 - a^2}$ **13.** $\sqrt[3]{0.1x^2}$ **15.** $\sqrt[4]{x^3 - 1}$ **17.** $\sqrt{\dfrac{3x}{5y}}$

19. $\sqrt[7]{\dfrac{5x - 15}{4x + 8}}$ **21.** $\sqrt[6]{5400}$ **23.** $\sqrt[6]{49x^3y^2}$

25. $\sqrt[6]{x^5 - 4x^4 + 4x^3}$ **27.** $\sqrt[10]{x^9y^7}$ **29.** $\sqrt[12]{x^{11}y^{10}}$

31. $\sqrt[20]{a^{18}b^{17}c^{14}}$ **33.** $3\sqrt{3}$ **35.** $2\sqrt{3}$ **37.** $2\sqrt{2}$

39. $2\sqrt{11}$ **41.** $6a^2\sqrt{b}$ **43.** $2x\sqrt[3]{y^2}$ **45.** $-2x^2\sqrt[3]{2}$

47. $f(x) = 5x\sqrt[3]{x^2}$ **49.** $f(x) = |7(x + 5)|,$ or $7|x + 5|$

51. $f(x) = |x - 1|\sqrt{5}$ **53.** $ab^2\sqrt{a}$ **55.** $xy^2z^3\sqrt[3]{x^2z}$

57. $-2ab^2\sqrt[5]{a^2b}$ **59.** $ab^2c\sqrt[5]{ab^2c^2}$ **61.** $3x^2\sqrt[4]{10x}$

63. $5\sqrt{7}$ **65.** $3\sqrt[3]{5}$ **67.** $13\sqrt[3]{y}$ **69.** $7\sqrt{2}$

71. $13\sqrt[3]{7} + \sqrt{3}$ **73.** $21\sqrt{3}$ **75.** $23\sqrt{5}$ **77.** $9\sqrt[3]{2}$

79. $(1 + 6a)\sqrt{5a}$ **81.** $(x + 2)\sqrt[3]{6x}$ **83.** $3\sqrt{a - 1}$

85. $(x + 3)\sqrt{x - 1}$ **87.** $f(x) = -6x\sqrt{5 + x}$

89. $f(x) = (x + 3x^2)\sqrt[4]{x - 1}$ **91.** 8

93. $9a^3b^2 + 5a^2b^2 - 8a^2b + 3ab$ **95.** $(2x + 7)(2x - 7)$

97. ◈   **99.** ◈

101. (a) $-3.3°C$; (b) $-16.6°C$; (c) $-25.5°C$; (d) $-54.0°C$

103. $25x^5\sqrt[3]{25x}$ **105.** $(7x^2 - 2y^2)\sqrt{x + y}$

107.   ; $\mathbb{R}$ **109.** ◈ 〰

$g(x) = \sqrt{(x + 3)^2}$

EXERCISE SET 7.4, PP. 401–403

1. $5\sqrt{2}$ **3.** $2\sqrt{21}$ **5.** 2   **7.** $a\sqrt[3]{10}$ **9.** $3x^4\sqrt{2}$

11. $s^2t^3\sqrt[3]{t}$ **13.** $(x + 5)^2$ **15.** $2ab^3\sqrt[4]{3a}$

17. $x(y + z)^2\sqrt[5]{x}$ **19.** $a\sqrt[4]{a}$ **21.** $b\sqrt[10]{b^9}$

23. $xy\sqrt[6]{xy^5}$ **25.** $3a^2b\sqrt[4]{ab}$ **27.** $xyz\sqrt[6]{x^5yz^2}$

29. $9a^2(b + 1)\sqrt[6]{243a^5(b + 1)^5}$ **31.** $\dfrac{5}{6}$ **33.** $\dfrac{4}{3}$ **35.** $\dfrac{7}{y}$

37. $\dfrac{5y\sqrt{y}}{x^2}$ **39.** $\dfrac{3a\sqrt[3]{a}}{2b}$ **41.** $\dfrac{2a}{bc^2}$ **43.** $\dfrac{ab^2}{c^2}\sqrt[4]{\dfrac{a}{c^2}}$

45. $\dfrac{2x}{y^2}\sqrt[5]{\dfrac{x}{y}}$ **47.** $\dfrac{xy}{z^2}\sqrt[6]{\dfrac{y^2}{z^3}}$ **49.** $\sqrt{5}$ **51.** 3

53. $y\sqrt{5y}$ **55.** $2\sqrt[3]{a^2b}$ **57.** $\sqrt{2ab}$ **59.** $2x^2y^3\sqrt[4]{y^3}$

61. $\sqrt[3]{x^2 + xy + y^2}$ **63.** $\sqrt[12]{a^5}$ **65.** $\sqrt[12]{x^2y^5}$

67. $\sqrt[10]{ab^9c^7}$ **69.** $\sqrt[20]{(3x - 1)^3}$ **71.** $\sqrt[15]{(2x + 1)^4}$

73. 8   **75.** Length: 20; width: 5

77. $a_1 = a_2 - \dfrac{m}{A},$ or $\dfrac{Aa_2 - m}{A}$ **79.** ◈   **81.** ◈

83. (a) 1.62 sec; (b) 1.99 sec; (c) 2.20 sec   **85.** $9\sqrt[3]{9n^2}$

87. $x - y$ **89.** 6   **91.** 〰

EXERCISE SET 7.5, PP. 408–409

1. $3\sqrt{7} - 7$ **3.** $\sqrt{6} - \sqrt{10}$ **5.** $2\sqrt{15} - 6\sqrt{3}$ **7.** -6

9. $a + 2a\sqrt[3]{3}$ **11.** 19   **13.** -19

15. $15 - 2\sqrt[3]{10} - \sqrt[3]{100}$ **17.** -6

19. $2\sqrt[3]{9} + 3\sqrt[3]{6} - 2\sqrt[3]{4}$ **21.** $3x + 2\sqrt{3xy} + y$

23. $x\sqrt[6]{xy^5} - \sqrt[15]{x^{13}y^{14}}$

25. $2m^2 + m\sqrt[4]{n} + 2m\sqrt[3]{n^2} + \sqrt[12]{n^{11}}$ **27.** $\sqrt[4]{2x^2} - x^3$

29. $x^2 - 7$ **31.** $2 - 3\sqrt{x} + x$ **33.** $27 - 10\sqrt{2}$

35. $8 + 2\sqrt{15}$ **37.** $15 - 10\sqrt{2}$ **39.** $\dfrac{\sqrt{35}}{7}$ **41.** $\dfrac{2\sqrt{15}}{5}$

43. $\dfrac{2\sqrt[3]{6}}{3}$ **45.** $\dfrac{\sqrt[3]{75ac^2}}{5c}$ **47.** $\dfrac{y\sqrt[3]{180x^2y}}{6x^2}$ **49.** $\dfrac{\sqrt[3]{2xy^2}}{xy}$

51. $\dfrac{\sqrt{14a}}{6}$ **53.** $\dfrac{3\sqrt{5y}}{10xy}$ **55.** $\dfrac{35 + 5\sqrt{2}}{47}$ **57.** $\dfrac{x - \sqrt{xy}}{x - y}$

59. $\dfrac{3 + 6\sqrt{2} + 4\sqrt{15} + 8\sqrt{30}}{-21}$ **61.** $\dfrac{3\sqrt{6} + 4}{2}$

63. $\dfrac{5}{\sqrt{35x}}$ **65.** $\dfrac{2}{\sqrt{6}}$ **67.** $\dfrac{52}{3\sqrt{91}}$ **69.** $\dfrac{7}{\sqrt[3]{98}}$

71. $\dfrac{7x}{\sqrt{21xy}}$ **73.** $\dfrac{2a^2}{\sqrt[3]{20ab}}$ **75.** $\dfrac{x^2y}{\sqrt{2xy}}$ **77.** $\dfrac{1}{6\sqrt{5} - 12}$

79. $\dfrac{-22}{\sqrt{6} + 5\sqrt{2} + 5\sqrt{3} + 25}$ **81.** $\dfrac{x - y}{x - 2\sqrt{xy} + y}$

83. $\dfrac{a^2b - c}{ab + \sqrt{bc} + a\sqrt{bc} + c}$ **85.** 6   **87.** $\dfrac{x - 2}{x + 3}$

89. ◈   **91.** $(\sqrt{x} + \sqrt{5})(\sqrt{x} - \sqrt{5})$

93. $(\sqrt{x} + \sqrt{a})(\sqrt{x} - \sqrt{a})$ **95.** $2x - 2\sqrt{x^2 - 4}$

97. $\dfrac{b^2 + \sqrt{b}}{1 + b + b^2}$ **99.** $\dfrac{x - 19}{x + 10\sqrt{x + 6} + 31}$

101. $\left(\dfrac{5x + 4y - 3}{xy}\right)\sqrt{xy}$ **103.** $\dfrac{7\sqrt{3}}{39}$ **105.**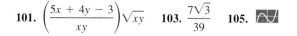

TECHNOLOGY CONNECTION 7.6, P. 411

1. x-coordinates of points of intersection should approximate the solutions of the examples.

EXERCISE SET 7.6, PP. 414–415

1. 7   **3.** 12   **5.** 168   **7.** 56   **9.** 3   **11.** 19   **13.** 0, 9
15. 64   **17.** -27 **19.** 125   **21.** No solution   **23.** 39
25. 88   **27.** -6 **29.** 5   **31.** $\frac{1}{2}$ **33.** 5   **35.** 7
37. $\frac{80}{9}$ **39.** -1 **41.** No solution   **43.** 1   **45.** 2
47. 4   **49.** 2   **51.**

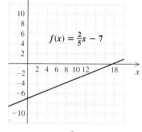

53. ◈   **55.** ◈   **57.** $h = \dfrac{v^2 r}{2gr - v^2}$

59. 22,500 ft   **61.** $-\dfrac{8}{9}$ **63.** $-8, 8$ **65.** $-1, 6$

67. $(2, 0)$ **69.** $(0, 0), \left(\dfrac{125}{4}, 0\right)$ **71.**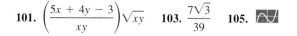

EXERCISE SET 7.7, PP. 420–422

1. $\sqrt{34}$; 5.831   **3.** $\sqrt{98}$; 9.899   **5.** 5   **7.** $\sqrt{31}$; 5.568
9. $\sqrt{12}$; 3.464   **11.** $\sqrt{n-1}$ **13.** $\sqrt{325}$ ft; 18.028 ft
15. $\sqrt{14,500}$ ft; 120.416 ft   **17.** 20 in.
19. $\sqrt{13,000}$ m; 114.018 m   **21.** $a = 5$; $c = 5\sqrt{2} \approx 7.071$
23. $a = 7$; $b = 7\sqrt{3} \approx 12.124$
25. $a = 5\sqrt{3} \approx 8.660$; $c = 10\sqrt{3} \approx 17.321$
27. $a = \dfrac{13\sqrt{2}}{2} \approx 9.192$; $b = \dfrac{13\sqrt{2}}{2} \approx 9.192$
29. $b = 14\sqrt{3} \approx 24.249$; $c = 28$ **31.** $3\sqrt{3} \approx 5.196$
33. $13\sqrt{2} \approx 18.385$ **35.** $\dfrac{19\sqrt{2}}{2} \approx 13.435$
37. $\sqrt{10561}$ ft ≈ 102.767 ft   **39.** $h = 2\sqrt{3}$ ft ≈ 3.464 ft
41. 8 ft   **43.** $(0, -4), (0, 4)$ **45.** 3, 8   **47.** $\left\{-\dfrac{2}{3}, 4\right\}$
49. ◈   **51.** $\sqrt{75}$ cm   **53.** 4 gal; the total area of the doors and windows is 164 ft² or more.

EXERCISE SET 7.8, PP. 428–429

1. $5i$ **3.** $i\sqrt{13}$, or $\sqrt{13}i$ **5.** $3i\sqrt{2}$, or $3\sqrt{2}i$
7. $i\sqrt{3}$, or $\sqrt{3}i$ **9.** $9i$ **11.** $10i\sqrt{3}$, or $10\sqrt{3}i$ **13.** $-7i$

15. $4 - 2\sqrt{15}i$ **17.** $(2 + 2\sqrt{3})i$ **19.** $9 + 5i$
21. $3 + 11i$ **23.** $4 + 5i$ **25.** $-1 - 5i$ **27.** $5 + 5i$
29. -30 **31.** 63   **33.** -35 **35.** $-\sqrt{42}$ **37.** $-5\sqrt{6}$
39. $-6 + 14i$ **41.** $-20 - 24i$ **43.** $-7 + 26i$
45. $40 + 13i$ **47.** $8 + 31i$ **49.** $7 - 61i$
51. $-15 - 30i$ **53.** $39 - 37i$ **55.** $21 - 20i$
57. $12 + 16i$ **59.** $21 + 20i$ **61.** $\frac{14}{5} + \frac{7}{5}i$ **63.** $\frac{6}{29} + \frac{15}{29}i$
65. $-\frac{8}{9}i$ **67.** $-\frac{1}{3} - \frac{7}{6}i$ **69.** $-\frac{23}{58} + \frac{43}{58}i$ **71.** $\frac{6}{25} - \frac{17}{25}i$
73. $-i$ **75.** 1   **77.** -1 **79.** i **81.** -1 **83.** $-125i$
85. 0   **87.** 0   **89.** 7   **91.** $-\frac{4}{3}, 7$ **93.** ◈
95. ◈   **97.** $-4 - 8i; -2 + 4i; 8 - 6i$ **99.** 0
101. $-1 - \sqrt{5}i$ **103.** $-\frac{2}{3}i$

REVIEW EXERCISES: CHAPTER 7, PP. 431–432

1. [7.1] $\frac{7}{6}$ **2.** [7.1] 0.5   **3.** [7.1] 5
4. [7.1] $\left\{x \mid x \geq \frac{7}{2}\right\}$, or $\left[\frac{7}{2}, \infty\right)$ **5.** [7.1] $7|a|$ **6.** [7.1] $|c + 8|$
7. [7.1] $|x - 3|$ **8.** [7.1] $|2x + 1|$ **9.** [7.1] -2
10. [7.4] $-\dfrac{4x^2}{3}$ **11.** [7.3] $|x^3 y^2|$, or $|x^3|y^2$ **12.** [7.3] $2x^2$
13. [7.2] $(5ab)^{4/3}$, or $5^{4/3}a^{4/3}b^{4/3}$ **14.** [7.2] $8a^6$
15. [7.2] $x^3 y^5$ **16.** [7.2] $a^3 b\sqrt[3]{ab^2}$ **17.** [7.2] $\dfrac{1}{x^{2/5}}$
18. [7.2] $7^{1/6}$ **19.** [7.1] $f(x) = 5|x - 3|$
20. [7.3] $\sqrt{15xy}$ **21.** [7.4] $3a\sqrt[3]{a^2 b^2}$ **22.** [7.3] $\sqrt[15]{a^{14}b^{11}}$
23. [7.4] $y\sqrt[3]{6}$ **24.** [7.4] $\dfrac{5\sqrt{x}}{2}$ **25.** [7.4] $\sqrt[12]{x^5}$
26. [7.3] $7\sqrt[3]{x}$ **27.** [7.3] $3\sqrt{3}$ **28.** [7.3] $(2x + y^2)\sqrt[3]{x}$
29. [7.3] $15\sqrt{2}$ **30.** [7.5] $-43 - 2\sqrt{10}$
31. [7.5] $a^2 - 2a\sqrt{2} + 2$ **32.** [7.5] $\dfrac{10a\sqrt{3} - 10\sqrt{3ab}}{a - b}$
33. [7.5] $\dfrac{30a}{\sqrt{3a}(\sqrt{a} + \sqrt{b})}$, or $\dfrac{30a}{\sqrt{3}(a + \sqrt{ab})}$ **34.** [7.6] 4
35. [7.6] 14   **36.** [7.7] 9 cm   **37.** [7.7] $\sqrt{24}$ ft; 4.899 ft
38. [7.7] $a = 10$; $b = 10\sqrt{3} \approx 17.321$
39. [7.8] $-2i\sqrt{2}$, or $-2\sqrt{2}i$ **40.** [7.8] $-2 - 9i$
41. [7.8] $1 + i$ **42.** [7.8] 29   **43.** [7.8] i
44. [7.8] $9 - 12i$ **45.** [7.8] $\frac{13}{25} - \frac{34}{25}i$
46. [5.8] 12, 13, 14   **47.** [6.4] $\frac{23}{7}$ **48.** [5.8] $-\frac{9}{2}, 3$
49. [6.1] $\dfrac{x(x + 2)}{(x + y)(x - 3)}$
50. [7.1] ◈ An absolute-value sign must be used to simplify $\sqrt[n]{x^n}$ when n is even, since x may be negative. If x is negative while n is even, the radical expression cannot be simplified to x, since $\sqrt[n]{x^n}$ represents the principal, or positive, root. When n is odd, there is only one root, and it will be positive or negative depending on the sign of x. Thus there is no absolute-value sign when n is odd.

51. [7.8]  Every real number is a complex number, but there are complex numbers that are not real. A complex number $a + bi$ is not real if $b \neq 0$.
52. [7.6] 3   **53.** [7.8] $-\frac{2}{5} + \frac{9}{10}i$

TEST: CHAPTER 7, P. 432

1. [7.3] $5\sqrt{3}$ **2.** [7.4] $-\frac{2}{x^2}$ **3.** [7.1] $10|a|$

4. [7.1] $|x - 4|$ **5.** [7.3] $x^2y\sqrt[5]{x^2y^3}$ **6.** [7.4] $\left|\frac{5x}{6y^2}\right|$

7. [7.3] $\sqrt[3]{10xy^2}$ **8.** [7.4] $\sqrt[5]{x^2y^2}$ **9.** [7.4] $xy\sqrt[4]{x}$
10. [7.4] $\sqrt[20]{a^3}$ **11.** [7.3] $5\sqrt{2}$ **12.** [7.3] $(x^2 + 3y)\sqrt{y}$
13. [7.5] $14 - 19\sqrt{x} - 3x$
14. [7.1], [7.3] $\{x \mid x \leq 2\}$, or $(-\infty, 2]$
15. [7.5] $27 + 10\sqrt{2}$ **16.** [7.5] $-\frac{13 + 8\sqrt{2}}{41}$

17. [7.6] 7   **18.** [7.7] Longer leg: $10\sqrt{3}$ cm ≈ 17.321 cm; hypotenuse: 20 cm   **19.** [7.7] $\sqrt{10{,}600}$ ft ≈ 102.956 ft
20. [7.8] $5i\sqrt{2}$, or $5\sqrt{2}i$ **21.** [7.8] $10 + 2i$
22. [7.8] -24 **23.** [7.8] $15 - 8i$ **24.** [7.8] $-\frac{13}{53} - \frac{19}{53}i$

25. [7.8] i **26.** [5.8] $-\frac{1}{3}, \frac{5}{2}$ **27.** [6.1] $\frac{x - 3}{x - 4}$

28. [6.4] No solution   **29.** [5.8] 16 and 18; -18 and -16
30. [7.6] 3   **31.** [7.8] $-\frac{17}{4}i$

CHAPTER 8

TECHNOLOGY CONNECTION 8.1, P. 440

1. The right-hand x-intercept should be an approximation of $4 + \sqrt{23}$.   **2.** x-intercepts should be approximations of $-3 + \sqrt{5}$ and $-3 - \sqrt{5}$ for Example 8; approximations of $(-5 + \sqrt{37})/2$ and $(-5 - \sqrt{37})/2$ for Example 10(b).
3. A grapher can give only rational-number approximations of the two irrational solutions. An *exact* solution cannot be found with a grapher.   **4.** The graph of $y = 4x^2 + 9$ has no x-intercepts.

EXERCISE SET 8.1, PP. 443–444

1. $\pm\sqrt{3}$ **3.** $\pm\frac{2}{5}i$ **5.** $\pm\sqrt{\frac{3}{2}}$, or $\pm\frac{\sqrt{6}}{2}$ **7.** $-9, 5$

9. $-5 \pm 2\sqrt{2}$ **11.** $7 \pm 2i$ **13.** $\frac{-3 \pm \sqrt{14}}{2}$ **15.** $-7, 13$

17. $3, 11$ **19.** $3 \pm \sqrt{13}$ **21.** $-13, -1$
23. $x^2 + 10x + 25, (x + 5)^2$ **25.** $x^2 - 6x + 9, (x - 3)^2$
27. $x^2 - 24x + 144, (x - 12)^2$ **29.** $x^2 + 9x + \frac{81}{4}$,
$\left(x + \frac{9}{2}\right)^2$ **31.** $x^2 - 3x + \frac{9}{4}, \left(x - \frac{3}{2}\right)^2$ **33.** $x^2 + \frac{2}{3}x + \frac{1}{9}$,
$\left(x + \frac{1}{3}\right)^2$ **35.** $x^2 - \frac{5}{6}x + \frac{25}{144}, \left(x - \frac{5}{12}\right)^2$

37. $x^2 + \frac{9}{5}x + \frac{81}{100}, \left(x + \frac{9}{10}\right)^2$ **39.** $-7, 1$ **41.** $5 \pm \sqrt{47}$
43. $-5, -1$ **45.** $3, 7$ **47.** $-2 \pm \sqrt{3}$ **49.** $3 \pm \sqrt{5}$
51. $-3 \pm 2i$ **53.** $-\frac{1}{2}, 3$ **55.** $-\frac{3}{2}, -\frac{1}{2}$ **57.** $-\frac{3}{2}, \frac{5}{3}$
59. $\frac{-2 \pm \sqrt{2}}{2}$ **61.** $\frac{5 \pm \sqrt{61}}{6}$ **63.** 10%   **65.** 18.75%
67. 4%   **69.** About 10.7 sec   **71.** About 6.3 sec
73.

$f(x) = 5 - 2x$

75. $3\sqrt[3]{10}$ **77.** 5   **79.** ◈   **81.** ◈   **83.** ± 18
85. $0, -\frac{7}{2}, 8, -\sqrt{5}, \sqrt{5}$ **87.** Barge: 8 km/h; fishing boat: 15 km/h   **89.** 〰   **91.** ◈ 〰

EXERCISE SET 8.2, PP. 449–450

1. $\frac{-7 \pm \sqrt{33}}{2}$ **3.** $-\frac{5}{3}, -1$ **5.** $\frac{1 \pm i\sqrt{7}}{2}$ **7.** $3 \pm 2i$

9. $3 \pm \sqrt{5}$ **11.** $\frac{-4 \pm \sqrt{19}}{3}$ **13.** $-1, 0$ **15.** $-\frac{9}{14}, 0$

17. $\frac{2}{5}$ **19.** $-2, -\frac{3}{4}$ **21.** $5, 10$ **23.** $\frac{13 \pm \sqrt{509}}{10}$

25. $2 \pm i\sqrt{5}$ **27.** $\frac{2}{3}, \frac{3}{2}$ **29.** $2, -1 \pm i\sqrt{3}$
31. $\frac{5 \pm \sqrt{37}}{6}$ **33.** $5 \pm \sqrt{53}$ **35.** $\frac{7 \pm \sqrt{85}}{2}$
37. $-5.31662479, 1.31662479$ **39.** 0.7639320225,
5.236067978 **41.** $-1.265564437, 2.765564437$

43. Kenyan: 30 lb; Peruvian: 20 lb   **45.** $\frac{3(x + 1)}{3x + 1}$

47. ◈   **49.** ◈   **51.** $(-5 - \sqrt{37}, 0), (-5 + \sqrt{37}, 0)$
53. $4 \pm 2\sqrt{2}$ **55.** $-1.1792101, 0.3392101$
57. $\sqrt{3}, \frac{3 - \sqrt{3}}{2}$ **59.** $\frac{1}{2}$ **61.** 〰

EXERCISE SET 8.3, PP. 454–457

1. First part: 60 mph; second part: 50 mph   **3.** 40 mph
5. Cessna: 150 mph, Beechcraft: 200 mph; or Cessna: 200 mph, Beechcraft: 250 mph
7. To Hillsboro: 10 mph; return trip: 4 mph
9. About 11 mph   **11.** 12 hr   **13.** 9.34 km/h

15. $r = \frac{1}{2}\sqrt{\frac{A}{\pi}}$ **17.** $r = \frac{-\pi h + \sqrt{\pi^2 h^2 + 2\pi A}}{2\pi}$

19. $s = \sqrt{\frac{kQ_1Q_2}{N}}$ **21.** $g = \frac{4\pi^2 l}{T^2}$

23. $c = \sqrt{d^2 - a^2 - b^2}$ **25.** $t = \dfrac{-v_0 + \sqrt{v_0^2 + 2gs}}{g}$

27. $n = \dfrac{1 + \sqrt{1 + 8N}}{2}$ **29.** $h = \dfrac{V^2}{12.25}$

31. $r = -1 + \dfrac{-P_2 + \sqrt{P_2^2 + 4AP_1}}{2P_1}$ **33.** **(a)** 10.1 sec;
(b) 7.49 sec; **(c)** 272.5 m   **35.** 2.9 sec   **37.** 0.87 sec
39. 2.5 m/sec   **41.** 7%   **43.** 1, 5   **45.** $2y\sqrt[3]{9x^2}$
47. ◆   **49.** ◆   **51.** \$2.50   **53.** $l = \dfrac{w + w\sqrt{5}}{2}$

55. $c = \dfrac{vm}{\sqrt{m^2 - m_0^2}}$ **57.** $A(S) = \dfrac{\pi S}{6}$

EXERCISE SET 8.4, PP. 460–461

1. Two rational   **3.** Two imaginary   **5.** Two irrational
7. One rational   **9.** Two imaginary   **11.** Two rational
13. Two rational   **15.** Two rational   **17.** Two irrational
19. Two imaginary   **21.** Two irrational
23. $x^2 + 4x - 21 = 0$ **25.** $x^2 - 6x + 9 = 0$
27. $x^2 + 7x + 10 = 0$ **29.** $3x^2 - 14x + 8 = 0$
31. $6x^2 - 5x + 1 = 0$ **33.** $x^2 - 0.8x - 0.84 = 0$
35. $x^2 - 7 = 0$ **37.** $x^2 - 18 = 0$ **39.** $x^2 + 9 = 0$
41. $x^2 - 10x + 29 = 0$ **43.** $x^2 - 4x - 6 = 0$
45. $x^3 - x^2 - 12x = 0$ **47.** $x^3 - 2x^2 - x + 2 = 0$
49. 6

51.

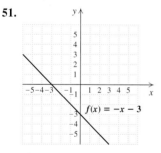

$f(x) = -x - 3$

53. ◆   **55.** ◆   **57.** $a = 1, b = 2, c = -3$
59. **(a)** 2; **(b)** $1 - i$
61. $\dfrac{-b + \sqrt{b^2 - 4ac}}{2a} + \dfrac{-b - \sqrt{b^2 - 4ac}}{2a} = \dfrac{-2b}{2a} = -\dfrac{b}{a}$
63. $h = -36, k = 15$
65. $x^4 - 14x^3 + 70x^2 - 126x + 29 = 0$
67. ◆   **69.** ◆

EXERCISE SET 8.5, P. 466

1. $\pm 1, \pm 2$ **3.** $\pm\sqrt{3}, \pm 3$ **5.** $\pm\dfrac{\sqrt{3}}{2}, \pm 2$
7. $9 + 4\sqrt{5}$ **9.** $\pm 2\sqrt{2}, \pm 3$ **11.** No solution
13. $-\dfrac{1}{2}, \dfrac{1}{3}$ **15.** $-\dfrac{4}{5}, 1$ **17.** $-27, 8$
19. 729   **21.** 1   **23.** No solution   **25.** $\dfrac{12}{5}$

27. $\left(\dfrac{4}{25}, 0\right)$ **29.** $\left(\dfrac{3 + \sqrt{33}}{2}, 0\right), \left(\dfrac{3 - \sqrt{33}}{2}, 0\right),$
$(4, 0), (-1, 0)$ **31.** $(-243, 0), (32, 0)$
33. $\left(\dfrac{9 + \sqrt{89}}{2}, 0\right), \left(\dfrac{9 - \sqrt{89}}{2}, 0\right), (-1 + \sqrt{3}, 0),$
$(-1 - \sqrt{3}, 0)$
35. $3x^2\sqrt{x}$ **37.** $\dfrac{x^3 + x^2 + 2x + 2}{x^3 - 1}$ **39.** ◆
41. $x = \pm\sqrt{\dfrac{-5 \pm \sqrt{37}}{6}}$ **43.** $-2, -1, 6, 7$
45. $\dfrac{100}{99}$ **47.** $-5, -3, -2, 0, 2, 3, 5$ **49.** 1, 3
51. 〰   **53.** ◆ 〰

EXERCISE SET 8.6, PP. 473–476

1. $k = 4; y = 4x$ **3.** $k = 1.7; y = 1.7x$ **5.** $k = \dfrac{15}{4}$;
$y = \dfrac{15}{4}x$ **7.** $k = 1.6; y = 1.6x$ **9.** 6 amperes
11. 241,920,000   **13.** 40 kg   **15.** 7,700,000 tons
17. $k = 60; y = \dfrac{60}{x}$ **19.** $k = 12; y = \dfrac{12}{x}$
21. $k = 36; y = \dfrac{36}{x}$ **23.** $k = 9; y = \dfrac{9}{x}$ **25.** 27 min
27. 160 cm³   **29.** 450 m   **31.** $y = \dfrac{2}{3}x^2$ **33.** $y = \dfrac{54}{x^2}$
35. $y = \dfrac{1}{2}xz$ **37.** $y = 0.3xz^2$ **39.** $y = \dfrac{4wx^2}{z}$
41. $y = \dfrac{xz}{5wp}$ **43.** 36 mph   **45.** 2.5 m   **47.** 1600 km
49. About 57.42 mph   **51.** $y = -\dfrac{2}{3}x - 5$
53. $\dfrac{c - 2a}{3c + 4a}$ **55.** 9   **57.** ◆   **59.** ◆
61. y is multiplied by 8.   **63.** W varies jointly as m_1 and
M_1 and inversely as the square of d.   **65.** **(a)** $k \approx 0.001$;
$N = \dfrac{0.001P_1P_2}{d^2}$; **(b)** 1137 km   **67.** $d(s) = \dfrac{28}{s}$; 70 yd

EXERCISE SET 8.7, PP. 484–485

1.   **3.**

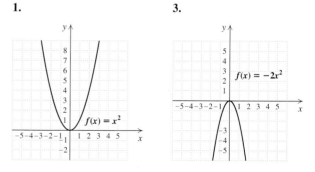

$f(x) = x^2$ $f(x) = -2x^2$

5.

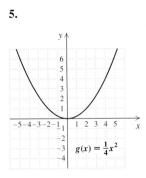

$g(x) = \frac{1}{4}x^2$

7.

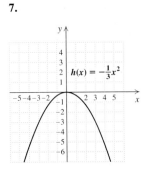

$h(x) = -\frac{1}{3}x^2$

21. Vertex: $(-1, 0)$;
axis of symmetry: $x = -1$

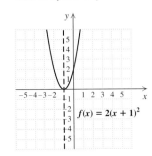

$f(x) = 2(x + 1)^2$

23. Vertex: $(3, 0)$;
axis of symmetry: $x = 3$

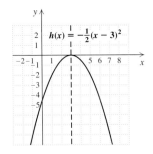

$h(x) = -\frac{1}{2}(x - 3)^2$

9.

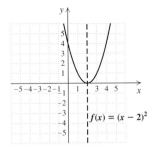

$f(x) = \frac{3}{2}x^2$

11. Vertex: $(-1, 0)$; axis
of symmetry: $x = -1$

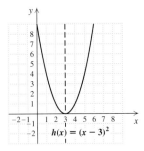

$g(x) = (x + 1)^2$

25. Vertex: $(1, 0)$;
axis of symmetry: $x = 1$

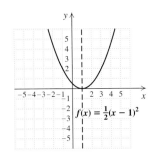

$f(x) = \frac{1}{2}(x - 1)^2$

27. Vertex: $(-5, 0)$;
axis of symmetry: $x = -5$

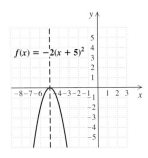

$f(x) = -2(x + 5)^2$

13. Vertex: $(2, 0)$;
axis of symmetry: $x = 2$

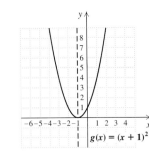

$f(x) = (x - 2)^2$

15. Vertex: $(3, 0)$;
axis of symmetry: $x = 3$

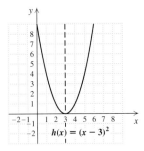

$h(x) = (x - 3)^2$

29. Vertex: $\left(\frac{1}{2}, 0\right)$;
axis of symmetry: $x = \frac{1}{2}$

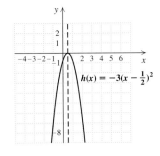

$h(x) = -3\left(x - \frac{1}{2}\right)^2$

31. Vertex: $(5, 1)$;
axis of symmetry: $x = 5$;
minimum: 1

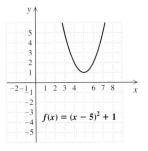

$f(x) = (x - 5)^2 + 1$

17. Vertex: $(-4, 0)$;
axis of symmetry: $x = -4$

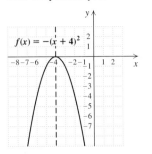

$f(x) = -(x + 4)^2$

19. Vertex: $(1, 0)$;
axis of symmetry: $x = 1$

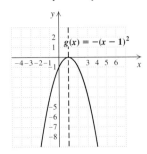

$g(x) = -(x - 1)^2$

33. Vertex: $(-1, -2)$;
axis of symmetry: $x = -1$;
minimum: -2

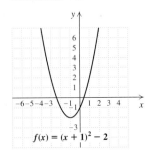

$f(x) = (x + 1)^2 - 2$

35. Vertex: $(-4, 1)$;
axis of symmetry: $x = -4$;
minimum: 1

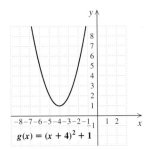

$g(x) = (x + 4)^2 + 1$

37. Vertex: $(1, -3)$;
axis of symmetry: $x = 1$;
maximum: -3

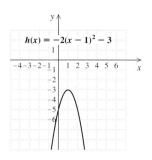

$h(x) = -2(x - 1)^2 - 3$

39. Vertex: $(-4, 1)$;
axis of symmetry: $x = -4$;
minimum: 1

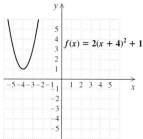

$f(x) = 2(x + 4)^2 + 1$

EXERCISE SET 8.8, PP. 490–491

1. (a) Vertex: $(2, 1)$;
axis of symmetry: $x = 2$;
(b)

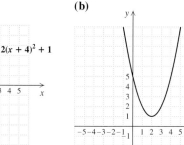

3. (a) Vertex: $(-3, 4)$;
axis of symmetry: $x = -3$;
(b)

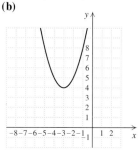

41. Vertex: $(1, 2)$;
axis of symmetry: $x = 1$;
maximum: 2

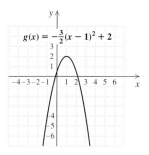

$g(x) = -\frac{3}{2}(x - 1)^2 + 2$

43. Vertex: $(9, 5)$;
axis of symmetry: $x = 9$;
minimum: 5
45. Vertex: $(-6, 11)$;
axis of symmetry: $x = -6$;
maximum: 11
47. Vertex: $\left(-\frac{1}{4}, -13\right)$;
axis of symmetry: $x = -\frac{1}{4}$;
minimum: -13
49. Vertex: $(-4.58, 65\pi)$;
axis of symmetry: $x = -4.58$; minimum: 65π

5. (a) Vertex: $(-4, 4)$;
axis of symmetry: $x = -4$;
(b)

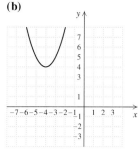

7. (a) Vertex: $(-4, -7)$;
axis of symmetry: $x = -4$;
(b)

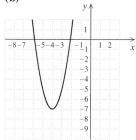

51. $(-5, -1)$ **53.** $x^2 + 5x + \frac{25}{4}$ **55.** ◈   **57.** ◈
59. $f(x) = 2(x - 5)^2$ **61.** $g(x) = -2(x + 4)^2$
63. $g(x) = -2(x - 3)^2 + 8$ **65.** $F(x) = 3(x - 5)^2 + 1$
67.   **69.**

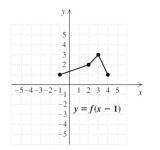

$y = f(x - 1)$

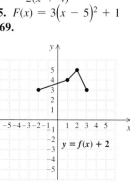

$y = f(x) + 2$

9. (a) Vertex: $(1, 6)$;
axis of symmetry: $x = 1$;
(b)

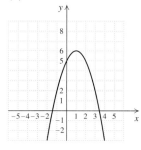

11. (a) Vertex: $\left(-\frac{3}{2}, -\frac{49}{4}\right)$;
axis of symmetry: $x = -\frac{3}{2}$;
(b)

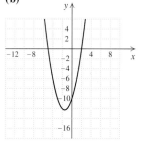

71.

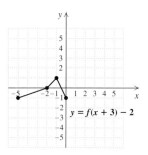

$y = f(x + 3) - 2$

73. 〰   **75.** ◈ 〰

13. (a) Vertex: $(4, 2)$;
axis of symmetry: $x = 4$;
(b)

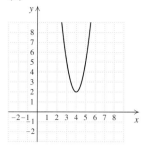

15. (a) Vertex: $\left(\frac{9}{2}, -\frac{81}{4}\right)$;
axis of symmetry: $x = \frac{9}{2}$;
(b)

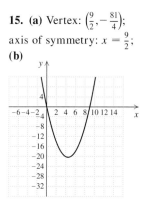

17. (a) Vertex: $(-1, -4)$; axis of symmetry: $x = -1$; **(b)**

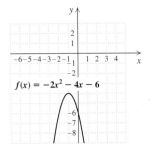

$f(x) = -2x^2 - 4x - 6$

19. (a) Vertex: $\left(\frac{7}{4}, -\frac{41}{8}\right)$; axis of symmetry: $x = \frac{7}{4}$; **(b)**

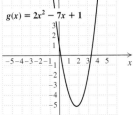

$g(x) = 2x^2 - 7x + 1$

21. (a) Vertex: $\left(\frac{5}{6}, \frac{1}{12}\right)$; axis of symmetry: $x = \frac{5}{6}$; **(b)**

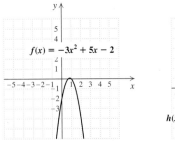

$f(x) = -3x^2 + 5x - 2$

23. (a) Vertex: $\left(-4, -\frac{5}{3}\right)$; axis of symmetry: $x = -4$; **(b)**

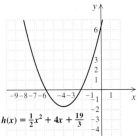

$h(x) = \frac{1}{2}x^2 + 4x + \frac{19}{3}$

25. $(3 - \sqrt{6}, 0)$, $(3 + \sqrt{6}, 0)$; $(0, 3)$ **27.** $(-1, 0)$, $(3, 0)$; $(0, 3)$ **29.** No x-intercepts; $(0, 4)$ **31.** $(2, 0)$; $(0, -4)$

33. $\left(\frac{3 - \sqrt{6}}{2}, 0\right)$, $\left(\frac{3 + \sqrt{6}}{2}, 0\right)$; $(0, 3)$ **35.** 5

37. $n = \dfrac{3Q}{A - T}$ **39.** ◈   **41.** ◈

43. (a) Maximum: 7.01412766; **(b)** $(-0.400174191, 0)$, $(0.821450787, 0)$; $(0, 6.18)$ **45. (a)** $-3, 1$; **(b)** $-4.5, 2.5$;

(c) $-5.5, 3.5$ **47.** $f(x) = 3\left[x - \left(-\dfrac{m}{6}\right)\right]^2 + \dfrac{11m^2}{12}$

49. $f(x) = -0.28x^2 - 0.56x + 6.72$

51.

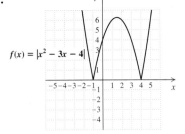

$f(x) = |x^2 - 3x - 4|$

53. ⌁

1. 14 ft by 14 ft; 196 ft^2 **3.** -9; -3 and 3
5. 169; 13 and 13   **7.** $-\frac{49}{4}$; $-\frac{7}{2}$ and $\frac{7}{2}$ **9.** 36; -6 and -6
11. 450 ft^2; 15 ft by 30 ft (The house serves as a 30-ft side.)   **13.** 3.5 in.   **15.** 3.5 hundred, or 350
17. $P(x) = -x^2 + 980x - 3000$; \$237,100 at $x = 490$
19. $f(x) = 2x^2 + 3x - 1$ **21.** $f(x) = -\frac{1}{4}x^2 + 3x - 5$
23. (a) $A(s) = \frac{3}{16}s^2 - \frac{135}{4}s + 1750$; **(b)** About 531
25. $h(d) = -0.0068d^2 + 0.8571d$ **27.** $\dfrac{x - 9}{(x + 9)(x + 7)}$
29. ◈   **31.** The radius of the circular portion of the window and the height of the rectangular portion should each be $\dfrac{24}{\pi + 4}$ ft.   **33.** 30   **35.** 78.4 ft   **37.** ⌁

1. $\{x|\ -0.78 \le x \le 1.59\}$, or $[-0.78, 1.59]$
2. $\{x|\ x \le -0.21\ or\ x \ge 2.47\}$, or $(-\infty, -0.21] \cup [2.47, \infty)$
3. $\{x|\ x < -1.26\ or\ x > 2.33\}$, or $(-\infty, -1.26) \cup (2.33, \infty)$
4. $\{x|\ x > -1.37\}$, or $(-1.37, \infty)$

1. $(-\infty, -4) \cup (3, \infty)$, or $\{x|\ x < -4\ or\ x > 3\}$
3. $[-7, 2]$, or $\{x|\ -7 \le x \le 2\}$ **5.** $(-1, 2)$, or $\{x|\ -1 < x < 2\}$ **7.** $(-\infty, -3] \cup [3, \infty)$, or $\{x|\ x \le -3\ or\ x \ge 3\}$ **9.** $(-\infty, \infty)$, or $\mathbb{R}$ **11.** $(-2, 6)$, or $\{x|\ -2 < x < 6\}$ **13.** $(-\infty, -2) \cup (0, 2)$, or $\{x|\ x < -2\ or\ 0 < x < 2\}$ **15.** $(-3, -1) \cup (2, \infty)$, or $\{x|\ -3 < x < -1\ or\ x > 2\}$ **17.** $(-\infty, -3) \cup (-2, 1)$, or $\{x|\ x < -3\ or\ -2 < x < 1\}$ **19.** $(-\infty, -7)$, or $\{x|\ x < -7\}$ **21.** $(-\infty, -1] \cup (3, \infty)$, or $\{x|\ x \le -1\ or\ x > 3\}$ **23.** $\left[-\frac{2}{3}, 3\right)$, or $\{x|\ -\frac{2}{3} \le x < 3\}$
25. $\left(\frac{3}{2}, 4\right)$, or $\{x|\ \frac{3}{2} < x < 4\}$ **27.** $(-\infty, -1] \cup [2, 5)$, or $\{x|\ x \le -1\ or\ 2 \le x < 5\}$ **29.** $(-\infty, -3) \cup [0, \infty)$, or $\{x|\ x < -3\ or\ x \ge 0\}$ **31.** $(0, \infty)$, or $\{x|\ x > 0\}$
33. $(-\infty, -4) \cup [1, 3)$, or $\{x|\ x < -4\ or\ 1 \le x < 3\}$
35. $\left(0, \frac{1}{4}\right)$, or $\{x|\ 0 < x < \frac{1}{4}\}$ **37.** $\sqrt[15]{a^{11}b^{13}}$ **39.** ◈
41. $(-\infty, -1 - \sqrt{5}) \cup (-1 + \sqrt{5}, \infty)$, or $\{x|\ x < -1 - \sqrt{5}\ or\ x > -1 + \sqrt{5}\}$ **43.** $\{0\}$
45. (a) $(10, 200)$, or $\{x|\ 10 < x < 200\}$;
(b) $[0, 10) \cup (200, \infty)$, or $\{x|\ 0 \le x < 10\ or\ x > 200\}$
47. $\{n|\ 12 \le n \le 25\}$ **49.** $f(x) = 0$ for $x = -2, 1, 3$; $f(x) < 0$ for $(-\infty, -2) \cup (1, 3)$, or $\{x|\ x < -2\ or\ 1 < x < 3\}$; $f(x) > 0$ for $(-2, 1) \cup (3, \infty)$, or $\{x|\ -2 < x < 1\ or\ x > 3\}$ **51.** $f(x)$ has no zeros; $f(x) < 0$ for $(-\infty, 0)$, or $\{x|\ x < 0\}$; $f(x) > 0$ for $(0, \infty)$, or $\{x|\ x > 0\}$ **53.** $f(x) = 0$ for $x = -2, 1, 2, 3$; $f(x) < 0$ for $(-2, 1) \cup (2, 3)$, or $\{x|\ -2 < x < 1\ or\ 2 < x < 3\}$; $f(x) > 0$ for $(-\infty, -2) \cup (1, 2) \cup (3, \infty)$, or $\{x|\ x < -2\ or\ 1 < x < 2\ or\ x > 3\}$ **55.** ⌁

REVIEW EXERCISES: CHAPTER 8, PP. 510–511

1. [8.1] $\pm\sqrt{\dfrac{7}{2}}$, or $\pm\dfrac{\sqrt{14}}{2}$ **2.** [8.1] $0, -\dfrac{5}{14}$

3. [8.1] 3, 9   **4.** [8.2] $\dfrac{5 \pm i\sqrt{11}}{2}$ **5.** [8.2] 3, 5

6. [8.2] $-0.3722813233,\ 5.372281323$ **7.** [8.2] $-\dfrac{1}{4},\ 1$

8. [8.1] $x^2 - 12x + 36;\ (x-6)^2$ **9.** [8.1] $x^2 + \dfrac{3}{5}x + \dfrac{9}{100}$; $\left(x + \dfrac{3}{10}\right)^2$ **10.** [8.1] $-4, 2$ **11.** [8.1] $3 \pm 2\sqrt{2}$

12. [8.1] 10%   **13.** [8.1] 6.7 sec

14. [8.3] About 2.89 mph   **15.** [8.3] 6 hr

16. [8.4] Two irrational   **17.** [8.4] Two imaginary

18. [8.4] $x^2 - 2x - 35 = 0$ **19.** [8.4] $x^2 + 8x + 16 = 0$

20. [8.5] $(-3, 0), (-2, 0), (2, 0), (3, 0)$ **21.** [8.5] $-5, 3$

22. [8.5] $\pm\sqrt{2}, \pm\sqrt{7}$ **23.** [8.6] 5 amperes

24. [8.6] 64 L   **25.** [8.6] $y = \dfrac{15xz}{2w}$

26. [8.7]
$x = -2$
$(-2, 4)$
$f(x) = -3(x+2)^2 + 4$
Maximum: 4

27. [8.8] **(a)** Vertex: (3, 5), axis of symmetry: $x = 3$;   **(b)**
$f(x) = 2x^2 - 12x + 23$

28. [8.8] $(2, 0), (7, 0);\ (0, 14)$ **29.** [8.3] $p = \dfrac{9\pi^2}{N^2}$

30. [8.3] $T = \dfrac{1 \pm \sqrt{1 + 24A}}{6}$ **31.** [8.9] -121; 11 and -11

32. [8.9] $f(x) = -x^2 + 6x - 2$

33. [8.10] $(-1, 0) \cup (3, \infty)$, or $\{x \mid -1 < x < 0\ or\ x > 3\}$

34. [8.10] $(-3, 5]$, or $\{x \mid -3 < x \le 5\}$ **35.** [3.3] 210 kg of A; 90 kg of B   **36.** [7.6] 2   **37.** [6.2] $\dfrac{x+1}{(x-3)(x-1)}$

38. [7.4] $3t^5 s\sqrt[3]{s}$ **39.** [8.8], [8.9] ◈ The x-coordinate of the maximum or minimum point lies halfway between the x-coordinates of the x-intercepts.   **40.** [8.2], [8.4] ◈ Yes; if the discriminant is a perfect square, then the solutions are rational numbers, p/q and r/s. (Note that if the discriminant is 0, then $p/q = r/s$.) Then the equation can be written in factored form, $(qx - p)(sx - r) = 0$.

41. [8.5] ◈ Four; let $u = x^2$. Then $au^2 + bu + c = 0$ has at most two solutions, $u = m$ or $u = n$. Now substitute x^2 for u and obtain $x^2 = m$ or $x^2 = n$. These equations yield the solutions $x = \pm\sqrt{m}$ and $x = \pm\sqrt{n}$. When $m \ne n$, the maximum number of solutions, four, occurs.

42. [8.1], [8.2], [8.8] ◈ Completing the square was used to solve quadratic equations and to graph quadratic functions by rewriting the function in the form $f(x) = a(x - h)^2 + k$.

43. [8.8] $f(x) = \dfrac{7}{15}x^2 - \dfrac{14}{15}x - 7$ **44.** [8.4] $h = 60$, $k = 60$ **45.** [8.5] 18, 324

TEST: CHAPTER 8, PP. 511–512

1. [8.1] $\pm\dfrac{4\sqrt{3}}{3}$ **2.** [8.2] 2, 9   **3.** [8.2] $\dfrac{-1 \pm i\sqrt{3}}{2}$

4. [8.5] $-2, \dfrac{2}{3}$ **5.** [8.2] $-4.192582404,\ 1.192582404$

6. [8.2] $-\dfrac{3}{4}, \dfrac{7}{3}$ **7.** [8.1] $x^2 + 14x + 49;\ (x+7)^2$

8. [8.1] $x^2 - \dfrac{2}{7}x + \dfrac{1}{49};\ \left(x - \dfrac{1}{7}\right)^2$ **9.** [8.1] $-6, 3$

10. [8.1] $-5 \pm \sqrt{10}$ **11.** [8.3] 16 km/h   **12.** [8.3] 2 hr

13. [8.4] Two imaginary   **14.** [8.4] $3x^2 + 5x - 2 = 0$

15. [8.5] $(-3, 0), (-1, 0), (-2 - \sqrt{5}, 0), (-2 + \sqrt{5}, 0)$

16. [8.6] $\dfrac{833}{125}$, or 6.664 in^2

17. [8.7]
$f(x) = 4(x-3)^2 + 5$
Minimum: 5

18. [8.8] **(a)** $(-1, -8)$, $x = -1$;   **(b)**
$f(x) = 2x^2 + 4x - 6$

19. [8.8] $(-2, 0), (3, 0);\ (0, -6)$

20. [8.3] $r = \sqrt{\dfrac{3V}{\pi} - R^2}$ **21.** [8.9] -16

22. [8.9] $f(x) = \dfrac{1}{5}x^2 - \dfrac{3}{5}x$ **23.** [8.10] $[-6, 1]$, or $\{x \mid -6 \le x \le 1\}$ **24.** [7.6] 6   **25.** [7.4] $ab\sqrt[4]{2a^2}$

26. [6.2] $\dfrac{x-8}{(x+6)(x+8)}$ **27.** [3.3] Each is 36.

28. [8.4] $\dfrac{1}{2}$ **29.** [8.4] $x^4 - 14x^3 + 67x^2 - 114x + 26 = 0$; answers may vary.   **30.** [8.4] $x^6 - 10x^5 + 20x^4 + 50x^3 - 119x^2 - 60x + 150 = 0$; answers may vary.

CHAPTER 9

TECHNOLOGY CONNECTION 9.1A, P. 517

1.

2.

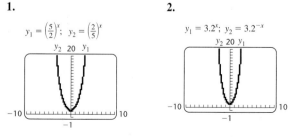

$y_1 = \left(\dfrac{5}{2}\right)^x;\ y_2 = \left(\dfrac{2}{5}\right)^x$

$y_1 = 3.2^x;\ y_2 = 3.2^{-x}$

3.
$y_1 = 9.34^x$; $y_2 = 9.34^{-x}$

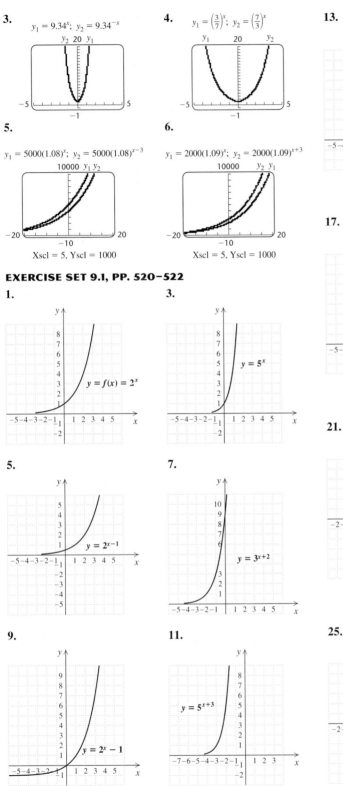

4.
$y_1 = \left(\frac{3}{7}\right)^x$; $y_2 = \left(\frac{7}{3}\right)^x$

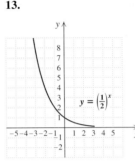

5.
$y_1 = 5000(1.08)^x$; $y_2 = 5000(1.08)^{x-3}$

Xscl = 5, Yscl = 1000

6.
$y_1 = 2000(1.09)^x$; $y_2 = 2000(1.09)^{x+3}$

Xscl = 5, Yscl = 1000

EXERCISE SET 9.1, PP. 520–522

1.

$y = f(x) = 2^x$

3.

$y = 5^x$

13.

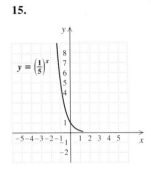
$y = \left(\frac{1}{2}\right)^x$

15.

$y = \left(\frac{1}{5}\right)^x$

5.

$y = 2^{x-1}$

7.

$y = 3^{x+2}$

17.

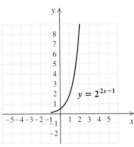
$y = 2^{2x-1}$

19.

$y = 2^{x-3} - 1$

9.

$y = 2^x - 1$

11.

$y = 5^{x+3}$

21.

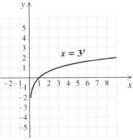
$x = 3^y$

23.

$x = \left(\frac{1}{2}\right)^y$

25.

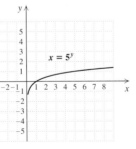
$x = 5^y$

27.

$x = \left(\frac{3}{2}\right)^y$

29.

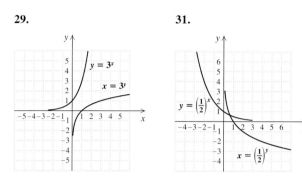

31.

33. (a) 384,160; (b) 2,066,105;
(c)

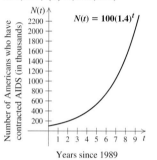

35. (a) 250,000; 166,667; 49,383; 4335;
(b)

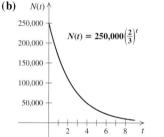

37. (a) 454,354,240 cm², 525,233,501,400 cm²;
(b)

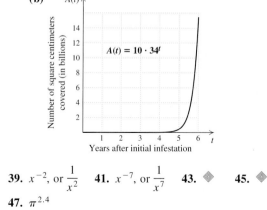

39. x^{-2}, or $\dfrac{1}{x^2}$ **41.** x^{-7}, or $\dfrac{1}{x^7}$ **43.** ◆   **45.** ◆

47. $\pi^{2.4}$

49.

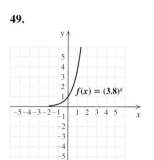

51.

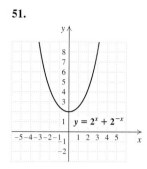

53.

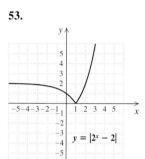

55.

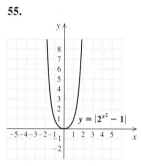

57.

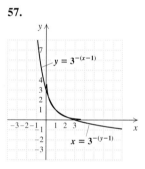

59. $y = 1.809071941(2.316985337)^t$, where t is the number of years since 1993 and y is in millions; 120.8 million

61. 19 wpm, 66 wpm, 110 wpm

TECHNOLOGY CONNECTION 9.2A, P. 525

1. A table shows that $y_2 = y_3$. $y_1 = 7x + 3;\ y_2 = y_1^2;$
$y_3 = (7x + 3)^2$

$y_2 = y_1^2;\ y_3 = (7x + 3)^2$

A graph can also be used:

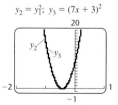

1. Graph each pair of functions in a square window along with the line $y = x$ and determine whether the first two functions are reflections of each other across $y = x$. For further verification, examine a table of values for each pair of functions. **2.** Yes; most graphers do not require that the inverse is a function.

EXERCISE SET 9.2, PP. 532–534

1. $f \circ g(x) = 12x^2 + 36x + 26$; $g \circ f(x) = 6x^2 + 1$

3. $f \circ g(x) = \dfrac{16}{x^2} - 1$; $g \circ f(x) = \dfrac{2}{4x^2 - 1}$

5. $f \circ g(x) = x^4 + 2x^2 - 2$;
$g \circ f(x) = x^4 - 6x^2 + 10$

7. $f(x) = x^2$; $g(x) = 7 - 5x$

9. $f(x) = x^5$; $g(x) = 3x^2 - 7$ **11.** $f(x) = \dfrac{2}{x}$;

$g(x) = x - 3$ **13.** $f(x) = \dfrac{1}{\sqrt{x}}$; $g(x) = 7x + 2$

15. $f(x) = \dfrac{x + 1}{x - 1}$; $g(x) = x^3$ **17.** Yes **19.** No

21. Yes **23.** No **25.** (a) Yes; (b) $f^{-1}(x) = x - 6$

27. (a) Yes; (b) $f^{-1}(x) = 3 - x$ **29.** (a) Yes;

(b) $g^{-1}(x) = x + 5$ **31.** (a) Yes; (b) $f^{-1}(x) = \dfrac{x}{4}$

33. (a) Yes; (b) $g^{-1}(x) = \dfrac{x - 3}{4}$ **35.** (a) No **37.** (a) Yes;

(b) $f^{-1}(x) = \dfrac{1}{x}$ **39.** (a) Yes; (b) $f^{-1}(x) = \dfrac{3x - 1}{2}$

41. (a) Yes; (b) $f^{-1}(x) = \sqrt[3]{x + 5}$ **43.** (a) Yes;

(b) $g^{-1}(x) = \sqrt[3]{x + 2}$ **45.** (a) Yes; (b) $f^{-1}(x) = x^2, x \geq 0$

47. (a) Yes; (b) $f^{-1}(x) = \sqrt{\dfrac{x - 1}{2}}$

49.

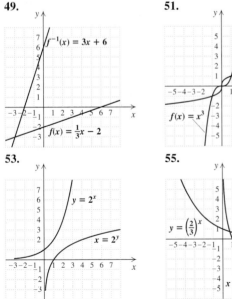

51.

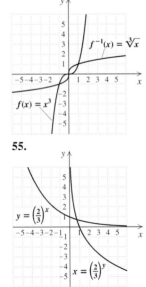

53.

55.

57.

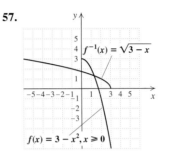

59. (1) $f^{-1} \circ f(x) = f^{-1}(f(x)) = f^{-1}\left(\frac{4}{5}x\right) = \frac{5}{4}\left(\frac{4}{5}x\right) = x$;
(2) $f \circ f^{-1}(x) = f(f^{-1}(x)) = f\left(\frac{5}{4}x\right) = \frac{4}{5}\left(\frac{5}{4}x\right) = x$

61. (1) $f^{-1} \circ f(x) = f^{-1}(f(x)) = f^{-1}\left(\dfrac{1 - x}{x}\right)$

$= \dfrac{1}{\left(\dfrac{1 - x}{x}\right) + 1}$

$= \dfrac{1}{\dfrac{1 - x + x}{x}}$

$= x$;

(2) $f \circ f^{-1}(x) = f(f^{-1}(x)) = f\left(\dfrac{1}{x + 1}\right)$

$= \dfrac{1 - \left(\dfrac{1}{x + 1}\right)}{\left(\dfrac{1}{x + 1}\right)}$

$= \dfrac{\dfrac{x + 1 - 1}{x + 1}}{\dfrac{1}{x + 1}} = x$

63. (a) 40, 42, 46, 50; (b) $f^{-1}(x) = x - 32$;
(c) 8, 10, 14, 18 **65.** $y = 9x$ **67.** $a^{17}b^{17}$

69. ◈ **71.** ◈ **73.** $g(x) = \dfrac{x}{2} + 20$ **75.** No

77. Yes **79.** (1) C; (2) A; (3) B; (4) D

EXERCISE SET 9.3, PP. 539–540

1.

3.

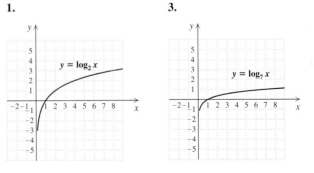

5.

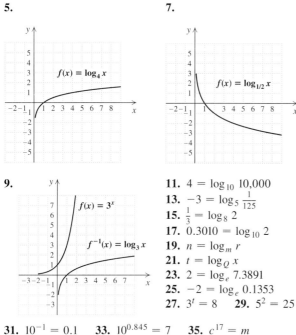

$f(x) = \log_4 x$

7.

$f(x) = \log_{1/2} x$

9.

$f(x) = 3^x$

$f^{-1}(x) = \log_3 x$

11. $4 = \log_{10} 10{,}000$
13. $-3 = \log_5 \frac{1}{125}$
15. $\frac{1}{3} = \log_8 2$
17. $0.3010 = \log_{10} 2$
19. $n = \log_m r$
21. $t = \log_Q x$
23. $2 = \log_e 7.3891$
25. $-2 = \log_e 0.1353$
27. $3^t = 8$ **29.** $5^2 = 25$

31. $10^{-1} = 0.1$ **33.** $10^{0.845} = 7$ **35.** $c^{17} = m$
37. $t^k = Q$ **39.** $e^{-1.3863} = 0.25$ **41.** $r^{-x} = T$ **43.** 81
45. 5   **47.** 4   **49.** 3   **51.** 8   **53.** 1   **55.** $\frac{1}{2}$ **57.** 4
59. 4   **61.** 0   **63.** 4   **65.** -2 **67.** 1   **69.** 0
71. 15   **73.** $\frac{2}{3}$ **75.** 7   **77.** $\frac{x(3y-2)}{2y+x}$ **79.** $\frac{1}{4096}$
81. $\frac{1}{\sqrt[3]{t}}$ **83.** ◈   **85.** ◈
87.

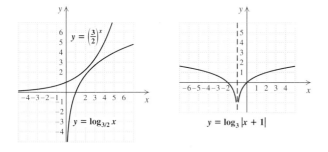

$y = \left(\frac{3}{2}\right)^x$

$y = \log_{3/2} x$

89.

$y = \log_3 |x+1|$

91. 25   **93.** $-\frac{7}{16}$ **95.** 3   **97.** 1   **99.** 1

EXERCISE SET 9.4, PP. 545–546

1. $\log_3 81 + \log_3 27$ **3.** $\log_4 64 + \log_4 16$
5. $\log_c x + \log_c y + \log_c z$ **7.** $\log_a (5 \cdot 14)$, or $\log_a 70$
9. $\log_c (t \cdot y)$ **11.** $7 \log_a t$ **13.** $6 \log_c y$
15. $-3 \log_b C$ **17.** $\log_2 64 - \log_2 16$
19. $\log_b m - \log_b n$ **21.** $\log_a \frac{15}{7}$
23. $2 \log_a x + 3 \log_a y + \log_a z$
25. $\log_b x + 2 \log_b y - 3 \log_b z$
27. $2 \log_a x - 3 \log_a y - \log_a z$

29. $\log_b x + 2 \log_b y - \log_b w - 3 \log_b z$
31. $\frac{1}{2}(6 \log_a x - 5 \log_a y - 8 \log_a z)$
33. $\frac{1}{3}(6 \log_a x + 3 \log_a y - 2 - 7 \log_a z)$
35. $\log_a x^4 y^3$ **37.** $\log_a x$ **39.** $\log_a \frac{y^3}{x^{3/2}}$
41. $\log_a (x+2)$ **43.** 2.708   **45.** -0.51 **47.** -1.609
49. $\frac{3}{2}$ **51.** Cannot be found   **53.** 4.317   **55.** 9
57. m **59.** $0 + i$, or i **61.** $5 + 10i$ **63.** $16 - 30i$
65. ◈   **67.** ◈   **69.** $\log_a (x^3 + y^3)$
71. $\frac{1}{2} \log_a (c - d) - \frac{1}{2} \log_a (c + d)$
73. -2 **75.** False
77. $y_1 = \log x^2$; $y_2 = \log x \cdot \log x$

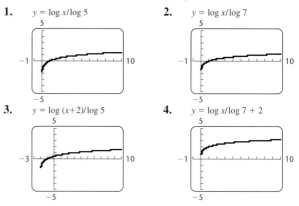

TECHNOLOGY CONNECTION 9.5, P. 551

1. $y = \log x/\log 5$

2. $y = \log x/\log 7$

3. $y = \log (x+2)/\log 5$

4. $y = \log x/\log 7 + 2$

EXERCISE SET 9.5, PP. 551–552

1. 0.6021   **3.** 1.7300   **5.** 2.6405   **7.** 4.1271
9. -0.2782 **11.** 199.5262   **13.** 1.4894
15. 0.0011   **17.** 1.6094   **19.** 4.1271   **21.** 8.3814
23. -5.0832 **25.** 15.0293   **27.** 0.0305
29. 109.9472   **31.** 2.5702   **33.** 6.6439   **35.** 2.1452
37. -2.3219 **39.** -2.3219 **41.** 3.5471
43.

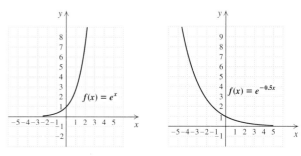

$f(x) = e^x$

45.

$f(x) = e^{-0.5x}$

TEST: CHAPTER 9, P. 573

1. [9.1]

$f(x) = 2^{x+3}$

2. [9.3]

$f(x) = \log_7 x$

3. [9.2] $f \circ g(x) = 25x^2 - 15x + 2$, $g \circ f(x) = 5x^2 + 5x - 2$

4. [9.2] No   **5.** [9.2] $f^{-1}(x) = \dfrac{x + 3}{4}$

6. [9.2] $f^{-1}(x) = \sqrt[3]{x} - 1$ **7.** [9.3] $\log_4 x = -3$

8. [9.3] $\log_{256} 16 = \dfrac{1}{2}$ **9.** [9.3] $4^2 = 16$

10. [9.3] $7^m = 49$ **11.** [9.4] $3 \log a + \dfrac{1}{2} \log b - 2 \log c$

12. [9.4] $\log_a x^{1/3} z^2$ **13.** [9.3] 3   **14.** [9.3] 23

15. [9.3] 1   **16.** [9.3] 0   **17.** [9.4] -0.544

18. [9.4] 1.079   **19.** [9.4] 1.322   **20.** [9.5] -1.9101

21. [9.5] 6309.5734   **22.** [9.5] -4.3949

23. [9.5] 107.7701   **24.** [9.5] 1.1881   **25.** [9.6] 5

26. [9.6] 2   **27.** [9.6] 10,000   **28.** [9.6] $\dfrac{1}{3}$

29. [9.6] $\dfrac{\log 1.2}{\log 7} \approx 0.0937$ **30.** [9.6] $e^{1/4} \approx 1.2840$

31. [9.6] 4   **32.** [9.7] **(a)** 2.05 ft/sec; **(b)** 984,262

33. [9.7] **(a)** $P(t) = 30e^{0.0062t}$, where $P(t)$ is in millions and t is the number of years after 1996; **(b)** 31.3 million; 32.7 million; **(c)** 2079; **(d)** 111.8 yr   **34.** [9.7] 4.6%

35. [9.7] 4684 yr   **36.** [9.7] $10^{-4.5}$ W/m²   **37.** [9.7] 7.0

38. [8.5] 1, 64   **39.** [8.3] $t = \dfrac{b \pm \sqrt{b^2 + 4aS}}{2a}$

40. [7.8] 29   **41.** [6.3] $\dfrac{1}{2x}$ **42.** [9.6] 316, -309

43. [9.4] 2

CUMULATIVE REVIEW: 1–9

1. [1.1], [1.6] 2   **2.** [1.2] 6   **3.** [1.6] $\dfrac{y^{12}}{16x^8}$

4. [1.6] $\dfrac{20x^6 z^2}{y}$ **5.** [1.6] $\dfrac{-y^4}{3z^5}$ **6.** [1.3] $-4x - 1$

7. [1.1] 25   **8.** [1.3] $\dfrac{11}{2}$ **9.** [3.2] $(3, -1)$

10. [3.4] $(1, -2, 0)$ **11.** [5.8] $-2, 5$ **12.** [6.4] $\dfrac{9}{2}$

13. [8.2] $\dfrac{5}{8}$ **14.** [7.6] 5   **15.** [7.6] $\dfrac{1}{2}$ **16.** [8.1] $\pm 5i$

17. [8.5] 9, 25   **18.** [8.5] $\pm 3, \pm 2$ **19.** [9.3] 8

20. [9.3] 7   **21.** [9.6] $\dfrac{3}{2}$ **22.** [9.6] $\dfrac{\log 7}{5 \log 3} \approx 0.3542$

23. [9.6] $\dfrac{80}{9}$ **24.** [8.10] $(-\infty, -5) \cup (1, \infty)$, or $\{x \mid x < -5$ or $x > 1\}$ **25.** [8.2] $-3 \pm 2\sqrt{5}$ **26.** [4.3] $\{x \mid x \leqslant -3$ or $x \geqslant 6\}$, or $(-\infty, -3] \cup [6, \infty)$ **27.** [6.8] $a = \dfrac{Db}{b - D}$

28. [6.8] $q = \dfrac{pf}{p - f}$ **29.** [1.5] $B = \dfrac{3M - 2A}{2}$, or $B = \dfrac{3}{2}M - A$ **30.** [3.7] 2   **31.** [3.7] 3

32. [5.8] $\left\{x \mid x \text{ is a real number and } x \neq -\dfrac{1}{3} \text{ and } x \neq 2\right\}$

33. [2.5] **(a)** $A(t) = -225t + 1400$, where t is the number of years after 1994; **(b)** 50 acres   **34.** [1.4] Length: 36 m; width: 20 m   **35.** [1.4] A: 15°; B: 45°; C: 120°

36. [6.5] $5\dfrac{5}{11}$ hr   **37.** [3.3] 60 L of Swim Clean; 40 L of Pure Swim   **38.** [6.5] $2\dfrac{7}{9}$ km/h

39. [8.9] -49; -7 and 7   **40.** [9.7] 78   **41.** [9.7] 67.5

42. [9.7] $P(t) = 52e^{0.002t}$, where P is in millions and t is the number of years after 1996

43. [9.7] 52.5 million, 52.9 million   **44.** [8.6] 18

45. [5.1] $7p^2 q^3 - 2p^3 q + pq - 6$

46. [5.1] $8x^2 - 11x - 1$ **47.** [5.2] $9x^4 - 12x^2 y + 4y^2$

48. [5.2] $10a^2 - 9ab - 9b^2$ **49.** [6.1] $\dfrac{(x + 4)(x - 3)}{2(x - 1)}$

50. [6.3] $\dfrac{1}{x - 4}$ **51.** [6.1] $\dfrac{a + 2}{6}$

52. [6.2] $\dfrac{7x + 4}{(x + 6)(x - 6)}$ **53.** [5.3] $x(y - 2z + w)$

54. [5.6] $(1 - 5x)(1 + 5x + 25x^2)$

55. [5.4] $2(3x - 2y)(x + 2y)$ **56.** [5.3] $(x^3 + 7)(x - 4)$

57. [5.5] $(a - 5 + 9b)(a - 5 - 9b)$ **58.** [5.5] $2(m + 3n)^2$

59. [5.5] $(x - 2y)(x + 2y)(x^2 + 4y^2)$ **60.** [2.2] -12

61. [6.6] $x^3 - 2x^2 - 4x - 12 + \dfrac{-42}{x - 3}$

62. [1.7] 1.8×10^{-1} **63.** [1.7] 5.0×10^{13}

64. [7.4] $2y^2 \sqrt[3]{y}$ **65.** [7.4] $14xy^2 \sqrt{x}$

66. [7.2] $81a^8 b \sqrt[3]{b}$ **67.** [7.5] $\dfrac{6 + \sqrt{y} - y}{4 - y}$

68. [7.4] $\sqrt[10]{(x + 5)^3}$ **69.** [7.8] $12 + 4\sqrt{3}\,i$

70. [7.8] $\dfrac{18}{25} + \dfrac{1}{25}i$ **71.** [9.2] $f^{-1}(x) = \dfrac{x - 7}{-2}$

72. [2.5] $f(x) = -5x - 3$ **73.** [2.5] $y = \dfrac{1}{2}x + \dfrac{13}{2}$

74. [2.4]

$5x = 15 + 3y$

75. [8.8]

$y = 2x^2 - 4x - 1$

76. [9.3]

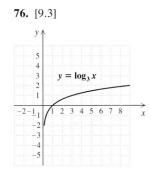

77. [9.1]

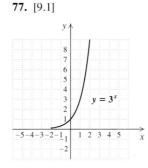

EXERCISE SET 10.1, PP. 586–588

1.

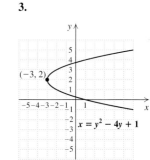

3.

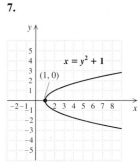

78. [4.4]

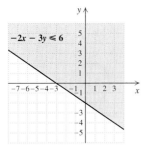

79. [8.7]

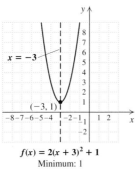

$f(x) = 2(x + 3)^2 + 1$
Minimum: 1

5.

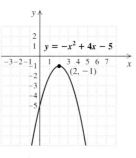

7.

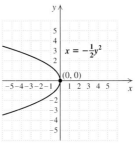

80. [9.4] $2 \log a + 3 \log c - \log b$

81. [9.4] $\log \left(\dfrac{x^3}{y^{1/2}z^2} \right)$ **82.** [9.3] $a^x = 5$

83. [9.3] $\log_x t = 3$ **84.** [9.5] -1.2545

85. [9.5] 776.2471 **86.** [9.5] 2.5479 **87.** [9.5] 0.2466

88. [6.4] All real numbers except 1 and -2

89. [9.6] $\dfrac{1}{3}, \dfrac{10,000}{3}$ **90.** [8.3] 35 mph

CHAPTER 10

TECHNOLOGY CONNECTION 10.1, P. 585

1. $x^2 + y^2 - 16 = 0$

2. $4x^2 + 4y^2 = 100$

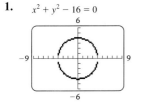

9.

11.

13.

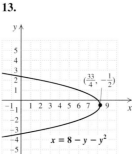

15.

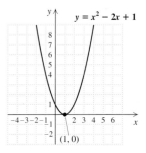

3. $x^2 + y^2 + 14x - 16y + 54 = 0$ **4.** $x^2 + y^2 - 10x - 11 = 0$

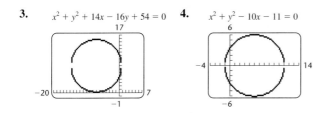

17.

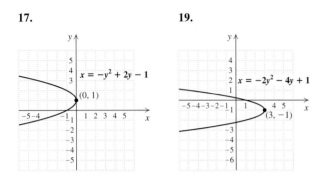

$x = -y^2 + 2y - 1$
$(0, 1)$

19.

$x = -2y^2 - 4y + 1$
$(3, -1)$

69. $(5, 0); \frac{1}{2}$

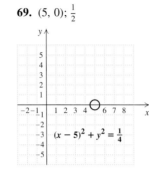

$(x - 5)^2 + y^2 = \frac{1}{4}$

71. $(-4, 3); \sqrt{40}, \text{ or } 2\sqrt{10}$

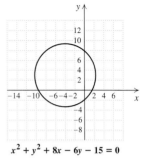

$x^2 + y^2 + 8x - 6y - 15 = 0$

21. 5   **23.** $\sqrt{18} \approx 4.243$ **25.** $\sqrt{200} \approx 14.142$

27. 17.8   **29.** $\dfrac{\sqrt{41}}{7} \approx 0.915$ **31.** $\sqrt{8} \approx 2.828$

33. $\sqrt{17 + 2\sqrt{14} + 2\sqrt{15}} \approx 5.677$ **35.** $\sqrt{s^2 + t^2}$

37. $(4, 1)$ **39.** $\left(\dfrac{7}{2}, \dfrac{7}{2}\right)$ **41.** $(-1, -3)$

43. $(-0.25, -0.3)$ **45.** $\left(-\dfrac{1}{12}, \dfrac{1}{24}\right)$

47. $\left(\dfrac{\sqrt{2} + \sqrt{3}}{2}, \dfrac{3}{2}\right)$ **49.** $x^2 + y^2 = 36$

51. $(x - 7)^2 + (y - 3)^2 = 5$
53. $(x + 4)^2 + (y - 3)^2 = 48$
55. $(x + 7)^2 + (y + 2)^2 = 50$ **57.** $x^2 + y^2 = 25$
59. $(x + 4)^2 + (y - 1)^2 = 20$

61. $(0, 0); 7$

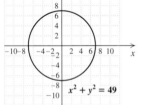

$x^2 + y^2 = 49$

63. $(-1, -3); 2$

$(x + 1)^2 + (y + 3)^2 = 4$

73. $(4, -1); 2$

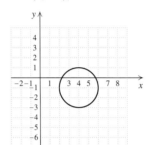

$x^2 + y^2 - 8x + 2y + 13 = 0$

75. $(0, -5); 10$

$x^2 + y^2 + 10y - 75 = 0$

77. $\left(-\dfrac{7}{2}, \dfrac{3}{2}\right);$ $\sqrt{\dfrac{98}{4}}, \text{ or } \dfrac{7\sqrt{2}}{2}$

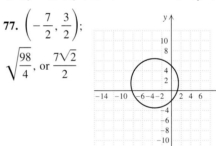

$x^2 + y^2 + 7x - 3y - 10 = 0$

79. $(0, 0); \dfrac{1}{6}$

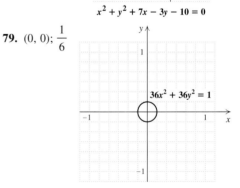

$36x^2 + 36y^2 = 1$

65. $(4, -3); \sqrt{10}$

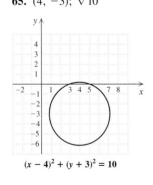

$(x - 4)^2 + (y + 3)^2 = 10$

67. $(0, 0); \sqrt{7}$

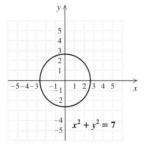

$x^2 + y^2 = 7$

81. 4 in.   **83.** $\left(\dfrac{35}{17}, \dfrac{5}{34}\right)$ **85.** ◈   **87.** ◈
89. $(x - 3)^2 + (y + 5)^2 = 9$ **91.** $(x - 3)^2 + y^2 = 25$
93. $(0, 4)$ **95. (a)** $(0, -8467.8);$ **(b)** 8487.3 mm
97. (a) $(0, -1522.8);$ **(b)** 1524.9 cm   **99.** 29 cm
101. (a) $-2.4, 3.4;$ **(b)** $-1.3, 2.3$

103. Let $P_1 = (x_1, y_1)$, $P_2 = (x_2, y_2)$, and
$M = \left(\dfrac{x_1 + x_2}{2}, \dfrac{y_1 + y_2}{2}\right)$. Let $d(AB)$ denote the distance
from point A to point B.

i) $d(P_1 M) = \sqrt{\left(\dfrac{x_1 + x_2}{2} - x_1\right)^2 + \left(\dfrac{y_1 + y_2}{2} - y_1\right)^2}$

$= \dfrac{1}{2}\sqrt{(x_2 - x_1)^2 + (y_2 - y_1)^2};$

$d(P_2 M) = \sqrt{\left(\dfrac{x_1 + x_2}{2} - x_2\right)^2 + \left(\dfrac{y_1 + y_2}{2} - y_2\right)^2}$

$= \dfrac{1}{2}\sqrt{(x_1 - x_2)^2 + (y_1 - y_2)^2}$

$= \dfrac{1}{2}\sqrt{(x_2 - x_1)^2 + (y_2 - y_1)^2} = d(P_1 M).$

ii) $d(P_1 M) + d(P_2 M) = \dfrac{1}{2}\sqrt{(x_2 - x_1)^2 + (y_2 - y_1)^2}$

$+ \dfrac{1}{2}\sqrt{(x_2 - x_1)^2 + (y_2 - y_1)^2}$

$= \sqrt{(x_2 - x_1)^2 + (y_2 - y_1)^2}$

$= d(P_1 P_2).$

105.

EXERCISE SET 10.2, PP. 593–594

1.

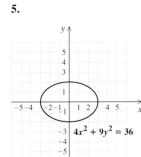

3.

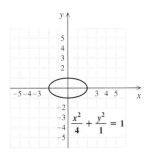

$\dfrac{x^2}{9} + \dfrac{y^2}{25} = 1$

5.

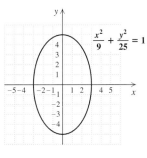

$4x^2 + 9y^2 = 36$

7.

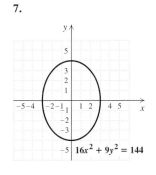

$16x^2 + 9y^2 = 144$

9.

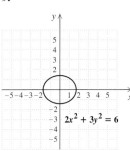

$2x^2 + 3y^2 = 6$

11.

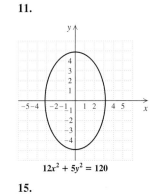

$12x^2 + 5y^2 = 120$

13.

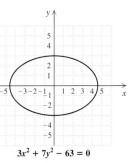

$3x^2 + 7y^2 - 63 = 0$

15.

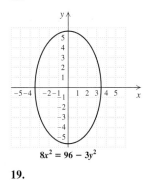

$8x^2 = 96 - 3y^2$

17.

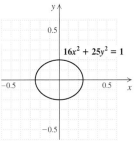

$16x^2 + 25y^2 = 1$

19.

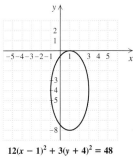

$\dfrac{(x - 2)^2}{9} + \dfrac{(y - 1)^2}{25} = 1$

21.

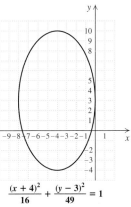

$\dfrac{(x + 4)^2}{16} + \dfrac{(y - 3)^2}{49} = 1$

23.

$12(x - 1)^2 + 3(y + 4)^2 = 48$

25.

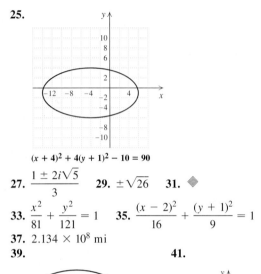

$$(x + 4)^2 + 4(y + 1)^2 - 10 = 90$$

27. $\dfrac{1 \pm 2i\sqrt{5}}{3}$ **29.** $\pm\sqrt{26}$ **31.**

33. $\dfrac{x^2}{81} + \dfrac{y^2}{121} = 1$ **35.** $\dfrac{(x - 2)^2}{16} + \dfrac{(y + 1)^2}{9} = 1$

37. 2.134×10^8 mi

39.

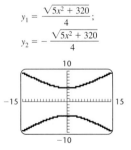

Seat   Center   Seat
of
office

Oval office

41.

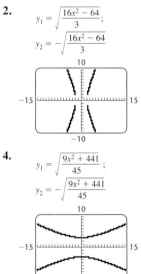

$$\dfrac{(x - 2)^2}{16} + \dfrac{(y + 1)^2}{4} = 1$$

43.

TECHNOLOGY CONNECTION 10.3, P. 597

1. $y_1 = \dfrac{\sqrt{15x^2 - 240}}{2}$;

$y_2 = -\dfrac{\sqrt{15x^2 - 240}}{2}$

2. $y_1 = \sqrt{\dfrac{16x^2 - 64}{3}}$;

$y_2 = -\sqrt{\dfrac{16x^2 - 64}{3}}$

3. $y_1 = \dfrac{\sqrt{5x^2 + 320}}{4}$;

$y_2 = -\dfrac{\sqrt{5x^2 + 320}}{4}$

4. $y_1 = \sqrt{\dfrac{9x^2 + 441}{45}}$;

$y_2 = -\sqrt{\dfrac{9x^2 + 441}{45}}$

EXERCISE SET 10.3, PP. 603–604

1.

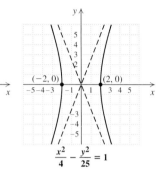

$$\dfrac{y^2}{9} - \dfrac{x^2}{9} = 1$$

3.

$$\dfrac{x^2}{4} - \dfrac{y^2}{25} = 1$$

5.

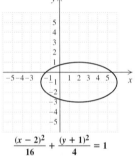

$$\dfrac{y^2}{36} - \dfrac{x^2}{9} = 1$$

7.

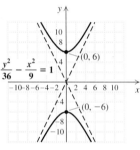

$$y^2 - x^2 = 25$$

9.

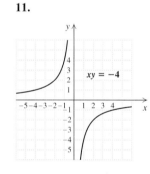

$$25x^2 - 16y^2 = 400$$

11.

$$xy = -4$$

13.

$$xy = 3$$

15.

$$xy = -2$$

17.

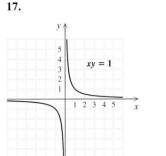

19. Circle   **21.** Ellipse
23. Hyperbola   **25.** Circle
27. Ellipse
29. Hyperbola
31. Parabola
33. Hyperbola   **35.** Circle
37. Ellipse   **39.** $5t^5$
41. $\dfrac{13 + 8\sqrt{6}}{10}$ **43.**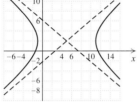

45. $\dfrac{y^2}{36} - \dfrac{x^2}{4} = 1$

47. C: $(5, 2)$; V: $(-1, 2)$, $(11, 2)$;
asymptotes: $y - 2 = \frac{5}{6}(x - 5)$,
$y - 2 = -\frac{5}{6}(x - 5)$

49. $\dfrac{(y + 3)^2}{4} - \dfrac{(x - 4)^2}{16} = 1$; C: $(4, -3)$; V: $(4, -5)$, $(4, -1)$;
asymptotes: $y + 3 = \frac{1}{2}(x - 4)$, $y + 3 = -\frac{1}{2}(x - 4)$

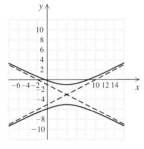

51. $\dfrac{(x + 3)^2}{1} - \dfrac{(y - 2)^2}{4} = 1$; C: $(-3, 2)$; V: $(-4, 2)$, $(-2, 2)$;
asymptotes: $y - 2 = 2(x + 3)$, $y - 2 = -2(x + 3)$

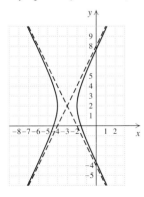

53.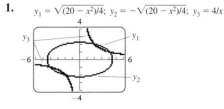

TECHNOLOGY CONNECTION 10.4A, P. 606

1. $(-1.50, -1.17)$; $(3.50, 0.50)$
2. $(-2.77, 2.52)$; $(-2.77, -2.52)$

TECHNOLOGY CONNECTION 10.4B, P. 609

1.

$y_1 = \sqrt{(20 - x^2)/4}$; $y_2 = -\sqrt{(20 - x^2)/4}$; $y_3 = 4/x$

EXERCISE SET 10.4, PP. 611–613

1. $(-8, -6)$, $(6, 8)$ **3.** $(2, 0)$, $(0, 3)$ **5.** $(2, 4)$, $(1, 1)$
7. $\left(\frac{11}{4}, -\frac{9}{8}\right)$, $(1, -2)$ **9.** $\left(-\frac{5}{3}, -\frac{13}{3}\right)$, $(3, 5)$
11. $\left(\dfrac{3 + \sqrt{7}}{2}, \dfrac{-1 + \sqrt{7}}{2}\right)$, $\left(\dfrac{3 - \sqrt{7}}{2}, \dfrac{-1 - \sqrt{7}}{2}\right)$
13. $\left(-3, \frac{5}{2}\right)$, $(3, 1)$ **15.** $(1, -7)$, $(-7, 1)$
17. $(3, 0)$, $\left(-\frac{9}{5}, \frac{8}{5}\right)$ **19.** $(1, 4)$, $(4, 1)$
21. $(0, 0)$, $(1, 1)$, $\left(-\dfrac{1}{2} + \dfrac{\sqrt{3}}{2}i, -\dfrac{1}{2} - \dfrac{\sqrt{3}}{2}i\right)$,
$\left(-\dfrac{1}{2} - \dfrac{\sqrt{3}}{2}i, -\dfrac{1}{2} + \dfrac{\sqrt{3}}{2}i\right)$ **23.** $(-3, 0)$, $(3, 0)$
25. $(-4, -3)$, $(-3, -4)$, $(3, 4)$, $(4, 3)$
27. $\left(\dfrac{6\sqrt{21}}{7}, \dfrac{4i\sqrt{35}}{7}\right)$, $\left(\dfrac{6\sqrt{21}}{7}, -\dfrac{4i\sqrt{35}}{7}\right)$,
$\left(-\dfrac{6\sqrt{21}}{7}, \dfrac{4i\sqrt{35}}{7}\right)$, $\left(-\dfrac{6\sqrt{21}}{7}, -\dfrac{4i\sqrt{35}}{7}\right)$
29. $(-\sqrt{2}, -\sqrt{14})$, $(-\sqrt{2}, \sqrt{14})$, $(\sqrt{2}, -\sqrt{14})$, $(\sqrt{2}, \sqrt{14})$
31. $(-2, -1)$, $(-1, -2)$, $(1, 2)$, $(2, 1)$
33. $(-3, -2)$, $(-2, -3)$, $(2, 3)$, $(3, 2)$
35. $(2, 5)$, $(-2, -5)$ **37.** $(3, 2)$, $(-3, -2)$
39. $(-3, 4)$, $(3, 4)$, $(0, -5)$
41. Length: 8 cm; width: 6 cm
43. Length: 5 in.; width: 4 in.
45. Length: 75 yd; width: 30 yd   **47.** 13 and 12
49. 24 ft, 16 ft   **51.** Length: $\sqrt{3}$ m; width: 1 m
53. $4\sqrt{3}$ **55.** 3.7 mph   **57.** ◈
59. 61.52 cm and 38.48 cm   **61.** $4x^2 + 3y^2 = 43$
63. 30   **65.** $\left(\frac{1}{3}, \frac{1}{2}\right)$, $\left(\frac{1}{2}, \frac{1}{3}\right)$

REVIEW EXERCISES: CHAPTER 10, P. 615

1. [10.1] 4   **2.** [10.1] 5   **3.** [10.1] $\sqrt{90.1} \approx 9.492$

4. [10.1] $\sqrt{9 + 4a^2}$ **5.** [10.1] $(4, 6)$ **6.** [10.1] $\left(-3, \frac{5}{2}\right)$
7. [10.1] $\left(\frac{3}{4}, \frac{\sqrt{3} - \sqrt{2}}{2}\right)$ **8.** [10.1] $\left(\frac{1}{2}, 2a\right)$
9. [10.1] $(-2, 3)$, $\sqrt{2}$ **10.** [10.1] $(5, 0)$, 7
11. [10.1] $(3, 1)$, 3 **12.** [10.1] $(-4, 3)$, $\sqrt{35}$
13. [10.1] $(x + 4)^2 + (y - 3)^2 = 48$
14. [10.1] $(x - 7)^2 + (y + 2)^2 = 20$
15. [10.3], [10.1] Circle   **16.** [10.3], [10.2] Ellipse

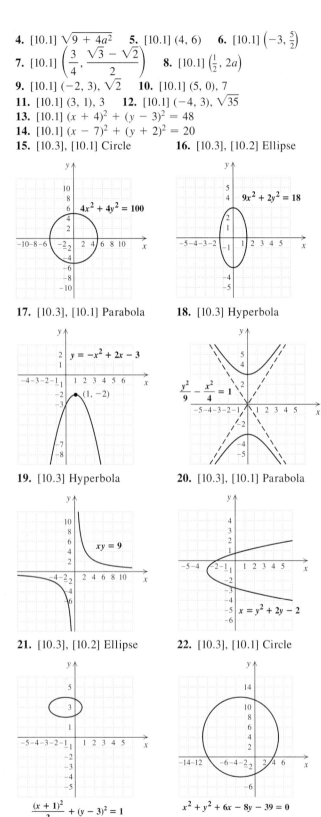

17. [10.3], [10.1] Parabola   **18.** [10.3] Hyperbola

19. [10.3] Hyperbola   **20.** [10.3], [10.1] Parabola

21. [10.3], [10.2] Ellipse   **22.** [10.3], [10.1] Circle

23. [10.4] $(7, 4)$ **24.** [10.4] $(2, 2)$, $\left(\frac{32}{9}, -\frac{10}{9}\right)$
25. [10.4] $(0, -3)$, $(2, 1)$
26. [10.4] $(4, 3)$, $(4, -3)$, $(-4, 3)$, $(-4, -3)$
27. [10.4] $(2, 1)$, $(\sqrt{3}, 0)$, $(-2, 1)$, $(-\sqrt{3}, 0)$
28. [10.4] $(3, -3)$, $\left(-\frac{3}{5}, \frac{21}{5}\right)$
29. [10.4] $(6, 8)$, $(6, -8)$, $(-6, 8)$, $(-6, -8)$
30. [10.4] $(2, 2)$, $(-2, -2)$, $(2\sqrt{2}, \sqrt{2})$, $(-2\sqrt{2}, -\sqrt{2})$
31. [10.4] 12 m by 7 m   **32.** [10.4] 4 and 8
33. [10.4] 32 cm, 20 cm   **34.** [10.4] 3 ft, 11 ft
35. [7.3] $3a^2b^3\sqrt[3]{3a^2b}$ **36.** [8.2] $-1 \pm 2i$
37. [7.5] $\dfrac{16 - a}{8 + 6\sqrt{a} + a}$ **38.** [6.5] 6 ft/sec
39. [10.1], [10.3] ◈ The graph of a parabola has one branch whereas the graph of a hyperbola has two branches. A hyperbola has asymptotes, but a parabola does not.
40. [10.1], [10.2], [10.3] ◈ Function notation is not used in this chapter because many of the relations are not functions. Function notation could be used for vertical parabolas and for hyperbolas that have the axes as asymptotes.
41. [10.4] $(-5, -4\sqrt{2})$, $(-5, 4\sqrt{2})$, $(3, -2\sqrt{2})$, $(3, 2\sqrt{2})$
42. [10.1] $(0, 6)$, $(0, -6)$
43. [10.1], [10.4] $(x - 2)^2 + (y + 1)^2 = 25$
44. [10.2] $\dfrac{x^2}{49} + \dfrac{y^2}{9} = 1$ **45.** [10.1] $\left(\frac{9}{4}, 0\right)$

TEST: CHAPTER 10, P. 616
1. [10.1] $9\sqrt{2} \approx 12.728$ **2.** [10.1] $2\sqrt{9 + a^2}$
3. [10.1] $\left(-\frac{1}{2}, \frac{7}{2}\right)$ **4.** [10.1] $(0, 0)$ **5.** [10.1] $(-2, 3)$, 8
6. [10.1] $(-2, 3)$, 3
7. [10.3], [10.1] Parabola   **8.** [10.3], [10.1] Circle

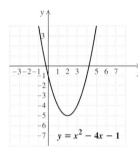

9. [10.3] Hyperbola

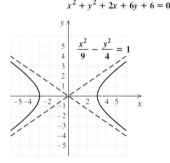

10. [10.3], [10.2] Ellipse

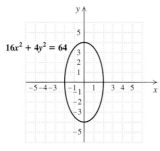

$16x^2 + 4y^2 = 64$

11. [10.3] Hyperbola   **12.** [10.3], [10.1] Parabola

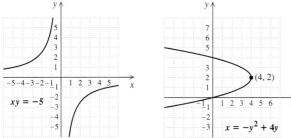

$xy = -5$

$x = -y^2 + 4y$

(4, 2)

13. [10.4] $(0, 3), (4, 0)$ **14.** [10.4] $(4, 0), (-4, 0)$
15. [10.4] 15 and 8 or 8 and 15   **16.** [10.4] 2 by 11
17. [10.4] $\sqrt{5}$ m, $\sqrt{3}$ m   **18.** [10.4] 16 ft by 12 ft
19. [8.2] $-1 \pm \sqrt{6}$ **20.** [7.3] $2ab^6\sqrt[3]{6a^2}$
21. [7.5] $\dfrac{8 - 6\sqrt{a} + a}{4 - a}$ **22.** [6.5] 6 mph
23. [10.2] $\dfrac{(x - 6)^2}{25} + \dfrac{(y - 3)^2}{9} = 1$
24. [10.1] $\left(0, -\dfrac{31}{4}\right)$ **25.** [10.4] 9

CHAPTER 11

EXERCISE SET 11.1, PP. 622–624

1. 2, 7, 12, 17; 47; 72   **3.** $\dfrac{1}{3}, \dfrac{1}{2}, \dfrac{3}{5}, \dfrac{2}{3}; \dfrac{5}{6}; \dfrac{15}{17}$
5. $-1, 0, 3, 8$; 80; 195   **7.** 2, $2\frac{1}{2}, 3\frac{1}{3}, 4\frac{1}{4}$; $10\frac{1}{10}$; $15\frac{1}{15}$
9. $-1, 4, -9, 16$; 100; -225
11. $-2, -1, 4, -7$; -25; 40   **13.** 16   **15.** 403
17. -23.5 **19.** -363 **21.** $\dfrac{441}{400}$ **23.** 43   **25.** $2n - 1$
27. $(-1)^n 2(3)^{n-1}$ **29.** $\dfrac{n}{n + 1}$ **31.** $3^{n/2}$
33. $-(3n - 2)$, or $2 - 3n$ **35.** 4   **37.** 30
39. $\dfrac{1}{2} + \dfrac{1}{4} + \dfrac{1}{6} + \dfrac{1}{8} + \dfrac{1}{10} = \dfrac{137}{120}$
41. $3^0 + 3^1 + 3^2 + 3^3 + 3^4 = 121$
43. $\dfrac{1}{2} + \dfrac{2}{3} + \dfrac{3}{4} + \dfrac{4}{5} + \dfrac{5}{6} + \dfrac{6}{7} + \dfrac{7}{8} + \dfrac{8}{9} = \dfrac{15,551}{2520}$
45. $(-1)^1 + (-1)^2 + (-1)^3 + (-1)^4 + (-1)^5 = -1$
47. $(-1)^2 2^1 + (-1)^3 2^2 + (-1)^4 2^3 + (-1)^5 2^4 + (-1)^6 2^5 +$
$(-1)^7 2^6 + (-1)^8 2^7 + (-1)^9 2^8 = -170$
49. $3 + 2 + 3 + 6 + 11 + 18 = 43$

51. $\dfrac{(-1)^3}{3 \cdot 4} + \dfrac{(-1)^4}{4 \cdot 5} + \dfrac{(-1)^5}{5 \cdot 6} = -\dfrac{1}{15}$ **53.** $\displaystyle\sum_{k=1}^{5} \dfrac{k + 1}{k + 2}$
55. $\displaystyle\sum_{k=1}^{6} k^2$ **57.** $\displaystyle\sum_{k=2}^{n} (-1)^k k^2$ **59.** $\displaystyle\sum_{k=1}^{\infty} 5k$
61. $\displaystyle\sum_{k=1}^{\infty} \dfrac{1}{k(k + 1)}$ **63.** 29   **65.** 1   **67.** 7   **69.**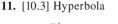
71. ◈   **73.** 0, 3, 12, 147, 21,612, 467,078,547
75. $5200, $3900, $2925, $2193.75, $1645.31, $1233.98,
$925.49, $694.12, $520.59, $390.44   **77.** $\dfrac{3}{2}, \dfrac{3}{2}, \dfrac{9}{8}, \dfrac{3}{4}, \dfrac{15}{32}, \dfrac{171}{32}$
79. $\{x \mid x = 4n - 1$, where n is a natural number$\}$
81. 1225

EXERCISE SET 11.2, PP. 630–632

1. $a_1 = 3, d = 5$ **3.** $a_1 = 6, d = -4$ **5.** $a_1 = \dfrac{3}{2}, d = \dfrac{3}{4}$
7. $a_1 = $2.12, d = 0.12 **9.** 47   **11.** -41
13. $-$1628.16   **15.** 27th   **17.** 102nd   **19.** 82   **21.** 5
23. 28   **25.** $a_1 = 8; d = -3$; 8, 5, 2, $-1, -4$ **27.** 780
29. 45,150   **31.** 2550   **33.** 918   **35.** 990
37. 42; 420   **39.** 1260   **41.** $31,000   **43.** 722
45. $a^k = P$ **47.** $x^2 + y^2 = 81$ **49.** ◈   **51.** $S_n = n^2$
53. $8760, $7961.77, $7163.54, $6365.31, $5567.08,
$4768.85, $3970.62, $3172.39, $2374.16, $1575.93
55. Let $d = $ the common difference. Since $p, m,$ and q form
an arithmetic sequence, $m = p + d$ and $q = p + 2d$. Then
$\dfrac{p + q}{2} = \dfrac{p + (p + 2d)}{2} = p + d = m.$

EXERCISE SET 11.3, PP. 639–641

1. 2   **3.** -1 **5.** $-\dfrac{1}{2}$ **7.** $\dfrac{1}{5}$ **9.** $\dfrac{3}{m}$ **11.** 320
13. 48   **15.** 648   **17.** $2331.64   **19.** $a_n = 3^{n-1}$
21. $a_n = (-1)^{n-1}$ **23.** $a_n = \dfrac{1}{x^n}$ **25.** 889   **27.** $\dfrac{547}{18}$
29. $\dfrac{1 - x^8}{1 - x}$, or $(1 + x)(1 + x^2)(1 + x^4)$ **31.** $5134.51
33. $\dfrac{27}{2}$ **35.** $\dfrac{49}{4}$ **37.** No   **39.** No   **41.** $\dfrac{43}{99}$
43. $25,000   **45.** $\dfrac{7}{9}$ **47.** $\dfrac{830}{99}$ **49.** $\dfrac{5}{33}$ **51.** $\dfrac{5}{1024}$ ft
53. 155,797   **55.** $43,318.94   **57.** 25 min
59. 3100.35 ft   **61.** 20.48 in.   **63.** $\left(-\dfrac{63}{29}, -\dfrac{114}{29}\right)$
65. ◈   **67.** ◈   **69.** $\dfrac{1 - x^n}{1 - x}$ **71.** ◈

EXERCISE SET 11.4, PP. 648–649

1. 40,320   **3.** 3,628,800   **5.** 210   **7.** 720   **9.** 28
11. 210   **13.** 190   **15.** 595
17. $m^5 + 5m^4 n + 10m^3 n^2 + 10m^2 n^3 + 5mn^4 + n^5$
19. $x^6 - 6x^5 y + 15x^4 y^2 - 20x^3 y^3 + 15x^2 y^4 - 6xy^5 + y^6$
21. $x^{10} - 15x^8 y + 90x^6 y^2 - 270x^4 y^3 + 405x^2 y^4 - 243y^5$
23. $729c^6 - 1458c^5 d + 1215c^4 d^2 - 540c^3 d^3 + 135c^2 d^4 -$
$18cd^5 + d^6$ **25.** $x^3 - 3x^2 y + 3xy^2 - y^3$

27. $x^9 + \dfrac{18x^8}{y} + \dfrac{144x^7}{y^2} + \dfrac{672x^6}{y^3} + \dfrac{2016x^5}{y^4} + \dfrac{4032x^4}{y^5} +$ $\dfrac{5376x^3}{y^6} + \dfrac{4608x^2}{y^7} + \dfrac{2304x}{y^8} + \dfrac{512}{y^9}$

29. $a^{10} - 5a^8b^3 + 10a^6b^6 - 10a^4b^9 + 5a^2b^{12} - b^{15}$

31. $9 - 12\sqrt{3}t + 18t^2 - 4\sqrt{3}t^3 + t^4$

33. $x^{-8} + 4x^{-4} + 6 + 4x^4 + x^8$ **35.** $15a^4b^2$

37. $-64,481,508a^3$ **39.** $1120x^{12}y^2$

41. $-1,959,552u^5v^{10}$ **43.** 4   **45.** 5.6348   **47.** ◆

49. $\dbinom{5}{2}(0.313)^3(0.687)^2 \approx 0.145$

51. $\dbinom{5}{2}(0.313)^3(0.687)^2 + \dbinom{5}{3}(0.313)^2(0.687)^3 +$ $\dbinom{5}{4}(0.313)(0.687)^4 + \dbinom{5}{5}(0.687)^5 \approx 0.964$

53. $\dbinom{n}{n-r} = \dfrac{n!}{[n-(n-r)]!\,(n-r)!}$
$\qquad = \dfrac{n!}{r!\,(n-r)!} = \dbinom{n}{r}$

55. $-4320x^6y^{9/2}$ **57.** $-\dfrac{35}{x^{1/6}}$

REVIEW EXERCISES: CHAPTER 11, PP. 650–651

1. [11.1] 1, 5, 9, 13; 29; 45   **2.** [11.1] $0, \frac{1}{5}, \frac{1}{5}, \frac{3}{17}; \frac{7}{65}; \frac{11}{145}$

3. [11.1] $a_n = -2n$ **4.** [11.1] $a_n = n^2$

5. [11.1] $-2 + 4 + (-8) + 16 + (-32) = -22$

6. [11.1] $-3 + (-5) + (-7) + (-9) + (-11) +$ $(-13) = -48$ **7.** [11.1] $\sum\limits_{k=1}^{5} 4k$ **8.** [11.1] $\sum\limits_{k=1}^{5} \dfrac{1}{(-2)^k}$

9. [11.2] 85   **10.** [11.2] $\frac{8}{3}$

11. [11.2] $d = 1.25, a_1 = 11.25$ **12.** [11.2] -544

13. [11.2] 8580   **14.** [11.3] $1024\sqrt{2}$ **15.** [11.3] $\frac{2}{3}$

16. [11.3] $a_n = 2(-1)^n$ **17.** [11.3] $a_n = 3\left(\dfrac{x}{4}\right)^{n-1}$

18. [11.3] 4095   **19.** [11.3] $-4095x$ **20.** [11.3] 12

21. [11.3] $\frac{49}{11}$ **22.** [11.3] No   **23.** [11.3] No

24. [11.3] \$40,000   **25.** [11.3] $\frac{5}{9}$ **26.** [11.3] $\frac{46}{33}$

27. [11.2] \$17.80   **28.** [11.2] 903   **29.** [11.3] \$22,521.92

30. [11.3] 6 m   **31.** [11.4] 40,320   **32.** [11.4] 56

33. [11.4] $190a^{18}b^2$

34. [11.4] $x^4 - 8x^3y + 24x^2y^2 - 32xy^3 + 16y^4$

35. [3.2] $\left(\frac{26}{11}, \frac{1}{11}\right)$ **36.** [9.6] $\frac{5}{9}$ **37.** [9.3] 13

38. [10.1] $(x-3)^2 + (y+1)^2 = 3$

39. [11.3] ◆ For a geometric sequence with $|r| < 1$, as n gets larger, the absolute value of the terms gets smaller, since $|r^n|$ gets smaller.

40. [11.4] ◆ The first form of the binomial theorem draws the coefficients from Pascal's triangle; the second form uses factorial notation. The second form avoids the need to compute all preceding rows of Pascal's triangle, and is generally easier to use when only one term of an expansion is needed. When several terms of an expansion are needed and n is not large (say, $n \le 8$), it is often easier to use Pascal's triangle.

41. [11.3] $\dfrac{1 - (-x)^n}{x + 1}$

42. [11.4] $x^{-15} + 5x^{-9} + 10x^{-3} + 10x^3 + 5x^9 + x^{15}$

TEST: CHAPTER 11, PP. 651–652

1. [11.1] 1, 7, 13, 19, 25; 91   **2.** [11.1] $a_n = 4\left(\frac{1}{3}\right)^n$

3. [11.1] $1 - 1 - 5 - 13 - 29 = -47$

4. [11.1] $\sum\limits_{k=1}^{5} k^3$ **5.** [11.2] -46 **6.** [11.2] $\frac{3}{8}$

7. [11.2] $a_1 = 31.2; d = -3.8$ **8.** [11.2] 2508

9. [11.3] $\frac{9}{128}$ **10.** [11.3] $\frac{2}{3}$ **11.** [11.3] $(-1)^{n+1}3^n$

12. [11.3] $511 + 511x$ **13.** [11.3] 1   **14.** [11.3] No

15. [11.3] $\dfrac{\$25,000}{23} \approx \$1086.96$ **16.** [11.3] $\frac{85}{99}$

17. [11.2] 63   **18.** [11.2] 1378   **19.** [11.3] \$8981.05

20. [11.3] 36 m   **21.** [11.4] 78

22. [11.4] $x^{10} - 15x^8y + 90x^6y^2 - 270x^4y^3 + 405x^2y^4 - 243y^5$ **23.** [11.4] $220a^9x^3$ **24.** [9.6] 3

25. [10.1] $(x-1)^2 + (y+2)^2 = 27$ **26.** [3.2] $(-1, 2)$

27. [9.3] $10^{1.5} = x$ **28.** [11.2] $n(n+1)$

29. [11.3] $\dfrac{1 - \left(\dfrac{1}{x}\right)^n}{1 - \dfrac{1}{x}}$, or $\dfrac{x^n - 1}{x^{n-1}(x-1)}$

CUMULATIVE REVIEW: 1–11

1. [1.6] $-45x^6y^{-4}$, or $\dfrac{-45x^6}{y^4}$ **2.** [1.2] 6.3

3. [1.3] $-3y + 17$ **4.** [1.1] 280   **5.** [1.1], [1.2] $\frac{7}{6}$

6. [5.1] $3a^2 - 8ab - 15b^2$ **7.** [5.1] $13x^3 - 7x^2 - 6x + 6$

8. [5.2] $6a^2 + 7a - 5$ **9.** [5.2] $9a^4 - 30a^2y + 25y^2$

10. [6.2] $\dfrac{4}{x+2}$ **11.** [6.1] $\dfrac{x-4}{3(x+2)}$

12. [6.1] $\dfrac{(x+y)(x^2 + xy + y^2)}{x^2 + y^2}$ **13.** [6.3] $x - a$

14. [5.5] $(2x-3)^2$ **15.** [5.6] $(3a-2)(9a^2 + 6a + 4)$

16. [5.3] $(a^2 - b)(a+3)$ **17.** [5.7] $3(y^2+3)(5y^2-4)$

18. [2.2] 20   **19.** [6.6] $7x^3 + 9x^2 + 19x + 38 + \dfrac{72}{x-2}$

20. [1.3] $\frac{2}{3}$ **21.** [6.4] $-\frac{6}{5}, 4$ **22.** [4.2] $\mathbb{R}$

23. [3.2] $(1, -1)$ **24.** [3.4] $(2, -1, 1)$ **25.** [7.6] 2
26. [8.5] $\pm 2, \pm 5$
27. [10.4] $(\sqrt{5}, \sqrt{3}), (\sqrt{5}, -\sqrt{3}), (-\sqrt{5}, \sqrt{3}),$
$(-\sqrt{5}, -\sqrt{3})$ **28.** [9.6] $\dfrac{\ln 8}{\ln 5} \approx 1.2920$ **29.** [9.6] 1005
30. [9.3] $\frac{1}{16}$ **31.** [9.6] $-\frac{1}{2}$ **32.** [4.3] $\{x|\ -2 \le x \le 3\}$,
or $[-2, 3]$ **33.** [8.1] $\pm i\sqrt{2}$ **34.** [8.2] $-2 \pm \sqrt{7}$
35. [8.10] $\{y|\ y < -5 \ or \ y > 2\}$, or $(-\infty, -5) \cup (2, \infty)$
36. [5.8] $-6, 8$ **37.** [10.4] 5 ft by 12 ft
38. [4.1] More than 4   **39.** [1.4] $65, 66, 67$ **40.** [3.3] $11\frac{3}{7}$
41. [3.3] 24 L of A; 56 L of B   **42.** [6.5] 350 mph
43. [6.5] $8\frac{2}{5}$ min or 8 min, 24 sec   **44.** [8.6] 20
45. [8.9] 1250 ft^2
46. [2.4]   **47.** [10.2]

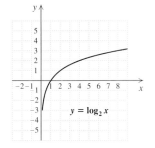

$3x - y = 6$

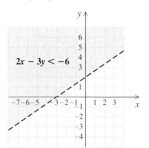

$\dfrac{x^2}{25} + \dfrac{y^2}{4} = 1$

48. [9.3]   **49.** [4.4]

$y = \log_2 x$

$2x - 3y < -6$

50. [8.7]

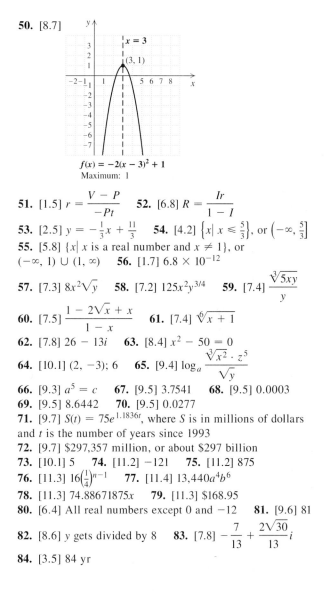

$f(x) = -2(x - 3)^2 + 1$
Maximum: 1

51. [1.5] $r = \dfrac{V - P}{-Pt}$ **52.** [6.8] $R = \dfrac{Ir}{1 - I}$
53. [2.5] $y = -\frac{1}{3}x + \frac{11}{3}$ **54.** [4.2] $\left\{x|\ x \le \frac{5}{3}\right\}$, or $\left(-\infty, \frac{5}{3}\right]$
55. [5.8] $\{x|\ x$ is a real number and $x \ne 1\}$, or
$(-\infty, 1) \cup (1, \infty)$ **56.** [1.7] 6.8×10^{-12}
57. [7.3] $8x^2\sqrt{y}$ **58.** [7.2] $125x^2y^{3/4}$ **59.** [7.4] $\dfrac{\sqrt[3]{5xy}}{y}$
60. [7.5] $\dfrac{1 - 2\sqrt{x} + x}{1 - x}$ **61.** [7.4] $\sqrt[6]{x + 1}$
62. [7.8] $26 - 13i$ **63.** [8.4] $x^2 - 50 = 0$
64. [10.1] $(2, -3)$; 6   **65.** [9.4] $\log_a \dfrac{\sqrt[3]{x^2} \cdot z^5}{\sqrt{y}}$
66. [9.3] $a^5 = c$ **67.** [9.5] 3.7541   **68.** [9.5] 0.0003
69. [9.5] 8.6442   **70.** [9.5] 0.0277
71. [9.7] $S(t) = 75e^{1.1836t}$, where S is in millions of dollars
and t is the number of years since 1993
72. [9.7] \$297,357 million, or about \$297 billion
73. [10.1] 5   **74.** [11.2] -121 **75.** [11.2] 875
76. [11.3] $16\left(\frac{1}{4}\right)^{n-1}$ **77.** [11.4] $13,440a^4b^6$
78. [11.3] $74.88671875x$ **79.** [11.3] \$168.95
80. [6.4] All real numbers except 0 and -12 **81.** [9.6] 81
82. [8.6] y gets divided by 8   **83.** [7.8] $-\dfrac{7}{13} + \dfrac{2\sqrt{30}}{13}i$
84. [3.5] 84 yr

Index

INDEX OF APPLICATIONS

In addition to the applications highlighted below, there are other applied problems and examples of problem solving in the text. An extensive list of their locations can be found under the heading "Applied problems" in the index at the back of the book.

(continued)

(continued)

Geometry

PLANE GEOMETRY

Rectangle
Area: $A = lw$
Perimeter: $P = 2l + 2w$

Square
Area: $A = s^2$
Perimeter: $P = 4s$

Triangle
Area: $A = \frac{1}{2}bh$

Triangle
Sum of Angle Measures:
$A + B + C = 180°$

Right Triangle
Pythagorean Theorem
(Equation):
$a^2 + b^2 = c^2$

Parallelogram
Area: $A = bh$

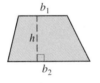

Trapezoid
Area: $A = \frac{1}{2}h(b_1 + b_2)$

Circle
Area: $A = \pi r^2$
Circumference:
$C = \pi d = 2\pi r$
$\left(\frac{22}{7} \text{ and } 3.14 \text{ are different}\right.$
approximations for π)

SOLID GEOMETRY

Rectangular Solid
Volume: $V = lwh$

Cube
Volume: $V = s^3$

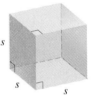

Right Circular Cylinder
Volume: $V = \pi r^2 h$
Total Surface Area:
$S = 2\pi rh + 2\pi r^2$

Right Circular Cone
Volume: $V = \frac{1}{3}\pi r^2 h$
Total Surface Area:
$S = \pi r^2 + \pi rs$
Slant Height:
$s = \sqrt{r^2 + h^2}$

Sphere
Volume: $V = \frac{4}{3}\pi r^3$
Surface Area: $S = 4\pi r^2$

Geometry

PLANE GEOMETRY

Rectangle
Area: $A = lw$
Perimeter: $P = 2l + 2w$

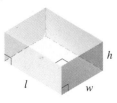

Square
Area: $A = s^2$
Perimeter: $P = 4s$

Triangle
Area: $A = \frac{1}{2}bh$

Triangle
Sum of Angle Measures:
$A + B + C = 180°$

Right Triangle
Pythagorean Theorem
(Equation):
$a^2 + b^2 = c^2$

Parallelogram
Area: $A = bh$

Trapezoid
Area: $A = \frac{1}{2}h(b_1 + b_2)$

Circle
Area: $A = \pi r^2$
Circumference:
$C = \pi d = 2\pi r$
$\left(\frac{22}{7} \text{ and } 3.14 \text{ are different}\right.$
approximations for $\left.\pi\right)$

SOLID GEOMETRY

Rectangular Solid
Volume: $V = lwh$

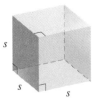

Cube
Volume: $V = s^3$

Right Circular Cylinder
Volume: $V = \pi r^2 h$
Total Surface Area:
$S = 2\pi rh + 2\pi r^2$

Right Circular Cone
Volume: $V = \frac{1}{3}\pi r^2 h$
Total Surface Area:
$S = \pi r^2 + \pi rs$
Slant Height:
$s = \sqrt{r^2 + h^2}$

Sphere
Volume: $V = \frac{4}{3}\pi r^3$
Surface Area: $S = 4\pi r^2$